Numerical Methods in Scientific Computing

Volume I

Numerical Methods in Scientific Computing

Volume I

Germund Dahlquist
Royal Institute of Technology
Stockholm, Sweden

Åke Björck
Linköping University
Linköping, Sweden

siam. Society for Industrial and Applied Mathematics • Philadelphia

10 9 8 7 6 5 4 3 2 1

Library of Congress Cataloging-in-Publication Data

Dahlquist, Germund.
Numerical methods in scientific computing / Germund Dahlquist, Åke Björck.
p.cm.
Includes bibliographical references and index.
ISBN 978-0-898716-44-3 (v. 1 : alk. paper)
1. Numerical analysis—Data processing. I. Björck, Åke, 1934- II. Title.

QA297.D335 2008
518—dc22 2007061806

To Marianne and Eva

Contents

List of Figures

List of Tables

List of Conventions

Besides the generally accepted mathematical abbreviations and notations (see, e.g., James and James, *Mathematics Dictionary* [1985, pp. 467–471]), the following notations are used in the book:

MATLAB® has been used for this book in testing algorithms. We also use its notations for array operations and the convenient colon notation.

.*	$A\,.*\,B$ element-by-element product $A(i,j)B(i,j)$
./	$A\,./B$ element-by-element division $A(i,j)/B(i,j)$
$i:k$	same as $i, i+1, \ldots, k$ and empty if $i > k$
$i:j:k$	same as $i, i+j, i+2j, \ldots, k$
$A(:,k)$, $A(i,:)$	the kth column, ith row of A, respectively
$A(i:k)$	same as $A(i), A(i+1), \ldots, A(k)$
$\lfloor x \rfloor$	floor, i.e., the largest integer $\le x$
$\lceil x \rceil$	roof, i.e., the smallest integer $\ge x$
e^x and $\exp(x)$	both denote the exponential function
$fl\,(x+y)$	floating-point operations; see Sec. 2.2.3
$\{x_i\}_{i=0}^n$	denotes the set $\{x_0, x_1, \ldots, x_n\}$
$[a,b]$	closed interval $(a \le x \le b)$
(a,b)	open interval $(a < x < b)$
$\text{sign}\,(x)$	$+1$ if $x \ge 0$, else -1
$\text{int}\,(a,b,c,\ldots,w)$	the smallest interval which contains $a, b, c, \ldots, w$
$f(x) = O(g(x)),\ x \to a$	$\lvert f(x)/g(x) \rvert$ is bounded as $x \to a$ (a can be finite, $+\infty$, or $-\infty$)
$f(x) = o(g(x)),\ x \to a$	$\lim_{x\to a} f(x)/g(x) = 0$
$f(x) \sim g(x),\ x \to a$	$\lim_{x\to a} f(x)/g(x) = 1$
$k \le i, j \le n$	means $k \le i \le n$ and $k \le j \le n$
$\mathcal{P}_k$	the set of polynomials of degree *less than* k
(f,g)	scalar product of functions f and g
$\lVert \cdot \rVert_p$	p-norm in a linear vector or function space; see Sec. 4.5.1–4.5.3 and Sec. A.3.3 in Online Appendix A
$E_n(f)$	$\text{dist}(f, \mathcal{P}_n)_{\infty,[a,b]}$; see Definition 4.5.6

The notations $a \approx b$, $a \lessapprox b$, and $a \gtrapprox b$ are defined in Sec. 2.1.2. Matrices and vectors are generally denoted by Roman letters A and b. A^T and b^T denote the transpose of the matrix A and the vector b, respectively. $(A,\ B)$ means a partitioned matrix; see Sec. A.2 in Online Appendix A. Notation for matrix computation can also be found in Online Appendix A. Notations for differences and difference operators, e.g., $\Delta^2 y_n$, $[x_0, x_1, x_2]f$, $\delta^2 y$, are defined in Chapters 3 and 4.

Preface

In 1974 the book by Dahlquist and Björck, *Numerical Methods*, was published in the Prentice–Hall Series in Automatic Computation, edited by George Forsythe. It was an extended and updated English translation of a Swedish undergraduate textbook used at the Royal Institute of Technology (KTH) in Stockholm. This book became one of the most successful titles at Prentice–Hall. It was translated into several other languages and as late as 1990 a Chinese edition appeared. It was reprinted in 2003 by Dover Publications.

In 1984 the authors were invited by Prentice–Hall to prepare a new edition of the book. After some attempts it soon became apparent that, because of the rapid development of the field, one volume would no longer suffice to cover the topics treated in the 1974 book. Thus a large part of the new book would have to be written more or less from scratch. This meant more work than we initially envisaged. Other commitments inevitably interfered, sometimes for years, and the project was delayed. The present volume is the result of several revisions worked out during the past 10 years.

Tragically, my mentor, friend, and coauthor Germund Dahlquist died on February 8, 2005, before this first volume was finished. Fortunately the gaps left in his parts of the manuscript were relatively few. Encouraged by his family, I decided to carry on and I have tried to the best of my ability to fill in the missing parts. I hope that I have managed to convey some of his originality and enthusiasm for his subject. It was a great privilege for me to work with him over many years. It is sad that he could never enjoy the fruits of his labor on this book.

Today mathematics is used in one form or another within most areas of science and industry. Although there has always been a close interaction between mathematics on the one hand and science and technology on the other, there has been a tremendous increase in the use of sophisticated mathematical models in the last decades. Advanced mathematical models and methods are now also used more and more within areas such as medicine, economics, and social sciences. Today, experiment and theory, the two classical elements of the scientific method, are supplemented in many areas by computations that are an equally important component.

The increased use of numerical methods has been caused not only by the continuing advent of faster and larger computers. Gains in problem-solving capabilities through better mathematical algorithms have played an equally important role. In modern scientific computing one can now treat more complex and less simplified problems through massive amounts of numerical calculations.

This volume is suitable for use in a basic introductory course in a graduate program in numerical analysis. Although short introductions to numerical linear algebra and differential equations are included, a more substantial treatment is deferred to later volumes. The book can also be used as a reference for researchers in applied sciences working in scientific computing. Much of the material in the book is derived from graduate courses given by the first author at KTH and Stanford University, and by the second author at Linköping University, mainly during the 1980s and 90s.

We have aimed to make the book as self-contained as possible. The level of presentation ranges from elementary in the first and second chapters to fairly sophisticated in some later parts. For most parts the necessary prerequisites are calculus and linear algebra. For some of the more advanced sections some knowledge of complex analysis and functional analysis is helpful, although all concepts used are explained.

The choice of topics inevitably reflects our own interests. We have included many methods that are important in large-scale computing and the design of algorithms. But the emphasis is on traditional and well-developed topics in numerical analysis. Obvious omissions in the book are wavelets and radial basis functions. Our experience from the 1974 book showed us that the most up-to-date topics are the first to become out of date.

Chapter 1 is on a more elementary level than the rest of the book. It is used to introduce a few general and powerful concepts and ideas that will be used repeatedly. An introduction is given to some basic methods in the numerical solution of linear equations and least squares problems, including the important singular value decomposition. Basic techniques for the numerical solution of initial value problems for ordinary differential equations is illustrated. An introduction to Monte Carlo methods, including a survey of pseudorandom number generators and variance reduction techniques, ends this chapter.

Chapter 2 treats floating-point number systems and estimation and control of errors. It is modeled after the same chapter in the 1974 book, but the IEEE floating-point standard has made possible a much more satisfactory treatment. We are aware of the fact that this aspect of computing is considered by many to be boring. But when things go wrong (and they do!), then some understanding of floating-point arithmetic and condition numbers may be essential. A new feature is a section on interval arithmetic, a topic which recently has seen a revival, partly because the directed rounding incorporated in the IEEE standard simplifies the efficient implementation.

In Chapter 3 different uses of infinite power series for numerical computations are studied, including ill-conditioned and semiconvergent series. Various algorithms for computing the coefficients of power series are given. Formal power series are introduced and their convenient manipulation using triangular Toeplitz matrices is described.

Difference operators are handy tools for the derivation, analysis, and practical application of numerical methods for many tasks such as interpolation, differentiation, and quadrature. A more rigorous treatment of operator series expansions and the use of the Cauchy formula and the fast Fourier transform (FFT) to derive the expansions are original features of this part of Chapter 3.

Methods for convergence acceleration of series (sequences) are covered in detail. For alternating series or series in a complex variable, Aitken extrapolation and Euler's transformation are the most important. Variants of Aitken, Euler–Maclaurin, and Richardson acceleration work for monotonic sequences. A partly new and more rigorous theoretical analysis given for completely monotonic sequences reflects Dahlquist's interest in analytic function theory. Although not intended for the novice, this has been included partly because it illustrates techniques that are of more general interest.

An exposition of continued fractions and Padé approximation, which transform a (formal) power series into a sequence of rational functions, concludes this chapter. This includes the ϵ-algorithm, the most important nonlinear convergence acceleration method, as well as the qd algorithm.

Chapter 4 treats several topics related to interpolation and approximation. Different bases for polynomial interpolation and related interpolation formulas are explained. The advantages of the barycentric form of Lagrange interpolation formula are stressed. Complex analysis is used to derive a general Lagrange–Hermite formula for polynomial interpolation in the complex plane. Algorithms for rational and multidimensional interpolation are briefly surveyed.

Interpolation of an analytic function at an infinite equidistant point set is treated from the point of view of complex analysis. Applications made to the Runge phenomenon and the Shannon sampling theorem. This section is more advanced than the rest of the chapter and can be skipped in a first reading.

Piecewise polynomials have become ubiquitous in computer aided design and computer aided manufacturing. We describe how parametric Bézier curves are constructed from piecewise Bernštein polynomials. A comprehensive treatment of splines is given and the famous recurrence relation of de Boor and Cox for B-splines is derived. The use of B-splines for representing curves and surfaces with given differentiability conditions is illustrated.

Function spaces are introduced in Chapter 4 and the concepts of linear operator and operator norm are extended to general infinite-dimensional vector spaces. The norm and distance formula, which gives a convenient error bound for general approximation problems, is presented. Inner product spaces, orthogonal systems, and the least squares approximation problem are treated next. The importance of the three-term recurrence formula and the Stieltjes procedure for numerical calculations is stressed. Chebyshev systems and theory and algorithms for approximation in maximum norm are surveyed.

Basic formulas and theorems for Fourier series and Fourier transforms are discussed next. Periodic continuation, sampled data and aliasing, and the Gibbs phenomenon are treated. In applications such as digital signal and image processing, and time-series analysis, the FFT algorithm (already used in Chapter 3) is an important tool. A separate section is therefore devoted to a matrix-oriented treatment of the FFT, including fast trigonometric transforms.

In Chapter 5 the classical Newton–Cotes rules for equidistant nodes and the Clenshaw–Curtis interpolatory rules for numerical integration are first treated. Next, extrapolation methods such as Romberg's method and the use of the ϵ-algorithm are described. The superconvergence of the trapezoidal rule in special cases and special Filon-type methods for oscillating integrands are discussed. A short section on adaptive quadrature follows.

Quadrature rules with both free and prescribed nodes are important in many contexts. A general technique of deriving formulas using the method of undetermined coefficients is given first. Next, Gauss–Christoffel quadrature rules and their properties are treated, and Gauss–Lobatto, Gauss–Radau, and Gauss–Kronrod rules are introduced. A more advanced exposition of relations between moments, tridiagonal matrices, and Gauss quadrature is included, but this part can be skipped at first reading.

Product rules for multidimensional integration formulas use simple generalizations of univariate rules and are applicable to rectangular domains. For more general domains, integration using irregular triangular grids is more suitable. The basic linear and quadratic interpolation formulas on such grids are derived. Together with a simple correction for curved boundaries these formulas are also very suitable for use in the finite element method. A discussion of Monte Carlo and quasi–Monte Carlo methods and their advantages for high-dimensional integration ends Chapter 5.

Chapter 6 starts with the bisection method. Next, fixed-point iterations are introduced and the contraction mapping theorem proved. Convergence order and the efficiency index are discussed. Newton's method is treated also for complex-valued equations and an interval Newton method is described. A discussion of higher-order methods, including the Schröder family of methods, is featured in this chapter.

Because of their importance for the matrix eigenproblem, algebraic equations are treated at length. The frequent ill-conditioning of roots is illustrated. Several classical methods are described, as well as an efficient and robust modified Newton method due to Madsen and Reid. Further, we describe the progressive qd algorithm and Sturm sequence methods, both of which are also of interest for the tridiagonal eigenproblem.

Three Online Appendices are available from the Web page of the book, `www.siam.org/books/ot103`. Appendix A is a compact survey of notations and some frequently used results in numerical linear algebra. Volume II will contain a full treatment of these topics. Online Appendix B describes Mulprec, a collection of MATLAB m-files for (almost) unlimited high precision calculation. This package can also be downloaded from the Web page. Online Appendix C is a more complete guide to literature, where advice is given on not only general textbooks in numerical analysis but also handbooks, encyclopedias, tables, software, and journals.

An important feature of the book is the large collection of problems and computer exercises included. This draws from the authors' 40+ year of experience in teaching courses in numerical analysis. It is highly recommended that a modern interactive system such as MATLAB is available to the reader for working out these assignments. The 1974 book also contained answers and solutions to most problems. It has not been possible to retain this feature because of the much greater number and complexity of the problems in the present book.

We have aimed to make and the bibliography as comprehensive and up-to-date as possible. A Notes and References section containing historical comments and additional references concludes each chapter. To remind the reader of the fact that much of the theory and many methods date one or several hundred years back in time, we have included more than 60 short biographical notes on mathematicians who have made significant contributions. These notes would not have been possible without the invaluable use of the biographies compiled at the School of Mathematics and Statistics, University of St Andrews, Scotland (`www-history.mcs.st.andrews.ac.uk`). Many of these full biographies are fascinating to read.

I am very grateful for the encouragement received from Marianne and Martin Dahlquist, who graciously allowed me to access computer files from Germund Dahlquist's personal computer. Without their support the completion of this book would not have been possible.

Many people read early drafts at various stages of the evolution of this book and contributed many corrections and constructive comments. I am particularly grateful to Nick Higham, Lothar Reichel, Zdenek Strakos, and several anonymous referees whose suggestions led to several major improvements. Other people who helped with proofreading include Bo Einarsson, Tommy Elfving, Pablo Guerrero-Garcia, Sven-Åke Gustafsson, and

Per Lötstedt. Thank you all for your interest in the book and for giving so much of your valuable time!

The book was typeset in LaTeX the references were prepared in BibTeX, and the index with MakeIndex. These are all wonderful tools and my thanks goes to Donald Knuth for his gift to mathematics. Thanks also to Cleve Moler for MATLAB, which was used in working out examples and for generating figures.

It is a pleasure to thank Elizabeth Greenspan, Sarah Murphy, and other staff at SIAM for their cheerful and professional support during all phases of the acquisition and production of the book.

Åke Björck
Linköping, July 2007

Chapter 1

Principles of Numerical Calculations

It is almost impossible to identify a mathematical theory no matter how "pure," that has never influenced numerical reasoning.
—B. J. C. Baxter and Arieh Iserles

1.1 Common Ideas and Concepts

Although numerical mathematics has been used for centuries in one form or another within many areas of science and industry,[1] modern scientific computing using electronic computers has its origin in research and developments during the Second World War. In the late 1940s and early 1950s the foundation of numerical analysis was laid as a separate discipline of mathematics. The new capability of performing billions of arithmetic operations cheaply has led to new classes of algorithms, which need careful analysis to ensure their accuracy and stability.

As a rule, applications lead to mathematical problems which in their complete form cannot be conveniently solved with exact formulas, unless one restricts oneself to special cases or simplified models. In many cases, one thereby reduces the problem to a sequence of linear problems—for example, linear systems of differential equations. Such an approach can quite often lead to concepts and points of view which, at least qualitatively, can be used even in the unreduced problems.

Recent developments have enormously increased the scope for using numerical methods. Gains in problem solving capabilities mean that today one can treat much more complex and less simplified problems through massive amounts of numerical calculations. This has increased the interaction of mathematics with science and technology.

In most numerical methods, one applies a small number of general and relatively simple ideas. These are then combined with one another in an inventive way and with such

[1]The Greek mathematician Archimedes (287–212 B.C.), Isaac Newton (1642–1727), and Carl Friedrich Gauss (1777–1883) were pioneering contributors to numerical mathematics.

knowledge of the given problem as one can obtain in other ways—for example, with the methods of mathematical analysis. Some knowledge of the background of the problem is also of value; among other things, one should take into account the orders of magnitude of certain numerical data of the problem.

In this chapter we shall illustrate the use of some general ideas behind numerical methods on some simple problems. These may occur as subproblems or computational details of larger problems, though as a rule they more often occur in a less pure form and on a larger scale than they do here. When we present and analyze numerical methods, to some degree we use the same approach which was first mentioned above: we study in detail special cases and simplified situations, with the aim of uncovering more generally applicable concepts and points of view which can guide us in more difficult problems.

It is important to keep in mind that the success of the methods presented depends on the smoothness properties of the functions involved. In this first survey we shall tacitly assume that the functions have as many well-behaved derivatives as are needed.

1.1.1 Fixed-Point Iteration

One of the most frequently occurring ideas in numerical calculations is **iteration** (from the Latin *iterare*, "to plow once again") or **successive approximation**. Taken generally, iteration means the repetition of a pattern of action or process. Iteration in this sense occurs, for example, in the repeated application of a numerical process—perhaps very complicated and itself containing many instances of the use of iteration in the somewhat narrower sense to be described below—in order to improve previous results. To illustrate a more specific use of the idea of iteration, we consider the problem of solving a (usually) nonlinear equation of the form

$$x = F(x), \tag{1.1.1}$$

where F is assumed to be a differentiable function whose value can be computed for any given value of a real variable x, within a certain interval. Using the method of iteration, one starts with an initial approximation x_0, and computes the sequence

$$x_1 = F(x_0), \qquad x_2 = F(x_1), \qquad x_3 = F(x_2), \dots. \tag{1.1.2}$$

Each computation of the type $x_{n+1} = F(x_n)$, $n = 0, 1, 2, \dots$, is called a **fixed-point iteration**. As n grows, we would like the numbers x_n to be better and better estimates of the desired root. If the sequence $\{x_n\}$ converges to a limiting value α, then we have

$$\alpha = \lim_{n\to\infty} x_{n+1} = \lim_{n\to\infty} F(x_n) = F(\alpha),$$

and thus $x = \alpha$ satisfies the equation $x = F(x)$. One can then stop the iterations when the desired accuracy has been attained.

A geometric interpretation of fixed point iteration is shown in Figure 1.1.1. A root of (1.1.1) is given by the abscissa (and ordinate) of an intersecting point of the curve $y = F(x)$ and the line $y = x$. Starting from x_0, the point $x_1 = F(x_0)$ on the x-axis is obtained by first drawing a horizontal line from the point $(x_0, F(x_0)) = (x_0, x_1)$ until it intersects the line $y = x$ in the point (x_1, x_1); from there we draw a vertical line to $(x_1, F(x_1)) = (x_1, x_2)$,

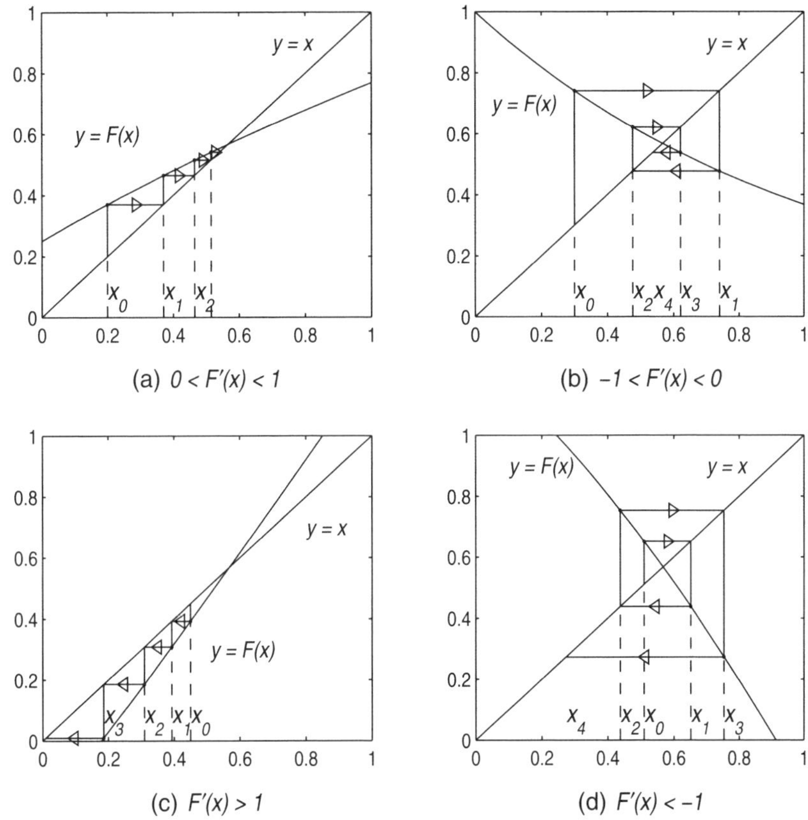

Figure 1.1.1. *Geometric interpretation of iteration* $x_{n+1} = F(x_n)$.

and so on in a "staircase" pattern. In Figure 1.1.1(a) it is obvious that the sequence $\{x_n\}$ converges monotonically to the root α. Figure 1.1.1(b) shows a case where F is a decreasing function. There we also have convergence, but not monotone convergence; the successive iterates x_n lie alternately to the right and to the left of the root α. In this case the root is bracketed by any two successive iterates.

There are also two divergent cases, exemplified by Figures 1.1.1(c) and (d). One can see geometrically that the quantity, which determines the rate of convergence (or divergence), is the slope of the curve $y = F(x)$ in the neighborhood of the root. Indeed, from the mean value theorem of calculus we have

$$\frac{x_{n+1} - \alpha}{x_n - \alpha} = \frac{F(x_n) - F(\alpha)}{x_n - \alpha} = F'(\xi_n),$$

where ξ_n lies between x_n and α. We see that if x_0 is chosen sufficiently close to the root (yet $x_0 \neq \alpha$), the iteration will converge if $|F'(\alpha)| < 1$. In this case α is called a **point of attraction**. The convergence is faster the smaller $|F'(\alpha)|$ is. If $|F'(\alpha)| > 1$, then α is a **point of repulsion** and the iteration diverges.

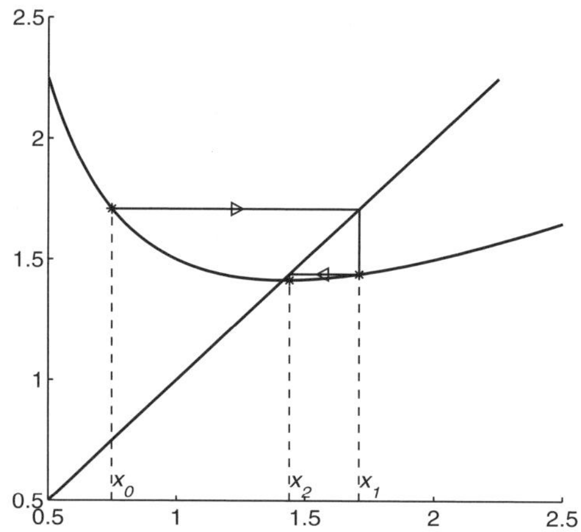

Figure 1.1.2. *The fixed-point iteration* $x_{n+1} = (x_n + c/x_n)/2$, $c = 2$, $x_0 = 0.75$.

Example 1.1.1.

The square root of $c > 0$ satisfies the equation $x^2 = c$, which can also be written $x = c/x$ or $x = (x + c/x)12$. This suggests the fixed-point iteration

$$x_{n+1} = \frac{1}{2}(x_n + c/x_n), \quad n = 1, 2, \ldots, \tag{1.1.3}$$

which is the widely used **Heron's rule**.[2] The curve $y = F(x)$ is in this case a hyperbola (see Figure 1.1.2).

From (1.1.3) follows

$$x_{n+1} \pm \sqrt{c} = \frac{1}{2}\left(x_n \pm 2\sqrt{c} + \frac{c}{x_n}\right) = \frac{(x_n \pm \sqrt{c})^2}{2x_n};$$

that is,

$$\frac{x_{n+1} - \sqrt{c}}{x_{n+1} + \sqrt{c}} = \left(\frac{x_n - \sqrt{c}}{x_n + \sqrt{c}}\right)^2. \tag{1.1.4}$$

We can take $e_n = \frac{x_n - \sqrt{c}}{x_n + \sqrt{c}}$ to be a measure of the error in x_n. Then (1.1.4) reads $e_{n+1} = e_n^2$ and it follows that $e_n = e_0^{2^n}$. If $|x_0 - \sqrt{c}| \neq |x_0 + \sqrt{c}|$, then $e_0 < 1$ and x_n converges to a square root of c when $n \to \infty$. Note that the iteration (1.1.3) can also be used for complex values of c.

For $c = 2$ and $x_0 = 1.5$, we get $x_1 = (1.5 + 2/1.5)12 = 15/12 = \mathbf{1.41}66666\ldots$, and

$$x_2 = \mathbf{1.414215}\,686274, \qquad x_3 = \mathbf{1.414213\,562375}$$

(correct digits shown in boldface). This can be compared with the exact value $\sqrt{2} = 1.414213\,562373\ldots$. As can be seen from Figure 1.1.2, a rough value for x_0 suffices. The

[2]Heron made important contributions to geometry and mechanics. He is believed to have lived in Alexandria, Egypt, during the first century A.D.

rapid convergence is due to the fact that for $\alpha = \sqrt{c}$ we have

$$F'(\alpha) = (1 - c/\alpha^2)/2 = 0.$$

One can in fact show that

$$\lim_{n\to\infty} \frac{|x_{n+1} - \sqrt{c}|}{|x_n - \sqrt{c}|^2} = C$$

for some constant $0 < C < \infty$, which is an example of what is known as **quadratic convergence**. Roughly, if x_n has t correct digits, then x_{n+1} will have at least $2t - 1$ correct digits.

The above iteration method is used quite generally on both pocket calculators and larger computers for calculating square roots.

Iteration is one of the most important aids for practical as well as theoretical treatment of both linear and nonlinear problems. One very common application of iteration is to the solution of *systems of equations*. In this case $\{x_n\}$ is a sequence of vectors, and F is a vector-valued function. When iteration is applied to *differential equations*, $\{x_n\}$ means a sequence of functions, and $F(x)$ means an expression in which integration or other operations on functions may be involved. A number of other variations on the very general idea of iteration will be given in later chapters.

The form of (1.1.1) is frequently called the **fixed-point form**, since the root α is a fixed point of the mapping F. An equation may not be given in this form originally. One has a certain amount of choice in the rewriting of an equation $f(x) = 0$ in fixed-point form, and the rate of convergence depends very much on this choice. The equation $x^2 = c$ can also be written, for example, as $x = c/x$. The iteration formula $x_{n+1} = c/x_n$ gives a sequence which alternates between x_0 (for even n) and c/x_0 (for odd n)—the sequence does not converge for any $x_0 \neq \sqrt{c}$!

1.1.2 Newton's Method

Let an equation be given in the form $f(x) = 0$, and for any $k \neq 0$, set

$$F(x) = x + kf(x).$$

Then the equation $x = F(x)$ is equivalent to the equation $f(x) = 0$. Since $F'(\alpha) = 1 + kf'(\alpha)$, we obtain the fastest convergence for $k = -1/f'(\alpha)$. Because α is not known, this cannot be applied literally. But if we use x_n as an approximation, this leads to the choice $F(x) = x - f(x)/f'(x)$, or the iteration

$$x_{n+1} = x_n - \frac{f(x_n)}{f'(x_n)}. \tag{1.1.5}$$

This is the celebrated **Newton's method**.[3] We shall derive it in another way below.

[3]Isaac Newton (1642–1727), English mathematician, astronomer, and physicist invented infinitesimal calculus independently of the German mathematician and philosopher Gottfried W. von Leibniz (1646–1716).

The equation $x^2 = c$ can be written in the form $f(x) = x^2 - c = 0$. Newton's method for this equation becomes

$$x_{n+1} = x_n - \frac{x_n^2 - c}{2x_n} = \frac{1}{2}\left(x_n + \frac{c}{x_n}\right), \quad n = 0, 1, 2, \dots, \tag{1.1.6}$$

which is the fast method in Example 1.1.1. More generally, Newton's method applied to the equation $f(x) = x^p - c = 0$ can be used to compute $c^{1/p}$, $p = \pm 1, \pm 2, \dots$, from the iteration

$$x_{n+1} = x_n - \frac{x_n^p - c}{p x_n^{p-1}}.$$

This can be written as

$$x_{n+1} = \frac{1}{p}\left((p-1)x_n + \frac{c}{x_n^{p-1}}\right) = \frac{x_n}{(-p)}[(1-p) - c x_n^{-p}]. \tag{1.1.7}$$

It is convenient to use the first expression in (1.1.7) when $p > 0$ and the second when $p < 0$. With $p = 2, 3$, and -2, respectively, this iteration formula is used for calculating $\sqrt{c}$, $\sqrt[3]{c}$, and $1/\sqrt{c}$. Also $1/c$, $(p = -1)$ can be computed by the iteration

$$x_{n+1} = x_n + x_n(1 - c x_n) = x_n(2 - c x_n),$$

using only multiplication and addition. In some early computers, which lacked division in hardware, this iteration was used to implement division, i.e., b/c was computed as $b(1/c)$.

Example 1.1.2.

We want to construct an algorithm based on Newton's method for the efficient calculation of the square root of any given floating-point number a. If we first shift the mantissa so that the exponent becomes even, $a = c \cdot 2^{2e}$, and $1/2 \le c < 2$; then

$$\sqrt{a} = \sqrt{c} \cdot 2^e.$$

We need only consider the reduced range $1/2 \le c \le 1$ since for $1 < c \le 2$ we can compute $\sqrt{1/c}$ and invert.[4]

To find an initial approximation x_0 to start the Newton iterations when $1/2 \le c < 1$, we can use linear interpolation of $x = \sqrt{c}$ between the endpoints $1/2, 1$, giving

$$x_0(c) = \sqrt{2}(1 - c) + 2(c - 1/2)$$

($\sqrt{2}$ is precomputed). The iteration then proceeds using (1.1.6).

For $c = 3/4$ ($\sqrt{c} = 0.86602540378444$) the result is $x_0 = (\sqrt{2} + 2)/4$ and (correct digits are in boldface)

$$x_0 = \mathbf{0.8}5355339059327, \qquad x_1 = \mathbf{0.866}11652351682,$$
$$x_2 = \mathbf{0.86602540}857756, \qquad x_3 = \mathbf{0.86602540378444},$$

[4]Since division is usually much slower than addition and multiplication, this may not be optimal.

The quadratic rate of convergence is apparent. Three iterations suffice to give about 16 digits of accuracy for all $x \in [1/2, 1]$.

Newton's method is based on **linearization**. This means that *locally*, i.e., in a small neighborhood of a point, *a more complicated function is approximated with a linear function*. In the solution of the equation $f(x) = 0$, this means geometrically that we seek the intersection point between the x-axis and the curve $y = f(x)$; see Figure 1.1.3. Assume

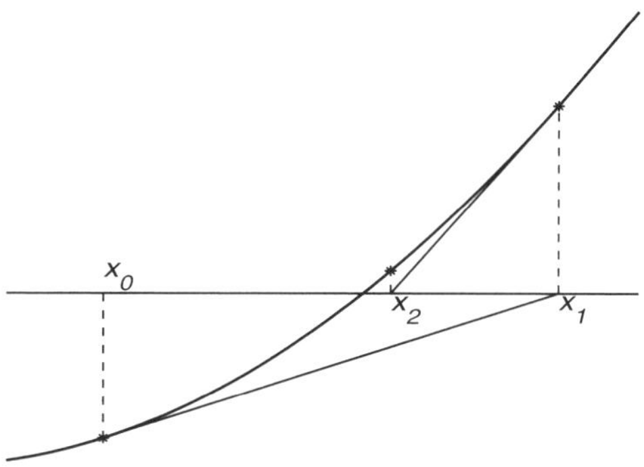

Figure 1.1.3. *Geometric interpretation of Newton's method.*

that we have an approximating value x_0 to the root. We then approximate the curve with its *tangent* at the point $(x_0, f(x_0))$. Let x_1 be the abscissa of the point of intersection between the x-axis and the tangent. Since the equation for the tangent reads

$$y - f(x_0) = f'(x_0)(x - x_0),$$

by setting $y = 0$ we obtain the approximation

$$x_1 = x_0 - f(x_0)/f'(x_0).$$

In many cases x_1 will have about twice as many correct digits as x_0. But if x_0 is a poor approximation and $f(x)$ far from linear, then it is possible that x_1 will be a worse approximation than x_0.

If we combine the ideas of iteration and linearization, that is, substitute x_n for x_0 and x_{n+1} for x_1, we rediscover Newton's method mentioned earlier. If x_0 is close enough to α, the iterations will converge rapidly (see Figure 1.1.3), but there are also cases of divergence.

An alternative to drawing the tangent to approximate a curve locally with a linear function is to choose two neighboring points on the curve and to approximate the curve with the *secant* which joins the two points; see Figure 1.1.4. The **secant method** for the solution of nonlinear equations is based on this approximation. This method, which preceded Newton's method, is discussed in more detail in Sec. 6.3.1.

Newton's method can be generalized to yield a quadratically convergent method for solving a **system of nonlinear equations**

$$f_i(x_1, x_2, \dots, x_n) = 0, \quad i = 1 : n. \tag{1.1.8}$$

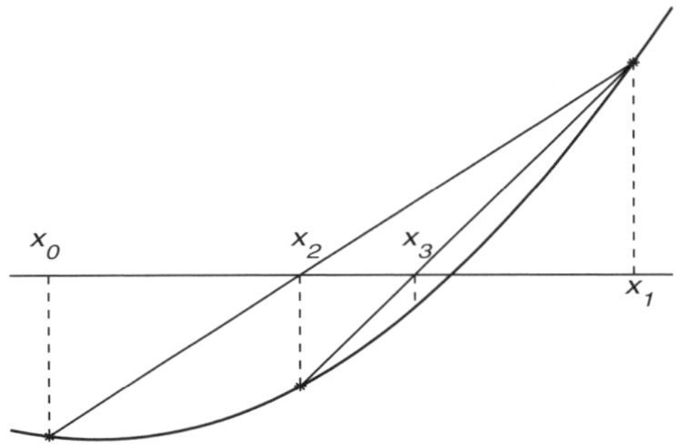

Figure 1.1.4. *Geometric interpretation of the secant method.*

Such systems arise in many different contexts in scientific computing. Important examples are the solution of systems of differential equations and optimization problems. We can write (1.1.8) as $\mathbf{f}(\mathbf{x}) = \mathbf{0}$, where $\mathbf{f}$ and $\mathbf{x}$ are vectors in $\mathbf{R}^n$. The vector-valued function $\mathbf{f}$ is said to be differentiable at the point $\mathbf{x}$ if each component is differentiable with respect to all the variables. The matrix of partial derivatives of $\mathbf{f}$ with respect to $\mathbf{x}$,

$$J(\mathbf{x}) = \mathbf{f}'(\mathbf{x}) = \begin{pmatrix} \frac{\partial f_1}{\partial x_1} & \cdots & \frac{\partial f_1}{\partial x_n} \\ \vdots & & \vdots \\ \frac{\partial f_n}{\partial x_1} & \cdots & \frac{\partial f_n}{\partial x_n} \end{pmatrix} \in \mathbf{R}^{n\times n}, \tag{1.1.9}$$

is called the **Jacobian** of $\mathbf{f}$.

Let $\mathbf{x}_k$ be the current approximate solution and assume that the matrix $\mathbf{f}'(\mathbf{x}_k)$ is nonsingular. Then in **Newton's method** for the system (1.1.8), the next iterate $\mathbf{x}_{n+1}$ is determined from the unique solution to the system of linear equations

$$J(\mathbf{x}_k)(\mathbf{x}_{n+1} - \mathbf{x}_k) = -\mathbf{f}(\mathbf{x}_k). \tag{1.1.10}$$

The linear system (1.1.10) can be solved by computing the LU factorization of the matrix $J(\mathbf{x})$; see Sec. 1.3.2.

Each step of Newton's method requires the evaluation of the n^2 entries of the Jacobian matrix $J(\mathbf{x}_k)$. This may be a time consuming task if n is large. If either the iterates or the Jacobian matrix are not changing too rapidly, it is possible to reevaluate $J(\mathbf{x}_k)$ only occasionally and use the same Jacobian in several steps. This has the further advantage that once we have computed the LU factorization of the Jacobian matrix, the linear system can be solved in only $O(n^2)$ arithmetic operations; see Sec. 1.3.2.

Example 1.1.3.

The following example illustrates the quadratic convergence of Newton's method for simple roots. The nonlinear system

$$\begin{aligned} x^2 + y^2 - 4x &= 0, \\ y^2 + 2x - 2 &= 0 \end{aligned}$$

has a solution close to $x_0 = 0.5$, $y_0 = 1$. The Jacobian matrix is

$$J(x, y) = \begin{pmatrix} 2x - 4 & 2y \\ 2 & 2y \end{pmatrix},$$

and Newton's method becomes

$$\begin{pmatrix} x_{k+1} \\ y_{k+1} \end{pmatrix} = \begin{pmatrix} x_k \\ y_k \end{pmatrix} - J(x_k, y_k)^{-1} \begin{pmatrix} x_k^2 + y_k^2 - 4x_n \\ y_k^2 + 2x_k - 2 \end{pmatrix}.$$

We get the following results:

k	x_k	y_k
1	**0.35**	**1.15**
2	**0.35424**528301887	**1.1365**2584085316
3	**0.35424868893**322	**1.13644297**217273
4	**0.35424868893541**	**1.13644296914943**

All digits are correct in the last iteration. The quadratic convergence is obvious; the number of correct digits approximately doubles in each iteration.

Often, the main difficulty in solving a nonlinear system is to find a sufficiently good starting point for the Newton iterations. Techniques for modifying Newton's method to ensure **global convergence** are therefore important in several dimensions. These must include techniques for coping with ill-conditioned or even singular Jacobian matrices at intermediate points. Such techniques will be discussed in Volume II.

1.1.3 Linearization and Extrapolation

The secant approximation is useful in many other contexts; for instance, it is generally used when one "reads between the lines" or interpolates in a table of numerical values. In this case the secant approximation is called **linear interpolation**. When the secant approximation is used in **numerical integration**, i.e., in the approximate calculation of a definite integral,

$$I = \int_a^b y(x)\, dx, \tag{1.1.11}$$

(see Figure 1.1.5) it is called the **trapezoidal rule**. With this method, the area between the curve $y = y(x)$ and the x-axis is approximated with the sum $T(h)$ of the areas of a series of parallel trapezoids. Using the notation of Figure 1.1.5, we have

$$T(h) = h\frac{1}{2}\sum_{i=0}^{n-1}(y_i + y_{i+1}), \qquad h = \frac{b-a}{n}. \tag{1.1.12}$$

(In the figure, $n = 4$.) We shall show in a later chapter that the error is very nearly proportional to h^2 when h is small. One can then, in principle, attain arbitrarily high accuracy by choosing h sufficiently small. But the computational work involved is roughly

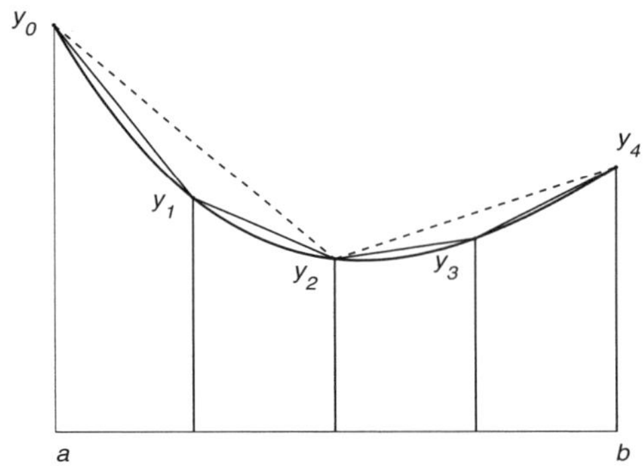

Figure 1.1.5. *Numerical integration by the trapezoidal rule* ($n = 4$).

proportional to the number of points where $y(x)$ must be computed, and thus inversely proportional to h. Hence the computational work grows rapidly as one demands higher accuracy (smaller h).

Numerical integration is a fairly common problem because only seldom can the "primitive" function be analytically calculated in a finite expression containing only elementary functions. It is not possible for such simple functions as e^{x^2} or $(\sin x)/x$. In order to obtain *higher accuracy* with significantly less work than the trapezoidal rule requires, one can use one of the following two important ideas:

(a) **local approximation** of the integrand with a polynomial of higher degree, or with a function of some other class, for which one knows the primitive function;

(b) computation with the trapezoidal rule for several values of h and then extrapolation to $h = 0$, the so-called **Richardson extrapolation**[5] or **deferred approach to the limit**, with the use of general results concerning the dependence of the error on h.

The technical details for the various ways of approximating a function with a polynomial, including Taylor expansions, interpolation, and the method of least squares, are treated in later chapters.

The extrapolation to the limit can easily be applied to numerical integration with the trapezoidal rule. As was mentioned previously, the trapezoidal approximation (1.1.12) to the integral has an error approximately proportional to the square of the step size. Thus, using two step sizes, h and $2h$, one has

$$T(h) - I \approx kh^2, \qquad T(2h) - I \approx k(2h)^2,$$

and hence $4(T(h) - I) \approx T(2h) - I$, from which it follows that

$$I \approx \frac{1}{3}(4T(h) - T(2h)) = T(h) + \frac{1}{3}(T(h) - T(2h)).$$

[5]Lewis Fry Richardson (1881–1953) studied mathematics, physics, chemistry, botany, and zoology. He graduated from King's College, Cambridge in 1903. He was the first (1922) to attempt to apply the method of finite differences to weather prediction, long before the computer age!

Thus, by adding the corrective term $\frac{1}{3}(T(h) - T(2h))$ to $T(h)$, one should get an estimate of I which is typically far more accurate than $T(h)$. In Sec. 3.4.6 we shall see that the improvement is in most cases quite striking. The result of the Richardson extrapolation is in this case equivalent to the classical **Simpson's rule** for numerical integration, which we shall encounter many times in this volume. It can be derived in several different ways. Section 3.4.5 also contains application of extrapolation to problems other than numerical integration, as well as a further development of the extrapolation idea, namely **repeated Richardson extrapolation**. In numerical integration this is also known as **Romberg's method**; see Sec. 5.2.2.

Knowledge of the behavior of the error can, together with the idea of extrapolation, lead to a powerful method for improving results. Such a line of reasoning is useful not only for the common problem of numerical integration, but also in many other types of problems.

Example 1.1.4.

The integral

$$\int_{10}^{12} f(x)\,dx$$

is computed for $f(x) = x^3$ by the trapezoidal method. With $h = 1$ we obtain $T(h) = 2695$, $T(2h) = 2728$, and extrapolation gives $T = 2684$, equal to the exact result.

Similarly, for $f(x) = x^4$ we obtain $T(h) = 30{,}009$, $T(2h) = 30{,}736$, and with extrapolation $T \approx 29{,}766.7$ (exact 29,766.4).

1.1.4 Finite Difference Approximations

The local approximation of a complicated function by a linear function leads to another frequently encountered idea in the construction of numerical methods, namely the approximation of a derivative by a difference quotient. Figure 1.1.6 shows the graph of a function

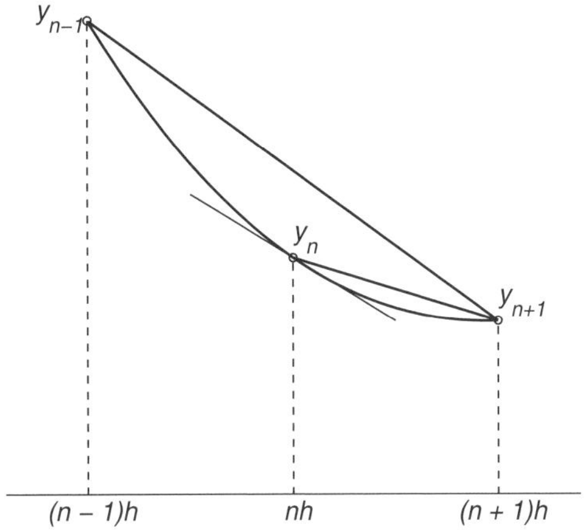

Figure 1.1.6. *Centered finite difference quotient.*

$y(x)$ in the interval $[x_{n-1}, x_{n+1}]$, where $x_{n+1} - x_n = x_n - x_{n-1} = h$; h is called the step size. If we set $y_i = y(x_i)$, $i = n-1, n, n+1$, then the derivative at x_n can be approximated by a **forward difference** quotient,

$$y'(x_n) \approx \frac{y_{n+1} - y_n}{h}, \tag{1.1.13}$$

or a similar backward difference quotient involving y_n and y_{n-1}. The error in the approximation is called a **discretization error**.

But it is conceivable that the **centered difference** approximation

$$y'(x_n) \approx \frac{y_{n+1} - y_{n-1}}{2h} \tag{1.1.14}$$

usually will be more accurate. It is in fact easy to motivate this. By Taylor's formula,

$$y(x+h) - y(x) = y'(x)h + y''(x)h^2/2 + y'''(x)h^3/6 + \cdots, \tag{1.1.15}$$

$$-y(x-h) + y(x) = y'(x)h - y''(x)h^2/2 + y'''(x)h^3/6 - \cdots. \tag{1.1.16}$$

Set $x = x_n$. Then, by the first of these equations,

$$y'(x_n) = \frac{y_{n+1} - y_n}{h} - \frac{h}{2}y''(x_n) - \cdots.$$

Next, add the two Taylor expansions and divide by $2h$. Then the first error term cancels and we have

$$y'(x_n) = \frac{y_{n+1} - y_{n-1}}{2h} - \frac{h^2}{6}y'''(x_n) - \cdots. \tag{1.1.17}$$

In what follows we call a formula (or a method), where a step size parameter h is involved, **accurate of order** p, if its error is approximately proportional to h^p. Since $y''(x)$ vanishes for all x if and only if y is a linear function of x, and similarly, $y'''(x)$ vanishes for all x if and only if y is a quadratic function, we have established the following important result.

Lemma 1.1.1.

The forward difference approximation (1.1.13) *is exact only for a linear function, and it is only first order accurate in the general case. The centered difference approximation* (1.1.14) *is exact also for a quadratic function, and is second order accurate in the general case.*

For the above reason the approximation (1.1.14) is, in most situations, preferable to (1.1.13). But there are situations when these formulas are applied to the approximate solution of differential equations where the forward difference approximation suffices, but where the centered difference quotient is entirely unusable, for reasons which have to do with how errors are propagated to later stages in the calculation. We shall not examine this phenomenon more closely here, but mention it only to intimate some of the surprising and fascinating mathematical questions which can arise in the study of numerical methods.

Higher derivatives can be approximated with **higher differences**, that is, differences of differences, another central concept in numerical calculations. We define

$$\begin{aligned}(\Delta y)_n &= y_{n+1} - y_n;\\ (\Delta^2 y)_n = (\Delta(\Delta y))_n &= (y_{n+2} - y_{n+1}) - (y_{n+1} - y_n)\\ &= y_{n+2} - 2y_{n+1} + y_n;\\ (\Delta^3 y)_n = (\Delta(\Delta^2 y))_n &= y_{n+3} - 3y_{n+2} + 3y_{n+1} - y_n;\end{aligned}$$

etc. For simplicity one often omits the parentheses and writes, for example, $\Delta^2 y_5$ instead of $(\Delta^2 y)_5$. The coefficients that appear here in the expressions for the higher differences are, by the way, the binomial coefficients. In addition, if we denote the step length by Δx instead of by h, we get the following formulas, which are easily remembered:

$$\frac{dy}{dx} \approx \frac{\Delta y}{\Delta x}, \qquad \frac{d^2 y}{dx^2} \approx \frac{\Delta^2 y}{(\Delta x)^2}, \tag{1.1.18}$$

etc. Each of these approximations is second order accurate for the value of the derivative at an x which equals the *mean value* of the largest and smallest x for which the corresponding value of y is used in the computation of the difference. (The formulas are only first order accurate when regarded as approximations to derivatives at other points between these bounds.) These statements can be established by arguments similar to the motivation for (1.1.13) and (1.1.14).

Taking the difference of the Taylor expansions (1.1.15)–(1.1.16) with one more term in each and dividing by h^2, we obtain the following important formula:

$$y''(x_n) = \frac{y_{n+1} - 2y_n + y_{n-1}}{h^2} - \frac{h^2}{12} y^{iv}(x_n) - \cdots.$$

Introducing the **central difference operator**

$$\delta y_n = y\left(x_n + \frac{1}{2}h\right) - y\left(x_n - \frac{1}{2}h\right) \tag{1.1.19}$$

and neglecting higher order terms we get

$$y''(x_n) \approx \frac{1}{h^2}\delta^2 y_n - \frac{h^2}{12} y^{iv}(x_n). \tag{1.1.20}$$

The approximation of (1.1.14) can be interpreted as an application of (1.1.18) with $\Delta x = 2h$, or as the mean of the estimates which one gets according to (1.1.18) for $y'((n + \frac{1}{2})h)$ and $y'((n - \frac{1}{2})h)$.

When the values of the function have errors (for example, when they are rounded numbers) the difference quotients become more and more uncertain the smaller h is. Thus if one wishes to compute the derivatives of a function one should be careful not to use too small a step length; see Sec. 3.3.4.

Example 1.1.5.

Assume that for $y = \cos x$, function values correct to six decimal digits are known at equidistant points:

x	y	Δy	$\Delta^2 y$
0.59	0.830941		
		−5605	
0.60	0.825336		−83
		−5688	
0.61	0.819648		

where the differences are expressed in units of 10^{-6}. This arrangement of the numbers is called a **difference scheme**. Using (1.1.14) and (1.1.18) one gets

$$y'(0.60) \approx (0.819648 - 0.830941)/0.02 = -0.56465,$$
$$y''(0.60) \approx -83 \cdot 10^{-6}/(0.01)^2 = -0.83.$$

The correct results are, with six decimals,

$$y'(0.60) = -0.564642, \qquad y''(0.60) = -0.825336.$$

In y'' we got only two correct decimal digits. This is due to **cancellation**, which is an important cause of loss of accuracy; see Sec. 2.3.4. Better accuracy can be achieved by *increasing* the step h; see Problem 1.1.5 at the end of this section.

A very important equation of mathematical physics is **Poisson's equation**:[6]

$$\frac{\partial^2 u}{\partial x^2} + \frac{\partial^2 u}{\partial y^2} = f(x, y), \quad (x, y) \in \Omega. \tag{1.1.21}$$

Here the function $f(x, y)$ is given together with some boundary condition on $u(x, y)$. Under certain conditions, gravitational, electric, magnetic, and velocity potentials satisfy **Laplace's equation**[7] which is (1.1.21) with $f(x, y) = 0$.

Finite difference approximations are useful for partial derivatives. Suppose that Ω is a rectangular region and introduce a **rectangular grid** that covers the rectangle. With grid spacing h and k, respectively, in the x and y directions, respectively, this consists of the points

$$x_i = x_0 + ih, \quad i = 0 : M, \qquad y_j = y_0 + jk, \quad j = 0 : N.$$

By (1.1.20), a second order accurate approximation of Poisson's equation is given by the **five-point operator**

$$\nabla_5^2 u = \frac{u_{i+1,j} - 2u_{i,j} + u_{i-1,j}}{h^2} + \frac{u_{i,j+1} - 2u_{i,j} + u_{i,j-1}}{k^2}.$$

For $k = h$

$$\nabla_5^2 u = \frac{1}{h^2}\big(u_{i,j+1} + u_{i-1,j} - 4u_{i,j} + u_{i+1,j} + u_{i,j-1}\big),$$

[6]Siméon Denis Poisson (1781–1840), professor at École Polytechnique. He has also given his name to the Poisson distribution in probability theory.

[7]Pierre-Simon, Marquis de Laplace (1749–1827), professor at École Militaire. Laplace was one of the most influential scientists of his time and did major work in probability and celestial mechanics.

which corresponds to the "computational molecule"

$$\frac{1}{h^2}\begin{bmatrix} & 1 & \\ 1 & -4 & 1 \\ & 1 & \end{bmatrix}.$$

If this is superimposed on each grid point we get one equation for the unknown values $u(x_i, y_j)$, $i = 1 : M-1$, $j = 1 : N-1$, at each interior point of the grid.

To get a solution we also need prescribed boundary conditions on u or $\partial u/\partial n$ on the boundary. The solution can then be obtained in the interior by solving a system of linear equations.

Review Questions

1.1.1 Make lists of the concepts and ideas which have been introduced. Review their use in the various types of problems mentioned.

1.1.2 Discuss the convergence condition and the rate of convergence of the fixed-point iteration method for solving a nonlinear equation $x = F(x)$.

1.1.3 What is meant by quadratic convergence of an iterative method for solving a nonlinear equation?

1.1.4 What is the trapezoidal rule? What is said about the dependence of its error on the step length?

1.1.5 How can Richardson extrapolation be used to improve the accuracy of the trapezoidal rule?

Problems and Computer Exercises

1.1.1 Calculate $\sqrt{10}$ to seven decimal places using the method in Example 1.1.1. Begin with $x_0 = 2$.

1.1.2 Consider $f(x) = x^3 - 2x - 5$. The cubic equation $f(x) = 0$ has been a standard test problem, since Newton used it in 1669 to demonstrate his method. By computing (say) $f(x)$ for $x = 1, 2, 3$, we see that $x = 2$ probably is a rather good initial guess. Iterate by Newton's method until you trust that the result is correct to six decimal places.

1.1.3 The equation $x^3 - x = 0$ has three roots, $-1, 0, 1$. We shall study the behavior of Newton's method on this equation, with the notations used in Sec. 1.1.1 and Figure 1.1.3.

(a) What happens if $x_0 = 1/\sqrt{3}$? Show that x_n converges to 1 for any $x_0 > 1/\sqrt{3}$. What is the analogous result for convergence to -1?

(b) What happens if $x_0 = 1/\sqrt{5}$? Show that x_n converges to 0 for any $x_0 \in (-1/\sqrt{5},\ 1/\sqrt{5})$.

Hint: Show first that if $x_0 \in (0,\ 1/\sqrt{5})$, then $x_1 \in (-x_0, 0)$. What can then be said about x_2?

(c) Find, by a drawing (with paper and pencil), $\lim x_n$ if x_0 is a little less than $1/\sqrt{3}$. Find by computation $\lim x_n$ if $x_0 = 0.46$.

(d) A complete discussion of the question in (c) is rather complicated, but there is an implicit recurrence relation that produces a decreasing sequence $\{a_1 = 1/\sqrt{3}, a_2, a_3, \dots\}$, by means of which one you can easily find $\lim_{n\to\infty} x_n$ for any $x_0 \in (1/\sqrt{5},\ 1/\sqrt{3})$. Try to find this recurrence.

Answer: $a_i - f(a_i)/f'(a_i) = -a_{i-1}$; $\lim_{n\to\infty} x_n = (-1)^i$ if $x_0 \in (a_i, a_{i+1})$; $a_1 = 0.577$, $a_2 = 0.462$, $a_3 = 0.450$, $a_4 \approx \lim_{i\to\infty} a_i = 1/\sqrt{5} = 0.447$.

1.1.4 Calculate $\int_0^{1/2} e^x\,dx$

(a) to six decimals using the primitive function.

(b) with the trapezoidal rule, using step length $h = 1/4$.

(c) using Richardson extrapolation to $h = 0$ on the results using step lengths $h = 1/2$ and $h = 1/4$.

(d) the ratio between the error in the result of (c) to that of (b).

1.1.5 In Example 1.1.5 we computed $y''(0.6)$ for $y = \cos x$, with step length $h = 0.01$. Make similar calculations using $h = 0.1$, $h = 0.05$, and $h = 0.001$. Which value of h gives the best result, using values of y to six decimal places? Discuss qualitatively the influences of both the rounding errors in the function values and the error in the approximation of a derivative with a difference quotient on the result for various values of h.

1.1.6 Give an approximate expression of the form $ah^b f^{(c)}(0)$ for the error of the estimate of the integral $\int_{-h}^{h} f(x)dx$ obtained by Richardson extrapolation (according to Sec. 1.1.3) from the trapezoidal values $T(h)$ and $T(2h)$.

1.2 Some Numerical Algorithms

For a given numerical problem one can consider many different algorithms. Even if they just differ in small details they can differ in efficiency and reliability and give approximate answers with widely varying accuracy. In the following we give a few examples of how algorithms can be developed to solve some typical numerical problems.

1.2.1 Solving a Quadratic Equation

An early example of pitfalls in computation studied by G. E. Forsythe [121] is the following. For computing the roots of the quadratic equation $ax^2 + bx + c = 0$, $a \neq 0$, elementary textbooks usually give the well-known formula

$$r_{1,2} = \big(-b \pm \sqrt{b^2 - 4ac}\big)/(2a).$$

Using this for the quadratic equation $x^2 - 56x + 1 = 0$, we get the two approximate real roots

$$r_1 = 28 + \sqrt{783} \approx 28 + 27.982 = 55.982 \pm \frac{1}{2}10^{-3},$$

$$r_2 = 28 - \sqrt{783} \approx 28 - 27.982 = 0.018 \pm \frac{1}{2}10^{-3}.$$

In spite of the fact that the square root used is given to five digits of accuracy, we get only two significant digits in r_2, while the relative error in r_1 is less than 10^{-5}. This shows that *there can be very poor relative accuracy in the difference between two nearly equal numbers*. This phenomenon is called **cancellation of terms**. It is a very common reason for poor accuracy in numerical calculations.

Notice that the subtraction in the calculation of r_2 was carried out exactly. *The cancellation in the subtraction only gives an indication of the unhappy consequence of a loss of information in previous steps, due to the rounding of one of the operands, and is not the cause of the inaccuracy.*

In numerical calculations, if possible one should try to avoid formulas that give rise to cancellation, as in the above example. For the quadratic equation this can be done by *rewriting of the formulas*. Comparing coefficients on both sides of

$$x^2 + (b/a)x + c/a = (x - r_1)(x - r_2) = x^2 - (r_1 + r_2)x + r_1 r_2,$$

we get the relation between coefficients and roots

$$r_1 + r_2 = -b/a, \qquad r_1 r_2 = c/a. \tag{1.2.1}$$

A more accurate value of the root of *smaller magnitude* is obtained by computing this root from the latter of these relations. We then get

$$r_2 = 1/55.982 = 0.0178629 \pm 0.0000002.$$

Five significant digits are now obtained also for this root.

1.2.2 Recurrence Relations

A common computational task is the evaluation of a polynomial

$$p(x) = a_0 x^n + a_1 x^2 + \cdots + a_{n-1} x + a_n$$

at a given point. This can be reformulated as

$$p(x) = (\cdots((a_0 x + a_1)x + a_2)x + \cdots + a_{n-1})x + a_n,$$

and written as a **recurrence relation**:

$$b_i(x) = b_{i-1}(x)x + a_i, \qquad i = 1 : n. \tag{1.2.2}$$

We note that this recurrence relation can be used in two different ways:

- it can be used *algebraically* to generate a sequence of Horner polynomials $b_i(x)$ such that $b_n(x) = p(x)$;
- it can be used *arithmetically* with a specific value $x = x_1$, which is **Horner's rule** for evaluating $p(x_1) = b_n(x_1)$.

Horner's rule requires n additions and multiplications for evaluating $p(x)$ for $x = x_1$. Note that if the powers are calculated recursively by $x_1^i = x_1 \cdot x_1^{i-1}$ and subsequently multiplied by a_{n-i}, this requires twice as many multiplications.

When a polynomial $p(x)$ is divided by $x - x_1$ the remainder equals $p(x_1)$; i.e., $p(x) = (x - x_1)q(x) + p(x_1)$. The quantities $b_i(x_1)$ from the Horner scheme (1.2.2) are of intrinsic interest because they are the coefficients of the quotient polynomial $q(x)$. This algorithm therefore performs the **synthetic division**

$$\frac{p(x) - p(x_1)}{x - x_1} = \sum_{i=0}^{n-1} b_i(x_1)x^{n-1-i}. \tag{1.2.3}$$

The proof of this result is left as an exercise.

Synthetic division is used, for instance, in the solution of algebraic equations when already computed roots are successively eliminated. After each elimination, one can deal with an equation of lower degree. This process is called **deflation**; see Sec. 6.5.4. As emphasized there, some care is necessary in the numerical application of this idea to prevent the propagation of roundoff errors.

The proof of the following useful relation is left as an exercise for the reader.

Lemma 1.2.1.

Let b_i be defined by (1.2.2) *and*

$$c_0 = b_0, \qquad c_i = b_i + xc_{i-1}, \quad i = 1 : n - 1. \tag{1.2.4}$$

Then $p'(x) = c_{n-1}$.

Due to their intrinsic constructive quality, recurrence relations are one of the basic mathematical tools of computation. There is hardly a computational task which does not use recursive techniques. One of the most important and interesting parts of the preparation of a problem for a computer is therefore to find a recursive description of the task. Often an enormous amount of computation can be described by a small set of recurrence relations.

Although recurrence relations are a powerful tool they are also susceptible to error growth. Each cycle of a recurrence relation not only generates its own errors but also inherits errors committed in all previous cycles. If conditions are unfavorable, the result may be disastrous. This aspect of recurrence relations and its prevention is therefore of great importance in computations and has been studied extensively; see [139].

Example 1.2.1.

Unless used in the right way, errors committed in a recurrence relation can grow exponentially and completely ruin the results. To compute the integrals

$$I_n = \int_0^1 \frac{x^n}{x+5}\,dx, \quad i = 1 : N,$$

one can use the recurrence relation

$$I_n + 5I_{n-1} = \frac{1}{n}, \tag{1.2.5}$$

which follows from

$$I_n + 5I_{n-1} = \int_0^1 \frac{x^n + 5x^{n-1}}{x+5}\,dx = \int_0^1 x^{n-1}\,dx = \frac{1}{n}.$$

Below we use this formula to compute I_8, using six decimals throughout. For $n = 0$ we have

$$I_0 = [\ln(x+5)]_0^1 \approx \ln 6 - \ln 5 = 0.182322.$$

Using the recurrence relation we get

$$\begin{aligned}
I_1 &= 1 - 5I_0 = 1 - 0.911610 = 0.088390,\\
I_2 &= 1/2 - 5I_1 = 0.500000 - 0.441950 = 0.058050,\\
I_3 &= 1/3 - 5I_2 = 0.333333 - 0.290250 = 0.043083,\\
I_4 &= 1/4 - 5I_3 = 0.250000 - 0.215415 = 0.034585,\\
I_5 &= 1/5 - 5I_4 = 0.200000 - 0.172925 = 0.027075,\\
I_6 &= 1/6 - 5I_5 = 0.166667 - 0.135375 = 0.031292,\\
I_7 &= 1/7 - 5I_6 = 0.142857 - 0.156460 = -0.013603.
\end{aligned}$$

It is strange that $I_6 > I_5$, and obviously absurd that $I_7 < 0$! The reason for the absurd result is that the roundoff error ϵ in $I_0 = 0.18232156\ldots$, whose magnitude is about $0.44 \cdot 10^{-6}$, is *multiplied* by (-5) in the calculation of I_1, which then has an error of -5ϵ. That error produces an error in I_2 of $5^2\epsilon$, and so forth. Thus the magnitude of the error in I_7 is $5^7\epsilon = 0.0391$, which is larger than the true value of I_7. On top of this are the roundoff errors committed in the various steps of the calculation. These can be shown, in this case, to be relatively unimportant.

If one uses higher precision, the absurd result will show up at a later stage. For example, a computer that works with a precision corresponding to about 16 decimal places gave a negative value to I_{22}, although I_0 had full accuracy. The above algorithm is an example of an unpleasant phenomenon, called **numerical instability**. In this simple case, one can avoid the numerical instability by reversing the direction of the recursion.

Example 1.2.2.

If we use the recurrence relation in the other direction,

$$I_{n-1} = (1/n - I_n)/5, \tag{1.2.6}$$

the errors will be *divided* by -5 in each step. But we need a starting value. We can directly see from the definition that I_n decreases as n increases. One can also surmise that I_n decreases slowly when n is large (the reader is encouraged to motivate this). Thus we try setting $I_{12} = I_{11}$. It then follows that

$$I_{11} + 5I_{11} \approx 1/12, \qquad I_{11} \approx 1/72 \approx 0.013889$$

(show that $0 < I_{12} < 1/72 < I_{11}$). Using the recurrence relation we get

$$I_{10} = (1/11 - 0.013889)/5 = 0.015404, \qquad I_9 = (1/10 - 0.015404)/5 = 0.016919$$

and, further,

$$\begin{aligned} I_8 &= 0.018838, & I_7 &= 0.021232, & I_6 &= 0.024325, & I_5 &= 0.028468, \\ I_4 &= 0.034306, & I_3 &= 0.043139, & I_2 &= 0.058039, & I_1 &= 0.088392, \end{aligned}$$

and finally $I_0 = 0.182322$. Correct!

If one instead simply takes $I_{12} = 0$ as the starting value, one gets $I_{11} = 0.016667$, $I_{10} = 0.018889$, $I_9 = 0, 016222$, $I_8 = 0.018978$, $I_7 = 0.021204$, $I_6 = 0.024331$, and $I_5, \dots, I_0$ have the same values as above. The difference in the values for I_{11} is 0.002778. The subsequent values of $I_{10}, I_9, \dots, I_0$ are quite close *because the error is divided by* -5 *in each step*. The results for I_n obtained above have errors which are less than 10^{-3} for $n \le 8$.

One should not draw erroneous conclusions from the above example. The use of a recurrence relation "backwards" is not a universal recipe, as will be seen later. Compare also Problems 1.2.7 and 1.2.8.

In Sec. 3.3.5 we will study the general linear homogeneous difference equation of kth order

$$y_{n+k} + a_1 y_{n+k-1} + \cdots + a_k y_n = 0, \tag{1.2.7}$$

with real or complex constant coefficients $a_1, \dots, a_k$. The stability properties of this type of equation are fundamental, since they arise in the numerical solution of ordinary and partial differential equations.

1.2.3 Divide and Conquer Strategy

A powerful strategy for solving large scale problems is the **divide and conquer** strategy (one of the oldest military strategies). This is one of the most powerful algorithmic paradigms for designing efficient algorithms. The idea is to split a high-dimensional problem into multiple problems (typically two for sequential algorithms) of lower dimension. Each of these is then again split into smaller subproblems, and so forth, until a number of sufficiently small problems are obtained. The solution of the initial problem is then obtained by combining the solutions of the subproblems working backward in the hierarchy.

We illustrate the idea on the computation of the sum $s = \sum_{i=1}^{n} a_i$. The usual way to proceed is to use the recursion

$$s_0 = 0, \qquad s_i = s_{i-1} + a_i, \qquad i = 1 : n.$$

Another order of summation is as illustrated below for $n = 2^3 = 8$:

$$\begin{array}{cccccccc} a_1 & a_2 & a_3 & a_4 & a_5 & a_6 & a_7 & a_8 \\ \searrow & \swarrow & \searrow & \swarrow & \searrow & \swarrow & \searrow & \swarrow \\ & s_{1:2} & & s_{3:4} & & s_{5:6} & & s_{7:8} \\ & \searrow & & \swarrow & & \searrow & & \swarrow \\ & & s_{1:4} & & & & s_{5:8} & \\ & & \searrow & & & \swarrow & & \\ & & & & s_{1:8} & & & \end{array},$$

where $s_{i:j} = a_i + \cdots + a_j$. In this table each new entry is obtained by adding its two neighbors in the row above. Clearly this can be generalized to compute an arbitrary sum of $n = 2^k$ terms in k steps. In the first step we perform $n/2$ sums of two terms, then $n/4$ partial sums each of four terms, etc., until in the kth step we compute the final sum.

This summation algorithm uses the same number of additions as the first one. But it has the advantage that it splits the task into *several subtasks that can be performed in parallel*. For large values of n this summation order can also be much more accurate than the conventional order (see Problem 2.3.5).

The algorithm can also be described in another way. Consider the summation algorithm

> $sum = s(i, j)$;
> **if** $j = i + 1$ **then** $sum = a_i + a_j$;
>   **else** $k = \lfloor (i + j)/2 \rfloor$; $sum = s(i, k) + s(k + 1, j)$;
> **end**

for computing the sum $s(i, j) = a_i + \cdots + a_j$, $j > i$. (Here and in the following $\lfloor x \rfloor$ denotes the **floor** of x, i.e., the largest integer $\leq x$. Similarly, $\lceil x \rceil$ denotes the **ceiling** of x, i.e., the smallest integer $\geq x$.) This function defines $s(i, j)$ in a recursive way; if the sum consists of only two terms, then we add them and return with the answer. Otherwise we split the sum in two and use the function again to evaluate the corresponding two partial sums. Espelid [114] gives an interesting discussion of such summation algorithms.

The function above is an example of a **recursive algorithm**—it calls itself. Many computer languages (for example, MATLAB) allow the definition of such recursive algorithms. The divide and conquer is a **top down** description of the algorithm in contrast to the **bottom up** description we gave first.

Example 1.2.3.

Sorting the items of a one-dimensional array in ascending or descending order is one of the most important problems in computer science. In numerical work, sorting is frequently needed when data need to be rearranged. One of the best known and most efficient sorting algorithms, **quicksort** by Hoare [202], is based on the divide and conquer paradigm. To sort an array of n items, $a[0 : n - 1]$, it proceeds as follows:

1. Select an element $a(k)$ to be the pivot. Commonly used methods are to select the pivot randomly or select the median of the first, middle, and last element in the array.

2. Rearrange the elements of the array a into a left and right subarray such that no element in the left subarray is larger than the pivot and no element in the right subarray is smaller than the pivot.

3. Recursively sort the left and right subarray.

The partitioning of a subarray $a[l : r]$, $l < r$, in step 2 can proceed as follows. Place the pivot in $a[l]$ and initialize two pointers $i = l$, $j = r + 1$. The pointer i is incremented until an element $a(i)$ is encountered which is larger than the pivot. Similarly, the pointer j is decremented until an element $a(j)$ is encountered which is smaller than the pivot. At this point the elements $a(i)$ and $a(j)$ are exchanged. The process continues until the pointers cross each other. Finally, the pivot element is placed in its correct position.

It is intuitively clear that this algorithm sorts the entire array and that no merging phase is needed.

There are many other examples of the power of the divide and conquer approach. It underlies the fast Fourier transform (Sec. 4.6.3) and is used in efficient automatic parallelization of many tasks, such as matrix multiplication; see [111].

1.2.4 Power Series Expansions

In many problems of applied mathematics, the solution of a given problem can be obtained as a power series expansion. Often the convergence of these series is quite fast. As an example we consider the task of computing, to five decimals, $y(0.5)$, where $y(x)$ is the solution to the differential equation

$$y'' = -xy,$$

with initial conditions $y(0) = 1$, $y'(0) = 0$. The solution cannot be simply expressed in terms of elementary functions. We shall use the **method of undetermined coefficients**. Thus we try substituting a series of the form:

$$y(x) = \sum_{n=0}^{\infty} c_n x^n = c_0 + c_1 x + c_2 x^2 + \cdots.$$

Differentiating twice we get

$$\begin{aligned} y''(x) &= \sum_{n=0}^{\infty} n(n-1)c_n x^{n-2} \\ &= 2c_2 + 6c_3 x + 12c_4 x^2 + \cdots + (m+2)(m+1)c_{m+2}x^m + \cdots, \\ -xy(x) &= -c_0 x - c_1 x^2 - c_2 x^3 - \cdots - c_{m-1}x^m - \cdots. \end{aligned}$$

Equating coefficients of x^m in these series gives

$$c_2 = 0, \qquad (m+2)(m+1)c_{m+2} = -c_{m-1}, \quad m \geq 1.$$

It follows from the initial conditions that $c_0 = 1$, $c_1 = 0$. Thus $c_n = 0$, if n is not a multiple of 3, and using the recursion we obtain

$$y(x) = 1 - \frac{x^3}{6} + \frac{x^6}{180} - \frac{x^9}{12\,960} + \cdots. \tag{1.2.8}$$

This gives $y(0.5) \approx 0.97925$. The x^9 term is ignored, since it is less than $2 \cdot 10^{-7}$. In this example also the first neglected term gives a rigorous bound for the error (i.e., for the remaining terms), since the absolute value of the term decreases, and the terms alternate in sign.

Since the calculation was based on a trial substitution, one should, strictly speaking, prove that the series obtained defines a function which satisfies the given problem. Clearly, the series converges at least for $|x| < 1$, since the coefficients are bounded. (In fact the series converges for all x.) Since a power series can be differentiated term by term in the interior of its interval of convergence, the proof presents no difficulty. Note, in addition, that the finite series obtained for $y(x)$ by breaking off after the x^9-term is the exact solution to the following *modified differential equation*:

$$y'' = -xy - \frac{x^{10}}{12\,960}, \qquad y(0) = 1, \qquad y'(0) = 0,$$

where the "perturbation term" $-x^{10}/12\,960$ has magnitude less than 10^{-7} for $|x| \leq 0.5$. It is possible to find rigorous bounds for the difference between the solutions of a differential system and a modified differential system.

The use of power series and rational approximations will be studied in depth in Chapter 3, where other more efficient methods than the Maclaurin series for approximation by polynomials will also be treated.

A different approximation problem, which occurs in many variants, is to approximate a function f specified at a one- or two-dimensional grid by a member f^* of a class of functions which is easy to work with mathematically. Examples are (piecewise) polynomials, rational functions, or trigonometric polynomials, where each particular function in the class is specified by the numerical values of a number of parameters.

In computer aided design (CAD) curves and surfaces have to be represented mathematically so that they can be manipulated and visualized easily. For this purpose **spline functions** are now used extensively with important applications in the aircraft and automotive industries; see Sec. 4.4. The name **spline** comes from a very old technique in drawing smooth curves, in which a thin strip of wood or rubber, called a draftsman's spline, is bent so that it passes through a given set of points. The points of interpolation are called **knots** and the spline is secured at the knots by means of lead weights called **ducks**. Before the computer age, splines were used in shipbuilding and other engineering designs.

Review Questions

1.2.1 What is a common cause of loss of accuracy in numerical calculations?

1.2.2 Describe Horner's rule and synthetic division.

1.2.3 Give a concise explanation why the algorithm in Example 1.2.1 did not work and why that in Example 1.2.2 did work.

1.2.4 Describe the basic idea behind the divide and conquer strategy. What is a main advantage of this strategy? How do you apply it to the task of summing n numbers?

Problems and Computer Exercises

1.2.1 (a) Use Horner's scheme to compute for $x = 2$

$$p(x) = x^4 + 2x^3 - 3x^2 + 2.$$

(b) Count the number of multiplications and additions required for the evaluation of a polynomial $p(z)$ of degree n by Horner's rule. Compare with the work needed when the powers are calculated recursively by $x^j = x \cdot x^{j-1}$ and subsequently multiplied by a_{n-j}.

1.2.2 $P(x) = 1 - \frac{1}{2}x^2 + \frac{1}{24}x^4$ is a polynomial approximation to $\cos x$ for small values of $|x|$. Estimate the errors of

$$P(x), \quad P'(x), \quad \frac{1}{x}\int_0^x P(t)\,dt,$$

and compare them for $x = 0.1$.

1.2.3 Show how repeated synthetic division can be used to move the origin of a polynomial; i.e., given $a_1, a_2, \dots, a_n$, and z, find $c_1, c_2, \dots, c_n$ so that

$$p_n(x) = \sum_{j=1}^{n} a_j x^{j-1} \equiv \sum_{j=1}^{n} c_j (x - z)^{j-1}.$$

Write a program for synthetic division (with this ordering of the coefficients) and apply it to this algorithm.

Hint: Apply synthetic division to $p_n(x)$, $p_{n-1}(x) = (p_n(x) - p_n(z))/(x - z)$, and so forth.

1.2.4 (a) Show that the transformation made in Problem 1.2.3 can also be expressed by means of the matrix-vector equation

$$c = \operatorname{diag}(1, z^{-1}, \dots, z^{1-n})\, P \operatorname{diag}(1, z, \dots, z^{n-1})\, a,$$

where $a = [a_1, a_2, \dots, a_n]^T$, $c = [c_1, c_2, \dots, c_n]^T$, and $\operatorname{diag}(1, z, \dots, z^{n-1})$ is a diagonal matrix with elements z^{j-1}, $j = 1 : n$. The matrix $P \in \mathbf{R}^{n\times n}$ has elements

$$p_{i,j} = \begin{cases} \binom{j-1}{i-1} & \text{if } j \ge i, \\ 0 & \text{otherwise.} \end{cases}$$

By convention, $\binom{0}{0} = 1$ here.

(b) Note the relation of P to the **Pascal triangle**, and show how P can be generated by a simple recursion formula. Also show how each element of P^{-1} can be expressed in terms of the corresponding element of P. How is the origin of the polynomial $p_n(x)$ moved, if you replace P by P^{-1} in the matrix-vector equation that defines c?

(c) If you reverse the order of the elements of the vectors a, c—this may sometimes be a more convenient ordering—how is the matrix P changed?

Comment: With terminology to be used much in this book (see Sec. 4.1.2), we can look upon a and c as different coordinate vectors for the same element in the n-dimensional linear space $\mathcal{P}_n$ of polynomials of degree *less than* n. The matrix P gives the coordinate transformation.

1.2.5 Derive recurrence relations and write a program for computing the coefficients of the *product* r of two polynomials p and q:

$$r(x) = p(x)q(x) = \left(\sum_{i=1}^{m} a_i x^{i-1}\right)\left(\sum_{j=1}^{n} b_j x^{j-1}\right) = \sum_{k=1}^{m+n-1} c_k x^{k-1}.$$

1.2.6 Let a, b be nonnegative integers, with $b \neq 0$. The division a/b yields the quotient q and the remainder r. Show that if a and b have a common factor, then that number is a divisor of r as well. Use this remark to derive the **Euclidean algorithm** for the determination of the greatest common divisor of a and b.

1.2.7 Derive a forward and a backward recurrence relation for calculating the integrals

$$I_n = \int_0^1 \frac{x^n}{4x+1}\,dx.$$

Why is the forward recurrence stable and the backward recurrence unstable in this case?

1.2.8 (a) Solve Example 1.2.1 on a computer, with the following changes: Start the recursion (1.2.5) with $I_0 = \ln 1.2$, and compute and print the sequence $\{I_n\}$ until I_n becomes negative for the first time.

(b) Start the recursion (1.2.6) first with the condition $I_{19} = I_{20}$, then with $I_{29} = I_{30}$. Compare the results you obtain and assess their approximate accuracy. Compare also with the results of part (a).

1.2.9 (a) Write a program (or study some library program) for finding the quotient $Q(x)$ and the remainder $R(x)$ of two polynomials $A(x)$, $B(x)$, i.e.,

$$A(x) = Q(x)B(x) + R(x), \qquad \deg R(x) < \deg B(x).$$

(b) Write a program (or study some library program) for finding the coefficients of a polynomial with given roots.

1.2.10 (a) Write a program (or study some library program) for finding the greatest common divisor of two *polynomials*. Test it on a number of polynomials of your own choice. Choose also some polynomials of a rather high degree, and do not choose only polynomials with small integer coefficients. Even if you have constructed the polynomials so that they should have a common divisor, rounding errors may disturb this, and some tolerance is needed in the decision whether a remainder is zero or not. One way of finding a suitable size of the tolerance is to make one or several runs where the coefficients are subject to some small random perturbations, and find out how much the results are changed.

(b) Apply the programs mentioned in the last two problems for finding and eliminating multiple zeros of a polynomial.

Hint: A multiple zero of a polynomial is a common zero of the polynomial and its derivative.

1.3 Matrix Computations

Matrix computations are ubiquitous in scientific computing. A survey of basic notation and concepts in matrix computations and linear vector spaces is given in Online Appendix A. This is needed for several topics treated in later chapters of this book. A fuller treatment of this topic will be given in Volume II.

In this section we focus on some important developments since the 1950s in the solution of linear systems. One is the systematic use of matrix notations and the interpretation of Gaussian elimination as matrix factorization. This **decompositional approach** has several advantages; e.g., a computed factorization can often be used with great savings to solve new problems involving the original matrix. Another is the rapid development of sophisticated iterative methods, which are becoming increasingly important as systems increase in size.

1.3.1 Matrix Multiplication

A **matrix**[8] A is a collection of $m \times n$ numbers ordered in m rows and n columns:

$$A = (a_{ij}) = \begin{pmatrix} a_{11} & a_{12} & \dots & a_{1n} \\ a_{21} & a_{22} & \dots & a_{2n} \\ \vdots & \vdots & \ddots & \vdots \\ a_{m1} & a_{m2} & \dots & a_{mn} \end{pmatrix}.$$

We write $A \in \mathbf{R}^{m \times n}$, where $\mathbf{R}^{m \times n}$ denotes the set of all real $m \times n$ matrices. If $m = n$, then the matrix A is said to be square and of order n. If $m \neq n$, then A is said to be rectangular.

The **product** of two matrices A and B is defined if and only if the number of columns in A equals the number of rows in B. If $A \in \mathbf{R}^{m \times p}$ and $B \in \mathbf{R}^{p \times n}$, then $C = AB \in \mathbf{R}^{m \times n}$, where

$$c_{ij} = \sum_{k=1}^{p} a_{ik} b_{kj}, \quad 1 \le i \le m, \quad 1 \le j \le n. \tag{1.3.1}$$

The product BA is defined only if $m = n$ and then $BA \in \mathbf{R}^{p \times p}$. Clearly, matrix multiplication is in general *not commutative*. In the exceptional case where $AB = BA$ holds, the matrices A and B are said to **commute**.

Matrix multiplication satisfies the associative and distributive rules:

$$A(BC) = (AB)C, \qquad A(B + C) = AB + AC.$$

However, the number of arithmetic operations required to compute the left- and right-hand sides of these equations can be very different!

Example 1.3.1.

Let the three matrices $A \in \mathbf{R}^{m \times p}$, $B \in \mathbf{R}^{p \times n}$, and $C \in \mathbf{R}^{n \times q}$ be given. Then computing the product ABC as $(AB)C$ requires $mn(p+q)$ multiplications, whereas $A(BC)$ requires $pq(m + n)$ multiplications.

[8]The first to use the term "matrix" was the English mathematician James Sylvester in 1850. Arthur Cayley then published *Memoir on the Theory of Matrices* in 1858, which spread the concept.

If A and B are square $n \times n$ matrices and $C = x \in \mathbf{R}^{n\times 1}$, a column vector of length n, then computing $(AB)x$ requires $n^2(n+1)$ multiplications, whereas $A(Bx)$ only requires $2n^2$ multiplications. When $n \gg 1$ this makes a great difference!

It is often useful to think of a matrix as being built up of blocks of lower dimensions. The great convenience of this lies in the fact that the operations of addition and multiplication can be performed by treating the blocks as *noncommuting scalars* and applying the definition (1.3.1). Of course the dimensions of the blocks must correspond in such a way that the operations can be performed.

Example 1.3.2.

Assume that the two $n \times n$ matrices are partitioned into 2×2 block form,

$$A = \begin{pmatrix} A_{11} & A_{12} \\ A_{21} & A_{22} \end{pmatrix}, \qquad B = \begin{pmatrix} B_{11} & B_{12} \\ B_{21} & B_{22} \end{pmatrix},$$

where A_{11} and B_{11} are square matrices of the same dimension. Then the product $C = AB$ equals

$$C = \begin{pmatrix} A_{11}B_{11} + A_{12}B_{21} & A_{11}B_{12} + A_{12}B_{22} \\ A_{21}B_{11} + A_{22}B_{21} & A_{21}B_{12} + A_{22}B_{22} \end{pmatrix}. \qquad (1.3.2)$$

Be careful to note that since matrix multiplication is not commutative the *order of the factors in the products cannot be changed.* In the special case of block upper triangular matrices this reduces to

$$\begin{pmatrix} R_{11} & R_{12} \\ 0 & R_{22} \end{pmatrix} \begin{pmatrix} S_{11} & S_{12} \\ 0 & S_{22} \end{pmatrix} = \begin{pmatrix} R_{11}S_{11} & R_{11}S_{12} + R_{12}S_{22} \\ 0 & R_{22}S_{22} \end{pmatrix}. \qquad (1.3.3)$$

Note that the product is again block upper triangular, and its block diagonal simply equals the products of the diagonal blocks of the factors.

It is important to know roughly how much work is required by different matrix algorithms. By inspection of (1.3.1) it is seen that computing the mp elements c_{ij} in the product $C = AB$ requires mnp additions and multiplications.

In matrix computations the number of multiplicative operations $(\times, /)$ is usually about the same as the number of additive operations $(+, -)$. Therefore, in older literature, a **flop** was defined to mean roughly the amount of work associated with the computation

$$s := s + a_{ik}b_{kj},$$

i.e., one addition *and* one multiplication (or division). In more recent textbooks (e.g., Golub and Van Loan [169])a flop is defined as one floating-point operation, doubling the older flop counts.[9] Hence, multiplication $C = AB$ of two square matrices of order n requires $2n^3$

[9]Stewart [335, p. 96] uses **flam** (floating-point addition and multiplication) to denote an "old" flop.

flops. The matrix-vector multiplication $y = Ax$, where $A \in \mathbf{R}^{n\times n}$ and $x \in \mathbf{R}^n$, requires $2mn$ flops.[10]

Operation counts are meant only as a rough appraisal of the work and one should not assign too much meaning to their precise value. On modern computer architectures the rate of transfer of data between different levels of memory often limits the actual performance. Also usually ignored is the fact that on many computers a division is five to ten times slower than a multiplication.

An operation count still provides useful information and can serve as an initial basis of comparison of different algorithms. It implies that the running time for multiplying two square matrices on a computer will increase roughly cubically with the dimension n. Thus, doubling n will approximately increase the work by a factor of eight; this is also apparent from (1.3.2).

A faster method for matrix multiplication would give more efficient algorithms for many linear algebra problems such as inverting matrices, solving linear systems, and solving eigenvalue problems. An intriguing question is whether it is possible to multiply two matrices $A, B \in \mathbf{R}^{n\times n}$ (or solve a linear system of order n) in less than n^3 (scalar) multiplications. The answer is yes! Strassen [341] developed a fast algorithm for matrix multiplication which, if used recursively to multiply two square matrices of dimension $n = 2^k$, reduces the number of multiplications from n^3 to $n^{\log_2 7} = n^{2.807\ldots}$. The key observation behind the algorithm is that the block matrix multiplication (1.3.2) can be performed with only *seven* block matrix multiplications and eighteen block matrix additions. Since for large dimensions matrix multiplication is much more expensive ($2n^3$ flops) than addition ($2n^2$ flops), this will lead to a savings in operations.

It is still an open (difficult!) question what the minimum exponent ω is such that matrix multiplication can be done in $O(n^\omega)$ operations. The fastest known algorithm, devised in 1987 by Coppersmith and Winograd [79], has $\omega < 2.376$. Many believe that an optimal algorithm can be found which reduces the number to essentially n^2. For a review of recent efforts in this direction using group theory, see Robinson [306]. (Note that for many of the theoretically "fast" methods large constants are hidden in the O notation.)

1.3.2 Solving Linear Systems by LU Factorization

The solution of **linear systems of equations** is the most frequently encountered task in scientific computing. One important source of linear systems is discrete approximations of continuous differential and integral equations.

A linear system can be written in matrix-vector form as

$$\begin{pmatrix} a_{11} & a_{12} & \cdots & a_{1n} \\ a_{21} & a_{22} & \cdots & a_{2n} \\ \vdots & \vdots & \ddots & \vdots \\ a_{m1} & a_{m2} & \cdots & a_{mn} \end{pmatrix} \begin{pmatrix} x_1 \\ x_2 \\ \vdots \\ x_n \end{pmatrix} = \begin{pmatrix} b_1 \\ b_2 \\ \vdots \\ b_m \end{pmatrix}, \tag{1.3.4}$$

where a_{ij} and b_i, $1 \le i \le m$, $1 \le j \le n$, are known input data and the task is to compute the unknowns x_j, $1 \le j \le n$. More compactly we write $Ax = b$, where $A \in \mathbf{R}^{m\times n}$ is a matrix and $x \in \mathbf{R}^n$ and $b \in \mathbf{R}^m$ are column vectors.

[10]To add to the confusion, in computer literature "flops" means floating-point operations per second.

Solving linear systems by **Gaussian elimination**[11] is taught in elementary courses in linear algebra. Although in theory this algorithm seems deceptively simple, the practical solution of large linear systems is far from trivial. In the 1940s, at the beginning of the computer age, there was a mood of pessimism among mathematicians about the possibility of accurately solving systems even of modest order, say $n = 100$. Today there is a deeper understanding of how Gaussian elimination performs in finite precision arithmetic. Linear systems with hundred of thousands of unknowns are now routinely solved in scientific computing.

Linear systems which (possibly after a permutation of rows and columns) are of triangular form are particularly simple to solve. Consider a square **upper triangular** linear system ($m = n$),

$$\begin{pmatrix} u_{11} & \dots & u_{1,n-1} & u_{1n} \\ & \ddots & \vdots & \vdots \\ & & u_{n-1,n-1} & u_{n-1,n} \\ & & & u_{nn} \end{pmatrix} \begin{pmatrix} x_1 \\ \vdots \\ x_{n-1} \\ x_n \end{pmatrix} = \begin{pmatrix} b_1 \\ \vdots \\ b_{n-1} \\ b_n \end{pmatrix}.$$

The matrix U is nonsingular if and only if

$$\det(U) = u_{11} \cdots u_{n-1,n-1} u_{nn} \neq 0.$$

If this is the case the unknowns can be computed by the following recursion:

$$x_n = b_n / u_{nn}, \qquad x_i = \left(b_i - \sum_{k=i+1}^{n} u_{ik} x_k \right) \Big/ u_{ii}, \quad i = n-1 : -1 : 1. \tag{1.3.5}$$

Since the unknowns are solved in reverse order this is called **back-substitution**. Thus the solution of a triangular system of order n can be computed in exactly n^2 flops; this is the same amount of work as required for *multiplying* a vector by a triangular matrix.

Similarly, a square linear system of **lower triangular** form $Lx = b$,

$$\begin{pmatrix} l_{11} & & & \\ l_{21} & l_{22} & & \\ \vdots & \vdots & \ddots & \\ l_{n1} & l_{n2} & \dots & l_{nn} \end{pmatrix} \begin{pmatrix} x_1 \\ x_2 \\ \vdots \\ x_n \end{pmatrix} = \begin{pmatrix} b_1 \\ b_2 \\ \vdots \\ b_n \end{pmatrix},$$

where L is nonsingular, can be solved by **forward-substitution:**

$$x_1 = b_1 / l_{11}, \qquad x_i = \left(b_i - \sum_{k=1}^{i-1} l_{ik} x_k \right) \Big/ l_{ii}, \quad i = 2 : n. \tag{1.3.6}$$

(Note that by reversing the order of the rows and columns an upper triangular system is transformed into a lower triangular and vice versa.)

[11]Named after the German mathematician Carl Friedrich Gauss (1777–1855), but known already in China as early as the first century B.C. Gauss was one of the greatest mathematician of the nineteenth century. He spent most of his life in Göttingen, where in his dissertation he gave the first proof of the fundamental theorem of algebra. He made fundamental contributions to number theory, differential geometry, celestial mechanics, and geodesy. He introduced the method of least squares and put it on a solid foundation.

When implementing a matrix algorithm on a computer, the *order of operations* in matrix algorithms may be important. One reason for this is the economizing of storage, since even matrices of moderate dimensions have a large number of elements. When the initial data are not needed for future use, computed quantities may overwrite data. To resolve such ambiguities in the description of matrix algorithms, it is important to be able to describe computations like those in (1.3.5) in a more precise form. For this purpose we will use an informal programming language, which is sufficiently precise for our purpose but allows the suppression of cumbersome details. We illustrate these concepts on the back-substitution algorithm given above. In the following back-substitution algorithm the solution x overwrites the data b.

ALGORITHM 1.1. *Back-Substitution.*

Given a nonsingular upper triangular matrix $U \in \mathbf{R}^{n\times n}$ and a vector $b \in \mathbf{R}^n$, the following algorithm computes $x \in \mathbf{R}^n$ such that $Ux = b$:

for $i = n : (-1) : 1$

$$s := \sum_{k=i+1}^{n} u_{ik} b_k;$$

$$b_i := (b_i - s)/u_{ii};$$

end

Here $x := y$ means that the value of y is evaluated and assigned to x. We use the convention that when the upper limit in a sum is smaller than the lower limit the sum is set to zero.

In the above algorithm the elements in U are accessed in a rowwise manner. In another possible sequencing of the operations the elements in U are accessed columnwise. This gives the following algorithm:

for $k = n : (-1) : 1$
  $b_k := b_k/u_{kk};$
  for $i = k - 1 : (-1) : 1$
    $b_i := b_i - u_{ik} b_k;$
  end
end

Such differences in the sequencing of the operations can influence the efficiency of the implementation depending on how the elements in the matrix U are stored.

Gaussian elimination uses the following elementary operations, which can be performed without changing the set of solutions:

- interchanging two equations,
- multiplying an equation by a nonzero scalar α,
- adding a multiple α of the ith equation to the jth equation, $j \neq i$.

These operations correspond in an obvious way to row operations carried out on the augmented matrix $(A,\ b)$. By performing a sequence of such elementary operations the system $Ax = b$ can be transformed into an upper triangular system which, as shown above, can be solved by recursive substitution.

In Gaussian elimination the unknowns are eliminated in a systematic way so that at the end an equivalent triangular system is produced, which can be solved by substitution. Consider the system (1.3.4) with $m = n$ and assume that $a_{11} \neq 0$. Then we can eliminate x_1 from the last $(n-1)$ equations by subtracting from the ith equation the multiple $l_{i1} = a_{i1}/a_{11}$, of the first equation. The last $(n-1)$ equations then become

$$\begin{pmatrix} a_{22}^{(2)} & \cdots & a_{2n}^{(2)} \\ \vdots & \ddots & \vdots \\ a_{n2}^{(2)} & \cdots & a_{nn}^{(2)} \end{pmatrix} \begin{pmatrix} x_2 \\ \vdots \\ x_n \end{pmatrix} = \begin{pmatrix} b_2^{(2)} \\ \vdots \\ b_n^{(2)} \end{pmatrix},$$

where the new elements are given by

$$a_{ij}^{(2)} = a_{ij} - \frac{a_{i1}a_{1j}}{a_{11}} = a_{ij} - l_{i1}a_{1j},$$

$$b_i^{(2)} = b_i - l_{i1}b_1, \quad i, j = 2 : n.$$

This is a system of $(n-1)$ equations in the $(n-1)$ unknowns $x_2, \dots, x_n$. All following steps are similar. In step k, $k = 1 : n-1$, if $a_{kk}^{(k)} \neq 0$, we eliminate x_k from the last $(n-k)$ equations giving a system containing only $x_{k+1}, \dots, x_n$. We take $l_{ik} = a_{ik}^{(k)}/a_{kk}^{(k)}$, and the elements of the new system are given by

$$a_{ij}^{(k+1)} = a_{ij}^{(k)} - \frac{a_{ik}^{(k)}a_{kj}^{(k)}}{a_{kk}^{(k)}} = a_{ij}^{(k)} - l_{ik}a_{kj}^{(k)}, \tag{1.3.7}$$

$$b_i^{(k+1)} = b_i^{(k)} - l_{ik}b_k^{(k)}, \quad i, j = k+1 : n. \tag{1.3.8}$$

The diagonal elements $a_{11}, a_{22}^{(2)}, \dots, a_{n,n}^{(n)}$, which appear in the denominator in (1.3.7) during the elimination are called **pivotal elements**. As long as these are nonzero, the elimination can be continued. After $(n-1)$ steps we get the single equation

$$a_{nn}^{(n)}x_n = b_n^{(n)}.$$

Collecting the first equation from each step we get

$$\begin{pmatrix} a_{11}^{(1)} & a_{12}^{(1)} & \cdots & a_{1n}^{(1)} \\ & a_{22}^{(2)} & \cdots & a_{2n}^{(2)} \\ & & \ddots & \vdots \\ & & & a_{nn}^{(n)} \end{pmatrix} \begin{pmatrix} x_1 \\ x_2 \\ \vdots \\ x_n \end{pmatrix} = \begin{pmatrix} b_1^{(1)} \\ b_2^{(2)} \\ \vdots \\ b_n^{(n)} \end{pmatrix}, \tag{1.3.9}$$

where we have introduced the notations $a_{ij}^{(1)} = a_{ij}$, $b_i^{(1)} = b_i$ for the coefficients in the original system. Thus (1.3.4) has been reduced to an equivalent nonsingular, upper triangular system (1.3.9), which can be solved by back-substitution.

We remark that no extra memory space is needed to store the multipliers. When $l_{ik} = a_{ik}^{(k)}/a_{kk}^{(k)}$ is computed the element $a_{ik}^{(k+1)}$ becomes equal to zero, so the multipliers can be stored in the lower triangular part of the matrix. Note also that if the multipliers l_{ik} are saved, then the operations on the vector b can be carried out at a later stage. *This observation is important in that it shows that when solving several linear systems with the same matrix A but different right-hand sides,*

$$AX = B, \quad X = (x_1, \dots, x_p), \quad B = (b_1, \dots, b_p),$$

the operations on A only have to be carried out once.

We now show another interpretation of Gaussian elimination. For notational convenience we assume that $m = n$ and that Gaussian elimination can be carried out without pivoting. Then Gaussian elimination can be interpreted as computing the factorization $A = LU$ of the matrix A into the product of a unit lower triangular matrix L and an upper triangular matrix U.

Depending on whether the element a_{ij} lies on, above, or below the principal diagonal we have

$$a_{ij}^{(n)} = \begin{cases} \dots = a_{ij}^{(i+1)} = a_{ij}^{(i)}, & i \le j; \\ \dots = a_{ij}^{(j+1)} = 0, & i > j. \end{cases}$$

Thus the elements a_{ij}, $1 \le i, j \le n$, are transformed according to

$$a_{ij}^{(k+1)} = a_{ij}^{(k)} - l_{ik}a_{kj}^{(k)}, \quad k = 1 : p, \quad p = \min(i-1, j). \tag{1.3.10}$$

If these equations are summed for $k = 1 : p$, we obtain

$$\sum_{k=1}^{p} \left(a_{ij}^{(k+1)} - a_{ij}^{(k)}\right) = a_{ij}^{(p+1)} - a_{ij} = -\sum_{k=1}^{p} l_{ik}a_{kj}^{(k)}.$$

This can also be written

$$a_{ij} = \begin{cases} a_{ij}^{(i)} + \sum_{k=1}^{i-1} l_{ik}a_{kj}^{(k)}, & i \le j, \\ 0 + \sum_{k=1}^{j} l_{ik}a_{kj}^{(k)}, & i > j, \end{cases}$$

or, if we define $l_{ii} = 1$, $i = 1 : n$,

$$a_{ij} = \sum_{k=1}^{r} l_{ik}u_{kj}, \quad u_{kj} = a_{kj}^{(k)}, \quad r = \min(i, j). \tag{1.3.11}$$

However, these equations are equivalent to the matrix equation

$$A = LU, \quad L = (l_{ik}), \quad U = (u_{kj}).$$

Here L and U are lower and upper triangular matrices, respectively. Hence Gaussian elimination computes a factorization of A into a product of a lower and an upper triangular matrix, the **LU factorization** of A. Note that since the unit diagonal elements in L need not be stored, it is possible to store the L and U factors in an array of the same dimensions as A.

ALGORITHM 1.2. *LU Factorization.*

Given a matrix $A = A^{(1)} \in \mathbf{R}^{n\times n}$ and a vector $b = b^{(1)} \in \mathbf{R}^n$, the following algorithm computes the elements of the reduced system of upper triangular form (1.3.9). It is assumed that $a_{kk}^{(k)} \neq 0$, $k = 1 : n$:

for $k = 1 : n-1$
  for $i = k+1 : n$
    $l_{ik} := a_{ik}^{(k)}/a_{kk}^{(k)}$; $a_{ik}^{(k+1)} := 0$;
    for $j = k+1 : n$
      $a_{ij}^{(k+1)} := a_{ij}^{(k)} - l_{ik}a_{kj}^{(k)}$;
    end
  end
end

Although the LU factorization is just a different interpretation of Gaussian elimination it turns out to have important conceptual advantages. It divides the solution of a linear system into two independent steps:

1. the factorization $A = LU$,
2. solution of the systems $Ly = b$ and $Ux = y$.

The LU factorization is a prime example of *the decompositional approach to matrix computation*. This approach came into favor in the 1950s and early 1960s and has been named as one of the ten algorithms having the most influence on science and engineering in the twentieth century. This interpretation of Gaussian elimination has turned out to be very fruitful. For example, it immediately follows that the inverse of A (if it exists) has the factorization

$$A^{-1} = (LU)^{-1} = U^{-1}L^{-1}.$$

This shows that the solution of linear system $Ax = b$,

$$x = A^{-1}b = U^{-1}(L^{-1}b),$$

can be computed by solving the two triangular systems $Ly = b$, $Ux = y$. Indeed it has been said (see Forsythe and Moler [124]) that "Almost anything you can do with A^{-1} can be done without it."

Another example is the problem of solving the transposed system $A^Tx = b$. Since

$$A^T = (LU)^T = U^TL^T,$$

we have that $A^T x = U^T(L^T x) = b$. It follows that x can be computed by solving the two triangular systems

$$U^T c = b, \qquad L^T x = c. \tag{1.3.12}$$

In passing we remark that Gaussian elimination is also an efficient algorithm for computing the **determinant** of a matrix A. It can be shown that the value of the determinant is unchanged if a row (column) multiplied by a scalar is added to another row (column) (see (A.2.4) in the Online Appendix). Further, if two rows (columns) are interchanged, the value of the determinant is multiplied by (-1). Since the determinant of a triangular matrix equals the product of the diagonal elements it follows that

$$\det(A) = \det(L)\det(U) = \det(U) = (-1)^q a_{11}^{(1)} a_{22}^{(2)} \cdots a_{nn}^{(n)}, \tag{1.3.13}$$

where q is the number of row interchanges performed.

From Algorithm 1.2 it follows that $(n-k)$ divisions and $(n-k)^2$ multiplications and additions are used in step k to transform the elements of A. A further $(n-k)$ multiplications and additions are used to transform the elements of b. Summing over k and neglecting low order terms, we find that the number of flops required for the reduction of $Ax = b$ to a triangular system by Gaussian elimination is

$$\sum_{k=1}^{n-1} 2(n-k)^2 \approx 2n^3/3, \qquad \sum_{k=1}^{n-1} 2(n-k) \approx n^2$$

for the transformation of A and the right-hand side b, respectively. Comparing this with the n^2 flops needed to solve a triangular system we conclude that, except for very small values of n, *the LU factorization of A dominates the work in solving a linear system.* If several linear systems with the same matrix A but different right-hand sides are to be solved, then the factorization needs to be performed only once.

Example 1.3.3.

Linear systems where the matrix A has only a few nonzero diagonals often arise. Such matrices are called **band matrices**. In particular, band matrices of the form

$$A = \begin{pmatrix} b_1 & c_1 & & & \\ a_1 & b_2 & c_2 & & \\ & \ddots & \ddots & \ddots & \\ & & a_{n-2} & b_{n-1} & c_{n-1} \\ & & & a_{n-1} & b_n \end{pmatrix} \tag{1.3.14}$$

are called **tridiagonal**. Tridiagonal systems of linear equations can be solved by Gaussian elimination with much less work than the general case. The following algorithm solves the tridiagonal system $Ax = g$ by Gaussian elimination without pivoting.

First compute the LU factorization $A = LU$, where

$$L = \begin{pmatrix} 1 & & & & \\ \gamma_1 & 1 & & & \\ & \gamma_2 & 1 & & \\ & & \ddots & \ddots & \\ & & & \gamma_{n-1} & 1 \end{pmatrix}, \qquad U = \begin{pmatrix} \beta_1 & c_1 & & & \\ & \beta_2 & c_2 & & \\ & & \ddots & \ddots & \\ & & & \beta_{n-1} & c_{n-1} \\ & & & & \beta_n \end{pmatrix}.$$

The new elements in L and U are obtained from the recursion: Set $\beta_1 = b_1$, and

$$\gamma_k = a_k/\beta_k, \qquad \beta_{k+1} = b_{k+1} - \gamma_k c_k, \qquad k = 1 : n-1. \tag{1.3.15}$$

(Check this by computing the product LU.) The solution to $Ax = L(Ux) = g$ is then obtained in two steps. First a forward-substitution to get $y = Ux$,

$$y_1 = g_1, \qquad y_{k+1} = g_{k+1} - \gamma_k y_k, \qquad k = 1 : n-1, \tag{1.3.16}$$

followed by a backward recursion for x,

$$x_n = y_n/\beta_n, \qquad x_k = (y_k - c_k x_{k+1})/\beta_k, \qquad k = n-1 : -1 : 1. \tag{1.3.17}$$

In this algorithm the LU factorization requires only about n divisions and n multiplications and additions. The solution of the lower and upper bidiagonal systems require about twice as much work.

Stability of Gaussian Elimination

If A is nonsingular, then Gaussian elimination can always be carried through provided row interchanges are allowed. In this more general case, Gaussian elimination computes an LU factorization of the matrix $\tilde{A}$ obtained by carrying out all row interchanges on A. In practice row interchanges are needed to ensure the numerical stability of Gaussian elimination. We now consider how the LU factorization has to be modified when such interchanges are incorporated.

Consider the case when in step k of Gaussian elimination a zero pivotal element is encountered, i.e., $a_{kk}^{(k)} = 0$. (The equations may have been reordered in previous steps, but we assume that the notations have been changed accordingly.) If A is nonsingular, then in particular its first k columns are linearly independent. This must also be true for the first k columns of the reduced matrix, and hence some element $a_{ik}^{(k)}$, $i = k : n$, must be nonzero, say $a_{rk}^{(k)} \neq 0$. *By interchanging rows k and r this element can be taken as pivot and it is possible to proceed with the elimination.* The important conclusion is that *any nonsingular system of equations can be reduced to triangular form by Gaussian elimination, if appropriate row interchanges are used.*

Note that when rows are interchanged in A the same interchanges must be made in the elements of the right-hand side b. Also, the computed factors L and U will be the same as if the row interchanges are first performed on A and the Gaussian elimination performed without interchanges. Row interchanges can be expressed as premultiplication with certain matrices, which we now introduce.

A **permutation matrix** $P \in \mathbf{R}^{n\times n}$ is a matrix whose columns are a permutation of the columns of the unit matrix, that is,

$$P = (e_{p_1}, \dots, e_{p_n}),$$

where $(p_1, \dots, p_n)$ is a permutation of $(1, \dots, n)$. Notice that in a permutation matrix every row and every column contains just one unity element. The transpose P^T of a permutation matrix is therefore again a permutation matrix. A permutation matrix is orthogonal $P^T P = I$, and hence P^T affects the reverse permutation.

A **transposition matrix** is a special permutation matrix,

$$I_{ij} = (\dots, e_{i-1}, e_j, e_{i+1}, \dots, e_{j-1}, e_i, e_{j+1}),$$

which equals the identity matrix except with columns i and j interchanged. By construction it immediately follows that I_{ij} is symmetric. $I_{ij}^2 = I$ and hence $I_{ij}^{-1} = I_{ij}$. If a matrix A is premultiplied by I_{ij}, this results in the interchange of *rows* i and j. Any permutation matrix can be expressed as a product of transposition matrices.

If P is a permutation matrix, then PA is the matrix A with its rows permuted. Hence, Gaussian elimination with row interchanges produces a factorization, which in matrix notations can be written

$$PA = LU,$$

where P is a permutation matrix. Note that P is uniquely represented by the integer vector $(p_1, \dots, p_n)$ and need never be stored as a matrix.

Assume that in the kth step, $k = 1 : n - 1$, we select the pivot element from row p_k, and interchange the rows k and p_k. Notice that in these row interchanges also, previously computed multipliers l_{ij} must take part. At completion of the elimination, we have obtained lower and upper triangular matrices L and U. We now make the important observation that these are the same triangular factors that are obtained if we *first* carry out the row interchanges $k \leftrightarrow p_k$, $k = 1 : n - 1$, on the *original matrix* A to get a matrix PA, where P is a permutation matrix, and then perform Gaussian elimination on PA *without any interchanges*. This means that Gaussian elimination with row interchanges computes the LU factors of the matrix PA. We now summarize the results and prove the uniqueness of the LU factorization.

To ensure the numerical stability in Gaussian elimination, except for special classes of linear systems, it will be necessary to perform row interchanges *not only when a pivotal element is exactly zero*. Usually it suffices to choose the pivotal element in step k as the element of largest magnitude in the unreduced part of the kth column. This is called **partial pivoting**.

Example 1.3.4.

The linear system

$$\begin{pmatrix} \epsilon & 1 \\ 1 & 1 \end{pmatrix} \begin{pmatrix} x_1 \\ x_2 \end{pmatrix} = \begin{pmatrix} 1 \\ 0 \end{pmatrix}$$

is nonsingular for any $\epsilon \neq 1$ and has the unique solution $x_1 = -x_2 = -1/(1 - \epsilon)$. But when $\epsilon = 0$ the first step in Gaussian elimination cannot be carried out. The remedy here is obviously to interchange the two equations, which directly gives an upper triangular system.

Suppose that in the system above $\epsilon = 10^{-4}$. Then the exact solution, rounded to four decimals, equals $x = (-1.0001, 1.0001)^T$. But if Gaussian elimination is carried through without interchanges, we obtain $l_{21} = 10^4$ and the triangular system

$$\begin{aligned} 0.0001x_1 + x_2 &= 1, \\ (1 - 10^4)x_2 &= -10^4. \end{aligned}$$

Suppose that the computation is performed using arithmetic with three decimal digits. Then in the last equation the coefficient $a_{22}^{(2)}$ will be rounded to -10^4 and the solution computed by back-substitution is $\bar{x}_2 = 1.000$, $\bar{x}_1 = 0$, which is a catastrophic result.

If before performing Gaussian elimination we interchange the two equations, then we get $l_{21} = 10^{-4}$ and the reduced system becomes

$$x_1 + x_2 = 0,$$
$$(1 - 10^{-4})x_2 = 1.$$

The coefficient $a_{22}^{(2)}$ is now rounded to 1, and the computed solution becomes $\bar{x}_2 = 1.000$, $\bar{x}_1 = -1.000$, which is correct to the precision carried.

In this simple example it is easy to see what went wrong in the elimination without interchanges. The problem is that *the choice of a small pivotal element gives rise to large elements in the reduced matrix* and the coefficient a_{22} in the original system is lost through rounding. Rounding errors which are small when compared to the large elements in the reduced matrix are unacceptable in terms of the original elements. When the equations are interchanged the multiplier is small and the elements of the reduced matrix are of the same size as in the original matrix.

In general an algorithm is said to be **backward stable** (see Definition 2.4.10) if the computed solution $\overline{w}$ equals *the exact solution* of a problem with "slightly perturbed data." The more or less final form of error analysis of Gaussian elimination was given by J. H. Wilkinson [375].[12]

Wilkinson showed that the computed triangular factors L and U of A obtained by Gaussian elimination *are the exact triangular factors of a slightly perturbed matrix* $A + E$. He further gave bounds on the elements of E. The essential condition for stability is that *no substantial growth occurs in the elements in* L *and* U; see Theorem 2.4.12. Although matrices can be constructed for which the element growth factor in Gaussian elimination with partial pivoting equals 2^{n-1}, quoting Kahan [219] we say that: "Intolerable pivot-growth (with partial pivoting) is a phenomenon that happens only to numerical analysts who are looking for that phenomenon." Why large element growth rarely occurs in practice with partial pivoting is a subtle and still not fully understood phenomenon. Trefethen and Schreiber [360] show that for certain distributions of random matrices the average element growth is close to $n^{2/3}$ for partial pivoting.

It is important to note that the fact that a problem has been solved by a backward stable algorithm *does not mean that the error in the computed solution is small.* If the matrix A is close to a singular matrix, then the solution is very sensitive to perturbations in the data. This is the case when the rows (columns) of A are almost linearly dependent. But this inaccuracy is intrinsic to the problem and cannot be avoided except by using higher precision in the calculations. Condition numbers for linear systems are discussed in Sec. 2.4.2.

An important special case of LU factorization is when the matrix A is symmetric, $A^T = A$, and **positive definite**, i.e.,

$$x^T Ax > 0 \quad \forall x \in \mathbf{R}^n, \qquad x \neq 0. \tag{1.3.18}$$

Similarly A is said to be **positive semidefinite** if $x^T Ax \geq 0$ for all $x \in \mathbf{R}^n$. Otherwise it is called **indefinite**.

[12]James Hardy Wilkinson (1919–1986), English mathematician who graduated from Trinity College, Cambridge. He became Alan Turing's assistant at the National Physical Laboratory in London in 1946, where he worked on the ACE computer project. He did pioneering work on numerical methods for solving linear systems and eigenvalue problems, and developed software and libraries of numerical routines.

For symmetric positive definite matrices there always exists a unique factorization

$$A = R^T R, \tag{1.3.19}$$

where R is an upper triangular matrix with positive diagonal elements. An important fact is that *no pivoting is needed for stability. Indeed, unless the pivots are chosen from the diagonal, pivoting is harmful since it will destroy symmetry.*

This is called the **Cholesky factorization.**[13] The elements in the Cholesky factor $R = (r_{ij})$ can be determined directly. The matrix equation $A = R^T R$ with R upper triangular can be written elementwise as

$$a_{ij} = \sum_{k=1}^{i} r_{ki} r_{kj} = \sum_{k=1}^{i-1} r_{ki} r_{kj} + r_{ii} r_{ij}, \quad 1 \le i \le j \le n. \tag{1.3.20}$$

These are $n(n+1)/2$ equations for the unknown elements in R. Solving for r_{ij} from the corresponding equation in (1.3.20), we obtain

$$r_{ij} = \left(a_{ij} - \sum_{k=1}^{i-1} r_{ki} r_{kj} \right) \Big/ r_{ii}, \quad i < j, \qquad r_{jj} = \left(a_{jj} - \sum_{k=1}^{j-1} r_{kj}^2 \right)^{1/2}. \tag{1.3.21}$$

If properly sequenced, these equations can be used in a recursive fashion to compute the elements in R, for example, one row at a time. The resulting algorithm requires n square roots and approximately $n^3/3$ flops, which is about half the work of an LU factorization.

We remark that the Cholesky factorization can be carried out also for a symmetric indefinite matrix, if at each step a positive pivot is chosen from the diagonal. However, for a symmetric *indefinite* matrix, such as the matrix in Example 1.3.4 with $\epsilon < 1$, no Cholesky factorization can exist.

1.3.3 Sparse Matrices and Iterative Methods

Following Wilkinson and Reinsch [381], a matrix A will be called **sparse** if the percentage of zero elements is large and its distribution is such that it is economical to take advantage of their presence. The nonzero elements of a sparse matrix may be concentrated on a narrow band centered on the diagonal. Alternatively, they may be distributed in a less systematic manner.

Large sparse linear systems arise in numerous areas of application, such as the numerical solution of partial differential equations, mathematical programming, structural analysis, chemical engineering, electrical circuits, and networks. Large could imply a value of n in the range 1000–1,000,000. Figure 1.3.1 shows a sparse matrix of order $n = 479$ with 1887 nonzero elements (or 0.9%) that arises from a model of an eight stage chemical distillation column.

The first task in solving a sparse system by Gaussian elimination is to permute the rows and columns so that not too many new nonzero elements are created during the elimination.

[13] André-Louis Cholesky (1875–1918), a French military officer involved in geodesy and surveying in Crete and North Africa just before World War One.

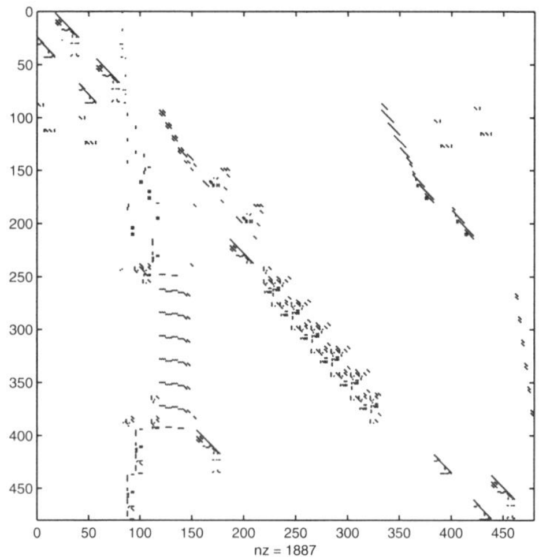

Figure 1.3.1. *Nonzero pattern of a sparse matrix from an eight stage chemical distillation column.*

Equivalently, we want to choose permutation matrices P and Q such that the LU factors of PAQ are as sparse as possible. Such a reordering will usually nearly minimize the number of arithmetic operations.

To find an *optimal* ordering which minimizes the number of nonzero elements in L and U is unfortunately an intractable problem, because the number of possible orderings of rows and columns are $(n!)^2$. Fortunately, there are heuristic ordering algorithms which do a good job. In Figure 1.3.2 we show the reordered matrix PAQ and its LU factors. Here L and U contain together 5904 nonzero elements or about 2.6%. The column ordering was obtained using a MATLAB version of the so-called column minimum degree ordering.

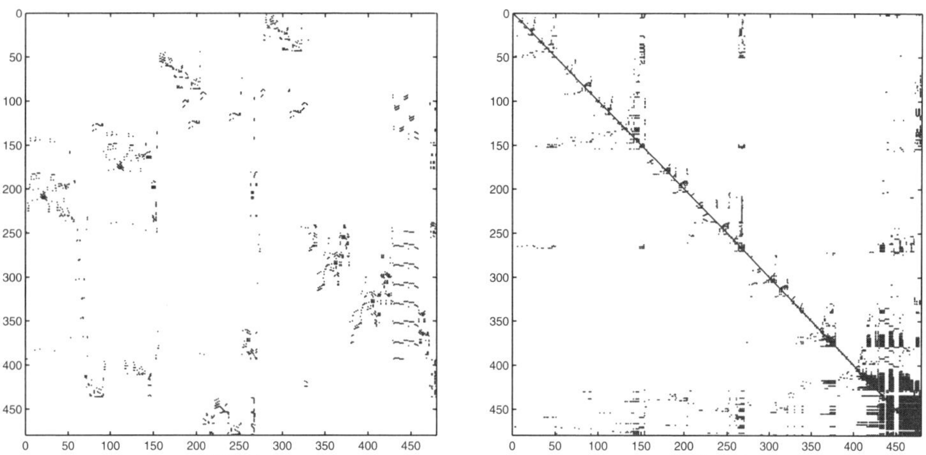

Figure 1.3.2. *Nonzero structure of the matrix A (left) and L + U (right).*

For the origin and details of this code we refer to Gilbert, Moler, and Schreiber [156]. We remark that, in general, some kind of stability check on the pivot elements must be performed during the factorization.

For many classes of sparse linear systems **iterative methods** are more efficient to use than **direct methods** such as Gaussian elimination. Typical examples are those arising when a differential equation in two or three dimensions is discretized. In iterative methods a sequence of approximate solutions is computed, which in the limit converges to the exact solution x. Basic iterative methods work *directly with the original matrix* A and therefore have the added advantage of requiring only extra storage for a few vectors.

In a classical iterative method due to Richardson [302], starting from $x^{(0)} = 0$, a sequence $x^{(k)}$ is defined by

$$x^{(k+1)} = x^{(k)} + \omega(b - Ax^{(k)}), \quad k = 0, 1, 2, \ldots, \tag{1.3.22}$$

where $\omega > 0$ is a parameter to be chosen. It follows easily from (1.3.22) that the error in $x^{(k)}$ satisfies $x^{(k+1)} - x = (I - \omega A)(x^{(k)} - x)$, and hence

$$x^{(k)} - x = (I - \omega A)^k (x^{(0)} - x).$$

It can be shown that, if all the eigenvalues λ_i of A are real and satisfy

$$0 < a \leq \lambda_i \leq b,$$

then $x^{(k)}$ will converge to the solution, when $k \to \infty$, for $0 < \omega < 2/b$.

Iterative methods are used most often for the solution of very large linear systems, which typically arise in the solution of boundary value problems of partial differential equations by finite difference or finite element methods. The matrices involved can be huge, sometimes involving several million unknowns. The LU factors of matrices arising in such applications typically contain orders of magnitude more nonzero elements than A itself. Hence, because of the storage and number of arithmetic operations required, Gaussian elimination may be far too costly to use. In a typical problem for the Poisson equation (1.1.21) the function is to be determined in a plane domain D, when the values of u are given on the boundary ∂D. Such **boundary value problems** occur in the study of steady states in most branches of physics, such as electricity, elasticity, heat flow, and fluid mechanics (including meteorology). Let D be a square grid with grid size h, i.e., $x_i = x_0 + ih$, $y_j = y_0 + jh$, $0 \leq i \leq N+1$, $0 \leq j \leq N+1$. Then the difference approximation yields

$$u_{i,j+1} + u_{i-1,j} + u_{i+1,j} + u_{i,j-1} - 4u_{i,j} = h^2 f(x_i, y_j)$$

$(1 \leq i, j \leq N)$. This is a huge system of linear algebraic equations; one equation for each interior gridpoint, altogether N^2 unknowns and equations. (Note that $u_{i,0}$, $u_{i,N+1}$, $u_{0,j}$, $u_{N+1,j}$ are known boundary values.) To write the equations in matrix-vector form we order the unknowns in a vector,

$$u = (u_{1,1}, \ldots, u_{1,N}, u_{2,1}, \ldots, u_{2,N}, \ldots, u_{N,1}, \ldots, u_{N,N}),$$

the so-called natural ordering. If the equations are ordered in the same order we get a system $Au = b$, where A is symmetric with all nonzero elements located in five diagonals; see Figure 1.3.3 (left).

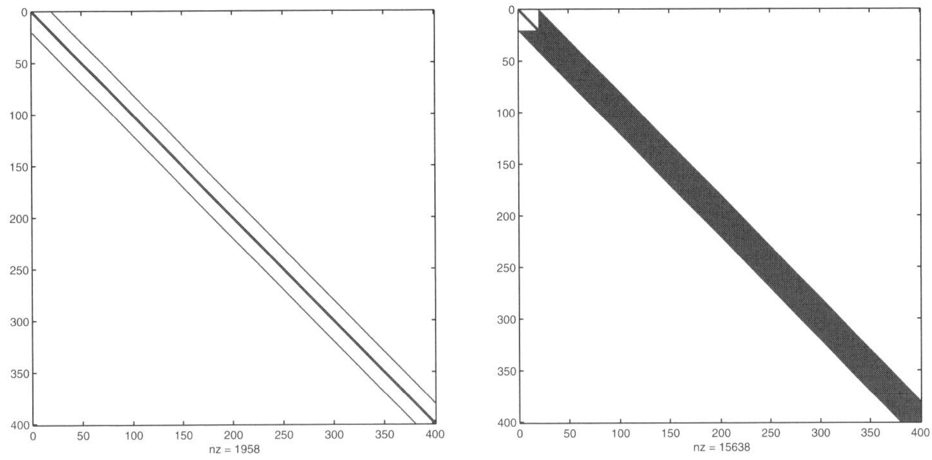

Figure 1.3.3. *Structure of the matrix A (left) and L + U (right) for the Poisson problem, N = 20 (rowwise ordering of the unknowns).*

In principle Gaussian elimination can be used to solve such systems. But even taking symmetry and the banded structure into account, this would require $\frac{1}{2} \cdot N^4$ multiplications, since in the LU factors the zero elements inside the outer diagonals will fill in during the elimination, as shown in Figure 1.3.3 (right).

The linear system arising from the Poisson equation has several features common to boundary value problems for other linear partial differential equations. One of these is that only a tiny fraction of the elements in each row of A are nonzero. Therefore, each iteration in Richardson's method requires only about kN^2 multiplications, i.e., k multiplications per unknown. Using iterative methods which take advantage of the sparsity and other features does allow the efficient solution of such systems. This becomes even more essential for three-dimensional problems.

As early as 1954, a simple atmospheric model was used for weather forecasting on an electronic computer. The net covered most of North America and Europe. During a 48 hour forecast, the computer solved (among other things) 48 Poisson equations (with different right-hand sides). This would have been impossible at that time if the special features of the system had not been used.

1.3.4 Software for Matrix Computations

In most computers in use today the key to high efficiency is to avoid as much as possible data transfers between memory, registers, and functional units, since these can be more costly than arithmetic operations on the data. This means that the operations have to be carefully structured. One observation is that Gaussian elimination consists of three nested loops, which can be ordered in $3 \cdot 2 \cdot 1 = 6$ ways. Disregarding the right-hand side vector b, each version does the operations

$$a_{ij}^{(k+1)} := a_{ij}^{(k)} - a_{kj}^{(k)} a_{ik}^{(k)} / a_{kk}^{(k)},$$

and only the ordering in which they are done differs. The version given above uses row operations and may be called the "kij" variant, where k refers to step number, i to row index, and j to column index. This version is not suitable for programming languages like Fortran 77, in which matrix elements are stored sequentially by columns. In such a language the form kji should be preferred, as well as a column oriented back-substitution rather than that in Algorithm 1.1.

The first collection of high quality linear algebra software was a series of algorithms written in Algol 60 that appeared in Wilkinson and Reinsch [381]. This contains 11 subroutines for linear systems, least squares, and linear programming, and 18 routines for the algebraic eigenvalue problem.

The Basic Linear Algebra Subprograms (BLAS) have become an important tool for structuring linear algebra computations. These are now commonly used to formulate matrix algorithms and have become an aid to clarity, portability, and modularity in modern software. The original set of BLAS [239], introduced in 1979, identified frequently occurring vector operations in matrix computation such as scalar product, adding of a multiple of one vector to another, etc. For example, the operation

$$y := \alpha x + y$$

in single precision is named SAXPY. By carefully optimizing them for each specific computer, performance was enhanced without sacrificing portability. These BLAS were adopted in the collections of Fortran subroutines LINPACK (see [101]) for linear systems and EISPACK (see [133]) for eigenvalue problems.

For modern computers it is important to avoid excessive data movements between different parts of memory hierarchy. To achieve this, so-called level 3 BLAS were introduced in the 1990s. These work on blocks of the full matrix and perform, for example, the operations

$$C := \alpha AB + \beta C, \qquad C := \alpha A^T B + \beta C, \qquad C := \alpha AB^T + \beta C.$$

Level 3 BLAS use $O(n^2)$ data but perform $O(n^3)$ arithmetic operations. This gives a surface-to-volume effect for the ratio of data movement to operations.

LAPACK (see [6]) is a linear algebra package initially released in 1992. LAPACK was designed to supersede and integrate the algorithms in both LINPACK and EISPACK. It achieves close to optimal performance on a large variety of computer architectures by expressing as much of the algorithm as possible as calls to level 3 BLAS. This is also an aid to clarity, portability, and modularity. LAPACK today is the backbone of the interactive matrix computing system MATLAB.

Example 1.3.5.

In 1974 the authors wrote in [89, Sec. 8.5.3] that "a full 1000×1000 system of equations is near the limit at what can be solved at a reasonable cost." Today systems of this size can easily be handled on a personal computer. The benchmark problem for the Japanese Earth Simulator, one of the world's fastest computers in 2004, was the solution of a system of size 1,041,216 on which a speed of 35.6×10^{12} operations per second was measured. This is a striking illustration of the progress in high speed matrix computing that has occurred in these 30 years!

Review Questions

1.3.1 How many operations are needed (approximately) for

(a) the multiplication of two square matrices $A, B \in \mathbf{R}^{n\times n}$?

(b) the LU factorization of a matrix $A \in \mathbf{R}^{n\times n}$?

(c) the solution of $Ax = b$, when the triangular factorization of A is known?

1.3.2 Show that if the kth diagonal entry of an upper triangular matrix is zero, then its first k columns are linearly dependent.

1.3.3 What is the LU factorization of an n by n matrix A, and how is it related to Gaussian elimination? Does it always exist? If not, give sufficient conditions for its existence.

1.3.4 (a) For what type of linear systems are iterative methods to be preferred to Gaussian elimination?

(b) Describe Richardson's method for solving $Ax = b$. What can you say about the error in successive iterations?

1.3.5 What does the acronym BLAS stand for? What is meant by level 3 BLAS, and why are they used in current linear algebra software?

Problems and Computer Exercises

1.3.1 Let A be a square matrix of order n and k a positive integer such that $2^p \le k < 2^{p+1}$. Show how A^k can be computed in at most $2pn^3$ multiplications.

Hint: Write k in the binary number system and compute $A^2, A^4, A^8, \dots$, by successive squaring; e.g., $13 = (1101)_2$ and $A^{13} = A^8A^4A$.

1.3.2 (a) Let A and B be square upper triangular matrices of order n. Show that the product matrix $C = AB$ is also upper triangular. Determine how many multiplications are needed to compute C.

(b) Show that if R is an upper triangular matrix with zero diagonal elements, then $R^n = 0$.

1.3.3 Show that there cannot exist an LU factorization

$$A = \begin{pmatrix} 0 & 1 \\ 1 & 1 \end{pmatrix} = \begin{pmatrix} l_{11} & 0 \\ l_{21} & l_{22} \end{pmatrix} \begin{pmatrix} u_{11} & u_{12} \\ 0 & u_{22} \end{pmatrix}.$$

Hint: Equate the $(1, 1)$-elements and deduce that either the first row or the first column in LU must be zero.

1.3.4 (a) Consider the special upper triangular matrix of order n,

$$U_n(a) = \begin{pmatrix} 1 & a & a & \cdots & a \\ & 1 & a & \cdots & a \\ & & 1 & \cdots & a \\ & & & \ddots & \vdots \\ & & & & 1 \end{pmatrix}.$$

Determine the solution x to the triangular system $U_n(a)x = e_n$, where $e_n = (0, 0, \dots, 0, 1)^T$ is the nth unit vector.

(b) Show that the inverse of an upper triangular matrix is also upper triangular. Determine for $n = 3$ the inverse of $U_n(a)$. Try also to determine $U_n(a)^{-1}$ for an arbitrary n. *Hint:* Note that $UU^{-1} = U^{-1}U = I$, the identity matrix.

1.3.5 A matrix H_n of order n such that $h_{ij} = 0$ whenever $i > j + 1$ is called an upper **Hessenberg** matrix. For $n = 5$ it has the structure

$$H_5 = \begin{pmatrix} h_{11} & h_{12} & h_{13} & h_{14} & h_{15} \\ h_{21} & h_{22} & h_{23} & h_{24} & h_{25} \\ 0 & h_{32} & h_{33} & h_{34} & h_{35} \\ 0 & 0 & h_{43} & h_{44} & h_{45} \\ 0 & 0 & 0 & h_{54} & h_{55} \end{pmatrix}.$$

(a) Determine the approximate number of operations needed to compute the LU factorization of H_n without pivoting.

(b) Determine the approximate number of operations needed to solve the linear system $H_n x = b$, when the factorization in (a) is given.

1.3.6 Compute the product $|L|\,|U|$ for the LU factors with and without pivoting of the matrix in Example 1.3.4. (Here $|A|$ denotes the matrix with elements $|a_{ij}|$.)

1.3.7 Let $A \in \mathbf{R}^{n\times n}$ be a given matrix. Show that if $Ax = y$ has *at least one* solution for any $y \in \mathbf{R}^n$, then it has *exactly one* solution for any $y \in \mathbf{R}^n$. (This is a useful formulation for showing uniqueness of approximation formulas.)

1.4 The Linear Least Squares Problem

A basic problem in science is to fit a mathematical model to given observations subject to errors. As an example, consider observations (t_i, y_i), $i = 1 : m$, to be fitted to a model described by a scalar function $y(t) = f(c, t)$, where $c \in \mathbf{R}^n$ is a parameter vector to be determined. There are two types of shortcomings to take into account: errors in the input data, and approximations made in the particular model (class of functions, form). We shall call these **measurement errors** and **errors in the model**, respectively.

Clearly the more observations that are available the more accurately will it be possible to determine the parameters in the model. One can also see this problem as analogous to the task of a communication engineer, to filter away *noise* from the *signal*. These questions are connected with both mathematical statistics and the mathematical discipline of approximation theory.

A simple example is when the model is *linear* in c and of the form

$$y(t) = \sum_{j=1}^{n} c_j \phi_j(t),$$

where $\phi_j(t)$ are given (possibly nonlinear) functions. Given $m > n$ measurements the resulting equations

$$y_i = \sum_{j=1}^{n} c_j \phi_j(t_i), \qquad i = 1 : m,$$

form an **overdetermined** linear system. If we set $A \in \mathbf{R}^{m\times n}$, $a_{ij} = \phi_j(t_i)$, then the system can be written in matrix form as $Ac = y$. In general such a system is inconsistent and has no solution. But we can try to find a vector $c \in \mathbf{R}^n$ such that Ac is the "best" approximation to y. This is equivalent to minimizing the size of the **residual vector** $r = y - Ac$.

There are many possible ways of defining the best solution to such an inconsistent linear system. A choice which can often be motivated for statistical reasons, and which also leads to a simple computational problem, is to take as the solution a vector c, which minimizes the sum of the squared residuals, that is,

$$\min_x \sum_{i=1}^m r_i^2 = \min_x \|y - Ac\|_2^2. \tag{1.4.1}$$

Here we have used the notation

$$\|x\|_2 = (|x_1|^2 + \cdots + |x_n|^2)^{1/2} = (x^T x)^{1/2}$$

for the Euclidean length of a vector x (see Online Appendix A). We call (1.4.1) a **linear least squares problem** and any minimizer x a **least squares solution** of the system $Ax = b$.

Gauss claims to have discovered the method of least squares in 1795 when he was 18 years old. He used it in 1801 to successively predict the orbit of the asteroid Ceres.

1.4.1 Basic Concepts in Probability and Statistics

We now introduce, without proofs, some basic concepts, formulas, and results from statistics which will be used later. Proofs may be found in most texts on these subjects.

The **distribution function** of a random variable X is denoted by the nonnegative and nondecreasing function

$$F(x) = Pr\{X \le x\}, \quad F(-\infty) = 0, \quad F(\infty) = 1.$$

If $F(x)$ is differentiable, the (probability) **density function**[14] is $f(x) = F'(x)$. Note that

$$f(x) \ge 0, \quad \int_{\mathbf{R}} f(x)\,dx = 1$$

and

$$Pr\{X \in [x, x + \Delta x]\} = f(x)\,\Delta x + o(\Delta x).$$

In the discrete case X can only take on discrete values x_i, $i = 1 : N$, and

$$Pr\{X = x_i\} = p_i, \quad i = 1 : N,$$

where $p_i \ge 0$ and $\sum_i p_i = 1$.

[14] In old literature a density function is often called a frequency function. The term *cumulative distribution* is also used as a synonym for distribution function. Unfortunately, distribution or probability distribution is sometimes used in the meaning of a density function.

The **mean** or the **expectation** of X is

$$E(X) = \begin{cases} \displaystyle\int_{-\infty}^{\infty} x f(x)\,dx, & \text{continuous case,} \\ \displaystyle\sum_{i=1}^{N} p_i x_i, & \text{discrete case.} \end{cases}$$

The **variance** of X equals

$$\sigma^2 = \operatorname{var}(X) = E((X-\mu)^2), \qquad \mu = E(X),$$

and $\sigma = \sqrt{\operatorname{var}(X)}$ is the **standard deviation**. The mean and standard deviation are frequently used as measures of the center and spread of a distribution.

Let X_k, $k = 1 : n$, be random variables with mean values μ_k. Then the **covariance matrix** is $V = \{\sigma_{jk}\}_{j,k=1}^{n}$, where

$$\begin{aligned} \sigma_{jk} = \operatorname{cov}(X_j, X_k) &= E((X_j - \mu_j)(X_k - \mu_k)) \\ &= E(X_j)E(X_k) - \mu_j\mu_k. \end{aligned}$$

If $\operatorname{cov}(X_j, X_k) = 0$, then X_j and X_k are said to be **uncorrelated**.

If the random variables X_k, $k = 1 : n$, are mutually uncorrelated, then V is a diagonal matrix.

Some formulas for the estimation of mean, standard deviation, etc. from results of simulation experiments or other statistical data are given in the computer exercises of Sec. 2.3.

1.4.2 Characterization of Least Squares Solutions

Let $y \in \mathbf{R}^m$ be a vector of observations that is related to a parameter vector $c \in \mathbf{R}^n$ by the linear relation

$$y = Ac + \epsilon, \qquad A \in \mathbf{R}^{m\times n}, \tag{1.4.2}$$

where A is a known matrix of full column rank and $\epsilon \in \mathbf{R}^m$ is a vector of random errors. We assume here that ϵ_i, $i = 1 : m$, has zero mean and that ϵ_i and ϵ_j, $i \neq j$, are uncorrelated, i.e.,

$$E(\epsilon) = 0, \qquad V(\epsilon) = \sigma^2 I.$$

The parameter c is then a random vector which we want to estimate in terms of the known quantities A and y. Let $y^T c$ be a linear functional of the parameter c in (1.4.2). We say that $\theta = \theta(A, y)$ is an unbiased linear estimator of $y^T c$ if $E(\theta) = y^T c$. It is a **best linear unbiased estimator** if θ has the smallest variance among all such estimators.

The following theorem[15] places the method of least squares on a sound theoretical basis.

[15]This theorem is originally due to C. F. Gauss (1821). His contribution was somewhat neglected until rediscovered by the Russian mathematician A. A. Markov in 1912.

Theorem 1.4.1 (*Gauss–Markov Theorem*).

Consider a linear model (1.4.2), *where ϵ is an uncorrelated random vector with zero mean and covariance matrix $V = \sigma^2 I$. Then the best linear unbiased estimator of any linear functional $y^T c$ is $y^T \hat{c}$, where*

$$\hat{c} = (A^T A)^{-1} A^T y \tag{1.4.3}$$

is the least squares estimator obtained by minimizing the sum of squares $\|f - Ac\|_2^2$. Furthermore, the covariance matrix of the least squares estimate $\hat{c}$ equals

$$V(\hat{c}) = \sigma^2 (A^T A)^{-1}, \tag{1.4.4}$$

and

$$s^2 = \|y - A\hat{c}\|_2^2/(m - n) \tag{1.4.5}$$

is an unbiased estimate of σ^2, i.e. $E(s^2) = \sigma^2$.

Proof. See Zelen [389, pp. 560–561]. □

The set of all least squares solutions can also be characterized geometrically. For this purpose we introduce two fundamental subspaces of $\mathbf{R}^m$, the **range** of A and the **null space** of A^T, defined by

$$\mathcal{R}(A) = \{z \in \mathbf{R}^m \mid z = Ax,\ x \in \mathbf{R}^n\}, \tag{1.4.6}$$

$$\mathcal{N}(A^T) = \{y \in \mathbf{R}^m \mid A^T y = 0\}. \tag{1.4.7}$$

If $z \in \mathcal{R}(A)$ and $y \in \mathcal{N}(A^T)$, then $z^T y = x^T A^T y = 0$, which shows that $\mathcal{N}(A^T)$ is the orthogonal complement of $\mathcal{R}(A)$.

By the Gauss–Markov theorem any least squares solution to an overdetermined linear system $Ax = b$ satisfies the **normal equations**

$$A^T A x = A^T b. \tag{1.4.8}$$

The normal equations are always consistent, since the right-hand side satisfies

$$A^T b \in \mathcal{R}(A^T) = \mathcal{R}(A^T A).$$

Therefore, a least squares solution always exists, although it may not be unique.

Theorem 1.4.2.

The vector x minimizes $\|b - Ax\|_2$ if and only if the residual vector $r = b - Ax$ is orthogonal to $\mathcal{R}(A)$ or, equivalently,

$$A^T (b - Ax) = 0. \tag{1.4.9}$$

Proof. Let x be a vector for which $A^T(b - Ax) = 0$. For any $y \in \mathbf{R}^n$, it holds that $b - Ay = (b - Ax) + A(x - y)$. Squaring this and using (1.4.9) we obtain

$$\|b - Ay\|_2^2 = \|b - Ax\|_2^2 + \|A(x - y)\|_2^2 \geq \|b - Ax\|_2^2,$$

where equality holds only if $A(x - y) = 0$.

Now assume that $A^T(b - Ax) = z \neq 0$. If $x - y = -\epsilon z$, we have for sufficiently small $\epsilon \neq 0$

$$\begin{aligned}\|b - Ay\|_2^2 &= \|b - Ax\|_2^2 + \epsilon^2\|Az\|_2^2 - 2\epsilon(Az)^T(b - Ax)\\ &= \|b - Ax\|_2^2 + \epsilon^2\|Az\|_2^2 - 2\epsilon\|z\|_2^2 < \|b - Ax\|_2^2,\end{aligned}$$

and thus x does not minimize $\|b - Ax\|_2$. □

From Theorem 1.4.2 it follows that a least squares solution x decomposes the right-hand side into two orthogonal components

$$b = Ax + r, \quad r = b - Ax \in \mathcal{N}(A^T), \quad Ax \in \mathcal{R}(A). \tag{1.4.10}$$

This geometric interpretation is illustrated in Figure 1.4.1. Note that although the solution x to the least squares problem may not be unique, the decomposition (1.4.10) always is unique.

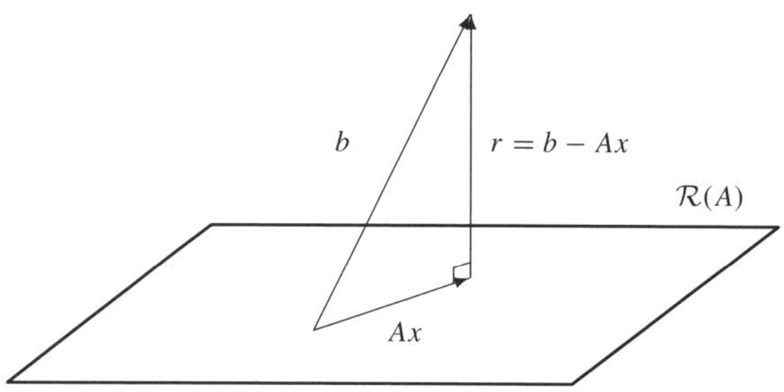

Figure 1.4.1. *Geometric characterization of the least squares solution.*

We now give a necessary and sufficient condition for the least squares solution to be unique.

Theorem 1.4.3.

The matrix A^TA is positive definite and hence nonsingular if and only if the columns of A are linearly independent, that is, when rank $(A) = n$. *In this case the least squares solution x is unique and given by*

$$x = (A^TA)^{-1}A^Tb. \tag{1.4.11}$$

Proof. If the columns of A are linearly independent, then $x \neq 0 \Rightarrow Ax \neq 0$. Therefore, $x \neq 0 \Rightarrow x^TA^TAx = \|Ax\|_2^2 > 0$, and hence A^TA is positive definite.

On the other hand, if the columns are linearly dependent, then for some $x_0 \neq 0$ we have $Ax_0 = 0$. Then $x_0^TA^TAx_0 = 0$, and therefore A^TA is not positive definite. When A^TA is positive definite it is also nonsingular and (1.4.11) follows. □

When A has full column rank A^TA is symmetric and positive definite and the normal equations can be solved by computing the Cholesky factorization $A^TA = R^TR$. The normal equations then become $R^T Rx = A^T b$, which decomposes as

$$R^T z = A^T b, \qquad Rx = z.$$

The first system is lower triangular and z is computed by forward-substitution. Then x is computed from the second upper triangular system by back-substitution. For many practical problems this **method of normal equations** is an adequate solution method, although its numerical stability is not the best.

Example 1.4.1.

The comet Tentax, discovered in 1968, is supposed to move within the solar system. The following observations of its position in a certain polar coordinate system have been made:

r	2.70	2.00	1.61	1.20	1.02
ϕ	48°	67°	83°	108°	126°

By Kepler's first law the comet should move in a plane orbit of elliptic or hyperbolic form, if the perturbations from planets are neglected. Then the coordinates satisfy

$$r = p/(1 - e\cos\phi),$$

where p is a parameter and e the eccentricity. We want to estimate p and e by the method of least squares from the given observations.

We first note that if the relationship is rewritten as

$$1/p - (e/p)\cos\phi = 1/r,$$

it becomes linear in the parameters $x_1 = 1/p$ and $x_2 = e/p$. We then get the linear system $Ax = b$, where

$$A = \begin{pmatrix} 1.0000 & -0.6691 \\ 1.0000 & -0.3907 \\ 1.0000 & -0.1219 \\ 1.0000 & 0.3090 \\ 1.0000 & 0.5878 \end{pmatrix}, \qquad b = \begin{pmatrix} 0.3704 \\ 0.5000 \\ 0.6211 \\ 0.8333 \\ 0.9804 \end{pmatrix}.$$

The least squares solution is $x = (0.6886 \quad 0.4839)^T$ giving $p = 1/x_1 = 1.4522$ and finally $e = px_2 = 0.7027$.

By (1.4.10), if x is a least squares solution, then Ax is the orthogonal projection of b onto $\mathcal{R}(A)$. Thus orthogonal projections play a central role in least squares problems. In general, a matrix $P_1 \in \mathbf{R}^{m\times m}$ is called a **projector** onto a subspace $S \subset \mathbf{R}^m$ if and only if it holds that

$$P_1 v = v \quad \forall v \in S, \qquad P_1^2 = P_1. \tag{1.4.12}$$

An arbitrary vector $v \in \mathbf{R}^m$ can then be decomposed as $v = P_1 v + P_2 v \equiv v_1 + v_2$, where $P_2 = I - P_1$. In particular, if P_1 is symmetric, $P_1 = P_1^T$, we have

$$P_1^T P_2 v = P_1^T (I - P_1) v = (P_1 - P_1^2) v = 0 \quad \forall v \in \mathbf{R}^m,$$

and it follows that $P_1^T P_2 = 0$. Hence $v_1^T v_2 = v^T P_1^T P_2 v = 0$ for all $v \in \mathbf{R}^m$, i.e., $v_2 \perp v_1$. In this case P_1 is the **orthogonal projector** onto S, and $P_2 = I - P_1$ is the orthogonal projector onto $S^\perp$.

In the full column rank case, rank $(A) = n$, of the least squares problem, the residual $r = b - Ax$ can be written $r = b - P_{\mathcal{R}(A)} b$, where

$$P_{\mathcal{R}(A)} = A(A^T A)^{-1} A^T \tag{1.4.13}$$

is the orthogonal projector onto $\mathcal{R}(A)$. If rank $(A) < n$, then A has a nontrivial null space. In this case if $\hat{x}$ is any vector that minimizes $\|Ax - b\|_2$, then the set of all least squares solutions is

$$\mathcal{S} = \{x = \hat{x} + y \mid y \in \mathcal{N}(A)\}. \tag{1.4.14}$$

In this set there is a unique solution of minimum norm characterized by $x \perp \mathcal{N}(A)$, which is called the **pseudoinverse solution**.

1.4.3 The Singular Value Decomposition

In the past the conventional way to determine the rank of a matrix A was to compute the row echelon form by Gaussian elimination. This would also show whether a given linear system is consistent or not. However, in floating-point calculations it is difficult to decide if a pivot element, or an element in the transformed right-hand side, should be considered as zero or nonzero. Such questions can be answered in a more satisfactory way by using the **singular value decomposition** (SVD), which is of great theoretical and computational importance.[16]

The geometrical significance of the SVD is as follows: The rectangular matrix $A \in \mathbf{R}^{m\times n}$, $m \geq n$, represents a mapping $y = Ax$ from $\mathbf{R}^n$ to $\mathbf{R}^m$. The image of the unit sphere $\|x\|_2 = 1$ is a hyper ellipse in $\mathbf{R}^m$ with axes equal to $\sigma_1 \geq \sigma_2 \cdots \geq \sigma_n \geq 0$. In other words, the SVD gives orthogonal bases in these two spaces such that the mapping is represented by the generalized diagonal matrix $\Sigma \in \mathbf{R}^{m\times n}$. This is made more precise in the following theorem, a constructive proof of which will be given in Volume II.

Theorem 1.4.4 (*Singular Value Decomposition*).

Any matrix $A \in \mathbf{R}^{m\times n}$ of rank r can be decomposed as

$$A = U\Sigma V^T, \qquad \Sigma = \begin{pmatrix} \Sigma_r & 0 \\ 0 & 0 \end{pmatrix} \in \mathbf{R}^{m\times n}, \tag{1.4.15}$$

where $\Sigma_r = \operatorname{diag}(\sigma_1, \sigma_2, \ldots, \sigma_r)$ is diagonal and

$$U = (u_1, \ldots, u_m) \in \mathbf{R}^{m\times m}, \qquad V = (v_1, \ldots, v_n) \in \mathbf{R}^{n\times n} \tag{1.4.16}$$

[16]The SVD was published by Eugenio Beltrami in 1873 and independently by Camille Jordan in 1874. Its use in numerical computations is much more recent, since a stable algorithm for computing the SVD did not become available until the publication of Golub and Reinsch [167] in 1970.

are square orthogonal matrices, $U^TU = I_m$, $V^TV = I_n$. *Here*

$$\sigma_1 \geq \sigma_2 \geq \cdots \geq \sigma_r > 0$$

are the $r \leq \min(m, n)$ *nonzero* **singular values** *of* A. *The vectors* u_i, $i = 1 : m$, *and* v_j, $j = 1 : n$, *are left and right* **singular vectors**. *(Note that if* $r = n$ *and/or* $r = m$, *some of the zero submatrices in* Σ *disappear.)*

The singular values of A are uniquely determined. For any distinct singular value $\sigma_j \neq \sigma_i$, $i \neq j$, the corresponding singular vector v_j is unique (up to a factor ± 1). For a singular value of multiplicity p the corresponding singular vectors can be chosen as any orthonormal basis for the unique subspace of dimension p that they span. Once the singular vectors v_j, $1 \leq j \leq r$, have been chosen, the vectors u_j, $1 \leq j \leq r$, are uniquely determined, and vice versa, by

$$u_j = \frac{1}{\sigma_j} A v_j, \qquad v_j = \frac{1}{\sigma_j} A^T u_j, \quad j = 1 : r. \tag{1.4.17}$$

By transposing (1.4.15) we obtain $A^T = V\Sigma^T U^T$, which is the SVD of A^T. Expanding (1.4.15), the SVD of the matrix A can be written as a sum of r matrices of rank one,

$$A = \sum_{i=1}^{r} \sigma_i u_i v_i^T. \tag{1.4.18}$$

The SVD gives orthogonal bases for the range and null space of A and A^T. Suppose that the matrix A has rank $r < \min(m, n)$. It is easy to verify that

$$\mathcal{R}(A) = \text{span}\,(u_1, \dots, u_r), \qquad \mathcal{N}(A^T) = \text{span}\,(u_{r+1}, \dots, u_m), \tag{1.4.19}$$

$$\mathcal{R}(A^T) = \text{span}\,(v_1, \dots, v_r), \qquad \mathcal{N}(A) = \text{span}\,(v_{r+1}, \dots, v_n). \tag{1.4.20}$$

It immediately follows that

$$\mathcal{R}(A)^\perp = \mathcal{N}(A^T), \qquad \mathcal{N}(A)^\perp = \mathcal{R}(A^T); \tag{1.4.21}$$

i.e., $\mathcal{N}(A^T)$ is the orthogonal complement to $\mathcal{R}(A)$, and $\mathcal{N}(A)$ is the orthogonal complement to $\mathcal{R}(A^T)$. This result is sometimes called the fundamental theorem of linear lgebra.

We remark that the SVD generalizes readily to complex matrices. The SVD of a matrix $A \in \mathbf{C}^{m\times n}$ is

$$A = (U_1\ U_2)\Sigma \begin{pmatrix} V_1^T \\ V_2^T \end{pmatrix}, \quad \Sigma = \begin{pmatrix} \Sigma_r & 0 \\ 0 & 0 \end{pmatrix} \in \mathbf{R}^{m\times n}, \tag{1.4.22}$$

where the singular values $\sigma_1, \sigma_2, \dots, \sigma_r$ are real and nonnegative, and U and V are square unitary matrices, $U^HU = I_m$, $V^HV = I_n$. (Here A^H denotes the conjugate transpose of A.)

The SVD can also be used to solve linear least squares problems in the case when the columns in A are linearly dependent. Then there is a vector $c \neq 0$ such that $Ac = 0$ and the least squares solution is not unique, then there exists a *unique least squares solution of minimum Euclidean length*, which solves the least squares problem

$$\min_{x\in S} \|x\|_2, \qquad S = \{x \in \mathbf{R}^n \mid \|b - Ax\|_2 = \min\}. \tag{1.4.23}$$

In terms of the SVD (1.4.22) of A the solution to (1.4.23) can be written $x = A^\dagger b$, where the matrix $A^\dagger$ is

$$A^\dagger = (V_1\ V_2)\Sigma^\dagger \begin{pmatrix} U_1^T \\ U_2^T \end{pmatrix}, \quad \Sigma^\dagger = \begin{pmatrix} \Sigma_r^{-1} & 0 \\ 0 & 0 \end{pmatrix} \in \mathbf{R}^{n\times m}. \tag{1.4.24}$$

The matrix $A^\dagger$ is unique and called the **pseudoinverse** of A, and $x = A^\dagger b$ is the pseudoinverse solution. Note that problem (1.4.23) includes as special cases the solution of both overdetermined and underdetermined linear systems.

The pseudoinverse $A^\dagger$ is often called the **Moore–Penrose inverse**. Moore developed the concept of the general reciprocal in 1920. In 1955 Penrose [288] gave an elegant algebraic characterization and showed that $X = A^\dagger$ is uniquely determined by the four **Penrose conditions**:

$$(1)\quad AXA = A, \qquad (2)\quad XAX = X, \tag{1.4.25}$$
$$(3)\quad (AX)^T = AX, \qquad (4)\quad (XA)^T = XA. \tag{1.4.26}$$

It can be directly verified that $X = A^\dagger$ given by (1.4.24) satisfies these four conditions. In particular this shows that $A^\dagger$ does not depend on the particular choices of U and V in the SVD.

The orthogonal projections onto the four fundamental subspaces of A have the following simple expressions in terms of the SVD:

$$\begin{aligned} P_{\mathcal{R}(A)} &= AA^\dagger = U_1U_1^T, & P_{\mathcal{N}(A^T)} &= I - AA^\dagger = U_2U_2^T, \\ P_{\mathcal{R}(A^T)} &= A^\dagger A = V_1V_1^T, & P_{\mathcal{N}(A)} &= I - A^\dagger A = V_2V_2^T. \end{aligned} \tag{1.4.27}$$

These first expressions are easily verified using the definition of an orthogonal projection and the Penrose conditions.

1.4.4 The Numerical Rank of a Matrix

Let A be a matrix of rank $r < \min(m, n)$, and E a matrix of small random elements. Then it is most likely that the perturbed matrix $A + E$ has maximal rank $\min(m, n)$. However, since $A + E$ is close to a rank deficient matrix, it should be considered as having **numerical rank** equal to r. In general, the numerical rank assigned to a matrix should depend on some tolerance δ, which reflects the error level in the data and/or the precision of the arithmetic used.

It can be shown that perturbations of an element of a matrix A result in perturbations of the same, or smaller, magnitude in its singular values. This motivates the following definition of numerical rank.

Definition 1.4.5.

A matrix $A \in \mathbf{R}^{m\times n}$ is said to have numerical δ-rank equal to k if

$$\sigma_1 \geq \cdots \geq \sigma_k > \delta \geq \sigma_{k+1} \geq \cdots \geq \sigma_p, \quad p = \min(m, n), \tag{1.4.28}$$

where σ_i are the singular values of A. Then the right singular vectors $(v_{k+1}, \ldots, v_n)$ form an orthogonal basis for the numerical null space *of A.*

Definition 1.4.5 assumes that there is a well-defined gap between σ_{k+1} and σ_k. When this is not the case the numerical rank of A is not well defined.

Example 1.4.2.

Consider an integral equation of the first kind,

$$\int_0^1 k(s,t)f(s)\,ds = g(t), \qquad k(s,t) = e^{-(s-t)^2},$$

on $-1 \le t \le 1$. In order to solve this equation numerically it must first be discretized. We introduce a uniform mesh for s and t on $[-1,1]$ with step size $h = 2/n$, $s_i = -1 + ih$, $t_j = -1 + jh$, $i, j = 0 : n$. Approximating the integral with the trapezoidal rule gives

$$h \sum_{i=0}^{n} w_i k(s_i, t_j) f(t_i) = g(t_j), \qquad j = 0 : n,$$

where $w_i = 1$, $i \neq 0, n$, and $w_0 = w_n = 1/2$. These equations form a linear system

$$Kf = g, \qquad K \in \mathbf{R}^{(n+1)\times(n+1)}, \qquad f, g \in \mathbf{R}^{n+1}.$$

For $n = 100$ the singular values σ_k of the matrix K were computed in IEEE double precision with a unit roundoff level of $1.11 \cdot 10^{-16}$ (see Sec. 2.2.3). They are displayed in logarithmic scale in Figure 1.4.2. Note that for $k > 30$ all σ_k are close to roundoff level, so the numerical rank of K certainly is smaller than 30. This means that the linear system $Kf = g$ is *numerically underdetermined* and has a meaningful solution only for special right-hand sides g.

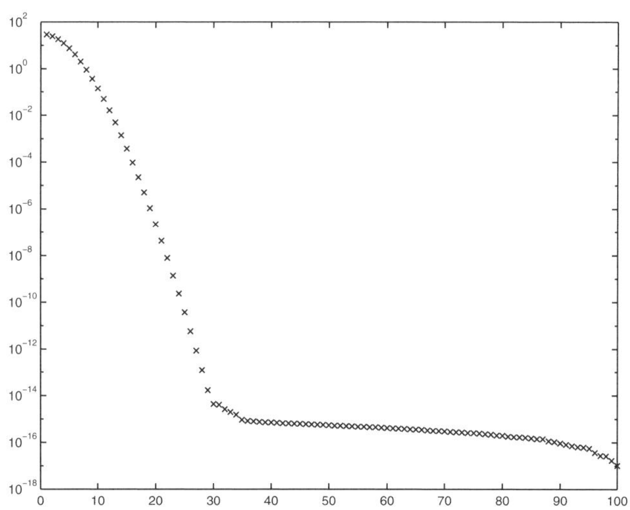

Figure 1.4.2. *Singular values of a numerically singular matrix.*

Equation (1.4.28) is a **Fredholm integral equation** of the first kind. It is known that such equations are **ill-posed** in the sense that the solution f does not depend continuously on the right-hand side g. This example illustrate how this inherent difficulty in the continuous problem carries over to the discretized problem.

Review Questions

1.4.1 State the Gauss–Markov theorem.

1.4.2 Show that the matrix $A^TA \in \mathbf{R}^{n\times n}$ of the normal equations is symmetric and positive semidefinite, i.e., $x^T(A^TA)x \geq 0$, for all $x \neq 0$.

1.4.3 Give two geometric conditions which are necessary and sufficient conditions for x to be the pseudoinverse solution of $Ax = b$.

1.4.4 (a) Which are the four fundamental subspaces of a matrix? Which relations hold between them?

(b) Show, using the SVD, that $P_{\mathcal{R}(A)} = AA^\dagger$ and $P_{\mathcal{R}(A^T)} = A^\dagger A$.

1.4.5 (a) Construct an example where $(AB)^\dagger \neq B^\dagger A^\dagger$.

(b) Show that if A is an $m\times r$ matrix, B is an $r\times n$ matrix, and rank $(A) =$ rank $(B) = r$, then $(AB)^\dagger = B^\dagger A^\dagger$.

Problems and Computer Exercises

1.4.1 In order to estimate the height above sea level for three points A, B, and C, the difference in altitude was measured between these points and points D, E, and F at sea level. The measurements obtained form a linear system in the heights x_A, x_B, and x_C of A, B, and C:

$$\begin{pmatrix} 1 & 0 & 0 \\ 0 & 1 & 0 \\ 0 & 0 & 1 \\ -1 & 1 & 0 \\ 0 & -1 & 1 \\ -1 & 0 & 1 \end{pmatrix} \begin{pmatrix} x_A \\ x_B \\ x_C \end{pmatrix} = \begin{pmatrix} 1 \\ 2 \\ 3 \\ 1 \\ 2 \\ 1 \end{pmatrix}.$$

Determine the least squares solution and verify that the residual vector is orthogonal to all columns in A.

1.4.2 Consider the least squares problem $\min_x \|Ax - b\|_2^2$, where A has full column rank. Partition the problem as

$$\min_{x_1,x_2} \left\| (A_1\ A_2) \begin{pmatrix} x_1 \\ x_2 \end{pmatrix} - b \right\|_2^2.$$

By a geometric argument show that the solution can be obtained as follows. First compute x_2 as the solution to the problem

$$\min_{x_2} \| P_{A_1}^{\perp}(A_2x_2 - b)\|_2^2,$$

where $P_{A_1}^{\perp} = I - P_{A_1}$ is the orthogonal projector onto $\mathcal{N}(A_1^T)$. Then compute x_2 as the solution to the problem

$$\min_{x_1} \|A_1x_1 - (b - A_2x_2)\|_2^2.$$

1.5 Numerical Solution of Differential Equations

1.5.1 Euler's Method

The approximate solution of *differential equations* is a very important task in scientific computing. Nearly all the areas of science and technology contain mathematical models which lead to systems of ordinary (or partial) differential equations. For the **step by step simulation** of such a system a **mathematical model** is first set up; i.e., **state variables** are set up which describe the essential features of the state of the system. Then the laws are formulated, which govern *the rate of change of the state variables*, and other *mathematical relations* between these variables. Finally, these equations are programmed for a computer to calculate approximately, step by step, the development in time of the system.

The reliability of the results depends primarily on the quality of the mathematical model and on the size of the time step. The choice of the time step is partly a question of economics. Small time steps may give you good accuracy, but also long computing time. More accurate numerical methods are often a good alternative to the use of small time steps.

The construction of a mathematical model is not trivial. Knowledge of numerical methods and programming helps in that phase of the job, but more important is a good understanding of the fundamental processes in the system, and that is beyond the scope of this text. It is, however, important to realize that if the mathematical model is bad, no sophisticated numerical techniques or powerful computers can stop the results from being unreliable, or even harmful.

A mathematical model can be studied by analytic or computational techniques. Analytic methods do not belong to this text. We want, though, to emphasize that the comparison of results obtained by applying analytic methods, in the special cases when they can be applied, can be very useful when numerical methods and computer programs are tested. We shall now illustrate these general comments using a particular example.

An **initial value problem** for an ordinary differential equation is to find $y(t)$ such that

$$\frac{dy}{dt} = f(t, y), \quad y(0) = c.$$

The differential equation gives, at each point (t, y), the direction of the tangent to the solution curve which passes through the point in question. The direction of the tangent changes continuously from point to point, but the simplest approximation (which was proposed as early as the eighteenth century by Euler[17]) is that one studies the solution for only certain

[17] Leonhard Euler (1707–1783), incredibly prolific Swiss mathematician. He gave fundamental contributions to many branches of mathematics and to the mechanics of rigid and deformable bodies, as well as to fluid mechanics.

values of $t = t_n = nh$, $n = 0, 1, 2, \ldots$, (h is called the "time step" or "step length") and assumes that dy/dt is constant between the points. In this way the solution is approximated by a polygon (Figure 1.5.1) which joins the points (t_n, y_n), $n = 0, 1, 2, \ldots$, where

$$y_0 = c, \qquad \frac{y_{n+1} - y_n}{h} = f(t_n, y_n). \tag{1.5.1}$$

Thus we have the simple difference equation known as **Euler's method**:

$$y_0 = c, \qquad y_{n+1} = y_n + hf(t_n, y_n), \qquad n = 0, 1, 2, \ldots. \tag{1.5.2}$$

During the computation, each y_n occurs first on the left-hand side, then *recurs* later on the right-hand side of an equation. (One could also call (1.5.2) an iteration formula, but one usually reserves the word "iteration" for the special case where a recursion formula is used solely as a means of calculating a limiting value.)

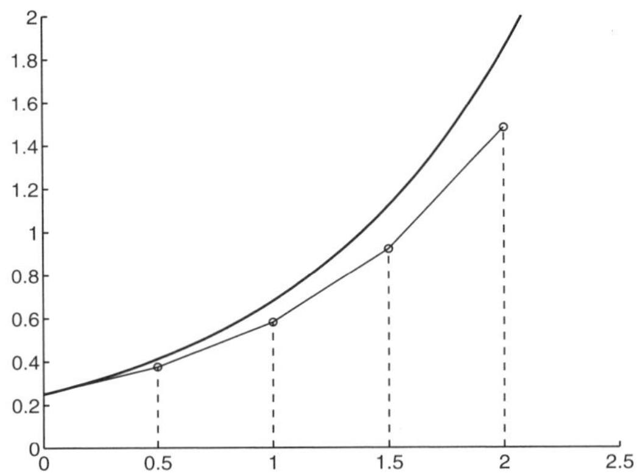

Figure 1.5.1. *Approximate solution of the differential equation* $dy/dt = y$, $y_0 = 0.25$, *by Euler's method with* $h = 0.5$.

1.5.2 An Introductory Example

Consider the motion of a ball (or a shot) under the influence of gravity and air resistance. It is well known that the trajectory is a parabola, when the air resistance is neglected and the force of gravity is assumed to be constant. We shall still neglect the variation of the force of gravity as well as the curvature and the rotation of the Earth. This means that we forsake serious applications to, for example, satellites. We shall, however, take the air resistance into account. We neglect the rotation of the shot around its own axis. Therefore, we can treat the problem as motion in a plane, but we have to forsake the application to, for example, table tennis, baseball, or a rotating projectile. Now we have introduced a number of assumptions, which define our **model** of reality.

The state of the ball is described by its position (x, y) and velocity (u, v), each of which has two Cartesian coordinates in the plane of motion. The x-axis is horizontal, and the y-axis is directed upward. Assume that the air resistance is a force P such that the direction is opposite to the velocity, and the strength z is proportional to the square of the speed and to the square of the radius R of the shot. If we denote by P_x and P_y the components of P along the x and y directions, respectively, we can then write

$$P_x = -mzu, \qquad P_y = -mzv, \qquad z = \frac{cR^2}{m}\sqrt{u^2 + v^2}, \tag{1.5.3}$$

where m is the mass of the ball.

For the sake of simplicity we assume that c is a constant. It actually depends on the density and the viscosity of the air. Therefore, we have to forsake the application to cannon shots, where the variation of the density with height is important. If one has access to a good model of the atmosphere, the variation of c would not make the numerical simulation much more difficult. This contrasts with analytic methods, where such a modification is likely to mean a considerable complication. In fact, even with a constant c, a purely analytic treatment offers great difficulties.

Newton's law of motion tells us that

$$m\,du/dt = P_x, \qquad m\,dv/dt = -mg + P_y, \tag{1.5.4}$$

where the term $-mg$ is the force of gravity. Inserting (1.5.3) into (1.5.4) and dividing by m we get

$$du/dt = -zu, \qquad dv/dt = -g - zv, \tag{1.5.5}$$

and by the definition of velocity,

$$dx/dt = u, \qquad dy/dt = v. \tag{1.5.6}$$

Equations (1.5.5) and (1.5.6) constitute a system of four differential equations for the four variables x, y, u, v. The initial state x_0, y_0 and u_0, v_0 at time $t_0 = 0$ is assumed to be given. A fundamental proposition in the theory of differential equations tells us that if initial values of the state variables u, v, x, y are given at some initial time $t = t_0$, then they will be uniquely determined for all $t > t_0$.

The simulation of the motion of the ball means that, at a sequence of time instances t_n, $n = 0, 1, 2, \ldots$, we determine the approximate values u_n, v_n, x_n, y_n. We first look at the simplest technique, using Euler's method with a constant time step h. Therefore, set $t_n = nh$. We replace the derivative du/dt by the forward difference quotient $(u_{n+1} - u_n)/h$, and similarly for the other variables. Hence after multiplication by h, the differential equations are replaced by the following system of **difference equations**:

$$\begin{aligned} x_{n+1} &= x_n + hu_n, \qquad y_{n+1} = y_n + hv_n, \\ u_{n+1} &= u_n - hz_n u_n, \\ v_{n+1} &= v_n - h(g + z_n v_n), \end{aligned} \tag{1.5.7}$$

where

$$z_n = \frac{cR^2}{m}\sqrt{u_n^2 + v_n^2}.$$

From this $x_{n+1}, y_{n+1}, u_{n+1}, v_{n+1}$, etc. are solved, step by step, for $n = 0, 1, 2, \ldots$, using the provided initial values x_0, y_0, u_0, and v_0.

We performed these computations until y_{n+1} became negative for the first time, with $g = 9.81$, $\phi = 60°$, and the initial values

$$x_0 = 0, \quad y_0 = 0, \quad u_0 = 100\cos\phi, \quad v_0 = 100\sin\phi.$$

Curves obtained for $h = 0.01$ and $cR^2/m = 0.25i \cdot 10^{-3}$, $i = 0 : 4$, are shown in Figure 1.5.2. There is, in this graphical representation, also an error due to the limited resolution of the plotting device.

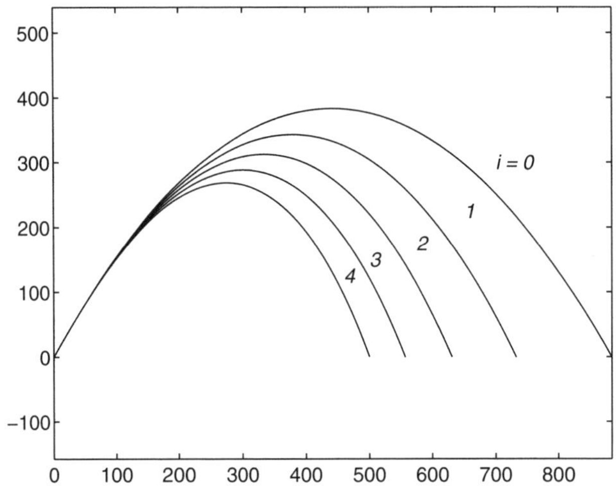

Figure 1.5.2. *Approximate trajectories computed with Euler's method with h* $= 0.01$.

In Euler's method the state variables are *locally approximated by linear functions* of time, one of the often recurrent ideas in numerical computation. We can use the same idea for computing the coordinate x^* of the point where the shot hits the ground. Suppose that y_{n+1} becomes negative for the first time when $n = N$. For $x_N \le x \le x_{N+1}$ we then approximate y by a linear function of x, represented by the secant through the points (x_N, y_N) and (x_{N+1}, y_{N+1}), i.e.,

$$y = y_N + (x - x_N)\frac{y_{N+1} - y_N}{x_{N+1} - x_N}.$$

By setting $y = 0$ we obtain

$$x^* = x_N - y_N\frac{x_{N+1} - x_N}{y_{N+1} - y_N}. \tag{1.5.8}$$

This is called (linear) **inverse interpolation**; see Sec. 4.2.2. The error from the linear approximation in (1.5.8) used for the computation of x^* is proportional to h^2. It is thus approximately equal to the error committed in *one single step* with Euler's method, and hence of less importance than the other error.

The case without air resistance ($i = 0$) can be solved exactly. In fact it can be shown that

$$x^* = 2u_0v_0/9.81 = 5000 \cdot \sqrt{3}/9.81 \approx 882.7986.$$

The computer produced $x^* \approx 883.2985$ for $h = 0.01$, and $x^* \approx 883.7984$ for $h = 0.02$. The error for $h = 0.01$ is therefore 0.4999, and for $h = 0.02$ it is 0.9998. The approximate proportionality to h is thus verified, actually more strikingly than could be expected!

It can be shown that *the error in the results obtained with Euler's method is also proportional to h (not h^2)*. Hence a disadvantage of the above method is that the step length h must be chosen quite small if reasonable accuracy is desired. In order to improve the method we can apply another idea mentioned previously, namely Richardson extrapolation. (The application differs a little from the one we saw previously, because now the error is approximately proportional to h, while for the trapezoidal rule it was approximately proportional to h^2.) For $i = 4$, the computer produced $x^* \approx 500.2646$ and $x^* \approx 500.3845$ for, respectively, $h = 0.01$ and $h = 0.02$. Now let x^* denote the *exact* horizontal coordinate of the landing point. Then

$$x^* - 500.2646 \approx 0.01k, \qquad x^* - 500.3845 \approx 0.02k.$$

By elimination of k we obtain

$$x^* \approx 2 \cdot 500.2646 - 500.3845 = 500.1447,$$

which should be a more accurate estimate of the coordinate. By a more accurate integration method we obtained 500.1440. Thus, in this case we gained more than two decimal digits by the use of Richardson extrapolation.

The simulations shown in Figure 1.5.2 required about 1500 time steps for each curve. This may seem satisfactory, but we must not forget that this is a very small task, compared with most serious applications. So we would like to have a method that allows *much larger time steps* than Euler's method.

1.5.3 Second Order Accurate Methods

In step by step computations we have to distinguish between the **local error**, the error that is committed at a single step, and the **global error**, the error of the final results. Recall that we say that a method is accurate of order p if its global error is approximately proportional to h^p. Euler's method is only first order accurate; we shall present a method that is second order accurate. To achieve the same accuracy as with Euler's method the number of steps can then be reduced to about the square root of the number of steps in Euler's method. In the above ball problem this means $\sqrt{1500} \approx 40$ steps. Since the amount of work is closely proportional to the number of steps this is an enormous savings!

Another question is how the step size h is to be chosen. It can be shown that even for rather simple examples (see below) it is adequate to use very *different step sizes* in different parts of the computation. Hence the automatic control of the step size (also called *adaptive* control) is an important issue.

Both requests can be met by an improvement of the Euler method (due to Runge[18]) obtained by applying the Richardson extrapolation in every second step. This is different from our previous application of the Richardson idea. We first introduce a better notation by writing a **system of differential equations** and the initial conditions in vector form

$$d\mathbf{y}/dt = \mathbf{f}(t, \mathbf{y}), \quad \mathbf{y}(a) = \mathbf{c}, \tag{1.5.9}$$

where $\mathbf{y}$ is a column vector that contains all the state variables.[19] With this notation methods for large systems of differential equations can be described as easily as methods for a single equation. The change of a system with time can then be thought of as a motion of the state vector in a multidimensional space, where the differential equation defines the **velocity field**. This is our first example of the central role of vectors and matrices in modern computing.

For the ball example, we have $a = 0$, and by (1.5.5) and (1.5.6),

$$\mathbf{y} = \begin{pmatrix} y_1 \\ y_2 \\ y_3 \\ y_4 \end{pmatrix} \equiv \begin{pmatrix} x \\ y \\ u \\ v \end{pmatrix}, \qquad \mathbf{f}(t, \mathbf{y}) = \begin{pmatrix} y_3 \\ y_4 \\ -zy_3 \\ -g - zy_4 \end{pmatrix}, \qquad \mathbf{c} = 10^2 \begin{pmatrix} 0 \\ 0 \\ \cos\phi \\ \sin\phi \end{pmatrix},$$

where

$$z = \frac{cR^2}{m}\sqrt{(y_3)^2 + (y_4)^2}.$$

The computations in the step which leads from t_n to t_{n+1} are then as follows:

i. One Euler step of length h yields the estimate:

$$\mathbf{y}^*_{n+1} = \mathbf{y}_n + h\mathbf{f}(t_n, \mathbf{y}_n).$$

ii. Two Euler steps of length $\frac{1}{2}h$ yield another estimate:

$$\mathbf{y}_{n+1/2} = \mathbf{y}_n + \frac{1}{2}h\mathbf{f}(t_n, \mathbf{y}_n), \qquad \mathbf{y}^{**}_{n+1} = \mathbf{y}_{n+1/2} + \frac{1}{2}h\mathbf{f}(t_{n+1/2}, \mathbf{y}_{n+1/2}),$$

where $t_{n+1/2} = t_n + h/2$.

iii. Then $\mathbf{y}_{n+1}$ is obtained by Richardson extrapolation:

$$\mathbf{y}_{n+1} = \mathbf{y}^{**}_{n+1} + (\mathbf{y}^{**}_{n+1} - \mathbf{y}^*_{n+1}).$$

It is conceivable that this yields a second order accurate method. It is left as an exercise (Problem 1.5.2) to verify that *this scheme is identical to the following somewhat simpler scheme* known as **Runge's second order method**:

$$\begin{aligned} \mathbf{k}_1 &= h_n\mathbf{f}(t_n, \mathbf{y}_n), \\ \mathbf{k}_2 &= h_n\mathbf{f}(t_n + h_n/2, \mathbf{y}_n + \mathbf{k}_1/2), \\ \mathbf{y}_{n+1} &= \mathbf{y}_n + \mathbf{k}_2, \end{aligned} \tag{1.5.10}$$

[18]Carle David Tolmé Runge (1856–1927), German mathematician. Runge was professor of applied mathematics at Göttingen from 1904 until his death.

[19]The boldface notation is temporarily used for vectors *in this section* only, not in the rest of the book.

where we have replaced h by h_n in order to include the use of variable step size. Another explanation of the second order accuracy of this method is that the displacement $\mathbf{k}_2$ equals the product of the step size and a sufficiently accurate estimate of the velocity at the *midpoint* of the time step. Sometimes this method is called the improved Euler method or Heun's method, but these names are also used to denote other second order accurate methods.

1.5.4 Adaptive Choice of Step Size

We shall now describe how the step size can be **adaptively** (or automatically) controlled by means of a tolerance tol, by which the user tells the program how large an error he tolerates in values of variables (relative to the values themselves).[20] Compute

$$\delta = \max_i \delta_i, \quad \delta_i = |k_{2,i} - k_{1,i}|/|3y_i|,$$

where δ_i is related to the *relative* error of the ith component of the vector $\mathbf{y}$ at the current step; see below.

A step size is *accepted* if $\delta \le \text{tol}$, and the next step should be

$$h_{next} = h \min\left\{1.5, \sqrt{\text{tol}/(1.2\delta)}\right\},$$

where 1.2 is a safety factor, since the future is never exactly like the past! The square root occurring here is due to the fact that this method is second order accurate; i.e., the global error is almost proportional to the square of the step size and δ is approximately proportional to h^2.

A step is *rejected* if $\delta > \text{tol}$, and *recomputed* with the step size

$$h_{next} = h \max\left\{0.1, \sqrt{\text{tol}/(1.2\delta)}\right\}.$$

The program needs a suggestion for the size of the first step. This can be a very rough guess, because the step size control described above will improve it automatically so that an adequate step size is found after a few steps (or recomputations, if the suggested step was too big). In our experience, a program of this sort can efficiently handle guesses that are wrong by several powers of 10. If $y(a) \ne 0$ and $y'(a) \ne 0$, you may try the initial step size

$$h = \frac{1}{4} \sum_i |y_i| \Big/ \sum_i |dy_i/dt|$$

evaluated at the initial point $t = a$. When you encounter the cases $y(a) = 0$ or $y'(a) = 0$ for the first time, you are likely to have gained enough experience to suggest something that the program can handle. More professional programs take care of this detail automatically.

The request for a certain *relative* accuracy may cause trouble when some components of y are close to zero. So, already in the first version of your program, you had better *replace y_i in the above definition of δ by*

$$\bar{y}_i = \max\{|y_i|, 0.001\}.$$

(You may sometimes have to replace the default value 0.001 by something else.)

[20] With the terminology that will be introduced in the next chapter, tol is, with the step size control described here, related to the *global relative errors*. At the time of writing, this contrasts to most codes for the solution of ordinary differential equations, in which the *local* errors per step are controlled by the tolerance.

It is a good habit to make a second run with a predetermined sequence of step sizes (if your program allows this) instead of adaptive control. Suppose that the sequence of time instances used in the first run is $t_0, t_1, t_2, \ldots$. Divide each subinterval $[t_n, t_{n+1}]$ into two steps of equal length. Thus, the second run still has variable step size and twice as many steps as the first run. The errors are therefore expected to be approximately $\frac{1}{4}$ of the errors of the first run. The first run can therefore use a tolerance that is four times as large than the error you can tolerate in the final result. Denote the results of the two runs by $y_I(t)$ and $y_{II}(t)$. You can plot $\frac{1}{3}(y_{II}(t) - y_I(t))$ versus t; this is an error curve for $y_{II}(t)$. Alternatively you can add $\frac{1}{3}(y_{II}(t) - y_I(t))$ to $y_{II}(t)$. This is another application of the Richardson extrapolation idea. The cost is only 50% more work than the plain result without an error curve.

If there are no singularities in the differential equation, $\frac{1}{3}(y_{II}(t) - y_I(t))$ *strongly overestimates the error of the extrapolated values*—typically by a factor such as $\text{tol}^{-1/2}$. It is, however, a nontrivial matter to find an error curve that strictly and realistically tells us how good the extrapolated results are. The reader is advised to test experimentally how this works on examples where the exact results are known.

An easier, though inferior, alternative is to run a problem with two different tolerances. One reason why this is inferior is that the two runs do not "keep in step," and then Richardson extrapolation cannot be easily applied.

If you request very high accuracy in your results, or if you are going to simulate a system over a very long time, you will need a method with a higher order of accuracy than two. The reduction of computing time, if you replace this method by a higher order method can be large, but the improvements are seldom as drastic as when you replace Euler's method by a second order accurate scheme like this. Runge's second order method is, however, no universal recipe. There are special classes of problems, notably the problems which are called "stiff," which need special methods.

One advantage of a second order accurate scheme when requests for accuracy are modest is that the quality of the computed results is normally not ruined by the use of *linear interpolation* at the graphical output, or in the postprocessing of numerical results. (After you have used a more than second order accurate integration method, it may be necessary to use more sophisticated interpolation at the graphical or numerical treatment of the results.) We suggest that you write or try to find a program that can be used for systems with (in principle) any number of equations; see the preface.

Example 1.5.1.

The differential equation

$$dy/dt = -\frac{1}{2}y^3,$$

with initial condition $y(1) = 1$, was treated by a program, essentially constructed as described above, with $\text{tol} = 10^{-4}$ until $t = 10^4$. When comparing the result with the exact solution $y(t) = t^{-1/2}$, it was found that the actual relative error remained at a little less than $1.5\,\text{tol}$ all the time when $t > 10$. The step size increased almost linearly with t from $h = 0.025$ to $h = 260$. The number of steps increased almost proportionally to $\log t$; the total number of steps was 374. Only one step had to be recomputed (except for the first step, where the program had to find an appropriate step size).

The computation was repeated with tol $= 4 \cdot 10^{-4}$. The experience was the same, except that the steps were about twice as long all the time. This is what can be expected, since the step sizes should be approximately proportional to $\sqrt{\text{tol}}$, for a second order accurate method. The total number of steps was 194.

Example 1.5.2.

The example of the motion of a ball was treated by Runge's second order method with the constant step size $h = 0.9$. The x-coordinate of the landing point became $x^* \approx 500.194$, which is more than twice as accurate than the result obtained by Euler's method (without Richardson extrapolation) with $h = 0.01$, which uses about 90 times as many steps.

We have now seen a variety of ideas and concepts which can be used in the development of numerical methods. A small warning is perhaps warranted here: it is not certain that the methods will work as well in practice as one might expect. This is because approximations and the restriction of numbers to a certain number of digits introduce errors which are propagated to later stages of a calculation. The manner in which errors are propagated is decisive for the practical usefulness of a numerical method. We shall examine such questions in Chapter 2. Later chapters will treat **propagation of errors** in connection with various typical problems.

The risk that error propagation may upstage the desired result of a numerical process should, however, not dissuade one from the use of numerical methods. It is often wise, though, to experiment with a proposed method on a simplified problem before using it in a larger context. Developments in hardware as well as software has created a far better environment for such work.

Review Questions

1.5.1 Explain the difference between the local and global error of a numerical method for solving a differential equation. What is meant by the order of accuracy of a method?

1.5.2 Describe how Richardson extrapolation can be used to increase the order of accuracy of Euler's method.

1.5.3 Discuss some strategies for the adaptive control of step length and estimation of global accuracy in the numerical solution of differential equations.

Problems and Computer Exercises

1.5.1 (a) Integrate numerically using Euler's method the differential equation $dy/dt = y$, with initial conditions $y(0) = 1$, to $t = 0.4$, with step length $h = 0.2$ and $h = 0.1$.

(b) Extrapolate to $h = 0$, using the fact that the error is approximately proportional to the step length. Compare the result with the exact solution of the differential equation and determine the ratio of the errors in the results in (a) and (b).

(c) How many steps would have been needed in order to attain, without using extrapolation, the same accuracy as was obtained in (b)?

1.5.2 (a) Write a program for the simulation of the motion of the ball using Euler's method and the same initial values and parameter values as above. Print only x, y at integer values of t and at the last two points (i.e., $n = N$ and $n = N + 1$) as well as the x-coordinate of the landing point. Take $h = 0.05$ and $h = 0.1$. As postprocessing, improve the estimates of x^* by Richardson extrapolation, and estimate the error by comparison with the results given in the text above.

(b) In (1.5.7), in the equations for x_{n+1} and y_{n+1}, replace the right-hand sides u_n and v_n by, respectively, u_{n+1} and v_{n+1}. Then proceed as in (a) and compare the accuracy obtained with that obtained in (a).

(c) Choose initial values which correspond to what you think is reasonable for shot put. Make experiments with several values of u_0, v_0 for $c = 0$. How much is x^* influenced by the parameter cR^2/m?

1.5.3 Verify that Runge's second order method, as described by (1.5.10), is equivalent to the scheme described a few lines earlier (with Euler steps and Richardson extrapolation).

1.5.4 Write a program for Runge's second order method with automatic step size control that can be applied to a system of differential equations. Store the results so that they can be processed afterward, for example, to make a table of the results; draw curves showing $y(t)$ versus t, or (for a system) y_2 versus y_1; or draw some other interesting curves.

Apply the program to Examples 1.5.1 and 1.5.2, and to the circle test, that is,

$$y_1' = -y_2, \qquad y_2' = y_1,$$

with initial conditions $y_1(0) = 1$, $y_2(0) = 0$. Verify that the exact solution is a uniform motion along the unit circle in the (y_1, y_2)-plane. Stop the computations after 10 revolutions ($t = 20\pi$). Make experiments with different tolerances, and determine how small the tolerance has to be in order that the circle on the screen does not become "thick."

1.6 Monte Carlo Methods

1.6.1 Origin of Monte Carlo Methods

In most of the applications of probability theory one makes a mathematical formulation of a stochastic problem (i.e., a problem where chance plays some part) and then solves the problem by using analytical or numerical methods. In the **Monte Carlo method** one does the opposite; a mathematical or physical problem is given, and one constructs a **numerical game of chance**, the mathematical analysis of which leads to the same equations as the given problem for, e.g., the probability of some event, or for the mean of some random variable in the game. One plays it N times and estimates the relevant quantities by traditional statistical methods. Here N is a large number, because the standard deviation of a statistical estimate typically *decreases only inversely proportionally to* $\sqrt{N}$.

The idea behind the Monte Carlo method was used by the Italian physicist Enrico Fermi to study neutron diffusion in the early 1930s. Fermi used a small mechanical adding machine for this purpose. With the development of computers larger problems could be tackled. At Los Alamos in the late 1940s the use of the method was pioneered by von Neumann,[21] Ulam,[22] and others for many problems in mathematical physics including approximating complicated multidimensional integrals. The picturesque name of the method was coined by Nicholas Metropolis.

The Monte Carlo method is now so popular that the definition is too narrow. For instance, in many of the problems where the Monte Carlo method is successful, there is already an element of chance in the system or process which one wants to study. Thus such games of chance can be considered numerical simulations of the most important aspects. In this wider sense the "Monte Carlo methods" also include techniques used by statisticians since around 1900, under names like *experimental* or *artificial sampling*. For example, statistical experiments were used to check the adequacy of certain theoretical probability laws that had been derived mathematically by the eminent scientist W. S. Gosset. (He used the pseudonym "Student" when he wrote on probability.)

Monte Carlo methods may be used when the changes in the system are described with a much more complicated type of equation than a system of ordinary differential equations. Note that there are many ways to combine analytical methods and Monte Carlo methods. An important rule is that *if a part of a problem can be treated with analytical or traditional numerical methods, then one should use such methods.*

The following are some areas where the Monte Carlo method has been applied:

(a) Problems in reactor physics; for example, a neutron, because it collides with other particles, is forced to make a random journey. In infrequent but important cases the neutron can go through a layer of (say) shielding material (see Figure 1.6.1).

(b) Technical problems concerning traffic (in telecommunication systems and railway networks; in the regulation of traffic lights, and in other problems concerning automobile traffic).

(c) Queuing problems.

(d) Models of conflict.

(e) Approximate computation of multiple integrals.

(f) Stochastic models in financial mathematics.

Monte Carlo methods are often used for the evaluation of high-dimensional (10 to 100) integrals over complicated regions. Such integrals occur in such diverse areas as

[21]John von Neumann was born János Neumann in Budapest 1903, and died in Washington D.C. 1957. He studied under Hilbert in Göttingen in 1926–27, was appointed professor at Princeton University in 1931, and in 1933 joined the newly founded Institute for Advanced Studies in Princeton. He built a framework for quantum mechanics, worked in game theory, and was one of the pioneers of computer science.

[22]Stanislaw Marcin Ulam, born in Lemberg, Poland (now Lwow, Ukraine) 1909, and died in Santa Fe, New Mexico, USA, 1984. Ulam obtained his Ph.D. in 1933 from the Polytechnic institute of Lwow, where he studied under Banach. He was invited to Harvard University by G. D. Birkhoff in 1935 and left Poland permanently in 1939. In 1943 he was asked by von Neumann to come to Los Alamos, where he remained until 1965.

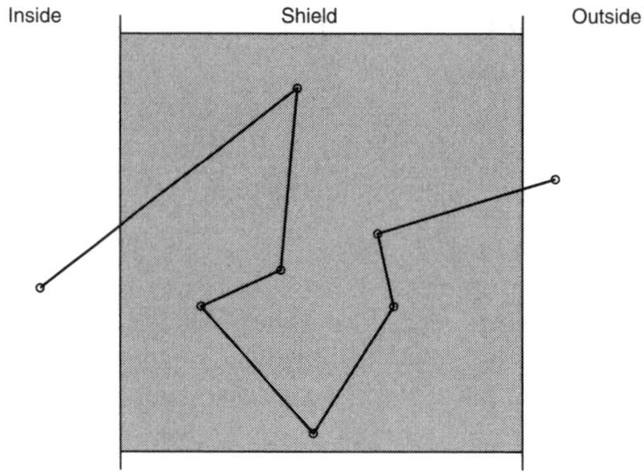

Figure 1.6.1. *Neutron scattering.*

quantum physics and mathematical finance. The integrand is then evaluated at random points uniformly distributed in the region of integration. The arithmetic mean of these function values is then used to approximate the integral; see Sec. 5.4.5.

In a simulation, one can study the result of various actions more cheaply, more quickly, and with less risk of organizational problems than if one were to take the corresponding actions on the actual system. In particular, for problems in applied operations research, it is quite common to take a shortcut from the actual system to a computer program for the game of chance, without formulating any mathematical equations. The game is then a model of the system. In order for the term "Monte Carlo method" to be correctly applied, however, **random choices** should occur in the calculations. This is achieved by using so-called **random numbers**; the values of certain variables are determined by a process comparable to dice throwing. Simulation is so important that several special programming languages have been developed exclusively for its use.[23]

1.6.2 Generating and Testing Pseudorandom Numbers

In the beginning, coins, dice, and roulette wheels were used for creating the randomness. For example, the sequence of 20 digits

$$11100\ 01001\ 10011\ 01100$$

is a record of 20 tosses of a coin where "heads" are denoted by 1 and "tails" by 0. Such digits are sometimes called (binary) **random digits**, assuming that we have a perfect coin—i.e., that heads and tails have the same probability of occurring. We also assume that the tosses of the coin are made in a statistically independent way.[24]

Similarly, decimal random digits could in principle be obtained by using a well-made icosahedral (20 sided) dice and assigning each decimal digit to two of its sides. Such

[23] One notable early example is the SIMULA programming language designed and built by Ole-Johan Dahl and Kristen Nygaard at the Norwegian Computing Center in Oslo 1962–67. It was originally built as a language for discrete event simulation, but was also influential because it introduced object-oriented programming concepts.

[24] Of course, these assumptions cannot be obtained in practice, as shown in theoretical and experimental studies by Persi Diaconis, Stanford University.

mechanical (or analogous electronic) devices have been used to produce **tables of random sampling digits**; the first one by Tippett was published in 1927 and was to be considered as a sequence of 40,000 independent observations of a random variable that equals one of the integer values $0, 1, 2, \ldots, 9$, each with probability $1/10$. In the early 1950s the Rand Corporation constructed a million-digit table of random numbers using an electrical "roulette wheel" ([295]). The wheel had 32 slots, of which 12 were ignored; the others were numbered from zero to nine twice. To test the quality of the randomness several tests were applied. Every block of a 1000 digits in the tables (and also the table as a whole) were tested. The *Handbook of Mathematical Functions* [1, Table 26.11][25] provides 2500 five-digit random numbers compiled from this set.

Example 1.6.1.

The random number generator, used for drawing of prizes of Swedish Premium Saving Bonds, was developed in 1962 by Dahlquist [86]. Speed is not a major concern for this application, since relatively few random decimal digits (about 50,000) are needed. Therefore, an algorithm which is easier to analyze was chosen. This uses a primary series of less than 240 decimal random digits produced by some other means. The length of this primary series is $n = p_1 + p_2 + \cdots + p_k$, where p_i are prime numbers and $p_i \neq p_j, i \neq j$. For the analysis it is assumed that the primary series is perfectly random.

The primary series is used to generate a much longer secondary series of prime numbers in a way that is best described by a mechanical analogy. Think of k cogwheels with p_i cogs, $i = 1 : k$, and place the digits from the primary series on the cogs of these. The first digit in the secondary series is obtained by adding the k digits (modulus 10) that are at the top position of each cogwheel. Then each wheel is turned one cog clockwise and the second digit is obtained in the same way as the first, etc. After $p_1 \cdot p_2 \cdots p_k$ steps we are back in the original position. This is the minimum period of the secondary series of random digits.

For the application mentioned above, $k = 7$ prime numbers, in the range $13 \leq p_i \leq 53$, are randomly selected. This gives a varying minimum period approximately equal to 10^8, which is much more than the number of digits used to produce the drawing list. Considering the public reaction, the primary series is generated by drawing from a tombola.

Random digits from a table can be packed together to give a sequence of equidistributed integers. For example, the sequence

$$55693 \quad 02945 \quad 81723 \quad 43588 \quad 81350 \quad 76302 \quad \ldots$$

can be considered as six five-digit random numbers, where each element in the sequence has a probability of 10^{-5} of taking on the value $0, 1, 2, \ldots, 99{,}999$. From the same digits one can also construct the sequence

$$0.556935, 0.029455, 0.817235, 0.435885, 0.813505, 0.763025, \ldots, \tag{1.6.1}$$

which can be considered a good approximation to a sequence of independent observations of a variable which is a sequence of uniform deviates in the interval $[0, 1)$. The 5 in the

[25]This classical *Handbook of Mathematical Functions*, edited by Milton Abramowitz and Irene A. Stegun, is used as a reference throughout this book. We will often refer to it as just "the Handbook."

sixth decimal place is added in order to get the correct mean (without this the mean would be 0.499995 instead of 0.5).

In a computer it is usually not appropriate to store a large table of random numbers. Several physical devices for random number generation have been proposed, using for instance electronic or radioactive noise, but very few seem to have been inserted in an actual computer system. Instead random numbers are usually produced by arithmetic methods, so-called **random number generators** (RNGs). The aim of a random number generator is to generate a sequence of numbers $u_1, u_2, u_3, \ldots$ that imitates the abstract mathematical concept of a sequence of mutually independent random variables uniformly distributed over the interval $[0, 1)$. Sequences obtained in this way are *uniquely determined* by one or more starting values called **seeds**, to be given by the user (or some default values). Random number generators should be analyzed theoretically and be backed by practical evidence from extensive statistical testing. According to a much quoted statement by D. H. Lehmer,[26]

> A random sequence is a vague notion ... in which each term is unpredictable to the uninitiated and whose digits pass a certain number of tests traditional with statisticians...

Because the set of floating-point numbers in [0, 1] is finite, although very large, there will eventually appear a number that has appeared before, say $u_{i+j} = u_i$ for some positive i, j. The sequence $\{u_n\}$ therefore repeats itself periodically for $n \geq i$; the length of the period is j. A truly random sequence is, of course, never periodic. For this and other reasons, a sequence generated like this is called a **pseudorandom** sequence. But the ability to repeat exactly the same sequence of numbers, which is needed for program verification and variance reduction, is a major advantage over generation by physical devices.

There are two popular myths about the making of a random number generator: first that it is impossible; second that it is trivial. We have seen that the first myth is correct, unless we add the prefix "pseudo."[27] The second myth, however, is completely false.

In a computer the fundamental concept is not a sequence of decimal random *digits*, but **uniform random deviates**, i.e., a sequence of *mutually independent observations of a random variable U* with a uniform distribution on [0, 1); the density function of U is thus (with a temporary notation)

$$f_1(u) = \begin{cases} 1 & \text{if } u \in [0, 1), \\ 0 & \text{otherwise.} \end{cases}$$

Random deviates for other distributions are generated by means of uniform deviates. For example, the variable $X = a+(b-a)U$ is a *uniform deviate on* $[a, b)$. Its density function is $f(x) = f_1((x-a)/(b-a))$. If $[a, b] = [0, 1]$, we usually write "uniform deviate" (without mentioning the interval). We often write "deviate" instead of "random deviate" when the meaning is evident from the context. Algorithms for generating deviates for several other distributions are given in Sec. 1.6.3.

---

[26] Some readers may think that Lehmer's definition is too vague. There have been many deep attempts for more precise formulation. See Knuth [230, pp. 149–179], who catches the flavor of the philosophical discussion of these matters and contributes to it himself.

[27] "Anyone who considers arithmetic methods of producing random numbers is, of course, in a state of sin."–John von Neumann (1951).

The most widely used generators for producing pseudorandom numbers are **multiple recursive generators**. These are based on a linear recurrence of order k,

$$x_n = \lambda_1 x_{n-1} + \cdots + \lambda_k x_{n-k} + c \mod P, \tag{1.6.2}$$

i.e., x_n is the remainder obtained when the right-hand side is divided by the modulus m. Here P is a positive integer and the coefficients $\lambda_1, \dots, \lambda_k$ belong to the set $\{0, 1, \dots, m-1\}$. The state at step n is $s_n = (x_{n-k+1}, \dots, x_n)$ and the generator is started from a seed $s_{k-1} = (x_0, \dots, x_{k-1})$. When m is large the output can be taken as the number $u_n = x_n/m$. For $k = 1$ we obtain the classical **mixed congruential method**

$$x_n = \lambda x_{n-1} + c \mod P.$$

An important characteristic of an RNG is its **period**, which is the maximum length of the sequence before it begins to repeat. Note that if the algorithm for computing x_n depends only on x_{n-1}, then the entire sequence repeats once the seed x_0 is duplicated. One can show that if $P = 2^t$ (which is natural on a binary computer) the period of the mixed congruential method is equal to 2^t, assuming that c is odd and that λ gives remainder 1 when divided by four. Also, if P is a prime number and if the coefficients λ_j satisfy certain conditions, then the generated sequence has the maximal period $m^k - 1$; see Knuth [230].

A good RNG should have a period that is guaranteed to be extremely long to make sure that no wrap-around can occur in practice. The linear congruential generator defined by

$$x_n = 16807 x_{n-1} \mod (2^{31} - 1), \tag{1.6.3}$$

with period $(2^{31} - 2)$, was proposed originally by Lewis, Goodman, and Miller (1969). It has been widely used in many software libraries for statistics, simulation, and optimization. In the survey by Park and Miller [283] this generator was proposed as a "minimal standard" against which other generators should be judged. A similar generator, but with the multiplier $7^7 = 823543$, was used in MATLAB 4.

Marsaglia [258] pointed out a theoretical weakness of all linear congruential generators. He showed that if k successive random numbers $(x_{i+1}, \dots, x_{i+k})$ at a time are generated and used to plot points in k-dimensional space, then they will lie on $(k-1)$-dimensional hyperplanes and will not fill up the space; see Figure 1.6.2 (left). More precisely, the values will lie on a set of at most $(k!m)^{1/k} \approx (k/e)m^{1/k}$ equidistant parallel hyperplanes in the k-dimensional hypercube $(0, 1)^k$. When the number of hyperplanes is too small, this obviously is a strong limitation to the k-dimensional uniformity. For example, for $m = 2^{31} - 1$ and $k = 3$, this is only about 1600 planes. This clearly may interfere with a simulation problem.

If the constants m, a, and c are not very carefully chosen, there will be many fewer hyperplanes than the maximum possible. One such infamous example is the linear congruential generator with $a = 65539$, $c = 0$, and $m = 2^{31}$ used by IBM mainframe computers for many years.

Another weakness of linear congruential generators is that their low order digits are much less random than their high order digits. Therefore, when only part of a generated random number is used, one should pick the high order digits.

One approach to better generators is to combine two RNGs. One possibility is to use a second RNG to shuffle the output of a linear congruential generator. In this way it is possible to get rid of some serial correlations in the output; see the generator ran1 described in Press et al. [294, Chapter 7.1].

A good generator should have been analyzed theoretically and be supported by practical evidence from extensive statistical and other tests. Knuth [230, Chapter 3] points out important ideas, concepts, and facts of the topic, but also mentions some scandalously poor RNGs that were in widespread daily use for decades as standard tools in computer libraries. Although the generators in daily use have improved, *many are still not satisfactory*. He ends this masterly chapter on random numbers with the following exercise: "Look at the subroutine library at your computer installation, and replace the random number generators by good ones. Try to avoid being too shocked at what you find."

L'Ecuyer [244] writes in 2001:

> Unfortunately, despite repeated warnings over the past years about certain classes of generators, and despite the availability of much better alternatives, simplistic and unsafe generators still abound in commercial software.

L'Ecuyer reports on tests of RNGs used in some popular software products. Microsoft Excel used the linear congruential generator

$$u_i = 9821.0u_{n-1} + 0.211327 \mod 1,$$

implemented directly for the u_i in floating-point arithmetic. Its period length depends on the precision of the arithmetic and it is not clear what it is. Microsoft Visual Basic used a linear congruential generator with period 2^{24}, defined by

$$x_i = 1140671485x_{i-1} + 12820163 \mod (2^{24}),$$

and takes $u_i = x_i/2^{24}$. The Unix standard library uses the recurrence

$$x_i = 25214903917x_{i-1} + 12820163 \mod (2^{48}),$$

with period 2^{48} and sets $u_i = x_i/2^{48}$. The Java standard library uses the same recurrence but constructs random deviates u_i from x_{2i} and x_{2i+1}.

In MATLAB 5 and later versions the previous linear congruential generator has been replaced with a much better generator, based on ideas of Marsaglia; see Figure 1.6.2 (right). This generator has a 35 element state vector and can generate all the floating-point numbers in the closed interval $[2^{-53}, 1 - 2^{-53}]$. Theoretically it can generate 2^{1492} values before repeating itself; see Moler [266]. If one generates one million random numbers a second it would take 10^{435} years before it repeats itself!

Some modern linear RNGs can generate huge samples of pseudorandom numbers very fast and reliably. The multiple recursive generator MRG32k3a proposed by L'Ecuyer has a period near 2^{191}. The **Mersenne twister** MT19937 by Matsumoto and Nishimura [261], the "world champion" of RNGs in the year 2000, has a period length of $2^{19,937} - 1$!

Many statistical tests have been adapted and extended for the examination of *arithmetic* methods of (pseudo)random number generation. In these, the observed frequencies

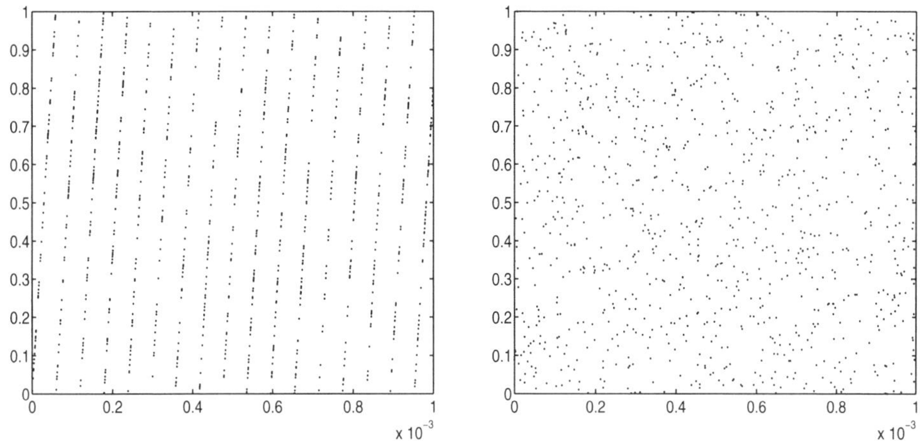

Figure 1.6.2. *Plots of pairs of* 10^6 *random uniform deviates* (U_i, U_{i+1}) *such that* $U_i < 0.0001$. *Left: MATLAB* 4; *Right: MATLAB* 5.

for some random variable associated with the test are compared with the theoretical frequencies on the hypothesis that the numbers are independent observations from a true sequence of random digits without bias. This is done by means of the famous χ^2-test of K. Pearson [287],[28] which we now describe.

Suppose that the space S of the random variable is divided into a finite number r of nonoverlapping parts $S_1, \dots, S_r$. These parts may be groups into which the sample values have been arranged for tabulation purposes. Let the corresponding group probabilities be

$$p_i = Pr(S_i), \quad i = 1, \dots, r, \quad \sum_{i=1}^{r} p_i = 1.$$

We now form a measure of the deviation of the observed frequencies $\nu_1, \dots, \nu_r$, $\sum_i \nu_i = n$, from the expected frequencies

$$\chi^2 = \sum_{i=1}^{r} \frac{(\nu_i - np_i)^2}{np_i} = \sum_{i=1}^{r} \frac{\nu_i^2}{np_i} - n. \tag{1.6.4}$$

It is known that as n tends to infinity the distribution of χ^2 tends to a limit independent of $Pr(S_i)$, which is the χ^2-distribution with $r - 1$ degrees of freedom.

Let χ_p^2 be a value such that $Pr(\chi^2 > \chi_p^2) = p\,\%$. Here p is chosen so small that we are practically certain that an event of probability $p\,\%$ will not occur in a single trial. The

[28]This paper, published in 1900 by the English mathematician Karl Pearson (1857–1936), is considered one of the foundations of modern statistics. In it he gave several examples, such as proving that some runs at roulette that he had observed during a visit to Monte Carlo were so far from expectations that the odds against an honest wheel were about 10^{29} to one.

hypothesis is rejected if the observed value of χ^2 is larger than χ_p^2. Often a rejection level of 5% or 1% is used.

Example 1.6.2.

In an experiment consisting of $n = 4040$ throws with a coin, $\nu_1 = 2048$ heads were obtained and hence $\nu_2 = n - \nu_1 = 1992$ tails. Is this consistent with the hypothesis that there is a probability of $p_1 = 1/2$ of throwing tails? Computing

$$\chi^2 = \frac{(\nu_1 - np_1)^2}{np_1} + \frac{(n - \nu_1 - np_1)^2}{np_1} = 2\frac{(2048 - 2020)^2}{2020} = 0.776$$

and using a rejection level of 5%, we find from a table of the χ^2-distribution with one degree of freedom that $\chi_5^2 = 3.841$. Hence the hypothesis is accepted at this level.

Several tests that have been used for testing RNGs are described in Knuth [230, Sec. 3.3]. Some of them are the following:

1. **Frequency test**. This test is to find out if the generated numbers are equidistributed. One divides the possible outcomes into equal nonoverlapping intervals and tallies the number of numbers in each interval.

2. **Poker test**. This test applies to generated digits, which are divided into nonoverlapping groups of five digits. Within the groups we study some (unordered) combinations of interest in poker. These are given below, together with their probabilities.

All different:	abcde	0.3024
One pair:	aabcd	0.5040
Two pairs:	aabbc	0.1080
Three of a kind:	aaabc	0.0720
Full house:	aaabb	0.0090
Four of a kind:	aaaab	0.0045
Five of a kind:	aaaaa	0.0001

3. **Gap test**. This test examines the length of "gaps" between occurrences of U_j in a certain range. If α and β are two numbers with $0 \le \alpha < \beta \le 1$, we consider the length of consecutive subsequences $U_j, U_{j+1}, \dots, U_{j+r}$ in which U_{j+r} lies between α and β but $U_j, U_{j+1}, \dots, U_{j+r-1}$ do not. These subsequence then represents a gap of length r.

   The special cases $(\alpha, \beta) = (0, 1/2)$ or $(1/2, 1)$ give rise to tests called "runs above the mean" and "runs below the mean," respectively.

   Working with single digits, we see that the gap equals the distance between two equal digits. The probability of a gap of length r in this case equals

$$p_r = 0.1(1 - 0.1)^r = 0.1(0.9)^r, \quad r = 0, 1, 2, \dots.$$

Example 1.6.3.

To test the two-dimensional behavior of an RNG we generated 10^6 pseudorandom numbers U_i. We then placed the numbers (U_i, U_{i+1}) in the unit square of the plot. A thin slice of the surface of the square, 0.0001 wide by 1.0 high, was then cut on its left side and stretched out horizontally. This corresponds to plotting only the pairs (U_i, U_{i+1}) such that $U_i < 0.0001$ (about 1000 points).

In Figure 1.6.2 we show the two plots from the generators in MATLAB 4 and MATLAB 5, respectively. The lattice structure is quite clear in the first plot. With the new generator no lattice structure is visible.

A statistical test studied by Knuth [230] is the **collision test**. In this test the interval useful $[0, 1)$ is first cut into n equal intervals, for some positive integer n. This partitions the hypercube $[0, 1)^d$ into $k = n^d$ cubic boxes. Then N random points are generated in $[0, 1)^d$ and we record the number of times C that a point falls in a box that already has a point in it. The expectation of the random number C is known to be of very good approximation when N is large. Indeed, C follows approximatively a Poisson distribution with mean equal to $N^2/(2k)$.

For this and other similar tests it has been observed that when the sample size N is increased the test starts to fail when N reaches a critical value N_0, and the failure is clear for all larger values of N. For the collision test it was observed by L'Ecuyer [244] that $N_0 \approx 16\rho^{1/2}$ for good linear congruential generators,where ρ is the period of the RNG. For another statistical test called the *birthday spacing* test the relation was $N_0 \approx 16\rho^{1/3}$.

From such tests is can be concluded that when large sample sizes are needed many RNGs are unsafe to use and can fail decisively. A period of 2^{24} or even 2^{48} may not be enough. Linear RNGs are also unsuitable for cryptographic applications, because the output is too predictable. For this reason, nonlinear generators have been developed, but these are in general much slower than the linear generators.

1.6.3 Random Deviates for Other Distributions

We have so far discussed how to generate sequences that behave as if they were random uniform deviates U on $[0, 1)$. By arithmetic operations one can form random numbers with other distributions. A simple example is that the random numbers

$$S = a + (b - a)U$$

will be uniformly distributed on $[a, b)$.

Monte Carlo methods often call for other kinds of distributions. We shall show here how to use uniform deviates to generating random deviates X for several other distributions. Many of the tricks used were originally suggested by John von Neumann in the early 1950s, but have since been improved and refined.

Discrete Distributions

Making a random choice from a finite number k of *equally probable* possibilities is equivalent to generating a random integer X between 1 and k. To do this we take a random deviate

U uniformly distributed on [0, 1), multiply it by k, and take the integer part

$$X = \lceil kU \rceil;$$

here $\lceil x \rceil$ denotes the smallest integer larger than or equal to x. There will be a small error because the set of floating-point numbers is finite, but this is usually negligible.

In a *more general situation*, we might want to give different probabilities to the values of a variable. Suppose we assign the values $X = x_i$, $i = 1 : k$ the probabilities p_i, $i = 1 : k$; note that $\sum p_i = 1$. We can then generate a uniform number U and let

$$X = \begin{cases} x_1 & \text{if } 0 \le U < p_1, \\ x_2 & \text{if } p_1 \le U < p_1 + p_2, \\ \vdots & \\ x_k & \text{if } p_1 + p_2 + \cdots p_{k-1} \le U < 1. \end{cases}$$

If k is large, and the sequence $\{p_i\}$ is irregular, it may require some thought how to find x quickly for a given u. See the analogous question of finding a first guess to the root of (1.6.5) below and the discussion in Knuth [230, Sec. 3.4.1].

A General Transformation from U to X

Suppose we want to generate numbers for a random variable X with a given continuous or discrete distribution function $F(x)$. (In the discrete case, the graph of the distribution function becomes a staircase; see the formulas above.) A general method for this is to solve the equation

$$F(X) = U \tag{1.6.5}$$

or, equivalently, $X = F^{-1}(U)$; see Figure 1.6.3. Because $F(x)$ is a nondecreasing function, and $Pr\{U \le u\} = u$ for all $u \in [0, 1]$, (1.6.5) is proved by the line

$$Pr\{X \le x\} = Pr\{F(X) \le F(x)\} = Pr\{U \le F(x)\} = F(x).$$

How to solve (1.6.5) efficiently is the main problem with this method. For some distributions we shall describe better methods below.

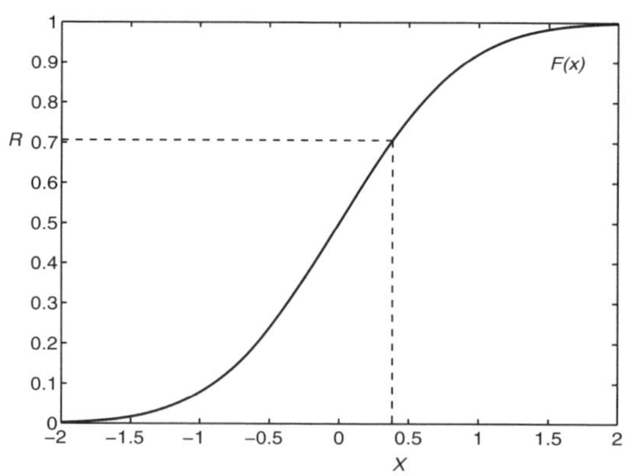

Figure 1.6.3. *Random number with distribution $F(x)$.*

Exponential Deviates

The exponential distribution with parameter $\lambda > 0$ occurs in queuing problems, for example, in telecommunications, to model arrival and service times. The important property is that the intervals of time between two successive events are a sequence of exponential deviates. The exponential distribution with mean $1/\lambda$ has density function $f(t) = \lambda e^{-\lambda t}$, $t > 0$, and distribution function

$$F(x) = \int_0^x \lambda e^{-\lambda t}\,dt = 1 - e^{-\lambda x}. \tag{1.6.6}$$

Using the general rule given above, exponentially distributed random numbers X can be generated as follows: Let U be a uniformly distributed random number in $[0, 1]$. Solving the equation $1 - e^{-\lambda X} = U$, we obtain

$$X = -\lambda^{-1}\ln(1 - U).$$

A drawback of this method is that the evaluation of the logarithm is relatively slow.

One important use of exponentially distributed random numbers is in the generation of so-called **Poisson processes**. Such processes are often fundamental in models of telecommunication systems and other service systems. A Poisson process with frequency parameter λ is a sequence of events characterized by the property that the probability of occurrence of an event in a short time interval $(t, t + \Delta t)$ is equal to $\lambda \cdot \Delta t + o(\Delta t)$, independent of the sequence of events previous to time t. An "event" can mean a call on a telephone line, the arrival of a customer to a store, etc. For simulating a Poisson process one can use the important property that the intervals of time between two successive events are independent exponentially distributed random numbers.

Normal Deviates

A normal deviate $N = N(0, 1)$ with zero mean and unit standard deviation has the density function

$$f(x) = \frac{1}{\sqrt{2\pi}} e^{-x^2/2}.$$

Then $\mu + \sigma N$ is a normal deviate with mean μ and standard deviation σ with density function $\frac{1}{\sigma} f((x - \mu)/\sigma)$. Since the normal distribution function

$$\Phi(x) = \frac{1}{\sqrt{2\pi}} \int_{-\infty}^x e^{-t^2/2}\,dt \tag{1.6.7}$$

is not an elementary function, solving equation (1.6.5) would be time consuming.

Fortunately, random normal deviates can be obtained in easier ways. In the **polar algorithm** a random point in the unit disk is first generated as follows. Let U_1, U_2 be two independent uniformly distributed random numbers on $[0, 1]$. Then the point (V_1, V_2), where $V_i = 2U_i - 1$, $i = 1, 2$, is uniformly distributed in the square $[-1, 1] \times [-1, 1]$. If we compute $S = V_1^2 + V_2^2$ and reject the point if it is outside the unit circle, i.e., if $S > 1$, remaining points will be uniformly distributed on the unit disk. For each accepted point we then form

$$N_1 = \tau V_1, \qquad N_2 = \tau V_2, \qquad \tau = \sqrt{\frac{-2\log S}{S}}. \tag{1.6.8}$$

It can be proved that N_1, N_2 are two independent normally distributed random numbers with zero mean and unit standard deviation.

We point out that N_1, N_2 can be considered to be rectangular coordinates of a point whose polar coordinates (r, ϕ) are determined by the equations

$$r^2 = N_1^2 + N_2^2 = -2 \ln S, \qquad \cos\phi = U_1/\sqrt{S}, \qquad \sin\phi = U_2/\sqrt{S}.$$

The correctness of the above procedure follows from the fact that the distribution function for a pair of independent normally distributed random variables is rotationally symmetric (uniformly distributed angle) and that their sum of squares is exponentially distributed with mean 2. For a proof of this, see Knuth [230, p. 123].

The polar algorithm (used previously in MATLAB 4) is not optimal. First, about $1 - \pi/4 \approx 21.5\%$ of the uniform numbers are rejected because the generated point falls outside the unit disk. Further, the calculation of the logarithm contributes significantly to the cost. From MATLAB version 5 and later, a more efficient table look-up algorithm developed by Marsaglia and Tsang [260] is used. This is called the "ziggurat" algorithm after the name of ancient Mesopotamian terraced temple mounds which look like two-dimensional step functions. A popular description of the ziggurat algorithm is given by Moler [267]; see also [220].

Example 1.6.4.

To simulate a two-dimensional Brownian motion, trajectories are generated as follows. Initially the particle is located at the origin $w_0 = (0, 0)^T$. At each time step the particle moves randomly,

$$w_{k+1} = w_k + h \begin{pmatrix} N_{1k} \\ N_{2k} \end{pmatrix}, \qquad k = 0 : n,$$

where N_{1k} and N_{2k} are normal random deviates generated according to (1.6.8). Figure 1.6.4 shows plots of 32 simulated paths with $h = 0.1$, each consisting of $n = 64$ time steps.

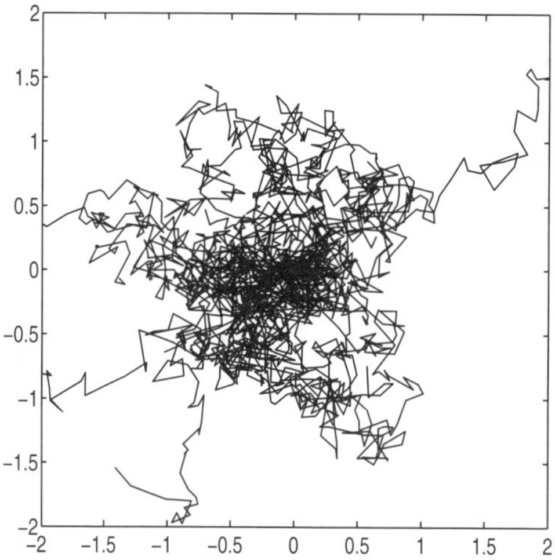

Figure 1.6.4. *Simulated two-dimensional Brownian motion. Plotted are* 32 *simulated paths with* $h = 0.1$, *each consisting of* 64 *steps.*

Chi-Square Distribution

The **chi-square distribution** function $P(\chi^2, n)$ is related to the incomplete gamma function (see Abramowitz and Stegun [1, Sec. 6.5]):

$$P(\chi^2, n) = \gamma(n/2, \chi^2/2). \tag{1.6.9}$$

Its complement $Q(\chi^2, n) = 1 - P(\chi^2, n)$ is the probability that the observed chi-square will exceed the value χ^2, even for a correct model. Subroutines for evaluating the χ^2-distribution function as well as other important statistical distribution functions are given in [294, Sec. 6.2–6.3].

Numbers belonging to the chi-square distribution can also be obtained by using the definition of the distribution. If $N_1, N_2, \dots, N_n$ are normal deviates with zero mean and unit variance, the number

$$Y_n = N_1^2 + N_2^2 + \cdots + N_n^2$$

is distributed as χ^2 with n degrees of freedom.

Other Methods

Several other methods to generate random deviates with Poisson, gamma, and binomial distribution are described in Knuth [230, Sec. 3.4]) and Press et al. [294, Chapter 7.3]. The **rejection method** is based on ideas of von Neumann (1951). A general method introduced by Marsaglia [257] is the **rectangle-wedge-tail method**; see references in Knuth [230]. Powerful combinations of rejection methods and the rectangle-wedge-tail method have been developed.

1.6.4 Reduction of Variance

From statistics, we know that if one makes n independent observations of a quantity whose standard deviation is σ, then the standard deviation of the mean is $\sigma/\sqrt{n}$. Hence, to increase the accuracy by a factor of 10 (say) we have to increase the number of experiments n by a factor 100.

Often a more efficient way than increasing the number of samples is to try to decrease the value of σ by redesigning the experiment in various ways. Assume that one has two ways (which require the same amount of work) of carrying out an experiment, and these experiments have standard deviations σ_1 and σ_2 associated with them. If one repeats the experiments n_1 and n_2 times (respectively), the same precision will be obtained if $\sigma_1/\sqrt{n_1} = \sigma_2/\sqrt{n_2}$, or

$$n_1/n_2 = \sigma_1^2/\sigma_2^2. \tag{1.6.10}$$

Thus if a variance reduction by a factor k can be achieved, then the number of experiments needed is also reduced by the same factor k.

Example 1.6.5.

In 1777 Buffon[29] carried out a probability experiment by throwing sticks over his shoulder onto a tiled floor and counting the number of times the sticks fell across the lines between the tiles. He stated that the favorable cases correspond "to the area of part of the cycloid whose generating circle has diameter equal to the length of the needle." To simulate Buffon's experiment we suppose a board is ruled with equidistant parallel lines and that a needle fine enough to be considered a segment of length l not longer than the distance d between consecutive lines is thrown on the board. The probability is then $2l/(\pi d)$ that it will hit one of the lines.

The Monte Carlo method and this game can be used to approximate the value of π. Take the distance δ between the center of the needle and the lines and the angle ϕ between the needle and the lines to be random numbers. By symmetry we can choose these to be rectangularly distributed on $[0, d/2]$ and $[0, \pi/2]$, respectively. Then the needle hits the line if $\delta < (l/2)\sin\phi$.

We took $l = d$. Let m be the number of hits in the first n throws in a Monte Carlo simulation with 1000 throws. The expected value of m/n is therefore $2/\pi$, and so $2n/m$ is an estimate of π after n throws. In the left part of Figure 1.6.5 we see how $2n/m$ varies with n in one simulation. The right part compares $|m/n - 2/\pi|$ with the standard deviation of m/n, which equals

$$\sqrt{\frac{2}{\pi}\left(1 - \frac{2}{\pi}\right)\frac{1}{n}}$$

and is, in the log-log diagram, represented by a straight line, the slope of which is $-1/2$. This can be taken as a test that the RNG in MATLAB is behaving correctly! (The spikes, directed downward in the figure, typically indicate where $m/n - 2/\pi$ changes sign.)

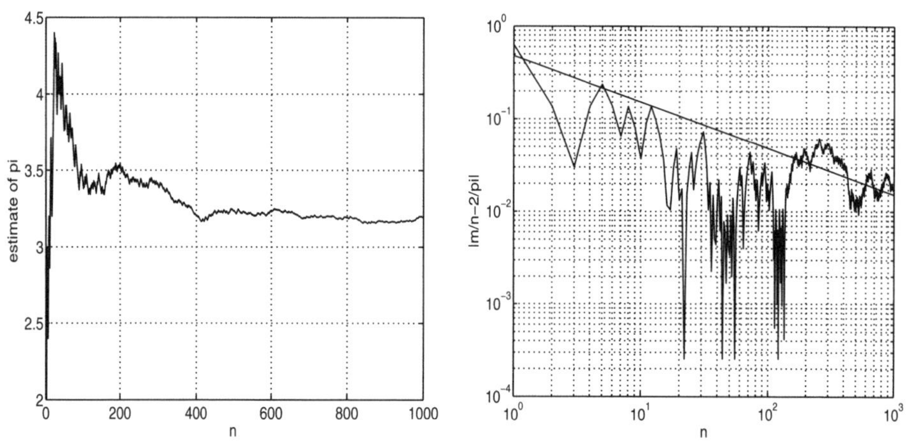

Figure 1.6.5. *The left part shows how the estimate of π varies with the number of throws. The right part compares $|m/n - 2/\pi|$ with the standard deviation of m/n.*

[29]Comte de Buffon (1707–1788), French natural scientist who contributed to the understanding of probability. He also computed the probability that the sun would continue to rise after having been observed to rise on n consecutive days.

An important means of reducing the variance of estimates obtained from the Monte Carlo method is to use **antithetic sequences**. If U_i, $i = 1 : n$, is a sequence of random uniform deviates on [0, 1], then $U_i' = 1 - U_i$, $i = 1 : n$, is an antithetic uniformly distributed sequence. From the sequence in (1.6.1) we get the antithetic sequence

$$0.443065, 0.970545, 0.182765, 0.564115, 0.186495, 0.236975, \dots. \tag{1.6.11}$$

Antithetic sequences of normally distributed numbers with zero mean are obtained simply by reversing the sign of the original sequence.

Roughly speaking, since the influence of chance has opposing effects in the two antithetic experiments, one can presume that the effect of chance on the *means* is much less than the effect of chance in the original experiments. In the following example we show how to make a quantitative estimate of the reduction of variance accomplished with the use of antithetic experiments.

Example 1.6.6.

Suppose the numbers x_i are the results of statistically independent measurements of a quantity with expected value μ, and standard deviation σ. Set

$$\bar{x} = \frac{1}{n}\sum_{i=1}^{n} x_i, \qquad s^2 = \frac{1}{n-1}\sum_{i=1}^{n}(x_i - \bar{x})^2.$$

Then $\bar{x}$ is an estimate of μ, and s is an estimate of σ.

In ten simulations and their antithetic experiments of a service system, the following values were obtained for the treatment time:

$$685 \quad 1\,045 \quad 718 \quad 615 \quad 1\,021 \quad 735 \quad 675 \quad 635 \quad 616 \quad 889\,.$$

From this experiment the mean for the treatment time is estimated as 763.4, and the standard deviation 51.5. Using an antithetic series, the following values were obtained:

$$731 \quad 521 \quad 585 \quad 710 \quad 527 \quad 574 \quad 607 \quad 698 \quad 761 \quad 532\,.$$

The series means are thus

$$708 \quad 783 \quad 651.5 \quad 662.5 \quad 774 \quad 654.5 \quad 641 \quad 666.5 \quad 688.5 \quad 710.5\,,$$

from which one gets the estimate 694.0 ± 15.9.

When one instead supplements the first sequence with 10 values using independent random numbers, the estimate 704 ± 36 using all 20 values is obtained. These results indicate that, in this example, using antithetical sequence produces the desired accuracy with $(15.9/36)^2 \approx 1/5$ of the work required if completely independent random numbers are used. This rough estimate of the work saved is uncertain, but indicates that it is very profitable to use the technique of antithetic series.

Example 1.6.7.

Monte Carlo methods have been used successfully to study queuing problems. A well-known example is a study by Bailey [12] to determine how to give appointment times to patients at a polyclinic. The aim is to find a suitable balance between the mean waiting times of both patients and doctors. This problem was in fact solved analytically—much later—after Bailey already had the results that he wanted; this situation is not uncommon when numerical methods (and especially Monte Carlo methods) have been used.

Suppose that k patients have been booked at the time $t = 0$ (when the clinic opens), and that the rest of the patients (altogether 10) are booked at intervals of 50 time units thereafter. The time of treatment is assumed to be exponentially distributed with mean 50. (Bailey used a distribution function which was based on empirical data.) We use the following numbers which are taken from a table of exponentially distributed random numbers with mean 100:

$$211 \quad 3 \quad 53 \quad 159 \quad 24 \quad 35 \quad 54 \quad 39 \quad 44 \quad 13\,.$$

Three alternatives, $k = 1, 2, 3$, are to be simulated. *By using the same random numbers for each k (hence the same treatment times) one gets a* **reduced variance** *in the estimate of the change in waiting times as k varies.* s

The computations are shown in Table 1.6.1. The following abbreviations are used in what follows: P = patient, D = doctor, T = treatment. An asterisk indicates that the patient did not need to wait. In the table, P_{arr} follows from the rule given previously for booking patients. The treatment time T_{time} equals $R/2$, where R are exponentially distributed numbers with mean 100 taken from a table, i.e., the mean treatment time is 50. T_{beg} equals the larger of the number P_{arr} (on the same row) and T_{end} (in the row just above), where $T_{end} = T_{beg} + T_{time}$.

Table 1.6.1. *Simulation of waiting times for patients at a polyclinic.*

	$k=1$					$k=2$	
P_{no}	P_{arr}	T_{beg}	R	T_{time}	T_{end}	P_{arr}	T_{end}
1	0*	0	211	106	106	0*	106
2	50	106	3	2	108	0	108
3	100	108	53	26	134	50	134
4	150*	150	159	80	230	100	214
5	200	230	24	12	242	150	226
6	250*	250	35	18	268	200	244
7	300*	300	54	27	327	250*	277
8	350*	350	39	20	370	300*	320
9	400*	400	44	22	422	350*	372
10	450*	450	13	6	456	400*	406
Σ	2250			319	2663	1800	2407

From the table we find that for $k = 1$ the doctor waited the time $D = 456-319 = 137$; the total waiting time for patients was $P = 2663 - 2250 - 319 = 94$. For $k = 2$ the corresponding waiting times were $D = 406-319 = 87$ and $P = 2407-1800-319 = 288$. Similar calculations for $k = 3$ gave $D = 28$ and $P = 553$ (see Figure 1.6.6). For $k \geq 4$ the doctor never needs to wait.

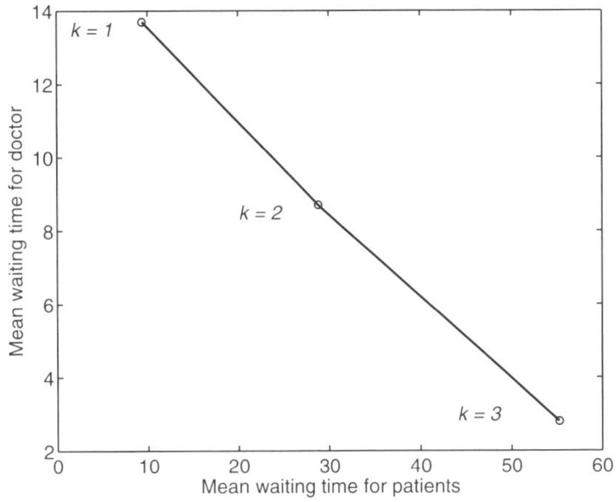

Figure 1.6.6. *Mean waiting times for doctor and patients at polyclinic.*

One cannot, of course, draw any tenable conclusions from one experiment. More experiments should be done in order to put the conclusions on statistically solid ground. Even isolated experiments, however, can give valuable suggestions for the planning of subsequent experiments, or perhaps suggestions of appropriate approximations to be made in the analytic treatment of the problem. The large-scale use of Monte Carlo methods requires careful planning to avoid drowning in enormous quantities of unintelligible results.

Two methods for **reduction of variance** have been introduced here: *antithetic sequence of random numbers* and the technique of using *the same random numbers in corresponding situations*. The latter technique is used when studying the changes in behavior of a system when a certain parameter is changed, for example, the parameter k in Example 1.6.7. Note that for this we need to be able to restart the RNG using the same seed. Other effective methods for reducing variance are **importance sampling** and **splitting techniques**; see Hammersley and Handscomb [183].

Review Questions

1.6.1 What is meant by the Monte Carlo method? Describe the origin of the method and give some typical applications. In general, how fast does the error decrease in estimates obtained with the Monte Carlo method?

1.6.2 Describe a linear congruential generator for generating a sequence of uniformly distributed pseudorandom numbers. What are some important properties of such a generator?

1.6.3 Describe a general method for obtaining random numbers with a given discrete or continuous distribution from uniformly distributed random numbers. Give examples of its use.

1.6.4 Describe some statistical tests which can be applied to an RNG.

1.6.5 What are the most important properties of a Poisson process? How can one generate a Poisson process with the help of random numbers?

1.6.6 What is the mixed congruential method for generating pseudorandom numbers? What important difference is there between the numbers generated by this method and "genuine" random numbers?

1.6.7 Explain what is meant by *reduction of variance* in estimates made with the Monte Carlo method. Give three methods for reduction of variance. What is the quantitative connection between reducing variance and decreasing the amount of computation needed in a given problem?

Problems and Computer Exercises

1.6.1 (C. Moler) Consider the toy RNG, $x_i = ax_i \bmod m$, with $a = 13$, $m = 31$, and start with $x_0 = 1$. Show that this generates a sequence consisting of a permutation of all integers from 1 to 30, and then repeats itself. Conclude that this generator has period $m - 1 = 30$, equal to the maximum possible.

1.6.2 Simulate (say) 360 throws with two standard dice. Denote the sum of the number of dots on the two dice on the nth throw by Y_n, $2 \le Y_n \le 12$. Tabulate or draw a histogram of the (absolute) frequency of the occurrence of j dots versus j, $j = 2:12$. Make a conjecture about the true value of $Pr(Y_n = j)$. Try to confirm it by repeating the experiment with fresh uniform random numbers. When you have found the right conjecture, you will find that it is not hard to prove.

1.6.3 (a) Let X, Y be independent uniform random numbers on the interval $[0, 1]$. Show that $Pr(X^2 + Y^2 \le 1) = \pi/4$, and estimate this probability by a Monte Carlo experiment with (say) 1000 pairs of random numbers. Produce a graphical output like in the Buffon needle problem.

(b) Conduct an antithetic experiment, and take the average of the two results. Is the average better than one could expect if the second experiment had been independent of the first one?

(c) Estimate similarly the volume of the four-dimensional unit ball. If you have enough time, use more random numbers. (The exact volume of the unit ball is $\pi^2/2$.)

1.6.4 A famous result by P. Diaconis asserts that it takes approximately $\frac{3}{2}\log_2 52 \approx 8.55$ riffle shuffles to randomize a deck of 52 cards, and that randomization occurs abruptly according to a "cutoff phenomenon." (After six shuffles the deck is still far from random!)

The following definition can be used for simulating a riffle shuffle. The deck of cards is first cut roughly in half according to a binomial distribution, i.e., the probability that ν cards are cut is $\binom{n}{\nu}2^{-n}$. The two halves are then riffled together by dropping cards roughly alternately from each half onto a pile, with the probability of a card being dropped from each half being proportional to the number of cards in it.

Write a program that uses uniform random numbers, and perhaps uses the formula $X = \lceil kR \rceil$, for several values of k, to simulate a random shuffle of a deck of 52 cards according to the above precise definition. This is for a *numerical* game; do not spend time drawing beautiful hearts, clubs, etc.

1.6.5 Brownian motion is the irregular motion of dust particles suspended in a fluid, being bombarded by molecules in a random way. Generate two sequences of random normal deviates a_i and b_i, and use these to simulate Brownian motion by generating a path defined by the points (x_i, y_i), where $x_0 = y_0 = 0$, $x_i = x_{i-1} + a_i$, $y_i = y_{i-1} + b_i$. Plot each point and connect the points with a straight line to visualize the path.

1.6.6 Repeat the simulation in the queuing problem in Example 1.6.7 for $k = 1$ and $k = 2$ using the sequence of exponentially distributed numbers R,

$$13 \quad 365 \quad 88 \quad 23 \quad 154 \quad 122 \quad 87 \quad 112 \quad 104 \quad 213,$$

antithetic to that used in Example 1.6.7. Compute the mean of the waiting times for the doctor and for all patients for this and the previous experiment.

1.6.7 A target with depth $2b$ and very large width is to be shot at with a cannon. (The assumption that the target is very wide makes the problem one-dimensional.) The distance to the center of the target is unknown, but estimated to be D. The difference between the actual distance and D is assumed to be a normally distributed random variable $X = N(0, \sigma_1)$.
One shoots at the target with a salvo of three shots, which are expected to travel a distance $D-a$, D, and $D+a$, respectively. The difference between the actual and the expected distance traveled is assumed to be a normally distributed random variable $N(0, \sigma_2)$; the resulting error component in the three shots is denoted by Y_{-1}, Y_0, Y_1. We further assume that these three variables are independent of each other and X.
One wants to know how the probability of at least one "hit" in a given salvo depends on a and b. Use normally distributed pseudorandom numbers to shoot 10 salvos and determine for each salvo the least value of b for which there is at least one hit in the salvo. Show that this is equal to

$$\min_k |X - (Y_k + ka)|, \quad k = -1, 0, 1.$$

Fire an antithetic salvo for each salvo.
Draw curves, for both $a = 1$ and $a = 2$, which give the probability of a hit as a function of the depth of the target. Use $\sigma_1 = 3$ and $\sigma_2 = 1$, and the same random numbers.

Notes and References

The methods and problems presented in this introductory chapter will be studied in greater detail later in this volume and in Volume II. In particular, numerical quadrature methods are studied in Chapter 5 and methods for solving a single nonlinear equation in Chapter 6. For a survey of sorting algorithms we refer to [294, Chapter 8]. A comprehensive treatment of sorting and searching is given in Knuth [231].

Although the history of Gaussian elimination goes back at least to Chinese mathematicians (about 250 B.C.), there was no practical experience of solving large linear systems until the advent of computers in the 1940s. Gaussian elimination was the first numerical algorithm to be subjected to a rounding error analysis. In 1946 there was a mood of pessimism about the stability of Gaussian elimination. Bounds had been produced showing that the error in the solution would be proportional to 4^n. This suggested that it would be impossible to solve even systems of modest order. A few years later J. von Neumann and H. H. Goldstein published more relevant error bounds. In 1948 A. M. Turing wrote a remarkable paper [362], in which he formulated the LU factorization and introduced matrix condition numbers.

Several of the great mathematicians at the turn of the nineteenth century worked on methods for solving overdetermined linear systems. In 1799, Laplace used the principle of minimizing the sum of absolute errors $|r_i|$, with the added conditions that the errors sum to zero. This leads to a solution x that satisfies at least n equations exactly. The method of least squares was first published as an algebraic procedure by Legendre in 1805 [245]. Gauss justified the least squares principle as a statistical procedure in [138], where he claimed to have used the method since 1795. This led to one of the most famous priority disputes in the history of mathematics. Gauss further developed the statistical aspects in 1821–1823. For an interesting account of the history of the invention of least squares, see Stigler [337].

For a comprehensive treatment of all aspects of random numbers we refer to Knuth [230]. Another good reference on the state of the art is the monograph by Niederreiter [275]. Guidelines for choosing a good RNG are given in Marsaglia [259], the monograph by Gentle [152], and in the two surveys L'Ecuyer [242, 243]. Hellekalek [191] explains how to access RNGs for practitioners. An introduction to Monte Carlo methods and their applications is given by Hammersley and Handscomb [183]. There is a close connection between random number generation and data encryptation; see Press et al. [294, Chapter 7.5].

Some later chapters in this book assume a working knowledge in numerical linear algebra. Online Appendix A gives a brief survey of matrix computations. A more in-depth treatment of direct and iterative methods for linear systems, least squares, and eigenvalue problems is planned for Volume II. Some knowledge of modern analysis including analytic functions is also needed for some more advanced parts of the book. The classical textbook by Apostol [7] is highly recommended as a suitable reference.

The *James & James Mathematics Dictionary* [210] is a high-quality general mathematics dictionary covering arithmetic to calculus, and it includes a multilingual index. *CRC Concise Encyclopedia of Mathematics* [370] is a comprehensive compendium of mathematical definitions, formulas, and references. A free Web encyclopedia containing surveys and references is Eric Weisstein's MathWorld at `mathworld.wolfram.com`.

The development of numerical analysis during the period when the foundation was laid in the sixteenth through the nineteenth century is traced in Goldstine [160]. Essays on the history of scientific computing can be found in Nash [274]. An interesting account of the developments in the twentieth century is given in [56]. An eloquent essay on the foundations of computational mathematics and its relation to other fields is given by Baxter and Iserles [21].

In 2000–2001, the *Journal of Computational and Applied Mathematics* published a series of papers on Numerical Analysis of the 20th Century, with the aim of presenting the historical development of numerical analysis and reviewing current research. The papers were arranged in seven volumes; see Online Appendix C3.

We give below a selection of textbooks and review papers on numerical methods. Even though the selection is by no means complete and reflects a subjective choice, we hope it can serve as a guide for a reader who, out of interest (or necessity!), wishes to deepen his or her knowledge. Both recent textbooks and older classics are included. Note that reviews of new books can be found in *Mathematical Reviews* as well as in the journals *SIAM Review* and *Mathematics of Computation*. A more complete guide to relevant literature and software is given in Online Appendix C.

Many outstanding textbooks in numerical analysis were originally published in the 1960s and 70s. The classical text by Hildebrand [201] can still be used as an introduction. Isaacson and Keller [208] give a rigorous mathematical treatment of classical topics, including differential equations and orthogonal polynomials. The present authors' textbook [89] was used at many universities in the USA and is still available.

Hamming [184] is a more applied text and aims at combining mathematical theory, heuristic analysis, and computing methods. It emphasizes the message that *"the purpose of computing is insight, not numbers."* An in-depth treatment of several areas such as numerical quadrature and approximation is found in the comprehensive book by Ralston and Rabinowitz [296]. This book also contains a large number of interesting and fairly advanced problems.

The book by Forsythe, Malcolm, and Moler [123] is notable in that it includes a set of Fortran subroutines of unusually high quality. Kahaner, Moler, and Nash [220] comes with a disk containing software. A good introduction to scientific computing is given by Golub and Ortega [166]. A matrix-vector approach is used in the MATLAB oriented text of Van Loan [367]. Analysis is complemented with computational experiments using a package of more than 200 m-files. Cheney and Kincaid [68] is an undergraduate text with many examples and exercises. Kincaid and Cheney [226] is a related textbook but more mathematically oriented. Two other good introductory texts are Eldén, Wittmeyer-Koch, and Nielsen [109] and Süli and Mayers [344].

Heath [190] is a popular, more advanced, and comprehensive text. Gautschi [147] is an elegant introductory text containing a wealth of computer exercises. Much valuable and hard-to-find information is included in notes after each chapter. The bestseller by Press et al. [294] surveys contemporary numerical methods for the applied scientist, but is weak on analysis.

Several good textbooks have been translated from German, notably the excellent book by Stoer and Bulirsch [338]. This is particularly suitable for a reader with a good mathematical background. Hämmerlin and Hoffmann [182] and Deuflhard and Hohmann [96] are less comprehensive but with a modern and careful treatment. Schwarz [318] is a mathematically oriented text which also covers ordinary and partial differential equations. Rutishauser [312] is an annotated translation of a highly original textbook by one of the pioneers of numerical analysis. Though brief, the book by Tyrtychnikov [363] is original and thorough. It also contains references to Russian literature unavailable in English.

Since numerical analysis is still in a dynamic stage it is important to keep track of new developments. An excellent source of survey articles on topics of current interest can be found in *Acta Numerica*, a Cambridge University Press Annual started in 1992. The journal *SIAM Review* also publishes high-quality review papers.

Another collection of outstanding survey papers on special topics is being published in a multivolume sequence in the *Handbook of Numerical Analysis* [70], edited by Philippe G. Ciarlet and Jacques-Louis Lions. It offers comprehensive coverage in all areas of numerical analysis as well as many actual problems of contemporary interest; see section C3 in Online Appendix C.

Chapter 2

How to Obtain and Estimate Accuracy

I always think I used computers for what
God had intended them for, to do arithmetic.
—Cleve Moler

2.1 Basic Concepts in Error Estimation

The main purpose of numerical analysis and scientific computing is to develop efficient and accurate methods to compute approximations to quantities that are difficult or impossible to obtain by analytic means. It has been convincingly argued (Trefethen [357]) that controlling rounding errors is just a small part of this, and that the main business of computing is the development of algorithms that converge rapidly. Even if we acknowledge the truth of this statement, it is still necessary to be able to control different sources of errors, including roundoff errors, so that these will not interfere with the computed results.

2.1.1 Sources of Error

Numerical results are affected by many types of errors. Some sources of error are difficult to influence; others can be reduced or even eliminated by, for example, rewriting formulas or making other changes in the computational sequence. Errors are propagated from their sources to quantities computed later, sometimes with a considerable amplification or damping. It is important to distinguish between the new error produced at the computation of a quantity (a source error), and the error inherited (propagated) from the data that the quantity depends on.

A. *Errors in Given Input Data.*
Input data can be the result of measurements which have been contaminated by different types of errors. In general one should be careful to distinguish between **systematic errors** and **random errors**. A systematic error can, for example, be produced by insufficiencies in the construction of an instrument of measurement;

such an error is the same in each trial. Random errors depend on the variation in the experimental environment which cannot be controlled.

B. *Rounding Errors During the Computations.*
A **rounding error** occurs whenever an irrational number, for example π, is shortened ("rounded off") to a fixed number of digits, or when a decimal fraction is converted to the binary form used in the computer. The limitation of floating-point numbers in a computer leads at times to a loss of information that, depending on the context, may or may not be important. Two typical cases are

(i) If the computer cannot handle numbers which have more than, say, s digits, then the exact product of two s-digit numbers (which contains $2s$ or $2s - 1$ digits) cannot be used in subsequent calculations; the product must be rounded off.

(ii) In a floating-point computation, if a relatively small term b is added to a, then some digits of b are "shifted out" (see Example 2.3.1), and they will not have any effect on future quantities that depend on the value of $a + b$.

The effect of such rounding can be quite noticeable in an extensive calculation, or in an algorithm which is numerically unstable.

C. *Truncation Errors.*
These are errors committed when a limiting process is truncated (broken off) before one has come to the limiting value. A **truncation error** occurs, for example, when an infinite series is broken off after a finite number of terms, or when a derivative is approximated with a difference quotient (although in this case the term **discretization error** is better). Another example is when a nonlinear function is approximated with a linear function, as in Newton's method. Observe the distinction between truncation error and rounding error.

D. *Simplifications in the Mathematical Model.*
In most of the applications of mathematics, one makes idealizations. In a mechanical problem one might assume that a string in a pendulum has zero mass. In many other types of problems it is advantageous to consider a given body to be homogeneously filled with matter, instead of being built of atoms. For a calculation in economics, one might assume that the rate of interest is constant over a given period of time. The effects of such sources of error are usually more difficult to estimate than the types named in A, B, and C.

E. *"Human" Errors and Machine Errors.*
In all numerical work, one must expect that clerical errors, errors in hand calculation, and misunderstandings will occur. One should even be aware that textbooks (!), tables, and formulas may contain errors. When one uses computers, one can expect errors in the program itself, typing errors in entering the data, operator errors, and (less frequently) pure machine errors.

Errors which are purely machine errors are responsible for only a very small part of the strange results which (occasionally with great publicity) are produced by computers. Most of the errors depend on the so-called human factor. As a rule, the effect of this type of error source cannot be analyzed with the help of the theoretical considerations of this chapter! We take up these sources of error in order to emphasize that both the person who carries out a calculation and the person who guides the work of others can plan so that such sources of error are not damaging. One can reduce the risk of such errors by suitable adjustments in working conditions and routines. Stress and fatigue are common causes of such errors.

Intermediate results that may reveal errors in a computation are not visible when using a computer. Hence the user must be able to verify the correctness of his results or be able to prove that his process cannot fail! Therefore, one should carefully consider what kind of checks can be made, either in the final result or in certain stages of the work, to prevent the necessity of redoing a whole project just because a small error has been made in an early stage. One can often discover whether calculated values are of the wrong order of magnitude or are not sufficiently regular, for example, using difference checks (see Sec. 3.3.1).

Occasionally one can check the credibility of several results at the same time by checking that certain relations are true. In linear problems, one often has the possibility of sum checks. In physical problems, one can check to see whether energy is conserved, although because of error sources A to D one cannot expect that it will be *exactly* conserved. In some situations, it can be best to treat a problem in two independent ways, although one can usually (as intimated above) check a result with less work than this.

Errors of type E do occur, sometimes with serious consequences. The first American Venus probe was lost due to a program fault caused by the inadvertent substitution of a statement in a Fortran program of the form `DO 3 I = 1.3` for one of the form `DO 3 I = 1,3`. Erroneously replacing the comma "," with a dot "." converts the intended loop statement into an assignment statement! A hardware error that got much publicity surfaced in 1994, when it was found that the INTEL Pentium processor gave wrong results for division with floating-point numbers of certain patterns. This was discovered during research on prime numbers (see Edelman [103]) and later fixed.

From a different point of view, one may distinguish between controllable and uncontrollable (or unavoidable) error sources. Errors of type A and D are usually considered to be uncontrollable in the numerical treatment (although feedback to the constructor of the mathematical model may sometimes be useful). Errors of type C are usually controllable. For example, the number of iterations in the solution of an algebraic equation, or the step size in a simulation, can be chosen either directly or by setting a tolerance.

The rounding error in the individual arithmetic operation (type B) is, in a computer, controllable only to a limited extent, mainly through the choice between single and double precision. A very important fact is, however, that it can often be controlled by appropriate rewriting of formulas or by other changes of the algorithm; see Example 2.3.3.

If it doesn't cost too much, a controllable error source should be controlled so that its effects are evidently negligible compared to the effects of the uncontrollable sources. A reasonable interpretation of "full accuracy" is that the controllable error sources should not increase the error of a result by more than about 20%. Sometimes, "full accuracy" may be expensive, for example, in terms of computing time, memory space, or programming efforts. Then it becomes important to estimate the relation between accuracy and these cost factors. One goal of the rest of this chapter is to introduce concepts and techniques useful for this purpose.

Many real-world problems contain some nonstandard features, where understanding the general principles of numerical methods can save much time in the preparation of a program as well as in the computer runs. Nevertheless, we strongly encourage the reader to use quality library programs whenever possible, since a lot of experience and profound theoretical analysis has often been built into these (sometimes far beyond the scope of this text). It is not practical to "reinvent the wheel."

2.1.2 Absolute and Relative Errors

Approximation is a central concept in almost all the uses of mathematics. One must often be satisfied with approximate values of the quantities with which one works. Another type of approximation occurs when one ignores some quantities which are small compared to others. Such approximations are often necessary to ensure that the mathematical and numerical treatment of a problem does not become hopelessly complicated.

We make the following definition.

Definition 2.1.1.

Let $\tilde{x}$ be an approximate value whose exact value is x. Then the **absolute error** *in $\tilde{x}$ is*

$$\Delta x = |\tilde{x} - x|,$$

and if $x \neq 0$ the **relative error** *is*

$$\Delta x / x = |(\tilde{x} - x)/x|.$$

In some books the error is defined with the opposite sign to what we use here. It makes almost no difference which convention one uses, as long as one is consistent. Note that $x - \tilde{x}$ is *the correction* which should be *added* to $\tilde{x}$ to get rid of the error. The correction and the absolute error then have the same magnitude but may have different signs.

In many situations one wants to compute a strict or approximate **bound** for the absolute or relative error. Since it is sometimes rather hard to obtain an error bound that is both strict and sharp, one sometimes prefers to use less strict but often realistic **error estimates**. These can be based on the first neglected term in some expansion, or on some other asymptotic considerations.

The notation $x = \tilde{x} \pm \epsilon$ means, in this book, $|\tilde{x} - x| \leq \epsilon$. For example, if $x = 0.5876 \pm 0.0014$, then $0.5862 \leq x \leq 0.5890$, and $|\tilde{x} - x| \leq 0.0014$. In other texts, the same plus–minus notation is sometimes used for the "standard error" (see Sec. 2.3.3) or some other measure of deviation of a statistical nature. If x is a vector $\|\cdot\|$, then the error bound and the relative error bound may be defined as bounds for

$$\|\tilde{x} - x\| \quad \text{and} \quad \|\tilde{x} - x\|/\|x\|,$$

respectively, where $\|\cdot\|$ denotes some vector norm (see Sec. A.3.3 in Online Appendix A). Then a bound $\|\tilde{x} - x\|/\|x\| \leq 1/2 \cdot 10^{-p}$ implies that components $\tilde{x}_i$ with $|\tilde{x}_i| \approx \|x\|$ have about p significant digits, but this is not true for components of smaller absolute value. An alternative is to use componentwise relative errors,

$$\max_i |\tilde{x}_i - x_i|/|x_i|, \tag{2.1.1}$$

but this assumes that $x_i \neq 0$ for all i.

We will distinguish between the terms accuracy and precision. By **accuracy** we mean the absolute or relative error of an approximate quantity. The term **precision** will be reserved for the accuracy with which the basic arithmetic operations $+$, $-$, $*$, $/$ are performed. For floating-point operations this is given by the unit roundoff; see (2.2.8).

Numerical results which are not followed by any error estimations should often, though not always, be considered as having an uncertainty of $\frac{1}{2}$ of a unit in the last decimal place. In presenting numerical results, it is a good habit, if one does not want to go through the difficulty of presenting an error estimate with each result, to give explanatory remarks such as

- "All the digits given are thought to be significant."
- "The data have an uncertainty of at most three units in the last digit."
- "For an ideal two-atom gas, $c_P/c_V = 1.4$ (exactly)."

We shall also introduce some notations, useful in practice, though their definitions are not exact in a mathematical sense:

> $a \ll b$ ($a \gg b$) is read "a is much smaller (much greater) than b." What is meant by "much smaller"(or "much greater") depends on the context—among other things, on the desired precision.
>
> $a \approx b$ is read "a is approximately equal to b" and means the same as $|a - b| \ll c$, where c is chosen appropriate to the context. We *cannot generally* say, for example, that $10^{-6} \approx 0$.
>
> $a \lesssim b$ (or $b \gtrsim a$) is read "a is less than or approximately equal to b" and means the same as "$a \le b$ or $a \approx b$."

Occasionally we shall have use for the following more precisely defined mathematical concepts:

> $f(x) = O(g(x))$, $x \to a$, means that $|f(x)/g(x)|$ is bounded as $x \to a$ (a can be finite, $+\infty$, or $-\infty$).
>
> $f(x) = o(g(x))$, $x \to a$, means that $\lim_{x\to a} f(x)/g(x) = 0$.
>
> $f(x) \sim g(x)$, $x \to a$, means that $\lim_{x\to a} f(x)/g(x) = 1$.

2.1.3 Rounding and Chopping

When one counts the *number of digits* in a numerical value one should not include zeros in the beginning of the number, as these zeros only help to denote where the decimal point should be. For example, the number 0.00147 has five decimals but is given with three digits. The number 12.34 has two decimals but is given with four digits.

If the magnitude of the error in a given numerical value $\tilde{a}$ does not exceed $\frac{1}{2} \cdot 10^{-t}$, then $\tilde{a}$ is said to have t **correct decimals**. The *digits* in $\tilde{a}$ which occupy positions where the unit is greater than or equal to 10^{-t} are then called **significant digits** (any initial zeros are not counted). Thus, the number 0.001234 ± 0.000004 has five correct decimals and three significant digits, while 0.001234 ± 0.000006 has four correct decimals and two significant digits. The number of correct decimals gives one an idea of the magnitude of the *absolute error*, while the number of significant digits gives a rough idea of the magnitude of the *relative error*.

We distinguish here between two ways of rounding off a number x to a given number t of decimals. In **chopping** (or round toward zero) one simply leaves off all the decimals to the right of the tth. That way is generally *not recommended* since the rounding error has, systematically, the opposite sign of the number itself. Also, the magnitude of the error can be as large as 10^{-t}.

In **rounding to nearest** (sometimes called "correct" or "optimal" rounding), one chooses a number with s decimals which is *nearest* to x. Hence if p is the part of the number which stands to the right of the sth decimal, one leaves the tth decimal unchanged if and only if $|p| < 0.5 \cdot 10^{-s}$. Otherwise one raises the sth decimal by 1. In the case of a tie, when x is equidistant to two s decimal digit numbers, then one raises the sth decimal if it is odd or leaves it unchanged if it is even (round to even). In this way, the error is positive or negative about equally often. The error in rounding a decimal number to s decimals will always lie in the interval $[-\frac{1}{2}10^{-s}, \frac{1}{2}10^{-s}]$.

Example 2.1.1.

Shortening to three decimals,

0.2397	rounds to 0.240	(is chopped to 0.239),
−0.2397	rounds to −0.240	(is chopped to −0.239),
0.23750	rounds to 0.238	(is chopped to 0.237),
0.23650	rounds to 0.236	(is chopped to 0.236),
0.23652	rounds to 0.237	(is chopped to 0.236).

Observe that when one rounds off a numerical value one produces an error; thus it is occasionally wise to give more decimals than those which are correct. Take $a = 0.1237 \pm 0.0004$, which has three correct decimals according to the definition given previously. If one rounds to three decimals, one gets 0.124; here the third decimal may not be correct, since the least possible value for a is 0.1233.

Suppose that you are tabulating a transcendental function and that a particular entry has been evaluated as 1.2845 correct to the digits given. You want to round the value to three decimals. Should the final digit be 4 or 5? The answer depends on whether there is a nonzero trailing digit. You compute the entry more accurately and find 1.28450, then 1.284500, then 1.2845000, etc. Since the function is transcendental, there clearly is no bound on the number of digits that has to be computed before distinguishing if to round to 1.284 or 1.285. This is called the **tablemaker's dilemma.**[30]

Example 2.1.2.

The difference between chopping and rounding can be important, as is illustrated by the following story. The index of the Vancouver Stock Exchange, founded at the initial value

[30]This can be used to advantage in order to protect mathematical tables from illegal copying by rounding a few entries incorrectly, where the error in doing so is insignificant due to several trailing zeros. An illegal copy could then be exposed simply by looking up these entries!

1000.000 in 1982, was hitting lows in the 500s at the end of 1983 even though the exchange apparently performed well. It was discovered (*The Wall Street Journal*, Nov. 8, 1983, p. 37) that the discrepancy was caused by a computer program which updated the index thousands of times a day and used chopping instead of rounding to nearest! The rounded calculation gave a value of 1098.892.

Review Questions

2.1.1 Clarify with examples the various types of error sources which occur in numerical work.

2.1.2 (a) Define "absolute error" and "relative error" for an approximation $\bar{x}$ to a scalar quantity x. What is meant by an error bound?

(b) Generalize the definitions in (a) to a vector x.

2.1.3 How is "rounding to nearest" performed?

2.1.4 Give π to four decimals using (a) chopping; (b) rounding.

2.1.5 What is meant by the "tablemaker's dilemma"?

2.2 Computer Number Systems

2.2.1 The Position System

In order to represent numbers in daily life, we use a **position system** with base 10 (the decimal system). Thus, to represent the numbers we use ten different characters, and the magnitude with which the digit a contributes to the value of a number depends on the digit's position in the number. If the digit stands n steps to the right of the decimal point, the value contributed is $a \cdot 10^{-n}$. For example, the sequence of digits 4711.303 means

$$4 \cdot 10^3 + 7 \cdot 10^2 + 1 \cdot 10^1 + 1 \cdot 10^0 + 3 \cdot 10^{-1} + 0 \cdot 10^{-2} + 3 \cdot 10^{-3}.$$

Every real number has a unique representation in the above way, except for the possibility of infinite sequences of nines—for example, the infinite decimal fraction 0.3199999... represents the same number as 0.32.

One can very well consider other position systems with bases other than 10. Any integer $\beta \geq 2$ (or $\beta \leq -2$) can be used as a base. One can show that every positive real number a has, with exceptions analogous to the nines sequences mentioned above, a unique representation of the form

$$a = d_n\beta^n + d_{n-1}\beta^{n-1} + \cdots + d_1\beta^1 + d_0\beta^0 + d_{-1}\beta^{-1} + d_{-2}\beta^{-2} + \cdots$$

or, more compactly, $a = (d_n d_{n-1} \ldots d_0.d_{-1}d_{-2}\ldots)_\beta$, where the coefficients d_i, the "digits" in the system with base β, are positive integers d_i such that $0 \leq d_i \leq \beta - 1$.

One of the greatest advantages of the position system is that one can give simple, general rules for the arithmetic operations. The smaller the base is, the simpler these rules become. This is just one reason why most computers operate in base 2, the **binary number system**. The addition and multiplication tables then take the following simple form:

$$0 + 0 = 0, \qquad 0 + 1 = 1 + 0 = 1, \qquad 1 + 1 = 10,$$
$$0 \cdot 0 = 0, \qquad 0 \cdot 1 = 1 \cdot 0 = 0, \qquad 1 \cdot 1 = 1.$$

In the binary system the number seventeen becomes 10001, since $1 \cdot 2^4 + 0 \cdot 2^3 + 0 \cdot 2^2 + 0 \cdot 2^1 + 1 \cdot 2^0 =$ sixteen + one = seventeen. Put another way $(10001)_2 = (17)_{10}$, where the index (in decimal representation) denotes the base of the number system. The numbers become longer written in the binary system; large integers become about 3.3 times as long, since N binary digits suffice to represent integers less than $2^N = 10^{N \log_{10} 2} \approx 10^{N/3.3}$.

Occasionally one groups together the binary digits in subsequences of three or four, which is equivalent to using 2^3 and 2^4, respectively, as the base. These systems are called the **octal** and **hexadecimal** number systems, respectively. The octal system uses the digits from 0 to 7; in the hexadecimal system the digits 0 through 9 and the letters A, B, C, D, E, F ("ten" through "fifteen") are used.

Example 2.2.1.

$$\begin{aligned}
(17)_{10} &= (10001)_2 = (21)_8 = (11)_{16}, \\
(13.25)_{10} &= (1101.01)_2 = (15.2)_8 = (D.4)_{16}, \\
(0.1)_{10} &= (0.000110011001\ldots)_2 = (0.199999\ldots)_{16}.
\end{aligned}$$

Note that *the finite decimal fraction* 0.1 *cannot be represented exactly by a finite fraction in the binary number system*! (For this reason some pocket calculators use base 10.)

Example 2.2.2.

In 1991 a Patriot missile in Saudi Arabia failed to track and interrupt an incoming Scud missile due to a precision problem. The Scud then hit an army barrack and killed 28 Americans. The computer used to control the Patriot missile was based on a design dating from the 1970s using 24-bit arithmetic. For the tracking computations, time was recorded by the system clock in tenths of a second but converted to a 24-bit floating-point number. Rounding errors in the time conversions caused an error in the tracking. After 100 hours of consecutive operations the calculated time in seconds was 359,999.6567 instead of the correct value 360,000, an error of 0.3433 seconds leading to an error in the calculated range of 687 meters; see Skeel [326]. Modified software was later installed.

In the binary system the "point" used to separate the integer and fractional part of a number (corresponding to the decimal point) is called the binary point. The digits in the binary system are called **bits** (**b**inary dig**its**).

We are so accustomed to the position system that we forget that it is built upon an ingenious idea. The reader can puzzle over how the rules for arithmetic operations would look if one used Roman numerals, a number system without the position principle described above.

Recall that rational numbers are precisely those real numbers which can be expressed as a quotient between two integers. Equivalently, rational numbers are those whose representation in a position system have a finite number of digits or whose digits are repeating.

We now consider the problem of conversion between two number systems with different bases. Since almost all computers use a binary system this problem arises as soon as one wants to input data in decimal form or print results in decimal form.

Algorithm 2.1. *Conversion between Number Systems.*

Let a be an integer given in number systems with base α. We want to determine its representation in a number system with base β:

$$a = b_n\beta^m + b_{m-1}\beta^{n-1} + \cdots + b_0, \quad 0 \le b_i < \beta. \tag{2.2.1}$$

The computations are to be done in the system with base α, and thus β also is expressed in this representation. The conversion is done by successive divisions of a with β: Set $q_0 = a$, and

$$q_k/\beta = q_{k+1} + b_k/\beta, \quad k = 0, 1, 2, \ldots \tag{2.2.2}$$

(q_{k+1} is the quotient and b_k is the remainder in the division).

If a is not an integer, we write $a = b + c$, where b is the integer part and

$$c = c_{-1}\beta^{-1} + c_{-2}\beta^{-2} + c_{-3}\beta^{-3} + \cdots \tag{2.2.3}$$

is the fractional part, where $c_{-1}, c_{-2}, \ldots$ are to be determined. These digits are obtained as the integer parts when successively multiplying c with β: Set $p_{-1} = c$, and

$$p_k \cdot \beta = c_k\beta + p_{k-1}, \quad k = -1, -2, -3, \ldots. \tag{2.2.4}$$

Since a finite fraction in a number system with base α usually does not correspond to a finite fraction in the number system with base β, rounding of the result is usually needed.

When converting by hand between the decimal system and the binary system, all computations are made in the decimal system ($\alpha = 10$ and $\beta = 2$). It is then more convenient to convert the decimal number first to octal or hexadecimal, from which the binary representation easily follows. If, on the other hand, the conversion is carried out on a binary computer, the computations are made in the binary system ($\alpha = 2$ and $\beta = 10$).

Example 2.2.3.

Convert the decimal number 176.524 to ternary form (base $\beta = 3$). For the integer part we get $176/3 = 58$ with remainder 2; $58/3 = 19$ with remainder 1; $19/3 = 6$ with remainder 1; $6/3 = 2$ with remainder 0; $2/3 = 0$ with remainder 2. It follows that $(176)_{10} = (20112)_3$.

For the fractional part we compute $.524 \cdot 3 = 1.572$, $.572 \cdot 3 = 1.716$, $.716 \cdot 3 = 2.148, \ldots$. Continuing in this way we obtain $(.524)_{10} = (.112010222\ldots)_3$. The finite decimal fraction does not correspond to a finite fraction in the ternary number system.

2.2.2 Fixed- and Floating-Point Representation

A computer is, in general, built to handle pieces of information of a fixed size called a **word**. The number of digits in a word (usually binary) is called the **word-length** of the computer.

Typical word-lengths are 32 and 64 bits. A real or integer number is usually stored in a word. Integers can be exactly represented, provided that the word-length suffices to store all the digits in its representation.

In the first generation of computers calculations were made in a **fixed-point** number system; i.e., real numbers were represented with a fixed number of t binary digits in the fractional part. If the word-length of the computer is $s+1$ bits (including the sign bit), then only numbers in the interval $I = [-2^{s-t}, 2^{s-t}]$ are permitted. Some common fixed-point conventions are $t = s$ (fraction convention) and $t = 0$ (integer convention). This limitation causes difficulties, since even when $x \in I$, $y \in I$, we can have $x - y \notin I$ or $x/y \notin I$.

In a fixed-point number system one must see to it that all numbers, even intermediate results, remain within I. This can be attained by multiplying the variables by appropriate **scale factors**, and then transforming the equations accordingly. This is a tedious process. Moreover it is complicated by the risk that if the scale factors are chosen carelessly, certain intermediate results can have many leading zeros which can lead to poor accuracy in the final results. As a consequence, current numerical analysis literature rarely deals with other than floating-point arithmetic. In scientific computing fixed-point representation is mainly limited to computations with integers, as in subscript expressions for vectors and matrices.

On the other hand, fixed-point computations can be much faster than floating-point, especially since modern microprocessors have superscalar architectures with several fixed-point units but only one floating-point unit. In computer graphics, fixed-point is used almost exclusively once the geometry is transformed and clipped to the visible window. Fixed-point square roots and trigonometric functions are also pretty quick and are easy to write.

By a **normalized floating-point representation** of a real number a, we mean a representation in the form

$$a = \pm m \cdot \beta^e, \quad \beta^{-1} \le m < 1, \quad e \text{ an integer}. \tag{2.2.5}$$

Such a representation is possible for all real numbers a and is unique if $a \neq 0$. (The number zero is treated as a special case.) Here the fraction part m is called the **mantissa**[31] or **significand**), e is the **exponent**, and β the **base** (also called the **radix**).

In a computer, the number of digits for e and m is limited by the word-length. Suppose that t digits are used to represent m. Then we can only represent floating-point numbers of the form

$$\bar{a} = \pm \overline{m} \cdot \beta^e, \quad \overline{m} = (0.d_1 d_2 \cdots d_t)_\beta, \quad 0 \le d_i < \beta, \tag{2.2.6}$$

where $\overline{m}$ is the mantissa m rounded to t digits, and the exponent is limited to a finite range

$$e_{\min} \le e \le e_{\max}. \tag{2.2.7}$$

A floating-point number system F is characterized by the base β, the precision t, and the numbers $e_{\min}$, $e_{\max}$. Only a finite set F of rational numbers can be represented in the form (2.2.6). The numbers in this set are called **floating-point numbers**. Since $d_1 \neq 0$ this set contains, including the number zero, precisely

$$2(\beta - 1)\beta^{t-1}(e_{\max} - e_{\min} + 1) + 1$$

[31] Strictly speaking, "mantissa" refers to the decimal part of a logarithm.

numbers. (Show this!) The limited number of digits in the exponent implies that a is limited in magnitude to an interval which is called the **range** of the floating-point system. If a is larger in magnitude than the largest number in the set F, then a cannot be represented at all (**exponent spill**). The same is true, in a sense, of numbers smaller than the smallest nonzero number in F.

Example 2.2.4.

Consider the floating-point number system for $\beta = 2, t = 3, e_{\min} = -1$, and $e_{\max} = 2$. The positive normalized numbers in the corresponding set F are shown in Figure 2.2.1. The set F contains exactly $2 \cdot 16 + 1 = 33$ numbers. In this example the nonzero numbers of smallest magnitude that can be represented are $(0.100)_2 \cdot 2^{-1} = \frac{1}{4}$ and the largest is $(0.111)_2 \cdot 2^2 = \frac{7}{2}$.

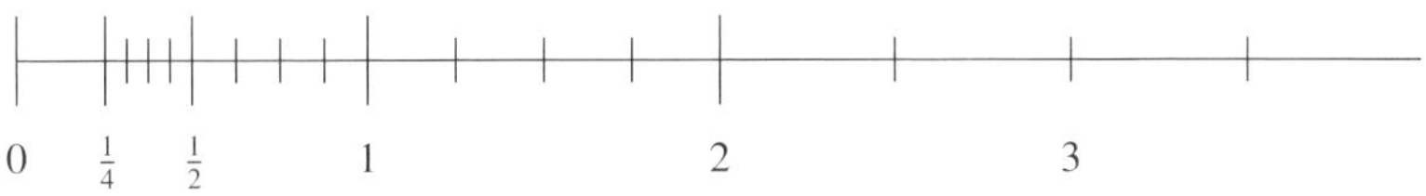

Figure 2.2.1. *Positive normalized numbers when $\beta = 2$, $t = 3$, and $-1 \le e \le 2$.*

Notice that floating-point numbers are not equally spaced; the spacing jumps by a factor of β at each power of β. This wobbling is smallest for $\beta = 2$.

Definition 2.2.1.

The spacing of floating-point numbers is characterized by the **machine epsilon**, *which is the distance ϵ_M from 1.0 to the next larger floating-point number.*

The leading significant digit of numbers represented in a number system with base β has been observed to closely fit a logarithmic distribution; i.e., the proportion of numbers with leading digit equal to n is $\ln_\beta(1 + 1/n)$ $(n = 0, 1, \ldots, \beta - 1)$. It has been shown that, under this assumption, taking the base equal to 2 will minimize the mean square representation error. A discussion of this intriguing fact, with historic references, is found in Higham [199, Sec. 2.7].

Even if the operands in an arithmetic operation are floating-point numbers in F, the *exact* result of the operation may not be in F. For example, the exact product of two floating-point t-digit numbers has $2t$ or $2t - 1$ digits.

If a real number a is in the range of the floating-point system, the obvious thing to do is to represent a by $\bar{a} = fl\,(a)$, where $fl\,(a)$ denotes a number in F which is nearest to a. This corresponds to rounding of the mantissa m, and according to Sec. 2.1.3, we have

$$|\overline{m} - m| \le \frac{1}{2}\beta^{-t}.$$

(There is one exception. If $|m|$ after rounding should be raised to 1, then $|\overline{m}|$ is set equal to 0.1 and e is raised by 1.) Since $m \ge 0.1$ this means that the magnitude of the relative error in $\bar{a}$ is at most equal to

$$\frac{\frac{1}{2}\beta^{-t} \cdot \beta^e}{m \cdot \beta^e} \leq \frac{1}{2}\beta^{-t+1}.$$

Even with the exception mentioned above this relative bound still holds. (If chopping is used, this doubles the error bound above.) This proves the following theorem.

Theorem 2.2.2.

In a floating-point number system $F = F(\beta, t, e_{\min}, e_{\max})$ every real number in the floating-point range of F can be represented with a relative error, which does not exceed the **unit roundoff** *u, which is defined by*

$$u = \begin{cases} \frac{1}{2}\beta^{-t+1} & \textit{if rounding is used,} \\ \beta^{-t+1} & \textit{if chopping is used.} \end{cases} \tag{2.2.8}$$

Note that in a floating-point system both large and small numbers are represented with nearly *the same relative precision*. The quantity u is, in many contexts, a natural unit for relative changes and relative errors. For example, termination criteria in iterative methods usually depend on the unit roundoff.

To measure the difference between a floating-point number and the real number it approximates we shall occasionally use "**unit in last place**" or **ulp**. We shall often say that "the quantity is perturbed by a few ulps." For example, if in a decimal floating-point system the number 3.14159 is represented as $0.3142 \cdot 10^1$, this has an error of 0.41 ulps.

Example 2.2.5.

Sometimes it is useful to be able to approximately determine the unit roundoff in a program at run time. This may be done using the observation that $u \approx \mu$, where μ is *the smallest floating-point number x for which $fl\,(1 + x) > 1$.* The following program computes a number μ which differs from the unit roundoff u by at most a factor of 2:

$$x := 1;$$
$$\textbf{while } 1 + x > 1 \quad x := x/2; \ \textbf{end};$$
$$\mu := x;$$

One reason why u does not exactly equal μ is that so-called double rounding may occur. This is when a result is first rounded to extended format and then to the target precision.

A floating-point number system can be extended by including **denormalized numbers** (also called subnormal numbers). These are numbers with the minimum exponent and with the most significant digit equal to zero. The three numbers

$$(.001)_2 2^{-1} = 1/16, \qquad (.010)_2 2^{-1} = 2/16, \qquad (.011)_2 2^{-1} = 3/16$$

can then also be represented (see Figure 2.2.2). Because the representations of denormalized numbers have initial zero digits these have fewer digits of precision than normalized numbers.

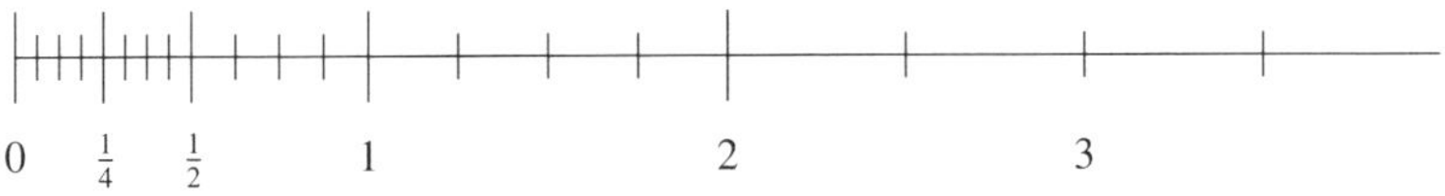

Figure 2.2.2. *Positive normalized and denormalized numbers when $\beta = 2$, $t = 3$, and $-1 \le e \le 2$.*

2.2.3 IEEE Floating-Point Standard

Actual computer implementations of floating-point representations may differ in detail from the one given above. Although some pocket calculators use a floating-point number system with base $\beta = 10$, almost all modern computers use base $\beta = 2$. Most current computers now conform to the IEEE 754 standard for binary floating-point arithmetic.[32] This standard from 1985 (see [205]), which is the result of several years' work by a subcommittee of the IEEE, is now implemented on almost all chips used for personal computers and workstations. There is also a standard IEEE 854 for radix independent floating-point arithmetic [206]. This is used with base 10 by several hand calculators.

The IEEE 754 standard specifies basic and extended formats for floating-point numbers, elementary operations and rounding rules available, conversion between different number formats, and binary–decimal conversion. The handling of exceptional cases like exponent overflow or underflow and division by zero are also specified.

Two main basic formats, single and double precision, are defined using 32 and 64 bits, respectively. In **single precision** a floating-point number a is stored as the sign s (one bit), the exponent e (8 bits), and the mantissa m (23 bits). In **double precision** 11 of the 64 bits are used for the exponent, and 52 bits are used for the mantissa. The value v of a is in the normal case

$$v = (-1)^s (1.m)_2 2^e, \qquad -e_{\min} \le e \le e_{\max}. \tag{2.2.9}$$

Note that the digit before the binary point is always 1 for a normalized number. Thus the normalization of the mantissa is different from that in (2.2.6). This bit is not stored (the hidden bit). In that way one bit is gained for the mantissa. A biased exponent is stored and no sign bit used for the exponent. In single precision $e_{\min} = -126$ and $e_{\max} = 127$, and $e + 127$ is stored.

The unit roundoff equals

$$u = \begin{cases} 2^{-24} \approx 5.96 \cdot 10^{-8} & \text{in single precision,} \\ 2^{-53} \approx 1.11 \cdot 10^{-16} & \text{in double precision.} \end{cases}$$

(The machine epsilon is twice as large.) The largest number that can be represented is approximately $2.0 \cdot 2^{127} \approx 3.4028 \times 10^{38}$ in single precision and $2.0 \cdot 2^{1023} \approx 1.7977 \times 10^{308}$ in double precision. The smallest normalized number is $1.0 \cdot 2^{-126} \approx 1.1755 \times 10^{-38}$ in single precision and $1.0 \cdot 2^{-1022} \approx 2.2251 \times 10^{-308}$ in double precision.

An exponent $e = e_{\min} - 1$ and $m \neq 0$ signifies the denormalized number

$$v = (-1)^s (0.m)_2 2^{e_{\min}}.$$

[32]W. Kahan, University of California, Berkeley, was given the Turing Award by the Association of Computing Machinery for his contribution to this standard.

The smallest denormalized number that can be represented is $2^{-126-23} \approx 1.4013 \cdot 10^{-45}$ in single precision and $2^{-1022-52} \approx 4.9407 \cdot 10^{-324}$ in double precision.

There are distinct representations for $+0$ and -0. ± 0 is represented by a sign bit, the exponent $e_{\min} - 1$, and a zero mantissa. Comparisons are defined so that $+0 = -0$. One use of a signed zero is to distinguish between positive and negative underflowed numbers. Another use occurs in the computation of complex elementary functions; see Sec. 2.2.4.

Infinity is also signed and $\pm\infty$ is represented by the exponent $e_{\max} + 1$ and a zero mantissa. When overflow occurs the result is set to $\pm\infty$. This is safer than simply returning the largest representable number, which may be nowhere near the correct answer. The result $\pm\infty$ is also obtained from the illegal operations $a/0$, where $a \neq 0$. The infinity symbol obeys the usual mathematical conventions such as $\infty + \infty = \infty$, $(-1) \times \infty = -\infty$, $a/\infty = 0$.

The IEEE standard also includes two extended precision formats that offer extra precision and exponent range. The standard only specifies a lower bound on how many extra bits it provides.[33] Most modern processors use 80-bit registers for processing real numbers and store results as 64-bit numbers according to the IEEE double precision standard. Extended formats simplify tasks such as computing elementary functions accurately in single or double precision. Extended precision formats are used also by hand calculators. These will often display 10 decimal digits but use 13 digits internally—"*the calculator knows more than it shows.*"

The characteristics of the IEEE formats are summarized in Table 2.2.1. (The hidden bit in the mantissa accounts for the $+1$ in the table. Note that double precision satisfies the requirements for the single extended format, so three different precisions suffice.)

Table 2.2.1. *IEEE floating-point formats.*

	Format	t	e	$e_{\min}$	$e_{\max}$
Single	32 bits	$23 + 1$	8 bits	-126	127
Single extended	$\geq$ 43 bits	≥ 32	$\geq$ 11 bits	≤ -1022	≥ 1023
Double	64 bits	$52 + 1$	11 bits	-1022	1023
Double extended	$\geq$ 79 bits	≥ 64	$\geq$ 15 bits	$\leq -16{,}382$	$\geq 16{,}383$

Example 2.2.6.

Although the exponent range of the floating-point formats seems reassuringly large, even simple programs can quickly give exponent spill. If $x_0 = 2$, $x_{n+1} = x_n^2$, then already $x_{10} = 2^{1024}$ is larger than what IEEE double precision permits. One should also be careful in computations with factorials, e.g., $171! \approx 1.24 \cdot 10^{309}$ is larger than the largest double precision number.

Four rounding modes are supported by the standard. The default rounding mode is round to the nearest representable number, with round to even in the case of a tie. (Some computers, in case of a tie, round away from zero, i.e., raise the absolute value of the

[33] Hardware implementation of extended precision normally does not use a hidden bit, so the double extended format uses 80 bits rather than 79.

number, because this is easier to realize technically.) Chopping is also supported as well as directed rounding to ∞ and to $-\infty$. The latter mode simplifies the implementation of interval arithmetic; see Sec. 2.5.3.

The standard specifies that all arithmetic operations should be performed as if they were first calculated to infinite precision and then rounded to a floating-point number according to one of the four modes mentioned above. This also includes the square root and conversion between an integer and a floating-point. The standard also requires that the conversion between internal formats and decimal be correctly rounded. This can be achieved using extra **guard digits** in the intermediate result of the operation before normalization and rounding. Using a single guard digit, however, will not always ensure the desired result. However, by introducing a second guard digit and a third sticky bit (the logical OR of all succeeding bits) the rounded exact result can be computed at only a slightly higher cost (Goldberg [158]). One reason for specifying precisely the results of arithmetic operations is to improve the portability of software. If a program is moved between two computers, both supporting the IEEE standard, intermediate results should be the same.

IEEE arithmetic is a closed system; that is, every operation, even mathematically invalid operations, such as $0/0$ or $\sqrt{-1}$, produces a result. To handle exceptional situations without aborting the computations some bit patterns (see Table 2.2.2) are reserved for special quantities like NaN ("Not a Number") and ∞. NaNs (there are more than one NaN) are represented by $e = e_{\max} + 1$ and $m \neq 0$.

Table 2.2.2. *IEEE* 754 *representation.*

Exponent	Mantissa	Represents
$e = e_{\min} - 1$	$m = 0$	± 0
$e = e_{\min} - 1$	$m \neq 0$	$\pm 0.m \cdot 2^{e_{\min}}$
$e_{\min} < e < e_{\max}$		$\pm 1.m \cdot 2^{e}$
$e = e_{\max} + 1$	$m = 0$	$\pm\infty$
$e = e_{\max} + 1$	$m \neq 0$	NaN

Note that the gap between zero and the smallest normalized number is $1.0 \times 2^{e_{\min}}$. This is much larger than for the spacing $2^{-t+1} \times 2^{e_{\min}}$ for the normalized numbers for numbers just larger than the underflow threshold; compare Figure 2.2.1. With denormalized numbers the spacing becomes more regular and permits what is called **gradual underflow**. This makes many algorithms well behaved close to the underflow threshold also. Another advantage of having gradual underflow is that it makes it possible to preserve the property

$$x = y \quad \Leftrightarrow \quad x - y = 0$$

as well as other useful relations. Several examples of how denormalized numbers make writing reliable floating-point code easier are analyzed by Demmel [94].

One illustration of the use of extended precision is in converting between IEEE 754 single precision and decimal. The converted single precision number should ideally be converted with enough digits so that when it is converted back the binary single precision number is recovered. It might be expected that since $2^{24} < 10^8$, eight decimal digits in the converted number would suffice. But it can be shown that nine decimal digits are needed to recover the binary number uniquely (see Goldberg [158, Theorem 15] and Problem 2.2.4). When converting back to binary form a rounding error as small as one ulp will give the wrong answer. To do this conversion efficiently, extended single precision is needed.[34]

A NaN is generated by operations such as $0/0$, $+\infty + (-\infty)$, $0 \times \infty$, and $\sqrt{-1}$. A NaN compares unequal with everything including itself. (Note that $x \neq x$ is a simple way to test if x equals a NaN.) When a NaN and an ordinary floating-point number are combined the result is the same as the NaN operand. A NaN is also often used for uninitialized or missing data.

Exceptional operations also raise a flag. The default is to set a flag and continue, but it is also possible to pass control to a trap handler. The flags are "sticky" in that they remain set until explicitly cleared. This implies that without a log file everything before the last setting is lost, which is why it is always wise to use a trap handler. There is one flag for each of the following five exceptions: underflow, overflow, division by zero, invalid operation, and inexact. For example, by testing the flags, it is possible to test if an overflow is genuine or the result of division by zero.

Because of cheaper hardware and increasing problem sizes, double precision is used more and more in scientific computing. With increasing speed and memory becoming available, bigger and bigger problems are being solved and actual problems may soon require more than IEEE double precision! When the IEEE 754 standard was defined no one expected computers able to execute more than 10^{12} floating-point operations per second.

2.2.4 Elementary Functions

Although the square root is included, the IEEE 754 standard does not deal with the implementation of other familiar elementary functions, such as the exponential function exp, the natural logarithm log, and the trigonometric and hyperbolic functions sin, cos, tan, sinh, cosh, tanh, and their inverse functions. With the IEEE 754 standard more accurate implementations are possible which in many cases give almost correctly rounded exact results. To always guarantee correctly rounded exact results sometimes requires computing many more digits than the target accuracy (cf. the tablemaker's dilemma) and therefore is in general too costly. It is also important to preserve monotonicity, e.g., $0 \le x \le y \le \pi/2 \Rightarrow \sin x \le \sin y$, and range restrictions, e.g., $\sin x \le 1$, but these demands may conflict with rounded exact results!

The first step in computing an elementary function is to perform a **range reduction**. To compute trigonometric functions, for example, $\sin x$, an additive range reduction is first performed, in which a reduced argument x^*, $-\pi/4 \le x^* \le \pi/4$, is computed by finding an integer k such that

$$x^* = x - k\pi/2 \quad (\pi/2 = 1.57079\,63267\,94896\,61923\ldots).$$

(Quantities such as $\pi/2$, $\log(2)$, that are often used in standard subroutines, are listed in decimal form to 30 digits and octal form to 40 digits in Hart et al. [187, Appendix C] and to

[34]It should be noted that some computer languages do not include input/output routines, but these are developed separately. This can lead to double rounding, which spoils the carefully designed accuracy in the IEEE 754 standard. (Some banks use separate routines with chopping even today—you may guess why!)

40 and 44 digits in Knuth [230, Appendix A].) Then $\sin x = \pm \sin x^*$ or $\sin x = \pm \cos x^*$, depending on if k mod 4 equals 0, 1, 2, or 3. Hence approximation for $\sin x$ and $\cos x$ need only be provided for $0 \le x \le \pi/4$. If the argument x is very large, then cancellation in the range reduction can lead to poor accuracy; see Example 2.3.7.

To compute $\log x$, $x > 0$, a multiplicative range reduction is used. If an integer k is determined such that

$$x^* = x/2^k, \quad x^* \in [1/2, 1],$$

then $\log x = \log x^* + k \cdot \log 2$.

To compute the exponential function $\exp(x)$ an integer k is determined such that

$$x^* = x - k \log 2, \quad x^* \in [0, \log 2] \quad (\log 2 = 0.69314\,71805\,59945\,30942\ldots).$$

It then holds that $\exp(x) = \exp(x^*) \cdot 2^k$ and hence we only need an approximation of $\exp(x)$ for the range $x \in [0, \log 2]$.

Coefficients of polynomial and rational approximations suitable for software implementations are tabulated in Hart et al. [187] and Cody and Waite [75]. But approximation of functions can now be simply obtained using software such as Maple [65]. For example, in Maple the commands

```
Digits = 40; minimax(exp(x), x = 0..1, [i,k],1,'err')
```

mean that we are looking for the coefficients of the minimax approximation of the exponential function on [0, 1] by a rational function with numerator of degree i and denominator of degree k with weight function 1, and that the variable `err` should be equal to the approximation error. The coefficients are to be computed to 40 decimal digits. A trend now is for elementary functions to be increasingly implemented in hardware. Hardware implementations are discussed by Muller [272]. Carefully implemented algorithms for elementary functions are available from `www.netlib.org/fdlibm` in the library package fdlibm (Freely Distributable Math. Library) developed by Sun Microsystems and used by MATLAB.

Example 2.2.7.

On a computer using IEEE double precision arithmetic the roundoff unit is $u = 2^{-53} \approx 1.1 \cdot 10^{-16}$. One wishes to compute $\sinh x$ with good *relative* accuracy, both for small and large $|x|$, at least moderately large. Assume that e^x is computed with a relative error less than u in the given interval. The formula $(e^x - e^{-x})/2$ for $\sinh x$ is sufficiently accurate except when $|x|$ is very small and cancellation occurs. Hence for $|x| \ll 1$, e^x and e^{-x} and hence $(e^x - e^{-x})/2$ can have *absolute* errors of order of magnitude (say) u. Then the *relative* error in $(e^x - e^{-x})/2$ can have magnitude $\approx u/|x|$; for example, this is more than 100% for $x \approx 10^{-16}$.

For $|x| \ll 1$ one can instead use (say) two terms in the series expansion for $\sinh x$,

$$\sinh x = x + x^3/3! + x^5/5! + \ldots .$$

Then one gets an absolute truncation error which is about $x^5/120$, and a roundoff error of the order of $2u|x|$. Thus the formula $x + x^3/6$ is better than $(e^x - e^{-x})/2$ if

$$|x|^5/120 + 2u|x| < u.$$

If $2u|x| \ll u$, we have $|x|^5 < 120u = 15 \cdot 2^{-50}$, or $|x| < 15^{1/5} \cdot 2^{-10} \approx 0.00168$ (which shows that $2u|x|$ really could be ignored in this rough calculation). Thus, if one switches from $(e^x - e^{-x})/2$ to $x + x^3/6$ for $|x| < 0.00168$, the relative error will nowhere exceed $u/0.00168 \approx 0.66 \cdot 10^{-13}$. If one needs higher accuracy, one can take more terms in the series, so that the switch can occur at a larger value of $|x|$.

For very large values of $|x|$ one must expect a relative error of order of magnitude $|xu|$ because of roundoff error in the argument x. Compare the discussion of range reduction in Sec. 2.2.4 and Problem 2.2.13.

For complex arguments the elementary functions have discontinuous jumps across when the argument crosses certain branch cuts in the complex plane. They are represented by functions which are single-valued excepts for certain straight lines called **branch cuts**. Where to put these branch cuts and the role of IEEE arithmetic in making these choices are discussed by Kahan [217].

Example 2.2.8.

The function $\sqrt{x}$ is multivalued and there is no way to select the values so the function is continuous over the whole complex plane. If a branch cut is made by excluding all real negative numbers from consideration the square root becomes continuous. Signed zero provides a way to distinguish numbers of the form $x + i(+0)$ and $x + i(-0)$ and to select one or the other side of the cut.

To test the implementation of elementary functions, a Fortran package ELEFUNT has been developed by Cody [73]. This checks the quality using identities like $\cos x = \cos(x/3)(4\cos^2(x/3) - 1)$. For complex elementary functions a package CELEFUNT serves the same purpose; see Cody [74].

2.2.5 Multiple Precision Arithmetic

Hardly any quantity in the physical world is known to an accuracy beyond IEEE double precision. A value of π correct to 20 decimal digits would suffice to calculate the circumference of a circle around the sun at the orbit of the Earth to within the width of an atom. There seems to be little need for multiple precision calculations. Occasionally, however, one may want to perform some calculations, for example, the evaluation of some mathematical constant (such as π and Euler's constant γ) or elementary functions, to very high precision.[35] Extremely high precision is sometimes needed in experimental mathematics when trying to discover new mathematical identities. Algorithms which may be used for these purposes include power series, continued fractions, solutions of equations with Newton's method, or other superlinearly convergent methods.

For performing such tasks it is convenient to use arrays to represent numbers in a floating-point form with a large base and a long mantissa, and to have routines for performing floating-point operations on such numbers. In this way it is possible to *simulate arithmetic of arbitrarily high precision* using standard floating-point arithmetic.

[35] In October 1995 Yasumasa Kanada of the University of Tokyo computed π to 6,442,458,938 decimals on a Hitachi supercomputer; see [11].

Brent [46, 45] developed the first major such multiple precision package in Fortran 66. His package represents multiple precision numbers as arrays of integers and operates on them with integer arithmetic. It includes subroutines for multiple precision evaluation of elementary functions. A more recent package called MPFUN, written in Fortran 77 code, is that of Bailey [9]. In MPFUN a multiple precision number is represented as a vector of single precision floating-point numbers with base 2^{24}. Complex multiprecision numbers are also supported. There is also a Fortran 90 version of this package [10], which is easy to use.

A package of MATLAB m-files called **Mulprec** for computations in, principally, unlimited precision floating-point, has been developed by the first-named author. Documentation of Mulprec and the m-files can be downloaded from the homepage of this book at `www.siam.org/books/ot103` together with some examples of its use.

Fortran routines for high precision computation are also provided in Press et al. [294, Sec. 20.6], and are also supported by symbolic manipulation systems such as Maple [65] and Mathematica [382]; see Online Appendix C.

Review Questions

2.2.1 What base β is used in the binary, octal, and hexadecimal number systems?

2.2.2 Show that any *finite* decimal fraction corresponds to a binary fraction that eventually is periodic.

2.2.3 What is meant by a normalized floating-point representation of a real number?

2.2.4 (a) How large can the maximum relative error be in representation of a real number a in the floating-point system $F = F(\beta, p, e_{\min}, e_{\max})$? It is assumed that a is in the range of F.

(b) How are the quantities "machine epsilon" and "unit roundoff" defined?

2.2.5 What are the characteristics of the IEEE single and double precision formats?

2.2.6 What are the advantages of including denormalized numbers in the IEEE standard?

2.2.7 Give examples of operations that give NaN as their result.

Problems and Computer Exercises

2.2.1 Which rational numbers can be expressed with a finite number of binary digits to the right of the binary point?

2.2.2 (a) Prove the algorithm for conversion between number systems given in Sec. 2.2.1.

(b) Give the hexadecimal form of the decimal numbers 0.1 and 0.3. What error is incurred in rounding these numbers to IEEE 754 single and double precision?

(c) What is the result of the computation $0.3/0.1$ in IEEE 754 single and double precision?

2.2.3 (Kahan) An (over)estimate of u can be obtained for almost any computer by evaluating $|3 \times (4/3 - 1) - 1|$ using rounded floating-point for every operation. Test this on a calculator or computer available to you.

2.2.4 (Goldberg [158]) The binary single precision numbers in the half-open interval $[10^3, 1024)$ have 10 bits to the left and 14 bits to the right of the binary point. Show that there are $(2^{10} - 10^3) \cdot 2^{14} = 393{,}216$ such numbers, but only $(2^{10} - 10^3) \cdot 10^4 = 240{,}000$ decimal numbers with eight decimal digits in the same interval. Conclude that eight decimal digits are not enough to uniquely represent single precision binary numbers in the IEEE 754 standard.

2.2.5 Suppose one wants to compute the power A^n of a square matrix A, where n is a positive integer. To compute $A^{k+1} = A \cdot A^k$ for $k = 1 : n-1$ requires $n-1$ matrix multiplications. Show that the number of multiplications can be reduced to less than $2\lfloor \log_2 n \rfloor$ by converting n into binary form and successively squaring $A^{2k} = (A^k)^2$, $k = 1 : \lfloor \log_2 n \rfloor$.

2.2.6 Give in decimal representation: (a) $(10000)_2$; (b) $(100)_8$; (c) $(64)_{16}$; (d) $(FF)_{16}$; (e) $(0.11)_8$; (f) the largest positive integer which can be written with 31 binary digits (answer with one significant digit).

2.2.7 (a) Show how the following numbers are stored in the basic single precision format of the IEEE 754 standard: 1.0; -0.0625; 250.25; 0.1.

(b) Give in decimal notation the largest and smallest positive numbers which can be stored in this format.

2.2.8 (Goldberg [158, Theorem 7]) When $\beta = 2$, if m and n are integers with $m < 2^{p-1}$ (p is the number of bits in the mantissa) and n has the special form $n = 2^i + 2^j$, then $fl((m/n) \times n) = m$ provided that floating-point operations are exactly rounded to nearest. The sequence of possible values of n starts with 1, 2, 3, 4, 5, 6, 8, 9, 10, 12, 16, 17. Test the theorem on your computer for these numbers.

2.2.9 Let pi be the closest floating-point number to π in double precision IEEE 754 standard. Find a sufficiently accurate approximation to π from a table and show that $\pi - pi \approx 1.2246 \cdot 10^{-16}$. What value do you get on your computer for $\sin \pi$?

2.2.10 (Edelman) Let x, $1 \le x < 2$, be a floating-point number in IEEE double precision arithmetic. Show that $fl(x \cdot fl(1/x))$ is either 1 or $1 - \epsilon_M/2$, where $\epsilon_M = 2^{-52}$ (the machine epsilon).

2.2.11 (N. J. Higham) Let a and b be floating-point numbers with $a \le b$. Show that the inequalities $a \le fl((a+b)/2) \le b$ can be violated in base 10 arithmetic. Show that $a \le fl(a + (b-a)/2) \le b$ in base β arithmetic.

2.2.12 (Muller) A rational approximation of $\tan x$ in $[-\pi/4, \pi/4]$ is

$$r(x) = \frac{(0.99999\,99328 - 0.09587\,5045x^2)x}{1 - (0.42920\,9672 + 0.00974\,3234x^2)x^2}.$$

Determine the approximate maximum error of this approximation by comparing with the function on your system on 100 equidistant points in $[0, \pi/4]$.

2.2.13 (a) Show how on a binary computer the exponential function can be approximated by first performing a range reduction based on the relation $e^x = 2^y$, $y = x/\log 2$, and then approximating 2^y on $y \in [0, 1/2]$.

(b) Show that since 2^y satisfies $2^{-y} = 1/2^y$ a rational function $r(y)$ approximating 2^y should have the form

$$r(y) = \frac{q(y^2) + ys(y^2)}{q(y^2) - ys(y^2)},$$

where q and s are polynomials.

(c) Suppose the $r(y)$ in (b) is used for approximating 2^y with

$$q(y) = 20.81892\,37930\,062 + y,$$
$$s(y) = 7.21528\,91511\,493 + 0.05769\,00723\,731y.$$

How many additions, multiplications, and divisions are needed in this case to evaluate $r(y)$? Investigate the accuracy achieved for $y \in [0, 1/2]$.

2.3 Accuracy and Rounding Errors

2.3.1 Floating-Point Arithmetic

It is useful to have a model of how the basic floating-point operations are carried out. If x and y are two floating-point numbers, we denote by

$$fl\,(x + y), \quad fl\,(x - y), \quad fl\,(x \times y), \quad fl\,(x/y)$$

the results of floating addition, subtraction, multiplication, and division, which the machine stores in memory (after rounding or chopping). In what follows we will assume that underflow or overflow does not occur, and that the following **standard model** for the arithmetic holds.

Definition 2.3.1.

Assume that $x, y \in F$. Then in the **standard model** *it holds that*

$$fl\,(x \text{ op } y) = (x \text{ op } y)(1 + \delta), \quad |\delta| \le u, \tag{2.3.1}$$

where u is the unit roundoff and "op" stands for one of the four elementary operations: $+$, $-$, $\times$, and $/$.

The standard model holds with the default rounding mode for computers implementing the IEEE 754 standard. In this case we also have

$$fl\,(\sqrt{x}) = \sqrt{x}(1 + \delta), \quad |\delta| \le u. \tag{2.3.2}$$

If a guard digit is lacking, then instead of (2.3.1) only the weaker model

$$fl\,(x \text{ op } y) = x(1 + \delta_1) \text{ op } y(1 + \delta_2), \quad |\delta_i| \le u, \tag{2.3.3}$$

holds for addition/subtraction. The lack of a guard digit is a serious drawback and can lead to damaging inaccuracy caused by cancellation. Many algorithms can be proved to work satisfactorily only if the standard model (2.3.1) holds. We remark that on current computers multiplication is as fast as addition/subtraction. Division usually is five to ten times slower than a multiply, and a square root about twice as slow as division.

Some earlier computers lack a guard digit in addition/subtraction. Notable examples are several pre-1995 models of Cray computers (Cray 1,2, X-MP,Y-MP, and C90), which were designed to have the highest possible floating-point performance. The IBM 360, which used a hexadecimal system, lacked a (hexadecimal) guard digit between 1964 and 1967. The consequences turned out to be so intolerable that a guard digit had to be retrofitted.

Sometimes the floating-point computation is more precise than what the standard model assumes. An obvious example is that when the exact value x op y can be represented as a floating-point number there is no rounding error at all.

Some computers can perform a *fused multiply–add* operation, i.e., an expression of the form $a \times x + y$ can be evaluated with just one instruction and there is only *one rounding error* at the end:

$$fl\,(a \times x + y) = (a \times x + y)(1 + \delta), \qquad |\delta| \le u.$$

Fused multiply–add can be used to advantage in many algorithms. For example, Horner's rule to evaluate the polynomial $p(x) = a_0x^n + a_1x^{n-1} + \cdots + a_{n-1}x + a_n$, which uses the recurrence relation $b_0 = a_0$, $b_i = b_{i-1} \cdot x + a_i$, $i = 1 : n$, needs only n fused multiply–add operations.

It is important to realize that floating-point operations have, to some degree, other properties than the exact arithmetic operations. Floating-point addition and multiplication are commutative but not associative, and the distributive law also fails for them. This makes the analysis of floating-point computations quite difficult.

Example 2.3.1.

To show that associativity does not, in general, hold for floating addition, consider adding the three numbers

$$a = 0.1234567 \cdot 10^0, \qquad b = 0.4711325 \cdot 10^4, \qquad c = -b$$

in a decimal floating-point system with $t = 7$ digits in the mantissa. The following scheme indicates how floating-point addition is performed:

$$fl\,(b + c) = 0, \qquad fl\,(a + fl\,(b + c)) = a = 0.1234567 \cdot 10^0$$

$$\begin{array}{rcr|r} a & = & 0.0000123 & 4567 \cdot 10^4 \\ +b & = & 0.4711325 & \cdot 10^4 \\ \hline fl\,(a+b) & = & 0.4711448 & \cdot 10^4 \\ c & = & -0.4711325 & \cdot 10^4 \end{array}.$$

The last four digits to the right of the vertical line are lost by **outshifting**, and

$$fl\,(fl\,(a + b) + c) = 0.0000123 \cdot 10^4 = 0.1230000 \cdot 10^0 \ne fl\,(a + fl\,(b + c)).$$

An interesting fact is that, assuming a guard digit is used, *floating-point subtraction of two sufficiently close numbers is always exact.*

Lemma 2.3.2 (*Sterbenz* [333]).

Let the floating-point numbers x and y satisfy

$$y/2 \le x \le 2y.$$

Then $fl(x - y) = x - y$, *unless* $x - y$ *underflows.*

Proof. By the assumption the exponent of x and y in the floating-point representations of x and y can differ by at most one unit. If the exponent is the same, then the exact result will be computed. Therefore, assume the exponents differ by one. After scaling and, if necessary, interchanging x and y, it holds that $x/2 \le y \le x < 2$ and the exact difference $z = x - y$ is of the form

$$\begin{array}{rcl} x & = & x_1.x_2 \ldots x_t \\ y & = & 0.y_1 \ldots y_{t-1}y_t \\ \hline z & = & z_1.z_2 \ldots z_t z_{t+1} \end{array}$$

But from the assumption, $x/2 - y \le 0$ or $x - y \le y$. Hence we must have $z_1 = 0$, and thus after shifting the exact result is also obtained in this case. □

With gradual underflow, as in the IEEE 754 standard, the condition that $x - y$ does not underflow can be dropped.

Example 2.3.2.

A corresponding result holds for any base β. For example, using four-digit floating decimal arithmetic we get with guard digit

$$fl\,(0.1000 \cdot 10^1 - 0.9999) = 0.0001 = 1.000 \cdot 10^{-4}$$

(exact), but without guard digit

$$fl\,(0.1000 \cdot 10^1 - 0.9999) = (0.1000 - 0.0999)10^1 = 0.0001 \cdot 10^1 = 1.000 \cdot 10^{-3}.$$

The last result satisfies (2.3.3) with $|\delta_i| \le 0.5 \cdot 10^{-3}$ since $0.10005 \cdot 10^1 - 0.9995 = 10^{-3}$.

Outshiftings are common causes of loss of information that may lead to **catastrophic cancellation** later, in the computations of a quantity that one would have liked to obtain with good relative accuracy.

Example 2.3.3.

An example where the result of Lemma 2.3.2 can be used to advantage is in computing compounded interest. Consider depositing the amount c every day in an account with an interest rate i compounded daily. With the accumulated capital, the total at the end of the year equals

$$c[(1 + x)^n - 1]/x, \quad x = i/n \ll 1,$$

and $n = 365$. Using this formula does not give accurate results. The reason is that a rounding error occurs in computing $fl(1 + x) = 1 + \bar{x}$ and low order bits of x are lost. For example, if $i = 0.06$, then $i/n = 0.0001643836$; in decimal arithmetic using six digits when this is added to one we get $fl(1+i/n) = 1.000164$, and thus four low order digits are lost.

The problem then is to accurately compute $(1+x)^n = \exp(n \log(1+x))$. The formula

$$\log(1+x) = \begin{cases} x & \text{if } fl\,(1+x) = 1, \\ x\dfrac{\log(1+x)}{(1+x)-1} & \text{otherwise} \end{cases} \tag{2.3.4}$$

can be shown to yield accurate results when $x \in [0, 3/4]$ and the computed value of $\log(1+x)$ equals the exact result rounded; see Goldberg [158, p. 12].

To check this formula we recall that the base e of the natural logarithm can be defined by the limit

$$e = \lim_{n\to\infty} (1 + 1/n)^n.$$

In Figure 2.3.1 we show computed values, using double precision floating-point arithmetic, of the sequence $|(1+1/n)^n - e|$ for $n = 10^p$, $p = 1 : 14$. More precisely, the expression was computed as

$$|\exp(n \log(1+1/n)) - \exp(1)|.$$

The smallest difference $3 \cdot 10^{-8}$ occurs for $n = 10^8$, for which about half the number of bits in $x = 1/n$ are lost. For larger values of n rounding errors destroy the convergence. But using (2.3.4) we obtain correct results for all values of n! (The Maclaurin series

$$\log(1+x) = x - x^2/2 + x^3/3 - x^4/4 + \cdots$$

will also give good results.)

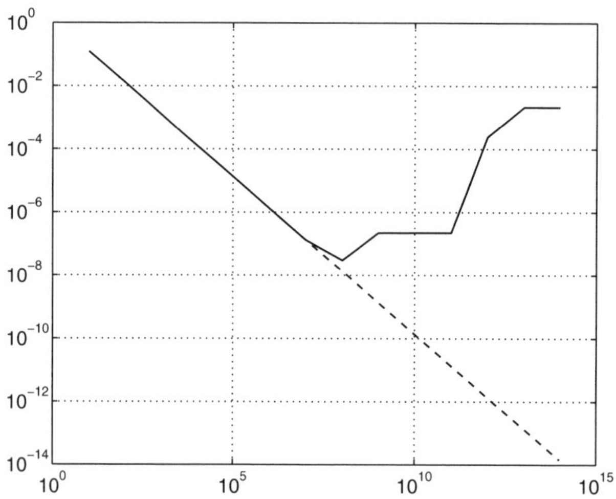

Figure 2.3.1. *Computed values for* $n = 10^p$, $p = 1 : 14$, *of the sequences: solid line* $|(1+1/n)^n - e|$; *dashed line* $|\exp(n \log(1+1/n)) - e|$ *using* (2.3.4).

A fundamental insight from the above examples can be expressed in the following way:

> "mathematically equivalent" formulas or algorithms are not in general "numerically equivalent."

This adds a new dimension to calculations in finite precision arithmetic and it will be a recurrent theme in the analysis of algorithms in this book.

By **mathematical equivalence** of two algorithms we mean here that the algorithms give exactly the same results from the same input data, if the computations are made without rounding error ("with infinitely many digits"). One algorithm can then, as a rule, formally be derived from the other using the rules of algebra for real numbers, and with the help of mathematical identities. Two algorithms are **numerically equivalent** if their respective floating-point results using the same input data are the same.

In error analysis for compound arithmetic expressions based on the standard model (2.3.1), one often needs an upper bound for quantities of this form:

$$\epsilon \equiv |(1+\delta_1)(1+\delta_2)\cdots(1+\delta_n) - 1|, \quad |\delta_i| \le u, \quad i = 1:n.$$

Then $\epsilon \le (1+u)^n - 1$. Assuming that $nu < 1$, an elementary calculation gives

$$\begin{aligned}(1+u)^n - 1 &= nu + \frac{n(n-1)}{2!}u^2 + \cdots + \binom{n}{k}u^k + \cdots \\ &< nu\left(1 + \frac{nu}{2} + \cdots + \left(\frac{nu}{2}\right)^{k-1} + \cdots\right) = \frac{nu}{1-nu/2}.\end{aligned} \tag{2.3.5}$$

Similarly, it can be shown that $(1-u)^{-n} - 1 < nu/(1-nu)$, and the following useful result follows (Higham [199, Lemma 3.1]).

Lemma 2.3.3.

Let $|\delta_i| \le u$, $\rho_i = \pm 1$, $i = 1{:}n$, and set

$$\prod_{i=1}^{n}(1+\delta_i)^{\rho_i} = 1 + \theta_n. \tag{2.3.6}$$

If $nu < 1$, then $|\theta_n| < \gamma_n$, where

$$\gamma_n = nu/(1-nu). \tag{2.3.7}$$

Complex arithmetic can be reduced to real arithmetic. Let $x = a + ib$ and $y = c + id$ be two complex numbers. Then we have

$$\begin{aligned} x \pm y &= a \pm c + i(b \pm d), \\ x \times y &= (ac - bd) + i(ad + bc), \\ x/y &= \frac{ac+bd}{c^2+d^2} + i\frac{bc-ad}{c^2+d^2}. \end{aligned} \tag{2.3.8}$$

Using the above formula, complex addition (subtraction) needs two real additions, and multiplying two complex numbers requires four real multiplications.

Lemma 2.3.4.

Assume that the standard model (2.3.1) *for floating-point arithmetic holds. Then, provided that no overflow or underflow occurs, no denormalized numbers are produced and the complex operations computed according to* (2.3.8) *satisfy*

$$\begin{aligned} fl\,(x \pm y) &= (x \pm y)(1+\delta), \quad |\delta| \le u, \\ fl\,(x \times y) &= x \times y(1+\delta), \quad |\delta| \le \sqrt{5}u, \\ fl\,(x/y) &= x/y(1+\delta), \quad |\delta| \le \sqrt{2}\gamma_4, \end{aligned} \tag{2.3.9}$$

where δ is a complex number and γ_n is defined in (2.3.7).

Proof. See Higham [199, Sec. 3.6]. The result for complex multiplication is due to Brent et al. [48]. □

The square root of a complex number $u + iv = \sqrt{x+iy}$ is given by

$$u = \left(\frac{r+x}{2}\right)^{1/2}, \qquad v = \left(\frac{r-x}{2}\right)^{1/2}, \qquad r = \sqrt{x^2+y^2}. \tag{2.3.10}$$

When $x > 0$ there will be cancellation when computing v, which can be severe if also $|x| \gg |y|$ (cf. Sec. 2.3.4). To avoid this we note that $uv = \sqrt{r^2 - x^2}/2 = y/2$, and thus v can be computed from $v = y/(2u)$. When $x < 0$ we instead compute v from (2.3.10) and set $u = y/(2v)$.

Most rounding error analyses given in this book are formulated for real arithmetic. Since the bounds in Lemma 2.3.4 are of the same form as the standard model for real arithmetic, these can simply be extended to complex arithmetic.

In some cases it may be desirable to avoid complex arithmetic when working with complex matrices. This can be achieved in a simple way by replacing the complex matrices and vectors by real ones of twice the order. Suppose that a complex matrix $A \in \mathbf{C}^{n\times n}$ and a complex vector $z \in \mathbf{C}^n$ are given, where

$$A = B + iC, \qquad z = x + iy,$$

with real B, C, x, and y. Form the real matrix $\tilde{A} \in \mathbf{R}^{2n\times 2n}$ and real vector $\tilde{z} \in \mathbf{R}^{2n}$ defined by

$$\tilde{A} = \begin{pmatrix} B & -C \\ C & B \end{pmatrix}, \qquad \tilde{z} = \begin{pmatrix} x \\ y \end{pmatrix}.$$

It is easy to verify the following rules:

$$\widetilde{(Az)} = \tilde{A}\tilde{z}, \qquad \widetilde{(AB)} = \tilde{A}\tilde{B}, \qquad \widetilde{(A^{-1})} = (\tilde{A})^{-1}.$$

Thus we can solve complex-valued matrix problems using algorithms for the real case. But this incurs a penalty in storage and arithmetic operations.

2.3.2 Basic Rounding Error Results

We now use the notation of Sec. 2.3.1 and the standard model of floating-point arithmetic (Definition 2.3.1) to carry out rounding error analysis of some basic computations. Most, but not all, results are still true if only the weaker bound (2.3.3) holds for addition and subtraction. Note that $fl\,(x \text{ op } y) = (x \text{ op } y)(1+\delta)$, $|\delta| \le u$, can be interpreted for multiplication to mean that $fl\,(x \cdot y)$ is the *exact result* of $x \cdot y(1+\delta)$ for some δ, $|\delta| \le u$. In the same way, the results using the three other operations can be interpreted as *the result of exact operations where the operands have been perturbed by a relative amount which does not exceed* u. In **backward error analysis** (see Sec. 2.4.4) one applies the above interpretation step by step backward in an algorithm.

By repeated use of the formula (2.3.1) in the case of multiplication, one can show that

$$fl\,(x_1x_2\cdots x_n) = x_1x_2(1+\delta_2)x_3(1+\delta_3)\cdots x_n(1+\delta_n),$$
$$|\delta_i| \le u, \quad i = 2:n$$

holds; i.e., the computed **product** $fl\,(x_1x_2\cdots x_n)$ is *exactly* equal to a product of the factors

$$\tilde{x}_1 = x_1, \quad \tilde{x}_i = x_i(1+\delta_i), \quad i = 2:n.$$

Using the estimate and notation of (2.3.7) it follows from this analysis that

$$|fl\,(x_1x_2\cdots x_n) - x_1x_2\cdots x_n| < \gamma_{n-1}|x_1x_2\cdots x_n|, \tag{2.3.11}$$

which bounds the **forward error** of the computed result.

For a **sum** of n floating-point numbers similar results can be derived. If the sum is computed in the natural order, we have

$$fl\,(\cdots(((x_1+x_2)+x_3)+\cdots+x_n))$$
$$= x_1(1+\delta_1)+x_2(1+\delta_2)+\cdots+x_n(1+\delta_n),$$

where

$$|\delta_1| < \gamma_{n-1}, \quad |\delta_i| < \gamma_{n+1-i}, \quad i = 2:n,$$

and thus the computed sum is *the exact sum* of the numbers $x_i(1+\delta_i)$. This also gives an estimate of the forward error

$$|fl\,(\cdots(((x_1+x_2)+x_3)+\cdots+x_n)) - (x_1+x_2+x_3+\cdots+x_n)|$$
$$< \sum_{i=1}^{n}\gamma_{n+1-i}|x_i| \le \gamma_{n-1}\sum_{i=1}^{n}|x_i|, \tag{2.3.12}$$

where the last upper bound holds independent of the summation order.

Notice that to minimize the first upper bound in equation (2.3.12), the terms *should be added in increasing order of magnitude*. For large n an even better bound can be shown if the summation is done using the divide and conquer technique described in Sec. 1.2.3; see Problem 2.3.5.

Example 2.3.4.

Using a hexadecimal machine ($\beta = 16$), with $t = 6$ and chopping ($u = 16^{-5} \approx 10^{-6}$), we computed

$$\sum_{n=1}^{10{,}000} n^{-2} \approx 1.644834$$

in two different orders. Using the natural summation order $n = 1, 2, 3, \ldots$ the error was $1.317 \cdot 10^{-3}$. Summing in the opposite order $n = 10{,}000, 9{,}999, 9{,}998, \ldots$ the error was reduced to $2 \cdot 10^{-6}$. This was not unexpected. Each operation is an addition, where the partial sum s is increased by n^{-2}. Thus, in each operation, one commits an error of about $s \cdot u$, and all these errors are added. Using the first summation order, we have $1 \le s \le 2$ in every step, but using the other order of summation we have $s < 10^{-2}$ in 9,900 of the 10,000 additions.

Similar bounds for roundoff errors can easily be derived for basic vector and matrix operations; see Wilkinson [377, pp. 114–118]. For an **inner product** $x^T y$ computed in the natural order we have

$$fl\,(x^T y) = x_1 y_1(1 + \delta_1) + x_2 y_2(1 + \delta_2) + \cdots + x_n y_n(1 + \delta_n),$$

where

$$|\delta_1| < \gamma_n, \quad |\delta_r| < \gamma_{n+2-i}, \quad i = 2 : n.$$

The corresponding forward error bound becomes

$$|fl\,(x^T y) - x^T y| < \sum_{i=1}^{n} \gamma_{n+2-i} |x_i||y_i| < \gamma_n \sum_{i=1}^{n} |x_i||y_i|.$$

If we let $|x|$, $|y|$ denote vectors with elements $|x_i|$, $|y_i|$ the last estimate can be written in the simple form

$$|fl\,(x^T y) - x^T y| < \gamma_n |x^T||y|. \tag{2.3.13}$$

This bound is independent of the summation order and also holds for the weaker model (2.3.3) valid with no guard digit rounding.

The outer product of two vectors $x, y \in \mathbf{R}^n$ is the matrix $xy^T = (x_i y_j)$. In floating-point arithmetic we compute the elements $fl\,(x_i y_j) = x_i y_j(1 + \delta_{ij})$, $\delta_{ij} \le u$, and thus

$$|fl\,(xy^T) - xy^T| \le u\,|xy^T|. \tag{2.3.14}$$

This is a satisfactory result for many purposes, but the computed result is not in general a rank one matrix and it is not possible to find Δx and Δy such that $fl(xy^T) = (x + \Delta x)(x + \Delta y)^T$.

The product of two t-digit floating-point numbers can be exactly represented with at most $2t$ digits. This allows inner products to be computed in extended precision without much extra cost. If fl_e denotes computation with extended precision and u_e the corresponding unit roundoff, then the forward error bound for an inner product becomes

$$|fl\,(fl_e(x^T y)) - x^T y| < u|x^T y| + \frac{nu_e}{1 - nu_e/2}(1 + u)|x^T||y|, \tag{2.3.15}$$

where the first term comes from the final rounding. If $|x^T||y| \leq u|x^T y|$, then the computed inner product is almost as accurate as the correctly rounded exact result. These accurate inner products can be used to improve accuracy by so-called *iterative refinement* in many linear algebra problems. But since computations in extended precision are machine dependent it has been difficult to make such programs portable.[36] The recent development of Extended and Mixed Precision BLAS (Basic Linear Algebra Subroutines; see [247]) may now make this more feasible.

Similar error bounds can easily be obtained for matrix multiplication. Let $A \in \mathbf{R}^{m\times n}$, $B \in \mathbf{R}^{n\times p}$, and denote by $|A|$ and $|B|$ matrices with elements $|a_{ij}|$ and $|b_{ij}|$. Then it holds that

$$|fl\,(AB) - AB| < \gamma_n |A||B|, \tag{2.3.16}$$

where the inequality is to be interpreted elementwise. Often we shall need bounds for some norm of the error matrix. From (2.3.16) it follows that

$$\|fl\,(AB) - AB\| < \gamma_n \|\,|A|\,\|\,\|\,|B|\,\|. \tag{2.3.17}$$

Hence, for the 1-norm, ∞-norm, and the Frobenius norm we have

$$\|fl\,(AB) - AB\| < \gamma_n \|A\|\,\|B\|, \tag{2.3.18}$$

but unless A and B have only nonnegative elements, we get for the 2-norm only the weaker bound

$$\|fl\,(AB) - AB\|_2 < n\gamma_n \|A\|_2\,\|B\|_2. \tag{2.3.19}$$

To reduce the effects of rounding errors in computing a sum $\sum_{i=1}^{n} x_i$ one can use **compensated summation**. In this algorithm the rounding error in each addition is estimated and then compensated for with a correction term. Compensated summation can be useful when a large number of small terms are to be added as in numerical quadrature. Another example is the case in the numerical solution of initial value problems for ordinary differential equations. Note that in this application the terms have to be added in the order in which they are generated.

Compensated summation is based on the possibility to *simulate double precision floating-point addition in single precision arithmetic*. To illustrate the basic idea we take, as in Example 2.3.1,

$$a = 0.1234567 \cdot 10^0, \qquad b = 0.4711325 \cdot 10^4,$$

so that $s = fl\,(a + b) = 0.4711448 \cdot 10^4$. Suppose we form

$$c = fl\,(fl\,(b - s) + a) = -0.1230000 \cdot 10^0 + 0.1234567 \cdot 10^0 = 4567000 \cdot 10^{-3}.$$

Note that the variable c is computed without error and picks up the information that was lost in the operation $fl\,(a + b)$.

[36] It was suggested that the IEEE 754 standard should require inner products to be precisely specified, but that did not happen.

ALGORITHM 2.2. *Compensated Summation.*

The following algorithm uses compensated summation to accurately compute the sum $\sum_{i=1}^{n} x_i$:

$$
\begin{aligned}
& s := x_1; \quad c := 0; \\
& \textbf{for } i = 2 : n \\
& \quad y := c + x_i; \\
& \quad t := s + y; \\
& \quad c := (s - t) + y; \\
& \quad s := t; \\
& \textbf{end}
\end{aligned}
$$

It can be proved (see Goldberg [158]) that on binary machines with a guard digit the computed sum satisfies

$$s = \sum_{i=1}^{n} (1 + \xi_i) x_i, \qquad |\xi_i| < 2u + O(nu^2). \tag{2.3.20}$$

This formulation is a typical example of a backward error analysis; see Sec. 2.4.4. The first term in the error bound is independent of n.

2.3.3 Statistical Models for Rounding Errors

The bounds for the accumulated rounding error we have derived so far are estimates of the **maximal error**. These bounds ignore the sign of the errors and tend to be much too pessimistic when the number of variables is large. However, they can still give valuable insight into the behavior of a method and be used for the purpose of comparing different methods.

An alternative is a statistical analysis of rounding errors, which is based on the assumption that rounding errors are independent and have some statistical distribution. It was observed already in the 1950s that rounding errors occurring in the solution of differential equations *are not random and are often strongly correlated.* This does not in itself preclude that useful information can sometimes be obtained by modeling them by random uncorrelated variables! In many computational situations and scientific experiments, where the error can be considered to have arisen from the addition of a large number of *independent* error sources of about the same magnitude, an assumption that the errors are normally distributed is justified.

Example 2.3.5.

Figure 2.3.2 illustrates the effect of rounding errors on the evaluation of two different expressions for the polynomial $p(x) = (x - 1)^5$ for $x \in [0.999, 1.001]$, in IEEE double precision (unit roundoff $u = 1.1 \cdot 10^{-16}$). Among other things it shows that the monotonicity of a function can be lost due to rounding errors. The model of rounding errors as independent random variables works well in this example. It is obvious that it would be impossible to locate the zero of $p(x)$ to a precision better than about $(0.5 \cdot 10^{-14})^{1/6} \approx 0.0007$ using the

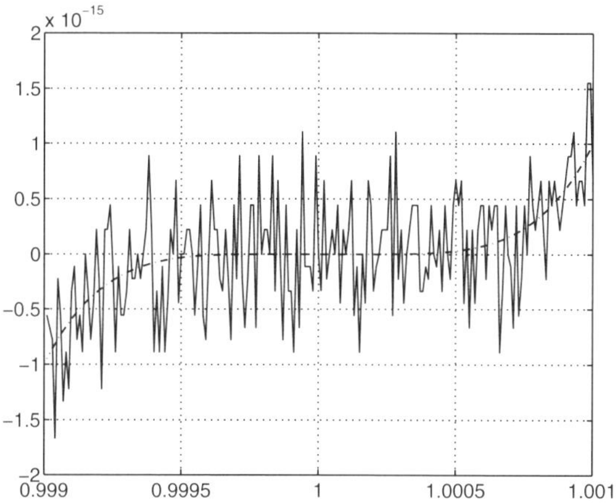

Figure 2.3.2. *Calculated values of a polynomial near a multiple root: solid line* $p(x) = x^5 - 5x^4 + 10x^3 - 10x^2 + 5x - 1 = 0$*; dashed line* $p(x) = (x - 1)^5$.

expanded form of $p(x)$. But using the expression $p(x) = (1 - x)^5$ function values can be evaluated with constant *relative* precision even close to $x = 1$, and the problem disappears!

This example shows that although multiple roots are in general ill-conditioned, an important exception is when the function $f(x)$ *is given in such a form that it can be computed with less absolute error as x approaches* α.

The theory of **standard error** is based on probability theory and will not be treated in detail here. The standard error of an estimate of a given quantity is the same as the *standard deviation of its sampling distribution.*

If in a sum $y = \sum_{i=1}^{n} x_i$ each x_i has error $|\Delta_i| \leq \delta$, then the maximum error bound for y is $n\delta$. Thus, *the maximal error grows proportionally to* n. If n is large—for example, $n = 1000$—then it is in fact highly improbable that the real error will be anywhere near $n\delta$, since that bound is attained only when every Δx_i has the same sign and the same maximal magnitude. Observe, though, that if positive numbers are added, each of which has been abridged to t decimals by chopping, then each Δx_i has the same sign and a magnitude which is on average $\frac{1}{2}\delta$, where $\delta = 10^{-t}$. Thus, the real error is often about 500δ.

If the numbers are rounded instead of chopped, and if one can assume that the errors in the various terms are stochastically independent with standard deviation ϵ, then the standard error in y becomes (see Theorem 2.4.5)

$$(\epsilon^2 + \epsilon^2 + \cdots + \epsilon^2)^{1/2} = \epsilon\sqrt{n}.$$

Thus the standard error of the sum grows only proportionally to $\sqrt{n}$. This supports the rule of thumb, suggested by Wilkinson [376, p. 26], that *if a rounding error analysis gives a bound* $f(n)u$ *for the maximum error, then one can expect the real error to be of size* $\sqrt{f(n)}u$.

If $n \gg 1$, then the error in y is, under the assumptions made above, approximately normally distributed with standard deviation $\sigma = \epsilon\sqrt{n}$. The corresponding frequency function,

$$f(t) = \frac{1}{\sqrt{2\pi}} e^{-t^2/2},$$

is illustrated in Figure 2.3.3; the curve shown there is also called the Gauss curve. The assumption that the error is normally distributed with standard deviation σ means, for example, that the statement "the magnitude of the error is greater than 2σ" (see the shaded area of Figure 2.3.3) is true in only about 5% of all cases (the unshaded area under the curve). More generally, the assertion that the magnitude of the error is larger than σ, 2σ, and 3σ respectively, is about 32%, 5%, and 0.27%.

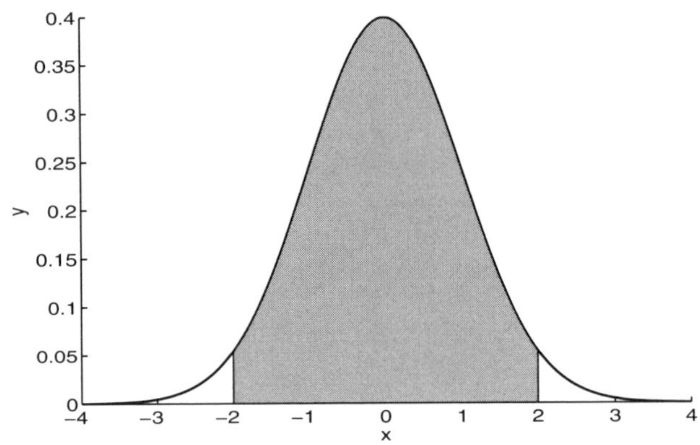

Figure 2.3.3. *The frequency function of the normal distribution for* $\sigma = 1$.

One can show that if the individual terms in a sum $y = \sum_{i=1}^{n} x_i$ have a **uniform** probability distribution in the interval $[-\frac{1}{2}\delta, \frac{1}{2}\delta]$, then the standard deviation of an individual term is $\delta/12$. Therefore, in only about 5% of cases is the error in the sum of 1000 terms greater than $2\delta\sqrt{1000/12} \approx 18\delta$, which can be compared to the maximum error 500δ. This shows that rounding can be far superior to chopping when a statistical interpretation (especially the assumption of independence) can be given to the principal sources of errors. Observe that, in the above, we have only considered the propagation of errors which were present in the original data, and have ignored the effect of possible roundoff errors in the additions themselves.

For rounding errors the formula for standard errors is used. For systematic errors, however, the formula for maximal error (2.4.5) should be used.

2.3.4 Avoiding Overflow and Cancellation

In the rare cases when input and output data are so large or small in magnitude that the range of the machine is not sufficient, one can use higher precision or else work with logarithms or some other transformation of the data. One should, however, keep in mind the risk that *intermediate results* in a calculation can produce an exponent which is too large (overflow) or too small (underflow) for the floating-point system of the machine. Different actions may be taken in such situations, as well for division by zero. Too small an exponent is usually, but not always, unproblematic. If the machine does not signal underflow, but simply sets the result equal to zero, there is a risk, however, of harmful consequences. Occasionally, "unexplainable errors" in output data are caused by underflow somewhere in the computations.

The **Pythagorean sum** $c = \sqrt{a^2 + b^2}$ occurs frequently, for example, in conversion to polar coordinates and in computing the complex modulus and complex multiplication. If the obvious algorithm is used, then damaging underflows and overflows may occur in the squaring of a and b even if a and b and the result c are well within the range of the floating-point system used. This can be avoided by using instead the algorithm: If $a = b = 0$, then $c = 0$; otherwise set $p = \max(|a|, |b|)$, $q = \min(|a|, |b|)$, and compute

$$\rho = q/p; \qquad c = p\sqrt{1 + \rho^2}. \tag{2.3.21}$$

Example 2.3.6.

The formula (2.3.8) for complex division suffers from the problem that intermediate results can overflow even if the final result is well within the range of the floating-point system. This problem can be avoided by rewriting the formula as for the Pythagorean sum: If $|c| > |d|$, then compute

$$\frac{a + ib}{c + id} = \frac{a + be}{r} + i\frac{b - ae}{r}, \qquad e = \frac{d}{c}, \qquad r = c + de.$$

If $|d| > |c|$, then $e = c/d$ is computed and a corresponding formula used.

Similar precautions are also needed for computing the Euclidean length (norm) of a vector $\|x\|_2 = \left(\sum_{i=1}^n x_i^2\right)^{1/2}$, $x \neq 0$. We could avoid overflows by first finding $x_{max} = \max_{1\le i\le n} |x_i|$ and then forming

$$s = \sum_{i=1}^{n} (x_i/x_{max})^2, \qquad \|x\|_2 = x_{max}\sqrt{s}. \tag{2.3.22}$$

This has the drawback of needing *two passes* through the data.

ALGORITHM 2.3.

The following algorithm, due to S. J. Hammarling, for computing the Euclidean length of a vector requires only one pass through the data. It is used in the level-1 BLAS routine xNRM2:

```
t = 0;  s = 1;
for i = 1 : n
  if |x_i| > 0
    if |x_i| > t
      s = 1 + s(t/x_i)^2;  t = |x_i|;
    else
      s = s + (x_i/t)^2;
    end
  end
end
||x||_2 = t*sqrt(s);
```

On the other hand this code does not vectorize and can therefore be slower if implemented on a vector computer.

One very common reason for poor accuracy in the result of a calculation is that somewhere a subtraction has been carried out in which the difference between the operands is considerably less than either of the operands. This causes a loss of relative precision. (Note that, on the other hand, relative precision is preserved in addition of nonnegative quantities, multiplication, and division.)

Consider the computation of $y = x_1 - x_2$, where $\tilde{x}_1 = x_1 + \Delta x_1$, $\tilde{x}_2 = x_2 + \Delta x_2$ are approximations to the exact values. If the operation is carried out exactly the result is $\tilde{y} = y + \Delta y$, where $\Delta y = \Delta x_1 - \Delta x_2$. But, since the errors Δx_1 and Δx_2 can have opposite sign, the best error bound for $\tilde{y}$ is

$$|\Delta y| \leq |\Delta x_1| + |\Delta x_2|. \tag{2.3.23}$$

Notice the plus sign! Hence for the relative error we have

$$\left|\frac{\Delta y}{y}\right| \leq \frac{|\Delta x_1| + |\Delta x_2|}{|x_1 - x_2|}. \tag{2.3.24}$$

This shows that *there can be very poor relative accuracy in the difference between two nearly equal numbers*. This phenomenon is called **cancellation of terms**.

In Sec. 1.2.1 it was shown that when using the well-known "textbook" formula

$$r_{1,2} = \left(-b \pm \sqrt{b^2 - 4ac}\right)/(2a)$$

for computing the real roots of the quadratic equation $ax^2 + bx + c = 0$ $(a \neq 0)$, cancellation could cause a loss of accuracy in the root of smallest magnitude. This can be avoided by computing the root of *smaller magnitude* from the relation $r_1 r_2 = c/a$ between coefficients and roots. The following is a suitable algorithm.

Algorithm 2.4. *Solving a Quadratic Equation.*

$d := b^2 - 4ac;$
if $d \geq 0$ % real roots
  $r_1 := -\text{sign}(b)\big(|b| + \sqrt{d}\big)/(2a);$
  $r_2 := c/(a \cdot r_1);$
else % complex roots $x + iy$
  $x := -b/(2a);$
  $y := \sqrt{-d}/(2a);$
end

Note that we define sign $(b) = 1$ if $b \geq 0$, else sign $(b) = -1$.[37] It can be proved that in IEEE arithmetic this algorithm computes *a slightly wrong solution to a slightly wrong problem*.

[37] In MATLAB sign $(0) = 0$, which can lead to failure of this algorithm.

Lemma 2.3.5.

Assume that Algorithm 2.4 is used to compute the roots $r_{1,2}$ of the quadratic equation $ax^2 + bx + c = 0$. Denote the computed roots by $\bar{r}_{1,2}$ and let $\tilde{r}_{1,2}$ be the exact roots of the nearby equation $ax^2 + bx + \tilde{c} = 0$, where $|\tilde{c} - c| \le \gamma_2 |\tilde{c}|$. Then $|\tilde{r}_i - \bar{r}_i| \le \gamma_5 |\tilde{r}_i|$, $i = 1, 2$.

Proof. See Kahan [216]. □

More generally, if $|\delta| \ll x$, then one should rewrite

$$\sqrt{x+\delta} - \sqrt{x} = \frac{x+\delta-x}{\sqrt{x+\delta}+\sqrt{x}} = \frac{\delta}{\sqrt{x+\delta}+\sqrt{x}}.$$

There are other exact ways of rewriting formulas which are as useful as the above; for example,

$$\cos(x+\delta) - \cos x = -2\sin(\delta/2)\sin(x+\delta/2).$$

If one cannot find an exact way of rewriting a given expression of the form $f(x+\delta) - f(x)$, it is often advantageous to use one or more terms in the Taylor series

$$f(x+\delta) - f(x) = f'(x)\delta + \frac{1}{2}f''(x)\delta^2 + \cdots$$

Example 2.3.7 (*Cody* [73]).

To compute $\sin 22$ we first find $\lfloor 22/(\pi/2) \rfloor = 14$. It follows that $\sin 22 = -\sin x^*$, where $x^* = 22 - 14(\pi/2)$. Using the correctly rounded 10-digit approximation $\pi/2 = 1.57079\,6327$ we obtain

$$x^* = 22 - 14 \cdot 1.57079\,6327 = 8.85142 \cdot 10^{-3}.$$

Here cancellation has taken place and the reduced argument has a maximal error of $7 \cdot 10^{-9}$. The actual error is slightly smaller since the correctly rounded value is $x^* = 8.85144\,8711 \cdot 10^{-3}$, which corresponds to a relative error in the computed $\sin 22$ of about $2.4 \cdot 10^{-6}$, in spite of using a 10-digit approximation to $\pi/2$.

For very large arguments the relative error can be much larger. Techniques for carrying out accurate range reductions without actually needing multiple precision calculations are discussed by Muller [272]; see also Problem 2.3.9.

In previous examples we got a warning that cancellation would occur, since x_2 was found as the difference between two nearly equal numbers each of which was, relatively, much larger than the difference itself. In practice, one does not always get such a warning, for two reasons: first, in using a computer one has no direct contact with the individual steps of calculation; second, cancellation can be spread over a great number of operations. This may occur in computing a partial sum of an infinite series. For example, in a series where the size of some terms are many orders of magnitude larger than the sum of the series, small relative errors in the computation of the large terms can then produce large errors in the result.

It has been emphasized here that calculations where cancellation occurs should be avoided. But there are cases where one has not been able to avoid it, and there is no time to wait for a better method. Situations occur in practice where (say) the first ten digits are lost, and we need a decent relative accuracy in what will be left.[38] Then, high accuracy is required in intermediate results. This is an instance where the high accuracy in IEEE double precision is needed!

Review Questions

2.3.1 What is the standard model for floating-point arithmetic? What weaker model holds if a guard digit is lacking?

2.3.2 Give examples to show that some of the axioms for arithmetic with real numbers do not always hold for floating-point arithmetic.

2.3.3 (a) Give the results of a backward and forward error analysis for computing $fl\,(x_1 + x_2 + \cdots + x_n)$. It is assumed that the standard model holds.

(b) Describe the idea in compensated summation.

2.3.4 Explain the terms "maximum error" and "standard error." What statistical assumption about rounding errors is often made when calculating the standard error in a sum due to rounding?

2.3.5 Explain what is meant by "cancellation of terms." Give an example of how this can be avoided by rewriting a formula.

Problems and Computer Exercises

2.3.1 Rewrite the following expressions to avoid cancellation of terms:
(a) $1 - \cos x$, $|x| \ll 1$; (b) $\sin x - \cos x$, $|x| \approx \pi/4$

2.3.2 (a) The expression $x^2 - y^2$ exhibits catastrophic cancellation if $|x| \approx |y|$. Show that it is more accurate to evaluate it as $(x + y)(x - y)$.

(b) Consider using the trigonometric identity $\sin^2 x + \cos^2 x = 1$ to compute $\cos x = (1 - \sin^2 x)^{1/2}$. For which arguments in the range $0 \le x \le \pi/4$ will this formula fail to give good accuracy?

2.3.3 The polar representation of a complex number is

$$z = x + iy = r(\sin\phi + \cos\phi) \equiv r \cdot e^{i\phi}.$$

Develop accurate formulas for computing this polar representation from x and y using real operations.

2.3.4 (Kahan) Show that with the use of fused multiply–add the algorithm

$$w = fl\,(b \times c); \quad y := fl\,(a \times d - w); \quad e := fl\,(b \times c - w); \quad z = fl\,(y - e);$$

[38] G. Dahlquist has encountered just this situation in a problem of financial mathematics.

computes with high relative accuracy

$$z = \det\begin{pmatrix} a & b \\ c & d \end{pmatrix} = ad - bc.$$

2.3.5 Suppose that the sum $s = \sum_{i=1}^{n} x_i, n = 2^k$, is computed using the divide and conquer technique described in Sec. 1.2.3. Show that this summation algorithm computes an exact sum

$$\bar{s} = \sum_{i=1}^{n} x_i(1 + \delta_i), \quad |\delta_i| \le \tilde{u} \log_2 n.$$

Hence for large values of n this summation order can be much more accurate than the conventional order.

2.3.6 Show that for the evaluation of a polynomial $p(x) = \sum_{i=0}^{n} a_i x^i$ by Horner's rule the following roundoff error estimate holds:

$$|fl\,(p(x)) - p(x)| < \gamma_1 \sum_{i=0}^{n} (2i + 1)|a_i|\,|x|^i, \quad (2nu \le 0.1).$$

2.3.7 In solving linear equations by Gaussian elimination expressions of the form $s = (c - \sum_{i=1}^{n-1} a_i b_i)/d$ often occur. Show that, by a slight extension of the result in the previous problem, that the computed $\bar{s}$ satisfies

$$\left|\bar{s}d - c + \sum_{i=1}^{n-1} a_i b_i\right| \le \gamma_n \left(|\bar{s}d| + \sum_{i=1}^{n-1} |a_i||b_i|\right),$$

where the inequality holds independent of the summation order.

2.3.8 The zeros of the reduced cubic polynomial $z^3 + 3qz - 2r = 0$ can be found from the Cardano–Tartaglia formula:

$$z = \left(r + \sqrt{q^3 + r^2}\right)^{1/3} + \left(r - \sqrt{q^3 + r^2}\right)^{1/3},$$

where the two cubic roots are to be chosen so that their product equals $-q$. One real root is obtained if $q^3 + r^2 \ge 0$, which is the case unless all three roots are real and distinct.

The above formula can lead to cancellation. Rewrite it so that it becomes more suitable for numerical calculation and requires the calculation of only one cubic root.

2.3.9 (Eldén and Wittmeyer-Koch) In the interval reduction for computing $\sin x$ there can be a loss of accuracy through cancellation in the computation of the reduced argument $x^* = x - k \cdot \pi/2$ when k is large. A way to avoid this without reverting to higher precision has been suggested by Cody and Waite [75]). Write

$$\pi/2 = \pi_0/2 + r,$$

where $\pi_0/2$ is *exactly representable* with a few digits in the (binary) floating-point system. The reduced argument is now computed as $x^* = (x - k \cdot \pi_0/2) - kr$. Here, unless k is very large, the first term can be computed without rounding error. The rounding error in the second term is bounded by $k|r|\,u$, where u is the unit roundoff. In IEEE single precision one takes

$$\pi_0/2 = 201/128 = 1.573125 = (10.1001001)_2, \qquad r = 4.838267949 \cdot 10^{-4}.$$

Estimate the relative error in the computed reduced argument x^* when $x = 1000$ and r is represented in IEEE single precision.

2.3.10 (Kahan [218]) The area A of a triangle with sides equal to a, b, c is given by Heron's formula:

$$A = \sqrt{s(s-a)(s-b)(s-c)}, \quad s = (a+b+c)/2.$$

Show that this formula fails for needle-shaped triangles, using five-digit decimal floating arithmetic and $a = 100.01$, $b = 99.995$, $c = 0.025$.
The following formula can be proved to work if addition/subtraction satisfies (2.3.21): Order the sides so that $a \geq b \geq c$, and use

$$A = \frac{1}{4}\sqrt{(a+(b+c))(c-(a-b))(c+(a-b))(a+(b-c))}.$$

Compute a correct result for the data above using this modified formula. If a person tells you that this gives an imaginary result if $a - b > c$, what do you answer him?

2.3.11 As is well known, $f(x) = (1+x)^{1/x}$ has the limit $e = 2.71828\ 18284\ 59045\ldots$ when $x \to \infty$. Study the sequences $f(x_n)$ for $x_n = 10^{-n}$ and $x_n = 2^{-n}$, for $n = 1, 2, 3, \ldots$. Stop when $x_n < 10^{-10}$ (or when $x_n < 10^{-20}$ if you are using double precision). Give your results as a table of n, x_n, and the relative error $g_n = (f(x_n) - e)/e$. Also plot $\log(|g_n|)$ against $\log(|x_n|)$. Comment on and explain your observations.

2.3.12 (a) Compute the derivative of the exponential function e^x at $x = 0$ by approximating with the difference quotients $(e^{x+h} - e^x)/h$, for $h = 2^{-i}$, $i = 1 : 20$. Explain your results.

(b) Repeat (a), but approximate with the central difference approximation $(e^{x+h} - e^{x-h})/(2h)$.

2.3.13 The hyperbolic cosine is defined by $\cosh t = (e^t + e^{-t})/2$, and its inverse function $t = \text{arccosh}\,(x)$ is the solution to

$$x = (e^t + e^{-t})/2.$$

Solving the quadratic equation $(e^t)^2 - 2xe^t + 1$, we find $e^t = x \pm (x^2-1)^{1/2}$ and

$$\text{arccos}\, x = \log(x \pm (x^2-1)^{1/2}).$$

(a) Show that this formula suffers from serious cancellation when the minus sign is used and x is large. Try, e.g., $x = \cosh(10)$ using double precision IEEE. (Using the plus sign will just transfer the problem to negative x.)

(b) A better formula is

$$\text{arccos}\, x = 2\log\left(((x+1)/2)^{1/2} + ((x-1)/2)^{1/2}\right).$$

This also avoids the squaring of x which can lead to overflow. Derive this formula and show that it is well behaved!

2.3.14 (Gautschi) Euler's constant $\gamma = 0.57721566490153286\ldots$ is defined as the limit

$$\gamma = \lim_{n\to\infty} \gamma_n, \quad \text{where} \quad \gamma_n = 1 + 1/2 + 1/3 + \cdots + 1/n - \log n.$$

Assuming that $\gamma - \gamma_n \sim cn^{-d}$, $n \to \infty$, for some constants c and $d > 0$, try to determine c and d experimentally on your computer.

2.3.15 In the statistical treatment of data, one often needs to compute the quantities

$$\bar{x} = \frac{1}{n}\sum_{i=1}^{n} x_i, \qquad s^2 = \frac{1}{n}\sum_{i=1}^{n}(x_i - \bar{x})^2.$$

If the numbers x_i are the results of statistically independent measurements of a quantity with expected value m, then $\bar{x}$ is an estimate of m, whose standard deviation is estimated by $s/\sqrt{n-1}$.

(a) The computation of $\bar{x}$ and m using the formulas above have the drawback that they require two passes through the data x_i. Let α be *a provisional mean*, chosen as an approximation to $\bar{x}$, and set $x_i' = x_i - \alpha$. Show that the formulas

$$\bar{x} = \alpha + \frac{1}{n}\sum_{i=1}^{n} x_i', \qquad s^2 = \frac{1}{n}\sum_{i=1}^{n}(x_i')^2 - (\bar{x} - \alpha)^2$$

hold for an arbitrary α.

(b) In 16 measurements of a quantity x one got the following results:

i	x_i	i	x_i	i	x_i	i	x_i
1	546.85	5	546.81	9	546.96	13	546.84
2	546.79	6	546.82	10	546.94	14	546.86
3	546.82	7	546.88	11	546.84	15	546.84
4	546.78	8	546.89	12	546.82	16	546.84

Compute $\bar{x}$ and s^2 to two significant digits using $\alpha = 546.85$.

(c) In the computations in (b), one never needed more than three digits. If one uses the value $\alpha = 0$, how many digits are needed in $(x_i')^2$ in order to get two significant digits in s^2? If one uses five digits throughout the computations, why is the cancellation in the s^2 more fatal than the cancellation in the subtraction $x_i' - \alpha$? (One can even get negative values for s^2!)

(d) If we define

$$m_k = \frac{1}{k}\sum_{i=1}^{k} x_i, \qquad q_k = \sum_{i=1}^{k}(x_i - m_k)^2 = \sum_{i=1}^{k} x_i^2 - \frac{1}{k}\Big(\sum_{i=1}^{k} x_i\Big)^2,$$

then it holds that $\bar{x} = m_n$, and $s^2 = q_n/n$. Show the recursion formulas

$$m_1 = x_1, \qquad m_k = m_{k-1} + (x_k - m_{k-1})/k$$
$$q_1 = 0, \qquad q_k = q_{k-1} + (x_k - m_{k-1})^2(k-1)/k.$$

2.3.16 Compute the sum in Example 2.3.4 using the natural summation ordering in IEEE 754 double precision. Repeat the computations using compensated summation (Algorithm 2.3).

2.4 Error Propagation

2.4.1 Numerical Problems, Methods, and Algorithms

By a **numerical problem** we mean here a clear and unambiguous description of the *functional connection* between **input data**—that is, the "independent variables" in the problem—and **output data**—that is, the desired results. Input data and output data consist of a finite number of real (or complex) quantities and are thus representable by finite dimensional vectors. The functional connection can be expressed in either explicit or implicit form. We require for the following discussion also that *the output data should be uniquely determined and depend continuously on the input data.*

By an **algorithm**[39] for a given numerical problem we mean a *complete description of well-defined operations* through which each permissible input data vector is transformed into an output data vector. By "operations" we mean here arithmetic and logical operations, which a computer can perform, together with references to previously defined algorithms. It should be noted that, as the field of computing has developed, more and more complex functions (for example, square root, circular, and hyperbolic functions) are built into the hardware. In many programming environments operations such as matrix multiplication, solution of linear systems, etc. are considered as "elementary operations" and for the user appear as black boxes.

(The concept algorithm can be analogously defined for problems completely different from numerical problems, with other types of input data and fundamental operations—for example, inflection, merging of words, and other transformations of words in a given language.)

Example 2.4.1.

To determine the largest real root of the cubic equation

$$p(z) = a_0z^3 + a_1z^2 + a_2z + a_3 = 0,$$

with real coefficients a_0, a_1, a_2, a_3, is a numerical problem. The input data vector is (a_0, a_1, a_2, a_3). The output is the desired root x; it is an implicitly defined function of the input data.

An algorithm for this problem can be based on Newton's method, supplemented with rules for how the initial approximation should be chosen and how the iteration process is to be terminated. One could also use other iterative methods, or algorithms based on the formula by Cardano–Tartaglia for the exact solution of the cubic equation (see Problem 2.3.8). Since this uses square roots and cube roots, one needs to assume that algorithms for the computation of these functions have been specified previously.

[39]The term "algorithm" is a Latinization of the name of the Arabic ninth century mathematician Al-Khowârizmî. He also introduced the word "algebra" (Al-jabr). Western Europe became acquainted with the Hindu positional number system from a Latin translation of his book entitled *Algorithmi de Numero Indorum.*

One often begins the construction of an algorithm for a given problem by breaking down the problem into subproblems in such a way that the output data from one subproblem are the input data to the next subproblem. Thus the distinction between problem and algorithm is not always so clear-cut. The essential point is that, in the formulation of the problem, one is only concerned with the initial state and the final state. In an algorithm, however, one should clearly define each step along the way, from start to finish.

We use the term **numerical method** in this book to mean a procedure either to approximate a mathematical problem with a numerical problem or to solve a numerical problem (or at least to transform it to a simpler problem). A numerical method should be more generally applicable than an algorithm, and set lesser emphasis on the completeness of the computational details. The transformation of a differential equation problem to a system of nonlinear equations can be called a numerical method—even without instructions as to how to solve the system of nonlinear equations. Newton's method is a numerical method for determining a root of a large class of nonlinear equations. In order to call it an algorithm, conditions for starting and stopping the iteration process should be added.

For a given numerical problem one can consider many different algorithms. As we have seen in Sec. 2.3 these can, in floating-point arithmetic, give approximations of widely varying accuracy to the exact solution.

Example 2.4.2.

The problem of solving the differential equation

$$\frac{d^2y}{dx^2} = x^2 + y^2$$

with boundary conditions $y(0) = 0$, $y(5) = 1$ is not a numerical problem according to the definition stated above. This is because the output is the *function* y, which cannot in any conspicuous way be specified by a finite number of parameters. The above mathematical problem can be *approximated with a numerical problem* if one specifies the output data to be the values of y for $x = h, 2h, 3h, \dots, 5-h$. Also, the domain of variation of the unknowns must be restricted in order to show that the problem has a unique solution. This can be done, however, and there are a number of different algorithms for solving the problem approximately, which have different properties with respect to the number of arithmetic operations needed and the accuracy obtained.

Before an algorithm can be used it has to be implemented in an algorithmic program language in a reliable and efficient manner. We leave these aspects aside for the moment, but *this is far from a trivial task. Most amateur algorithm writers seem to think that an algorithm is ready at the point where a professional realizes that the hard and tedious work is just beginning* (George E. Forsythe [120]).

2.4.2 Propagation of Errors and Condition Numbers

In scientific computing the given input data are usually imprecise. The errors in the input will propagate and give rise to errors in the output. In this section we develop some general tools for studying the propagation of errors. Error-propagation formulas are also of great interest in the *planning and analysis of scientific experiments.*

Note that rounding errors from each step in a calculation are also propagated to give errors in the final result. For many algorithms a rounding error analysis can be given, which shows that the computed result always equals the exact (or slightly perturbed) result of a nearby problem, where the input data have been slightly perturbed (see, e.g, Lemma 2.3.5). The effect of rounding errors on the final result can then be estimated using the tools of this section.

We first consider two simple special cases of error propagation. For a sum of an arbitrary number of terms we get the following lemma by induction from (2.3.23).

Lemma 2.4.1.

In addition (and subtraction) a bound for the absolute errors in the result is given by the sum of the bounds for the absolute errors of the operands:

$$y = \sum_{i=1}^{n} x_i, \qquad |\Delta y| \le \sum_{i=1}^{n} |\Delta x_i|. \tag{2.4.1}$$

To obtain a corresponding result for the error propagation in multiplication and division, we start with the observations that for $y = \log x$ we have $\Delta(\log x) \approx \Delta(x)/x$. In words, *the relative error in a quantity is approximately equal to the absolute error in its natural logarithm.* This is related to the fact that displacements of the same length at different places on a logarithmic scale mean the same relative change of the value. From this we obtain the following result.

Lemma 2.4.2.

In multiplication and division, an approximate bound for the relative error is obtained by adding the relative errors of the operands. More generally, for $y = x_1^{m_1} x_2^{m_2} \cdots x_n^{m_n}$,

$$\left|\frac{\Delta y}{y}\right| \lessapprox \sum_{i=1}^{n} |m_i| \left|\frac{\Delta x_i}{x_i}\right|. \tag{2.4.2}$$

Proof. The proof follows by differentiating $\log y = m_1 \log x_1 + m_2 \log x_2 + \cdots + m_n \log x_n$. □

Example 2.4.3.

In Newton's method for solving a nonlinear equation a correction is to be calculated as a quotient $\Delta x = f(x_k)/f'(x_k)$. Close to a root the relative error in the computed value of $f(x_k)$ can be quite large due to cancellation. How accurately should one compute $f'(x_k)$, assuming that the work grows as one demands higher accuracy? Since the limit for the relative error in Δx is equal to the sum of the bounds for the relative errors in $f(x_k)$ and $f'(x_k)$, there is no gain in making the relative error in $f'(x_k)$ very much less than the relative error in $f(x_k)$. This observation is of great importance, particularly in the generalization of Newton's method to *systems* of nonlinear equations.

We now study the propagation of errors in more general nonlinear expressions. Consider first the case when we want to compute a function $y = f(x)$ of a single real variable

x. How is the error in x propagated to y? Let $\tilde{x} - x = \Delta x$. Then, a natural way is to approximate $\Delta y = \tilde{y} - y$ with the differential of y. By the mean value theorem, $\Delta y = f(x + \Delta x) - f(x) = f'(\xi)\Delta x$, where ξ is a number between x and $x + \Delta x$. Suppose that $|\Delta x| \le \epsilon$. Then it follows that

$$|\Delta y| \le \max_{\xi} |f'(\xi)|\epsilon, \quad \xi \in [x - \epsilon, x + \epsilon]. \tag{2.4.3}$$

In practice, it is usually sufficient to replace ξ by the available estimate of x. *Even if high precision is needed in the value of $f(x)$, one rarely needs a high relative precision in an error bound or an error estimate.* (In the neighborhood of zeros of the first derivative $f'(x)$ one has to be more careful.)

By the implicit function theorem a similar result holds if y is an implicit function of x defined by $g(x, y) = 0$. If $g(x, y) = 0$ and $\frac{\partial g}{\partial y}(x, y) \ne 0$, then in a neighborhood of x, y there exists a unique function $y = f(x)$ such that $g(x, f(x)) = 0$ and it holds that

$$f'(x) = -\frac{\partial g}{\partial x}(x, f(x)) \Big/ \frac{\partial g}{\partial y}(x, f(x)).$$

Example 2.4.4.

The result in Lemma 2.3.5 does not say whether the computed roots of the quadratic equation are close to the exact roots r_1, r_2. To answer that question we must determine how sensitive the roots are to a relative perturbation in the coefficient c. Differentiating $ax^2 + bx + c = 0$, where $x = x(c)$ with respect to c, we obtain $(2ax + b)dx/dc + 1 = 0$, $dx/dc = -1/(2ax + b)$. With $x = r_1$ and using $r_1 + r_2 = -b/a$, $r_1 r_2 = c/a$ this can be written as

$$\frac{dr_1}{r_1} = -\frac{dc}{c}\frac{r_2}{r_1 - r_2}.$$

This shows that when $|r_1 - r_2| \ll |r_2|$ the roots can be very sensitive to small relative perturbations in c.

When $r_1 = r_2$, that is, when there is a double root, this linear analysis breaks down. Indeed, it is easy to see that the equation $(x - r)^2 - \Delta c = 0$ has roots $x = r \pm \sqrt{\Delta c}$.

To analyze error propagation in a function of several variables $f = f(x_1, \dots, x_n)$, we need the following generalization of the mean value theorem.

Theorem 2.4.3.

Assume that the real-valued function f is differentiable in a neighborhood of the point $x = (x_1, x_2, \dots, x_n)$, and let $x = x + \Delta x$ be a point in this neighborhood. Then there exists a number θ such that

$$\Delta f = f(x + \Delta x) - f(x) = \sum_{i=1}^{n} \frac{\partial f}{\partial x_i}(x + \theta \Delta x)\Delta x_i, \quad 0 \le \theta \le 1.$$

Proof. The proof follows by considering the function $F(t) = f(x + t\Delta x)$, and using the mean value theorem for functions of one variable and the chain rule. □

From Theorem 2.4.3 it follows that the perturbation Δf is approximately equal to the total differential. The use of this approximation means that the function $f(x)$ is, in a neighborhood of x that contains the point $x + \Delta x$, approximated by a linear function. All the techniques of differential calculus, such as logarithmic differentiation and implicit differentiation, may be useful for the calculation of the total differential; see the examples and the problems at the end of this section.

Theorem 2.4.4 (*General Formula for Error Propagation*).

Let the real-valued function $f = f(x_1, x_2, \dots, x_n)$ be differentiable in a neighborhood of the point $x = (x_1, x_2, \dots, x_n)$ with errors $\Delta x_1, \Delta x_2, \dots, \Delta x_n$. Then it holds that

$$\Delta f \approx \sum_{i=1}^{n} \frac{\partial f}{\partial x_i} \Delta x_i. \tag{2.4.4}$$

Then for the maximal error in $f(x_1, \dots, x_n)$ we obtain the approximate upper bound

$$|\Delta f| \lessapprox \sum_{i=1}^{n} \left| \frac{\partial f}{\partial x_i} \right| |\Delta x_i|, \tag{2.4.5}$$

where the partial derivatives are evaluated at x.

In order to get a *strict* bound for $|\Delta f|$, one should use in (2.4.5) the maximum absolute values of the partial derivatives in a neighborhood of the known point x. In most practical situations it suffices to calculate $|\partial f/\partial x_i|$ at x and then add a certain marginal amount (5 to 10 percent, say) for safety. Only if the Δx_i are large or if the derivatives have a large relative variation in the neighborhood of x need the maximal values be used. (The latter situation occurs, for example, in a neighborhood of an extremal point of $f(x)$.)

The bound in Theorem 2.4.4 is the best possible, unless one knows some dependence between the errors of the terms. Sometimes it can, for various reasons, be a gross overestimate of the real error.

Example 2.4.5.

Compute error bounds for $f = x_1^2 - x_2$, where $x_1 = 1.03 \pm 0.01$, $x_2 = 0.45 \pm 0.01$. We obtain

$$\left| \frac{\partial f}{\partial x_1} \right| = |2x_1| \le 2.1, \qquad \left| \frac{\partial f}{\partial x_2} \right| = |-1| = 1,$$

and find $|\Delta f| \le 2.1 \cdot 0.01 + 1 \cdot 0.01 = 0.031$, or $f = 1.061 - 0.450 \pm 0.032 = 0.611 \pm 0.032$. The error bound has been raised 0.001 because of the rounding in the calculation of x_1^2.

One is seldom asked to give mathematically guaranteed error bounds. More often it is satisfactory to give an estimate of the *order of magnitude* of the anticipated error. The bound for $|\Delta f|$ obtained with Theorem 2.4.3 estimates the maximal error, i.e., covers the worst possible cases, where the sources of error Δx_i contribute with the same sign and magnitudes equal to the error bounds for the individual variables.

In practice, the trouble with formula (2.4.5) is that it often gives bounds which are too coarse. More realistic estimates are often obtained using the standard error introduced in

Sec. 2.3.3. Here we give without proof the result for the general case, which can be derived using probability theory and (2.4.4). (Compare with the result for the standard error of a sum given in Sec. 2.3.3.)

Theorem 2.4.5.

Assume that the errors $\Delta x_1, \Delta x_2, \dots, \Delta x_n$ are independent random variables with mean zero and standard deviations $\epsilon_1, \epsilon_2, \dots, \epsilon_n$. Then the standard error ϵ for $f(x_1, x_2, \dots, x_n)$ is given by the formula

$$\epsilon \approx \left(\sum_{i=1}^{n} \left(\frac{\partial f}{\partial x_i} \right)^2 \epsilon_i^2 \right)^{1/2}. \tag{2.4.6}$$

Analysis of error propagation is more than just a means for judging the reliability of calculated results. As remarked above, it has an equally important function as a means for *the planning of a calculation or scientific experiment.* It can help in the choice of algorithm, and in making certain decisions during a calculation. An example of such a decision is the choice of step length during a numerical integration. Increased accuracy often has to be bought at the price of more costly or complicated calculations. One can also shed some light on the degree to which it is advisable to obtain a new apparatus to improve the measurements of a given variable when the measurements of other variables are subject to error as well.

It is useful to have a measure of how sensitive the output data are to small changes in the input data. In general, if "small" changes in the input data can result in "large" changes in the output data, we call the problem **ill-conditioned**; otherwise it is called **well-conditioned**. (The definition of large may differ from problem to problem depending on the accuracy of the data and the accuracy needed in the solution.)

Definition 2.4.6.

Consider a numerical problem $y = f(x) \in R^m$, $x \in R^n$, or in component form $y_j = f_j(x_1, \dots, x_n)$, $j = 1 : m$. Let $\hat{x}$ be fixed and assume that neither $\hat{x}$ nor $\hat{y}0f(\hat{x})$ is zero. The sensitivity of y with respect to small changes in x can be measured by the relative condition number

$$\kappa(f; \hat{x}) = \lim_{\epsilon \to 0} \sup_{\|h\|=\epsilon} \left\{ \frac{\|f(x+h) - f(x)\|}{\|f(x)\|} \Big/ \frac{\|h\|}{\|x\|} \right\}. \tag{2.4.7}$$

We have used a vector norm $\|\cdot\|$ to measure the size of a vector; see Sec. A.4.3 in Online Appendix A. Common vector norms are the p-norms defined by

$$\|x\|_p = (x_1|^p + |x_2|^p + \cdots + |x_n|^p)^{1/p}, \quad 1 \le p < \infty,$$

where one usually takes $p = 1, 2$, or $p = \infty$.

The condition number (2.4.7) is a function of the input data $\hat{x}$ and also depends on the choice of norms in the data space and the solution space. It measures the maximal amount which a given relative perturbation is magnified by the function f, in the limit of infinitely small perturbations. For perturbations of sufficiently small size we have the estimate

$$\|\tilde{y} - y\| \le \kappa\epsilon\|y\| + O(\epsilon^2).$$

We can expect to have roughly $s = log_{10}\kappa$ less significant decimal digits in the solution than in the input data. However, *this may not hold for all components of the output.*

Assume that f has partial derivatives with respect to x_i, $i = 1 : n$, and let $J(x)$ be the **Jacobian matrix**

$$J_{ij}(x) = \frac{\partial f_j(x)}{\partial x_i}, \quad j = 1 : m, \quad i = 1 : n. \tag{2.4.8}$$

Then, for any matrix norm subordinate to the vector norm (see Online Appendix A.3.3), the condition number defined above can be expressed as

$$\kappa(f; \hat{x}) = \frac{\|J(\hat{x})\| \|\hat{x}\|}{\|f(\hat{x})\|}. \tag{2.4.9}$$

For a composite function $g \circ f$ the chain rule for derivatives can be used to show that

$$\kappa(g \circ f; \hat{x}) \leq \kappa(g; \hat{y})\kappa(f; \hat{x}). \tag{2.4.10}$$

If the composite function is ill-conditioned we can infer from this that at least one of the functions g and f must be ill-conditioned.

If $y = f(x)$ is a linear (bounded) function $y = Mx$, where $M \in \mathbf{R}^{m\times n}$, then according to (2.4.9)

$$\kappa(M; x) = \|M\| \frac{\|x\|}{\|y\|}.$$

This inequality is sharp in the sense that for any matrix norm and for any M and x there exists a perturbation δb such that equality holds.

If M is a square and invertible matrix, then from $x = M^{-1}y$ we conclude that $\|x\| \leq \|M^{-1}\| \, \|y\|$. This gives the upper bound

$$\kappa(M; x) \leq \|M\| \|M^{-1}\|, \tag{2.4.11}$$

which is referred to as the condition number of M. For given x (or y), this upper bound may not be achievable for any perturbation of x. The inequality (2.4.11) motivates the following definition.

Theorem 2.4.7.

The **condition number** *for a square nonsingular matrix* $M \in \mathbf{R}^{n\times n}$ *equals* $\kappa(M) = \|M\| \, \|M^{-1}\|$, *where* $\| \cdot \|$ *is a subordinate matrix norm. In particular, for the Euclidean norm*

$$\kappa(M) = \kappa_2(M) = \|M\|_2 \, \|M^{-1}\|_2 = \sigma_1/\sigma_n, \tag{2.4.12}$$

where σ_1 *and* σ_n *are the largest and smallest singular value of* M.

The last expression in (2.4.12) follows by the observation that if M has singular values σ_i, $i = 1 : n$, then M^{-1} has singular values $1/\sigma_i$, $i = 1 : n$; see Theorem 1.4.4.

We note some simple properties of $\kappa(M)$. From $(\alpha M)^{-1} = M^{-1}/\alpha$ it follows that $\kappa(\alpha M) = \kappa(M)$; i.e., the condition number is invariant under multiplication of M by a scalar. Matrix norms are submultiplicative, i.e., $\|KM\| \leq \|K\| \, \|M\|$. From the definition and the identity $MM^{-1} = I$ it follows that

$$\kappa(M) = \|M\|_2 \|M^{-1}\|_2 \geq \|I\| = 1,$$

i.e., the condition number κ_2 is always greater than or equal to one. The composite mapping of $z = Ky$ and $y = Mx$ is represented by the matrix product KY, and we have

$$\kappa(KM) \leq \kappa(K)\kappa(M).$$

It is important to note that the condition number is a property of the mapping $x \to y$ and does not depend on the algorithm used to evaluate y! An ill-conditioned problem is *intrinsically* difficult to solve accurately using *any* numerical algorithm. Even if the input data are exact, rounding errors made during the calculations in floating-point arithmetic may cause large perturbations in the final result. Hence, in some sense an ill-conditioned problem is not well posed.

Example 2.4.6.

If we get an inaccurate solution to an ill-conditioned problem, then often nothing can be done about the situation. (If you ask a stupid question you get a stupid answer!) But sometimes the difficulty depends on the form one has chosen to represent the input and output data of the problem.

The polynomial

$$P(x) = (x - 10)^4 + 0.200(x - 10)^3 + 0.0500(x - 10)^2 - 0.00500(x - 10) + 0.00100$$

is identical with a polynomial Q which, if the coefficients are rounded to six digits, becomes

$$\tilde{Q}(x) = x^4 - 39.8000x^3 + 594.050x^2 - 3941.00x + 9805.05.$$

One finds that $P(10.11) = 0.0015 \pm 10^{-4}$, where only three digits are needed in the computation, while $\tilde{Q}(10.11) = -0.0481 \pm \frac{1}{2} \cdot 10^{-4}$, in spite of the fact that eight digits were used in the computation. The rounding to six digits of the coefficients of Q has thus caused an error in the polynomial's value at $x = 10.11$; the erroneous value is more than 30 times larger than the correct value and has the wrong sign. When the coefficients of Q are input data, the problem of computing the value of the polynomial for $x \approx 10$ is far more ill-conditioned than when the coefficients of P are input data.

The conditioning of a problem can to some degree be illustrated geometrically. A numerical problem P means a mapping of the space X of possible input data onto the space Y of the output data. The dimensions of these spaces are usually quite large. In Figure 2.4.1 we picture a mapping in two dimensions. Since we are considering relative changes, we take the coordinate axis to be logarithmically scaled. A small circle of radius r is mapped onto an ellipse whose ratio of major to minor axis is κr, where κ is the condition number of the problem P.

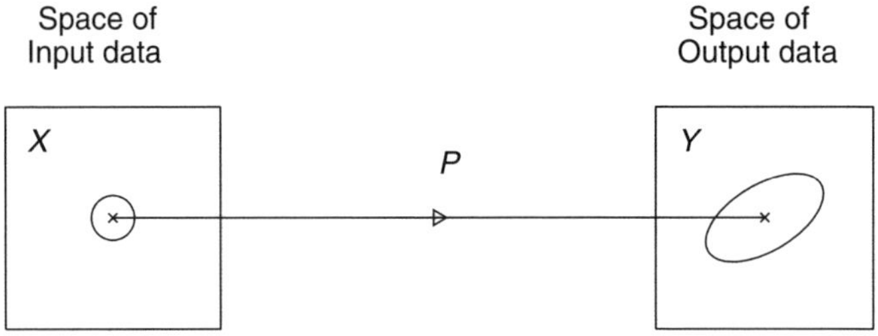

Figure 2.4.1. *Geometrical illustration of the condition number.*

2.4.3 Perturbation Analysis for Linear Systems

Consider the linear system $y = Ax$, where A is nonsingular and $y \neq 0$. From the analysis in the previous section we know that the condition number of the inverse mapping $x = A^{-1}y \neq 0$ is bounded by the condition number

$$\kappa(A^{-1}) = \kappa(A) = \|A^{-1}\| \, \|A\|.$$

Assume that the elements of the matrix A are given data and subject to perturbations δA. The perturbed solution $x + \delta x$ satisfies the linear system

$$(A + \delta A)(x + \delta x) = y.$$

Subtracting $Ax = y$ we obtain $(A + \delta A)\delta x = -\delta A x$. Assuming also that the matrix $(A + \delta A) = A(I + A^{-1}\delta A)$ is nonsingular and solving for δx yields

$$\delta x = -(I + A^{-1}\delta A)^{-1} A^{-1} \delta A x, \tag{2.4.13}$$

which is the basic identity for the analysis. Taking norms gives

$$\|\delta x\| \leq \|(I + A^{-1}\delta A)^{-1}\| \, \|A^{-1}\| \|\delta A\| \, \|x\|.$$

It can be shown (see Problem 2.4.9) that if $\|A^{-1}\delta A\| < 1$, then $A + \delta A$ is nonsingular and

$$\|(I + A^{-1}\delta A)^{-1}\| < 1/(1 - \|A^{-1}\delta A\|).$$

Neglecting second order terms,

$$\frac{\|\delta x\|}{\|x\|} \lessapprox \kappa(A) \frac{\|\delta A\|}{\|A\|}. \tag{2.4.14}$$

This shows that $\kappa(A)$ is also the condition number of $x = A^{-1}y$ with respect to perturbations in A.

For any real, orthogonal matrix Q we have

$$\kappa_2(Q) = \|Q\|_2 \|Q^{-1}\|_2 = 1,$$

so Q is perfectly conditioned. By Lemma A.4.1 (see Online Appendix A) we have $\|QAP\|_2 = \|A\|_2$ for any orthogonal P and Q. It follows that

$$\kappa_2(PAQ) = \kappa_2(A),$$

i.e., *the condition number of a matrix A is invariant under orthogonal transformations*. This important fact is one reason why orthogonal transformations play a central role in numerical linear algebra.

How large may κ be before we consider the problem to be ill-conditioned? That depends on the accuracy of the data and the accuracy desired in the solution. If the data have a relative error of 10^{-7}, then we can guarantee a (normwise) relative error in the solution $\leq 10^{-3}$ if $\kappa \leq 0.5 \cdot 10^4$. But to guarantee a (normwise) relative error in the solution $\leq 10^{-6}$ we need to have $\kappa \leq 5$.

Example 2.4.7.

The Hilbert matrix H_n of order n with elements

$$H_n(i, j) = h_{ij} = 1/(i + j - 1), \quad 1 \leq i, j \leq n,$$

is a notable example of an ill-conditioned matrix. In Table 2.4.1 approximate condition numbers of Hilbert matrices of order ≤ 12, computed in IEEE double precision, are given. For $n > 12$ the Hilbert matrices are too ill-conditioned even for IEEE double precision! From a result by G. Szegö (see Gautschi [147, p. 34]) it follows that

$$\kappa_2(H_n) \approx \frac{(\sqrt{2} + 1)^{4(n+1)}}{2^{15/4}\sqrt{\pi n}} \sim e^{3.5n},$$

i.e., the condition numbers grow exponentially with n. Although the severe ill-conditioning exhibited by the Hilbert matrices is rare, moderately ill-conditioned linear systems do occur regularly in many practical applications!

Table 2.4.1. *Condition numbers of Hilbert matrices of order* ≤ 12.

n	$\kappa_2(H_n)$	n	$\kappa_2(H_n)$
1	1	7	$4.753 \cdot 10^8$
2	19.281	8	$1.526 \cdot 10^{10}$
3	$5.241 \cdot 10^2$	9	$4.932 \cdot 10^{11}$
4	$1.551 \cdot 10^4$	10	$1.602 \cdot 10^{13}$
5	$4.766 \cdot 10^5$	11	$5.220 \cdot 10^{14}$
6	$1.495 \cdot 10^7$	12	$1.678 \cdot 10^{16}$

The normwise condition analysis in the previous section usually is satisfactory when the linear system is "well scaled." If this is not the case, then a **componentwise** analysis may give sharper bounds. We first introduce some notations. The absolute values $|A|$ and $|b|$ of a matrix A and vector b are interpreted componentwise

$$|A|_{ij} = (|a_{ij}|), \qquad |b|_i = (|b_i|).$$

The **partial ordering** "$\leq$" for the absolute values of matrices $|A|$, $|B|$ and vectors $|b|$, $|c|$ is to be interpreted componentwise:[40]

$$|A| \leq |B| \iff |a_{ij}| \leq |b_{ij}|, \qquad |b| \leq |c| \iff |b_i| \leq |c_i|.$$

[40]Note that $A \leq B$ in other contexts means that $B - A$ is positive semidefinite.

It follows easily that $|AB| \le |A|\,|B|$ and a similar rule holds for matrix-vector multiplication.

Taking absolute values in (2.4.13) gives componentwise error bounds for the corresponding perturbations in x,

$$|\delta x| \le |(I + A^{-1}\delta A)^{-1}|\,|A^{-1}|(|\delta A||x| + |\delta b|).$$

The matrix $(I - |A^{-1}||\delta A|)$ is guaranteed to be nonsingular if $\|\,|A^{-1}|\,|\delta A|\,\| < 1$.

Assume now that we have componentwise bounds for the perturbations in A and b, say

$$|\delta A| \le \omega|A|, \qquad |\delta b| \le \omega|b|. \tag{2.4.15}$$

Neglecting second order terms in ω and using (2.4.15) gives

$$|\delta x| \lessapprox |A^{-1}|(|\delta A||x| + |\delta b|) \le \omega|A^{-1}|(|A|\,|x| + |b|). \tag{2.4.16}$$

Taking norms in (2.4.16) we get

$$\|\delta x\| \lessapprox \omega\|\,|A^{-1}|(|A|\,|x| + |b|)\,\| + O(\omega^2). \tag{2.4.17}$$

The scalar quantity

$$\kappa_{|A|}(A) = \|\,|A^{-1}|\,|A|\,\| \tag{2.4.18}$$

is called the **Bauer–Skeel condition number** of the matrix A.

A different way to examine the sensitivity of various matrix problems is the differentiation of a parametrized matrix. Suppose that λ is a scalar and that $A(\lambda)$ is a matrix with elements $a_{ij}(\lambda)$ that are differentiable functions of λ. Then by the derivative of the matrix $A(\lambda)$ we mean the matrix

$$A'(\lambda) = \frac{d}{d\lambda}A(\lambda) = \left(\frac{da_{ij}}{d\lambda}\right). \tag{2.4.19}$$

Many of the rules for differentiation of scalar functions are easily generalized to differentiation of matrices. For differentiating a product of two matrices there holds

$$\frac{d}{d\lambda}[A(\lambda)B(\lambda)] = \frac{d}{d\lambda}[A(\lambda)]B(\lambda) + A(\lambda)\frac{d}{d\lambda}[B(\lambda)]. \tag{2.4.20}$$

Assuming that $A^{-1}(\lambda)$ exists, using this rule on the identity $A^{-1}(\lambda)A(\lambda) = I$ we obtain

$$\frac{d}{d\lambda}[A^{-1}(\lambda)]A(\lambda) + A^{-1}(\lambda)\frac{d}{d\lambda}[A(\lambda)] = 0,$$

or, solving for the derivative of the inverse,

$$\frac{d}{d\lambda}[A^{-1}(\lambda)] = -A^{-1}(\lambda)\frac{d}{d\lambda}[A(\lambda)]A^{-1}(\lambda). \tag{2.4.21}$$

2.4.4 Error Analysis and Stability of Algorithms

One common reason for poor accuracy in the computed solution is that the problem is ill-conditioned. But poor accuracy can also be caused by a poorly constructed algorithm. We say in general that an algorithm is **unstable** if it can introduce large errors in the computed solutions to a well-conditioned problem.

We consider in the following a finite algorithm with input data $(a_1, \dots, a_r)$, which by a sequence of arithmetic operations is transformed into the output data $(w_1, \dots, w_s)$, There are two basic forms of roundoff error analysis for such an algorithm, which are both useful:

(i) In **forward** error analysis, one attempts to find bounds for the errors in the solution $|\overline{w}_i - w_i|$, $i = 1 : s$, where $\overline{w}_i$ denotes the computed value of w_i. The main tool used in forward error analysis is the propagation of errors, as studied in Sec. 2.4.2.

(ii) In **backward** error analysis, one attempts to determine a modified set of data $a_i + \Delta a_i$ such that the computed solution $\overline{w}_i$ is the *exact solution*, and give bounds for $|\Delta a_i|$. There may be an infinite number of such sets; in this case we seek to minimize the size of$|\Delta a_i|$. However, it can also happen, even for very simple algorithms, that no such set exists.

Sometimes, when a pure backward error analysis cannot be achieved, one can show that the computed solution is a slightly perturbed solution to a problem with slightly modified input data. An example of such a **mixed error analysis** is the error analysis given in Lemma 2.3.5 for the solution of a quadratic equation.

In backward error analysis no reference is made to the exact solution for the original data. In practice, when the data are known only to a certain accuracy, the "exact" solution may not be well defined. Then any solution whose backward error is smaller than the domain of uncertainty of the data may be considered to be a satisfactory result.

A frequently occurring backward error problem is the following. Suppose we are given an approximate solution y to a linear system $Ax = b$. We want to find out if y is the *exact solution to a nearby perturbed system* $(A + \Delta A)y = b + \Delta b$. To this end we define the normwise backward error of y as

$$\eta(y) = \min\{\epsilon \mid (A + \Delta A)y = b + \Delta b,\ \|\Delta A\| \le \epsilon\|A\|,\ \|\Delta b\| \le \epsilon\|b\|\}. \qquad (2.4.22)$$

The following theorem tells us that the normwise backward error of y is small if the residual vector $b - Ay$ is small.

Theorem 2.4.8 (*Rigal and Gaches* [305]).
The normwise backward error of y is given by

$$\eta(y) = \frac{\|r\|}{\|A\|\,\|y\| + \|b\|}, \qquad (2.4.23)$$

where $r = b - Ay$, and $\|\cdot\|$ is any consistent norm.

Similarly, we define the **componentwise backward error** $\omega(y)$ of y by

$$\omega(y) = \min\{\epsilon \mid (A + \Delta A)y = b + \Delta b,\ |\Delta A| \le \epsilon\|A\|,\ |\Delta b| \le \epsilon|b|\}. \qquad (2.4.24)$$

As the following theorem shows, there is a simple expression also for $\omega(y)$.

Theorem 2.4.9 (*Oettli and Prager* [277]).

Let $r = b - A\bar{x}$, E and f be nonnegative, and set

$$\omega(y) = \max_i \frac{|r_i|}{(E|\bar{x}| + f)_i}, \tag{2.4.25}$$

where $\xi/0$ is interpreted as zero if $\xi = 0$ and infinity otherwise.

By means of backward error analysis it has been shown, even for many quite complicated matrix algorithms, that the computed results which the algorithm produces under the influence of roundoff error are the *exact* output data of a problem of the same type in which the relative change in data only are of the order of the unit roundoff u.

Definition 2.4.10.

An algorithm is **backward stable** *if the computed solution $\overline{w}$ for the data a is the exact solution of a problem with slightly perturbed data $\bar{a}$ such that for some norm $\|\cdot\|$ it holds that*

$$\|\bar{a} - a\|/\|a\| < c_1 u, \tag{2.4.26}$$

where c_1 is a not too large constant and u is the unit roundoff.

We are usually satisfied if we can prove normwise forward or backward stability for some norm, for example, $\|\cdot\|_2$ or $\|\cdot\|_\infty$. Occasionally we may like the estimates to hold componentwise,

$$|\bar{a}_i - a_i|/|a_i| < c_2 u, \quad i = 1:r. \tag{2.4.27}$$

For example, by equation (2.3.16) the usual algorithm for computing an inner product $x^T y$ is backward stable for elementwise relative perturbations.

We would like stability to hold for some *class of input data*. For example, a numerical algorithm for solving systems of linear equations $Ax = b$ is backward stable for a class of matrices $\mathcal{A}$ if for each $A \in \mathcal{A}$ and for each b the computed solution $\bar{x}$ satisfies $\bar{A}\bar{x} = \bar{b}$, where $\bar{A}$ and $\bar{b}$ are close to A and b.

To yield error bounds for $\overline{w}_i$, a backward error analysis has to be complemented with a perturbation analysis. For this the error propagation formulas in Sec. 2.4.2 can often be used. If the condition number of the problem is κ, then it follows that

$$\|\overline{w} - w\| \le c_1 u \kappa \|w\| + O(u^2). \tag{2.4.28}$$

Hence the error in the solution may still be large if the problem is ill-conditioned. But we have obtained an answer which is the exact mathematical solution to a problem with data close to the one we wanted to solve. If the perturbations $\bar{a} - a$ are within the uncertainties of the given data, *the computed solution is as good as our data warrant*!

A great advantage of backward error analysis is that, when it applies, it tends to give much sharper results than a forward error analysis. Perhaps more important, it usually also gives a better insight into the stability (or lack of it) of the algorithm.

By the definition of the condition number κ it follows that backward stability implies forward stability, but *the converse is not true*. Many important direct algorithms for solving linear systems are known to be backward stable. The following result for the Cholesky factorization is an important example.

Theorem 2.4.11 (*Wilkinson* [378]).
Let $A \in \mathbf{R}^{n\times n}$ be a symmetric positive definite matrix. Provided that

$$2n^{3/2}u\kappa(A) < 0.1, \tag{2.4.29}$$

the Cholesky factor of A can be computed without breakdown, and the computed factor $\bar{R}$ satisfies

$$\bar{R}^T\bar{R} = A + E, \quad \|E\|_2 < 2.5n^{3/2}u\|A\|_2, \tag{2.4.30}$$

and hence is the exact Cholesky factor of a matrix close to A.

For the LU factorization of matrix A the following componentwise backward error result is known.

Theorem 2.4.12.
If the LU factorization of the matrix $A \in \mathbf{R}^{n\times n}$ runs to completion, then the computed factors $\bar{L}$ and $\bar{U}$ satisfy

$$A + E = \bar{L}\bar{U}, \quad |E| \le \gamma_n|\bar{L}|\,|\bar{U}|, \tag{2.4.31}$$

where $\gamma_n = nu/(1 - nu)$, and u is the unit roundoff.

This shows that unless the elements in the computed factors $|\bar{L}|$ and $|\bar{U}|$ become large, LU factorization is backward stable.

Example 2.4.8.
For $\epsilon = 10^{-6}$ the system

$$\begin{pmatrix} \epsilon & 1 \\ 1 & 1 \end{pmatrix}\begin{pmatrix} x_1 \\ x_2 \end{pmatrix} = \begin{pmatrix} 1 \\ 0 \end{pmatrix}$$

is well-conditioned and has the exact solution $x_1 = -x_2 = -1/(1-\epsilon) \approx -1$. In Gaussian elimination we multiply the first equation by 10^6, and subtract from the second, giving $(1 - 10^6)x_2 = -10^6$. Rounding this to $x_2 = 1$ is correct to six digits. In the back-substitution to obtain x_1, we then get $10^{-6}x_1 = 1 - 1$, or $x_1 = 0$, which is a completely wrong result. This shows that Gaussian elimination can be an unstable algorithm unless row (and/or column) interchanges are performed to limit element growth.

Some algorithms, including most iterative methods, are not backward stable. Then it is necessary to weaken the definition of stability. In practice an algorithm can be considered stable if *it produces accurate solutions for well-conditioned problems*. Such an algorithm can be called **weakly stable**. Weak stability may be sufficient for giving confidence in an algorithm.

Example 2.4.9.

In the method of normal equations for computing the solution of a linear least squares problem one first forms the matrix $A^T A$. This product matrix can be expressed in outer form as

$$A^T A = \sum_{i=1}^{m} a_i a_i^T,$$

where a_i^T is the ith row of A, i.e., $A^T = (a_1 \quad a_2 \quad \dots \quad a_m)$. From (2.3.14) it follows that this computation is not backward stable; i.e., it is not true that $fl(A^T A) = (A+E)^T(A+E)$ for some small error matrix E. In order to avoid loss of significant information higher precision needs to be used.

Backward stability is easier to prove when there is a sufficiently large set of input data compared to the number of output data. When computing the outer product xy^T (as in Example 2.4.9) there are $2n$ data and n^2 results. This is not a backward stable operation. When the input data are structured rather than general, backward stability often does not hold.

Example 2.4.10.

Many algorithms for solving a linear system $Ax = b$ are known to be backward stable; i.e., the computed solution is the exact solution of a system $(A+E)x = b$, where the normwise relative error $\|E\|/\|A\|$ is not much larger than the machine precision. In many applications the system matrix is **structured**. An important example is **Toeplitz matrices** T, whose entries are constant along every diagonal:

$$T = (t_{i-j})_{1\le i,j\le n} = \begin{pmatrix} t_0 & t_1 & \dots & t_{n-1} \\ t_{-1} & t_0 & \dots & t_{n-2} \\ \vdots & \vdots & \ddots & \vdots \\ t_{-n+1} & t_{-n+2} & \dots & t_0 \end{pmatrix} \in \mathbf{R}^{n\times n}. \tag{2.4.32}$$

Note that a Toeplitz matrix is completely specified by its first row and column, i.e., the $2n-1$ quantities $t = (t_{-n+1}, \dots, t_0, \dots, t_{n-1})$.

Ideally, in a strict backward error analysis, we would like to show that a solution algorithm always computes *an exact solution to a nearby Toeplitz system* defined by $T + S$, where S is small. It has been shown that no such algorithm can exist! We have to be content with algorithms that (at best) compute the exact solution of $(T + E)x = b$, where $\|E\|$ is small but E unstructured.

In the construction of an algorithm for a given problem, one often breaks down the problem into a chain of subproblems, $P_1, P_2, \dots, P_k$, for which algorithms $A_1, A_2, \dots, A_k$ are known, in such a way that the output data from P_{i-1} are the input data to P_i. *Different ways of decomposing the problem give different algorithms with, as a rule, different stability properties.* It is dangerous if the last subproblem in such a chain is ill-conditioned. On the other hand, it need not be dangerous if the first subproblem is ill-conditioned, if the problem itself is well-conditioned. *Even if the algorithms for all the subproblems are stable, we cannot conclude that the composed algorithm is stable.*

Example 2.4.11.

The problem of computing the eigenvalues λ_i of a symmetric matrix A, given its elements (a_{ij}), is always a well-conditioned numerical problem with absolute condition number equal to 1. Consider an algorithm which breaks down this problem into two subproblems:

- P_1: compute the coefficients of the characteristic polynomial of the matrix A $p(\lambda) = \det(A - \lambda I)$ of the matrix A.
- P_2: compute the roots of the polynomial $p(\lambda)$ obtained from P_1.

It is well known that the second subproblem P_2 can be very ill-conditioned. For example, for a symmetric matrix A with eigenvalues $\pm 1, \pm 2, \ldots, \pm 20$ the condition number for P_2 is 10^{14} in spite of the fact that the origin lies exactly between the largest and smallest eigenvalues, so that one cannot blame the high condition number on a difficulty of the same type as that encountered in Example 2.4.7.

The important conclusion that eigenvalues should not be computed as outlined above is further discussed in Sec. 6.5.2.

On the other hand, as the next example shows, it need not be dangerous if the first subproblem of a decomposition is ill-conditioned, even if the problem itself is well-conditioned.

Example 2.4.12.

The first step in many algorithms for computing the eigenvalues λ_i of a symmetric matrix A is to use orthogonal similarity transformations to symmetric tridiagonal form,

$$Q^T A Q = T = \begin{pmatrix} \alpha_1 & \beta_2 & & & \\ \beta_2 & \alpha_2 & \beta_3 & & \\ & \ddots & \ddots & \ddots & \\ & & \beta_{n-1} & \alpha_{n-1} & \beta_n \\ & & & \beta_n & \alpha_n \end{pmatrix}.$$

In the second step the eigenvalues of T, which coincide with those of A, are computed.

Wilkinson [377, Sec. 5.28] showed that the computed tridiagonal matrix can differ significantly from the matrix corresponding to exact computation. Hence here the first subproblem is ill-conditioned. (This fact is not as well known as it should be and still alarms many users!) But the second subproblem is well-conditioned and the combined algorithm is known to be backward stable, i.e., the computed eigenvalues are the exact eigenvalues of a matrix $A + E$, where $\|E\|_2 < c(n)u\|A\|_2$. This is a more complex example of a calculation, where rounding errors cancel!

It should be stressed that *the primary purpose of a rounding error analysis is to give insight into the properties of the algorithm*. In practice we can usually expect a much smaller backward error in the computed solutions than the bounds derived in this section. It is appropriate to recall here a remark by J. H. Wilkinson:

> All too often, too much attention is paid to the precise error bound that has been established. The main purpose of such an analysis is either to establish

the essential numerical stability of an algorithm or to show why it is unstable and in doing so expose what sort of change is necessary to make it stable. The precise error bound is not of great importance.

The treatment in this section is geared toward matrix problems and is not very useful, for example, for time-dependent problems in ordinary and partial differential equations. In Sec. 1.5 some methods for the numerical solution of an initial value problem

$$y'' = -y, \quad y(0) = 0, \quad y'(0) = 1$$

were studied. As will be illustrated in Example 3.3.15, catastrophic error growth can occur in such processes. The notion of stability is here related to the stability of linear difference equations.

Review Questions

2.4.1 The maximal error bounds for addition and subtraction can for various reasons be a coarse overestimate of the real error. Give two reasons, preferably with examples.

2.4.2 How is the condition number $\kappa(A)$ of a matrix A defined? How does $\kappa(A)$ relate to perturbations in the solution x to a linear system $Ax = b$, when A and b are perturbed?

2.4.3 Define the condition number of a numerical problem P of computing output data $y_1, \dots, y_m$ given input data $x_1, \dots, x_n$.

2.4.4 Give examples of well-conditioned and ill-conditioned problems.

2.4.5 What is meant by (a) a forward error analysis; (b) a backward error analysis; (c) a mixed error analysis?

2.4.6 What is meant by (a) a backward stable algorithm; (b) a forward stable algorithm; (c) a mixed stable algorithm; (d) a weakly stable algorithm?

Problems and Computer Exercises

2.4.1 (a) Determine the maximum error for $y = x_1 x_2^2/\sqrt{x_3}$, where $x_1 = 2.0 \pm 0.1$, $x_2 = 3.0 \pm 0.2$, and $x_3 = 1.0 \pm 0.1$. Which variable contributes most to the error?

(b) Compute the standard error using the same data as in (a), assuming that the error estimates for the x_i indicate standard deviations.

2.4.2 One wishes to compute $f = (\sqrt{2} - 1)^6$, using the approximate value 1.4 for $\sqrt{2}$. Which of the following mathematically equivalent expressions gives the best result?

$$\frac{1}{(\sqrt{2}+1)^6}; \quad (3-2\sqrt{2})^3; \quad \frac{1}{(3+2\sqrt{2})^3}; \quad 99 - 70\sqrt{2}; \quad \frac{1}{99+70\sqrt{2}}$$

2.4.3 Analyze the error propagation in x^α

(a) if x is exact and α in error; (b) if α is exact and x in error.

2.4.4 One is observing a satellite in order to determine its speed. At the first observation, $R = 30{,}000 \pm 10$ miles. Five seconds later, the distance has increased by $r = 125.0 \pm 0.5$ miles and the change in the angle is $\phi = 0.00750 \pm 0.00002$ radians. What is the speed of the satellite, assuming that it moves in a straight line and with constant speed in the interval?

2.4.5 One has measured two sides and the included angle of a triangle to be $a = 100.0 \pm 0.1$, $b = 101.0 \pm 0.1$, and angle $C = 1.00^\circ \pm 0.01^\circ$. Then the third side is given by the cosine theorem

$$c = (a^2 + b^2 - 2ab\cos C)^{1/2}.$$

(a) How accurately is it possible to determine c from the given data?

(b) How accurately does one get c if one uses the value $\cos 1^\circ = 0.9998$, which is correct to four decimal places?

(c) Show that by rewriting the cosine theorem as

$$c = ((a - b)^2 + 4ab\sin^2(C/2))^{1/2}$$

it is possible to compute c to full accuracy using only a four-decimal table for the trigonometric functions.

2.4.6 Consider the linear system

$$\begin{pmatrix} 1 & \alpha \\ \alpha & 1 \end{pmatrix} \begin{pmatrix} x \\ y \end{pmatrix} = \begin{pmatrix} 1 \\ 0 \end{pmatrix},$$

where $\alpha \neq 1$. What is the relative condition number for computing x? Using Gaussian elimination and four decimal digits, compute x and y for $\alpha = 0.9950$ and compare with the exact solution $x = 1/(1 - \alpha^2)$, $y = -\alpha/(1 - \alpha^2)$.

2.4.7 (a) Let two vectors u and v be given with components (u_1, u_2) and (v_1, v_2). The angle ϕ between u and v is given by the formula

$$\cos\phi = \frac{u_1 v_1 + u_2 v_2}{(u_1^2 + u_2^2)^{1/2}(v_1^2 + v_2^2)^{1/2}}.$$

Show that computing the angle ϕ from the components of u and v is a well-conditioned problem.

Hint: Take the partial derivative of $\cos\phi$ with respect to u_1, and from this compute $\partial\phi/\partial u_1$. The other partial derivatives are obtained by symmetry.

(b) Show that the formula in (a) is *not* stable for small angles ϕ.

(c) Show that the following algorithm is stable. First normalize the vectors $\tilde{u} = u/\|u\|_2$, $\tilde{v} = v/\|v\|_2$. Then compute $\alpha = \|\tilde{u} - \tilde{v}\|_2$, $\beta = \|\tilde{u} + \tilde{v}\|_2$ and set

$$\phi = \begin{cases} 2\arctan(\alpha/\beta) & \text{if } \alpha \leq \beta, \\ \pi - 2\arctan(\beta/\alpha) & \text{if } \alpha > \beta. \end{cases}$$

2.4.8 For the integral

$$I(a, b) = \int_0^1 \frac{e^{-bx}}{a + x^2} dx,$$

the physical quantities a and b have been measured to be $a = 0.4000 \pm 0.003$,

$b = 0.340 \pm 0.005$. When the integral is computed for various perturbed values of a and b, one obtains

a	b	I
0.39	0.34	1.425032
0.40	0.32	1.408845
0.40	0.34	1.398464
0.40	0.36	1.388198
0.41	0.34	1.372950

Estimate the uncertainty in $I(a, b)$!

2.4.9 (a) Let $B \in \mathbf{R}^{n\times n}$ be a matrix for which $\|B\| < 1$. Show that the infinite sum and product

$$(I - B)^{-1} = \begin{cases} I + B + B^2 + B^3 + B^4 \cdots, \\ (I + B)(I + B^2)(I + B^4)(I + B^8) \cdots \end{cases}$$

both converge to the indicated limit.

Hint: Use the identity $(I - B)(I + B + \cdots + B^k) = I - B^{k+1}$.

(b) Show that the matrix $(I - B)$ is nonsingular and that

$$\|(I - B)^{-1}\| \le 1/(1 - \|B\|).$$

2.4.10 Solve the linear system in Example 2.4.8 with Gaussian elimination after exchanging the two equations. Do you now get an accurate result?

2.4.11 Derive forward and backward recursion formulas for calculating the integrals

$$I_n = \int_0^1 \frac{x^n}{4x + 1}\, dx.$$

Why is one algorithm stable and the other unstable?

2.4.12 (a) Use the results in Table 2.4.1 to determine constants c and q such that $\kappa(H_n) \approx c \cdot 10^q$.

(b) Compute the Bauer–Skeel condition number $\operatorname{cond}(H_n) = \|\,|H_n^{-1}||H_n|\,\|_2$, of the Hilbert matrices for $n = 1 : 12$. Compare the result with the values of $\kappa(H_n)$ given in Table 2.4.1.

2.4.13 Vandermonde matrices are structured matrices of the form

$$V_n = \begin{pmatrix} 1 & 1 & \cdots & 1 \\ \alpha_1 & \alpha_2 & \cdots & \alpha_n \\ \vdots & \vdots & \cdots & \vdots \\ \alpha_1^{n-1} & \alpha_2^{n-1} & \cdots & \alpha_n^{n-1} \end{pmatrix}.$$

Let $\alpha_j = 1 - 2(j - 1)/(n - 1)$, $j = 1 : n$. Compute the condition numbers $\kappa_2(V_n)$ for $n = 5, 10, 15, 20, 25$. Is the growth in $\kappa_2(V_n)$ exponential in n?

2.5 Automatic Control of Accuracy and Verified Computing

2.5.1 Running Error Analysis

A different approach to rounding error analysis is to perform the analysis automatically, for *each particular computation*. This gives an *a posteriori* error analysis in contrast to the *a priori* error analysis discussed above.

A simple form of a posteriori analysis, called running error analysis, was used in the early days of computing; see Wilkinson [380]. To illustrate his idea we rewrite the basic model for floating-point arithmetic as

$$x \text{ op } y = fl\,(x \text{ op } y)(1+\epsilon).$$

This holds for most implementations of floating-point arithmetic. Then, the actual error can be estimated $|fl\,(x \text{ op } y) - x \text{ op } y| \le u|fl\,(x \text{ op } y)|$. Note that the error is now given in terms of the *computed* result and is available in the computer at the time the operation is performed. This running error analysis can often be easily implemented. We just take an existing program and modify it, so that as each arithmetic operation is performed, the absolute value of the computed results is added into the accumulating error bound.

Example 2.5.1.

The inner product $fl\,(x^T y)$ is computed by the program

$s = 0$; $\eta = 0$;
for $i = 1, 2, \dots, n$
  $t = fl\,(x_i y_i)$; $\eta = \eta + |t|$;
  $s = fl\,(s + t)$; $\eta = \eta + |s|$;
end

For the final error we have the estimate $|fl\,(x^T y) - x^T y| \le \eta u$. Note that a running error analysis takes advantage of cancellations in the sum. This is in contrast to the previous estimates, which we call a priori error analysis, where the error estimate is the same for all distribution of signs of the elements x_i and y_i.

Efforts have been made to design the computational unit of a computer so that it gives, in every arithmetic operation, only those digits of the result which are judged to be significant (possibly with a fixed number of extra digits), so-called *unnormalized floating arithmetic*. This method reveals poor construction in algorithms, but in many other cases it gives a significant and unnecessary loss of accuracy. The mechanization of the rules, which a knowledgeable and experienced person would use for control of accuracy in hand calculation, is not as free from problems as one might expect. As a complement to arithmetical operations of conventional type, the above type of arithmetic is of some interest, but it is doubtful that it will ever be widely used.

A fundamental difficulty in automatic control of accuracy is that to decide how many digits are needed in a quantity to be used in later computation, one needs to consider the *entire context of the computations*. It can in fact occur that the errors in many operands depend on each other in such a way that they cancel each other. Such a **cancellation of error**, which is a completely different phenomenon from the previously discussed cancellation of terms, is most common in larger problems, but will be illustrated here with a simple example.

Example 2.5.2.

Suppose we want to compute $y = z_1 + z_2$, where $z_1 = \sqrt{x^2+1}$, $z_2 = 200 - x$, $x = 100 \pm 1$, with a rounding error which is negligible compared to that resulting from the errors in z_1 and z_2. The best possible error bounds in the intermediate results are $z_1 = 100 \pm 1$, $z_2 = 100 \pm 1$. It is then tempting to be satisfied with the result $y = 200 \pm 2$.

But the errors in z_1 and z_2 due to the uncertainty in x will, to a large extent, cancel each other! This becomes clear if we rewrite the expression as

$$y = 200 + \left(\sqrt{x^2+1} - x\right) = 200 + \frac{1}{\sqrt{x^2+1} + x}.$$

It follows that $y = 200 + z$, where $1/202 \lessapprox z \le 1/198$. Thus y can be computed with an absolute error less than about $2/(200)^2 = 0.5 \cdot 10^{-4}$. Therefore, using the expression $y = z_1 + z_2$ the intermediate results z_1 and z_2 should be computed to four decimals even though the last integer in these is uncertain. The result is $y = 200.0050 \pm \frac{1}{2}10^{-4}$.

In larger problems, such a cancellation of errors can occur even though one cannot easily give a way to rewrite the expressions involved. The authors have seen examples where the final result, a sum of seven terms, was obtained correctly to eight decimals even though the terms, which were complicated functions of the solution to a system of nonlinear equations with 14 unknowns, were correct only to three decimals! Another nontrivial example is given in Example 2.4.12.

2.5.2 Experimental Perturbations

In many practical problems, the functional dependence between input data and output data is so complicated that it is difficult to directly apply the general formulas for error propagation derived in Sec. 2.4.3. One can then investigate the sensitivity of the output data for perturbations in the input data by means of an **experimental perturbational calculation**. One performs the calculations many times with perturbed input data and studies the perturbations in the output data. For example, instead of using the formula for standard error of a function of many variables, given in Theorem 2.4.5, it is often easier to compute the function a number of times with randomly perturbed arguments and to use them to estimate the standard deviation of the function numerically.

Important data, such as the step length in a numerical integration or the parameter which determines when an iterative process is going to be broken off, should be varied with all the other data left unchanged. If one can easily *vary the precision of the machine* in the arithmetic operations one can get an idea of the influence of rounding errors. It is generally not necessary to make a perturbational calculation for each and every data component; one can instead *perturb many input data simultaneously*—for example, by using random numbers.

A perturbational calculation often gives not only an error estimate but also greater insight into the problem. Occasionally, it can be difficult to interpret the perturbational data correctly, since the disturbances in the output data depend not only on the mathematical problem but also on the choice of numerical method and the details in the design of the algorithm. The rounding errors during the computation are not the same for the perturbed and unperturbed problem. Thus if the output data react more sensitively than one had anticipated, it can be difficult to immediately point out the source of the error. It can then be profitable to plan a series of perturbation experiments with the help of which one can separate the effects of the various sources of error. If the dominant source of error is the method or the algorithm, then one should try another method or another algorithm. It is beyond the scope of this book to give further comments on the planning of such experiments. Imagination and the general insights regarding error analysis, which this chapter is meant to give, play a large role.

2.5.3 Interval Arithmetic

In **interval arithmetic** one assumes that all input values are given as intervals and systematically calculates an inclusion interval for each intermediate result. It is partly an automation of calculation with maximal error bounds. The importance of interval arithmetic is that it provides a tool for computing validated answers to mathematical problems.

The modern development of interval arithmetic was initiated by the work of R. E. Moore [271]. Interval arithmetic has since been developed into a useful tool for many problems in scientific computing and engineering. A noteworthy example of its use is the verification of the existence of a Lorenz attractor by W. Tucker [361]. Several extensive surveys on interval arithmetic are available; see [3, 4, 225]. Hargreaves [186] gives a short tutorial on INTLAB and also a good introduction to interval arithmetic.

The most frequently used representations for the intervals are the **infimum-supremum** representations

$$I = [a, b] := \{x \mid a \le x \le b\}, \quad (a \le b), \tag{2.5.1}$$

where x is a real number. The absolute value or the **magnitude** of an interval is defined as

$$|\,[a, b]\,| = \operatorname{mag}([a, b]) = \max\{|x| \mid x \in [a, b]\}, \tag{2.5.2}$$

and the **mignitude** of an interval is defined as

$$\operatorname{mig}([a, b]) = \min\{|x| \mid x \in [a, b]\}. \tag{2.5.3}$$

In terms of the endpoints we have

$$\operatorname{mag}([a, b]) = \max\{|a|, |b|\},$$
$$\operatorname{mig}([a, b]) = \begin{cases} \min\{|a|, |b|\} & \text{if } 0 \notin [a, b], \\ 0 & \text{otherwise.} \end{cases}$$

The result of an interval operation equals the range of the corresponding real operation. For example, the difference between the intervals $[a_1, a_2]$ and $[b_1, b_2]$ is defined as the shortest interval which contains all the numbers $x_1 - x_2$, where $x_1 \in [a_1, a_2]$, $x_2 \in [b_1, b_2]$, i.e., $[a_1, a_2]-[b_1, b_2] := [a_1-b_2, a_2-b_1]$. Other elementary interval arithmetic operations are similarly defined:

$$[a_1, a_2]\,\mathrm{op}\,[b_1, b_2] := \{x_1\,\mathrm{op}\,x_2 \mid x_1 \in [a_1, a_2],\ x_2 \in [b_1, b_2]\}, \tag{2.5.4}$$

where op $\in \{+, -, \times, \mathrm{div}\}$. The interval value of a function ϕ (for example, the elementary functions sin, cos, exp, log) evaluated on an interval is defined as

$$\phi([a, b]) = \Big[\inf_{x\in[a,b]} \phi(x),\ \sup_{x\in[a,b]} \phi(x) \Big].$$

Operational Definitions

Although (2.5.4) characterizes interval arithmetic operations, we also need **operational definitions**. We take

$$\begin{aligned}
&[a_1, a_2] + [b_1, b_2] = [a_1 + b_1, a_2 + b_2],\\
&[a_1, a_2] - [b_1, b_2] = [a_1 - b_2, a_2 - b_1],\\
&[a_1, a_2] \times [b_1, b_2] = \big[\min\{a_1b_1, a_1b_2, a_2b_1, a_2b_2\}, \max\{a_1b_1, a_1b_2, a_2b_1, a_2b_2\}\big],\\
&1/[a_1, a_2] = [1/a_2, 1/a_1] \quad \text{for} \quad a_1a_2 > 0,\\
&[a_1, a_2]/[b_1, b_2] = [a_1, a_2] \cdot (1/[b_1, b_2]).
\end{aligned} \tag{2.5.5}$$

It is easy to prove that in exact interval arithmetic the operational definitions above give the exact range (2.5.4) of the interval operations. Division by an interval containing zero is not defined and may cause an interval computation to come to a premature end.

A degenerate interval with radius zero is called a point interval and can be identified with a single number $a \equiv [a, a]$. In this way the usual arithmetic is recovered as a special case. The intervals $0 = [0, 0]$ and $1 = [1, 1]$ are the neutral elements with respect to interval addition and interval multiplication, respectively. A nondegenerate interval has no inverse with respect to addition or multiplication. For example, we have

$$[1, 2] - [1, 2] = [-1, 1], \qquad [1, 2]/[1, 2] = [0.5, 2].$$

For interval operations the **commutative law**

$$[a_1, a_2]\,\mathrm{op}\,[b_1, b_2] = [b_1, b_2]\,\mathrm{op}\,[a_1, a_2]$$

holds. But the distributive law has to be replaced by so-called **subdistributivity**:

$$[a_1, a_2]([b_1, b_2] + [c_1, c_2]) \subseteq [a_1, a_2][b_1, b_2] + [a_1, a_2][c_1, c_2]. \tag{2.5.6}$$

This unfortunately means that expressions, which are equivalent in real arithmetic, differ in exact interval arithmetic. The simple example

$$[-1, 1]([1, 1] + [-1, -1]) = 0 \subset [-1, 1][1, 1] + [-1, 1][-1, -1] = [-2, 2]$$

shows that $-[-1, 1]$ is not the additive inverse to $[-1, 1]$ and also illustrates (2.5.6).

The operations introduced are **inclusion monotonic**, i.e.,

$$[a_1, a_2] \subseteq [c_1, c_2],\ [b_1, b_2] \subseteq [d_1, d_2] \quad \Rightarrow \quad [a_1, a_2]\,\mathrm{op}\,[b_1, b_2] \subseteq [c_1, c_2]\,\mathrm{op}\,[d_1, d_2]. \tag{2.5.7}$$

An alternative representation for an interval is the **midpoint-radius** representation, for which we use brackets,

$$\langle c, r\rangle := \{x \bigm| |x - c| \le r\} \quad (0 \le r), \tag{2.5.8}$$

where the midpoint and radius of the interval $[a, b]$ are defined as

$$c = \mathrm{mid}\,([a, b]) = \frac{1}{2}(a + b), \qquad r = \mathrm{rad}\,([a, b]) = \frac{1}{2}(b - a). \tag{2.5.9}$$

For intervals in the midpoint-radius representation we take as operational definitions

$$\begin{aligned} &\langle c_1, r_1\rangle + \langle c_2, r_2\rangle = \langle c_1 + c_2, r_1 + r_2\rangle,\\ &\langle c_1, r_1\rangle - \langle c_2, r_2\rangle = \langle c_1 - c_2, r_1 + r_2\rangle,\\ &\langle c_1, r_1\rangle \times \langle c_2, r_2\rangle = \langle c_1c_2, |c_1|r_2 + r_1|c_2| + r_1r_2\rangle,\\ &1/\langle c, r\rangle = \langle c/(|c|^2 - r^2), r/(|c|^2 - r^2)\rangle, \quad (|c| > r),\\ &\langle c_1, r_1\rangle/\langle c_2, r_2\rangle = \langle c_1, r_1\rangle \times (1/\langle c_2, r_2\rangle). \end{aligned} \tag{2.5.10}$$

For addition, subtraction, and inversion, these give the exact range, but for multiplication and division they overestimate the range (see Problem 2.5.2). For multiplication we have for any $x_1 \in \langle c_1, r_1\rangle$ and $x_2 \in \langle c_2, r_2\rangle$

$$\begin{aligned} |x_1x_2 - c_1c_2| &= |c_1(x_2 - c_2) + c_2(x_1 - c_1) + (x_1 - c_1)(x_2 - c_2)|\\ &\le |c_1|r_2 + |c_2|r_1 + r_1r_2. \end{aligned}$$

In implementing interval arithmetic using floating-point arithmetic, the operational interval results may not be exactly representable as floating-point numbers. Then if the lower bound is rounded down to the nearest smaller machine number and the upper bound rounded up, the exact result must be contained in the resulting interval. Recall that these rounding modes (rounding to $-\infty$ and $+\infty$) are supported by the IEEE 754 standard. For example, using five significant decimal arithmetic, we would like to get

$$[1, 1] + [-10^{-10}, 10^{-10}] = [0.99999, 1.0001]$$

or, in midpoint-radius representation,

$$\langle 1, 0\rangle + \langle 0, 10^{-10}\rangle = \langle 1, 10^{-10}\rangle.$$

Note that in the conversion between decimal and binary representation rounding the appropriate rounding mode must also be used where needed. For example, converting the point interval 0.1 to binary IEEE double precision we get an interval with radius $1.3878 \cdot 10^{-17}$. The conversion between the infimum-supremum representation is straightforward but the infimum-supremum and midpoint may not be exactly representable.

Interval arithmetic applies also to complex numbers. A complex interval in the infimum-supremum representation is

$$[z_1, z_2] = \{z = x + iy \mid x \in [x_1, x_2],\ y \in [y_1, y_2]\}.$$

This defines a closed *rectangle in the complex plane* defined by the two real intervals,

$$[z_1, z_2] = [x_1, x_2] + i[y_1, y_2], \quad x_1 \le x_2, \quad y_1 \le y_2.$$

This can be written more compactly as $[z_1, z_2] := \{z \mid z_1 \le z \le z_2\}$, where we use the partial ordering

$$z \le w \iff \Re z \le \Re w \quad \text{and} \quad \Im z \le \Im w.$$

Complex interval operations in the infimum-supremum arithmetic are defined in terms of the real intervals in the same way as the complex operations are defined for complex numbers $z = x + iy$. For addition and subtraction the result coincides with the exact range. This is not the case for complex interval multiplication, where the result is a rectangle in the complex plane, whereas the actual range is not of this shape. Therefore, for complex intervals, multiplication will result in an overestimation.

In the complex case the midpoint-radius representation is

$$\langle c, r\rangle := \{z \in \mathbf{C} \mid |z - c| \le r\}, \quad 0 \le r,$$

where the midpoint c is now a complex number. This represents a *closed circular disk in the complex plane*. The operational definitions (2.5.10) are still valid, except that some operations now are complex operations, and that inversion becomes

$$1/\langle c, r\rangle = \big\langle \bar{c}/(|c|^2 - r^2), r/(|c|^2 - r^2)\big\rangle \quad \text{for} \quad |c| > r,$$

where $\bar{c}$ is the complex conjugate of c. Complex interval midpoint-radius arithmetic is also called **circular arithmetic**. For complex multiplications it generates less overestimation than the infimum-supremum notation.

Although the midpoint-radius arithmetic seems more appropriate for complex intervals, most older implementations of interval arithmetic use infimum-supremum arithmetic. One reason for this is the overestimation also caused for real intervals by the operational definitions for midpoint-radius multiplication and division. Rump [307] has shown that the overestimation is bounded by a factor 1.5 in radius and that midpoint-radius arithmetic allows for a much faster implementation for modern vector and parallel computers.

2.5.4 Range of Functions

One use of interval arithmetic is to enclose the range of a real-valued function. This can be used, for example, for localizing and enclosing global minimizers and global minima of a real function of one or several variables in a region. It can also be used for verifying the nonexistence of a zero of $f(x)$ in a given interval.

Let $f(x)$ be a real function composed of a finite number of elementary operations and standard functions. If one replaces the variable x by an interval $[\underline{x}, \overline{x}]$ and evaluates the resulting interval expression, one gets as the result an interval denoted by $f([\underline{x}, \overline{x}])$. (It is assumed that all operations can be carried out.) A fundamental result in interval arithmetic is that this evaluation is inclusion monotonic, i.e.,

$$[\underline{x}, \overline{x}] \subseteq [\underline{y}, \overline{y}] \quad \Rightarrow \quad f([\underline{x}, \overline{x}]) \subseteq f([\underline{y}, \overline{y}]).$$

In particular this means that

$$x \subseteq [\underline{x}, \overline{x}] \Rightarrow f(x) \subseteq f([\underline{x}, \overline{x}]),$$

i.e., $f([x])$ contains the range of $f(x)$ over the interval $[\underline{x}, \overline{x}]$. A similar result holds also for functions of several variables $f(x_1, \dots, x_n)$.

An important case when interval evaluation gives the exact range of a function is when $f(x_1, \dots, x_n)$ is a rational expression, where each variable x_i occurs only once in the function.

Example 2.5.3.

In general narrow bounds cannot be guaranteed. For example, if $f(x) = x/(1-x)$, then

$$f([2,3]) = [2,3]/(1-[2,3]) = [2,3]/[-2,-1] = [-3,-1].$$

The result contains but does not coincide with the exact range $[-2, -3/2]$. But if we rewrite the expression as $f(x) = 1/(1/x - 1)$, where x only occurs once, then we get

$$f([2,3]) = 1/(1/[2,3]-1) = 1/[-2/3, -1/2] = [-2, -3/2],$$

which is the exact range.

When interval analysis is used in a naive manner as a simple technique for simulating forward error analysis, it does not usually give sharp bounds on the total computational error. To get useful results the computations in general need to be arranged so that overestimation is minimized as much as possible. Often a refined design of the algorithm is required in order to prevent the bounds for the intervals from becoming unacceptably coarse. The answer $[-\infty, \infty]$ is of course always correct but not at all useful!

The remainder term in Taylor expansions includes a variable ξ, which is known to lie in an interval $\xi \in [a, b]$. This makes it suitable to treat the remainder term with interval arithmetic.

Example 2.5.4.

Evaluate for $[x] = [2, 3]$ the polynomial

$$p(x) = 1 - x + x^2 - x^3 + x^4 - x^5.$$

Using exact interval arithmetic we find

$$p([2,3]) = [-252, 49]$$

(verify this!). This is an overestimate of the exact range, which is $[-182, -21]$. Rewriting $p(x)$ in the form $p(x) = (1-x)(1+x^2+x^4)$ we obtain the correct range. In the first example there is a **cancellation of errors** in the intermediate results (cf. Example 2.5.2), which is not revealed by the interval calculations.

Sometimes it is desired to compute a tiny interval that is guaranteed to enclose a real simple root x^* of $f(x)$. This can be done using an interval version of Newton's method. Suppose that the function $f(x)$ is continuously differentiable. Let $f'([x_0])$ denote an interval containing $f'(x)$ for all x in a finite interval $[x] := [a, b]$. Define the Newton operator $N([x])$, $[x] = [a, b]$, by

$$N([x]) := m - \frac{f(m)}{f'([x])}, \qquad m = \text{mid}\,[x]. \tag{2.5.11}$$

For the properties of the **interval Newton's method**

$$[x_{k+1}] = N([x_k]), \quad k = 0, 1, 2, \dots;$$

see Sec. 6.3.3.

Another important application of interval arithmetic is to initial value problems for ordinary differential equations

$$y' = f(x, y), \quad y(x_0) = y_0, \quad y \in \mathbf{R}^n.$$

Interval techniques can be used to provide for errors in the initial values, as well as truncation and rounding errors, so that at each step intervals are computed that contain the actual solution. But it is a most demanding task to construct an interval algorithm for the initial value problem, and such algorithms tend to be significantly slower than corresponding point algorithms. One problem is that a wrapping effect occurs at each step and causes the computed interval widths to grow exponentially. This is illustrated in the following example.

Example 2.5.5.

The recursion formulas

$$x_{n+1} = (x_n - y_n)/\sqrt{2}, \qquad y_{n+1} = (x_n + y_n)/\sqrt{2}$$

mean a series of 45-degree rotations in the xy-plane (see Figure 2.5.1). By a two-dimensional interval one means a rectangle whose sides are parallel to the coordinate axes.

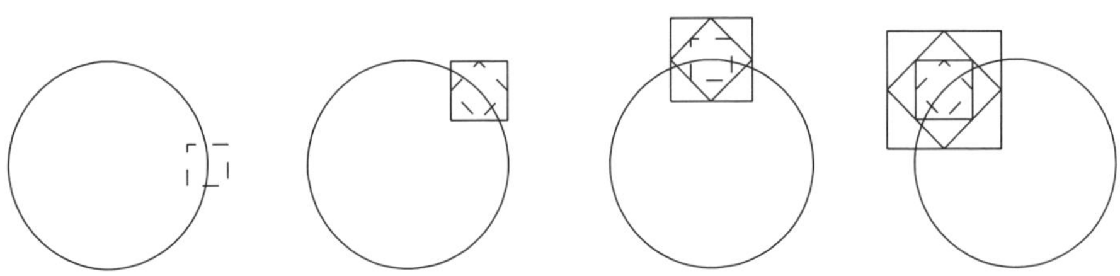

Figure 2.5.1. *Wrapping effect in interval analysis.*

If the initial value (x_0, y_0) is given as an interval $[x_0] = [1-\epsilon, 1+\epsilon]$, $[y_0] = [-\epsilon, \epsilon]$ (see the dashed square, in the leftmost portion of Figure 2.5.1), then (x_n, y_n) will, with *exact* performance of the transformations, also be a square with side 2ϵ, for all n (see the other squares in Figure 2.5.1). If the computations are made using interval arithmetic, rectangles with sides parallel to the coordinate axis will, in each step, be circumscribed about the exact image of the interval one had in the previous step. Thus the interval is multiplied by $\sqrt{2}$ in each step. After 40 steps, for example, the interval has been multiplied by $2^{20} > 10^6$. This phenomenon, intrinsic to interval computations, is called the **wrapping effect**. (Note that if one uses disks instead of rectangles, there would not be any difficulties in this example.)

2.5.5 Interval Matrix Computations

In order to treat multidimensional problems we introduce interval vectors and matrices. An interval vector is denoted by $[x]$ and has interval components $[x_i] = [\underline{x_i}, \overline{x_i}])$, $i = 1 : n$. Likewise an interval matrix $[A] = ([a_{ij}])$ has interval elements

$$[a_{ij}] = [\underline{a}_{ij}, \overline{a}_{ij}], \quad i = 1 : m, \quad j = 1 : n.$$

Operations between interval matrices and interval vectors are defined in an obvious manner. The interval matrix-vector product $[A][x]$ is the smallest interval vector, which contains the set

$$\{Ax \mid A \in [A],\ x \in [x]\}$$

but normally does not coincide with it. By the inclusion property it holds that

$$\{Ax \mid A \in [A],\ x \in [x]\} \subseteq [A][x] = \left(\sum_{j=1}^{n} [a_{ij}][x_j]\right).$$

In general, there will be an overestimation in enclosing the image with an interval vector, caused by the fact that the image of an interval vector under a linear transformation in general is not an interval vector. This phenomenon, intrinsic to interval computations, is similar to the wrapping effect described in Example 2.5.5.

Example 2.5.6.

We have

$$A = \begin{pmatrix} 1 & 1 \\ -1 & 1 \end{pmatrix}, \quad [x] = \begin{pmatrix} [0, 1] \\ [0, 1] \end{pmatrix} \quad \Rightarrow \quad A[x] = \begin{pmatrix} [0, 2] \\ [-1, 1] \end{pmatrix}.$$

Hence $b = (2 \quad -1)^T \in A[x]$, but there is no $x \in [x]$ such that $Ax = b$. (The solution to $Ax = b$ is $x = (3/2 \quad 1/2)^T$.)

The magnitude of an interval vector or matrix is interpreted componentwise and defined by

$$|\,[x]\,| = (|\,[x_1]\,|, |\,[x_2]\,|, \dots, |\,[x_n]\,|)^T,$$

where the magnitude of the components is defined by

$$|\,[a, b]\,| = \max\{|x| \mid x \in [a, b]\}. \tag{2.5.12}$$

The ∞-norm of an interval vector or matrix is defined as the ∞-norm of its magnitude,

$$\| [x] \|_\infty = \| \, | [x] | \, \|_\infty, \qquad \| [A] \|_\infty = \| \, | [A] | \, \|_\infty. \tag{2.5.13}$$

In implementing matrix multiplication it is important to avoid case distinctions in the inner loops, because that would make it impossible to use fast vector and matrix operations. Using interval arithmetic it is possible to compute strict enclosures of the product of two matrices. Note that this is also needed in the case of the product of two point matrices since rounding errors will usually occur.

We assume that the command

$$\text{setround}(i), \quad i = -1, 0, 1,$$

sets the rounding mode to $-\infty$, to nearest, and to $+\infty$, respectively. (Recall that these rounding modes are supported by the IEEE standard.) Let A and B be point matrices and suppose we want to compute an interval matrix $[C]$ such that

$$fl(A \cdot B) \subset [C] = [C_{\text{inf}}, C_{\text{sup}}].$$

Then the following simple code, using two matrix multiplications, does that:

$$\begin{aligned} &\text{setround}(-1); \quad C_{\text{inf}} = A \cdot B; \\ &\text{setround}(1); \quad C_{\text{sup}} = A \cdot B; \end{aligned}$$

We next consider the product of a point matrix A and an interval matrix $[B] = [B_{\text{inf}}, B_{\text{sup}}]$. The following code performs this using four matrix multiplications:

$$\begin{aligned} &A_- = \min(A, \ 0); \quad A_+ = \max(A, \ 0); \\ &\text{setround}(-1); \\ &C_{\text{inf}} = A_+ \cdot B_{\text{inf}} + A_- \cdot B_{\text{sup}}; \\ &\text{setround}(1); \\ &C_{\text{sup}} = A_- \cdot B_{\text{inf}} + A_+ \cdot B_{\text{sup}}; \end{aligned}$$

(Note that the commands $A_- = \min(A, \ 0)$ and $A_+ = \max(A, \ 0)$ act componentwise.) An algorithm for computing the product of two interval matrices using eight matrix multiplications is given by Rump [308].

Fast portable codes for interval matrix computations that make use of the Basic Linear Algebra Subroutines (BLAS) and IEEE 754 standard are now available. This makes it possible to efficiently use high-performance computers for interval computation. INTLAB (INTerval LABoratory) by Rump [308, 307] is based on MATLAB, and it is particularly easy to use. It includes many useful subroutines, for example, one to compute an enclosure of the difference between the solution and an approximate solution $x_m = C\text{mid}\,[b]$. Verified solutions of linear least squares problems can also be computed.

Review Questions

2.5.1 (a) Define the magnitude and mignitude of an interval $I = [a, b]$.

(b) How is the ∞-norm of an interval vector defined?

2.5.2 Describe the two different ways of representing intervals used in real and complex interval arithmetic. Mention some of the advantages and drawbacks of each of these.

2.5.3 An important property of interval arithmetic is that the operations are inclusion monotonic. Define this term.

2.5.4 What is meant by the "wrapping effect" in interval arithmetic and what are its implications? Give some examples of where it occurs.

2.5.5 Assume that the command

$$\text{setround}(i), \quad i = -1, 0, 1,$$

sets the rounding mode to $-\infty$, to nearest, and to $+\infty$, respectively. Give a simple code that, using two matrix multiplications, computes an interval matrix $[C]$ such that for point matrices A and B

$$fl(A \cdot B) \subset [C] = [C_{\text{inf}}, C_{\text{sup}}].$$

Problems and Computer Exercises

2.5.1 Carry out the following calculations in exact interval arithmetic:

(a) $[0, 1] + [1, 2]$; (b) $[3, 3.1] - [0, 0, 2]$; (c) $[-4. - 1] \cdot [-6, 5]$;

(d) $[2, 2] \cdot [-1, 2]$; (e) $[-1, 1]/[-2, -0.5]$; (f) $[-3, 2] \cdot [-3.1, 2.1]$.

2.5.2 Show that using the operational definitions (2.5.5) the product of the disks $\langle c_1, r_1\rangle$ and $\langle c_2, r_2\rangle$ contains zero if $c_1 = c_2 = 1$ and $r_1 = r_2 = \sqrt{2} - 1$.

2.5.3 (Stoer) Using Horner's rule and exact interval arithmetic, evaluate the cubic polynomial

$$p(x) = ((x - 3)x + 3)x, \quad [x] = [0.9, 1.1].$$

Compare the result with the exact range, which can be determined by observing that $p(x) = (x - 1)^3 + 1$.

2.5.4 Treat Example 1.2.2 using interval analysis and four decimal digits. Starting with the inclusion interval $I_{10} = [0, 1/60] = [0, 0.01667]$ generate successively intervals I_k, $k = 9 : -1 : 5$, using interval arithmetic and the recursion

$$I_{n-1} = 1/(5n) - I_n/5.$$

Notes and References

Many different aspects of number systems and floating-point computations are treated in Knuth [230, Chapter 4], including the historical development of number representation. Leibniz (1703) seems to have been the first to discuss binary arithmetic. He did not advocate it for practical calculations, but stressed its importance for number-theoretic investigations. King Charles XII of Sweden came upon the idea of radix 8 arithmetic in 1717. He felt this to be more convenient than the decimal notation and considered introducing octal arithmetic into Sweden. He died in battle before decreeing such a change.

In the early days of computing, floating-point computations were not built into the hardware but implemented in software. The earliest subroutines for floating-point arithmetic were probably those developed by J. H. Wilkinson at the National Physical Laboratory, England, in 1947. A general source on floating-point computation is Sterbenz [333]. Goldberg [158] is an excellent tutorial on the IEEE 754 standard for floating-point arithmetic defined in [205]. A self-contained, accessible, and easy to read introduction with many illustrating examples is the monograph by Overton [280]. An excellent treatment on floating-point computation, rounding error analysis, and related topics is given in Higham [199, Chapter 2]. Different aspects of accuracy and reliability are discussed in [108].

The fact that thoughtless use of mathematical formulas and numerical methods can lead to disastrous results are exemplified by Stegun and Abramowitz [331] and Forsythe [122]. Numerous examples in which incorrect answers are obtained from plausible numerical methods can be found in Fox [125].

Statistical analysis of rounding errors goes back to an early paper of Goldstine and von Neumann [161]. Barlow and Bareiss [15] have investigated the distribution of rounding errors for different modes of rounding under the assumption that the mantissas of the operands are from a logarithmic distribution.

Conditioning numbers of general differentiable functions were first studied by Rice [301]. Backward error analysis was developed and popularized by J. H. Wilkinson in the 1950s and 1960s, and the classic treatise on rounding error analysis is [376]. The more recent survey [380] gives a good summary and a historical background. Kahan [216] gives an in-depth discussion of rounding error analysis with examples of how flaws in the design of hardware and software in computer systems can have undesirable effects on accuracy. The normwise analysis is natural for studying the effect of orthogonal transformations in matrix computations; see Wilkinson [376]. The componentwise approach, used in perturbation analysis for linear systems by Bauer [19], can give sharper results and has gained in popularity.

Condition numbers are often viewed pragmatically as the coefficients of the backward errors in bounds on forward errors. Wilkinson in [376] avoids a precise definition of condition numbers in order to use them more freely. The more precise limsup definition in Definition 2.4.6 is usually attributed to Rice [301].

Even in the special literature, the discussion of planning of experimental perturbations is surprisingly meager. An exception is the collection of software tools called PRECISE, developed by Chaitin-Chatelin et al.; see [63, 64]. These are designed to help the user set up computer experiments to explore the impact of the quality of convergence of numerical methods. PRECISE involves a statistical analysis of the effect on a computed solution of random perturbations in data.

Chapter 3

Series, Operators, and Continued Fractions

No method of solving a computational problem
is really available to a user until it is completely
described in an algebraic computer language
and made completely reliable.
—George E. Forsythe

3.1 Some Basic Facts about Series

3.1.1 Introduction

Series expansions are a very important aid in numerical calculations, especially for quick estimates made in hand calculation—for example, in evaluating functions, integrals, or derivatives. Solutions to differential equations can often be expressed in terms of series expansions. Since the advent of computers it has, however, become more common to treat differential equations directly using, for example, finite difference or finite element approximations instead of series expansions. Series have some advantages, especially in problems containing parameters. As well, automatic methods for formula manipulation and some new numerical methods provide new possibilities for series.

In this section we will discuss general questions concerning the use of infinite series for numerical computations including, for example, the estimation of remainders, power series, and various algorithms for computing their coefficients. Often a series expansion can be derived by simple operations with a known series. We also give an introduction to formal power series. The next section treats perturbation expansions, ill-conditioned expansions, and semiconvergent expansions, from the point of view of computing.

Methods and results will sometimes be formulated in terms of *series*, sometimes in terms of *sequences*. These formulations are equivalent, since the sum of an infinite series is defined as the limit of the sequence $\{S_n\}$ of its partial sums

$$S_n = a_1 + a_2 + \cdots + a_n.$$

Conversely, any sequence $S_1, S_2, S_3, \dots$ can be written as the partial sums of a series,

$$S_1 + (S_2 - S_1) + (S_3 - S_2) + \cdots.$$

In practice, one is seldom seriously concerned about a rigorous error bound when the computed terms decrease rapidly, and it is "obvious" that the terms will continue to decrease equally quickly. One can then break off the series and use either the last included term or a coarse estimate of the **first neglected term** as an estimate of the remainder.

This rule is not very precise. *How rapidly is "rapidly"?* Questions like this occur everywhere in scientific computing. If mathematical rigor costs little effort or little extra computing time, then it should, of course, be used. Often, however, an error bound that is both rigorous and realistic may cost more than what is felt reasonable for (say) a one-off problem.

In problems where guaranteed error bounds are not asked for, when it is enough to obtain a feeling for the reliability of the results, one can handle these matters in the same spirit as one handles risks in everyday life. It is then a matter of experience to formulate a simple and *sufficiently* reliable **termination criterion** based on the automatic inspection of the successive terms.[41]

The inexperienced scientific programmer may, however, find such questions hard, even in simple cases. In the production of general purpose mathematical software, or in a context where an inaccurate numerical result can cause a disaster, such questions are serious and sometimes hard for the experienced scientific programmer also. For this reason, we shall formulate a few theorems with which one can often transform the feeling that "the remainder is negligible" to a mathematical proof. There are, in addition, actually numerically useful *divergent* series; see Sec. 3.2.6. When one uses such series, estimates of the remainder are clearly essential.

Assume that we want to compute a quantity S, which can be expressed in a series expansion, $S = \sum_{j=0}^{\infty} a_j$, and set

$$S_n = \sum_{j=0}^{n} a_j, \qquad R_n = S - S_n.$$

We call $\sum_{j=n+1}^{\infty} a_j$ the **tail** of the series; a_n is the "last included term" and a_{n+1} is the "first neglected term." The remainder R_n with reversed sign is called the **truncation error**.[42]

The tail of a convergent series can often be compared to a series with a known sum, such as a geometric series, or with an integral which can be computed directly.

Theorem 3.1.1 (*Comparison with a Geometric Series*).
If $|a_{j+1}| \le k|a_j|$ for all $j \ge n$, where $k < 1$, then

$$|R_n| \le \frac{|a_{n+1}|}{1-k} \le \frac{k|a_n|}{1-k}.$$

In particular if $k < 1/2$, then it is true that the absolute value of the remainder is less than the last included term.

[41] Termination criteria for iterative methods will be discussed in Sec. 6.1.3.

[42] In this terminology the remainder is the *correction* one has to make in order to eliminate the error.

Proof. By induction, one finds that $|a_j| \le k^{j-1-n}|a_{n+1}|$, $j \ge n+1$, since

$$|a_j| \le k^{j-1-n}|a_{n+1}| \Rightarrow |a_{j+1}| \le k|a_j| \le k^{j-n}|a_{n+1}|.$$

Thus

$$|R_n| \le \sum_{j=n+1}^{\infty} |a_j| \le \sum_{j=n+1}^{\infty} k^{j-1-n}|a_{n+1}| = \frac{|a_{n+1}|}{1-k} \le \frac{k|a_n|}{1-k},$$

according to the formula for the sum of an infinite geometric series. The last statement follows from the inequality $k/(1-k) < 1$, when $k < 1/2$. □

Example 3.1.1.

In a power series with slowly varying coefficients, $a_j = j^{1/2}\pi^{-2j}$. Then $a_6 < 2.45 \cdot 0.0000011 < 3 \cdot 10^{-6}$, and

$$\frac{|a_{j+1}|}{|a_j|} \le \frac{(j+1)^{1/2}}{j^{1/2}} \frac{\pi^{-2j-2}}{\pi^{-2j}} \le \left(1+\frac{1}{6}\right)^{1/2} \pi^{-2} < 0.11$$

for $j \ge 6$. Thus, by Theorem 3.1.1, $|R_6| < 3 \cdot 10^{-6} \frac{0.11}{1-0.11} < 4 \cdot 10^{-7}$.

Theorem 3.1.2 (*Comparison of a Series with an Integral*).

If $|a_j| \le f(j)$ for all $j \ge n+1$, where $f(x)$ is a nonincreasing function for $x \ge n$, then

$$|R_n| \le \sum_{j=n+1}^{\infty} |a_j| \le \int_n^{\infty} f(x)\,dx,$$

which yields an upper bound for $|R_n|$, if the integral is finite.

If $a_{j+1} \ge g(j) > 0$ for all $j \ge n$, we also obtain a lower bound for the error, namely

$$R_n = \sum_{j=n+1}^{\infty} a_j > \int_n^{\infty} g(x)\,dx.$$

Proof. See Figure 3.1.1. □

Example 3.1.2.

When a_j is slowly decreasing, the two error bounds are typically rather close to each other; hence, they are rather realistic bounds, much larger than the first neglected term a_{n+1}. Let $a_j = 1/(j^3+1)$, $f(x) = x^{-3}$. It follows that

$$0 < R_n \le \int_n^{\infty} x^{-3}\,dx = n^{-2}/2.$$

In addition this bound gives an asymptotically correct estimate of the remainder, as $n \to \infty$, which shows that R_n is here significantly larger than the first neglected term.

For alternating series the situation is typically quite different.

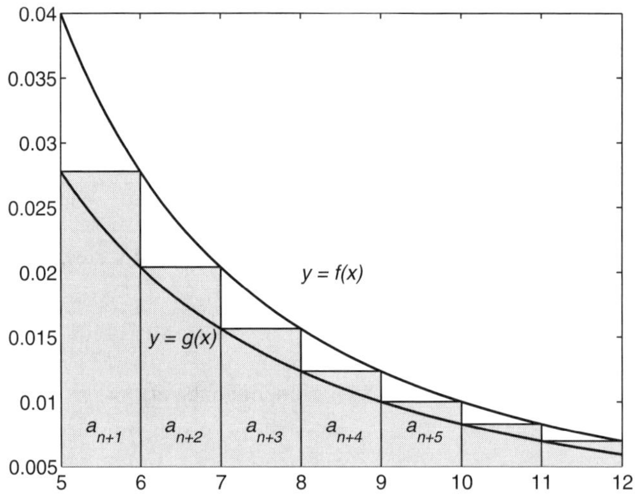

Figure 3.1.1. *Comparison of a series with an integral,* $(n = 5)$.

Definition 3.1.3.

A series is **alternating** *for* $j \geq n$ *if, for all* $j \geq n$, a_j *and* a_{j+1} *have opposite signs or, equivalently,* $\operatorname{sign} a_j \operatorname{sign} a_{j+1} \leq 0$, *where* $\operatorname{sign} x$ *(read "signum" of* x*) is defined by*

$$\operatorname{sign}\ x = \begin{cases} +1 & \text{if } x > 0, \\ 0 & \text{if } x = 0, \\ -1 & \text{if } x < 0. \end{cases}$$

Theorem 3.1.4.

If R_n *and* R_{n+1} *have opposite signs, then* S *lies between* S_n *and* S_{n+1}. *Furthermore*

$$S = \frac{1}{2}(S_n + S_{n+1}) \pm \frac{1}{2}|a_{n+1}|.$$

We also have the weaker results:

$$|R_n| \leq |a_{n+1}|, \qquad |R_{n+1}| \leq |a_{n+1}|, \qquad \operatorname{sign} R_n = \operatorname{sign} a_{n+1}.$$

This theorem has nontrivial applications to practically important divergent sequences; see Sec. 3.2.6.

Proof. The fact that R_{n+1} and R_n have opposite signs means, quite simply, that one of S_{n+1} and S_n is too large and the other is too small, i.e., S lies between S_{n+1} and S_n. Since $a_{n+1} = S_{n+1} - S_n$, one has for positive values of a_{n+1} the situation shown in Figure 3.1.2. From this figure, and an analogous one for the case of $a_{n+1} < 0$, the remaining assertions of the theorem clearly follow. □

The actual error of the average $\frac{1}{2}(S_n + S_{n+1})$ is, for slowly convergent alternating series, usually much smaller than the error bound $\frac{1}{2}|a_{n+1}|$. For example, if $S_n = 1 - \frac{1}{2} +$

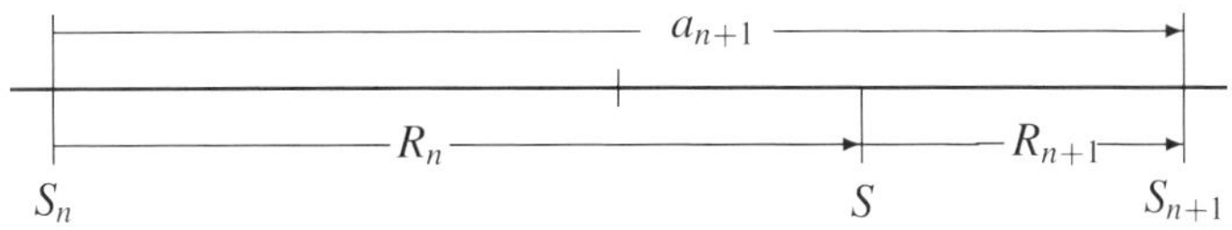

Figure 3.1.2. *A series where R_n and R_{n+1} have different signs.*

$\frac{1}{3} - \cdots \pm \frac{1}{n}$, $\lim S_n = \ln 2 \approx 0.6931$, the error bound for $n = 4$ is 0.1, while the actual error is less than 0.01. A systematic exploration of this observation, by means of repeated averaging, is carried out in Sec. 3.4.3.

Theorem 3.1.5.

For an alternating series, the absolute values of whose terms approach zero monotonically, the remainder has the same sign as the first neglected term a_{n+1}, and the absolute value of the remainder does not exceed $|a_{n+1}|$. (It is well known that such a series is convergent.)

Proof. Sketch: That the theorem is true is almost clear from Figure 3.1.3. The figure shows how S_j depends on j when the premises of the theorem are fulfilled. A formal proof is left to the reader. □

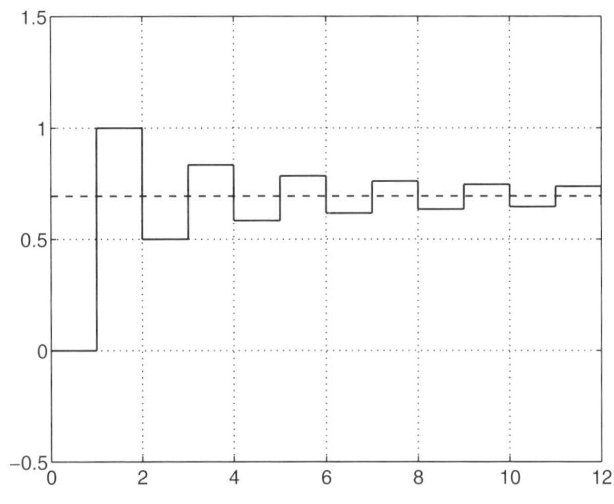

Figure 3.1.3. *Successive sums of an alternating series.*

The use of this theorem will be illustrated in Example 3.1.3. An important generalization is given as Problem 3.3.2 (g).

In the preceding theorems the ideas of well-known convergence criteria are extended to bounds or estimates of the error of a truncated expansion. In Sec. 3.4, we shall see a further extension of these ideas, namely for *improving the accuracy* obtained from a sequence of truncated expansions. This is known as *convergence acceleration.*

3.1.2 Taylor's Formula and Power Series

Consider an expansion into powers of a complex variable z, and suppose that it is convergent for some $z \neq 0$, and denote its sum by $f(z)$,

$$f(z) = \sum_{j=0}^{\infty} a_j z^j, \quad z \in \mathbf{C}. \tag{3.1.1}$$

It is then known from complex analysis that the series (3.1.1) either converges for all z, or it has a **circle of convergence** with radius ρ such that it converges for all $|z| < \rho$, and diverges for $|z| > \rho$. (For $|z| = \rho$ convergence or divergence is possible.) The radius of convergence is determined by the relation

$$\rho = \limsup |a_n|^{-1/n}. \tag{3.1.2}$$

Another formula is $\rho = \lim |a_n|/|a_{n+1}|$, *if this limit exists.*

The function $f(z)$ can be expanded into powers of $z - a$ around any point of analyticity,

$$f(z) = \sum_{j=0}^{\infty} a_j (z-a)^j, \quad z \in \mathbf{C}. \tag{3.1.3}$$

By **Taylor's formula** the coefficients are given by

$$a_0 = f(a), \qquad a_j = f^{(j)}(a)/j!, \quad j \geq 1. \tag{3.1.4}$$

In the general case this infinite series is called a Taylor series; in the special case, $a = 0$, it is by tradition called a **Maclaurin series**.[43]

The function $f(z)$ is analytic inside its circle of convergence and has at least one singular point on its boundary. The singularity of f, which is closest to the origin, can often be found easily from the expression that defines $f(z)$; thus the radius of convergence of a Maclaurin series can often be easily found.

Note that these Taylor coefficients are *uniquely determined* for the function f. This is true also for a nonanalytic function, for example, if $f \in C^p[a, b]$, although in this case the coefficient a_j exists only for $j \leq p$. In Figure 3.1.4 the partial sums of the Maclaurin expansions for the functions $f(x) = \cos x$ and $f(x) = 1/(1 + x^2)$ are shown. The series for $\cos x$ converges for all x, but rounding errors cause trouble for large values of x; see Sec. 3.2.5. For $1/(1 + x^2)$ the radius of convergence is 1.

There are several expressions for the remainder $R_n(z)$ when the expansion for $f(z)$ is truncated after the term that contains z^{n-1}. In order to simplify the notation, *we put $a = 0$ and consider the Maclaurin series.* The following *integral form* can be obtained by the application of repeated integration by parts to the integral $z \int_0^1 f'(zt)\,dt$:

$$R_n(z) = z^n \int_0^1 \frac{(1-t)^{n-1}}{(n-1)!} f^{(n)}(zt)\,dt; \tag{3.1.5}$$

[43]Brook Taylor (1685–1731), who announced his theorem in 1712, and Colin Maclaurin (1698–1746), were British mathematicians.

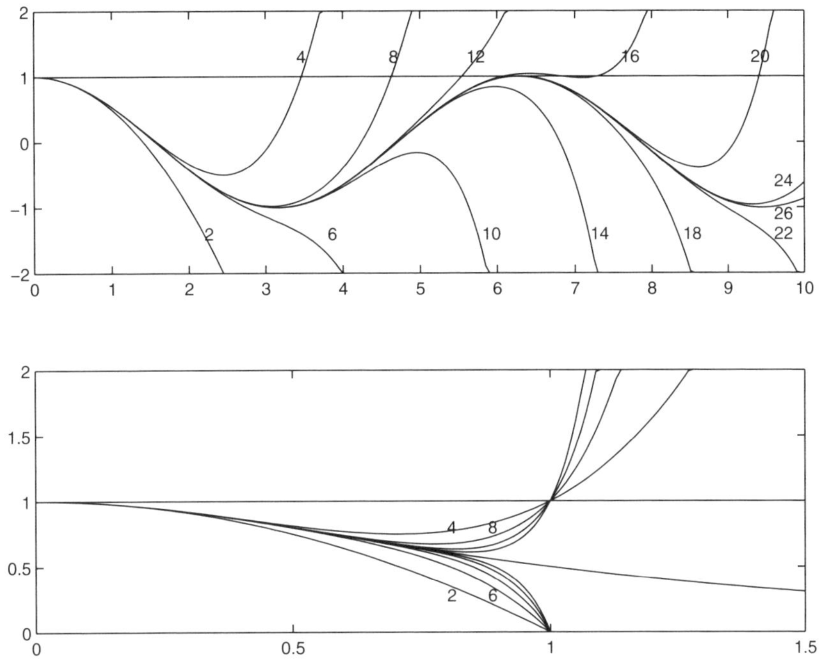

Figure 3.1.4. *Partial sums of the Maclaurin expansions for two functions. The upper curves are for* $\cos x$, $n = 0 : 2 : 26$, $0 \le x \le 10$. *The lower curves are for* $1/(1+x^2)$, $n = 0 : 2 : 18$, $0 \le x \le 1.5$.

the details are left for Problem 3.2.10 (b). From this follows the upper bound

$$|R_n(z)| \le \frac{1}{n!}|z|^n \max_{0 \le t \le 1} |f^{(n)}(zt)|. \tag{3.1.6}$$

This holds also in the complex case: if f is analytic on the segment from 0 to z, one integrates along this segment, i.e., for $0 \le t \le 1$; otherwise another path is to be chosen. The remainder formulas (3.1.5), (3.1.6) require only that $f \in C^n$. It is thus not necessary that the infinite expansion converges or even exists.

For a real-valued function, Lagrange's[44] formula for the remainder term

$$R_n(x) = \frac{f^{(n)}(\xi)x^n}{n!}, \quad \xi \in [0, x], \tag{3.1.7}$$

is obtained by the mean value theorem of integral calculus. For complex-valued functions and, more generally, for vector-valued functions, the mean value theorem and Lagrange's remainder term are not valid with a single ξ. (Sometimes componentwise application with different ξ is possible.) A different form (3.2.11) for the remainder, valid in the complex

[44]Joseph Louis Lagrange (1736–1813) was born in Turin, Italy. When Euler returned from Berlin to St. Petersburg in 1766, Lagrange accepted a position in the Berlin Academy. He stayed there until 1787, when he moved to Paris, where he remained until his death. Lagrange made fundamental contributions to most branches of mathematics and mechanics.

plane, is given in Sec. 3.2.2 in terms of the **maximum modulus** $M(r) = \max_{|z|=r} |f(z)|$, which may sometimes be easier to estimate than the nth derivative. A power series is uniformly convergent in any closed bounded region strictly inside its circle of convergence. Roughly speaking, the series can be manipulated like a polynomial, as long as z belongs to such a region:

- It can be integrated or differentiated term by term.
- Substitutions can be performed, and terms can be rearranged.

A power series can also be multiplied by another power series, as shown in the next theorem.

Theorem 3.1.6 (*Cauchy Product*).
If $f(z) = \sum_{i=0}^{\infty} a_i z^i$ *and* $g(z) = \sum_{j=0}^{\infty} b_j z^j$, *then* $f(z)g(z) = \sum_{n=0}^{\infty} c_n z^n$, *where*

$$c_n = a_0 b_n + a_1 b_{n-1} + \cdots + a_n b_0 = \sum_{i=0}^{n} a_i b_{n-i}. \tag{3.1.8}$$

The expression on the right side of (3.1.8) *is called the* **convolution** *or the* **Cauchy product** *of the coefficient sequences of* f *and* g.

Example 3.1.3.

Many important functions in applied mathematics cannot be expressed in finite terms of elementary functions and must be approximated by numerical methods. One such function is the **error function** defined by

$$\operatorname{erf}(x) = \frac{2}{\sqrt{\pi}} \int_0^x e^{-t^2}\, dt. \tag{3.1.9}$$

This function is encountered, for example, in computing the distribution function of a normal deviate. It takes the values $\operatorname{erf}(0) = 0$, $\operatorname{erf}(\infty) = 1$ and is related to the incomplete gamma functions (see the Handbook [1, Sec. 6.5]) by $\operatorname{erf}(x) = \gamma(1/2, x^2)$.

Suppose one wishes to compute $\operatorname{erf}(x)$ for $x \in [-1, 1]$ with a relative error less than 10^{-10}. One can then approximate the function by a power series. Setting $z = -t^2$ in the well-known Maclaurin series for e^z, truncating after $n+1$ terms, and integrating term by term we obtain

$$\operatorname{erf}(x) \approx \frac{2}{\sqrt{\pi}} \int_0^x \sum_{j=0}^{n} (-1)^j \frac{t^{2j}}{j!}\, dt = \frac{2}{\sqrt{\pi}} \sum_{j=0}^{n} a_j x^{2j+1} =: p_{2n+1}(x), \tag{3.1.10}$$

where

$$a_0 = 1, \quad a_j = \frac{(-1)^j}{j!(2j+1)}.$$

(Note that $\operatorname{erf}(x)$ is an odd function of x.) This series converges for all x, but is suitable for numerical computations only for values of x which are not too large. To evaluate the series we note that the coefficients a_j satisfy the recurrence relation

$$a_j = -a_{j-1}\frac{(2j-1)}{j(2j+1)}.$$

This recursion shows that for $x \in [0, 1]$ the absolute values of the terms $t_j = a_j x^{2j+1}$ decrease monotonically. By Theorem 3.1.5 this implies that the absolute error in a partial sum is bounded by the absolute value of the first neglected term $a_n x^n$.

A possible algorithm for evaluating the sum in (3.1.10) is as follows: Set $s_0 = t_0 = x$; for $j = 1, 2, \ldots,$ compute

$$t_j = -t_{j-1}\frac{(2j-1)}{j(2j+1)}x^2, \quad s_j = s_{j-1} + t_j \tag{3.1.11}$$

until $|t_j| \le 10^{-10}s_j$. Here we have estimated the error by the last term added in the series. Since we have to compute this term for the error estimate we might as well use it! Note also that in this case, where the number of terms is not fixed in advance, Horner's rule is not suitable for the evaluation. Figure 3.1.5 shows the graph of the relative error in the computed approximation $p_{2n+1}(x)$. At most 12 terms in the series were needed.

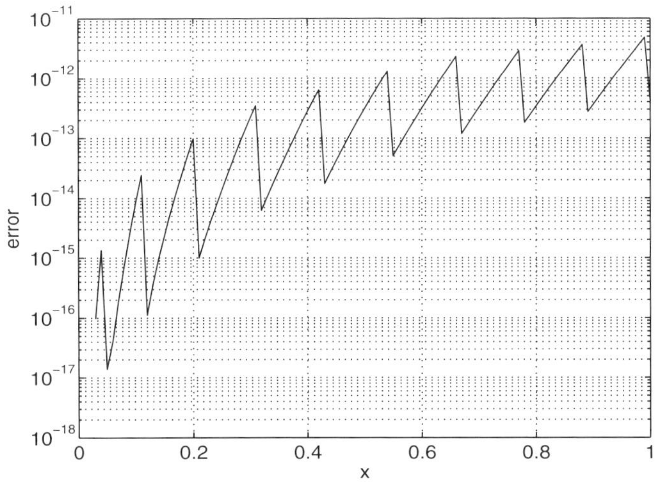

Figure 3.1.5. *Relative error in approximations of the error function by a Maclaurin series truncated after the first term that satisfies the condition in* (3.1.11).

The use of the Taylor coefficient formula and Lagrange's form of the remainder may be inconvenient, and it is often easier to obtain an expansion by manipulating some known expansions. The geometric series

$$\frac{1}{1-z} = 1 + z + z^2 + z^3 + \cdots + z^{n-1} + \frac{z^n}{1-z}, \quad z \ne 1, \tag{3.1.12}$$

is of particular importance; note that the remainder $z^n/(1-z)$ is valid even when the expansion is divergent.

Example 3.1.4.

Set $z = -t^2$ in the geometric series, and integrate:

$$\int_0^x (1+t^2)^{-1}\,dt = \sum_{j=0}^{n-1} \int_0^x (-t^2)^j\,dt + \int_0^x (-t^2)^n (1+t^2)^{-1}\,dt.$$

Using the mean value theorem of integral calculus on the last term we get

$$\arctan x = \sum_{j=0}^{n-1} \frac{(-1)^j x^{2j+1}}{2j+1} + \frac{(1+\xi^2)^{-1}(-1)^n x^{2n+1}}{2n+1} \tag{3.1.13}$$

for some $\xi \in \operatorname{int}[0, x]$. Both the remainder term and the actual derivation are much simpler than what one would get by using Taylor's formula with Lagrange's remainder term. Note also that Theorem 3.1.4 is applicable to the series obtained above for all x and n, even for $|x| > 1$, when the infinite power series is divergent.

Some useful expansions are collected in Table 3.1.1. These formulas will be used often without a reference; the reader is advised to memorize the expansions. "Remainder ratio" denotes the *ratio* of the remainder to the first neglected term, if $x \in \mathbf{R}$; ξ means a number between 0 and x. Otherwise these expansions are valid in the unit circle of $\mathbf{C}$ or in the whole of $\mathbf{C}$.

The binomial coefficients are, also for noninteger k, defined by

$$\binom{k}{n} = \frac{k(k-1)\cdots(k-n+1)}{1\cdot 2\cdots n}.$$

For example, setting $k = 1/2$ gives

$$(1+x)^{1/2} = 1 + \frac{x}{2} - \frac{x^2}{8} + \frac{x^3}{16} - \cdots \quad \text{if } |x| < 1.$$

Depending on the context, the binomial coefficients may be computed by one of the following well-known recurrences:

$$\binom{k}{n+1} = \binom{k}{n}\frac{(k-n)}{(n+1)} \quad \text{or} \quad \binom{k+1}{n} = \binom{k}{n} + \binom{k}{n-1}, \tag{3.1.14}$$

with appropriate initial conditions. The latter recurrence follows from the matching of the coefficients of t^n in the equation $(1+t)^{k+1} = (1+t)(1+t)^k$. (Compare the Pascal triangle; see Problem 1.2.4.) The explicit formula $\binom{k}{n} = k!/(n!(k-n)!)$, for integers k, n, is to be avoided if k can become large, because $k!$ has overflow for $k > 170$ even in IEEE double precision arithmetic.

The exponent k in $(1+x)^k$ is not necessarily an integer; it can even be an irrational or a complex number. This function may be defined as $(1+x)^k = e^{k\ln(1+x)}$. Since $\ln(1+x)$

Table 3.1.1. *Maclaurin expansions for some elementary functions.*

Function	Expansion ($x \in \mathbf{C}$)	Remainder ratio ($x \in \mathbf{R}$)
$(1-x)^{-1}$	$1 + x + x^2 + x^3 + \cdots$ if $\lvert x\rvert < 1$	$(1-x)^{-1}$ if $x \neq 1$
$(1+x)^k$	$1 + kx + \binom{k}{2}x^2 + \cdots$ if $\lvert x\rvert < 1$	$(1+\xi)^{k-n}$ if $x > -1$
$\ln(1+x)$	$x - \frac{x^2}{2} + \frac{x^3}{3} - \frac{x^4}{4} + \cdots$ if $\lvert x\rvert < 1$	$(1+\xi)^{-1}$ if $x > -1$
e^x	$1 + x + \frac{x^2}{2!} + \frac{x^3}{3!} + \cdots$ all x	e^{ξ}, all x
$\sin x$	$x - \frac{x^3}{3!} + \frac{x^5}{5!} - \frac{x^7}{7!} + \cdots$ all x	$\cos\xi$, all x, n odd
$\cos x$	$1 - \frac{x^2}{2!} + \frac{x^4}{4!} - \frac{x^6}{6!} + \cdots$ all x	$\cos\xi$, all x, n even
$\frac{1}{2}\ln\left(\frac{1+x}{1-x}\right)$	$x + \frac{x^3}{3} + \frac{x^5}{5} + \cdots$ if $\lvert x\rvert < 1$	$\frac{1}{1-\xi^2}$, $\lvert x\rvert < 1$, n even
$\arctan x$	$x - \frac{x^3}{3} + \frac{x^5}{5} + \cdots$ if $\lvert x\rvert < 1$	$\frac{1}{1+\xi^2}$, all x

is **multivalued**, $(1+x)^k$ is multivalued too, unless k is an integer. We can, however, make them single-valued by forbidding the complex variable x to take real values less than -1. In other words, we make a **cut** along the real axis from -1 to $-\infty$ that the complex variable must not cross. (The cut is outside the circle of convergence.) We obtain the **principal branch** by requiring that $\ln(1+x) > 0$ if $x > 0$. Let $1 + x = re^{i\phi}$, $r > 0$, $\phi \to \pm\pi$. Note that

$$1 + x \to -r, \qquad \ln(1+x) \to \ln r + \begin{cases} +i\pi & \text{if } \phi \to \pi, \\ -i\pi & \text{if } \phi \to -\pi. \end{cases} \tag{3.1.15}$$

Two important power series, not given in Table 3.1.1, are the following.

The Gauss hypergeometric function[45]

$$\begin{aligned} F(a,b,c;z) = 1 &+ \frac{ab}{c}\frac{z}{1!} + \frac{a(a+1)b(b+1)}{c(c+1)}\frac{z^2}{2!} \\ &+ \frac{a(a+1)(a+2)b(b+1)(b+2)}{c(c+1)(c+2)}\frac{z^3}{3!} + \dots, \end{aligned} \tag{3.1.16}$$

where a and b are complex constants and $c \neq -1, -2, -3, \dots$. The radius of convergence for this series equals unity; see [1, Chap. 15].

[45]Gauss presented his paper on this series in 1812.

Kummer's confluent hypergeometric function[46]

$$M(a, b; z) = 1 + \frac{a}{b}\frac{z}{1!} + \frac{a(a+1)}{b(b+1)}\frac{z^2}{2!} + \frac{a(a+1)(a+2)}{b(b+1)(b+2)}\frac{z^3}{3!} + \dots, \tag{3.1.17}$$

converges for all z (see [1, Chap. 13]). It is named "confluent" because

$$M(a, c; z) = \lim_{b\to\infty} F(a, b, c, z/b).$$

The coefficients of these series are easily computed and the functions are easily evaluated by recurrence relations. (You also need some criterion for the truncation of the series, adapted to your demands of accuracy.) In Sec. 3.5, these functions are also expressed in terms of infinite *continued fractions* that typically converge faster and in larger regions than the power series do.

Example 3.1.5.

The following procedure can generally be used in order to find the *expansion of the quotient of two expansions*. We illustrate it in a case where the result is of interest to us later.

The **Bernoulli**[47] **numbers** B_n are defined by the Maclaurin series

$$\frac{x}{e^x - 1} \equiv \sum_{j=0}^{\infty} \frac{B_j x^j}{j!}. \tag{3.1.18}$$

For $x = 0$ the left-hand side is defined by l'Hôpital's rule; the value is 1. If we multiply this equation by the denominator, we obtain

$$x \equiv \left(\sum_{i=1}^{\infty} \frac{x^i}{i!}\right)\left(\sum_{j=0}^{\infty} \frac{B_j x^j}{j!}\right).$$

By matching the coefficients of x^n, $n \geq 1$, on both sides, we obtain a recurrence relation for the Bernoulli numbers, which can be written in the form

$$B_0 = 1, \quad \sum_{j=0}^{n-1} \frac{1}{(n-j)!}\frac{B_j}{j!} = 0, \quad n \geq 2, \quad \text{i.e.,} \sum_{j=0}^{n-1} \binom{n}{j} B_j = 0. \tag{3.1.19}$$

The last equation is a recurrence that determines B_{n-1} in terms of Bernoulli numbers with smaller subscripts; hence $B_0 = 1$, $B_1 = -\frac{1}{2}$, $B_2 = \frac{1}{6}$, $B_3 = 0$, $B_4 = -\frac{1}{30}$, $B_5 = 0$, $B_6 = \frac{1}{42}, \dots$.

[46]Ernst Eduard Kummer (1810–1893), a German mathematician, was a professor in Berlin from 1855 to his death. He extended Gauss's work on hypergeometric series. Together with Weierstrass and Kronecker, he made Berlin into one of the leading centers of mathematics at that time.

[47]Jacob (or James) Bernoulli (1654–1705), a Swiss mathematician, was one of the earliest to realize the power of infinitesimal calculus. The Bernoulli numbers were published posthumously in 1713, in his fundamental work Ars Conjectandi (on probability). The notation for Bernoulli numbers varies in the literature. Our notation seems to be the most common in modern texts. Several members of the Bernoulli family enriched mathematics by their teaching and writing. Their role in the history of mathematics resembles the role of the Bach family in the history of music.

We see that the Bernoulli numbers are rational. We shall now demonstrate that $B_n = 0$, *when n is odd, except for n = 1*:

$$\frac{x}{e^x - 1} + \frac{x}{2} = \frac{x}{2}\frac{e^x + 1}{e^x - 1} = \frac{x}{2}\frac{e^{x/2} + e^{-x/2}}{e^{x/2} - e^{-x/2}} = \sum_{n=0}^{\infty} \frac{B_{2n}x^{2n}}{(2n)!}. \tag{3.1.20}$$

Since the next-to-last term is an even function its Maclaurin expansion contains only even powers of x, and hence the last expansion is also true.

The recurrence obtained for the Bernoulli numbers by the matching of coefficients in the equation

$$(e^{x/2} - e^{-x/2})\left(\sum_{n=0}^{\infty} B_{2n}x^{2n}/(2n)!\right) = \frac{1}{2}x\left(e^{x/2} + e^{-x/2}\right)$$

is not the same as the one we found above. It turns out to have better properties of numerical stability. We shall look into this experimentally in Problem 3.1.10 (g).

The singularities of the function $x/(e^x - 1)$ are poles at $x = 2n\pi i$, $n = \pm 1, \pm 2, \pm 3, \ldots$; hence the radius of convergence is 2π. Further properties of Bernoulli numbers and the related Bernoulli polynomials and periodic functions are presented in Sec. 3.4.5, where they occur as coefficients in the important Euler–Maclaurin formula.

If r is large, the following formula is very efficient; the series on its right-hand side then converges rapidly:

$$B_{2r}/(2r)! = (-1)^{r-1}2(2\pi)^{-2r}\left(1 + \sum_{n=2}^{\infty} n^{-2r}\right). \tag{3.1.21}$$

This is a particular case ($t = 0$) of a Fourier series for the Bernoulli functions that we shall encounter in Lemma 3.4.9(c). In fact, you obtain IEEE double precision accuracy for $r > 26$, even if the infinite sum on the right-hand side is totally ignored. Thanks to (3.1.21) we do not need to worry much over the instability of the recurrences. When r is *very* large, however, we must be careful about underflow and overflow.

The **Euler numbers** E_n, which will be used later, are similarly defined by the generating function

$$\frac{1}{\cosh z} \equiv \sum_{n=0}^{\infty} \frac{E_n z^n}{n!}, \qquad |z| < \frac{\pi}{2}. \tag{3.1.22}$$

Obviously $E_n = 0$ for all odd n. It can be shown that the Euler numbers are integers, $E_0 = 1$, $E_2 = -1$, $E_4 = 5$, $E_6 = -61$; see Problem 3.1.7(c).

Example 3.1.6.

Let $f(x) = (x^3 + 1)^{-\frac{1}{2}}$. Compute $\int_{10}^{\infty} f(x)\,dx$ to nine decimal places, and $f'''(10)$, with at most 1% error. Since x^{-1} is fairly small, we expand in powers of x^{-1}:

$$\begin{aligned} f(x) &= x^{-3/2}(1 + x^{-3})^{-1/2} = x^{-3/2}\left(1 - \frac{1}{2}x^{-3} + \frac{1\cdot 3}{8}x^{-6} - \cdots\right) \\ &= x^{-1.5} - \frac{1}{2}x^{-4.5} + \frac{3}{8}x^{-7.5} - \cdots. \end{aligned}$$

By integration,

$$\int_{10}^{\infty} f(x)\,dx = 2\cdot 10^{-0.5} - \frac{1}{7}10^{-3.5} + \frac{3}{52}10^{-6.5} + \cdots = 0.632410375.$$

Each term is less than 0.001 of the previous term.

By differentiating the series three times, we similarly obtain

$$f'''(x) = -\frac{105}{8}x^{-4.5} + \frac{1\,287}{16}x^{-7.5} + \cdots.$$

For $x = 10$ the second term is less than 1% of the first; the terms after the second decrease quickly and are negligible. One can show that the magnitude of each term is less than $8\,x^{-3}$ of the previous term. We get $f'''(10) = -4.12\cdot 10^{-4}$ to the desired accuracy. The reader is advised to carry through the calculation in more detail.

Example 3.1.7.

One wishes to compute the exponential function e^x with full accuracy in IEEE double precision arithmetic (unit roundoff $u = 2^{-53} \approx 1.1\cdot 10^{-16}$). The method of **scaling and squaring** is based on the following idea. If we let $m \geq 1$ be an integer and set $y = x/2^m$, then

$$e^x = (e^y)^{2^m}.$$

Here the right-hand side can be computed by squaring e^y m times. By choosing m large enough, e^y can be computed by a truncated Taylor expansion with k terms; see Sec. 3.1.2.

The integers m and k should be chosen so that the bound

$$\frac{1}{k!}y^k \leq \frac{1}{k!}\left(\frac{\log 2}{2^m}\right)^k$$

for the truncation error, multiplied by 2^m to take into account the propagation of error due to squaring e^{x^*}, is bounded by the unit roundoff u. Subject to this constraint, m and k are determined to minimize the computing time. If the Taylor expansion is evaluated by Horner's rule this is approximately proportional to $(m + 2k)$. In IEEE double precision arithmetic with $u = 2^{-53}$ we find that $(k, m) = (7, 7)$ and $(8, 5)$ are good choices. Note that to keep the rounding error sufficiently small, part of the computations must be done in extended precision.

We remark that rational approximations often give much better accuracy than polynomial approximations. This is related to the fact that continued fraction expansions often converge much faster than those based on power series; see Sec. 3.5.3, where Padé approximations for the exponential function are given.

In numerical computation a series should be regarded as a finite expansion together with a remainder. Taylor's formula with the remainder (3.1.5) is valid for any function $f \in C^n[a, a+x]$, but *the infinite series is valid only if the function is analytic in a complex neighborhood of a.*

If a function is not analytic at zero, it can happen that the Maclaurin expansion converges to a wrong result. A classical example (see the Appendix to Chapter 6 in Courant [82]) is

$$f(x) = \begin{cases} e^{-1/x^2} & \text{if } x \neq 0, \\ 0 & \text{if } x = 0. \end{cases}$$

It can be shown that all its Maclaurin coefficients are zero. This trivial Maclaurin expansion converges for all x, *but the sum is wrong for* $x \neq 0$. There is nothing wrong with the use of Maclaurin's formula as a finite expansion with a remainder. Although the remainder that in this case equals $f(x)$ itself does not tend to 0 *as* $n \to \infty$ *for a fixed* $x \neq 0$, it tends to 0 faster than any power of x, *as* $x \to 0$, *for any fixed* n. The "expansion" gives, for example, an *absolute* error less than 10^{-43} for $x = 0.1$, but the *relative* error is 100%. Also note that this function can be added to any function without changing its Maclaurin expansion.

From the point of view of complex analysis, however, the origin is a singular point for this function. Note that $|f(z)| \to \infty$ as $z \to 0$ along the imaginary axis, and this prevents the application of any theorem that would guarantee that the infinite Maclaurin series represents the function. This trouble does not occur for a truncated Maclaurin expansion around a point, where the function under consideration is analytic. The size of the first nonvanishing neglected term then gives a good hint about the truncation error, when $|z|$ is a small fraction of the radius of convergence.

The above example may sound like a purely theoretical matter of curiosity. We emphasize this *distinction between the convergence and the validity of an infinite expansion* in this text as a background to other expansions of importance in numerical computation, such as the Euler–Maclaurin expansion in Sec. 3.4.5, which may converge to the wrong result, and in the application to a well-behaved analytic function. On the other hand, we shall see in Sec. 3.2.6 that divergent expansions can sometimes be very useful. The universal recipe *in numerical computation* is to consider an infinite series as a finite expansion plus a remainder term. But a more algebraic point of view on a series is often useful *in the design of a numerical method*; see Sec. 3.1.5 (Formal Power Series) and Sec. 3.3.2 (The Calculus of Operators). Convergence of an expansion is neither necessary nor sufficient for its success in practical computation.

3.1.3 Analytic Continuation

Analytic functions have many important properties that you may find in any text on complex analysis. A good summary for the purpose of numerical mathematics is found in the first chapter of Stenger [332]. Two important properties are contained in the following lemma.

We remark that the region of analyticity of a function $f(z)$ is an *open* set. If we say that $f(z)$ is analytic on a closed real interval, it means that there exists an open set in $\mathbf{C}$ that contains this interval, where $f(z)$ is analytic.

Lemma 3.1.7.

An analytic function can only have a finite number of zeros in a compact subset of the region of analyticity, unless the function is identically zero.

Suppose that two functions f_1 *and* f_2 *are analytic in regions* D_1 *and* D_2*, respectively. Suppose that* $D_1 \cap D_2$ *contains an interval throughout which* $f_1(z) = f_2(z)$*. Then* $f_1(z) = f_2(z)$ *in the intersection* $D_1 \cap D_2$.

Proof. We refer, for the first part, to any text on complex analysis. Here we closely follow Titchmarsh [351]. The second part follows by the application of the first part to the function $f_1 - f_2$. □

A consequence of this is known as *the permanence of functional equations*. That is, in order to prove the validity of a functional equation (or "a formula for a function") in a region of the complex plane, it may be sufficient to prove its validity in (say) an interval of the real axis, under the conditions specified in the lemma.

Example 3.1.8 (*The Permanence of Functional Equations*).

We know from elementary real analysis that the functional equation

$$e^{(p+q)z} = e^{pz}e^{qz}, \quad (p, q \in \mathbf{R}),$$

holds for all $z \in \mathbf{R}$. We also know that all three functions involved are analytic for all $z \in \mathbf{C}$. Set $D_1 = D_2 = \mathbf{C}$ in the lemma, and let "the interval" be any compact interval of $\mathbf{R}$. The lemma then tells us that the displayed equation holds for all complex z.

The right- and left-hand sides then have identical power series. Applying the convolution formula and matching the coefficients of z^n, we obtain

$$\frac{(p+q)^n}{n!} = \sum_{j=0}^{n} \frac{p^j}{j!} \frac{q^{n-j}}{(n-j)!}, \quad \text{i.e.,} \quad (p+q)^n = \sum_{j=0}^{n} \frac{n!}{j!(n-j)!} p^j q^{n-j}.$$

This is not a very sensational result. It is more interesting to start from the following functional equation:

$$(1+z)^{p+q} = (1+z)^p(1+z)^q.$$

The same argumentation holds, except that—by the discussion around Table 3.1.1—D_1, D_2 should be equal to the complex plane with a cut from -1 to $-\infty$, and that the Maclaurin series is convergent in the unit disk only. We obtain the equations

$$\binom{p+q}{n} = \sum_{j=0}^{n} \binom{p}{j}\binom{q}{n-j}, \quad n = 0, 1, 2, \ldots. \tag{3.1.23}$$

(They can also be proved by induction, but it is not needed.) This sequence of algebraic identities, where *each identity contains a finite number of terms*, is equivalent to the above functional equation.

We shall see that this observation is useful for motivating certain "*symbolic computations*" with power series, which can provide elegant derivations of useful formulas in numerical mathematics.

Now we may consider the aggregate of values of $f_1(z)$ and $f_2(z)$ at points interior to D_1 or D_2 as a single analytic function f. Thus f is analytic in the union $D_1 \cup D_2$, and $f(z) = f_1(z)$ in D_1, $f(z) = f_2(z)$ in D_2.

The function f_2 may be considered as extending the domain in which f_1 is defined, and it is called a (single-valued) **analytic continuation** of f_1. In the same way, f_1 is an analytic continuation of f_2. Analytic continuation denotes both this process of extending the definition of a given function and the result of the process. We shall see examples of this, e.g., in Sec. 3.1.4. Under certain conditions the analytic continuation is unique.

Theorem 3.1.8.

Suppose that a region D is overlapped by regions D_1, D_2, and that $(D_1 \cap D_2) \cap D$ contains an interval. Let f be analytic in D, let f_1 be an analytic continuation of f to D_1, and let f_2 be an analytic continuation of f to D_2 so that

$$f(z) = f_1(z) = f_2(z) \quad \text{in} \quad (D_1 \cap D_2) \cap D.$$

Then either of these functions provides a single-valued analytic continuation of f to $D_1 \cap D_2$. The results of the two processes are the same.

Proof. Since $f_1 - f_2$ is analytic in $D_1 \cap D_2$, and $f_1 - f_2 = 0$ in the set $(D_1 \cap D_2) \cap D$, which contains an interval, it follows from Lemma 3.1.7 that $f_1(z) = f_2(z)$ in $D_1 \cap D_2$, which proves the theorem. □

If the set $(D_1 \cap D_2) \cap D$ is *void*, the conclusion in the theorem *may not be valid*. We may still consider the aggregate of values as a single analytic function, but *this function can be multivalued in $D_1 \cap D_2$*.

Example 3.1.9.

For $|x| < 1$ the important formula

$$\arctan x = \frac{1}{2i} \ln\left(\frac{1+ix}{1-ix}\right)$$

easily follows from the expansions in Table 3.1.1. The function $\arctan x$ has an analytic continuation as single-valued functions in the complex plane with cuts along the imaginary axis from i to ∞ and from $-i$ to $-\infty$. It follows from the theorem that "the important formula" is valid in this set.

3.1.4 Manipulating Power Series

In some contexts, algebraic recurrence relations can be used for the computation of the coefficients in Maclaurin expansions, particularly if only a moderate number of coefficients are wanted. We shall study a few examples.

Example 3.1.10 (*Expansion of a Composite Function*).

Let $g(x) = b_0 + b_1 x + b_2 x^2 + \cdots$, $f(z) = a_0 + a_1 z + a_2 z^2 + \cdots$ be given functions, analytic at the origin. Find the power series

$$h(x) = f(g(x)) \equiv c_0 + c_1 x + c_2 x^2 + \cdots.$$

In particular, we shall study the case $f(z) = e^z$.

The first idea we may think of is to substitute the expansion $b_0 + b_1 x + b_2 x^2 + \cdots$ for z into the power series for $f(z)$. This is, however, *no good unless* $g(0) = b_0 = 0$, because

$$(g(x))^k = b_0^k + k b_0^{k-1} b_1 x + \cdots$$

gives a contribution to $c_0, c_1, \cdots$ for every k, and thus we cannot successively compute the c_j by *finite* computation.

Now suppose that $b_0 = 0$, $b_1 = 1$, i.e., $g(x) = x + b_2 x^2 + b_3 x^3 + \cdots$. (The assumption that $b_1 = 1$ is not important, but it simplifies the writing.) Then c_j depends only on $b_k, a_k, k \le j$, since $(g(x))^k = x^k + k b_2 x^{k+1} + \cdots$. We obtain

$$h(x) = a_0 + a_1 x + (a_1 b_2 + a_2) x^2 + (a_1 b_3 + 2a_2 b_2 + a_3) x^3 + \cdots,$$

and the coefficients of $h(x)$ come out recursively,

$$c_0 = a_0, \quad c_1 = a_1, \quad c_2 = a_1 b_2 + a_2, \quad c_3 = a_1 b_3 + 2a_2 b_2 + a_3, \ldots.$$

Now consider the case $f(z) = e^z$, i.e., $a_n = 1/n!$. We first see that it is then also easy to handle the case that $b_0 \neq 0$, since

$$e^{g(x)} = e^{b_0} e^{b_1 x + b_2 x^2 + b_3 x^3 + \cdots}.$$

But there exists a more important simplification if $f(z) = e^z$. Note that h satisfies the differential equation $h'(x) = g'(x) h(x)$, $h(0) = e^{b_0}$. Hence

$$\sum_{n=0}^{\infty} (n+1) c_{n+1} x^n \equiv \sum_{j=0}^{\infty} (j+1) b_{j+1} x^j \sum_{k=0}^{\infty} c_k x^k.$$

Set $c_0 = e^{b_0}$, apply the convolution formula (3.1.8), and match the coefficients of x^n on the two sides:

$$(n+1) c_{n+1} = b_1 c_n + 2 b_2 c_{n-1} + \cdots + (n+1) b_{n+1} c_0, \quad (n = 0, 1, 2, \ldots).$$

This recurrence relation is more easily programmed than the general procedure indicated above. Other functions that satisfy appropriate differential equations can be treated similarly; see Problem 3.1.8. More information is found in Knuth [230, Sec. 4.7].

Formulas like these are often used in packages for **symbolic differentiation** and for **automatic** or **algorithmic differentiation**. Expanding a function into a Taylor series is equivalent to finding the sequence of derivatives of the function at a given point. The goal of *symbolic* differentiation is to obtain analytic *expressions* for derivatives of functions given in analytic form. This is handled by computer algebra systems, for example, Maple or Mathematica.

In contrast, the goal of **automatic** or **algorithmic differentiation** is to extend an algorithm (a program) for the computation of the *numerical values* of a few functions to an algorithm that also computes *the numerical values* of a few derivatives of these functions, without truncation errors. A simple example, Horner's rule for computing values and derivatives for a polynomial, was given in Sec. 1.2.1. At the time of this writing, lively and active research is being performed on theory, software development, and applications of automatic differentiation. Typical applications are in the solution of ordinary differential equations by Taylor expansion; see the example in Sec. 1.2.4. Such techniques are also used in optimization for partial derivatives of low order for the computation of Jacobian and Hessian matrices.

Sometimes power series are needed with many terms, although rarely more than, say 30. (The ill-conditioned series are exceptions; see Sec. 3.2.5.) The determination of the coefficients can be achieved by the **Toeplitz matrix method** using floating-point computation and an interactive matrix language. Computational details will be given in Problems 3.1.10–3.1.13 for MATLAB. These problems are also available on the home page of the book, `www.siam.org/books/ot103`. (Systems like Maple and Mathematica that include exact arithmetic and other features are evidently also useful here.) An alternative method, the **Cauchy–FFT method**, will be described in Sec. 3.2.2.

Both methods will be applied later in the book. See in particular Sec. 3.3.4, where they are used for deriving approximation formulas in the form of *expansions in powers of elementary difference or differential operators*. In such applications, the coefficient vector, v (say), is obtained in floating-point arithmetic (usually in a very short time).

Very accurate rational approximations to v, often even the exact values, can be obtained (again in a very short time) by applying the MATLAB function `[Nu,De] = rat(v,Tol)`, which returns two integer vectors so that `abs(Nu./De - v) <= Tol*abs(v)` results, with a few different values of the tolerance. This function is based on a continued fraction algorithm, given in Sec. 3.5.1, for finding the best rational approximation to a real number. This can be used for the *"cleaning"* of numerical results which have, for practical reasons, been computed by floating-point arithmetic, although the exact results are known to be (or strongly believed to be) rather simple rational numbers. The algorithm attempts to remove the "dirt" caused by computational errors. In Sec. 3.5.1 you will also find some comments of importance for the interpretation of the results, for example, for judging whether the rational numbers are exact results or only good approximations.

Let $f(z)$ be a function analytic at $z = 0$ with power series

$$f(z) = \sum_{j=0}^{\infty} a_j z^j.$$

We can associate this power series with an infinite upper triangular **semicirculant matrix**

$$C_f = \begin{pmatrix} a_0 & a_1 & a_2 & a_3 & \dots \\ & a_0 & a_1 & a_2 & \dots \\ & & a_0 & a_1 & \dots \\ & & & a_0 & \dots \\ & & & & \ddots \end{pmatrix}. \tag{3.1.24}$$

This matrix has constant entries along each diagonal in C_f and is therefore also a **Toeplitz matrix**.[48] A truncated power series $f_N(z) = \sum_{j=0}^{N-1} a_j z^j$ is represented by the finite leading

[48]Otto Toeplitz (1881–1940), German mathematician. While in Göttingen 1906–1913, influenced by Hilbert's work on integral equations, he studied summation processes and discovered what is now known as Toeplitz operators.

principal $N \times N$ submatrix of C_f (see Definition A.2.1 in the online Appendix), which can be written as

$$f_N(S_N) = \sum_{j=0}^{N-1} a_j S_N^j, \tag{3.1.25}$$

where S_N is a **shift matrix**. For example, with $N = 4$,

$$f_N(S_N) = \begin{pmatrix} a_0 & a_1 & a_2 & a_3 \\ 0 & a_0 & a_1 & a_2 \\ 0 & 0 & a_0 & a_1 \\ 0 & 0 & 0 & a_0 \end{pmatrix}, \qquad S_N = \begin{pmatrix} 0 & 1 & 0 & 0 \\ 0 & 0 & 1 & 0 \\ 0 & 0 & 0 & 1 \\ 0 & 0 & 0 & 0 \end{pmatrix}.$$

The following properties of S_N explain the term "shift matrix":

$$S_N \begin{pmatrix} x_1 \\ x_2 \\ x_3 \\ x_4 \end{pmatrix} = \begin{pmatrix} x_2 \\ x_3 \\ x_4 \\ 0 \end{pmatrix}, \qquad (x_1, x_2, x_3, x_4)S_N = (0, x_1, x_2, x_3).$$

What do the powers of S_N look like? Note that $S_N^N = 0$, i.e., S_N is a **nilpotent** matrix. This is one of the reasons why the Toeplitz matrix representation is convenient for work with truncated power series, since it follows that

$$f(S_N) = \sum_{j=0}^{\infty} a_j S_N^j = \sum_{j=0}^{N-1} a_j S_N^j = f_N(S_N).$$

It is easily verified that a **product** of upper triangular Toeplitz matrices is of the same type. Also note that the multiplication of such matrices is *commutative*. It is also evident that a **linear combination** of such matrices is of the same type. Further, it holds that

$$(f \cdot g)(S_N) = f(S_N)g(S_N) = f_N(S_N)g_N(S_N),$$
$$(\alpha f + \beta g)(S_N) = \alpha f_N(S_N) + \beta g_N(S_N).$$

Similarly the **quotient** of two upper triangular Toeplitz matrices, say,

$$Q(S_N) = f(S_N) \cdot g(S_N)^{-1}, \tag{3.1.26}$$

is also a matrix of the same type. (A hint for a proof is given in Problem 3.1.10.)[49] Note that $Q(S_N) \cdot g(S_N) = f(S_N)$. (In general, Toeplitz matrices are not nilpotent, and the product of two *nontriangular* Toeplitz matrices is not a Toeplitz matrix; this also holds for the inverse. In this section we shall deal only with upper triangular Toeplitz matrices.)

[49] In the terminology of algebra, the set of upper triangular $N \times N$ Toeplitz matrices, i.e., $\{\sum_{j=0}^{N-1} \alpha_j S_N^j\}$, $\alpha_j \in \mathbf{C}$, is a **commutative integral domain**, i.e., isomorphic with the set of polynomials $\sum_{j=0}^{N-1} \alpha_j x^j$ modulo x^N, where x is an indeterminate.

ALGORITHM 3.1. *Expand Row to Toeplitz Matrix.*

An upper triangular Toeplitz matrix of order N is uniquely determined by its first row r. The following MATLAB function expands this row to a triangular Toeplitz matrix:

```
function T = toep(r,N);
% toep expands the row vector r into an upper triangular
% Toeplitz matrix T; N is an optional argument.
lr= length(r);
if (nargin==1 | lr > N), N = lr; end;
if lr < N,   r = [r, zeros(1,N-lr)]; end;
gs = zeros(N,N);
for i = 1:N
   gs(i,i:N) = r(1:N-i+1);
end
T = gs;
```

CPU time and memory space can be saved by working with the first rows of the Toeplitz matrices instead of with the full triangular matrices. We shall *denote by* $f1$, $g1$, *the row vectors with the first N coefficients of the Maclaurin expansions of* $f(z)$, $g(z)$. They are equal to the first rows of the matrices $f(S_N)$, $g(S_N)$, respectively.

Suppose that $f1$, $g1$ are given. We shall compute the first row of $f(S_N)\cdot g(S_N)$ in a similar notation. Then since

$$e_1^T(f(S_N)\cdot g(S_N)) = (e_1^T f(S_N))\cdot g(S_N),$$

this can be written `f1*toep(g1,N)`. Notice that you never have to multiply two triangular matrices if you work with the first rows only. Thus, only about N^2 flops and (typically) an application of the toep(r, N) algorithm are needed.

With similar notations as above, the computation of the quotient in (3.1.26) can be neatly written in MATLAB as

$$\texttt{q1} = \texttt{f1/toep(g1,N)}.$$

Note that this is the *vector by matrix* division of MATLAB. Although the discussion in Sec. 1.3.2 is concerned with linear systems with a *column* as the unknown (instead of a row), we draw from it the conclusion that only about N^2 scalar flops are needed, instead of the $N^3/6$ needed in the solution of the matrix equation $Q\cdot g(S_N) = f(S_N)$.

A library called `toeplib` is given in Problem 3.1.10(a), which consists of short MATLAB scripts mainly based on Table 3.1.1. In the following problems the series of the library are combined by elementary operations to become interesting examples of the Toeplitz matrix method. The convenience, the accuracy, and the execution time are probably much better than you would expect; even the authors were surprised.

Next we shall study how a **composite function** $h(z) = f(g(z))$ can be expanded in powers of z. *Suppose that* $f(z)$ and $g(z)$ are analytic at $z = 0$, $f(z) = \sum_{j=1}^{\infty} f1(j)z^{j-1}$.

An important assumption is that $g(0) = 0$. Then we can set $g(z) = z\bar{g}(z)$, hence $(g(z))^n = z^n(\bar{g}(z))^n$ and, because $S_N^n = 0$, $n \geq N$, we obtain

$$(g(S_N))^n = S_N^n \cdot (\bar{g}(S_N))^n = 0 \quad \text{if } n \geq N \text{ and } g(0) = 0,$$

$$h(S_N) \equiv f(g(S_N)) = \sum_{j=1}^{N} f1(j)(g(S_N))^{j-1} \quad \text{if } g(0) = 0. \tag{3.1.27}$$

This matrix polynomial can be computed by a matrix version of Horner's rule. The row vector version of this equation is written $h1 = \text{comp}(f1, g1, N)$.

If $g(0) \neq 0$, (3.1.27) still provides an "expansion," but it is **wrong**; see Problem 3.1.12 (c). Suppose that $|g(0)|$ is less than the radius of convergence of the Maclaurin expansion of $f(x)$. Then a correct expansion is obtained by a different decomposition. Set $\tilde{g}(z) = g(z) - g(0)$, $\tilde{f}(x) = f(x + g(0))$. Then $\tilde{f}$, $\tilde{g}$ are analytic at $z = 0$, $\tilde{g}(0) = 0$, and $\tilde{f}(\tilde{g}(z)) = f(g(z)) = h(z)$. Thus, (3.1.27) and its row vector implementations can be used if $\tilde{f}$, $\tilde{g}$ are substituted for f, g.

ALGORITHM 3.2. *Expansion of Composite Function.*

The following MATLAB function uses the Horner recurrence for the expansion of a composite function and evaluates the matrix polynomial $f(g(S_N))$ according to (3.1.27). If $g(0) = 0$, the following MATLAB function evaluates the first row of $h(S_N) = f(g(S_N))$.

```
function h1 = comp(f1,g1,N);
% INPUT: the integer N and the rows f1, g1, with
% the first N Maclaurin coefficients for the analytic
% functions f(z), g(z).
% OUTPUT: The row h1 with the first N Maclaurin coeffi-
% cients for the composite function h(z) = f(g(z)),
% where g1(1) = g(0) = 0.
% Error message if g(0) \ne 0.
if g1(1) ~= 0,
  error('g(0) ~= 0 in a composite function f(g(z))')
end
e1t = zeros(1,N);  e1t(1)=1;
r = f1(N)*e1t;
gs = toep(g1,N);
for j = N-1:-1:1,
  r = r*gs + f1(j)*e1t;
end
h1 = r;
```

Analytic functions of matrices are defined by their Taylor series. For example, the series

$$e^A = I + A + \frac{A^2}{2!} + \frac{A^3}{3!} + \cdots$$

converges elementwise for any matrix A. There exist several algorithms for computing e^A, $\sqrt{A}$, $\log A$, where A is a square matrix. One can form linear combinations, products, quotients, and composite functions of them. For example, a "principal matrix value" of $Y = (I + A)^\alpha$ is obtained by

$$B = \log(I + A), \quad Y = e^{\alpha B}.$$

For a composite matrix function $f(g(A))$, it is *not* necessary that $g(0) = 0$, but it is *important* that $g(z)$ and $f(g(z))$ are analytic when z is an eigenvalue of A. We obtain *truncated power series if* $A = S_N$; note that S_N has a multiple eigenvalue at zero. The coding, and the manual handling in interactive computing, are convenient with matrix functions, but the computer has to perform more operations on full triangular matrices than with the row vector level algorithms described above. Therefore, for *very* long expansions the earlier algorithms are notably faster.

If the given power series, $f(x)$, $g(x)$, ... have *rational coefficients*, then the exact results of a sequence of additions, multiplications, divisions, compositions, differentiations, and integrations will have rational coefficients, because the algorithms are all formed by a finite number of scalar additions, multiplications, and divisions. As mentioned above, very accurate rational approximations, often even the exact values, can be quickly obtained by applying a continued fraction algorithm (presented in Sec. 3.5.1) to the results of a floating-point computation.

If $f(x)$ is an even function, its power series contains only even powers of x. You gain space and time by letting the shift matrix S_N correspond to x^2 (instead of x). Similarly, if $f(x)$ is an odd function, you can instead work with the even function $f(x)/x$, and let S_N correspond to x^2.

Finally, we consider a classical problem of mathematics, known as **power series reversion**. The task is to find the power series for the **inverse function** $x = g(y)$ of the function $y = f(x) = \sum_{j=0}^{\infty} a_j x^j$, in the particular case where $a_0 = 0$, $a_1 = 1$. Note that even if the series for $f(x)$ is finite, the series for $g(y)$ is in general infinite!

The following simple cases of power series reversion are often sufficient and useful in low order computations with paper and pencil.

$$\begin{aligned} & y = x + ax^k + \cdots, \quad (k > 1), \\ \Rightarrow\ & x = y - ax^k - \cdots = y - ay^k - \cdots; \end{aligned} \tag{3.1.28}$$

$$\begin{aligned} & y = f(x) \equiv x + a_2x^2 + a_3x^3 + a_4x^4 + \cdots, \\ \Rightarrow\ & x = g(y) \equiv y - a_2y^2 + (2a_2^2 - a_3)y^3 - (5a_2^3 - 5a_2a_3 + a_4)y^4 + \cdots. \end{aligned} \tag{3.1.29}$$

An application of power series reversion occurs in the derivation of a family of iterative methods of arbitrary high order for solving scalar nonlinear equations; see Sec. 6.2.3.

The radius of convergence depends on the singularities of $g(y)$, which are typically related to the singularities of $f(x)$ and to the zeros of $f'(x)$ (Why?). There are other cases, for example, if $f'(x) \to 0$ as $x \to \infty$, then $\lim f(x)$ may be a singularity of $g(y)$.

Knuth [230, Sec 4.7] presents several algorithms for power series reversion, including a classical algorithm due to Lagrange (1768) that requires $O(N^3)$ operations to compute the first N terms. An algorithm due to Brent and Kung [47] is based on an adaptation to formal power series of Newton's method (1.2.3) for solving a numerical algebraic equation. For power series reversion, the equation to be solved reads

$$f(g(y)) = y, \tag{3.1.30}$$

where the coefficients of g are the unknowns. The number of correct terms is roughly doubled in each iteration, as long as N is not exceeded. In the usual numerical application of Newton's method to a scalar nonlinear equation (see Secs. 1.2 and 6.3) it is the number of significant digits that is (approximately) doubled, so-called quadratic convergence. Brent and Kung's algorithm can be implemented in about $150\,(N \log N)^{3/2}$ scalar flops.

We now develop a convenient Toeplitz matrix implementation of the Brent and Kung algorithm. It requires about $cN^3 \log N$ scalar flops with a moderate value of c. It is thus much inferior to the original algorithm if N is very large. In some interesting interactive applications, however, N rarely exceeds 30. In such cases our implementation is satisfactory, unless (say) hundreds of series are to be reversed. Let

$$y = f(x) = \sum_{j=1}^{\infty} f1(j)x^{j-1},$$

where $f1(1) = f(0) = 0$, $f1(2) = f'(0) = 1$ (with the notation used previously). Power series reversion is to find the power series for the inverse function

$$x = g(y) = \sum_{j=1}^{\infty} g1(j)y^{j-1},$$

where $g1(1) = g(0) = 0$. We work with truncated series with N terms in the Toeplitz matrix representation. The inverse function relationship gives the matrix equation $f(g(S_N)) = S_N$. Because $g(0) = 0$, we have, by (3.1.27),

$$f(g(S_N)) = \sum_{j=1}^{N} f1(j)g(S_N)^{j-1}.$$

Now Horner's rule can be used for computing the polynomial *and its derivative*, the latter being obtained by algorithmic differentiation; see Sec. 1.2.1.

ALGORITHM 3.3. *Power Series Reversion.*

The first row of this matrix equation is treated by Newton's method in the MATLAB function `breku`[50] listed below. The Horner algorithms are adapted to the first row. The notations in the code are almost the same as in the theoretical description, although lower case letters are used, e.g., the matrix $g(S_N)$ is denoted gs, and $fgs1$ is the first row of the matrix $f(g(S_N))$.

[50]The name "breku" comes from Brent and Kung, who were probably the first mathematicians to apply Newton's method to series reversion, although with a different formulation of the equation than ours (no Toeplitz matrices).

The equation reads $fgs1 - s1 = 0$.

```
function g1 = breku(f1,N);
% INPUT: The row vector f1 that represents a (truncated)
% Maclaurin series. N is optional input; by default
% N = length(f1). If length(f1) < N, f1 is extended to
% length N by zeros.
% OUTPUT: The row g1, i.e., the first N terms of the series
% x = g(y), where y = f(x).
% Note that f1(1) = 0, f1(2) = 1; if not, there will
% be an error message.
if ~(f1(1) ~= 0 | f1(2) ~= 1),
   error('wrong f1(1) or f1(2)');
end
lf1 = length(f1);
if (nargin == 1|lf1 > N), N = lf1; end
if lf1 < N, f1 = [f1 zeros(1, N-lf1)]  end
maxiter = floor(log(N)/log(2));
e1t = [1, zeros(1,N-1)];
s1 = [0 1 zeros(1,N-2)];  gs1 = s1;
for iter = 0:maxiter
    gs   = toep(gs1,N);
% Horner's scheme for computing the first rows
% of f(gs) and f'(g(s)):
    fgs1 = f1(N)*e1t;   der1 = zeros(1,N);
    for j  =  N-1:-1:1
        ofgs1 = fgs1;    %ofgs1 means "old" fgs1
        fgs1 = ofgs1*gs + f1(j)*e1t ;
        der1 = ofgs1 + der1*gs ;
    end
    % A Newton iteration for the equation fgs1 - s1 = 0:
    gs1 = gs1 - (fgs1 - s1)/toep(der1,N);
end
g1 = gs1;
```

3.1.5 Formal Power Series

A power series is not only a means for numerical computation; it is also an aid for deriving formulas in numerical mathematics and in other branches of applied mathematics. Then one has another, more algebraic, aspect of power series that we shall briefly introduce. A more rigorous and detailed treatment is found in Henrici [196, Chapter 1], and in the literature quoted there.

The set $\mathcal{P}$ of **formal power series** consists of all expressions of the form

$$\mathbf{P} = a_0 + a_1\mathbf{x} + a_2\mathbf{x}^2 + \cdots,$$

where the coefficients a_j may be real or complex numbers (or elements in some other field), while $\mathbf{x}$ is an algebraic **indeterminate**; $\mathbf{x}$ and its powers can be viewed as place keepers.

The sum of $\mathbf{P}$ and another formal power series, $\mathbf{Q} = b_0 + b_1\mathbf{x} + b_2\mathbf{x}^2 + \cdots$, is *defined* as

$$\mathbf{P} + \mathbf{Q} = (a_0 + b_0) + (a_1 + b_1)\mathbf{x} + (a_2 + b_2)\mathbf{x}^2 + \cdots.$$

Similarly, the *Cauchy product* is *defined* as

$$\mathbf{PQ} = c_0 + c_1\mathbf{x} + c_2\mathbf{x}^2 + \cdots, \qquad c_n = \sum_{j=0}^{n} a_j b_{n-j},$$

where the coefficients are given by the convolution formula (3.1.8). The multiplicative identity element is the series $\mathbf{I} := 1 + 0\mathbf{x} + 0\mathbf{x}^2 + \cdots$. The division of two formal power series is defined by a recurrence, as indicated in Example 3.1.5, if and only if the first coefficient of the denominator is not zero. In algebraic terminology, the set $\mathcal{P}$ together with the operations of addition and multiplication is an **integral domain**.

No real or complex values are assigned to $\mathbf{x}$ and $\mathbf{P}$. Convergence, divergence, and remainder term have no relevance for formal power series. The coefficients of a formal power series may even be such that the series diverges for any nonzero complex value that you substitute for the indeterminate, for example, the series

$$\mathbf{P} = 0!\mathbf{x} - 1!\mathbf{x}^2 + 2!\mathbf{x}^3 - 3!\mathbf{x}^4 + \cdots. \tag{3.1.31}$$

Other operations are defined without surprises, for example, the derivative of $\mathbf{P}$ is *defined* as $\mathbf{P}' = 1a_1 + 2a_2\mathbf{x} + 3a_3\mathbf{x}^2 + \cdots$. The limit process, by which the derivative is defined in calculus, does not exist for formal power series. The usual rules for differentiation are still valid, and as an exercise you may verify that the formal power series defined by (3.1.31) satisfies the formal differential equation $\mathbf{x}^2\mathbf{P}' = \mathbf{x} - \mathbf{P}$.

Formal power series can be used for deriving identities. In most applications in this book difference operators or differential operators are substituted for the indeterminates, and the identities are then used in the derivation of approximation formulas, and for interpolation, numerical differentiation, and integration.

The formal definitions of the Cauchy product, (i.e., convolution) and division are rarely used in practical calculation. It is easier to work with upper triangular $N \times N$ Toeplitz matrices, as in Sec. 3.1.4, where N is any natural number. Algebraic calculations with these matrices are isomorphic with calculations with formal power series modulo $\mathbf{x}^N$.

If you perform operations on matrices $f_M(S), g_M(S), \ldots$, where $M < N$, the results are equal to the principal $M \times M$ submatrices of the results obtained with the matrices $f_N(S), g_N(S), \ldots$. This fact follows directly from the equivalence with power series manipulations. It is related to the fact that in the multiplication of block upper triangular matrices, the diagonal blocks of the product equal the products of the diagonal blocks, and no new off-diagonal blocks enter; see Example A.2.1 in Online Appendix A.

So, we can easily *define the product of two infinite upper triangular matrices*, $C = AB$, by stating that if $i \le j \le n$, then c_{ij} has the same value that it has in the $N \times N$ submatrix $C_N = A_N B_N$ for every $N \ge n$. In particular C is upper triangular, and note that there are no conditions on the behavior of the elements a_{ij}, b_{ij} as $i, j \to \infty$. One can show that this product is associative and distributive. For the infinite triangular Toeplitz matrices it is commutative too.[51]

[51] For infinite *nontriangular* matrices the definition of a product generally contains conditions on the behavior of the elements as $i, j \to \infty$, but we shall not discuss this here.

The mapping of formal power series onto the set of infinite semicirculant matrices is *an isomorphism.* (see Henrici [196, Sec. 1.3]). If the formal power series $a_0 + a_1\mathbf{x} + a_2\mathbf{x}^2 + \cdots$ and its reciprocal series, which exists if and only if $a_0 \neq 0$, are represented by the semicirculants A and B, respectively, Henrici proves that $AB = BA = I$, where I is the unit matrix of infinite order. This indicates how to define the inverse of any infinite upper triangular matrix if all diagonal elements $a_{ii} \neq 0$.

If a function f of a complex variable z is analytic at the origin, then *we define*[52] $f(\mathbf{x})$ as the formal power series with the same coefficients as the Maclaurin series for $f(z)$. In the case of a multivalued function we take the principal branch.

There is a kind of "permanence of functional equations" also for the generalization from a function $g(z)$ of a complex variable that is analytic at the origin, to the formal power series $g(\mathbf{x})$. We illustrate a *general principle* on an important special example that we formulate as a lemma, since we shall need it in the next section.

Lemma 3.1.9.

$$(e^{\mathbf{x}})^\theta = e^{\theta\mathbf{x}}, \quad (\theta \in \mathbf{R}). \tag{3.1.32}$$

Proof. Let the coefficient of $\mathbf{x}^j$ in the expansion of the left-hand side be $\phi_j(\theta)$. The corresponding coefficient for the right-hand side is $\theta^j/j!$. If we replace $\mathbf{x}$ by a complex variable z, the power series coefficients are the same and we know that $(e^z)^\theta = e^{\theta z}$, hence $\phi_j(\theta) = \theta^j/j!$, $j = 1, 2, 3 \ldots$, and therefore

$$\sum_0^\infty \phi_j(\theta)\mathbf{x}^j = \sum_0^\infty (\theta^j/j!)\mathbf{x}^j,$$

and the lemma follows. □

Example 3.1.11.

Find (if possible) a formal power series $\mathbf{Q} = 0 + b_1\mathbf{x} + b_2\mathbf{x}^2 + b_3\mathbf{x}^3 + \cdots$ that satisfies

$$e^{-\mathbf{Q}} = 1 - \mathbf{x}, \tag{3.1.33}$$

where $e^{-\mathbf{Q}} = 1 - \mathbf{Q} + \mathbf{Q}^2/2! - \cdots$.

We can, in principle, determine an arbitrarily long sequence $b_1, b_2, b_3, \ldots, b_k$, by matching the coefficients of $\mathbf{x}, \mathbf{x}^2, \mathbf{x}^3, \ldots, \mathbf{x}^k$, in the two sides of the equation. We display the first three equations.

$$\begin{aligned} 1 - (b_1\mathbf{x} + b_2\mathbf{x}^2 + b_3\mathbf{x}^3 + \cdots) + (b_1\mathbf{x} + b_2\mathbf{x}^2 + \cdots)^2/2 - (b_1\mathbf{x} + \cdots)^3/6 + \cdots \\ = 1 - 1\mathbf{x} + 0\mathbf{x}^2 + 0\mathbf{x}^3 + \cdots. \end{aligned}$$

For any natural number k, the matching condition is of the form

$$-b_k + \phi_k(b_{k-1}, b_{k-2}, \ldots, b_1) = 0.$$

[52] Henrici (see reference above) does not use this concept—it may not be established.

This shows that the coefficients are *uniquely* determined.

$$-b_1 = -1 \Rightarrow b_1 = 1,$$
$$-b_2 + b_1^2/2 = 0 \Rightarrow b_2 = 1/2,$$
$$-b_3 + b_1 b_2 - b_1/6 = 0 \Rightarrow b_3 = 1/3.$$

There exists, however, *a much easier way* to determine the coefficients. For the analogous problem with a complex variable z, we know that the solution is unique,

$$q(z) = -\ln(1-z) = \sum_{j=1}^{\infty} z^j/j$$

(the principal branch, where $b_0 = 0$), and hence $\sum_1^\infty \mathbf{x}^j/j$ is the unique formal power series that solves the problem, and we can use the notation $\mathbf{Q} = -\ln(1-\mathbf{x})$ for it.[53]

The theory of formal power series can in a similar way justify many elegant "symbolic" applications of power series for deriving mathematical formulas.

Review Questions

3.1.1 (a) Formulate three general theorems that can be used for estimating the remainder term in numerical series.

(b) What can you say about the remainder term, if the nth term is $O(n^{-k})$, $k > 1$? Suppose in addition that the series is alternating. What further condition should you add, in order to guarantee that the remainder term will be $O(n^{-k})$?

3.1.2 Give, with convergence conditions, the Maclaurin series for $\ln(1+x)$, e^x, $\sin x$, $\cos x$, $(1+x)^k$, $(1-x)^{-1}$, $\ln \frac{1+x}{1-x}$, $\arctan x$.

3.1.3 Describe the main features of a few methods to compute the Maclaurin coefficients of, e.g., $\sqrt{2e^x - 1}$.

3.1.4 Give generating functions of the Bernoulli and the Euler numbers. Describe generally how to derive the coefficients in a quotient of two Maclaurin series.

3.1.5 If a functional equation, for example, $4(\cos x)^3 = \cos 3x + 3\cos x$, is known to be valid for real x, how do you know that it holds also for all complex x? Explain what is meant by the statement that it holds also for formal power series, and why this is true.

3.1.6 (a) Show that multiplying two arbitrary upper triangular matrices of order N uses $\sum_{k=1}^{N} k(N-k) \approx N^3/6$ flops, compared to $\sum_{k=1}^{N} k \approx N^2/2$ for the product of a row vector and an upper triangular matrix.

(b) Show that if $g(x)$ is a power series and $g(0) = 0$, then $g(S_N)^n = 0$, $n \geq N$. Make an operation count for the evaluation of the matrix polynomial $f(g(S_N))$ by the matrix version of Horner's scheme.

[53] The three coefficients b_j computed above agree, of course, with $1/j$, $j = 1:3$.

(c) Consider the product $f(S_N)g(S_N)$, where $f(x)$ and $g(x)$ are two power series. Show, using rules for matrix multiplication, that for any $M < N$ the leading $M \times M$ block of the product matrix equals $f(S_M)g(S_M)$.

3.1.7 Consider a power series $y = f(x) = \sum_{j=0}^{\infty} a_j x^j$, where $a_0 = 0$, $a_1 = 1$. What is meant by reversion of this power series? In the Brent–Kung method the problem of reversion of a power series is formulated as a nonlinear equation. Write this equation for the Toeplitz matrix representation of the series.

3.1.8 Let $\mathbf{P} = a_0 + a_1\mathbf{x} + a_2\mathbf{x}^2 + \cdots$ and $\mathbf{Q} = b_0 + b_1\mathbf{x} + b_2\mathbf{x}^2 + \cdots$ be two formal power series. Define the sum $\mathbf{P} + \mathbf{Q}$ and the Cauchy product $\mathbf{PQ}$.

Problems and Computer Exercises

3.1.1 In how large a neighborhood of $x = 0$ does one get, respectively, four and six correct decimals using the following approximations?

(a) $\sin x \approx x$; (b) $(1 + x^2)^{-1/2} \approx 1 - x^2/2$; (c) $(1 + x^2)^{-1/2}e^{\sqrt{\cos x}} \approx e(1 - \frac{3}{4}x^2)$.

Comment: The truncation error is asymptotically qx^p where you know p.

An alternative to an exact algebraic calculation of q is a numerical estimation of q, by means of the actual error for a suitable value of x—neither too big nor too small (!). (Check the estimate of q for another value of x.)

3.1.2 (a) Let a, b be the lengths of the two smaller sides of a right angle triangle, $b \ll a$. Show that the hypotenuse is approximately $a + b^2/(2a)$ and estimate the error of this approximation. If $a = 100$, how large is b allowed to be, in order that the absolute error should be less than 0.01?

(b) How large a relative error do you commit when you approximate the length of a small circular arc by the length of the chord? How big is the error if the arc is 100 km on a great circle of the Earth? (Approximate the Earth by a ball of radius $40{,}000/(2\pi)$ km.)

(c) How accurate is the formula $\arctan x \approx \pi/2 - 1/x$ for $x \gg 1$?

3.1.3 (a) Compute $10 - (999.999)^{1/3}$ to nine significant digits by the use of the binomial expansion. Compare your result with the result obtained by a computer in IEEE double precision arithmetic, directly from the first expression.

(b) How many terms of the Maclaurin series for $\ln(1 + x)$ would you need in order to compute $\ln 2$ with an error less than 10^{-6} ? How many terms do you need if you use instead the series for $\ln((1 + x)/(1 - x))$, with an appropriate choice of x?

3.1.4 It is well known that $\text{erf}(x) \to 1$ as $x \to \infty$. If $x \gg 1$ the relative accuracy of the complement $1 - \text{erf}(x)$ is of interest. But the series expansion used in Example 3.1.3 for $x \in [0, 1]$ is not suitable for large values of x. Why?

Hint: Derive an approximate expression for the largest term.

3.1.5 Compute by means of appropriate expansions, not necessarily in powers of t, the following integrals to (say) five correct decimals.

(This is for paper, pencil, and a pocket calculator.)

$$\text{(a)} \int_0^{0.1} (1 - 0.1 \sin t)^{1/2}\, dt; \qquad \text{(b)} \int_{10}^{\infty} (t^3 - t)^{-1/2}\, dt.$$

3.1.6 (a) Expand $\arcsin x$ in powers of x by the integration of the expansion of $(1-x^2)^{-1/2}$.

(b) Use the result in (a) to prove the expansion

$$x = \sinh x - \frac{1}{2}\frac{\sinh^3 x}{3} + \frac{1\cdot 3}{2\cdot 4}\frac{\sinh^5 x}{5} - \frac{1\cdot 3\cdot 5}{2\cdot 4\cdot 6}\frac{\sinh^7 x}{7} + \cdots.$$

3.1.7 (a) Consider the power series for

$$(1+x)^{-\alpha}, \quad x > 0, \quad 0 < \alpha < 1.$$

Show that it is equal to the hypergeometric function $F(\alpha, 1, 1, -x)$. Is it true that the expansion is alternating? Is it true that the remainder has the same sign as the first neglected term, for $x > 1$, where the series is divergent? What do the Theorems 3.1.4 and 3.1.5 tell you in the cases $x < 1$ and $x > 1$?

Comment: An application of the divergent case for $\alpha = \frac{1}{2}$ is found in Problem 3.2.9 (c).

(b) Express the coefficients of the power series expansions of $y \cot y$ and $\ln(\sin y/y)$ in terms of the Bernoulli numbers.

Hint: Set $x = 2iy$ into (3.1.20). Differentiate the second function.

(c) Find a recurrence relation for the Euler numbers E_n (3.1.22) and use it for showing that these numbers are integers.

(d) Show that

$$\frac{1}{2}\ln\left(\frac{z+1}{z-1}\right) = \frac{1}{z} + \frac{1}{3z^3} + \frac{1}{5z^5} + \cdots, \quad |z| > 1.$$

Find a recurrence relation for the coefficients of the expansion

$$\left(\ln\left(\frac{z+1}{z-1}\right)\right)^{-1} = \frac{1}{2}z - \mu_1 z^{-1} - \mu_3 z^{-3} - \mu_5 z^{-5} - \cdots, \quad |z| > 1.$$

Compute μ_1, μ_3, μ_5 and determine $\sum_0^\infty \mu_{2j+1}$ by letting $z \downarrow 1$. (Full rigor is not required.)

Hint: Look at Example 3.1.5.

3.1.8 The power series expansion $g(x) = b_1 x + b_2 x^2 + \cdots$ is given. Find recurrence relations for the coefficients of the expansion for $h(x) \equiv f(g(x)) = c_0 + c_1 x + c_2 x^2 + \cdots$ in the following cases:

(a) $h(x) = \ln(1 + g(x))$, $f(x) = \ln(1+x)$.

Hint: Show that $h'(x) = g'(x) - h'(x)g(x)$. Then proceed analogously to Example 3.1.10.

Answer:

$$c_0 = 0, \quad c_n = b_n - \frac{1}{n}\sum_{j=1}^{n-1}(n-j)c_{n-j}b_j.$$

(b) $h(x) = (1+g(x))^k$, $f(x) = (1+x)^k$, $k \in \mathbf{R}$, $k \neq 1$.
Hint: Show that $g(x)h'(x) = kh(x)g'(x) - h'(x)$. Then proceed analogously to Example 3.1.10.
Answer:

$$c_0 = 1, \quad c_n = \frac{1}{n}\sum_{j=1}^{n}\big((k+1)j - n\big)c_{n-j}b_j,$$

$n = 1, 2, \ldots$. The recurrence relation is known as the J. C. P. Miller's formula.

(c) $h_1(y) = \cos g(x)$, $h_2(y) = \sin g(x)$, simultaneously.

Hint: Consider instead $h(y) = e^{ig(x)}$, and separate real and imaginary parts afterward.

3.1.9 (a) If you want $N > 3$ terms in the power series expansion of the function $f(x) = (1+x+x^2)/(1-x+x^2)$, you must augment the expansions for the numerator and denominator by sequences of zeros, so that the order of Toeplitz matrix becomes N. Show experimentally and theoretically that the first row of

$$(I_N + S_N + S_N^2)/(I_N - S_N + S_N^2)$$

is obtained by the statement

```
[1, 1, 1, zeros(1,N-3)]/toep([1, -1, 1, zeros(1,N-3)])
```

(b) Let $f(z) = -z^{-1}\ln(1-z)$. Compute the first six coefficients of the Maclaurin series for the functions $f(z)^k$, $k = 1:5$, in floating-point, and convert them to rational form. (The answer is given in (3.3.22) and an application to numerical differentiation in Example 3.3.6.)
If you choose an appropriate tolerance in the MATLAB function rat you will obtain an accurate rational approximation, but it is not necessarily exact. Try to judge which of the coefficients are exact.

(c) Compute in floating-point the coefficients μ_{2j-1}, $j = 1:11$, defined in Problem 3.1.7 (d), and convert them to rational form.

Hint: First seek an equivalent problem for an expansion in ascending powers.

(d) Prove that $Q = f(S_N)g(S_N)^{-1}$ is an upper triangular Toeplitz matrix.

Hint: Define $Q = toep(q1, N)$, where $q1$ is defined by (3.1.26), and show that each row of the equation $Q \cdot g(S_N) = f(S_N)$ is satisfied.

3.1.10 (a) Study the following library of MATLAB lines for common applications of the Toeplitz matrix method for arbitrary given values of N. All series are truncated to N terms. The shift matrix S_N corresponds to the variable x. You are welcome to add new "cases," e.g., for some of the exercises below.

```
function y = toeplib(cas,N,par);
% cas is a string parameter; par is an optional real
% or complex scalar with default value 1.
%
if nargin == 2, par = 1; end
if cas == 'bin',
  y = [1  cumprod(par:-1:par-N+2)./cumprod(1:N-1)];
% y = 1st row of binomial series (1+x)^par, par in R;
elseif  cas == 'con',
  y = cumprod([1  par*ones(1,N-1)]);
% The array multiplication y.*f1 returns the first
% row of f(par*S_N);
% sum(y.*f1) evaluates f(par). See also Problem~(b).
elseif cas == 'exp',
  y = [1  cumprod(par./[1:(N-1)])];
% y = 1st row of exponential \exp(par*x).
% Since par can be complex, trigonometric functions
% can also be expanded.
elseif cas == 'log',
  y = [0  1./[1:(N-1)]].*cumprod ([-1 -par*ones(1:N-1)]);
% y = 1st row of logarithm \ln(1+par*x).
elseif cas == 'e1t',
  y = [1  zeros(1,N-1)];  % y = e_1^T
elseif cas == 'SN1', y = [0  1  zeros(1,N-2)];
% y = 1st row of S_N.
elseif cas == 'dif', y = [0  1:(N-1)];
% y.*f1 returns xf'(x).
else  cas == 'int', y =1./[1:N];
% y.*f1 returns {1\over x}\int_0^x f(t) dt.
end
```

(b) *Evaluation of $f(x)$* Given `N` and `f1` of your own choice, set

```
fterms = toeplib('con',N,x).*f1.
```

What is `sum(fterms)` and `cumsum(fterms)`? When can `sum(fliplr(fterms))` be useful?

(c) Write a code that, for arbitrary given N, returns the first rows of the Toeplitz matrices for $\cos x$ and $\sin x$, with S_N corresponding to x, and then transforms them to first rows for Toeplitz matrices with S_N corresponding to x^2. Apply this for (say) $N = 36$, to determine the errors of the coefficients of $4(\cos x)^3 - 3\cos x - \cos 3x$.

(d) Find out how a library "`toeplib2`" designed for Toeplitz matrices for *even* functions, where S_N corresponds to x^2, must be different from `toeplib`. For example, how are `cas == 'dif'` and `cas == 'int'` to be changed?

(e) Unfortunately, a `toeplib` "case" has at most one parameter, namely `par`. Write a code that calls `toeplib` twice for finding the Maclaurin coefficients of the three parameter function $y = (a + bx)^\alpha$, $\quad a > 0, b, \alpha$ real. Compute the coefficients in two different ways for $N = 24$, $a = 2$, $b = -1$, $\alpha = \pm 3$, and compare the results for estimating the accuracy of the coefficients.

(f) Compute the Maclaurin expansions for $(1 - x^2)^{-1/2}$ and $\arcsin x$, and for $y = 2\text{arcsinh}\,(x/2)$. Expand also dy/dx and y^2. Convert the coefficients to rational numbers, as long as they seem to be reliable. Save the results, or make it easy to reproduce them, for comparisons with the results of Problem 3.1.12 (a).

Comment: The last three series are fundamental for the expansions of differential operators in powers of central difference operators, which lead to highly accurate formulas for numerical differentiation.

(g) Two power series that generate the Bernoulli numbers are given in Example 3.1.5, namely

$$x \equiv \left(\sum_{i=1}^{\infty} \frac{x^i}{i!} \right) \left(\sum_{j=0}^{\infty} \frac{B_j x^j}{j!} \right), \qquad \frac{x}{2} \frac{e^{x/2} + e^{-x/2}}{e^{x/2} - e^{-x/2}} = \sum_{j=0}^{\infty} \frac{B_{2j} x^{2j}}{(2j)!}.$$

Compute B_{2j} for (say) $j \leq 30$ in floating-point using each of these formulas, and compute the differences in the results, which are influenced by rounding errors. Try to find whether one of the sequences is more accurate than the other by means of the formula in (3.1.21) for (say) $j > 4$. Then convert the results to rational numbers. Use several tolerances in the function `rat` and compare with [1, Table 23.2]. Some of the results are likely to disagree. Why?

(h) The Kummer confluent hypergeometric function $M(a, b, x)$ is defined by the power series (3.1.17). Kummer's first identity,

$$M(a, b, -x) = e^{-x} M(b - a, b, x),$$

is important, for example, because the series on the left-hand side is ill-conditioned if $x \gg 1, a > 0, b > 0$, while the expression on the right-hand side is well-conditioned. Check the identity experimentally by computing the difference between the series on the left-hand side and on the right for a few values of a, b, The computed coefficients are afflicted by rounding errors. Are the differences small enough to convince you of the validity of the formula?

3.1.11 (a) *Matrix functions in* MATLAB. For $h(z) = e^{g(z)}$ it is convenient to use the matrix function `expm(g(SN))` or, on the vector level, `h1 = e1t*expm(g(SN))`, rather than to use `h1 = comp(f1,g1)`. If $f(0) \neq 0$, you can analogously use the functions `logm` and `sqrtm`. They may be slower and less accurate than `h1 = comp(f1,g1)`, but they are typically fast and accurate enough.

Compare computing times and accuracy in the use of `expm(k * logm(eye(N)) + SN)` and `toeplib('bin',N,k)` for a few values of N and k.

Comment: Note that for triangular Toeplitz matrices the diagonal elements are multiple eigenvalues.

(b) Expand $e^{\sin(z)}$ in powers of z in two ways: *first* using the function in Problem 3.1.12 (a); *second* using the matrix functions of MATLAB. Show that the latter can be written

```
HN = expm(imag(expm(i*SN))).
```

Do not be surprised if you find a dirty imaginary part of H_N. Kill it!
Compare the results of the two procedures. If you have done the runs appropriately, the results should agree excellently.

(c) Treat the series $h(z) = \sqrt{(1+e^z)}$ in three different ways and compare the results with respect to validity, accuracy, and computing time.

(i) Set $ha(z) = h(z)$, and determine $f(z)$, $g(z)$, analytic at $z = 0$, so that $g(0) = 0$. Compute `ha1 = comp(f1,g1,N)`. Do you trust the result?

(ii) Set $h(z) = H(z)$. Compute `HN = sqrtm(eye(N) + expm(SN))`.
In the first test, i.e., for $N = 6$, display the matrix H_N and check that H_N is an upper triangular Toeplitz matrix. For larger values of N, display the first row only and compare it to $ha1$. If you have done all this correctly, the agreement should be extremely good, and we can practically conclude that both are very accurate.

(iii) Try the "natural," although "illegal," decomposition $hb(z) = f(g(z))$, with $f(x) = (1+x)^{0.5}$, $g(z) = e^z$. Remove temporarily the error stop. Demonstrate by numerical experiment that $hb1$ is very wrong. If this is a surprise, read Sec. 3.1.4 once more.

3.1.12 (a) Apply the function `breku` for the reversion of power series to the computation of $g(y)$ for $f(x) = \sin x$ and for $f(x) = 2\sinh(x/2)$. Compare with the results of Problem 3.1.10 (f). Then reverse the two computed series $g(y)$, and study how you return to the original expansion of $f(x)$, more or less accurately. Use "tic" and "toc" to take the time for a few values of N.

(b) Compute $g(y)$ for $f(x) = \ln(1+x)$, $f(x) = e^x - 1$, $f(x) = x + x^2$, and $f(x) = x + x^2 + x^3$. If you know an analytic expression for $g(y)$, find the Maclaurin expansion for this, and compare with the expansions obtained from `breku`.

(c) Set $y = f(x)$ and suppose that $y(0) \neq 0$, $y'(0) \neq 0$. Show how the function `breku` can be used for expanding the inverse function in powers of $(y - y(0))/y'(0)$. Construct some good test examples.

(d) For the equation $\sin x - (1-y)x = 0$, express $x^2 = g(y)$ (why x^2?), with $N = 12$. Then express x in the form $x \approx \pm y^{1/2} P(y)$, where $P(y)$ is a truncated power series with (say) 11 terms.

3.1.13 The inverse function $w(y)$ of $y(w) = we^w$ is known as the Lambert W function.[54] The power series expansion for $w(y)$ is

$$w(y) = y + \sum_{n=2}^{\infty} \frac{(-1)^{n-1} n^{n-2}}{(n-1)!} y^n$$
$$= y - y^2 + \frac{3}{2}y^3 - \frac{8}{3}y^4 + \frac{125}{24}y^5 - \frac{54}{5}y^6 + \frac{16807}{720}y^7 - \cdots.$$

Estimate the radius of convergence for $f(x) = xe^x$ approximately by means of the ratios of the coefficients computed in (d), and exactly.

[54] Johann Heinrich Lambert (1728–1777), a German mathematician, physicist, and astronomer, was a colleague of Euler and Lagrange at the Berlin Academy of Sciences. He is best known for his illumination laws and for the continued fraction expansions of elementary functions; see Sec. 3.5.1. His W function was "rediscovered" a few years ago; see [81].

3.2 More about Series

3.2.1 Laurent and Fourier Series

A **Laurent series** is a series of the form

$$\sum_{n=-\infty}^{\infty} c_n z^n. \tag{3.2.1}$$

Its convergence region is the intersection of the convergence regions of the expansions

$$\sum_{n=0}^{\infty} c_n z^n \quad \text{and} \quad \sum_{m=1}^{\infty} c_{-m} z^{-m},$$

the interior of which are determined by conditions of the form $|z| < r_2$ and $|z| > r_1$. The convergence region can be void, for example, if $r_2 < r_1$.

If $0 < r_1 < r_2 < \infty$, then the convergence region is an *annulus*, $r_1 < |z| < r_2$. The series defines an analytic function in the annulus. Conversely, if $f(z)$ is a **single-valued analytic function** in this annulus, it is represented by a Laurent series, which converges uniformly in every closed subdomain of the annulus.

The coefficients are determined by the following formula, due to Cauchy:[55]

$$c_n = \frac{1}{2\pi i} \int_{|z|=r} z^{-n-1} f(z) dz, \quad r_1 < r < r_2, -\infty < n < \infty \tag{3.2.2}$$

and

$$|c_n| \le r^{-n} \max_{|z|=r} |f(z)|. \tag{3.2.3}$$

The extension to the case when $r_2 = \infty$ is obvious; the extension to $r_1 = 0$ depends on whether there are any terms with negative exponents or not. In the extension of *formal* power series to *formal Laurent series*, however, only a finite number of terms with negative indices are allowed to be different from zero; see Henrici [196, Sec. 1.8]. If you substitute z for z^{-1} an infinite number of negative indices is allowed, if the number of positive indices is finite.

Example 3.2.1.

A function may have several Laurent expansions (with different regions of convergence), for example,

$$(z-a)^{-1} = \begin{cases} -\sum_{n=0}^{\infty} a^{-n-1} z^n & \text{if } |z| < |a|, \\ \sum_{m=1}^{\infty} a^{m-1} z^{-m} & \text{if } |z| > |a|. \end{cases}$$

The function $1/(z-1) + 1/(z-2)$ has three Laurent expansions, with validity conditions $|z| < 1$, $1 < |z| < 2$, $2 < |z|$, respectively. The series contains both positive and negative powers of z in the middle case only. The details are left for Problem 3.2.4 (a).

[55]Augustin Cauchy (1789–1857) is the father of modern analysis. He is the creator of complex analysis, in which this formula plays a fundamental role.

Remark 3.2.1. The restriction to *single-valued* analytic functions is important in this subsection. In this book we cannot entirely avoid working with **multivalued** functions such as $\sqrt{z}$, $\ln z$, z^α, (α noninteger). We always work with such a function, however, in some region where one branch of it, determined by some convention, is single-valued. In the examples mentioned, the natural conventions are to require the function to be positive when $z > 1$, and to forbid z to cross the negative real axis. In other words, the complex plane has a **cut** along the negative real axis. The annulus mentioned above is incomplete in these cases; its intersection with the negative real axis is missing, and we cannot use a Laurent expansion.

For a function like $\ln(\frac{z+1}{z-1})$, we can, depending on the context, cut out either the interval $[-1, 1]$ or the complement of this interval with respect to the real axis. We then use an expansion into negative or into positive powers of z, respectively.

If $r_1 < 1 < r_2$, we set $F(t) = f(e^{it})$. Note that $F(t)$ is a periodic function; $F(t + 2\pi) = F(t)$. By (3.2.1) and (3.2.2), the Laurent series then becomes for $z = e^{it}$ a **Fourier series**:

$$F(t) = \sum_{n=-\infty}^{\infty} c_n e^{int}, \qquad c_n = \frac{1}{2\pi}\int_{-\pi}^{\pi} e^{-int} F(t)\,dt. \tag{3.2.4}$$

Note that $c_{-m} = O(r_1^m)$ for $m \to +\infty$, and $c_n = O(r_2^{-n})$ for $n \to +\infty$. *The formulas in* (3.2.4), *however, are valid in much more general situations*, where $c_n \to 0$ much more slowly, and where $F(t)$ cannot be continued to an analytic function $f(z)$, $z = re^{it}$, in an annulus. (Typically, in such a case $r_1 = 1 = r_2$.)

A Fourier series is often written in the following form:

$$F(t) = \frac{1}{2}a_0 + \sum_{k=1}^{\infty}(a_k \cos kt + b_k \sin kt). \tag{3.2.5}$$

Consider $c_k e^{ikt} + c_{-k}e^{-ikt} \equiv a_k \cos kt + b_k \sin kt$. Since $e^{\pm ikt} = \cos kt \pm i \sin kt$, we obtain for $k \geq 0$

$$a_k = c_k + c_{-k} = \frac{1}{\pi}\int_{-\pi}^{\pi} F(t)\cos kt\,dt, \qquad b_k = i(c_k - c_{-k}) = \frac{1}{\pi}\int_{-\pi}^{\pi} F(t)\sin kt\,dt. \tag{3.2.6}$$

Also note that $a_k - ib_k = 2c_k$. If $F(t)$ is real for $t \in \mathbf{R}$, then $c_{-k} = \bar{c}_k$.

We mention without proof the important **Riemann–Lebesgue theorem**,[56, 57] by which the Fourier coefficients c_n *tend to zero as* $n \to \infty$ *for any function that is integrable* (in the sense of Lebesgue), a fortiori for any periodic function that is continuous everywhere. A finite number of finite jumps in each period are also allowed.

A function $F(t)$ is said to be of **bounded variation** in an interval if, in this interval, it can be expressed in the form $F(t) = F_1(t) - F_2(t)$, where F_1 and F_2 are nondecreasing

[56] George Friedrich Bernhard Riemann (1826–1866), a German mathematician, made fundamental contributions to analysis and geometry. In his habilitation lecture 1854 in Göttingen, Riemann introduced the curvature tensor and laid the groundwork for Einstein's general theory of relativity.

[57] Henri Léon Lebesgue (1875–1941), a French mathematician, created path-breaking general concepts of measure and integral.

bounded functions. A finite number of jump discontinuities are allowed. The variation of F over the interval $[a, b]$ is denoted $\int_a^b |dF(t)|$. If F is differentiable the variation of F equals $\int_a^b |F'(t)|\,dt$.

Another classical result in the theory of Fourier series reads as follows: *If $F(t)$ is of bounded variation in the closed interval $[-\pi, \pi]$, then $c_n = O(n^{-1})$*; see Titchmarsh [351, Secs. 13.21, 13.73]. This result can be generalized as the following theorem.

Theorem 3.2.1.

Suppose that $F^{(p)}$ is of bounded variation on $[-\pi, \pi]$, and that $F^{(j)}$ is continuous everywhere for $j < p$. Denote the Fourier coefficients of $F^{(p)}(t)$ by $c_n^{(p)}$. Then

$$c_n = (in)^{-p} c_n^{(p)} = O(n^{-p-1}). \tag{3.2.7}$$

Proof. The theorem follows from the above classical result, after the integration of the formula for c_n in (3.2.2) by parts p times. □

Bounds for the truncation error of a Fourier series can also be obtained from this. The details are left for Problem 3.2.4 (d), together with a further generalization. A similar result is that $c_n = o(n^{-p})$ if $F^{(p)}$ is integrable, hence a fortiori if $F \in C^p$.

In particular, we find for $p = 1$ (since $\sum n^{-2}$ is convergent) that the Fourier series (3.2.2) *converges absolutely and uniformly* in **R**. It can also be shown that *the Fourier series is valid*, i.e., the sum is equal to $F(t)$.

3.2.2 The Cauchy–FFT Method

An alternative method for deriving coefficients of power series when many terms are needed is based on the following classic result. Suppose that the value $f(z)$ of an analytic function can be computed at any point inside and on the circle $C_r = \{z : |z - a| = r\}$, and set

$$M(r) = \max |f(z)|, \quad z = a + re^{i\theta} \in C_r.$$

Then the coefficients of the Taylor expansion around a are determined by Cauchy's formula,

$$a_n = \frac{1}{2\pi i} \int_{C_r} \frac{f(z)}{(z-a)^{(n+1)}}\,dz = \frac{r^{-n}}{2\pi} \int_0^{2\pi} f(a + re^{i\theta}) e^{-ni\theta}\,d\theta. \tag{3.2.8}$$

For a derivation, multiply the Taylor expansion (3.1.3) by $(z - a)^{-n-1}$, integrate term by term over C_r, and note that

$$\frac{1}{2\pi i} \int_{C_r} (z-a)^{j-n-1}\,dz = \frac{1}{2\pi} \int_0^{2\pi} r^{j-n} e^{(j-n)i\theta}\,d\theta = \begin{cases} 1 & \text{if } j = n, \\ 0 & \text{if } j \neq n. \end{cases} \tag{3.2.9}$$

From the definitions and (3.2.8) it follows that

$$|a_n| \le r^{-n} M(r). \tag{3.2.10}$$

Further, with $z' = a + r'e^{i\theta}$, $0 \le r' < r$, we have

$$|R_n(z')| \le \sum_{j=n}^{\infty} |a_j(z'-a)^j| \le \sum_{j=n}^{\infty} r^{-j} M(r)(r')^j = \frac{M(r)(r'/r)^n}{1-r'/r}. \tag{3.2.11}$$

This form of the remainder term of a Taylor series is useful in theoretical studies, and also for practical purpose, if the maximum modulus $M(r)$ is easier to estimate than the nth derivative.

Set $\Delta\theta = 2\pi/N$, and apply the trapezoidal rule (see (1.1.12)) to the second integral in (3.2.8). Note that the integrand has the same value for $\theta = 2\pi$ as for $\theta = 0$. The terms $\frac{1}{2}f_0$ and $\frac{1}{2}f_N$ that appear in the general trapezoidal rule can therefore in this case be replaced by f_0. Then

$$a_n \approx \tilde{a}_n \equiv \frac{1}{Nr^n} \sum_{k=0}^{N-1} f(a + re^{ik\Delta\theta})e^{-ink\Delta\theta}, \quad n = 0 : N-1. \tag{3.2.12}$$

The approximate Taylor coefficients $\tilde{a}_n$, or rather the numbers $a_n^\star = \tilde{a}_n N r^n$, are here expressed as a case of the (direct) **discrete Fourier transform (DFT)**. More generally, this transform maps an *arbitrary* sequence $\{\alpha_k\}_0^{N-1}$ to a sequence $\{a_n^\star\}_0^{N-1}$, by the following equations:

$$a_n^\star = \sum_{k=0}^{N-1} \alpha_k e^{-ink\Delta\theta}, \quad n = 0 : N-1. \tag{3.2.13}$$

It will be studied more systematically in Sec. 4.6.2.

If N is a power of 2, it is shown in Sec. 4.7 that, given the N values α_k, $k = 0 : N-1$, and $e^{-i\Delta\theta}$, *no more than $N \log_2 N$ complex multiplications and additions are needed for the computation of all the N coefficients $a_n^\star$*, if an implementation of the DFT known as the **fast Fourier transform (FFT)** is used. This makes our theoretical considerations very practical.

It is also shown in Sec. 4.7 that the **inverse** of the DFT (3.2.13) is given by the formulas

$$\alpha_k = (1/N) \sum_{n=0}^{N-1} a_n^\star e^{ink\Delta\theta}, \quad k = 0 : N-1. \tag{3.2.14}$$

This looks almost like the direct DFT (3.2.13), except for the sign of i and the factor $1/N$. It can therefore also be performed by means of $O(N \log N)$ elementary operations, instead of the $O(N^3)$ operations that the most obvious approach to this task would require (i.e., by solving the linear system (3.2.13)).

In our context, i.e., the computation of Taylor coefficients, we have, by (3.2.12) and the line after that equation,

$$\alpha_k = f(a + re^{ik\Delta\theta}), \qquad a_n^\star = \tilde{a}_n N r^n. \tag{3.2.15}$$

Set $z_k = a + re^{ik\Delta\theta}$. Using (3.2.15), the inverse transformation then becomes[58]

$$f(z_k) = \sum_{n=0}^{N-1} \tilde{a}_n (z_k - a)^n, \quad k = 0 : N-1. \tag{3.2.16}$$

[58]One interpretation of these equations is that the polynomial $\sum_{n=0}^{N-1} \tilde{a}_n (z-a)^n$ is the solution of a special, although important, interpolation problem for the function f, analytic inside a circle in $\mathbf{C}$.

Since the Taylor coefficients are equal to $f^{(n)}(a)/n!$, this is de facto a method for the accurate *numerical differentiation of an analytic function*.[59] If r and N are chosen appropriately, it is more well-conditioned than most methods for *numerical differentiation*, such as the difference approximations mentioned in Chapter 1; see also Sec. 3.3. It requires, however, complex arithmetic for a convenient implementation. We call this the **Cauchy–FFT method** for Taylor coefficients and differentiation.

The question arises of how to choose N and r. Theoretically, any r less than the radius of convergence ρ would do, but there may be trouble with cancellation if r is small. On the other hand, the truncation error of the numerical integration usually increases with r. *Scylla and Charybdis situations*[60] like this are very common with numerical methods.

Typically it is the rounding error that sets the limit for the accuracy; it is usually not expensive to choose r and N such that the truncation error becomes much smaller. A rule of thumb for this situation is to guess a value of $\hat{N}$, i.e., how many terms will be needed in the expansion, and then to try two values for N (powers of 2) larger than $\hat{N}$. *If ρ is finite* try $r = 0.9\rho$ and $r = 0.8\rho$, and compare the results. They may or may not indicate that some other values of N and r should also be tried. On the other hand, *if $\rho = \infty$* try, for example, $r = 1$ and $r = 3$, and compare the results. Again, the results indicate whether or not more experiments should be done.

One can also combine numerical experimentation with a theoretical analysis of a more or less simplified model, including a few elementary optimization calculations. The authors take the opportunity to exemplify below this type of "hard analysis" on this question.

We first derive two lemmas, which are important also in many other contexts. First we have a discrete analogue of (3.2.9).

Lemma 3.2.2.

Let p and N be integers. Then

$$\sum_{k=0}^{N-1} e^{2\pi i p k/N} = 0,$$

unless $p = 0$ or p is a multiple of N. In these exceptional cases every term equals 1, *and the sum equals N.*

Proof. If p is neither zero nor a multiple of N, the sum is a geometric series, the sum of which is equal to

$$(e^{2\pi i p} - 1)/(e^{2\pi i p/N} - 1) = 0.$$

The rest of the statement is obvious. □

We next show an error estimate for the approximation provided by the trapezoidal rule (3.2.12).

---

[59] The idea of using Cauchy's formula and FFT for numerical differentiation seems to have been first suggested by Lyness and Moler [252]; see Henrici [194, Sec. 3].

[60] According to the *American Heritage Dictionary*, Scylla is a rock on the Italian side of the Strait of Messina, opposite to the whirlpool Charybdis, personified by Homer (*Ulysses*) as a female sea monster who devoured sailors. The problem is to navigate safely between them.

Lemma 3.2.3.

Suppose that $f(z) = \sum_0^\infty a_n(z-a)^n$ *is analytic in the disk* $|z-a| < \rho$. *Let* $\tilde{a}_n$ *be defined by* (3.2.12), *where* $0 < r < \rho$. *Then*

$$\tilde{a}_n - a_n = a_{n+N}\, r^N + a_{n+2N}\, r^{2N} + a_{n+3N}\, r^{3N} + \cdots, \quad 0 \le n < N. \tag{3.2.17}$$

Proof. Since $\Delta\theta = 2\pi/N$,

$$\begin{aligned}\tilde{a}_n &= \frac{1}{Nr^n}\sum_{k=0}^{N-1} e^{-2\pi i nk/N} \sum_{m=0}^{\infty} a_m \left(re^{2\pi i k/N}\right)^m \\ &= \frac{1}{Nr^n}\sum_{m=0}^{\infty} a_m r^m \sum_{k=0}^{N-1} e^{2\pi i(-n+m)k/N}.\end{aligned}$$

By the previous lemma, the inner sum of the last expression is zero, unless $m-n$ is a multiple of N. Hence (recall that $0 \le n < N$),

$$\tilde{a}_n = \frac{1}{Nr^n}\left(a_n r^n N + a_{n+N}\, r^{n+N} N + a_{n+2N}\, r^{n+2N} N + \cdots\right),$$

from which (3.2.17) follows. □

Lemma 3.2.3 can, with some modifications, be generalized to Laurent series (and to complex Fourier series); for example, (3.2.17) becomes

$$\tilde{c}_n - c_n = \cdots c_{n-2N} r^{-2N} + c_{n-N} r^{-N} + c_{n+N} r^N + c_{n+2N} r^{2N} \cdots. \tag{3.2.18}$$

Let $M(r)$ be the maximum modulus for the function $f(z)$ on the circle C_r, and denote by $M(r)U$ an upper bound for the error of a computed function value $f(z)$, $|z| = r$, where $U \ll 1$. Assume that rounding errors during the computation of $\tilde{a}_n$ are of minor importance. Then, by (3.2.12), $M(r)U/r^n$ is a bound for the *rounding error* of $\tilde{a}_n$. (The rounding errors during the computation can be included by a redefinition of U.)

Next we shall consider the *truncation error* of (3.2.12). First we *estimate* the coefficients that occur in (3.2.17) by means of $\max |f(z)|$ on a circle with radius r'; $r' > r$, where r is the radius of the circle used in the *computation* of the first N coefficients. Thus, in (3.2.8) we substitute r', j for r, n, respectively, and obtain the inequality

$$|a_j| \le M(r')(r')^{-j}, \quad 0 < r < r' < \rho.$$

The actual choice of r' strongly depends on the function f. (In rare cases we may choose $r' = \rho$.) Put this inequality into (3.2.17), where we shall choose $r < r' < \rho$. Then

$$\begin{aligned}|\tilde{a}_n - a_n| &\le M(r')\left((r')^{-n-N} r^N + (r')^{-n-2N} r^{2N} + (r')^{-n-3N} r^{3N} + \cdots\right) \\ &= M(r')(r')^{-n}\left((r/r')^N + (r/r')^{2N} + (r/r')^{3N} + \cdots\right) \\ &= \frac{M(r')(r')^{-n}}{(r'/r)^N - 1}.\end{aligned}$$

We make a digression here, because *this is an amazingly good result*. The trapezoidal rule that was used in the calculation of the Taylor coefficients is typically expected to have an error that is $O((\Delta\theta)^2) = O(N^{-2})$. (As before, $\Delta\theta = 2\pi/N$.) This application is, however, *a very special situation: a periodic analytic function is integrated over a full period*. We shall return to results like this in Sec. 5.1.4. In this case, for fixed values of r, r', the truncation error is

$$O\big((r/r')^N\big) = O\big(e^{-\eta/\Delta\theta}\big), \quad \eta > 0, \quad \Delta\theta \to 0+. \tag{3.2.19}$$

This tends to zero faster than any power of $\Delta\theta$!

It follows that a bound for the total error of $\tilde{a}_n$, i.e., the sum of the bounds for the rounding and the truncation errors, is given by

$$UM(r)r^{-n} + \frac{M(r')(r')^{-n}}{(r'/r)^N - 1}, \quad r < r' < \rho. \tag{3.2.20}$$

Example 3.2.2 (*Scylla and Charybdis in the Cauchy–FFT*).

We shall discuss how to choose the parameters r and N so that the *absolute error bound* of a_n, given in (3.2.20) becomes uniformly small for (say) $n = 0 : \hat{n}$. $1 + \hat{n} \gg 1$ is thus the number of Taylor coefficients requested. The parameter r' does not belong to the Cauchy–FFT method, but it has to be chosen well in order to make the *bound* for the truncation error realistic.

The discussion is rather technical, and you may omit it at a first reading. It may, however, be useful to study this example later, because similar technical subproblems occur in many serious discussions of numerical methods that contain parameters that should be appropriately chosen.

First consider the *rounding error*. By the maximum modulus theorem, $M(r)$ is an increasing function; hence, for $r > 1$, $\max_n M(r)r^{-n} = M(r) > M(1)$. On the other hand, for $r \le 1$, $\max_n M(r)r^{-n} = M(r)r^{-\hat{n}}$; $\hat{n}$ was introduced in the beginning of this example. Let r^* be the value of r, for which this maximum is minimal. Note that $r^* = 1$ unless $M'(r)/M(r) = \hat{n}/r$ for some $r \le 1$.

Then try to determine N and $r' \in [r^*, \rho)$ so that, for $r = r^*$, the bound for the second term of (3.2.20) becomes much smaller than the first term, i.e., *the truncation error is made negligible compared to the rounding error. This works well if* $\rho \gg r^*$. *In such cases, we may therefore choose* $r = r^*$, and the total error is then just a little larger than $UM(r^*)(r^*)^{-\hat{n}}$.

For example, if $f(z) = e^z$, then $M(r) = e^r$, $\rho = \infty$. In this case $r^* = 1$ (since $\hat{n} \gg 1$). Then we shall choose N and $r' = N$ so that $e^{r'}/((r')^N - 1) \ll eU$. One can show that it is sufficient to choose $N \gg |\ln U/\ln|\ln U||$. For instance, if $U = 10^{-16}$, this is satisfied with a wide margin by $N = 32$. In IEEE double precision arithmetic, the choice $r = 1, N = 32$ gave an error less than $2{\cdot}10^{-16}$. The results were much worse for $r = 10$ and for $r = 0.1$; the maximum error of the first 32 coefficients became $4 \cdot 10^{-4}$ and $9 \cdot 10^{13}$(!), respectively. In the latter case the errors of the first eight coefficients did not exceed 10^{-10}, but the rounding error of a_n, due to cancellations, increased rapidly with n.

If ρ *is not much larger than* r^*, the procedure described above may lead to a value of N that is much larger than $\hat{n}$. In order to avoid this, we set $\hat{n} = \alpha N$. We now confine the discussion to the case that $r < r' < \rho \le 1$, $n = 0 : \hat{n}$. Then, with all other parameters fixed, the bound in (3.2.20) is maximal for $n = \hat{n}$. We simplify this bound; $M(r)$ is replaced by the larger quantity $M(r')$, and the denominator is replaced by $(r'/r)^N$. Then, for given r', α, N, we set $x = (r/r')^N$ and determine x so that

$$M(r')(r')^{-\alpha N}(Ux^{-\alpha} + x)$$

is minimized. The minimum is obtained for $x = (\alpha U)^{1/(1+\alpha)}$, i.e., for $r = r'x^{1/N}$, and the minimum is equal to[61]

$$M(r')(r')^{-n}U^{1/(1+\alpha)}c(\alpha), \quad \text{where} \quad c(\alpha) = (1+\alpha)\alpha^{-\alpha/(1+\alpha)}.$$

We see that the error bound contains the factor $U^{1/(1+\alpha)}$. This is proportional to $2\,U^{1/2}$ for $\alpha = 1$, and to $1.65\,U^{4/5}$ for $\alpha = \frac{1}{4}$. The latter case is thus much more accurate, but for the same $\hat{n}$ one has to choose N four times as large, which leads to more than four times as many arithmetic operations. In practice, $\hat{n}$ is usually given, and the order of magnitude of U can be estimated. Then α is to be chosen to make a compromise between the requirements for a good accuracy and for a small volume of computation. *If ρ is not much larger than r^**, we may choose

$$N = \hat{n}/\alpha, \quad x = (\alpha U)^{1/(1+\alpha)}, \quad r = r'x^{1/N}.$$

Experiments were conducted with

$$f(z) = \ln(1-z),$$

for which $\rho = 1$, $M(1) = \infty$. Take $\hat{n} = 64$, $U = 10^{-15}$, $r' = 0.999$. Then $M(r') = 6.9$. For $\alpha = 1,\ 1/2,\ 1/4$, we have $N = 64,\ 128,\ 256$, respectively. The above theory suggests $r = 0.764,\ 0.832,\ 0.894$, respectively. The theoretical estimates of the absolute errors become 10^{-9}, $2.4\cdot 10^{-12}$, $2.7\cdot 10^{-14}$, respectively. The smallest errors obtained in experiments with these three values of α are $6\cdot 10^{-10}$, $1.8\cdot 10^{-12}$, $1.8\cdot 10^{-14}$, which were obtained for $r = 0.766,\ 0.838,\ 0.898$, respectively. So, the theoretical predictions of these experimental results are very satisfactory.

3.2.3 Chebyshev Expansions

The **Chebyshev**[62] **polynomials** of the first kind are defined by

$$T_n(z) = \cos(n \arccos z), \quad n \geq 0, \tag{3.2.21}$$

that is, $T_n(z) = \cos(n\phi)$, where $z = \cos\phi$. From the well-known trigonometric formula

$$\cos(n+1)\phi + \cos(n-1)\phi = 2\cos\phi\cos n\phi$$

[61]This is a rigorous upper bound of the error for this value of r, in spite of simplifications in the formulation of the minimization.

[62]Pafnuty Lvovich Chebyshev (1821–1894), Russian mathematician, pioneer in approximation theory and the constructive theory of functions. His name has many different transcriptions, for example, Tschebyscheff. This may explain why the polynomials that bear his name are denoted $T_n(x)$. He also made important contributions to probability theory and number theory.

follows, by induction, the important **recurrence relation**: $T_0(z) = 1$, $T_1(z) = z$,

$$T_{n+1}(z) = 2zT_n(z) - T_{n-1}(z), \quad (n \geq 1). \tag{3.2.22}$$

Using this recurrence relation we obtain

$$T_2(z) = 2z^2 - 1, \quad T_3(z) = 4z^3 - 3z, \quad T_4(z) = 8z^4 - 8z^2 + 1,$$

$$T_5(z) = 16z^5 - 20z^3 + 5z, \quad T_6(z) = 32z^6 - 48z + 18z^2 - 1, \dots.$$

Clearly $T_n(z)$ is the nth degree polynomial,

$$T_n(z) = z^n - \binom{n}{2} z^{n-2}(1 - z^2) + \binom{n}{4} z^{n-4}(1 - z^2)^2 - \cdots.$$

The Chebyshev polynomials of the second kind,

$$U_{n-1}(z) = \frac{1}{n+1} T'(z) = \frac{\sin(n\phi)}{\sin \phi}, \quad \phi = \arccos z, \tag{3.2.23}$$

satisfy the same recurrence relation, with the initial conditions $U_{-1}(z) = 0$, $U_0(z) = 1$; its degree is $n - 1$. (When we write just "Chebyshev polynomial," we refer to the first kind.)

The Chebyshev polynomial $T_n(x)$ has n zeros in $[-1, 1]$ given by

$$x_k = \cos\left(\frac{2k-1}{n}\frac{\pi}{2}\right), \quad k = 1 : n, \tag{3.2.24}$$

the **Chebyshev points**, and $n + 1$ *extrema*

$$x'_k = \cos\left(\frac{k\pi}{n}\right), \quad k = 0 : n. \tag{3.2.25}$$

These results follow directly from the fact that $\cos(n\phi) = 0$ for $\phi = (2k + 1)\pi/(2n)$, and that $\cos(n\phi) = \pm 1$ for $\phi = k\pi/n$.

Note that from (3.2.21) it follows that $|T_n(x)| \leq 1$ for $x \in [-1, 1]$, even though its leading coefficient is as large as 2^{n-1}.

Example 3.2.3.

Figure 3.2.1 shows a plot of the Chebyshev polynomial $T_{20}(x)$ for $x \in [-1, 1]$. Setting $z = 1$ in the recurrence relation (3.2.22) and using $T_0(1) = T_1(1) = 1$, it follows that $T_n(1) = 1$, $n \geq 0$. From $T'_0(1) = 0$, $T'_1(1) = 1$ and differentiating the recurrence relation we get

$$T'_{n+1}(z) = 2(zT'_n(z) + T_n(z)) - T'_{n-1}(z), \quad (n \geq 1).$$

It follows easily by induction that $T'_n(1) = n^2$, i.e., *outside the interval* $[-1, 1]$ *the Chebyshev polynomials grow rapidly.*

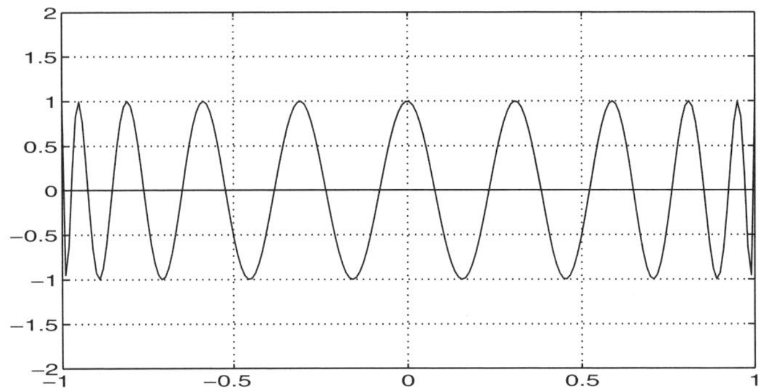

Figure 3.2.1. *Graph of the Chebyshev polynomial* $T_{20}(x)$, $x \in [-1, 1]$.

The Chebyshev polynomials have a unique **minimax property**. (For a use of this property, see Example 3.2.4.)

Lemma 3.2.4 (*Minimax Property*).

The Chebyshev polynomials have the following **minimax property**: *Of all nth degree polynomials with leading coefficient* 1, *the polynomial* $2^{1-n}T_n(x)$ *has the smallest magnitude* 2^{1-n} *in* $[-1, 1]$.

Proof. Suppose there were a polynomial $p_n(x)$, with leading coefficient 1 such that $|p_n(x)| < 2^{1-n}$ for all $x \in [-1, 1]$. Let x_k', $k = 0 : n$, be the abscissae of the extrema of $T_n(x)$. Then we would have

$$p_n(x_0') < 2^{1-n}T_n(x_0'), \qquad p_n(x_1') > 2^{1-n}T_n(x_1'), \qquad p_n(x_2') < 2^{1-n}T_n(x_2'), \quad \dots,$$

etc., up to x_n'. From this it follows that the polynomial

$$p_n(x) - 2^{1-n}T_n(x)$$

changes sign in each of the n intervals (x_k', x_{k+1}'), $k = 0 : n-1$. This is impossible, since the polynomial is of degree $n - 1$. This proves the minimax property. □

The Chebyshev expansion of a function $f(z)$,

$$f(z) = \sum_{j=0}^{\infty} c_j T_j(z), \tag{3.2.26}$$

is an important aid in studying functions on the interval $[-1, 1]$. If one is working with a function $f(t)$, $t \in [a, b]$, then one should make the substitution

$$t = \tfrac{1}{2}(a + b) + \tfrac{1}{2}(b - a)x, \tag{3.2.27}$$

which maps the interval $[-1, 1]$ onto $[a, b]$.

Consider the approximation to the function $f(x) = x^n$ on $[-1, 1]$ by a polynomial of lower degree. From the minimax property of Chebyshev polynomials it follows that the maximum magnitude of the error is minimized by the polynomial

$$p(x) = x^n - 2^{1-n}T_n(x). \tag{3.2.28}$$

From the symmetry property $T_n(-x) = (-1)^n T_n(x)$, it follows that this polynomial has in fact degree $n - 2$. The error $2^{1-n}T_n(x)$ assumes its extrema 2^{1-n} in a sequence of $n + 1$ points, $x_i = \cos(i\pi/n)$. The sign of the error alternates at these points.

Suppose that one has obtained, for example, by Taylor series, a truncated power series approximation to a function $f(x)$. By repeated use of (3.2.28), the series can be replaced by a polynomial of lower degree with a moderately increased bound for the truncation error. This process, called **economization of power series** often yields a useful polynomial approximation to $f(x)$ with a considerably smaller number of terms than the original power series.

Example 3.2.4.

If the series expansion $\cos x = 1 - x^2/2 + x^4/24 - \cdots$ is truncated after the x^4-term, the maximum error is 0.0014 in $[-1, 1]$. Since $T_4(x) = 8x^4 - 8x^2 + 1$, it holds that

$$x^4/24 \approx x^2/24 - 1/192$$

with an error which does not exceed $1/192 = 0.0052$. Thus the approximation

$$\cos x = (1 - 1/192) - x^2(1/2 - 1/24) = 0.99479 - 0.45833x^2$$

has an error whose magnitude does not exceed $0.0052 + 0.0014 < 0.007$. This is less than one-sixth of the error 0.042, which is obtained if the power series is truncated after the x^2-term.

Note that for the economized approximation $\cos(0)$ is not approximated by 1. It may not be acceptable that such an exact relation is lost. In this example one could have asked for a polynomial approximation to $(1 - \cos x)/x^2$ instead.

If a Chebyshev expansion converges rapidly, the truncation error is, by and large, determined by the first few neglected terms. As indicated by Figures 3.2.1 and 3.2.5 (see Problem 3.2.3), the error curve is oscillating with slowly varying amplitude in $[-1, 1]$. In contrast, the truncation error of a power series is proportional to a power of x. Note that $f(z)$ is allowed to have a singularity arbitrarily close to the interval $[-1, 1]$, and the convergence of the Chebyshev expansion will still be exponential, although the exponential rate deteriorates, as $R \downarrow 1$.

Important properties of trigonometric functions and Fourier series can be reformulated in the terminology of Chebyshev polynomials. For example, they satisfy certain orthogonality relations; see Example 4.5.10. Also, results like (3.2.7), concerning how the rate of decrease of the coefficients or the truncation error of a Fourier series is related to the smoothness properties of its sum, can be translated to Chebyshev expansions. So, even *if f is not analytic, its Chebyshev expansion converges under amazingly general conditions* (unlike a power series), but the convergence is much slower than exponential. A typical result reads as follows: *if* $f \in C^k[-1,1]$, $k>0$, *there exists a bound for the truncation error that decreases uniformly like* $O(n^{-k}\log n)$. Sometimes convergence acceleration can be successfully applied to such series.

Set $w = e^{i\phi} = \cos\phi + i\sin\phi$, where ϕ and $z = \cos\phi$ may be complex. Then

$$w = z \pm \sqrt{z^2-1}, \qquad z = \cos\phi = \frac{1}{2}(w + w^{-1}),$$

and

$$T_n(z) = \cos n\phi = \frac{1}{2}(w^n + w^{-n}), \tag{3.2.29}$$

$$\left(z + \sqrt{z^2-1}\right)^n = T_n(z) + U_{n-1}(z)\sqrt{z^2-1},$$

where $U_{n-1}(z)$ is the Chebyshev polynomials of the second kind; see (3.2.23). It follows that the Chebyshev expansion (3.2.26) formally corresponds to a symmetric Laurent expansion,

$$g(w) = f\left(\tfrac{1}{2}(w+w^{-1})\right) = \sum_{-\infty}^{\infty} a_j w^j, \qquad a_{-j} = a_j = \begin{cases} \frac{1}{2}c_j & \text{if } j>0, \\ c_0 & \text{if } j=0. \end{cases}$$

It can be shown by the parallelogram law that $|z+1| + |z-1| = |w| + |w|^{-1}$. Hence, if $R > 1$, $z = \frac{1}{2}(w+w^{-1})$ maps the annulus $\{w : R^{-1} < |w| < R\}$, twice onto an ellipse $\mathcal{E}_R$, determined by the relation

$$\mathcal{E}_R = \{z : |z-1| + |z+1| \le R + R^{-1}\}, \tag{3.2.30}$$

with foci at 1 and -1. The axes are, respectively, $R + R^{-1}$ and $R - R^{-1}$, and hence R *is the sum of the semiaxes.*

Note that as $R \to 1$, the ellipse degenerates into the interval $[-1,1]$. As $R \to \infty$, it becomes close to the circle $|z| < \frac{1}{2}R$. It follows from (3.2.29) that this family of confocal ellipses are level curves of $|w| = |z \pm \sqrt{z^2-1}|$. In fact, we can also write

$$\mathcal{E}_R = \left\{z : 1 \le |z + \sqrt{z^2-1}| \le R\right\}. \tag{3.2.31}$$

Theorem 3.2.5 (*Bernštein's Approximation Theorem*).

Let $f(z)$ *be real-valued for* $z \in [-1,1]$, *analytic and single-valued for* $z \in \mathcal{E}_R$, $R>1$. *Assume that* $|f(z)| \le M$ *for* $z \in \mathcal{E}_R$. *Then*[63]

$$\left| f(x) - \sum_{j=0}^{n-1} c_j T_j(x) \right| \le \frac{2MR^{-n}}{1-1/R} \quad \textit{for } x \in [-1,1].$$

Proof. Set as before $z = \frac{1}{2}(w+w^{-1})$, $g(w) = f(\frac{1}{2}(w+w^{-1}))$. Then $g(w)$ is analytic in the annulus $R^{-1} + \epsilon \le |w| \le R - \epsilon$, and hence the Laurent expansion (3.2.1) converges there. In particular it converges for $|w| = 1$, hence the Chebyshev expansion for $f(x)$ converges when $x \in [-1,1]$.

[63] A generalization to complex values of x is formulated in Problem 3.2.11.

Set $r = R - \epsilon$. By Cauchy's formula we obtain, for $j > 0$,

$$|c_j| = 2|a_j| = \left|\frac{2}{2\pi i}\int_{|w|=r} g(w)w^{-(j+1)}dw\right| \le \frac{2}{2\pi}\int_0^{2\pi} Mr^{-j-1}r\,d\phi = 2Mr^{-j}.$$

We then obtain, for $x \in [-1, 1]$,

$$\left|f(x) - \sum_{j=0}^{n-1} c_j T_j(x)\right| = \left|\sum_n^{\infty} c_j T_j(x)\right| \le \sum_n^{\infty} |c_j| \le 2M \sum_n^{\infty} r^{-j} \le 2M \frac{r^{-n}}{1 - 1/r}.$$

This holds for any $\epsilon > 0$. We can here let $\epsilon \to 0$ and thus replace r by R. □

The Chebyshev polynomials are perhaps the most important example of a family of **orthogonal polynomials**; see Sec. 4.5.5. The numerical value of a truncated Chebyshev expansion can be computed by means of **Clenshaw's algorithm**; see Theorem 4.5.21.

3.2.4 Perturbation Expansions

In the equations of applied mathematics it is often possible to identify a small dimensionless parameter (say) ϵ, $\epsilon \ll 1$. The case when $\epsilon = 0$ is called the *reduced problem* or the unperturbed case, and one asks for a **perturbation expansion**, i.e., an expansion of the solution of the perturbed problem into powers of the perturbation parameter ϵ. In many cases it can be proved that the expansion has the form $c_0 + c_1\epsilon + c_2\epsilon^2 + \cdots$, but there are also important cases where the expansion contains fractional or a few negative powers.

In this subsection, we consider an analytic equation $\phi(z, \epsilon) = 0$ and seek expansions for the roots $z_i(\epsilon)$ in powers of ϵ. This has some practical interest in its own right, but it is mainly to be considered as a preparation for more interesting applications of perturbation methods to more complicated problems. A simple perturbation example for a *differential equation* is given in Problem 3.2.9.

If $z_i(0)$ is a simple root, i.e., if $\partial\phi/\partial z \neq 0$, for $(z, \epsilon) = (z_i(0), 0)$, then a theorem of complex analysis tells us that $z_i(\epsilon)$ is an analytic function in a neighborhood of the origin. Hence the expansion

$$z_i(\epsilon) - z_i(0) = c_1\epsilon + c_2\epsilon^2 + \cdots$$

has a positive (or infinite) radius of convergence. We call this a **regular perturbation problem**. The techniques of power series reversion, presented in Sec. 3.1.4, can often be applied after some preparation of the equation. Computer algebra systems are also used in perturbation problems, if expansions with many terms are needed.

Example 3.2.5.

We shall expand the roots of

$$\phi(z, \epsilon) \equiv \epsilon z^2 - z + 1 = 0$$

into powers of ϵ. The reduced problem $-z + 1 = 0$ has only one finite root, $z_1(0) = 1$. Set $z = 1 + x\epsilon$, $x = c_1 + c_2\epsilon + c_3\epsilon^2 + \cdots$. Then $\phi(1 + x\epsilon, \epsilon)/\epsilon = (1 + x\epsilon)^2 - x = 0$, i.e.,

$$(1 + c_1\epsilon + c_2\epsilon^2 + \cdots)^2 - (c_1 + c_2\epsilon + c_3\epsilon^2 + \cdots) = 0.$$

Matching the coefficients of $\epsilon^0,\ \epsilon^1, \epsilon^2$, we obtain the system

$$\begin{aligned}
1 - c_1 = 0 &\Rightarrow c_1 = 1, \\
2c_1 - c_2 = 0 &\Rightarrow c_2 = 2, \\
2c_2 + c_1^2 - c_3 = 0 &\Rightarrow c_3 = 5;
\end{aligned}$$

hence $z_1(\epsilon) = 1 + \epsilon + 2\epsilon^2 + 5\epsilon^3 + \cdots$.

Now, the easiest way to obtain the expansion for the second root, $z_2(\epsilon)$, is to use the fact that the sum of the roots of the quadratic equation equals ϵ^{-1}; hence $z_2(\epsilon) = \epsilon^{-1} - 1 - \epsilon - 2\epsilon^2 + \cdots$.

Note the appearance of the term ϵ^{-1}. This is due to a characteristic feature of this example. The degree of the polynomial is lower for the reduced problem than it is for $\epsilon \neq 0$; one of the roots escapes to ∞ as $\epsilon \to 0$. This is an example of a **singular perturbation** problem, an important type of problem for differential equations; see Problem 3.2.7.

If $\partial\phi/\partial z = 0$, for some z_i, the situation is more complicated; z_i is a multiple root, and the expansions look different. If $z_i(0)$ is a k-fold root, then there may exist an expansion of the form

$$z_i(\epsilon) = c_0 + c_1\epsilon^{1/k} + c_2(\epsilon^{1/k})^2 + \cdots$$

for each of the k roots of ϵ, but this is not always the case. See (3.2.32) below, where the expansions are of a different type. *If one tries to determine the coefficients in an expansion of the wrong form, one usually runs into contradictions,* but the question about the right form of the expansions still remains.

The answers are given by the classical theory of *algebraic functions*, where Riemann surfaces and Newton polygons are two of the key concepts; see, e.g., Bliss [35]. We shall, for several reasons, not use this theory here. One reason is that it seems hard to generalize some of the methods of algebraic function theory to more complicated equations, such as differential equations. We shall instead use a general **balancing procedure**, recommended in Lin and Segel [246, Sec. 9.1], where it is applied to singular perturbation problems for differential equations too.

The basic idea is very simple: each term in an equation behaves like some power of ϵ. *The equation cannot hold unless there is a β such that a pair of terms of the equation behave like $A\epsilon^\beta$ (with different values of A), and the ϵ-exponents of the other terms are larger than or equal to β.* (Recall that larger exponents make smaller terms.)

Let us return to the previous example. Although we have already determined the expansion for $z_2(\epsilon)$ (by a trick that may not be useful for problems other than single analytic equations), we shall use this task to illustrate the balancing procedure. Suppose that

$$z_2(\epsilon) \sim A\epsilon^\alpha, \quad (\alpha < 0).$$

The three terms of the equation $\epsilon z^2 - z + 1 = 0$ then get the exponents

$$1 + 2\alpha, \quad \alpha, \quad 0.$$

Try the first two terms as the candidates for being the dominant pair. Then $1 + 2\alpha = \alpha$, hence $\alpha = -1$. The three exponents become $-1,\ -1,\ 0$. Since the third exponent is larger than the exponent of the candidates, this choice of pair seems possible, but we have not shown that it is the only possible choice.

Now try the first and the third terms as candidates. Then $1 + 2\alpha = 0$, hence $\alpha = -\frac{1}{2}$. The exponent of the noncandidate is $-\frac{1}{2} \leq 0$; this candidate pair is thus impossible. Finally, try the second and the third terms. Then $\alpha = 0$, but we are only interested in negative values of α.

The conclusion is that we can try coefficient matching in the expansion $z_2(\epsilon) = c_{-1}\epsilon^{-1} + c_0 + c_1\epsilon + \cdots$. We don't need to do it, since we know the answer already, but it indicates how to proceed in more complicated cases.

Example 3.2.6.

First consider the equation $z^3 - z^2 + \epsilon = 0$. The reduced problem $z^3 - z^2 = 0$ has a single root, $z_1 = 1$, and a double root, $z_{2,3} = 0$. No root has escaped to ∞. By a similar coefficient matching as in the previous example we find that $z_1(\epsilon) = 1 - \epsilon - 2\epsilon^2 + \cdots$. For the double root, set $z = A\epsilon^{\beta}$, $\beta > 0$. The three terms of the equation obtain the exponents $3\beta,\ 2\beta,\ 1$. Since 3β is dominated by 2β we conclude that $2\beta = 1$, i.e., $\beta = 1/2$,

$$z_{2,3}(\epsilon) = c_0\epsilon^{1/2} + c_1\epsilon + c_2\epsilon^{3/2} + \cdots.$$

By matching the coefficients of $\epsilon,\ \epsilon^{3/2},\ \epsilon^2$, we obtain the system

$$\begin{aligned} -c_0^2 + 1 = 0 &\Rightarrow c_0 = \pm 1, \\ -2c_0c_1 + c_0^3 = 0 &\Rightarrow c_1 = \frac{1}{2}, \\ -2c_0c_2 - c_1^2 + 2c_0^2c_1 + c_1c_0^2 = 0 &\Rightarrow c_2 = \pm\frac{5}{8}; \end{aligned}$$

hence $z_{2,3}(\epsilon) = \pm\epsilon^{1/2} + \frac{1}{2}\epsilon \pm \frac{5}{8}\epsilon^{3/2} + \cdots$.

There are, however, equations with a double root, where the perturbed pair of roots do not behave like $\pm c_0\epsilon^{1/2}$ as $\epsilon \to 0$. In such cases the balancing procedure may help. Consider the equation

$$(1 + \epsilon)z^2 + 4\epsilon z + \epsilon^2 = 0. \tag{3.2.32}$$

The reduced problem is $z^2 = 0$, with a double root. Try $z \sim A\epsilon^{\alpha}$, $\alpha > 0$. The exponents of the three terms become $2\alpha,\ \alpha + 1,\ 2$. We see that $\alpha = 1$ makes the three exponents all equal to 2; this is fine. So, set $z = \epsilon y$. The equation reads, after division by ϵ^2, $(1 + \epsilon)y^2 + 4y + 1 = 0$, hence $y(0) = a \equiv -2 \pm \sqrt{3}$. Coefficient matching yields the result

$$z = \epsilon y = a\epsilon + (-a^2/(2(a + 2)))\epsilon^2 + \cdots,$$

where all exponents are natural numbers.

If ϵ is small enough, the last term included can serve as an error estimate. A more reliable error estimate (or even an error bound) can be obtained by inserting the truncated expansion into the equation. It shows that *the truncated expansion satisfies a modified equation exactly*. The same idea can be applied to equations of many other types; see Problem 3.2.9.

3.2.5 Ill-Conditioned Series

Slow convergence is not the only numerical difficulty that occurs in connection with infinite series. There are also series with oscillating terms and a complicated type of catastrophic cancellation. The size of some terms is many orders of magnitude larger than the sum of the series. Small relative errors in the computation of the large terms lead to a large relative error in the result. We call such a series **ill-conditioned.**

Such series have not been subject to many systematic investigations. One simply tries to avoid them. For the important "special functions" of applied mathematics, such as Bessel functions, confluent hypergeometric functions, etc., there usually exist *expansions into descending powers of* z that can be useful, when $|z| \gg 1$ and the usual series, in *ascending* powers, are divergent or ill-conditioned. Another possibility is to use *multiple precision* in computations with ill-conditioned power series; this is relatively expensive and laborious (but the difficulties should not be exaggerated). There are, however, also other, less well known possibilities that will now be exemplified. The subject is still open for fresh ideas, and we hope that the following pages and the related problems at the end of the section will stimulate some readers to think about it.

First, we shall consider power series of the form

$$\sum_{n=0}^{\infty} \frac{(-x)^n c_n}{n!}, \tag{3.2.33}$$

where $x \gg 1$, although not so large that there is risk for overflow. We assume that the coefficients c_n are positive and slowly varying (relative to $(-x)^n/n!$). The ratio of two consecutive terms is

$$\frac{c_{n+1}}{c_n}\frac{-x}{n+1} \approx \frac{-x}{n+1}.$$

We see that the series converges for all x, and that the magnitude increases if and only if $n+1 < |x|$. *The term of largest magnitude is thus obtained for* $n \approx |x|$. Denote its magnitude by $M(x)$. Then, for $x \gg 1$, the following type of approximations can be used for crude estimates of the number of terms needed and the arithmetic precision that is to be used in computations related to ill-conditioned power series: $M(x) \approx c_x e^x (2\pi x)^{-1/2}$; i.e.,

$$\log_{10} M(x)/c_0 \approx 0.43x - \tfrac{1}{2}\log_{10}(2\pi x). \tag{3.2.34}$$

This follows from the classical **Stirling's formula,**

$$x! \sim \left(\frac{x}{e}\right)^x \sqrt{2\pi x}\left[1 + \frac{1}{12x} + \frac{1}{288x^2} + \cdots\right], \quad x \gg 1, \tag{3.2.35}$$

that gives $x!$ with a relative error that is about $1/(12x)$. You find a proof of this in most textbooks on calculus. It will be used often in the rest of this book. A more accurate and general version is given in Example 3.4.12 together with a few more facts about the gamma function, $\Gamma(z)$, an analytic function that interpolates the factorial $\Gamma(n+1) = n!$ if n is a natural number. Sometimes the notation $z!$ is used instead of $\Gamma(z+1)$ even if z is not an integer.

There exist **preconditioners**, i.e., transformations that can convert classes of ill-conditioned power series (with accurately computable coefficients) to more well-conditioned problems. One of the most successful preconditioners known to the authors is the following:

$$\sum_{n=0}^{\infty} \frac{(-x)^n c_n}{n!} = e^{-x} \sum_{n=0}^{\infty} \frac{x^n b_n}{n!}, \qquad b_n = (-\Delta)^n c_0. \tag{3.2.36}$$

A hint for proving this identity is given in Problem 3.3.22. The notation $\Delta^n c_n$ for high order differences was introduced in Sec. 1.1.4.

For the important class of sequences $\{c_n\}$ which are **completely monotonic**, $(-\Delta)^n c_0$ is positive and smoothly decreasing; see Sec. 3.4.4.

Example 3.2.7.

Consider the function

$$F(x) = \frac{1}{x} \int_0^x \frac{1 - e^{-t}}{t} \, dt = 1 - \frac{x}{2^2 \cdot 1!} + \frac{x^2}{3^2 \cdot 2!} - \cdots,$$

i.e., $F(x)$ is a particular case of (3.2.33) with $c_n = (n+1)^{-2}$. We shall look at three methods of computing $F(x)$ for $x = 10 : 10 : 50$, named A, B, C. $F(x)$ decreases smoothly from 0.2880 to 0.0898. The computed values of $F(x)$ are denoted $FA(x)$, $FB(x)$, $FC(x)$.

The coefficients c_n, $n = 0 : 119$, are given in IEEE floating-point, double precision. The results in Table 3.2.1 show that (except for $x = 50$) 120 terms is much more than necessary for the rounding of the coefficients to become the dominant error source.

Table 3.2.1. *Results of three ways to compute* $F(x) = (1/x) \int_0^x (1/t)(1 - e^{-t}) \, dt$.

x	10	20	30	40	50
$F(x) \approx$	0.2880	0.1786	0.1326	0.1066	0.0898
lasttermA	$1 \cdot 10^{-82}$	$8 \cdot 10^{-47}$	$7 \cdot 10^{-26}$	$6 \cdot 10^{-11}$	$2 \cdot 10^{1}$
$M(x; A)$	$3 \cdot 10^{1}$	$1 \cdot 10^{5}$	$9 \cdot 10^{8}$	$1 \cdot 10^{13}$	$1 \cdot 10^{17}$
$\lvert FA(x) - F(x) \rvert$	$2 \cdot 10^{-15}$	$5 \cdot 10^{-11}$	$2 \cdot 10^{-7}$	$3 \cdot 10^{-3}$	$2 \cdot 10^{1}$
lasttermB	$4 \cdot 10^{-84}$	$1 \cdot 10^{-52}$	$4 \cdot 10^{-36}$	$2 \cdot 10^{-25}$	$2 \cdot 10^{-18}$
$M(x; B)$	$4 \cdot 10^{-2}$	$2 \cdot 10^{-2}$	$1 \cdot 10^{-2}$	$7 \cdot 10^{-3}$	$5 \cdot 10^{-3}$
$\lvert FC(x) - FB(x) \rvert$	$7 \cdot 10^{-9}$	$2 \cdot 10^{-14}$	$6 \cdot 10^{-17}$	0	$1 \cdot 10^{-16}$

(A) We use (3.2.33) without preconditioner. $M(x; A)$ is the largest magnitude of the terms of the expansion. $M(x; A) \cdot 10^{-16}$ gives the order of magnitude of the effect of the rounding errors on the computed value $FA(x)$. Similarly, the truncation error is crudely estimated by lasttermA. See Figure 3.2.2. Since the largest term is 10^{13}, it is no surprise that the relative error of the sum is not better than 0.03, in spite of double precision floating-point being used. Note the scale, and look also in the table.

(B) We use the preconditioner (3.2.36). In this example $c_n = (n + 1)^{-2}$. In Problem 3.3.3 (c) we find the following explicit expressions, related to the series on the right-hand

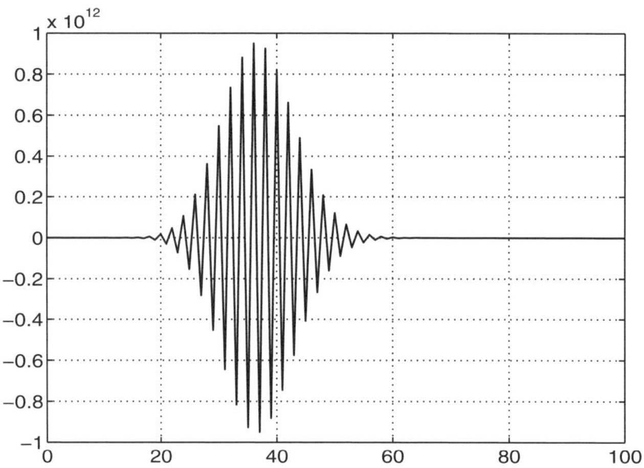

Figure 3.2.2. *Example* 3.2.7(A)*: Terms of* (3.2.33), $c_n = (n+1)^{-2}$, $x = 40$, *no preconditioner.*

side of the preconditioner for this example:

$$(-\Delta)^n c_0 = (-\Delta)^n c_m|_{m=0} = c_0(-\Delta)^n x^{-2}|_{x=1} = \frac{c_0}{n+1}\sum_{k=0}^{n}\frac{1}{k+1},$$

$$F(x) = c_0 e^{-x}\sum_{n=0}^{\infty}\frac{x^n}{(n+1)!}\sum_{k=0}^{n}\frac{1}{k+1}. \tag{3.2.37}$$

Note that $(-\Delta)^m c_0$ is positive and smoothly decreasing.

The largest term is thus smaller than the sum, and the series (3.2.37) is **well-conditioned**. The largest term is now about $7 \cdot 10^{-3}$ and the computed sum is correct to 16 decimal places. Multiple precision is not needed here. It can be shown that if $x \gg 1$, the mth term is approximately proportional to the value at m of the normal probability density with mean x and standard deviation equal to $\sqrt{x}$; note the resemblance to a Poisson distribution. The terms of the right-hand side, including the factor e^{-x}, become a so-called **bell sum**; see Figure 3.2.3.

$M(x; B)$ and lasttermB are defined analogously to $M(x; A)$ and lasttermA. The B-values are very different from the A-values. In fact they indicate that *all values of* $FB(x)$ *referred to in Table* 3.2.1 *give* $F(x)$ *to full accuracy*.

(C) The following expression for $F(x)$,

$$xF(x) \equiv \sum_{n=1}^{\infty}\frac{(-x)^n}{nn!} = -\gamma - \ln x - E_1(x), \quad E_1(x) = \int_x^{\infty}\frac{e^{-t}}{t}\,dt, \tag{3.2.38}$$

is valid for all $x > 0$; see [1, Sec. 5.1.11]. $E_1(x)$ is known as the **exponential integral**, and

$$\gamma = 0.57721\,56649\,01532\,86061\ldots$$

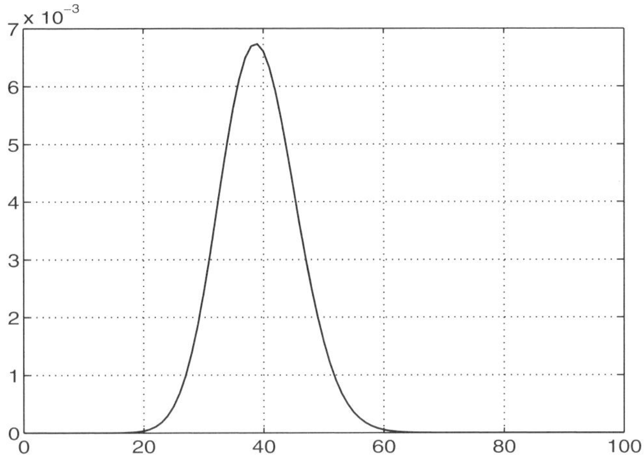

Figure 3.2.3. *Example* 3.2.7(B)*:* $c_n = (n+1)^{-2}$, $x = 40$, *with preconditioner in* (3.2.36).

is the well-known **Euler's constant**. In the next section, an asymptotic expansion for $E_1(x)$ for $x \gg 1$ is derived, the first two terms of which are used here in the computation of $F(x; C)$ for Table 3.2.1.

$$E_1(x) \approx e^{-x}(x^{-1} - x^{-2}), \quad x \gg 1.$$

This approximation is the dominant part of the error of $F(x; C)$; it is less than $e^{-x}2x^{-4}$. $F(x; C)$ gives full accuracy for (say) $x > 25$.

More examples of sequences, for which rather simple explicit expressions for the high order differences are known, are given in Problem 3.3.3. Kummer's confluent hypergeometric function $M(a, b, x)$ was defined in (3.1.17). We have

$$M(a, b, -x) = 1 + \sum_{n=1}^{\infty} \frac{(-x)^n c_n}{n!}, \qquad c_n = c_n(a, b) = \frac{a(a+1)\dots(a+n-1)}{b(b+1)\dots(b+n-1)}.$$

In our context $b > a > 0$, $n > 0$. The oscillatory series for $M(a, b, -x)$, $x > 0$, is ill-conditioned if $x \gg 1$.

By Problem 3.3.3, $(-\Delta)^n c_0(a, b) = c_n(b-a, b) > 0, n > 0$; hence the preconditioner (3.2.36) yields the equation

$$M(a, b, -x) = e^{-x} M(b-a, b, x), \tag{3.2.39}$$

where the series on the right-hand side has positive terms, because $b - a > 0$, $x > 0$, and is a well-conditioned *bell sum*. The mth term has typically a sharp maximum for $m \approx x$; compare Figure 3.2.3. Equation (3.2.39), is in the theory of the confluent hypergeometric functions, known as **Kummer's first identity**. It is emphasized here because several functions with famous names of their own are particular cases of the Kummer function. (Several other particular cases are presented in Sec. 3.5.1 together with continued fractions.) These share the numerous useful properties of Kummer's function, for example, the above identity; see the theory in Lebedev [240, Secs. 9.9–9.14][64] and the formulas in [1, Chap. 13, particularly Table 13.6 of special cases]. An important example is the error function (see Example 3.1.3) that can be expressed in terms of Kummer's confluent hypergeometric as

$$\operatorname{erf}(x) = \frac{2}{\sqrt{\pi}} \int_0^x e^{-t^2}\,dt = \frac{2x}{\sqrt{\pi}} M\left(\frac{1}{2}, \frac{3}{2}, -x^2\right). \tag{3.2.40}$$

If we cannot find explicit expressions for high-order differences, we can make a *difference scheme* by the recurrence $\Delta^{m+1}c_n = \Delta^m c_{n+1} - \Delta^m c_n$. Unfortunately the computation of a difference scheme suffers from numerical instability. Suppose that the absolute errors of the c_n are bounded by ϵ. Then the absolute errors can become as large as 2ϵ in the first differences, 4ϵ in the second differences, etc. More generally, the absolute errors of $(-\Delta)^m c_n$ can become as large as $2^m\epsilon$. (You will find more about this in Examples 3.3.2 and 3.3.3.) In connection with ill-conditioned series, this instability is much more disturbing than in the traditional applications of difference schemes to interpolation where m is seldom much larger than 10. Recall that $m \approx x$ for the largest term of the preconditioned series. Thus, if $x > 53$ even this term may not have any correct bit if IEEE double precision arithmetic is used, and many terms are needed after this.

Therefore, during the computation of the new coefficients $(-\Delta)^m c_n$ (only once for the function F, and with double accuracy in the results), the old coefficients c_n must be available with multiple accuracy, and multiple precision must be used in the computation of their difference scheme. Otherwise, we cannot evaluate the series with decent accuracy for much larger values of x than we could have done without preconditioning. Note, however, that if satisfactory coefficients have been obtained for the preconditioned series, double precision is sufficient when the series is evaluated for large values of x. (It is different for method A above.)

Let $F(x)$ be the function that we want to compute for $x \gg 1$, where it is defined by an ill-conditioned power series $F_1(x)$. A more general preconditioner can be described as follows. Try to find a power series $P(x)$ with positive coefficients such that the power series $P(x)F_1(x)$ has less severe cancellations than $F_1(x)$.

In order to distinguish between the algebraic manipulation and the numerical evaluation of the functions defined by these series, we introduce the indeterminate $\mathbf{x}$ and describe a **more general preconditioner** as follows:

$$\mathbf{F}_2^*(\mathbf{x}) = \mathbf{P}(\mathbf{x}) \cdot \mathbf{F}_1(\mathbf{x}), \qquad F_2(x) = F_2^*(x)/P(x). \tag{3.2.41}$$

The second statement is a usual scalar evaluation (no boldface). Here $P(x)$ may be evaluated by some other method than the power series, if it is more practical. If $P(x) = e^x$ and $F_1(x)$ is the series defined by (3.2.33), then it can be shown that $F_2(x)$ is mathematically equivalent to the right-hand side of (3.2.36). In these cases $F_2(x)$ has positive coefficients.

If, however, $F_1(x)$ has a positive zero, this is also a zero of $F_2^*(x)$, and hence it is impossible that all coefficients of the series $\mathbf{F}_2^*(\mathbf{x})$ have the same sign. Nevertheless, the following example shows that the preconditioner (3.2.41) can sometimes be successfully used in such a case too.

[64]Unfortunately, the formulation of Kummer's first identity in [240, eq. (9.11.2)] contains a serious sign error.

Example 3.2.8.

The two functions

$$J_0(x) = \sum_{n=0}^{\infty}(-1)^n \frac{(x^2/4)^n}{(n!)^2}, \qquad I_0(x) = \sum_{n=0}^{\infty} \frac{(x^2/4)^n}{(n!)^2}$$

are examples of Bessel functions of the first kind; I_0 is nowadays called a modified Bessel function. $J_0(x)$ is oscillatory and bounded, while $I_0(x) \sim e^x/\sqrt{2\pi x}$ for $x \gg 1$. Since all coefficients of I_0 are positive, we shall set $P = I_0$, $F_1 = J_0$, and try

$$\mathbf{F_2^*(x) = IJ(x) \equiv I_0(x) \cdot J_0(x)}, \qquad F_2(x) = F_2^*(x)/I_0(x)$$

as a preconditioner for the power series for $J_0(x)$, which is ill-conditioned if $x \gg 1$. In Table 3.2.2, lines 2 and 7 are obtained from the fully accurate built-in functions for $J_0(x)$ and $I_0(x)$. $J(x; N1)$ is computed in IEEE double precision arithmetic from $N1$ terms of the above power series for $J_0(x)$. $N1 = N1(x)$ is obtained by a termination criterion that should give full accuracy or, if the estimate of the effect of the rounding error is bigger than 10^{-16}, the truncation error should be smaller than this estimate. We omit the details; see Problem 3.2.9 (d).

Table 3.2.2. *Evaluation of some Bessel functions.*

1	x	10	20	30	40	50
2	$J_0(x) \approx$	$-2 \cdot 10^{-1}$	$2 \cdot 10^{-1}$	$-9 \cdot 10^{-2}$	$7 \cdot 10^{-3}$	$6 \cdot 10^{-2}$
3	$N1(x)$	26	41	55	69	82
4	$J(x; N1) - J_0(x)$	$9 \cdot 10^{-14}$	$3 \cdot 10^{-10}$	$-2 \cdot 10^{-6}$	$-1 \cdot 10^{-1}$	$-2 \cdot 10^{2}$
5	$N2(x)$	16	26	36	46	55
6	$IJ(x; N2) \approx$	$-7 \cdot 10^{2}$	$7 \cdot 10^{6}$	$-7 \cdot 10^{10}$	$1 \cdot 10^{14}$	$2 \cdot 10^{19}$
7	$I_0(x) \approx$	$3 \cdot 10^{3}$	$4 \cdot 10^{7}$	$8 \cdot 10^{11}$	$1 \cdot 10^{16}$	$3 \cdot 10^{20}$
8	$IJ(x)/I_0(x) - J_0(x)$	$3 \cdot 10^{-17}$	$2 \cdot 10^{-14}$	$3 \cdot 10^{-13}$	$-5 \cdot 10^{-12}$	$2 \cdot 10^{-10}$

The coefficients of $\mathbf{IJ(x)}$ are obtained from the second expression for γ_m given in Problem 3.2.9 (c). $N2 = N2(x)$ is the number of terms used in the expansion of $\mathbf{IJ(x)}$, by a termination criterion similar to the one described for $J(x; N1)$. Compared to line 4, line 8 is a remarkable improvement, obtained without the use of multiple precision.

For series of the form

$$\sum_{n=0}^{\infty} a_n \frac{(-x^2)^n}{(2n)!}$$

one can generate a preconditioner from $P(x) = \cosh x$. This can also be applied to $J_0(x)$ and other Bessel functions; see Problem 3.2.9 (e).

There are several procedures for *transforming a series into an integral* that can then be computed by numerical integration or be expanded in another series that may have better convergence or conditioning properties. An integral representation may also provide an analytic continuation of the function represented by the original series. Integral representations may be obtained in several different ways; we mention two of these. Either there exist integral representations of the coefficients,[65] or one can use general procedures in complex analysis that transform series into integrals. They are due to Cauchy, Plana, and Lindelöf; see Dahlquist [87].

3.2.6 Divergent or Semiconvergent Series

That a series is convergent is no guarantee that it is numerically useful. In this section, we shall see examples of the reverse situation: a divergent series can be of use in numerical computations. This sounds strange, but it refers to series where the size of the terms decreases rapidly at first and increases later, and where an error bound (see Figure 3.2.4), can be obtained in terms of the first neglected term. Such series are sometimes called **semiconvergent**.[66] An important subclass are the **asymptotic series**; see below.

Example 3.2.9.

We shall derive a semiconvergent series for the computation of Euler's function

$$f(x) = e^x E_1(x) = e^x \int_x^\infty e^{-t} t^{-1}\, dt = \int_0^\infty e^{-u}(u+x)^{-1}\, du$$

for large values of x. (The second integral was obtained from the first by the substitution $t = u + x$.) The expression $(u+x)^{-1}$ should first be expanded in a geometric series with remainder term, valid even for $u > x$,

$$(u+x)^{-1} = x^{-1}(1 + x^{-1}u)^{-1} = x^{-1}\sum_{j=0}^{n-1}(-1)^j x^{-j} u^j + (-1)^n (u+x)^{-1}(x^{-1}u)^n.$$

We shall frequently use the well-known formula

$$\int_0^\infty u^j e^{-u}\, du = j! = \Gamma(j+1).$$

We write $f(x) = S_n(x) + R_n(x)$, where

$$S_n(x) = x^{-1}\sum_{j=0}^{n-1}(-1)^j x^{-j}\int_0^\infty u^j e^{-u} du = \frac{1}{x} - \frac{1!}{x^2} + \frac{2!}{x^3} - \cdots + (-1)^{n-1}\frac{(n-1)!}{x^n},$$

$$R_n(x) = (-1)^n \int_0^\infty (u+x)^{-1}\left(\frac{u}{x}\right)^n e^{-u} du.$$

The terms in $S_n(x)$ qualitatively behave as in Figure 3.2.4. The ratio between the last term in S_{n+1} and the last term in S_n is

$$-\frac{n!}{x^{n+1}}\frac{x^n}{(n-1)!} = -\frac{n}{x}, \tag{3.2.42}$$

[65] For hypergeometric or confluent hypergeometric series see Lebedev [240, Secs. 9.1 and 9.11] or [1, Secs. 15.3 and 13.2].

[66] A rigorous theory of semiconvergent series was developed by Stieltjes and Poincaré in 1886.

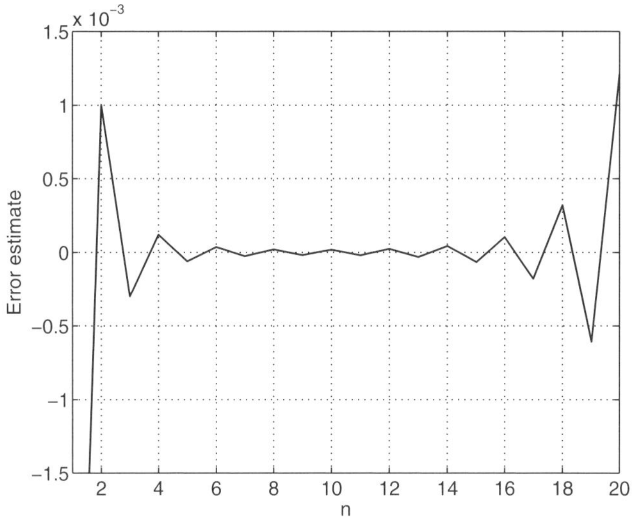

Figure 3.2.4. *Error estimates of the semiconvergent series of Example* 3.2.9 *for* $x = 10$; *see* (3.2.43).

and since the absolute value of that ratio for fixed x is unbounded as $n \to \infty$, the sequence $\{S_n(x)\}_{n=1}^{\infty}$ *diverges for every positive* x. But since sign $R_n(x) = (-1)^n$ for $x > 0$, it follows fromTheorem 3.1.4 that

$$f(x) = \frac{1}{2}\Big(S_n(x) + S_{n+1}(x)\Big) \pm \frac{1}{2}\frac{n!}{x^{n+1}}. \tag{3.2.43}$$

The idea is now to choose n so that the estimate of the remainder is as small as possible. According to (3.2.42), this happens when n is equal to the integer part of x. For $x = 5$ we choose $n = 5$,

$$S_5(5) = 0.2 - 0.04 + 0.016 - 0.0096 + 0.00768 = 0.17408,$$
$$S_6(5) = S_5(5) - 0.00768 = 0.16640,$$

which gives $f(5) = 0.17024 \pm 0.00384$. The correct value is 0.17042, so the actual error is only 5% of the error bound. For $n = x = 10$, the error estimate is $1.0144 \cdot 10^{-5}$.

For larger values of x the accuracy attainable increases. One can show that the bound for the *relative* error using the above computational scheme decreases approximately as $(\pi{\cdot}x/2)^{1/2}e^{-x}$, an extremely good accuracy for large values of x, if one stops at the smallest term. It can even be improved further, by the use of the convergence acceleration techniques presented in Sec. 3.4, notably the *repeated averages* algorithm, also known as the **Euler transformation**; see Sec. 3.4.3. The algorithms for the transformation of a power series into a rapidly convergent continued fraction, mentioned in Sec. 3.5.1, can also be successfully applied to this example and to many other divergent expansions.

One can derive the same series expansion as above by repeated integration by parts. This is often a good way to derive numerically useful expansions, convergent or semiconvergent, with a remainder in the form of an integral. For convenient reference, we formulate this as a lemma that is easily proved by induction and the mean value theorem of integral calculus. See Problem 3.2.10 for applications.

Lemma 3.2.6 (*Repeated Integration by Parts*).

Let $F \in C^p(a,b)$, let G_0 be a piecewise continuous function, and let $G_0, G_1, \dots$ be a sequence of functions such that $G'_{j+1}(x) = G_j(x)$ with suitably chosen constants of integration. Then

$$\int_a^b F(t)G_0(t)\,dt = \sum_{j=0}^{p-1}(-1)^j F^{(j)}(t)G_{j+1}(t)\Big|_{t=a}^{b} + (-1)^p \int_a^b F^{(p)}(t)G_p(t)\,dt.$$

The sum is the "expansion," and the last integral is the "remainder." If $G_p(t)$ has a constant sign in (a,b), the remainder term can also be written in the form

$$(-1)^p F^{(p)}(\xi)(G_{p+1}(b) - G_{p+1}(a)), \quad \xi \in (a,b).$$

The expansion in Lemma 3.2.6 is valid as an *infinite* series, if and only if the remainder tends to 0 as $p \to \infty$. Even if the sum converges as $p \to \infty$, it may converge to the wrong result.

The series in Example 3.2.9 is an expansion in *negative* powers of x, with the property that for all n, the remainder, when $x \to \infty$, approaches zero faster than the last included term. Such an expansion is said to **represent** $f(x)$ **asymptotically** as $x \to \infty$. Such an **asymptotic series** can be either convergent or divergent (semiconvergent). In many branches of applied mathematics, divergent asymptotic series are an important aid, though they are often needlessly surrounded by an air of mysticism.

It is important to appreciate that *an asymptotic series does not define a sum uniquely*. For example $f(x) = e^{-x}$ is asymptotically represented by the series $\sum j = 0^\infty 0 \cdot x^{-j}$, as $x \to \infty$. Thus e^{-x} (and many other functions) can be added to the function for which the expansion was originally obtained.

Asymptotic expansions are not necessarily expansions into negative powers of x. An expansion into *positive* powers of $x - a$,

$$f(x) \sim \sum_{\nu=0}^{n-1} c_\nu (x-a)^\nu + R_n(x),$$

represents $f(x)$ asymptotically when $x \to a$ if

$$\lim_{x\to a}(x-a)^{-(n-1)}R_n(x) = 0.$$

Asymptotic expansions of the error of a numerical method into positive powers of a step length h are of great importance in the more advanced study of numerical methods. Such expansions form the basis of simple and effective acceleration methods for improving numerical results; see Sec. 3.4.

Review Questions

3.2.1 Give the Cauchy formula for the coefficients of Taylor and Laurent series, and describe the Cauchy–FFT method. Give the formula for the coefficients of a Fourier series. For which of the functions in Table 3.1.1 does another Laurent expansion also exist?

3.2.2 Describe by an example the balancing procedure that was mentioned in the subsection about perturbation expansions.

3.2.3 Define the Chebyshev polynomials, and tell some interesting properties of these and of Chebyshev expansions. For example, what do you know about the speed of convergence of a Chebyshev expansion for various classes of functions? (The detailed expressions are not needed.)

3.2.4 Describe and exemplify what is meant by an ill-conditioned power series and a preconditioner for such a series.

3.2.5 (a) Define what is meant when one says that the series $\sum_0^\infty a_n x^{-n}$

- converges to a function $f(x)$ for $x \geq R$;
- represents a function $f(x)$ asymptotically as $x \to \infty$.

(b) Give an example of a series that represents a function asymptotically as $x \to \infty$, although it diverges for every finite positive x.

(c) What is meant by semiconvergence? Say a few words about termination criteria and error estimation.

Problems and Computer Exercises

3.2.1 Some of the functions appearing in Table 3.1.1 and in other examples and problems are *not single-valued* in the complex plane. Brush up your complex analysis and find out how to define the branches, where these expansions are valid, and (if necessary) define cuts in the complex plane that must not be crossed. It turns out not to be necessary for these expansions. Why?

(a) If you have access to programs for functions of complex variables (or to commands in some package for interactive computation), find out the conventions used for functions like square root, logarithm, powers, arc tangent, etc. If the manual does not give enough detail, invent numerical tests, both with strategically chosen values of z and with random complex numbers in some appropriate domain around the origin. For example, do you obtain

$$\ln\left(\frac{z+1}{z-1}\right) - \ln(z+1) + \ln(z-1) = 0 \quad \forall z?$$

Or, what values of $\sqrt{z^2-1}$ do you obtain for $z = \pm i$? What values should you obtain, if you want the branch which is positive for $z > 1$?

(b) What do you obtain if you apply Cauchy's coefficient formula or the Cauchy–FFT method to find a Laurent expansion for $\sqrt{z}$? Note that $\sqrt{z}$ is analytic everywhere in an annulus, but that does not help. The expansion is likely to become weird. Why?

3.2.2 Apply (on a computer) the Cauchy–FFT method to find the Maclaurin coefficients a_n of (say) e^z, $\ln(1-z)$, and $(1+z)^{1/2}$. Conduct experiments with different values of r and N, and compare with the exact coefficients. This presupposes that you have access to good programs for complex arithmetic and FFT.

Try to summarize your experiences of how the error of a_n depends on r, N. You may find some guidance in Example 3.2.2.

3.2.3 (a) Suppose that r is located inside the unit circle; t is real. Show that

$$\frac{1-r^2}{1-2r\cos t+r^2} = 1+2\sum_{n=1}^{\infty} r^n \cos nt,$$

$$\frac{2r\sin t}{1-2r\cos t+r^2} = 2\sum_{n=1}^{\infty} r^n \sin nt.$$

Hint: First suppose that r is real. Set $z = re^{it}$. Show that the two series are the real and imaginary parts of $(1+z)/(1-z)$. Finally, make an analytic continuation of the results.

(b) Let a be positive, $x \in [-a, a]$, while w is complex, $w \notin [-a, a]$. Let $r = r(w)$, $|r| < 1$ be a root of the quadratic $r^2 - (2w/a)r + 1 = 0$. Show that (with an appropriate definition of the square root)

$$\frac{1}{w-x} = \frac{1}{\sqrt{w^2-a^2}} \cdot \left(1+2\sum_{n=1}^{\infty} r^n T_n\left(\frac{x}{a}\right)\right), \qquad (w \notin [-a,\ a],\ x \in [-a, a]).$$

(c) Find the expansion of $1/(1+x^2)$ for $x \in [-1.5, 1.5]$ into the polynomials $T_n(x/1.5)$. Explain the order of magnitude of the error and the main features of the error curve in Figure 3.2.5.

Hint: Set $w = i$, and take the imaginary part. Note that r becomes imaginary.

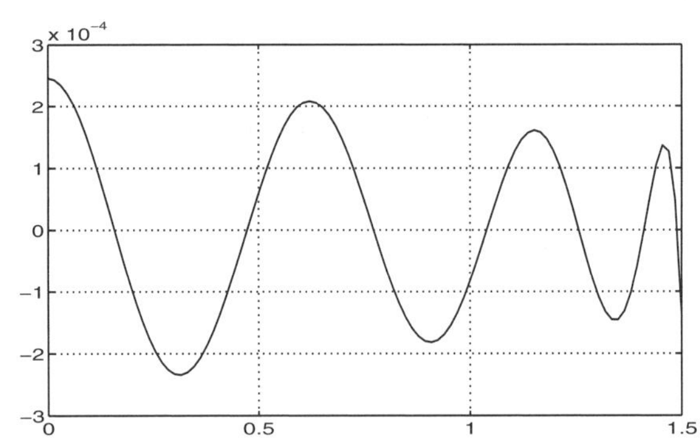

Figure 3.2.5. *The error of the expansion of* $f(x) = 1/(1+x^2)$ *in a sum of Chebyshev polynomials* $\{T_n(x/1.5)\}$, $n \leq 12$.

3.2.4 (a) Find the Laurent expansions for

$$f(z) = 1/(z-1) + 1/(z-2).$$

(b) How do you use the Cauchy–FFT method for finding Laurent expansions? Test your ideas on the function in the previous subproblem (and on a few other functions). There may be some pitfalls with the interpretation of the output from the FFT program, related to so-called **aliasing**; see Sec. 4.6.6 and Strang [339].

(c) As in Sec. 3.2.1, suppose that $F^{(p)}$ is of bounded variation in $[-\pi, \pi]$ and denote the Fourier coefficients of $F^{(p)}$ by $c_n^{(p)}$. Derive the following generalization of (3.2.7):

$$c_n = \frac{(-1)^{n-1}}{2\pi} \sum_{j=0}^{p-1} \frac{F^{(j)}(\pi) - F^{(j)}(-\pi)}{(in)^{j+1}} + \frac{c_n^{(p)}}{(in)^p}.$$

Show that if we add the condition that $F \in C^j[-\infty, \infty]$, $j < p$, then the asymptotic results given in (and after) (3.2.7) hold.

(d) Let $z = \frac{1}{2}(w + w^{-1})$. Show that $|z-1| + |z+1| = |w| + |w|^{-1}$.
Hint: Use the parallelogram law, $|p-q|^2 + |p+q|^2 = 2(|p|^2 + |q|^2)$.

3.2.5 (a) The expansion of $\operatorname{arcsinh} t$ into powers of t, truncated after t^7, is obtained from Problem 3.1.6 (b). Using economization of a power series, construct from this a polynomial approximation of the form $c_1 t + c_3 t^3$ for the interval $t \in [-\frac{1}{2}, \frac{1}{2}]$. Give bounds for the truncation error for the original truncated expansion and for the economized expansion.

(b) The graph of $T_{20}(x)$ for $x \in [-1, 1]$ is shown in Figure 3.2.1. Draw the graph of $T_{20}(x)$ for (say) $x \in [-1.1, 1.1]$.

3.2.6 Compute a few terms of the expansions into powers of ϵ or k of each of the roots of the following equations, so that the error is $O(\epsilon^2)$ or $O(k^{-2})$ (ϵ is small and positive; k is large and positive). Note that some terms may have fractional or negative exponents. Also try to fit an expansion of the wrong form in some of these examples, and see what happens.

(a) $(1+\epsilon)z^2 - \epsilon = 0$; (b) $\epsilon z^3 - z^2 + 1 = 0$; (c) $\epsilon z^3 - z + 1 = 0$;
(d) $z^4 - (k^2+1)z^2 - k^2 = 0$, $(k^2 \gg 1)$.

3.2.7 The solution of the boundary value problem

$$(1+\epsilon)y'' - \epsilon y = 0, \quad y(0) = 0, \quad y(1) = 1,$$

has an expansion of the form $y(t;\epsilon) = y_0(t) + y_1(t)\epsilon + y_2(t)\epsilon^2 + \cdots$.

(a) By coefficient matching, set up differential equations and boundary conditions for y_0, y_1, y_2, and solve them. You naturally use the boundary conditions of the original problem for y_0. Make sure you use the right boundary conditions for y_1, y_2.

(b) Set $R(t) = y_0(t) + \epsilon y_1(t) - y(t;\epsilon)$. Show that $R(t)$ satisfies the (modified) differential equation

$$(1+\epsilon)R'' - \epsilon R = \epsilon^2(7t - t^3)/6, \quad R(0) = 0, \quad R(1) = 0.$$

3.2.8 (a) Apply Kummer's first identity (3.2.39) to the error function $\operatorname{erf}(x)$, to show that

$$\operatorname{erf}(x) = \frac{2x}{\sqrt{\pi}} e^{-x^2} M\Big(1, \frac{3}{2}, x^2\Big) = \frac{2x}{\sqrt{\pi}} e^{-x^2} \Big(1 + \frac{2x^2}{3} + \frac{(2x^2)^2}{3 \cdot 5} + \frac{(2x^2)^3}{3 \cdot 5 \cdot 7} + \cdots\Big).$$

Why is this series well-conditioned? (Note that it is a bell sum; compare Figure 3.2.3.) Investigate the largest term, rounding errors, truncation errors, and termination criterion.

(b) $\operatorname{erfc}(x)$ has a semiconvergent expansion for $x \gg 1$ that begins

$$\operatorname{erfc}(x) = 1 - \operatorname{erf}(x) = \frac{2}{\sqrt{\pi}} \int_x^\infty e^{-t^2}\, dt = \frac{e^{-x^2}}{x\sqrt{\pi}} \Big(1 - \frac{1}{2x^2} + \frac{3}{4x^4} - \frac{15}{8x^6} + \cdots\Big).$$

Give an explicit expression for the coefficients, and show that the series diverges for every x. Where is the smallest term? Estimate its size.

Hint: Set $t^2 = x^2 + u$, and proceed analogously to Example 3.2.8. See Problem 3.1.7 (c), $\alpha = \frac{1}{2}$, about the remainder term. Alternatively, apply repeated integration by parts; it may be easier to find the remainder in this way.

3.2.9 *Other notations for series, with application to Bessel functions.*

(a) Set

$$f(x) = \sum_{n=0}^{\infty} \frac{a_n x^n}{n!}, \qquad g(x) = \sum_{n=0}^{\infty} \frac{b_n x^n}{n!}, \qquad h(x) = \sum_{n=0}^{\infty} \frac{c_n x^n}{n!},$$

$$\phi(w) = \sum_{n=0}^{\infty} \frac{\alpha_n w^n}{n!n!}, \qquad \psi(w) = \sum_{n=0}^{\infty} \frac{\beta_n w^n}{n!n!}, \qquad \chi(w) = \sum_{n=0}^{\infty} \frac{\gamma_n w^n}{n!n!}.$$

Let $h(x) = f(x) \cdot g(x)$, $\chi(w) = \phi(w) \cdot \psi(w)$. Show that

$$c_n = \sum_{j=0}^{n} \binom{n}{j} a_j b_{n-j}, \qquad \gamma_n = \sum_{j=0}^{n} \binom{n}{j}^2 \alpha_j \beta_{n-j}.$$

Derive analogous formulas for series of the form $\sum_{n=0}^{\infty} a_n w^n/(2n)!$.
Suggest how to *divide* two power series in these notations.

(b) Let $a_j = (-1)^j a'_j$, $g(x) = e^x$. Show that

$$c_n = \sum_{j=0}^{n} \binom{n}{j} (-1)^j a'_j.$$

Comment: By (3.2.1), this can also be written $c_n = (-1)^n \Delta^n a_0$. This proves the mathematical equivalence of the preconditioners (3.1.55) and (3.1.59) if $P(x) = e^x$.

(c) Set, according to Example 3.2.8 and part (a) of this problem, $w = -x^2/4$,

$$J_0(x) = \sum_{n=0}^{\infty} \frac{(-1)^n w^n}{n!n!}, \qquad I_0(x) = \sum_{n=0}^{\infty} \frac{w^n}{n!n!}, \qquad IJ(x) \equiv I_0(x) J_0(x) = \sum_{n=0}^{\infty} \frac{\gamma_n w^n}{n!n!}.$$

Show that

$$\gamma_n = \sum_{j=0}^{n} (-1)^j \binom{n}{j} \binom{n}{n-j} = \begin{cases} (-1)^m \binom{2m}{m} & \text{if } n = 2m, \\ 0 & \text{if } n = 2m+1. \end{cases}$$

Hint: The first expression for γ_n follows from (a). It can be interpreted as the coefficient of t^n in the product $(1-t)^n(1+t)^n$. The second expression for γ_n is the same coefficient in $(1-t^2)^n$.

(d) The second expression for γ_n in (c) is used in Example 3.2.8.[67] Reconstruct and extend the results of that example. Design a termination criterion. Where is the largest modulus of a term of the preconditioned series, and how large is it approximately? Make a crude guess in advance of the rounding error in the preconditioned series.

(e) Show that the power series of $J_0(x)$ can be written in the form

$$\sum_{n=0}^{\infty} a_n \frac{(-x^2)^n}{(2n)!},$$

where a_n is positive and decreases slowly and smoothly.

Hint: Compute a_{n+1}/a_n.

(f) It is known (see Lebedev [240, eq. (9.13.11)]) that

$$J_0(x) = e^{-ix} M\left(\frac{1}{2}, 1; 2ix\right),$$

where $M(a, b, c)$ is Kummer's confluent hypergeometric function, this time with an imaginary argument. Show that Kummer's first identity is unfortunately of no use here for preconditioning the power series.

Comment: Most of the formulas and procedures in this problem can be generalized to the series for the Bessel functions of the first kind of general integer order, $(z/2)^{-n}J_n(x)$. These belong to the most studied functions of applied mathematics, and there exist more efficient methods for computing them; see, e.g., Press et al. [294, Chapter 6]. This problem shows, however, that *preconditioning can work well* for a nontrivial power series, and it is worth being tried.

3.2.10. (a) Derive the expansion of Example 3.2.5 by repeated integration by parts.

(b) Derive the Maclaurin expansion with the remainder according to (3.1.5) by the application of repeated integration by parts to the equation

$$f(z) - f(0) = z \int_0^1 f'(zt)\, d(t-1).$$

3.2.11. Show the following generalization of Theorem 3.2.5. Assume that $|f(z)| \le M$ for $z \in \mathcal{E}_R$. Let $|\zeta| \in \mathcal{E}_\rho$, $1 < \rho < r \le R - \epsilon$. Then the Chebyshev expansion of $f(\zeta)$

[67] It is much better conditioned than the first expression. This may be one reason why multiple precision is not needed here.

satisfies the inequality

$$\left| f(\zeta) - \sum_{j=0}^{n-1} c_j T_j(\zeta) \right| \le \frac{2M(\rho/R)^n}{1-\rho/R}.$$

Hint: Set $\omega = \zeta + \sqrt{\zeta^2 - 1}$, and show that $|T_j(\zeta)| = |\frac{1}{2}(\omega^j + \omega^{-j})| \le \rho^j$.

3.3 Difference Operators and Operator Expansions

3.3.1 Properties of Difference Operators

Difference operators are handy tools for the derivation, analysis, and practical application of numerical methods for many problems for interpolation, differentiation, and quadrature of a function in terms of its values at equidistant arguments. The simplest notations for difference operators and applications to derivatives were mentioned in Sec. 1.1.4.

Let y denote a sequence $\{y_n\}$. Then we define the **shift operator** E (or translation operator) and the **forward difference operator** Δ by the relations

$$Ey = \{y_{n+1}\}, \qquad \Delta y = \{y_{n+1} - y_n\};$$

E and Δ are thus operators which map one sequence to another sequence. Note, however, that if y_n is defined for $a \le n \le b$ only, then Ey_b is not defined, and the sequence Ey has fewer elements than the sequence y. (It is therefore sometimes easier to extend the sequences to infinite sequences, for example, by adding zeros in both directions outside the original range of definition.)

These operators are **linear**, i.e., if α, β are real or complex constants and if y, z are two sequences, then $E(\alpha y + \beta z) = \alpha Ey + \beta Ez$, and similarly for Δ.

Powers of E and Δ are defined recursively, i.e.,

$$E^k y = E(E^{k-1}y), \qquad \Delta^k y = \Delta(\Delta^{k-1}y).$$

By induction, the first relation yields $E^k y = \{y_{n+k}\}$. We extend the validity of this relation to $k = 0$ by setting $E^0 y = y$ and to negative values of k. $\Delta^k y$ is called the kth difference of the sequence y. We make the convention that $\Delta^0 = 1$. There will be little use of Δ^k for negative values of k in this book, although Δ^{-1} can be interpreted as a summation operator.

Note that $\Delta y = Ey - y$, and $Ey = y + \Delta y$ for any sequence y. It is therefore convenient to express these as equations between operators:

$$\Delta = E - 1, \qquad E = 1 + \Delta.$$

The identity operator is in this context traditionally denoted by 1. It can be shown that all formulas derived from the axioms of commutative algebra can be used for these operators, for example, the binomial theorem for positive integral k,

$$\Delta^k = (E-1)^k = \sum_{j=0}^{k} (-1)^{k-j} \binom{k}{j} E^j, \qquad E^k = (1+\Delta)^k = \sum_{j=0}^{k} \binom{k}{j} \Delta^j, \tag{3.3.1}$$

giving

$$(\Delta^k y)_n = \sum_{j=0}^{k} (-1)^{k-j} \binom{k}{j} y_{n+j}, \qquad y_{n+k} = (E^k y)_n = \sum_{j=0}^{k} \binom{k}{j} (\Delta^j y)_n. \tag{3.3.2}$$

We abbreviate the notation further and write, for example, $Ey_n = y_{n+1}$ instead of $(Ey)_n = y_{n+1}$, and $\Delta^k y_n$ instead of $(\Delta^k y)_n$. But it is important to remember that Δ *operates on sequences* and not on elements of sequences. Thus, strictly speaking, this abbreviation is incorrect, though convenient. The formula for E^k will, in the next subsection, be extended to an infinite series for nonintegral values of k, but that is beyond the scope of algebra.

A **difference scheme** consists of a sequence and its difference sequences, arranged in the following way:

$$\begin{array}{ccccc}
y_0 & & & & \\
 & \Delta y_0 & & & \\
y_1 & & \Delta^2 y_0 & & \\
 & \Delta y_1 & & \Delta^3 y_0 & \\
y_2 & & \Delta^2 y_1 & & \Delta^4 y_0 \\
 & \Delta y_2 & & \Delta^3 y_1 & \\
y_3 & & \Delta^2 y_2 & & \\
 & \Delta y_3 & & & \\
y_4 & & & &
\end{array}$$

A difference scheme is best computed by successive subtractions; the formulas in (3.3.1) are used mostly in theoretical contexts.

In many applications the quantities y_n are computed in increasing order $n = 0, 1, 2, \ldots,$ and it is natural that a difference scheme is constructed by means of the quantities previously computed. One therefore introduces the **backward difference operator**

$$\nabla y_n = y_n - y_{n-1} = (1 - E^{-1}) y_n.$$

For this operator we have

$$\nabla^k = (1 - E^{-1})^k, \qquad E^{-k} = (1 - \nabla)^k. \tag{3.3.3}$$

Note the **reciprocity** in the relations between ∇ and E^{-1}.

Any linear combination of the elements $y_n, y_{n-1}, \ldots, y_{n-k}$ *can also be expressed as a linear combination of* $y_n, \nabla y_n, \ldots, \nabla^k y_n$, and vice versa.[68] For example,

$$y_n + y_{n-1} + y_{n-2} = 3y_n - 3\nabla y_n + \nabla^2 y_n,$$

because $1 + E^{-1} + E^{-2} = 1 + (1 - \nabla) + (1 - \nabla)^2 = 3 - 3\nabla + \nabla^2$. By reciprocity, we also obtain $y_n + \nabla y_n + \nabla^2 y_n = 3y_n - 3y_{n-1} + y_{n-2}$.

[68] An analogous statement holds for the elements $y_n, y_{n+1}, \ldots, y_{n+k}$ and forward differences.

In this notation the difference scheme reads as follows.

$$
\begin{array}{ccccccccc}
y_0 & & & & & & & & \\
 & \nabla y_1 & & & & & & & \\
y_1 & & \nabla^2 y_2 & & & & & & \\
 & \nabla y_2 & & \nabla^3 y_3 & & & & & \\
y_2 & & \nabla^2 y_3 & & \nabla^4 y_4 & & & & \\
 & \nabla y_3 & & \nabla^3 y_4 & & & & & \\
y_3 & & \nabla^2 y_4 & & & & & & \\
 & \nabla y_4 & & & & & & & \\
y_4 & & & & & & & &
\end{array}
$$

In the backward difference scheme the subscripts are constant along diagonals directed upward (backward) to the right, while in the forward difference scheme subscripts are constant along diagonals directed downward (forward). Note, for example, that $\nabla^k y_n = \Delta^k y_{n-k}$. In a computer, a backward difference scheme is preferably stored as a lower triangular matrix.

Example 3.3.1.

Part of the difference scheme for the sequence $y = \{\ldots, 0, 0, 0, 1, 0, 0, 0, \ldots\}$ is given below.

$$
\begin{array}{ccccccccc}
 & & & 0 & & 1 & & -7 & \\
 & & 0 & & 1 & & -6 & & 28 \\
 & 0 & & 1 & & -5 & & 21 & \\
0 & & 1 & & -4 & & 15 & & -56 \\
 & 1 & & -3 & & 10 & & -35 & \\
1 & & -2 & & 6 & & -20 & & 70 \\
 & -1 & & 3 & & -10 & & 35 & \\
0 & & 1 & & -4 & & 15 & & -56 \\
 & 0 & & -1 & & 5 & & -21 & \\
 & & 0 & & 1 & & -6 & & 28 \\
 & & & 0 & & -1 & & 7 &
\end{array}
$$

This example shows the *effect of a disturbance in one element* on the sequence of the higher differences. Because the effect broadens out and grows quickly, difference schemes are useful in the investigation and correction of computational and other errors, so-called **difference checks**. Notice that, since the differences are *linear* functions of the sequence, a **superposition principle** holds. The effect of errors can thus be estimated by studying simple sequences such as the one above.

Example 3.3.2.

The following is a difference scheme for a five-decimal table of the function $f(x) = \tan x$, $x \in [1.30, 1.36]$, with step $h = 0.01$. The differences are given with 10^{-5} as unit.

x	y	∇y	$\nabla^2 y$	$\nabla^3 y$	$\nabla^4 y$	$\nabla^5 y$	$\nabla^6 y$
1.30	3.60210						
		14498					
1.31	3.74708		1129				
		15627		140			
1.32	3.90335		1269		26		
		16896		166		2	
1.33	4.07231		1435		28		9
		18331		194		11	
1.34	4.25562		1629		39		
		19960		233			
1.35	4.45522		1862				
		21822					
1.36	4.67344						

We see that the differences decrease roughly by a factor of 0.1—that indicates that the step size has been chosen suitably for the purposes of interpolation, numerical quadrature, etc. until the last two columns, where the rounding errors of the function values have a visible effect.

Example 3.3.3.

For the sequence $y_n = (-1)^n$ one finds easily that

$$\nabla y_n = 2y_n, \quad \nabla^2 y_n = 4y_n, \dots, \qquad \nabla^k y_n = 2^k y_n.$$

If the errors in the elements of the sequence are bounded by ϵ, it follows that the errors of the kth differences are bounded by $2^k \epsilon$. A rather small reduction of this bound is obtained if the errors are assumed to be independent random variables (cf. Problem 3.4.24).

It is natural also to consider *difference operations on functions* not just on sequences. E and Δ map the function f onto functions whose values at the point x are

$$E\, f(x) = f(x+h), \qquad \Delta f(x) = f(x+h) - f(x),$$

where h is the *step size*. Of course, Δf depends on h; in some cases this should be indicated in the notation. One can, for example, write $\Delta_h f(x)$, or $\Delta f(x; h)$. If we set $y_n = f(x_0 + nh)$, the difference scheme of the function with step size h is the same as for the sequence $\{y_n\}$. Again it is important to realize that, in this case, the operators act on *functions*, not on the values of functions. It would be more correct to write $f(x_0 + h) = (Ef)(x_0)$. Actually, the notation $(x_0)Ef$ would be even more logical, since the insertion of the value of the argument x_0 is the last operation to be done, and the convention for the order of execution of operators proceeds from right to left.[69]

[69] The notation $[x_0]f$ occurs, however, naturally in connection with divided differences; see Sec. 4.2.1.

Note that *no new errors are introduced during the computation of the differences, but the effects of the original* irregular errors, *for example, rounding errors in y, grow exponentially*. Note that systematic errors, for example, truncation errors in the numerical solution of a differential equation, often have a smooth difference scheme. For example, if the values of y have been produced by the iterative solution of an equation, where x is a parameter, with the same number of iterations for every x and y and the same algorithm for the first approximation, then the truncation error of y is likely to be a smooth function of x.

Difference operators are in many respects similar to differentiation operators. Let f be a polynomial. By Taylor's formula,

$$\Delta f(x) = f(x+h) - f(x) = hf'(x) + \frac{1}{2}h^2 f''(x) + \cdots.$$

We see from this that $\deg \Delta f = \deg f - 1$. Similarly, for differences of higher order, *if f is a polynomial of degree less than k, then*

$$\Delta^{k-1} f(x) = constant, \qquad \Delta^p f(x) = 0 \; \forall p \geq k.$$

The same holds for backward differences.

The following important result can be derived directly from Taylor's theorem with the integral form of the remainder. Assume that all derivatives of f up to kth order are continuous. If $f \in C^k$,

$$\Delta^k f(x) = h^k f^{(k)}(\zeta), \quad \zeta \in [x, x+kh]. \tag{3.3.4}$$

Hence $h^{-k}\Delta^k f(x)$ is an approximation to $f^{(k)}(x)$; the error of this approximation approaches zero as $h \to 0$ (i.e., as $\zeta \to x$). As a rule, the error is approximately proportional to h. We postpone the proof to Sec. 4.2.1, where it appears as a particular case of a theorem concerning divided differences.

Even though difference schemes do not have the same importance today that they had in the days of hand calculations or calculation with desk calculators, they are still important conceptually, and we shall also see how they are still useful in practical computing. In a computer it is more natural to store a difference scheme as an array, with $y_n, \nabla y_n, \nabla^2 y_n, \dots, \nabla^k y_n$ in a row (instead of along a diagonal).

Many formulas for differences are analogous to formulas for derivatives, though usually more complicated. The following results are among the most important.

Lemma 3.3.1.

It holds that

$$\Delta^k(a^x) = (a^h - 1)^k a^x, \qquad \nabla^k(a^x) = (1 - a^{-h})^k a^x. \tag{3.3.5}$$

For sequences, i.e., if $h = 1$,

$$\Delta^k\{a^n\} = (a-1)^k\{a^n\}, \qquad \Delta^k\{2^n\} = \{2^n\}. \tag{3.3.6}$$

Proof. Let c be a given constant. For $k = 1$ we have

$$\Delta(ca^x) = ca^{x+h} - ca^x = ca^x a^h - ca^x = c(a^h - 1)a^x.$$

The general result follows easily by induction. The backward difference formula is derived in the same way. □

Lemma 3.3.2 (*Difference of a Product*).

$$\Delta(u_n v_n) = u_n \Delta v_n + \Delta u_n \, v_{n+1}. \tag{3.3.7}$$

Proof. We have

$$\begin{aligned}\Delta(u_n v_n) &= u_{n+1} v_{n+1} - u_n v_n \\ &= u_n (v_{n+1} - v_n) + (u_{n+1} - u_n) v_{n+1}.\end{aligned}$$

Compare the above result with the formula for differentials, $d(uv) = udv + vdu$. Note that we have v_{n+1} (not v_n) on the right-hand side. □

Lemma 3.3.3 (*Summation by Parts*).

$$\sum_{n=0}^{N-1} u_n \Delta v_n = u_N v_N - u_0 v_0 - \sum_{n=0}^{N-1} \Delta u_n \, v_{n+1}. \tag{3.3.8}$$

Proof. (Compare with the rule for integration by parts and its proof!) Notice that

$$\begin{aligned}\sum_{n=0}^{N-1} \Delta w_n &= (w_1 - w_0) + (w_2 - w_1) + \cdots + (w_N - w_{N-1}) \\ &= w_N - w_0.\end{aligned}$$

Use this on $w_n = u_n v_n$. From the result in Lemma 3.3.1 one gets after summation

$$u_N v_N - u_0 v_0 = \sum_{n=0}^{N-1} u_n \Delta v_n + \sum_{n=0}^{N-1} \Delta u_n v_{n+1},$$

and the result follows. (For an extension, see Problem 3.3.2 (d).) □

3.3.2 The Calculus of Operators

Formal calculations with operators, using the rules of algebra and analysis, are often an elegant means of assistance in *finding approximation formulas that are exact for all polynomials of degree less than (say) k*, and they should therefore be useful for functions that can be accurately approximated by such a polynomial. Our calculations often lead to divergent (or semiconvergent) series, but the way we handle them can usually be justified by means of the theory of formal power series, of which a brief introduction was given at the end of Sec. 3.1.5. The operator calculations also provide error estimates, asymptotically valid as the step size $h \to 0$. Rigorous error bounds can be derived by means of Peano's remainder theorem in Sec. 3.3.3.

Operator techniques are sometimes successfully used (see Sec. 3.3.4) in a way that is hard, or even impossible, to justify by means of formal power series. It is then not trivial to formulate appropriate conditions for the success and to derive satisfactory error bounds and error estimates, but it can sometimes be done.

We make a digression about terminology. More generally, *the word* **operator** *is in this book used for a function that maps a linear space $\mathcal{S}$ into another linear space $\mathcal{S}'$.* $\mathcal{S}$ can, for example, be a space of functions, a coordinate space, or a space of sequences. The dimension of these spaces can be finite or infinite. For example, the differential operator D maps the infinite-dimensional space $C^1[a, b]$ of functions with a continuous derivative, defined on the interval $[a, b]$, into the space $C[a, b]$ of continuous functions on the same interval.

In the following we denote by $\mathcal{P}_n$ the set of polynomials of degree *less than* n.[70] Note that $\mathcal{P}_n$ is an n-dimensional linear space for which $\{1, x, x^2, \dots, x^{n-1}\}$ is a basis called the *power basis*; the coefficients $(c_1, c_2, \dots, c_n)$ are then the *coordinates* of the polynomial p defined by $p(x) = \sum_{i=1}^{n} c_i x^{i-1}$.

For simplicity, we shall assume that the space of functions on which the operators are defined is $C^\infty(-\infty, \infty)$, i.e., the functions are infinitely differentiable on $(-\infty, \infty)$. This sometimes requires (theoretically) a modification of a function outside the bounded interval, where it is interesting. There are techniques for achieving this, but they are beyond the scope of this book. Just imagine that they have been applied.

We define the following operators:

$$\begin{aligned}
&Ef(x) = f(x+h) && \text{Shift (or translation) operator,}\\
&\Delta f(x) = f(x+h) - f(x) && \text{Forward difference operator,}\\
&\nabla f(x) = f(x) - f(x-h) && \text{Backward difference operator,}\\
&Df(x) = f'(x) && \text{Differentiation operator,}\\
&\delta f(x) = f(x+\tfrac{1}{2}h) - f(x-\tfrac{1}{2}h) && \text{Central difference operator,}\\
&\mu f(x) = \tfrac{1}{2}\big(f(x+\tfrac{1}{2}h) + f(x-\tfrac{1}{2}h)\big) && \text{Averaging operator.}
\end{aligned}$$

Suppose that the values of f are given on an equidistant grid only, e.g., $x_j = x_0 + jh$, $j = -M : N$ (j is an integer). Set $f_j = f(x_j)$. Note that $\delta f_j, \delta^3 f_j, \dots$ (odd powers) and μf_j *cannot* be exactly computed; they are available halfway between the grid points. (A way to get around this is given later; see (3.3.45).) The even powers $\delta^2 f_j, \delta^4 f_j, \dots$ and $\mu\delta f_j, \mu\delta^3 f_j, \dots$ *can* be exactly computed. This follows from the formulas

$$\mu\delta f(x) = \frac{1}{2}\big(f(x+h) - f(x-h)\big), \quad \mu\delta = \frac{1}{2}(\Delta + \nabla), \quad \delta^2 = \Delta - \nabla. \qquad (3.3.9)$$

Several other notations are in use. For example, in the study of difference methods for partial differential equations D_{+h}, D_{0h}, and D_{-h} are used instead of Δ, $\mu\delta$, and ∇, respectively.

An operator P is said to be a **linear operator** if

$$P(\alpha f + \beta g) = \alpha P f + \beta P g$$

holds for arbitrary complex constants α, β and arbitrary functions f, g. The above six operators are all linear. The operation of multiplying by a constant α is also a linear operator.

[70] Some authors use similar notations to denote the set of polynomials of degree less than or equal to n.

If P and Q are two operators, then their sum and product can be defined in the following way:

$$\begin{aligned}(P+Q)f &= Pf + Qf,\\ (P-Q)f &= Pf - Qf,\\ (PQ)f &= P(Qf),\\ (\alpha P)f &= \alpha(Pf),\\ P^n f &= P\cdot P \cdots Pf, \quad n \text{ factors}.\end{aligned}$$

Two operators are equal, $P = Q$, if $Pf = Qf$, for all f in the space of functions considered. Notice that $\Delta = E - 1$. One can show that the following rules hold for all linear operators:

$$P + Q = Q + P, \qquad P + (Q + R) = (P + Q) + R,$$
$$P(Q + R) = PQ + PR, \qquad P(QR) = (PQ)R.$$

The above six operators, E, Δ, ∇, hD, δ, and μ, and the combinations of them by these algebraic operations make a *commutative ring*. Thus, $PQ = QP$ holds for these operators, and any algebraic identity that is generally valid in such rings can be used.

If $\mathcal{S} = \mathbf{R^n}$, $\mathcal{S}' = \mathbf{R^m}$, and the elements are *column* vectors, then the linear operators are matrices of size $[m, n]$. They generally do not commute.

If $\mathcal{S}' = \mathbf{R}$ or $\mathbf{C}$, the operator is called a **functional**. Examples of functionals are, if x_0 denotes a fixed (though arbitrary) point,

$$Lf = f(x_0), \quad Lf = f'(x_0), \quad Lf = \int_0^1 e^{-x} f(x)\,dx, \quad \int_0^1 |f(x)|^2\,dx;$$

all except the last one are **linear functionals**.

There is a subtle distinction here. For example, E is a linear operator that maps a function to a function. Ef is the function whose value at the point x is $f(x+h)$. If we consider a fixed point x_0, then $(Ef)(x_0)$ is a scalar. This is therefore a *linear functional*. We shall allow ourselves to simplify the notation and to write $Ef(x_0)$, but it must be understood that E operates on the function f, not on the function value $f(x_0)$. This was just one example; simplifications like this will be made with other operators than E, and similar simplifications in notation were suggested earlier in this chapter. There are, however, situations where it is, for the sake of clarity, advisable to return to the more specific notation with a larger number of parentheses.

If we represent the vectors in $\mathbf{R}^n$ by *columns* y, the linear functionals in $\mathbf{R}^n$ are the scalar products $a^T x = \sum_{i=1}^n a_i y_i$; every *row* a^T thus defines a linear functional.

Examples of linear functionals in $\mathcal{P}_k$ are linear combinations of a finite number of function values, $Lf = \sum a_j f(x_j)$. If $x_j = x_0 + jh$ the same functional can be expressed in terms of differences, e.g., $\sum a'_j \Delta^j f(x_0)$; see Problem 3.3.4. The main purpose of this section is to show how operator methods can be used for finding approximations of this form to linear functionals in more general function spaces. First, we need a general theorem.

Theorem 3.3.4.

Let $x_1, x_2, \ldots, x_k$ be k distinct real (or complex) numbers. Then no nontrivial relation of the form

$$\sum_{j=1}^{k} a_j f(x_j) = 0 \tag{3.3.10}$$

can hold for all $f \in \mathcal{P}_k$. If we add one more point (x_0), there exists only one nontrivial relation of the form $\sum_{j=0}^{k} a'_j f(x_j) = 0$ (except that it can be multiplied by an arbitrary constant). In the equidistant case, i.e., if $x_j = x_0 + jh$, then

$$\sum_{j=0}^{k} a'_j f(x_j) \equiv c\Delta^k f(x_0), \quad c \neq 0.$$

Proof. If (3.3.10) were valid for all $f \in \mathcal{P}_k$, then the linear system $\sum_{j=1}^{k} x_j^{i-1} a_j = 0$, $i = 1 : k$, would have a nontrivial solution $(a_1, a_2, \ldots, a_k)$. The matrix of the system, however, is a **Vandermonde matrix**,[71]

$$V = [x_j^{i-1}]_{i,j=1}^{k} = \begin{pmatrix} 1 & 1 & \cdots & 1 \\ x_1 & x_2 & \cdots & x_k \\ \vdots & \vdots & \cdots & \vdots \\ x_1^{k-1} & x_2^{k-1} & \cdots & x_k^{k-1} \end{pmatrix} \tag{3.3.11}$$

(see Problem 3.3.1). Its determinant can be shown to equal the product of all differences, i.e.,

$$\det(V) = \prod_{1 \le i < j \le k} (x_i - x_j). \tag{3.3.12}$$

This is nonzero if and only if the points are distinct.

Now we add the point x_0. Suppose that there exist two relations,

$$\sum_{j=0}^{k} b_j f(x_j) = 0, \qquad \sum_{j=0}^{k} c_j f(x_j) = 0,$$

with linearly independent coefficient vectors. Then we can find a (nontrivial) linear combination, where x_0 has been eliminated, but this contradicts the result that we have just proved. Hence the hypothesis is wrong; the two coefficient vectors must be proportional.

We have seen above that, in the equidistant case, $\Delta^k f(x_0) = 0$ is such a relation. More generally, we shall see in Chapter 4 that, for $k + 1$ arbitrary distinct points, the kth order *divided difference* is zero for all $f \in \mathcal{P}_k$. □

[71]Alexandre Théophile Vandermonde (1735–1796), member of the French Academy of Sciences, is regarded as the founder of the theory of determinants. What is now referred to as the Vandermonde matrix does not seem to appear in his writings!

Corollary 3.3.5.

Suppose that a formula for interpolation, numerical differentiation, or integration has been derived by an operator technique. If it is a linear combination of the values of $f(x)$ *at* k given *distinct points* x_j, $j = 1 : k$, *and is exact for all* $f \in \mathcal{P}_k$, *this formula is unique. (If it is exact for all* $f \in \mathcal{P}_m$, $m < k$, *only, it is not unique.)*

In particular, for any $\{c_j\}_{j=1}^k$, *a unique polynomial* $P \in \mathcal{P}_k$ *is determined by the interpolation conditions* $P(x_j) = c_j$, $j = 1 : k$.

Proof. The difference between two formulas that use the same function values would lead to a relation that is impossible, by the theorem. □

Now we shall go outside of polynomial algebra and consider also *infinite series of operators*. The Taylor series

$$f(x+h) = f(x) + hf'(x) + \frac{h^2}{2!}f''(x) + \frac{h^3}{3!}f'''(x) + \cdots$$

can be written symbolically as

$$Ef = \left(1 + hD + \frac{(hD)^2}{2!} + \frac{(hD)^3}{3!} + \cdots\right)f.$$

We can here treat hD like an algebraic indeterminate, and consider the series inside the parenthesis (without the operand) as a *formal power series*.[72]

For a formal power series the concepts of convergence and divergence do not exist. When the operator series acts on a function f, and is evaluated at a point c, we obtain an ordinary numerical series, related to the linear functional $Ef(c) = f(c+h)$. We know that this Taylor series may converge or diverge, depending on f, c, and h.

Roughly speaking, the last part of Sec. 3.1.5 tells us that, with some care, "analytic functions" of one indeterminate can be handled with the same rules as analytic functions of one complex variable.

Theorem 3.3.6.

$$\begin{aligned} e^{hD} &= E = 1 + \Delta, \qquad e^{-hD} = E^{-1} = 1 - \nabla, \\ 2\sinh\frac{1}{2}hD &= e^{hD/2} - e^{-hD/2} = \delta, \\ (1+\Delta)^\theta &= (e^{hD})^\theta = e^{\theta hD}, \quad (\theta \in \mathbf{R}). \end{aligned}$$

Proof. The first formula follows from the previous discussion. The second and the third formulas are obtained in a similar way. (Recall the definition of δ.) The last formula follows from the first formula together with Lemma 3.1.9 (in Sec. 3.1.5). □

It follows from the power series expansion that

$$(e^{hD})^\theta f(x) = e^{\theta hD} f(x) = f(x + \theta h),$$

[72]We now abandon the bold face notation for indeterminates and formal power series used in Sec. 3.1.5 for the function e^{hD}, which is defined by this series. The reader is advised to take a look again at the last part of Sec. 3.1.5.

when it converges. Since $E = e^{hD}$ it is natural to *define*

$$E^\theta f(x) = f(x + \theta h),$$

and we extend this definition to such values of θ that the power series for $e^{\theta hD} f(x)$ is divergent. Note that, for example, the formula

$$E^{\theta_2} E^{\theta_1} f(x) = E^{\theta_2+\theta_1} f(x)$$

follows from this definition.

When one works with operators or functionals it is advisable to avoid notations like Δx^n, $De^{\alpha x}$, where the variables appear in the operands. For two important functions we therefore set

$$F_\alpha : F_\alpha(x) = e^{\alpha x}, \qquad f_n : f_n(x) = x^n. \tag{3.3.13}$$

Let P be any of the operators mentioned above. When applied to F_α it acts like a scalar that we shall call **the scalar of the operator**[73] and denote by $\mathrm{sc}(P)$:

$$PF_\alpha = \mathrm{sc}(P) F_\alpha.$$

We may also write $\mathrm{sc}(P; h\alpha)$ if it is desirable to emphasize its dependence on $h\alpha$. (We normalize the operators so that this is true; for example, we work with hD instead of D.) Note that

$$\mathrm{sc}(\beta P + \gamma Q) = \beta \mathrm{sc}(P) + \gamma \mathrm{sc}(Q), \quad (\beta, \gamma \in \mathbf{C}),$$
$$\mathrm{sc}(PQ) = \mathrm{sc}(P)\mathrm{sc}(Q).$$

For our most common operators we obtain

$$\begin{aligned}
\mathrm{sc}(E^\theta) &= e^{\theta h\alpha},\\
\mathrm{sc}(\nabla) &= \mathrm{sc}(1 - E^{-1}) = 1 - e^{-h\alpha},\\
\mathrm{sc}(\Delta) &= \mathrm{sc}(E - 1) = e^{h\alpha} - 1,\\
\mathrm{sc}(\delta) &= \mathrm{sc}(E^{1/2} - E^{-1/2}) = e^{h\alpha/2} - e^{-h\alpha/2}.
\end{aligned}$$

Let Q_h be one of the operators hD, Δ, δ, ∇. It follows from the last formulas that

$$\mathrm{sc}(Q_h) \sim h\alpha, \quad (h \to 0); \qquad |\mathrm{sc}(Q_h)| \le |h\alpha| e^{|h\alpha|}.$$

The main reason for grouping these operators together is that each of them has the important property (3.3.4), i.e., $Q_h^k f(c) = h^k f^{(k)}(\zeta)$, where ζ lies in the smallest interval that contains all the arguments used in the computation of $Q_h^k f(c)$. Hence,

$$f \in \mathcal{P}_k \quad \Rightarrow \quad Q_h^n f = 0 \quad \forall n \ge k. \tag{3.3.14}$$

This property[74] makes each of these four operators well suited to be the indeterminate in a formal power series that, hopefully, will be able to generate a sequence of approximations,

[73] In applied Fourier analysis this scalar is, for $\alpha = i\omega$, often called the *symbol of the operator*.

[74] The operators E and μ do *not* possess this property.

$L_1, L_2, L_3 \ldots$, to a given linear operator L. L_n is the nth partial sum of a formal power series for L. Then

$$f \in \mathcal{P}_k \quad \Rightarrow \quad L_n f = L_k f \quad \forall n \geq k. \tag{3.3.15}$$

We shall see in the next theorem that, for expansion into powers of Q_h,

$$\lim_{n\to\infty} L_n f(x) = Lf(x)$$

if f is a polynomial. This is not quite self-evident because it is not true for all functions f, and we have seen in Sec. 3.1.5 that it can happen that an expansion converges to a "wrong result." We shall see more examples of that later. *Convergence does not necessarily imply validity.*

Suppose that z is a complex variable, and that $\phi(z)$ is analytic at the origin, i.e., $\phi(z)$ is equal to its Maclaurin series, (say)

$$\phi(z) = a_0 + a_1 z + a_2 z^2 + \cdots,$$

if $|z| < \rho$ for some $\rho > 0$. For multivalued functions we always refer to the principal branch. The operator function $\phi(Q_h)$ is usually defined by the *formal* power series,

$$\phi(Q_h) = a_0 + a_1 Q_h + a_2 Q_h^2 + \cdots,$$

where Q_h is treated like an algebraic indeterminate.

The operators $E, hD, \Delta, \delta, \nabla$, and μ are related to each others. See Table 3.3.1, which is adapted from an article by the eminent blind British mathematician W. G. Bickley [25]. Some of these formulas follow almost directly from the definitions; others are derived in this section. We find the value sc$(\cdot)$ for each of these operators by *substituting α for D in the last column of the table.* (Why?)

Table 3.3.1. *Bickley's table of relations between difference operators.*

	E	Δ	δ	∇	hD
E	E	$1+\Delta$	$1+\frac{1}{2}\delta^2+\delta\sqrt{1+\frac{1}{4}\delta^2}$	$\dfrac{1}{1-\nabla}$	e^{hD}
Δ	$E-1$	Δ	$\delta\sqrt{1+\frac{1}{4}\delta^2}+\frac{1}{2}\delta^2$	$\dfrac{\nabla}{1-\nabla}$	$e^{hD}-1$
δ	$E^{1/2}-E^{-1/2}$	$\Delta(1+\Delta)^{-1/2}$	δ	$\nabla(1-\nabla)^{-1/2}$	$2\sinh\frac{1}{2}hD$
∇	$1-E^{-1}$	$\dfrac{\Delta}{1+\Delta}$	$\delta\sqrt{1+\frac{1}{4}\delta^2}-\frac{1}{2}\delta^2$	∇	$1-e^{-hD}$
hD	$\ln E$	$\ln(1+\Delta)$	$2\sinh^{-1}\frac{1}{2}\delta$	$-\ln(1-\nabla)$	hD
μ	$\frac{1}{2}(E^{1/2}+E^{-1/2})$	$\dfrac{1+\frac{1}{2}\Delta}{(1+\Delta)^{1/2}}$	$\sqrt{1+\frac{1}{4}\delta^2}$	$\dfrac{1-\frac{1}{2}\nabla}{(1-\nabla)^{1/2}}$	$\cosh\frac{1}{2}hD$

Example 3.3.4.

The definition of ∇ reads in operator form $E^{-1} = 1-\nabla$. This can be looked upon as a formal power series (with only two nonvanishing terms) for the reciprocal of E with ∇ as the indeterminate. By the rules for formal power series mentioned in Sec. 3.1.5, we obtain *uniquely*

$$E = (E^{-1})^{-1} = (1 - \nabla)^{-1} = 1 + \nabla + \nabla^2 + \cdots.$$

We find in Table 3.3.1 an equivalent expression containing a fraction line. Suppose that we have proved the last column of the table. Thus, $\mathrm{sc}(\nabla) = 1 - e^{-h\alpha}$, hence

$$\mathrm{sc}((1 - \nabla)^{-1}) = (e^{-h\alpha})^{-1} = e^{h\alpha} = \mathrm{sc}(E).$$

Example 3.3.5.

Suppose that we have proved the first and the last columns of Bickley's table (except for the equation $hD = \ln E$). We shall prove one of the formulas in the second column, namely the equation

$$\delta = \Delta(1 + \Delta)^{-1/2}.$$

By the first column, the right-hand side is equal to $(E - 1)E^{-1/2} = E^{1/2} - E^{-1/2} = \delta$.

We shall also compute $\mathrm{sc}(\Delta(1 + \Delta)^{-1/2})$. Since $\mathrm{sc}(\Delta) = e^{h\alpha} - 1$ we obtain

$$\begin{aligned} \mathrm{sc}(\Delta(1 + \Delta)^{-1/2}) &= (e^{h\alpha} - 1)(e^{h\alpha})^{-1/2} = e^{h\alpha/2} - e^{-h\alpha/2} \\ &= 2 \sinh \frac{1}{2} h\alpha = \mathrm{sc}(\delta). \end{aligned}$$

With the aid of Bickley's table, we are in a position to transform L into the form $\phi(Q_h)R_h$. (A sum of several such expressions with different indeterminates can also be treated.)

- Q_h is the one of the four operators, hD, Δ, δ, ∇, which we have chosen to be the "indeterminate."

  $$Lf \simeq \phi(Q_h)f = (a_0 + a_1 Q_h + a_2 Q_h^2 + \cdots)f. \tag{3.3.16}$$

  The coefficients a_j are the same as the Maclaurin coefficients of $\phi(z)$, $z \in \mathbf{C}$, if $\phi(z)$ is analytic at the origin. They can be determined by the techniques described in Sec. 3.1.4 and Sec. 3.1.5. The meaning of the relation $\simeq$ will hopefully be clear from the following theorem.

- R_h is, e.g., $\mu\delta$ or E^k, k integer, or more generally any linear operator with the properties that $R_h F_\alpha = \mathrm{sc}(R_h)F_\alpha$, and that the values of $R_h f(x_n)$ on the grid $x_n = x_0 + nh$, n integer, are determined by the values of f on the same grid.

Theorem 3.3.7.

Recall the notation Q_h for either of the operators Δ, δ£, ∇, hD, and the notations $F_\alpha(x) = e^{\alpha x}$, $f_n(x) = x^n$. Note that

$$F_\alpha(x) = \sum_{n=0}^{\infty} \frac{\alpha^n}{n!} f_n(x). \tag{3.3.17}$$

Also recall the scalar of an operator and its properties, for example,

$$LF_\alpha = \mathrm{sc}(L)F_\alpha, \qquad Q_h^j F_\alpha = (\mathrm{sc}(Q_h))^j F_\alpha;$$

for the operators under consideration the scalar depends on $h\alpha$.

We make the following assumptions:

(i) *A formal power series equation* $L = \sum_{j=0}^{\infty} a_j Q_h^j$ *has been derived.*[75] *Furthermore,* $|\mathrm{sc}(Q_h)| < \rho$, *where* ρ *is the radius of convergence of the series* $\sum a_j z^j$, $z \in \mathbf{C}$, *and*

$$\mathrm{sc}(L) = \sum_{j=0}^{\infty} a_j (\mathrm{sc}(Q_h))^j. \tag{3.3.18}$$

(ii) *At* $\alpha = 0$ *it holds that*

$$L \frac{\partial^n}{\partial \alpha^n} F_\alpha(x) = \frac{\partial^n}{\partial \alpha^n} (L F_\alpha)(x)$$

or, equivalently,

$$L \int_C \frac{F_\alpha(x)\, d\alpha}{\alpha^{n+1}} = \int_C \frac{(L F_\alpha)(x)\, d\alpha}{\alpha^{n+1}}, \tag{3.3.19}$$

where C *is any circle with the origin as center.*

(iii) *The domain of* x *is a bounded interval* I_1 *in* $\mathbf{R}$.

Then it holds that

$$L F_\alpha = \left(\sum_{j=0}^{\infty} a_j Q_h^j \right) F_\alpha \quad \text{if } |\mathrm{sc}(Q_h)| < \rho, \tag{3.3.20}$$

$$L f(x) = \sum_{j=0}^{k-1} a_j Q_h^j f(x) \quad \text{if } f \in \mathcal{P}_k, \tag{3.3.21}$$

for any positive integer k.

A rigorous error bound for (3.3.21), *if* $f \notin \mathcal{P}_k$, *is obtained in Peano's theorem* (3.3.8).

An asymptotic error estimate (as $h \to 0$ *for fixed* k*) is given by the first neglected nonvanishing term* $a_r Q_h^r f(x) \sim a_r (hD)^r f(x)$, $r \ge k$, *if* $f \in C^r[I]$, *where the interval* I *must contain all the points used in the evaluation of* $Q_h^r f(x)$.

Proof. By assumption (i),

$$L F_\alpha = \mathrm{sc}(L) F_\alpha = \lim_{J \to \infty} \sum_{j=0}^{J-1} a_j \mathrm{sc}(Q_h^j) F_\alpha = \lim_{J \to \infty} \sum_{j=0}^{J-1} a_j Q_h^j F_\alpha = \lim_{J \to \infty} \left(\sum_{j=0}^{J-1} a_j Q_h^j \right) F_\alpha,$$

hence $L F_\alpha = (\sum_{j=0}^{\infty} Q_h^j) F_\alpha$. This proves the first part of the theorem.

By (3.3.17), Cauchy's formula (3.2.8), and assumption (ii),

$$\begin{aligned} \frac{2\pi i}{n!} L f_n(x) &= L \int_C \frac{F_\alpha(x)\, d\alpha}{\alpha^{n+1}} = \int_C \frac{(L F_\alpha)(x)\, d\alpha}{\alpha^{n+1}} \\ &= \int_C \sum_{j=0}^{J-1} \frac{a_j Q_h^j F_\alpha(x)\, d\alpha}{\alpha^{n+1}} + \int_C \sum_{j=J}^{\infty} \frac{a_j \mathrm{sc}(Q_h)^j F_\alpha(x)\, d\alpha}{\alpha^{n+1}}. \end{aligned}$$

[75] To simplify the writing, the operator R_h is temporarily neglected. See one of the comments below.

Let ϵ be any positive number. Choose J so that the modulus of the last term becomes $\epsilon\theta_n 2\pi/n!$, where $|\theta_n| < 1$. This is possible, since $|\text{sc}(Q_h)| < \rho$; see assumption (i). Hence, for every $x \in I_1$,

$$Lf_n(x) - \epsilon\theta_n = \frac{n!}{2\pi i}\sum_{j=0}^{J-1} a_j Q_h^j \int_C \frac{F_\alpha(x)\,d\alpha}{\alpha^{n+1}} = \sum_{j=0}^{J-1} a_j Q_h^j f_n(x) = \sum_{j=0}^{k-1} a_j Q_h^j f_n(x).$$

The last step holds if $J \geq k > n$ because, by (3.3.14), $Q_h^j f_n = 0$ for $j > n$. It follows that

$$\left|Lf_n(x) - \sum_{j=0}^{k-1} a_j Q_h^j f_n(x)\right| < \epsilon \quad \forall\epsilon > 0,$$

and hence $Lf_n = \sum_{j=0}^{k-1} a_j Q_h^j f_n$.

If $f \in \mathcal{P}_k$, f is a linear combination of f_n, $n = 0 : k-1$. Hence $Lf = \sum_{j=0}^{k-1} a_j Q_h^j f$ if $f \in \mathcal{P}_k$. This proves the second part of the theorem.

The error bound is derived in Sec. 3.3.1. Recall the important formula (3.3.4) that expresses the kth difference as the value of the kth derivative in a point located in an interval that contains all the points used in the computation of the kth difference; i.e., the ratio of the error estimate $a_r(hD)^r f(x)$ to the true truncation error tends to 1, as $h \to 0$. □

Remark 3.3.1. This theorem is concerned with series of powers of the four operators collectively denoted Q_h. One may try to use operator techniques also to *find* a formula involving, for example, an infinite expansion into powers of the operator E. Then one should try afterward to find sufficient conditions for the validity of the result. This procedure will be illustrated in connection with Euler–Maclaurin's formula in Sec. 3.4.5.

Sometimes, operator techniques which are not covered by this theorem can, after appropriate restrictions, be justified (or even replaced) by *transform methods*, for example, z-, Laplace, or Fourier transforms.

The operator R_h that was introduced just before the theorem was neglected in the proof in order to simplify the writing. We now have to multiply the operands by R_h in the proof and in the results. This changes practically nothing for F_α, since $R_h F_\alpha = \text{sc}(R_h)F_\alpha$. In (3.3.21) there is only a trivial change, because the polynomials f and $R_h f$ may not have the same degree. For example, if $R_h = \mu\delta$ and $f \in P_k$, then $R_h f \in P_{k-1}$. The verification of the assumptions typically offers no difficulties.

It follows from the linearity of (3.3.20) that *it is satisfied also if F_α is replaced by a linear combination of exponential functions F_α with different* α, provided that $|\text{sc}(Q_h)| < \rho$ for all the occurring α. With some care, one can let the linear combination be an infinite series or an integral.

There are two things to note in connection with the asymptotic error estimates. First, the step size should be small enough; this means in practice that, in the beginning, the magnitude of the differences should decrease rapidly, as their order increases. When the order of the differences becomes large, it often happens that the moduli of the differences also increase. This can be due to two causes: semiconvergence (see the next comment) and/or rounding errors.

The *rounding errors* of the data may have such large effects on the high-order differences (recall Example 3.3.2) that the error estimation does not make sense. One should then use a smaller value of the order k, where the rounding errors have a smaller influence. An advantage with the use of a difference scheme is that it is relatively easy to choose the order k adaptively, and sometimes the step size h also.

This comment is of particular importance for numerical differentiation. Numerical illustrations and further comments are given below in Example 3.3.6 and Problem 3.3.7 (b), and in several other places.

The sequence of approximations to Lf may converge or diverge, depending on f and h. It is also often *semiconvergent* (recall Sec. 3.2.6), but in practice the rounding errors mentioned in the previous comment have often, though not always, taken over already, when the truncation error passes its minimum; see Problem 3.3.7 (b).

By Theorem 3.3.6, $e^{-hD} = 1 - \nabla$. We look upon this as a formal power series; the indeterminate is $Q_h = \nabla$. By Example 3.1.11,

$$L = hD = -\ln(1 - \nabla) = \nabla + \frac{1}{2}\nabla^2 + \frac{1}{3}\nabla^3 + \cdots. \tag{3.3.22}$$

Now we present verification of the assumptions of Theorem 3.3.7:[76]

(i) $\mathrm{sc}(\nabla) = 1 - e^{-h\alpha}$; the radius of convergence is $\rho = 1$.

$$\mathrm{sc}(L) = \mathrm{sc}(hD) = h\alpha; \qquad \sum_{j=1}^{\infty} \mathrm{sc}(\nabla)^j/j = -\ln(1 - (1 - e^{-h\alpha})) = h\alpha.$$

The convergence condition $|\mathrm{sc}(\nabla)| < 1$ reads $h\alpha > -\ln 2 = -0.69$ if α is real, $|h\omega| < \pi/3$ if $\alpha = i\omega$.

(ii) For $\alpha = 0$, $D\dfrac{\partial^n}{\partial\alpha^n}(e^{\alpha x}) = Dx^n = nx^{n-1}$. By Leibniz' rule

$$\frac{\partial^n}{\partial\alpha^n}(\alpha e^{\alpha x}) = 0x^n + nx^{n-1}.$$

By the theorem, we now obtain the **backward differentiation formula** that is exact for all $f \in \mathcal{P}_k$:

$$hf'(x) = \left(\nabla + \frac{1}{2}\nabla^2 + \frac{1}{3}\nabla^3 + \cdots + \frac{1}{k-1}\nabla^{k-1}\right) f(x). \tag{3.3.23}$$

By Theorem 3.3.4, this is the *unique* formula of this type that uses the values of $f(x)$ at the k points $x_n : -h : x_{n-k+1}$. The same approximation can be derived in many other ways, perhaps with a different appearance; see Chapter 4. This derivation has several advantages; the same expansion yields approximation formulas for every k, and *if* $f \in C^k$, $f \notin \mathcal{P}_k$, *the first neglected term, i.e.,* $\frac{1}{k}\nabla_h^k f(x_n)$, *provides an* **asymptotic error estimate** if $f^{(k)}(x_n) \neq 0$.

[76] Recall the definition of the scalar sc(·), given after (3.3.13).

Example 3.3.6.

We now apply formula (3.3.23) to the table in Example 3.3.2, where $f(x) = \tan x$, $h = 0.01$, $k = 6$,

$$0.01 f'(1.35) \approx 0.1996 + \frac{0.0163}{2} + \frac{0.0019}{3} + \frac{0.0001}{4} - \frac{0.0004}{5};$$

i.e., we obtain a sequence of approximate results,

$$f'(1.35) \approx 19.96, \quad 20.78, \quad 20.84, \quad 20.84, \quad 20.83.$$

The correct value to 3D is $(\cos 1.35)^{-2} = 20.849$. Note that the last result is worse than the next to last. Recall the last comments on the theorem. In this case this is due to the rounding errors of the data. Upper bounds for their effect of the sequence of approximate values of $f'(1.35)$ are, by Example 3.3.3, shown in the series

$$10^{-2}\left(1 + \frac{2}{2} + \frac{4}{3} + \frac{8}{4} + \frac{16}{5} + \cdots\right).$$

A larger version of this problem was run on a computer with the machine unit $2^{-53} \approx 10^{-16}$; $f(x) = \tan x$, $x = 1.35 : -0.01 : 1.06$. In the beginning the error decreases rapidly, but after 18 terms the rounding errors take over, and the error then grows almost exponentially (with constant sign). The eighteenth term and its rounding error have almost the same modulus (but opposite sign). The smallest error equals $5 \cdot 10^{-10}$, and is obtained after 18 terms; after 29 terms the actual error has grown to $2 \cdot 10^{-6}$. Such a large number of terms is seldom used in practice, unless a very high accuracy is demanded; see also Problem 3.3.7 (b), a computer exercise that offers both similar and different experiences.

Equation (3.3.22)—or its variable step size variant in Chapter 4—is the basis of the important **backward differentiation formula (BDF) method** for the numerical integration of ordinary differential equations.

Coefficients for backward differentiation formulas for higher derivatives are obtained from the equations

$$(hD/\nabla)^k = (-\ln(1-\nabla)/\nabla)^k.$$

The following formulas were computed by means of the matrix representation of a truncated power series:

$$\begin{pmatrix} hD/\nabla \\ (hD/\nabla)^2 \\ (hD/\nabla)^3 \\ (hD/\nabla)^4 \\ (hD/\nabla)^5 \end{pmatrix} = \begin{pmatrix} 1 & 1/2 & 1/3 & 1/4 & 1/5 \\ 1 & 1 & 11/12 & 5/6 & 137/180 \\ 1 & 3/2 & 7/4 & 15/8 & 29/15 \\ 1 & 2 & 17/6 & 7/2 & 967/240 \\ 1 & 5/2 & 25/6 & 35/6 & 1069/144 \end{pmatrix} \cdot \begin{pmatrix} 1 \\ \nabla \\ \nabla^2 \\ \nabla^3 \\ \nabla^4 \end{pmatrix}. \qquad (3.3.24)$$

The rows of the matrix are the first rows taken from the matrix representation of each of the expansions $(hD/\nabla)^k$, $k = 1 : 5$.

When the effect of the *irregular errors* of the data on a term becomes larger in magnitude than the term itself, the term should, of course, be neglected; it does more harm than good. This happens relatively early for the derivatives of high order; see Problem 3.3.7. When these formulas are to be used inside a program (rather than during an interactive post-processing of results of an automatic computation), some rules for *automatic truncation* have to be designed, an interesting kind of detail in scientific computing.

The *forward differentiation formula*, which is analogously based on the operator series

$$hD = \ln(1+\Delta) = \Delta - \frac{1}{2}\Delta^2 + \frac{1}{3}\Delta^3 - \cdots \tag{3.3.25}$$

is sometimes useful too. We obtain the coefficients for derivatives of higher order by inserting minus signs in the second and fourth columns of the matrix in (3.3.24).

A straightforward solution to this problem is to use the derivative of the corresponding interpolation polynomial as the approximation to the derivative of the function. This can also be done for higher-order derivatives.

A grid (or a table) may be too sparse to be useful for numerical differentiation and for the computation of other linear functionals. For example, we saw above that the successive backward differences of $e^{i\omega x}$ increase exponentially if $|\omega h| > \pi/3$. In such a case the grid, where the values are given, gives insufficient information about the function. One also says that "the grid does not *resolve* the function." This is often indicated by a strong variation in the higher differences. But even this indication can sometimes be absent. An extreme example is $f(x) = \sin(\pi x/h)$, on the grid $x_j = jh$, $j = 0, \pm 1, \pm 2, \dots$. All the higher differences, and thus the estimates of $f'(x)$ at all grid points, are zero, but the correct values of $f'(x_j)$ are certainly not zero. Therefore, this is an example where the expansion (trivially) converges, but it is not valid! (Recall the discussion of a Maclaurin expansion for a nonanalytic function at the end of Sec. 3.1.2. Now a similar trouble can also occur for an analytic function.)

A less trivial example is given by the functions

$$f(x) = \sum_{n=1}^{20} a_n \sin(2\pi n x), \qquad g(x) = \sum_{n=1}^{10} (a_n + a_{10+n}) \sin(2\pi n x).$$

On the grid $f(x) = g(x)$, hence they have the same difference scheme, but $f'(x) \neq g'(x)$ on the grid, and typically $f(x) \neq g(x)$ between the grid points.

3.3.3 The Peano Theorem

One can often, by a combination of theoretical and numerical evidence, rely on asymptotic error estimates. Since there are exceptions, it is interesting that there are two general methods for deriving strict error bounds. We call one of them the **norms and distance formula**. This is not restricted to polynomial approximation, and it is typically easy to use, but it requires some advanced concepts and often overestimates the error. We therefore postpone the presentation of that method to Sec. 4.5.2.

We shall now give another method, due to Peano.[77] Consider a linear functional

$$\tilde{L}f = \sum_{j=1}^{p} b_j f(x_j)$$

for the approximate computation of another linear functional, for example,

$$Lf = \int_0^1 \sqrt{x} f(x)\,dx.$$

Suppose that it is exact when it is applied to any polynomial of degree less than k: In other words, $\tilde{L}f = Lf$ for all $f \in \mathcal{P}_k$. The remainder is then itself a linear functional, $R = L - \tilde{L}$, with the special property that

$$Rf = 0 \quad \text{if} \quad f \in \mathcal{P}_k.$$

The next theorem gives a representation for such functionals which provides a universal device for deriving error bounds for approximations of the type that we are concerned with. Let $f \in C^n[a,b]$. In order to make the discussion less abstract we confine it to functionals of the following form, $0 \le m < n$,

$$Rf = \int_a^b \phi(x) f(x)\,dx + \sum_{j=1}^{p} \big(b_{j,0} f(x_j) + b_{j,1} f'(x_j) + \cdots + b_{j,m} f^{(m)}(x_j)\big), \tag{3.3.26}$$

where the function ϕ is integrable, the points x_j lie in the bounded real interval $[a,b]$, and $b_{j,m} \ne 0$ for at least one value of j. Moreover, *we assume that*

$$Rp = 0 \quad \forall\, p \in \mathcal{P}_k. \tag{3.3.27}$$

We define the function[78]

$$t_+ = \max(t, 0), \qquad t_+^j = (t_+)^j, \quad t_+^0 = \frac{1 + \operatorname{sign} t}{2}. \tag{3.3.28}$$

The function t_+^0 is often denoted $H(t)$ and is known as the **Heaviside unit step function.**[79] The function sign is defined as in Definition 3.1.3, i.e., sign $x = 0$, if $x = 0$. Note that $t_+^j \in C^{j-1}$, $(j \ge 1)$.

The **Peano kernel** $K(u)$ of the functional R is defined by the equation

$$K(u) = \frac{1}{(k-1)!} R_x (x-u)_+^{k-1}, \quad x \in [a,b], \quad u \in (-\infty, \infty). \tag{3.3.29}$$

The subscript in R_x indicates that R acts on the variable x (not u).

---

[77] Giuseppe Peano (1858–1932) was an Italian mathematician and logician.

[78] We use the neutral notation t here for the variable, to avoid tying up the function too closely with the variables x and u, which play a special role in the following.

[79] Oliver Heaviside (1850–1925), English physicist.

The function $K(u)$ *vanishes outside* $[a, b]$ because

- if $u > b$, then $u > x$; hence $(x-u)_+^{k-1} = 0$ and $K(u) = 0$.
- if $u < a$, then $x > u$. It follows that $(x-u)_+^{k-1} = (x-u)^{k-1} \in \mathcal{P}_k$; hence $K(u) = 0$, by (3.3.29) and (3.3.27).

If $\phi(x)$ is a polynomial, then $K(u)$ becomes a piecewise polynomial; the points x_j are the joints of the pieces. In this case $K \in C^{k-m-2}$; the order of differentiability may be lower, if ϕ has singularities.

We are now in a position to prove an important theorem.

Theorem 3.3.8 (*Peano's Remainder Theorem*).

Suppose that $Rp = 0$ for all $p \in \mathcal{P}_k$. Then,[80] *for all $f \in C^k[a, b]$,*

$$Rf = \int_{-\infty}^{\infty} f^{(k)}(u)K(u)\,du. \tag{3.3.30}$$

The definition and some basic properties of the Peano kernel $K(u)$ were given above.

Proof. By Taylor's formula,

$$f(x) = \sum_{j=1}^{k-1} \frac{f^{(j)}(a)}{j!}(x-a)^j + \int_a^x \frac{f^{(k)}(u)}{(k-1)!}(x-u)^{k-1}\,du.$$

This follows from putting $n = k$, $z = x - a$, $t = (u-a)/(x-u)$ into (3.1.5). We rewrite the last term as $\int_a^\infty f^{(k)}(u)(x-u)_+^{k-1}du$. Then apply the functional $R = R_x$ to both sides. Since we can allow the interchange of the functional R with the integral, for the class of functionals that we are working with, this yields

$$Rf = 0 + R\int_a^\infty \frac{f^{(k)}(u)(x-u)_+^{k-1}}{(k-1)!}\,du = \int_a^\infty \frac{f^{(k)}(u)R_x(x-u)_+^{k-1}}{(k-1)!}\,du.$$

The theorem then follows from (3.3.29). ☐

Corollary 3.3.9.

Suppose that $Rp = 0$ for all $p \in \mathcal{P}_k$. Then

$$R_x(x-a)^k = k!\int_{-\infty}^{\infty} K(u)\,du. \tag{3.3.31}$$

For any $f \in C^k[a, b]$, $Rf = \frac{f^{(k)}(\xi)}{k!}R_x((x-a)^k)$ holds for some $\xi \in (a, b)$ if and only if $K(u)$ does not change its sign.

If $K(u)$ changes its sign, the best possible error bound *reads*

$$|Rf| \le \sup_{u\in[a,b]} |f^{(k)}(u)| \int_{-\infty}^{\infty} |K(u)|\,du;$$

a formula with $f^{(k)}(\xi)$ is not generally true in this case.

[80]The definition of $f^{(k)}(u)$ for $u \notin [a, b]$ is arbitrary.

Proof. First suppose that $K(u)$ does not change sign. Then, by (3.3.30) and the mean value theorem of integral calculus, $Rf = f^{(k)}(\xi)\int_{-\infty}^{\infty} K(u)du$, $\xi \in [a,b]$. For $f(x) = (x-a)^k$ this yields (3.3.31). The "if" part of the corollary follows from the combination of these formulas for Rf and $R(x-a)^k$.

If $K(u)$ changes its sign, the "best possible bound" is approached by a sequence of functions f chosen so that (the continuous functions) $f^{(k)}(u)$ approach (the discontinuous function) sign $K(u)$. The "only if" part follows. □

Example 3.3.7.

The remainder of the *trapezoidal rule* (one step of length h) reads

$$Rf = \int_0^h f(x)\,dx - \frac{h}{2}(f(h) + f(0)).$$

We know that $Rp = 0$ for all $p \in \mathcal{P}_2$. The Peano kernel is zero for $u \notin [0,h]$, while for $u \in [0,h]$

$$K(u) = \int_0^h (x-u)_+\,dx - \frac{h}{2}((h-u)_+ + 0) = \frac{(h-u)^2}{2} - \frac{h(h-u)}{2} = \frac{-u(h-u)}{2} < 0.$$

We also compute

$$\frac{Rx^2}{2!} = \int_0^h \frac{x^2}{2}\,dx - \frac{h\cdot h^2}{2\cdot 2} = \frac{h^3}{6} - \frac{h^3}{4} = -\frac{h^3}{12}.$$

Since the Peano kernel does not change sign, we conclude that

$$Rf = -\frac{h^3}{12}f''(\xi), \quad \xi \in (0,h).$$

Example 3.3.8 (*Peano Kernels for Difference Operators*).

Let $Rf = \Delta^3 f(a)$, and set $x_i = a + ih$, $i = 0:3$. Note that $Rp = 0$ for $p \in \mathcal{P}_3$. Then

$$\begin{aligned} Rf &= f(x_3) - 3f(x_2) + 3f(x_1) - f(x_0),\\ 2K(u) &= (x_3-u)_+^2 - 3(x_2-u)_+^2 + 3(x_1-u)_+^2 - (x_0-u)_+^2; \end{aligned}$$

i.e.,

$$2K(u) = \begin{cases} 0 & \text{if } u > x_3,\\ (x_3-u)^2 & \text{if } x_2 \le u \le x_3,\\ (x_3-u)^2 - 3(x_2-u)^2 & \text{if } x_1 \le u \le x_2,\\ (x_3-u)^2 - 3(x_2-u)^2 + 3(x_1-u)^2 \equiv (u-x_0)^2 & \text{if } x_0 \le u \le x_1,\\ (x_3-u)^2 - 3(x_2-u)^2 + 3(x_1-u)^2 - (x_0-u)^2 \equiv 0 & \text{if } u < x_0. \end{cases}$$

For the simplification of the last two lines we used that $\Delta_u^3(x_0-u)^2 \equiv 0$. Note that $K(u)$ is a piecewise polynomial in $\mathcal{P}_3$ and that $K''(u)$ is discontinuous at $u = x_i$, $i = 0:3$.

It can be shown (numerically or analytically) that $K(u) > 0$ in the interval (u_0, u_3). This is no surprise because, by (3.3.4), $\Delta^n f(x) = h^n f^{(n)}(\xi)$ for any integer n, and, by the above corollary, this could not be generally true if $K(u)$ changes its sign. These calculations can be generalized to $\Delta^k f(a)$ for an arbitrary integer k. This example will be generalized in Sec. 4.4.2 to divided differences of nonequidistant data.

In general it is rather laborious to determine a Peano kernel. Sometimes one can show that the kernel is a piecewise polynomial, that it has a symmetry, and that it has a simple form in the intervals near the boundaries. All this can simplify the computation, and might have been used in these examples.

It is usually much easier to compute $R((x-a)^k)$, and an *approximate error estimate* is often given by

$$Rf \sim \frac{f^{(k)}(a)}{k!} R\big((x-a)^k\big), \quad f^{(k)}(a) \neq 0. \tag{3.3.32}$$

For example, suppose that $x \in [a, b]$, where $b - a$ is of the order of magnitude of a step size parameter h, and that f is analytic in $[a, b]$. By Taylor's formula,

$$f(x) = p(x) + \frac{f^{(k)}(a)}{k!}(x-a)^k + \frac{f^{(k+1)}(a)}{(k+1)!}(x-a)^{k+1} + \cdots, \qquad f^{(k)}(a) \neq 0,$$

where $p \in \mathcal{P}_k$; hence $Rp = 0$. Most of the common functionals can be applied term by term. Then

$$Rf = 0 + \frac{f^{(k)}(a)}{n!} R_x(x-a)^k + \frac{f^{(k+1)}(a)}{(k+1)!} R_x(x-a)^{k+1} + \cdots.$$

Assume that, for some c, $R_x(x-a)^k = O(h^{k+c})$ for $k = 1, 2, 3, \ldots$. (This is often the case.) Then (3.3.32) becomes an **asymptotic error estimate** as $h \to 0$. It was mentioned above that for formulas derived by operator methods, an asymptotic error estimate is directly available anyway, but if a formula is derived by other means (see Chapter 4) this error estimate is important.

Asymptotic error estimates are frequently used in computing, because they are often much easier to derive and apply than strict error bounds. The question is, however, how to know that "the computation is in the asymptotic regime," where an asymptotic estimate is practically reliable. Much can be said about this central question of applied mathematics. Let us here just mention that a difference scheme displays well the quantitative properties of a function needed to make the judgment.

If $Rp = 0$ for $p \in \mathcal{P}_k$, then a fortiori $Rp = 0$ for $p \in \mathcal{P}_{k-i}$, $i = 0 : k$. We may thus obtain a Peano kernel for each i, which is temporarily denoted by $K_{k-i}(u)$. They are obtained by integration by parts,

$$R_k f = \int_{-\infty}^{\infty} K_k(u) f^{(k)}(u)\, du = \int_{-\infty}^{\infty} K_{k-1}(u) f^{(k-1)}(u)\, du \tag{3.3.33}$$

$$= \int_{-\infty}^{\infty} K_{k-2}(u) f^{(k-2)}(u)\, du = \ldots, \tag{3.3.34}$$

where $K_{k-i} = (-D)^i K_k$, $i = 1, 2, \ldots,$ *as long as* K_{k-i} *is integrable. The lower-order kernels are useful, e.g., if the actual function* f *is not as smooth as the usual remainder formula requires.*

For the trapezoidal rule we obtained in Example 3.3.7

$$K_1(u) = \frac{h}{2}u_+^0 + \frac{h}{2} - u + \frac{h}{2}(u-h)_+^0.$$

A second integration by parts can only be performed within the framework of Dirac's delta functions (distributions); K_0 is not integrable. A reader who is familiar with these generalized functions may enjoy the following formula:

$$Rf = \int_{-\infty}^{\infty} K_0(u)f(u)du \equiv \int_{-\infty}^{\infty}\Big(-\frac{h}{2}\delta(u) + 1 - \frac{h}{2}\delta(u-h)\Big)f(u)du.$$

This is for one step of the trapezoidal rule, but many functionals can be expressed analogously.

3.3.4 Approximation Formulas by Operator Methods

We shall now demonstrate how operator methods are very useful for deriving approximation formulas. For example, in order to find interpolation formulas we consider the operator expansion

$$f(b-\gamma h) = E^{-\gamma}f(b) = (1-\nabla)^{\gamma}f(b) = \sum_{j=0}^{\infty}\binom{\gamma}{j}(-\nabla)^j f(b).$$

The verification of the assumptions of Theorem 3.3.7 offers no difficulties, and we omit the details. Truncate the expansion before $(-\nabla)^k$. By the theorem we obtain, for every γ, an approximation formula for $f(b-\gamma h)$ that uses the function values $f(b-jh)$ for $j = 0 : k-1$; it is exact if $f \in \mathcal{P}_k$ and is unique in the sense of Theorem 3.3.4. We also obtain an asymptotic error estimate if $f \notin \mathcal{P}_k$, namely the first neglected term of the expansion, i.e.,

$$\binom{\gamma}{k}(-\nabla)^k f(b) \sim \binom{\gamma}{k}(-h)^k f^{(k)}(b).$$

Note that the binomial coefficients are polynomials in the variable γ, and hence also in the variable $x = b - \gamma h$.

It follows that the approximation formula yields **a unique polynomial** $P_B \in \mathcal{P}_k$ that solves the **interpolation problem**: $P_B(b-hj) = f(b-hj)$, $j = 0 : k-1$ (B stands for backward). If we set $x = b - \gamma h$, we obtain

$$\begin{aligned} P_B(x) = E^{-\gamma}f(b) = (1-\nabla)^{\gamma}f(a) &= \sum_{j=0}^{k-1}\binom{\gamma}{j}(-\nabla)^j f(b) \qquad (3.3.35)\\ &= f(b-\gamma h) + O(h^k f^{(k)}). \end{aligned}$$

Similarly, the interpolation polynomial $P_F \in \mathcal{P}_k$ that uses *forward* differences based on the values of f at $a, a+h, \dots, a+(k-1)h$ reads, if we set $x = a + \theta h$,

$$\begin{aligned} P_F(x) = E^{\theta}f(a) = (1+\Delta)^{\theta}f(a) &\sum_{j=0}^{k-1}\binom{\theta}{j}\Delta^j f(a) \qquad (3.3.36)\\ &= f(a+\theta h) + O(h^k f^{(k)}). \end{aligned}$$

These formulas are known as **Newton's interpolation formulas** for constant step size, backward and forward. The generalization to variable step size will be found in Sec. 4.2.1.

There exists a similar expansion for *central differences*. Set

$$\phi_0(\theta) = 1, \quad \phi_1(\theta) = \theta, \quad \phi_j(\theta) = \frac{\theta}{j}\binom{\theta + \frac{1}{2}j - 1}{j-1}, \quad (j > 1). \tag{3.3.37}$$

ϕ_j is an even function if j is even, and an odd function if j is odd. It can be shown that $\delta^j \phi_k(\theta) = \phi_{k-j}(\theta)$ and $\delta^j \phi_k(0) = \delta_{j,k}$ (Kronecker's delta). The functions ϕ_k have thus an analogous relation to the operator δ as, for example, the functions $\theta^j/j!$ and $\binom{\theta}{j}$ have to the operators D and Δ, respectively. We obtain the following expansion, analogous to Taylor's formula and Newton's forward interpolation formula. The proof is left for Problem 3.3.5 (b). Then

$$E^\theta f(a) = \sum_{j=0}^{k-1} \phi_j(\theta)\delta^j f(a) = f(a + \theta h) + O(h^k f^{(k)}). \tag{3.3.38}$$

The direct practical importance of this formula is small, since $\delta^j f(a)$ cannot be expressed as a linear combination of the given data when j is odd. There are several formulas in which this drawback has been eliminated by various transformations. They were much in use before the computer age; each formula had its own group of fans. We shall derive only one of them, by a short break-neck application of the formal power series techniques.[81] Note that

$$\begin{aligned}
E^\theta &= e^{\theta h D} = \cosh \theta h D + \sinh \theta h D,\\
\delta^2 &= e^{hD} - 2 + e^{-hD}, \qquad e^{hD} - e^{-hD} = 2\mu\delta,\\
\cosh \theta h D &= \frac{1}{2}(E^\theta + E^{-\theta}) = \sum_{j=0}^{\infty} \phi_{2j}(\theta)\delta^{2j},\\
\sinh \theta h D &= \frac{1}{\theta}\frac{d(\cosh \theta h D)}{d(hD)} = \sum_{j=0}^{\infty} \phi_{2j}(\theta)\frac{1}{\theta}\frac{d\delta^{2j}}{d\delta^2}\frac{d\delta^2}{d(hD)}\\
&= \sum_{j=0}^{\infty} \phi_{2j}(\theta)\frac{j\delta^{2(j-1)}}{\theta}(e^{hD} - e^{-hD}) = \sum_{j=0}^{\infty} \phi_{2j}(\theta)\frac{2j}{\theta}\mu\delta^{2j-1}.
\end{aligned}$$

Hence,

$$f(x_0 + \theta h) = f_0 + \theta\mu\delta f_0 + \frac{\theta^2}{2!}\delta^2 f_0 + \sum_{j=2}^{\infty} \phi_{2j}(\theta)\Big(\frac{2j}{\theta}\mu\delta^{2j-1} f_0 + \delta^{2j} f_0\Big). \tag{3.3.39}$$

This is known as **Stirling's interpolation formula**.[82] The first three terms have been taken out from the sum, in order to show their simplicity and their resemblance to Taylor's formula. They yield the most practical formula for quadratic interpolation; it is easily remembered

[81] Differentiation of a formal power series with respect to an indeterminate has a purely algebraic definition. See the last part of Sec. 3.1.5.

[82] James Stirling (1692–1770), British mathematician perhaps most famous for his amazing approximation to $n!$.

and worth being remembered. An approximate error bound for this quadratic interpolation reads $|0.016\delta^3 f|$ if $|\theta| < 1$.

Note that

$$\phi_{2j}(\theta) = \theta^2(\theta^2 - 1)(\theta^2 - 4)\cdots(\theta^2 - (j-1)^2)/(2j)!.$$

The expansion yields a true interpolation formula if it is truncated after an *even* power of δ. For $k = 1$ you see that $f_0 + \theta\mu\delta f_0$ is not a formula for linear interpolation; it uses three data points instead of two. It is similar for all odd values of k.

Strict error bounds can be found by means of Peano's theorem, but the remainder given by Theorem 4.2.3 for Newton's general interpolation formula (that does not require equidistant data) typically give the answer easier. Both are typically of the form $c_{k+1} f^{(k+1)}(\xi)$ and require a bound for a derivative of high order. The assessment of such a bound typically costs much more work than performing interpolation in one point.

A more practical approach is to estimate a bound for this derivative by means of a bound for the differences of the same order. (Recall the important formula in (3.3.4).) This is not a rigorous *bound*, but it typically yields a quite reliable error *estimate*, in particular if you put a moderate safety factor on the top of it. There is much more to be said about the choice of step size and order; we shall return to these kinds of questions in later chapters.

You can make error estimates during the computations; it can happen sooner or later that it does not decrease when you increase the order. You may just as well stop there, and accept the most recent value as the result. This event is most likely due to the influence of irregular errors, but it can also indicate that the interpolation process is semiconvergent only.

The attainable accuracy of polynomial interpolation applied to a table with n equidistant values of an analytic function depends strongly on θ; the results are much poorer near the boundaries of the data set than near the center. This question will be illuminated in Sec. 4.7 by means of complex analysis.

Example 3.3.9.

The continuation of the difference scheme of a polynomial is a classical application of a difference scheme for obtaining a smooth extrapolation of a function outside its original domain. Given the values $y_{n-i} = f(x_n - ih)$ for $i = 1 : k$ and the backward differences, $\nabla^j y_{n-1}$, $j = 1 : k-1$. Recall that $\nabla^{k-1} y$ is a constant for $y \in \mathcal{P}_k$. Consider the algorithm

$$\begin{aligned} &\nabla^{k-1} y_n = \nabla^{k-1} y_{n-1};\\ &\textbf{for } j = k-1 : -1 : 1\\ &\quad \nabla^{j-1} y_n = \nabla^{j-1} y_{n-1} + \nabla^j y_n;\\ &\textbf{end}\\ &y_n = \nabla^0 y_n; \end{aligned} \tag{3.3.40}$$

It is left for Problem 3.3.2 (g) to show that the result y_n is the value at $x = x_n$ of the interpolation polynomial which is determined by y_{n-i}, $i = 1 : k$. This is a kind of inverse use of a difference scheme; there are additions from right to left along a diagonal, instead of subtractions from left to right.

This algorithm, which needs additions only, was used long ago for the production of mathematical tables, for example, for logarithms. Suppose that one knows, by means of a series expansion, a relatively complicated polynomial approximation to (say) $f(x) = \ln x$, that is accurate enough in (say) the interval $[a, b]$, and that this has been used for the computation of k very accurate values $y_0 = f(a)$, $y_1 = f(a+h), \dots, y_{k-1}$, needed for starting the difference scheme. The algorithm is then used for $n = k,\ k+1,\ k+2, \dots, (b-a)/h$. $k-1$ additions only are needed for each value y_n. Some analysis must have been needed for the choice of the step h to make the tables useful with (say) linear interpolation, and for the choice of k to make the basic polynomial approximation accurate enough over a substantial number of steps. The precision used was higher when the table was produced than when it was used. When $x = b$ was reached, a new approximating polynomial was needed for continuing the computation over another interval (at least a new value of $\nabla^{k-1} y_n$).[83]

The algorithm in (3.3.40) can be generalized to the case of nonequidistant with the use of divided differences; see Sec. 4.2.1.

We now derive some central difference formulas for numerical differentiation. From the definition and from Bickley's table (Table 3.3.1),

$$\delta \equiv E^{1/2} - E^{-1/2} = 2\sinh\Big(\frac{1}{2}hD\Big). \tag{3.3.41}$$

We may therefore put $x = \frac{1}{2}hD$, $\sinh x = \frac{1}{2}\delta$ into the expansion (see Problem 3.1.7)

$$x = \sinh x - \frac{1}{2}\frac{\sinh^3 x}{3} + \frac{1\cdot 3}{2\cdot 4}\frac{\sinh^5 x}{5} - \frac{1\cdot 3\cdot 5}{2\cdot 4\cdot 6}\frac{\sinh^7 x}{7} + \cdots,$$

with the result

$$hD = 2\operatorname{arcsinh}\frac{\delta}{2} = \delta - \frac{\delta^3}{24} + \frac{3\delta^5}{640} - \frac{5\delta^7}{7168} + \frac{35\delta^9}{294{,}912} - \frac{63\delta^{11}}{2{,}883{,}584} + \cdots. \tag{3.3.42}$$

The verification of the assumptions of Theorem 3.3.7 follows the pattern of the proof of (3.3.23), and we omit the details. Since $\operatorname{arcsinh} z$, $z \in \mathbf{C}$, has the same singularities as its derivative $(1+z^2)^{-1/2}$, namely $z = \pm i$, it follows that the expansion in (3.3.42), if $\operatorname{sc}(\delta/2)$ is substituted for $\delta/2$, converges if $\operatorname{sc}(\delta/2) < 1$; hence $\rho = 2$.

By squaring the above relation, we obtain

$$\begin{aligned}(hD)^2 &= \delta^2 - \frac{\delta^4}{12} + \frac{\delta^6}{90} - \frac{\delta^8}{560} + \frac{\delta^{10}}{3150} - \frac{\delta^{12}}{16{,}632} + \cdots,\\ f''(x_0) &\approx \Big(1 - \frac{\delta^2}{12} + \frac{\delta^4}{90} - \frac{\delta^6}{560} + \frac{\delta^8}{3150} - \frac{\delta^{10}}{16{,}632} + \cdots\Big)\frac{\delta^2 f_0}{h^2}.\end{aligned} \tag{3.3.43}$$

By Theorem 3.3.7 (3.3.43) holds for all polynomials. Since the first neglected nonvanishing term of (3.3.43) when applied to f is (asymptotically) $c\delta^{12} f''(x_0)$, the formula for $f''(x)$

[83] This procedure was the basis of the unfinished Difference Engine project of the great nineteenth century British computer pioneer Charles Babbage. He abandoned it after a while in order to spend more time on his huge Analytic Engine project, which was also unfinished. He documented a lot of ideas, where he was (say) 100 years ahead of his time. "Difference engines" based on Babbage's ideas were, however, constructed in Babbage's own time, by the Swedish inventors Scheutz (father and son) in 1834 and by Wiberg in 1876. They were applied to, among other things, the automatic calculation and printing of tables of logarithms; see Goldstine [159].

is exact if $f'' \in \mathcal{P}_{12}$, i.e., if $f \in \mathcal{P}_{14}$, although only 13 values of $f(x)$ are used. We thus gain one degree and, in the application to functions other than polynomials, one order of accuracy, compared to what we may have expected by counting unknowns and equations only; see Theorem 3.3.4. *This is typical for a problem that has a symmetry with respect to the hull of the data points.*

Suppose that the values $f(x)$ are given on the grid $x = x_0 + nh$, n integer. Since (3.3.42) contains odd powers of δ, it cannot be used to compute f'_n on the same grid, as pointed out in the beginning of Sec. 3.3.2. This difficulty can be overcome by means of another formula given in Bickley's table, namely

$$\mu = \sqrt{1 + \delta^2/4}. \tag{3.3.44}$$

This is derived as follows. The formulas

$$\mu = \cosh \frac{hD}{2}, \qquad \frac{\delta}{2} = \sinh \frac{hD}{2}$$

follow rather directly from the definitions; the details are left for Problem 3.3.6 (a). The formula $(\cosh hD)^2 - (\sinh hD)^2 = 1$ holds also for formal power series. Hence

$$\mu^2 - \frac{1}{4}\delta^2 = 1 \quad \text{or} \quad \mu^2 = 1 + \frac{1}{4}\delta^2,$$

from which the relation (3.3.44) follows.

If we now multiply the right-hand side of (3.3.42) by the expansion

$$1 = \mu\Big(1 + \frac{1}{4}\delta^2\Big)^{-1/2} = \mu\Big(1 - \frac{\delta^2}{8} + \frac{3\delta^4}{128} - \frac{5\delta^6}{1{,}024} + \frac{35\delta^8}{32{,}768} + \cdots\Big), \tag{3.3.45}$$

we obtain

$$hD = \Big(1 - \frac{\delta^2}{6} + \frac{\delta^4}{30} - \frac{\delta^6}{140} + \frac{\delta^8}{630} - \cdots\Big)\mu\delta. \tag{3.3.46}$$

This leads to a useful central difference formula for the first derivative (where we have used more terms than we displayed in the above derivation):

$$f'(x_0) = \Big(1 - \frac{\delta^2}{6} + \frac{\delta^4}{30} - \frac{\delta^6}{140} + \frac{\delta^8}{630} - \frac{\delta^{10}}{2772} + \cdots\Big)\frac{f_1 - f_{-1}}{2h}. \tag{3.3.47}$$

If you truncate the operator expansion in (3.3.47) after the δ^{2k} term, you obtain exactly the derivative of the interpolation polynomial of degree $2k+1$ for $f(x)$ that is determined by the $2k+2$ values f_i, $i = \pm 1, \pm 2, \ldots, \pm(k+1)$. Note that all the neglected terms in the expansion vanish when $f(x)$ is any polynomial of degree $2k+2$, independent of the value of f_0. (Check the statements first for $k = 0$; you will recognize a familiar property of the parabola.) So, although we search for a formula that is exact in $\mathcal{P}_{2k+2}$, we actually find a formula that is exact in $\mathcal{P}_{2k+3}$.

By the multiplication of the expansions in (3.3.43) and (3.3.46), we obtain the following formulas, which have applications in other sections:

$$\begin{aligned}
(hD)^3 &= \Big(1 - \frac{1}{4}\delta^2 + \frac{7}{120}\delta^4 + \cdots\Big)\mu\delta^3, \\
(hD)^5 &= \Big(1 - \frac{1}{3}\delta^2 + \cdots\Big)\mu\delta^5, \\
(hD)^7 &= \mu\delta^7 + \cdots.
\end{aligned} \tag{3.3.48}$$

Another valuable feature typical for expansions in powers of δ^2 is the rapid convergence. It was mentioned earlier that $\rho = 2$, hence $\rho^2 = 4$, (while $\rho = 1$ for the backward differentiation formula). The error constants of the differentiation formulas obtained by (3.3.43) and (3.3.47) are thus relatively small.

All this is typical for the symmetric approximation formulas which are based on central differences; see, for example, the above formula for $f''(x_0)$, or the next example. In view of this, can we forget the forward and backward difference formulas altogether? Well, this is not quite the case, since one must often deal with data that are unsymmetric with respect to the point where the result is needed. For example, given f_{-1}, f_0, f_1, how would you compute $f'(x_1)$? Asymmetry is also typical for the application to *initial value problems* for differential equations. In such applications methods based on symmetric rules for differentiation or integration have sometimes inferior properties of numerical stability.

We shall study the computation of $f'(x_0)$ using the operator expansion (3.3.47). The truncation error (called R_T) can be estimated by the first neglected term, where

$$\frac{1}{h}\mu\delta^{2k+1}f_0 \approx h^{2k} f^{(2k+1)}(x_0).$$

The irregular errors in the values of $f(x)$ are of much greater importance in numerical differentiation than in interpolation and integration. Suppose that the function values have errors whose magnitude does not exceed $\frac{1}{2}U$. Then the error bound on $\mu\delta f_0 = \frac{1}{2}(f_1 - f_{-1})$ is also equal to $\frac{1}{2}U$. Similarly, one can show that the error bounds in $\mu\delta^{(2k+1)}f_0$, for $k = 1 : 3$, are $1.5U$, $5U$, $417.5U$, respectively. Thus one gets the upper bounds $U/(2h)$, $3U/(4h)$, and $11U/(12h)$ for the roundoff error R_{XF} with one, two, and three terms in (3.3.47).

Example 3.3.10.

Assume that k terms in the formula above are used to approximate $f'(x_0)$, where $f(x) = \ln x$, $x_0 = 3$, and $U = 10^{-6}$. Then

$$f^{(2k+1)}(3) = (2k)!/3^{2k+1},$$

and for the truncation and roundoff errors we get

k	1	2	3
R_T	$0.0123h^2$	$0.00329h^4$	$0.00235h^6$
R_{XF}	$(1/2h)10^{-6}$	$(3/4h)10^{-6}$	$(11/12h)10^{-6}$

The plots of R_T and R_{XF} versus h in a log-log diagram in Figure 3.3.1 are straight lines that well illustrate quantitatively the conflict between truncation and roundoff errors. The truncation error increases, and the effect of the irregular error decreases with h. One sees how the choice of h, which minimizes the sum of the bounds for the two types of error, depends on U and k, and tells us what accuracy can be obtained. The optimal step lengths for $k = 1, 2, 3$ are $h = 0.0344$, $h = 0.1869$, and $h = 0.3260$, giving error bounds $2.91 \cdot 10^{-5}$, $8.03 \cdot 10^{-6}$, and $5.64 \cdot 10^{-6}$. Note that the optimal error bound with $k = 3$ is not much better than that for $k = 2$.

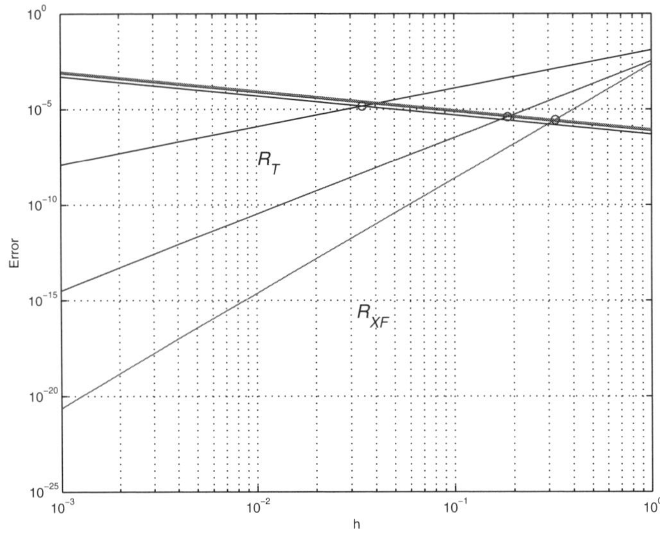

Figure 3.3.1. *Bounds for truncation error R_T and roundoff error R_{XF} in numerical differentiation as functions of h ($U = 0.5 \cdot 10^{-6}$).*

The effect of the pure rounding errors is important, though it should not be exaggerated. Using IEEE double precision with $u = 1.1 \cdot 10^{-16}$, one can obtain the first two derivatives very accurately by the optimal choice of h. The corresponding figures are $h = 2.08 \cdot 10^{-5}$, $h = 2.19 \cdot 10^{-3}$, and $h = 1.36 \cdot 10^{-2}$, giving the optimal error bounds $1.07 \cdot 10^{-11}$, $1.52 \cdot 10^{-13}$, and $3.00 \cdot 10^{-14}$, respectively.

It is left to the user (Problem 4.3.8) to check and modify the experiments and conclusions indicated in this example.

When a problem has a symmetry around some point x_0, you are advised to try to derive a δ^2-expansion. The first step is to express the relevant operator in the form $\Phi(\delta^2)$, where the function Φ is analytic at the origin.

To find a δ^2-expansion for $\Phi(\delta^2)$ is algebraically the same thing as expanding $\Phi(z)$ into powers of a complex variable z. Thus, the methods for the manipulation of power series mentioned in Sec. 3.1.4 and Problem 3.1.8 are available, and so is the Cauchy–FFT method. *For suitably chosen r, N you evaluate*

$$\Phi(re^{2\pi ik/N}), \quad k = 0 : N-1,$$

and obtain the coefficients of the δ^2-expansion by the FFT! You can therefore derive a long expansion, and later truncate it as needed. You also obtain error estimates for all these truncated expansions for free. By the assumed symmetry there will be even powers of δ only in the expansion. Some computation and storage can be saved by working with $F(\sqrt{z})$ instead.

Suppose that you have found a truncated δ^2-expansion, (say)

$$A(\delta^2) \equiv a_1 + a_2\delta^2 + a_3\delta^4 + \cdots + a_{k+1}\delta^{2k},$$

but you want instead an equivalent symmetric expression of the form

$$B(E) \equiv b_1 + b_2(E + E^{-1}) + b_3(E^2 + E^{-2}) + \cdots + b_{k+1}(E^k + E^{-k}).$$

Note that $\delta^2 = E - 2 + E^{-1}$. The transformation $A(\delta^2) \mapsto B(E)$ can be performed in several ways. Since it is linear it can be expressed by a matrix multiplication of the form $b = M_{k+1}a$, where a, b are column vectors for the coefficients, and M_{k+1} is the $(k+1) \times (k+1)$ upper triangular submatrix in the northwest corner of a matrix M that turns out to be

$$M = \begin{pmatrix} 1 & -2 & 6 & -20 & 70 & -252 & 924 & -3432 \\ & 1 & -4 & 15 & -56 & 210 & -792 & 3003 \\ & & 1 & -6 & 28 & -120 & 495 & -2002 \\ & & & 1 & -8 & 45 & -220 & 1001 \\ & & & & 1 & -10 & 66 & -364 \\ & & & & & 1 & -12 & 91 \\ & & & & & & 1 & -14 \\ & & & & & & & 1 \end{pmatrix}. \tag{3.3.49}$$

This 8×8 matrix is sufficient for a δ^2-expansion up to the term $a_8\delta^{14}$. Note that the matrix elements are binomial coefficients that can be generated recursively (Sec. 3.1.2). It is easy to extend by the recurrence that is mentioned in the theorem below. Also note that the matrix can be looked upon as the lower part of a thinned Pascal triangle.

Theorem 3.3.10.

The elements of M are

$$M_{ij} = \begin{cases} (-1)^{j-1}\binom{2j-2}{j-i} & \text{if } 1 \le i \le j, \\ 0 & \text{if } i > j. \end{cases} \tag{3.3.50}$$

We extend the definition by setting $M_{0,j} = M_{2,j}$. Then the columns of M are obtained by the recurrence

$$M_{i,j+1} = M_{i+1,j} - 2M_{i,j} + M_{i-1,j}. \tag{3.3.51}$$

Proof. Recall that $\delta = (1 - E^{-1})E^{1/2}$ and put $m - \nu = \mu$. Hence

$$\begin{aligned} \delta^{2m} &= (1 - E^{-1})^{2m} E^m = \sum_{\nu=0}^{2m} (-1)^\nu \binom{2m}{\nu} E^{m-\nu} \\ &= (-1)^m \binom{2m}{m} + \sum_{\mu=1}^{m} (-1)^{m-\mu} \binom{2m}{m-\mu} (E^\mu + E^{-\mu}). \end{aligned} \tag{3.3.52}$$

Since

$$(1 \quad \delta^2 \quad \delta^4 \quad \ldots) = (1 \quad (E - E^{-1}) \qquad (E^2 - E^{-2}) \quad \ldots)\, M,$$

we have in the result of (3.3.52) an expression for column $m+1$ of M. By putting $j = m+1$ and $i = \mu + 1$, we obtain (3.3.50). The proof of the recurrence is left to the reader. (Think of Pascal's triangle.) □

The integration operator D^{-1} is defined by the relation

$$(D^{-1}f)(x) = \int^x f(t)\,dt.$$

The lower limit is not fixed, so $D^{-1}f$ contains an arbitrary integration constant. Note that $DD^{-1}f = f$, while $D^{-1}Df = f + C$, where C is the integration constant. A difference expression like

$$D^{-1}f(b) - D^{-1}f(a) = \int_a^b f(t)\,dt$$

is uniquely defined. So is $\delta D^{-1}f$, but $D^{-1}\delta f$ has an integration constant.

A right-hand inverse can be also defined for the operators Δ, ∇, and δ. For example, $(\nabla^{-1}u)_n = \sum^{j=n} u_j$ has an arbitrary summation constant but, for example, $\nabla\nabla^{-1} = 1$, and $\Delta\nabla^{-1} = E\nabla\nabla^{-1} = E$ are uniquely defined.

One can make the inverses unique by restricting the class of sequences (or functions). For example, if we require that $\sum_{j=0}^{\infty} u_j$ is convergent, and make the convention that $(\Delta^{-1}u)_n \to 0$ as $n \to \infty$, then $\Delta^{-1}u_n = -\sum_{j=n}^{\infty} u_j$; notice the minus sign. Also notice that this is consistent with the following formal computation:

$$(1 + E + E^2 + \cdots)u_n = (1 - E)^{-1}u_n = -\Delta^{-1}u_n.$$

We recommend, however, some extra care with infinite expansions into powers of operators like E that is not covered by Theorem 3.3.7, but the finite expansion

$$1 + E + E^2 + \cdots + E^{n-1} = (E^n - 1)(E - 1)^{-1} \tag{3.3.53}$$

is valid.

In Chapter 5 we will use operator methods together with the Cauchy–FFT method for finding the **Newton–Cotes'** formulas for symmetric numerical integration. Operator techniques can also be extended to *functions of several variables*. The basic relation is again the operator form of Taylor's formula, which in the case of two variables reads

$$\begin{aligned} u(x_0 + h, y_0 + k) &= \exp\left(h\frac{\partial}{\partial x} + k\frac{\partial}{\partial y}\right) u(x_0, y_0) \\ &= \exp\left(h\frac{\partial}{\partial x}\right) \exp\left(k\frac{\partial}{\partial y}\right) u(x_0, y_0). \end{aligned} \tag{3.3.54}$$

3.3.5 Single Linear Difference Equations

Historically, the term **difference equation** was probably first used in connection with an equation of the form

$$b_0\Delta^k y_n + b_1\Delta^{k-1} y_n + \cdots + b_{k-1}\Delta y_n + b_k y_n = 0, \quad n = 0, 1, 2, \ldots,$$

which resembles a linear homogeneous differential equation. It follows, however, from the discussion after (3.3.1) and (3.3.3) that this equation can also be written in the form

$$y_{n+k} + a_1 y_{n+k-1} + \cdots + a_k y_n = 0, \tag{3.3.55}$$

and nowadays this is what one usually means by a single homogeneous linear difference equation of kth order with *constant coefficients*; a difference equation without differences. More generally, if we let the coefficients a_i depend on n we have a linear difference equation with *variable coefficients*. If we replace the zero on the right-hand side with some known quantity r_n, we have an *inhomogeneous* linear difference equation.

These types of equations are the main topic of this section. The coefficients and the unknown are real or complex numbers. We shall occasionally see examples of more general types of difference equations, e.g., a nonlinear difference equation

$$F(y_{n+k}, y_{n+k-1}, \ldots, y_n) = 0,$$

and *first order systems* of difference equations, i.e.,

$$y_{n+1} = A_n y_n + r_n,$$

where r_n and y_n are vectors while A_n is a square matrix. Finally, *partial difference equations*, where you have two (or more) subscripts in the unknown, occur often as numerical methods for partial differential equations, but they have many other important applications too.

A difference equation can be viewed as a *recurrence relation*. With given values of $y_0, y_1, \ldots, y_{k-1}$, called the **initial values** or the **seed** of the recurrence, we can successively compute $y_k, y_{k+1}, y_{k+2}, \ldots$; we see that *the general solution of a kth order difference equation contains k arbitrary constants*, just like the general solution of the kth order differential equation. There are other important similarities between difference and differential equations, for example, the following superposition result.

Lemma 3.3.11.

The general solution of a nonhomogeneous linear difference equation (also with variable coefficients) is the sum of one particular solution of it, and the general solution of the corresponding homogeneous difference equation.

In practical computing, the recursive computation of the solution of difference equations is most common. It was mentioned at the end of Sec. 3.2.3 that many important functions, e.g., Bessel functions and orthogonal polynomials, satisfy second order linear difference equations with variable coefficients (although this terminology was not used there). Other important applications are the multistep methods for ordinary differential equations.

In such an application you are usually interested in the solution for one particular initial condition, but due to rounding errors in the initial values you obtain another solution. It is therefore of interest to know the behavior of the solutions of the corresponding homogeneous difference equation. The questions are

- *Can we use a recurrence to find the desired solution accurately?*

- *How shall we use a recurrence, forward or backward?*

Forward recurrence is the type we described above. In backward recurrence we choose some large integer N, and give (almost) arbitrary values of y_{N+i}, $i = 0 : k-1$, as seeds, and compute y_n for $n = N-1 : -1 : 0$.

We have seen this already in Example 1.2.1 for an inhomogeneous first order recurrence relation. There it was found that the forward recurrence was useless, while backward recurrence, with a rather naturally chosen seed, gave satisfactory results. It is often like this, though not always. In Problem 1.2.7 it is the other way around: the forward recurrence is useful, and the backward recurrence is useless.

Sometimes **boundary values** are prescribed for a difference equation instead of initial values, (say) p values at the beginning and $q = k - p$ values at the end, e.g., the values of $y_0, y_1, \ldots, y_{p-1}$ and $y_{N-q}, \ldots, {}_{N-1}$, y_N are given. Then the difference equation can be treated as a *linear system* with $N - k$ unknown. This also holds for a difference equation with variable coefficients and for an inhomogeneous difference equation. *From the point of view of numerical stability, such a treatment can be better than either recurrence.* The amount of work is somewhat larger, not very much though, for the matrix is a band matrix. For a fixed number of bands *the amount of work to solve such a linear system is proportional to the number of unknowns.* An important particular case is when $k = 2$, $p = q = 1$; the linear system is then tridiagonal. An algorithm for tridiagonal linear systems is described in Example 1.3.3.

Another similarity for differential and difference equations is that the general solution of a linear equation with constant coefficients has a simple closed form. Although, in most cases real-world problems have variable coefficients (or are nonlinear), one can often formulate a class of model problems with constant coefficients with similar features. The analysis of such model problems can give hints, e.g., whether forward or backward recurrence should be used, or other questions related to the design and the analysis of the numerical stability of a numerical method for a more complicated problem.

We shall therefore now study how to solve a *single homogeneous linear difference equation with constant coefficients* (3.3.55), i.e.,

$$y_{n+k} + a_1 y_{n+k-1} + \cdots + a_k y_n = 0.$$

It is satisfied by the sequence $\{y_j\}$, where $y_j = cu^j$ ($u \neq 0$, $c \neq 0$) if and only if $u^{n+k} + a_1 u^{n+k-1} + \cdots + a_k u^n = 0$, i.e., when

$$\phi(u) \equiv u^k + a_1 u^{k-1} + \cdots + a_k = 0. \tag{3.3.56}$$

Equation (3.3.56) is called the **characteristic equation** of (3.3.55); $\phi(u)$ is called the **characteristic polynomial**.

Theorem 3.3.12.

If the characteristic equation has k different *roots,* $u_1, \dots, u_k$*, then the general solution of* (3.3.55) *is given by the sequences* $\{y_n\}$*, where*

$$y_n = c_1 u_1^n + c_2 u_2^n + \cdots + c_k u_k^n, \tag{3.3.57}$$

where $c_1, c_2, \dots, c_k$ *are arbitrary constants.*

Proof. That $\{y_n\}$ satisfies (3.3.55) follows from the previous comments and from the fact that the equation is linear. The parameters $c_1, c_2, \dots, c_k$ can be adjusted to arbitrary initial conditions $y_0, y_1, \dots, y_{k-1}$ by solving the system of equations

$$\begin{pmatrix} 1 & 1 & \cdots & 1 \\ u_1 & u_2 & \cdots & u_k \\ \vdots & \vdots & & \vdots \\ u_1^{k-1} & u_2^{k-1} & \cdots & u_k^{k-1} \end{pmatrix} \begin{pmatrix} c_1 \\ c_2 \\ \vdots \\ c_k \end{pmatrix} = \begin{pmatrix} y_0 \\ y_1 \\ \vdots \\ y_{k-1} \end{pmatrix}.$$

The matrix is a Vandermonde matrix and its determinant is thus equal to the product of all differences $(u_i - u_j)$, $i \ge j$, $1 < i \le k$, which is nonzero; see the proof of Theorem 3.3.4. □

Example 3.3.11.

Consider the difference equation $y_{n+2} - 5y_{n+1} + 6y_n = 0$ with initial conditions $y_0 = 0$, $y_1 = 1$. Forward recurrence yields $y_2 = 5$, $y_3 = 19$, $y_4 = 65, \dots$.

The characteristic equation $u^2 - 5u + 6 = 0$ has roots $u_1 = 3$, $u_2 = 2$; Hence, the general solution is $y_n = c_1 3^n + c_2 2^n$. The initial conditions give the system of equations

$$c_1 + c_2 = 0, \qquad 3c_1 + 2c_2 = 1,$$

with solution $c_1 = 1$, $c_2 = -1$; hence $y_n = 3^n - 2^n$.

As a check we find $y_2 = 5$, $y_3 = 19$ in agreement with the results found by using forward recurrence.

Example 3.3.12.

Consider the difference equation

$$T_{n+1}(x) - 2xT_n(x) + T_{n-1}(x) = 0, \quad n \ge 1, \quad -1 < x < 1,$$

with initial conditions $T_0(x) = 1$, $T_1(x) = x$. We obtain $T_2(x) = 2x^2 - 1$, $T_3(x) = 4x^3 - 3x$, $T_4(x) = 8x^4 - 8x^2 + 1, \dots$. By induction, $T_n(x)$ is an nth degree polynomial in x.

We can obtain a simple formula for $T_n(x)$ by solving the difference equation. The characteristic equation is $u^2 - 2xu + 1 = 0$, with roots $u = x \pm i\sqrt{1 - x^2}$. Set $x = \cos\phi$, $0 < x < \pi$. Then $u = \cos\phi \pm i \sin\phi$, and thus $u_1 = e^{i\phi}$, $u_2 = e^{-i\phi}$, $u_1 \ne u_2$. The general solution is $T_n(x) = c_1 e^{in\phi} + c_2 e^{-in\phi}$, and the initial conditions give

$$c_1 + c_2 = 1, \qquad c_1 e^{i\phi} + c_2 e^{-i\phi} = \cos\phi,$$

with solution $c_1 = c_2 = 1/2$. Hence, $T_n(x) = \cos(n\phi)$, $x = \cos\phi$. These polynomials are thus identical to the important Chebyshev polynomials $T_n(x)$ that were introduced in (3.2.21).

We excluded the cases $x = 1$ and $x = -1$, i.e., $\phi = 0$ and $\phi = \pi$, respectively. For the particular initial values of this example, there are no difficulties; the solution $T_n(x) = \cos n\phi$ depends continuously on ϕ, and as $\phi \to 0$ or $\phi \to \pi$, $T_n(x) = \cos n\phi$ converges to 1 for all n or $(-1)^n$ for all n, respectively.

When we ask for the general solution of the difference equation matters are a little more complicated, because the characteristic equation has in these cases a double root: $u = 1$ for $x = 1$, $u = -1$ for $x = -1$. Although they are thus covered by the next theorem, we shall look at them directly because they are easy to solve, and they are a good preparation for the general case.

If $x = 1$, the difference equation reads $T_{n+1} - 2T_n + T_{n-1} = 0$, i.e., $\Delta^2 T_n = 0$. We know from before (see, e.g., Theorem 3.3.4) that this is satisfied if and only if $T_n = an + b$. The solution is no longer built up by exponentials; a linear term is there too.

If $x = -1$, the difference equation reads $T_{n+1} + 2T_n + T_{n-1} = 0$. Set $T_n = (-1)^n V_n$. The difference equation becomes, after division by $(-1)^{n+1}$, $V_{n+1} - 2V_n + V_{n-1} = 0$, with the general solution $V_n = an + b$; hence $T_n = (-1)^n(an + b)$.

Theorem 3.3.13.

When u_i is an m_i-fold root of the characteristic equation, then the difference (3.3.55) *is satisfied by the sequence $\{y_n\}$, where*

$$y_n = P_i(n)u_i^n$$

and P_i is an arbitrary polynomial in $\mathcal{P}_{m_i}$. The general solution of the difference equation is a linear combination of solutions of this form using all the distinct roots of the characteristic equation.

Proof. We can write the polynomial $P \in \mathcal{P}_{m_i}$ in the form

$$P_i(n) = b_1 + b_2 n + b_3 n(n-1) + \cdots + b_{m_i} n(n-1)\cdots(n - m_i + 2).$$

Thus it is sufficient to show that (3.3.55) is satisfied when

$$y_n = n(n-1)\cdots(n-p+1)u_i^n = (u^p \partial^p(u^n)/\partial u^p)_{u=u_i}, \quad p = 1 : m_i - 1. \tag{3.3.58}$$

Substitute this in the left-hand side of (3.3.55):

$$u^p \frac{\partial^p}{\partial u^p}\left(u^{n+k} + a_1 u^{n+k-1} + \cdots + a_k u^n\right) = u^p \frac{\partial^p}{\partial u^p}(\phi(u)u^n)$$
$$= u^p \left(\phi^{(p)}(u)u^n + \binom{p}{1}\phi^{(p-1)}(u)nu^{n-1} + \cdots + \binom{p}{p}\phi(u)\frac{\partial^p}{\partial u^p}(u^n)\right).$$

The last manipulation was made using Leibniz's rule.

Now ϕ and all the derivatives of ϕ which occur in the above expression are zero for $u = u_i$, since u_i is an m_i-fold root. Thus the sequences $\{y_n\}$ in (3.3.58) satisfy the difference equation. We obtain a solution with $\sum m_i = k$ parameters by the linear combination of such solutions derived from the different roots of the characteristic equation.

It can be shown (see Henrici [192, p. 214]) that these solutions are linearly independent. (This also follows from a different proof, where a difference equation of higher order is transformed to a system of first order difference equations. This transformation also leads to other ways of handling inhomogeneous difference equations than those which are presented in this section.) □

Note that the double root cases discussed in the previous example are completely in accordance with this theorem. We look at one more example.

Example 3.3.13.
Consider the difference equation $y_{n+3} - 3y_{n+2} + 4y_n = 0$. The characteristic equation is $u^3 - 3u^2 + 4 = 0$ with roots $u_1 = -1$, $u_2 = u_3 = 2$. Hence, the general solution reads

$$y_n = c_1(-1)^n + (c_2 + c_3 n)2^n.$$

For a **nonhomogeneous** linear difference equation of order k, one can often find a *particular solution* by the use of an Ansatz[84] with undetermined coefficients; thereafter, by Lemma 3.3.11 one can get the general solution by adding the general solution of the homogeneous difference equation.

Example 3.3.14.
Consider the difference equation $y_{n+1} - 2y_n = a^n$, with initial condition $y_0 = 1$. Try the Ansatz $y_n = ca^n$. One gets

$$ca^{n+1} - 2ca^n = a^n, \quad c = 1/(a-2), \quad a \neq 2.$$

Thus the general solution is $y_n = a^n/(a-2) + c_1 2^n$. By the initial condition, $c_1 = 1 - 1/(a-2)$, hence

$$y_n = \frac{a^n - 2^n}{a-2} + 2^n. \tag{3.3.59}$$

When $a \to 2$, l'Hôpital's rule gives $y_n = 2^n + n2^{n-1}$. Notice how the Ansatz must be modified when a is a root of the characteristic equation.

The general rule when the right-hand side is of the form $P(n)a^n$ (or a sum of such terms), where P is a polynomial, is that the contribution of this term to y_n is $Q(n)a^n$, where Q is a polynomial. If a does not satisfy the characteristic equation, then $\deg Q = \deg P$; if a is a single or a double root of the characteristic equation, then $\deg Q = \deg P + 1$ or $\deg Q = \deg P + 2$, respectively, and so on. The coefficients of Q are determined by the insertion of $y_n = Q(n)a^n$ on the left-hand side of the equation and matching the coefficients with the right-hand side.

Another way to find a particular solution is based on the calculus of operators. Let an inhomogeneous difference equation be given in the form $\psi(Q)y_n = b_n$, where Q is one of the operators Δ, δ, and ∇, or an operator easily derived from these, for example, $\frac{1}{6}\delta^2$ (see Problem 3.3.27(d)). In Sec. 3.1.5 $\psi(Q)^{-1}$ was defined by the formal power series with the same coefficients as the Maclaurin series for the function $1/\psi(z)$, $z \in \mathbf{C}$, $\psi(0) \neq 0$. In simple cases, e.g., if $\psi(Q) = a_0 + a_1 Q$, these coefficients are usually easily found. Then

[84]An Ansatz (German term) is an assumed form for a mathematical statement that is not based on any underlying theory or principle.

$\psi(Q)^{-1}b_n$ *is a particular solution* of the difference equation $\psi(Q)y_n = b_n$; the truncated expansions approximate this. Note that if $Q = \delta$ or ∇, the infinite expansion demands that b_n is also defined if $n < 0$.

Note that a similar technique, with the operator D, can also be applied to linear differential equations. Today this technique has to a large extent been replaced by the Laplace transform,[85] which yields essentially the same algebraic calculations as operator calculus.

In some branches of applied mathematics it is popular to treat nonhomogeneous difference equations by means of a **generating function**, also called the **z-transform**, since both the definition and the practical computations are analogous to the Laplace transform. The z-transform of the sequence $y = \{y_n\}_0^\infty$ is

$$Y(z) = \sum_{n=0}^{\infty} y_n z^{-n}. \tag{3.3.60}$$

Note that the sequence $\{Ey\} = \{y_{n+1}\}$ has the z-transform $zY(z) - y_0$, $\{E^2y\} = \{y_{n+2}\}$ has the z-transform $z^2Y(z) - y_0 z - y_1$, etc.

If $Y(z)$ is available in *analytic* form, it can often be brought to a sum of functions whose inverse z-transforms are known by means of various analytic techniques, notably expansion into partial fractions if $Y(z)$ is a rational function. On the other hand, if *numerical values* of $Y(z)$ have been computed for complex values of z on some circle in $\mathbf{C}$ by means of an algorithm, then y_n can be determined by an obvious modification of the Cauchy–FFT method described in Sec. 3.2.2 (for expansions into negative powers of z). More information about the z-transform can be found in Strang [339, Sec. 6.3].

We are now in a position to exemplify in more detail the use of linear difference equations to studies of numerical stability, of the type mentioned above.

Theorem 3.3.14 (*Root Condition*).

Necessary and sufficient for boundedness (stability) of all solutions of the difference (3.3.55) *for all positive n is the following* **root condition**: *(We shall say either that a difference equation or that a characteristic polynomial satisfies the root condition; the meaning is the same.)*

i. *All roots of characteristic* (3.3.56) *are located inside or on the unit circle $|z| \le 1$;*

ii. *The roots on the unit circle are simple.*

Proof. The proof follows directly from Theorem 3.3.13. □

This root condition corresponds to cases where it is the absolute error that matters. It is basic in the theory of linear multistep methods for ordinary differential equations. Computer graphics and an algebraic criterion due to Schur are useful for investigations of the root condition, particularly, if the recurrence relation under investigation contains parameters.

[85] The Laplace transform is traditionally used for similar problems for linear differential equations, for example, in electrical engineering.

There are important applications of single linear difference equations to the study of the stability of numerical methods. When a recurrence is used one is usually interested in the solution for one particular initial condition. But a rounding error in an initial value produces a different solution, and it is therefore of interest to know the behavior of other solutions of the corresponding homogeneous difference equation. We have seen this already in Example 1.2.1 for an inhomogeneous first order recurrence relation, but it is even more important for recurrence relations of higher order.

The following example is based on a study done by Todd[86] in 1950 (see [352]).

Example 3.3.15.

Consider the initial value problem

$$y''(x) = -y, \quad y(0) = 0, \quad y'(0) = 1, \tag{3.3.61}$$

with the exact solution $y(x) = \sin x$. To compute an approximate solution $y_k = y(x_k)$ at equidistant points $x_k = kh$, where h is a step length, we approximate the second derivative according to (3.3.43):

$$h^2 y_k'' = \delta^2 y_k + \frac{\delta^4 y_k}{12} + \frac{\delta^6 y_k}{90} + \cdots. \tag{3.3.62}$$

We first use the first term only; the second term shows that the truncation error of this approximation of y_k'' is asymptotically $h^2 y^{(4)}/12$. We then obtain the difference equation $h^{-2}\delta^2 y_k = -y_k$ or, in other words,

$$y_{k+2} = (2 - h^2) y_{k+1} - y_k, \quad y_0 = 0, \tag{3.3.63}$$

where a suitable value of y_1 is to be assigned. In the third column of Table 3.3.2 we show the results obtained using this recursion formula with $h = 0.1$ and $y_1 = \sin 0.1$. All computations in this example were carried out using IEEE double precision arithmetic. We obtain about three digits of accuracy at the end of the interval $x = 1.2$.

Since the algorithm was based on a second order accurate approximation of y'' one may expect that the solution of the differential equation is also second order accurate. This turns out to be correct in this case; for example, if we divide the step size by two, the errors will approximately be divided by four. We shall, however, see that we cannot always draw conclusions of this kind; we also have to take the numerical stability into account.

In the hope of obtaining a more accurate solution, we shall now use one more term in the expansion (3.3.62); the third term then shows that the truncation error of this approximation is asymptotically $h^4 y^{(6)}/90$. The difference equation now reads

$$\delta^2 y_k - \frac{1}{12}\delta^4 y_k = -h^2 y_k, \tag{3.3.64}$$

or

$$y_{k+2} = 16 y_{k+1} - (30 - 12h^2) y_k + 16 y_{k-1} - y_{k-2}, \quad k \geq 2, \quad y_0 = 0, \tag{3.3.65}$$

[86] John Todd (1911–2007), born in Ireland, was a pioneer in computing and numerical analysis. During World War II he was head of the British Admiralty Computing Services. At the end of the war he earned his nickname "Savior of Oberwolfach" by protecting the Mathematical Research Institute at Oberwolfach in Germany from destruction by Moroccan troops. In 1947 he joined the National Bureau of Standards (NBS) in Washington, DC, where he became head of the Computation Laboratory and in 1954 Chief of the Numerical Analysis Section. In 1957 he took up a position as Professor of Mathematics at the California Institute of Technology.

Table 3.3.2. *Integrating $y'' = -y$, $y(0) = 0$, $y'(0) = 1$; the letters U and S in the headings of the last two columns refer to "Unstable" and "Stable."*

x_k	$\sin x_k$	2nd order	4th order U	4th order S
0.1	0.0998334166	0.0998334	0.0998334166	0.0998334166
0.2	0.1986693308	0.1986685	0.1986693307	0.1986693303
0.3	0.2955202067	0.2955169	0.2955202067	0.2955202050
0.4	0.3894183423	0.3894101	0.3894183688	0.3894183382
0.5	0.4794255386	0.4794093	0.4794126947	0.4794255305
0.6	0.5646424734	0.5646143	0.5643841035	0.5646424593
0.7	0.6442176872	0.6441732	0.6403394433	0.6442176650
0.8	0.7173560909	0.7172903	0.6627719932	0.7173560580
0.9	0.7833269096	0.7832346	0.0254286676	0.7833268635
1.0	0.8414709848	0.8413465	−9.654611899	0.8414709226
1.1	0.8912073601	0.8910450	−144.4011267	0.8912072789
1.2	0.9320390860	0.9318329	−2010.123761	0.9320389830

where starting values for y_1, y_2, and y_3 need to be assigned. We choose the correct values of the solution rounded to double precision. The results from this recursion are shown in the fourth column of Table 3.3.2. We see that disaster has struck—the recursion is severely unstable! Already for $x = 0.6$ the results are less accurate than the second order scheme. For $x \geq 0.9$ the errors completely dominate the unstable method.

We shall now look at these difference equations from the point of view of the root condition. The characteristic equation for (3.3.63) reads $u^2 - (2 - h^2)u + 1 = 0$, and since $|2 - h^2| < 2$, direct computation shows that it has simple roots of unit modulus. The root condition is satisfied. By Example 3.3.12, the solution of (3.3.63) is $y_n = T_n(1 - h^2/2)$. For the second order method the absolute error at $x = 1.2$ is approximately $2.1 \cdot 10^{-4}$, whereas for the stable fourth order method the error is $1.0 \cdot 10^{-7}$.

For (3.3.65) the characteristic equation reads $u^4 - 16u^3 + (30 - 12h^2)u^2 - 16u + 1 = 0$. We see immediately that *the root condition cannot be satisfied.* Since the sum of the roots equals 16, it is impossible that all roots are inside or on the unit circle. In fact, the largest root equals 13.94. So, a tiny error at $x = 0.1$ has been multiplied by $13.94^{14} \approx 10^{16}$ at the end.

A stable fourth order accurate method can easily be constructed. Using the differential equation we replace the term $\delta^4 y_k$ in (3.3.64) by $h^2\delta^2 y_k'' = -h^2\delta^2 y_k$. This leads to the recursion formula[87]

$$y_{k+1} = \left(2 - \frac{h^2}{1 + h^2/12}\right) y_k - y_{k-1}, \qquad y_0 = 0, \tag{3.3.66}$$

which can be traced back at least to B. Numerov (1924) (cf. Problem 3.4.27). This difference equation satisfies the root condition if $h^2 < 6$ (see Problem 3.3.25 (a)). It requires y_0, $y_1 \approx y(h)$ as the seed. The results using this recursion formula with $h = 0.1$ and $y_1 = \sin 0.1$

[87]Boris Vaisyevich Numerov (1891–1941) Russian astronomer and professor at the University of Leningrad.

are shown in the fifth column of Table 3.3.2. The error at the end is about $2 \cdot 10^{-7}$, which is much better than the $3.7 \cdot 10^{-4}$ obtained with the second order method.

Remark 3.3.2. If the solution of the original problem is itself strongly decreasing or strongly increasing, one should consider the location of the characteristic roots with respect to a circle in the complex plane that corresponds to the interesting solution. For example, if the interesting root is 0.8, then a root equal to −0.9 causes oscillations that may eventually become disturbing if one is interested in *relative* accuracy in a long run, even if the oscillating solution is small in the beginning.

Many problems contain homogeneous or nonhomogeneous linear difference equations with variable coefficients, for which the solutions are not known in a simple closed form.

We now confine the discussion to the cases where the original problem is to compute a particular solution of a *second order difference equation with variable coefficients*; several interesting problems of this type were mentioned above, and we formulated the questions of whether we *can* use a recurrence to find the desired solution accurately, and *how* we shall use a recurrence, forward or backward. Typically the original problem contains some parameter, and one usually wants to make a study for an interval of parameter values.

Such questions are sometimes studied with *frozen coefficients*, i.e., the model problems are in the class of difference equations with constant coefficients in the range of the actual coefficients of the original problem. If one of the types of recurrence is satisfactory (i.e., numerically stable in some sense) for all model problems, one would like to conclude that they are satisfactory also for the original problem, but *the conclusion is not always valid* without further restrictions on the coefficients—see a counterexample in Problem 3.3.27.

The technique with *frozen coefficients provides just a hint* that should always be *checked by numerical experiments* on the original problem. It is beyond the scope of this text to discuss what restrictions are needed. *If the coefficients of the original problem are slowly varying, however, there is a good chance that the numerical tests will confirm the hint*—but again, how slowly is "slowly"? A warning against the use of one of the types of recurrence may also be a valuable result of a study, although it is negative.

The following lemma exemplifies a type of tool that may be useful in such cases. The proof is left for Problem 3.3.24 (a). Another useful tool is presented in Problem 3.3.26 (a) and applied in Problem 3.3.26 (b).

Lemma 3.3.15.

Suppose that the wanted sequence y_n^* satisfies a difference equation (with constant coefficients),

$$\alpha y_{n+1} + \beta y_n - \gamma y_{n-1} = 0, \quad (\alpha > \gamma > 0,\ \beta > 0),$$

and that y_n^* is known to be positive for all sufficiently large n. Then the characteristic roots can be written $0 < u_1 < 1$, $u_2 < 0$, and $|u_2| > u_1$. Then y_n^* is unique apart from a positive factor c; $y_n^* = cu_1^n$, $c > 0$.

A solution $\bar{y}_n$, called the trial solution, *that is approximately of this form can be computed for $n = N : -1 : 0$ by* backward *recurrence starting with the "seed" $y_{N+1} = 0$, $y_N = 1$. If an accurate value of y_0^* is given, the desired solution is*

$$y_n^* = \bar{y}_n y_0^* / \bar{y}_0,$$

with a relative error approximately proportional to $(u_2/u_1)^{n-N}$ *(neglecting a possible error in* y_0^**). (If* y_n^* *is defined by some other condition, one can proceed analogously.)*

The *forward* recurrence is not recommended for finding y_n^* in this case, since the positive term $c_1 u_1^n$ will eventually be drowned by the oscillating term $c_2 u_2^n$ that will be introduced by the rounding errors. The proof is left for Problem 3.3.27. Even if y_0 (in the use of the forward recurrence) has no rounding errors, such errors committed at later stages will yield similar contributions to the numerical results.

Example 3.3.16.

The "original problem" is to compute the parabolic cylinder function $U(a, x)$ which satisfies the difference equation

$$\left(a + \frac{1}{2}\right) U(a+1, x) + xU(a, x) - U(a-1, x) = 0;$$

see Handbook [1, Chap. 19, in particular Example 19.28.1].

To be more precise, we consider the case $x = 5$. Given $U(3, 5) = 5.2847 \cdot 10^{-6}$ (obtained from a table in [1, p. 710]), we want to determine $U(a, 5)$ for integer values of a, $a > 3$, as long as $|U(a, 5)| > 10^{-15}$. We guess (a priori) that the discussion can be restricted to the interval (say) $a = [3, 15]$. The above lemma then gives the hint of a backward recurrence, for $a = a' - 1 : -1 : 3$ for some appropriate a' (see below), in order to obtain a trial solution $\bar{U}_a$ with the seed $\bar{U}_{a'} = 1$, $\bar{U}_{a'+1} = 0$. Then the wanted solution becomes, by the lemma (with changed notation),

$$U(a, 5) = \bar{U}_a U(3, 5)/\bar{U}_3.$$

The positive characteristic root of the frozen difference equation varies from 0.174 to 0.14 for $a = 5 : 15$, while the modulus of the negative root is between 6.4 and 3.3 times as large. This motivates a choice of $a' \approx 4 + (-9 - \log 5.3)/\ln 0.174 \approx 17$ for the backward recursion; it seems advisable to choose a' (say) four units larger than the value where U becomes negligible.

Forward recurrence with correctly rounded starting values $U(3, 5) = 5.2847 \cdot 10^{-6}$, $U(4, 5) = 9.172 \cdot 10^{-7}$ gives oscillating (absolute) errors of relatively slowly decreasing amplitude, approximately 10^{-11}, that gradually drown the exponentially decreasing true solution. The estimate of $U(a, 5)$ itself became negative for $a = 10$, and then the results oscillated with approximate amplitude 10^{-11}, while the correct results decrease from the order of 10^{-11} to 10^{-15} as $a = 10 : 15$. The details are left for Problem 3.3.25 (b).

It is conceivable that this procedure can be used for all x in some interval around five, but we refrain from presenting the properties of the parabolic cylinder function needed for determining the interval.

If the problem is nonlinear, one can instead solve the original problem with two seeds, (say) y'_N, y''_N, and study how the results deviate. The seeds should be so close that a linearization like $f(y'_n) - f(y''_n) \approx r_n(y'_n - y''_n)$ is acceptable, but $y'_n - y''_n$ should be well above the rounding error level. A more recent and general treatment of these matters is found in [96, Chapter 6].

Review Questions

3.3.1 Give expressions for the shift operator E^k in terms of Δ, ∇, and hD, and expressions for the central difference operator δ^2 in terms of E and hD.

3.3.2 Derive the best upper bound for the error of $\Delta^n y_0$, if we only know that the absolute value of the error of y_i, $i = 0, \dots, n$ does not exceed ϵ.

3.3.3 There is a theorem (and a corollary) about existence and uniqueness of approximation formulas of a certain type that are exact for polynomials of certain class. Formulate these results, and sketch the proofs.

3.3.4 What bound can be given for the kth difference of a function in terms of a bound for the kth derivative of the same function?

3.3.5 Formulate the basic theorem concerning the use of operator expansions for deriving approximation formulas for linear operators.

3.3.6 Discuss how various sources of error influence the choice of step length in numerical differentiation.

3.3.7 Formulate Peano's remainder theorem, and compute the Peano kernel for a given symmetric functional (with at most four subintervals).

3.3.8 Express polynomial interpolation formulas in terms of forward and backward difference operators.

3.3.9 Give Stirling's interpolation formula for quadratic interpolation with approximate bounds for truncation error and irregular error.

3.3.10 Derive central difference formulas for $f'(x_0)$ and $f''(x_0)$ that are exact for $f \in \mathcal{P}_4$. They should only use function values at x_j, $j = 0, \pm 1, \pm 2, \dots$, as many as needed. Give asymptotic error estimates.

3.3.11 Derive the formula for the general solution of the difference equation $y_{n+k} + a_1 y_{n+k-1} + \cdots + a_k y_n = 0$, when the characteristic equation has simple roots only. What is the general solution when the characteristic equation has multiple roots?

3.3.12 What is the general solution of the difference equation $\Delta^k y_n = an + b$?

Problems and Computer Exercises

3.3.1 Prove the formula (3.3.12) for the determinant of the Vandermonde matrix $V = V(x_1, \dots, x_k)$. For definition and properties of a determinant, Section A.3 in Online Appendix A.

Hint: Considered as a function of x_1, det V is a polynomial of degree $k - 1$. Since the determinant is zero if two columns are identical, this polynomial has the roots $x_1 = x_j$, $j = 2 : k$. Hence

$$\det V = c(x_2, \dots, x_k)(x_1 - x_2) \cdots (x_1 - x_k),$$

where c does not depend on x_1. Similarly, viewed as a polynomial of x_2 the determinant must contain the factor $(x_2 - x_1)(x_2 - x_3)\cdots(x_2 - x_k)$, etc.

3.3.2 (a) Show that $(1+\Delta)(1-\nabla) = 1$, $\Delta - \nabla = \Delta\nabla = \delta^2 = E - 2 + E^{-1}$, and that $\delta^2 y_n = y_{n+1} - 2y_n + y_{n-1}$.

(b) Let $\Delta^p y_n$, $\nabla^p y_m$, $\delta^p y_k$ all denote the same quantity. How are n, m, k connected? Along which lines in the difference scheme are the subscripts constant?

(c) Given the values of $y_n, \nabla y_n, \dots, \nabla^k y_n$, for a particular value of n, find a recurrence relation for computing $y_n, y_{n-1}, \dots, y_{n-k}$, by simple additions only. On the way you obtain the full difference scheme of this sequence.

(d) *Repeated summation by parts.* Show that if $u_1 = u_N = v_1 = v_N = 0$, then

$$\sum_{n=1}^{N-1} u_n \Delta^2 v_{n-1} = -\sum_{n=1}^{N-1} \Delta u_n \Delta v_n = \sum_{n=1}^{N-1} v_n \Delta^2 u_{n-1}.$$

(e) Show that if $\Delta^k v_n \to 0$, as $n \to \infty$, then $\sum_{n=m}^{\infty} \Delta^k v_n = -\Delta^{k-1} v_m$.

(f) Show that $(\mu\delta^3 + 2\mu\delta) f_0 = f_2 - f_{-2}$.

(g) Show the validity of the algorithm in (3.3.40). Babbage's favorite example was $f(x) = x^2 + x + 41$. Given $f(x)$ for $x = 0, 1, 2$, compute the backward differences for $x = 2$ and use the algorithm to obtain $f(3)$. Then compute $f(x)$ for (say) $x = 4 : 10$, by repeated use of the algorithm. (This is simple enough for paper and pencil, since the algorithm contains only additions.)

3.3.3 (a) Prove by induction, the following two formulas:

$$\Delta_x^j \binom{x}{k} = \binom{x}{k-j}, \quad j \le k,$$

where Δ_x means differencing with respect to x, with $h = 1$, and

$$\Delta^j x^{-1} = \frac{(-h)^j j!}{x(x+h)\cdots(x+jh)}.$$

Find the analogous expression for $\nabla^j x^{-1}$.

(b) What formulas with derivatives instead of differences are these formulas analogous to?

(c) Show the following formulas if x, a are integers:

$$\sum_{n=a}^{x-1} \binom{n}{k-1} = \binom{x}{k} - \binom{a}{k},$$

$$\sum_{n=x}^{\infty} \frac{1}{n(n+1)\cdots(n+j)} = \frac{1}{j} \cdot \frac{1}{x(x+1)\cdots(x+j-1)}.$$

Modify these results for noninteger x; $x - a$ is still an integer.

(d) Suppose that $b \neq 0, -1, -2, \dots$, and set

$$c_0(a,b) = 1, \quad c_n(a,b) = \frac{a(a+1)\dots(a+n-1)}{b(b+1)\dots(b+n-1)}, \quad n = 1,2,3,\dots.$$

Show by induction that

$$(-\Delta)^k c_n(a,b) = c_k(b-a,b)c_n(a,b+k),$$

and that hence $(-\Delta)^n c_0(a,b) = c_n(b-a,b)$.

(e) Compute for $a = e$, $b = \pi$ (say), $c_n(a,b)$, $n = 1 : 100$. How do you avoid overflow? Compute $\Delta^n c_0(a,b)$, both numerically by the difference scheme and according to the formula in (d). Compare the results and formulate your experiences. Do the same with $a = e$, $b = \pi^2$.

Do the same with $\Delta^j x^{-1}$ for various values of x, j, and h.

3.3.4 Set

$$\begin{aligned} Y_{ord} &= (y_{n-k}, y_{n-k+1}, \dots, y_{n-1}, y_n), \\ Y_{dif} &= (\nabla^k y_n, \nabla^{k-1} y_n, \dots, \nabla y_n, y_n). \end{aligned}$$

Note that the results of this problem also hold if the y_j are column vectors.

(a) Find a matrix P such that $Y_{dif} = Y_{ord} P$. Show that

$$Y_{ord} = Y_{dif} P, \quad \text{hence} \quad P^{-1} = P.$$

How do you generate this matrix by means of a simple recurrence relation?

Hint: P is related to the Pascal matrix, but do not forget the minus signs in this triangular matrix. Compare Problem 1.2.4.

(b) Suppose that $\sum_{j=0}^{k} \alpha_j E^{-j}$ and $\sum_{j=0}^{k} a_j \nabla^j$ represent the same operator. Set $\alpha = (\alpha_k, \alpha_{k-1}, \dots, \alpha_0)^T$ and $a = (a_k, a_{k-1}, \dots, a_0)^T$, i.e., $Y_{ord} \cdot \alpha \equiv Y_{dif} \cdot a$. Show that $Pa = \alpha$, $P\alpha = a$.

(c) The matrix P depends on the integer k. Is it true that the matrix which is obtained for a certain k is a submatrix of the matrix you obtain for a larger value of k?

(d) Compare this method of performing the mapping $Y_{ord} \mapsto Y_{dif}$ with the ordinary construction of a difference scheme. Consider the number of arithmetic operations, the kind of arithmetic operations, rounding errors, convenience of programming in a language with matrix operations as primary operations, etc. In the same way, compare this method of performing the inverse mapping with the algorithm in Problem 3.3.2 (c).

3.3.5 (a) Set $f(x) = \tan x$. Compute by using the table of $\tan x$ (in Example 3.3.2) and the interpolation and differentiation formulas given in the above examples (almost) as accurately as possible the quantities

$$f'(1.35), \quad f(1.322), \quad f'(1.325), \quad f''(1.32).$$

Estimate the influence of rounding errors of the function values and estimate the truncation errors.

(b) Write a program for computing a difference scheme. Use it for computing the difference scheme for more accurate values of $\tan x$, $x = 1.30 : 0.01 : 1.35$, and calculate improved values of the functionals in (a). Compare the error estimates with the true errors.

(c) Verify the assumptions of Theorem 3.3.7 for one of the three interpolation formulas in Sec. 3.3.4.

(d) It is rather easy to find the values at $\theta = 0$ of the first two derivatives of Stirling's interpolation formula. You find thus explicit expressions for the coefficients in the formulas for $f'(x_0)$ and $f''(x_0)$ in (3.3.47) and (3.3.43), respectively. Check numerically a few coefficients in these equations, and explain why they are reciprocals of integers. Also note that each coefficient in (3.3.47) has a simple relation to the corresponding coefficient in (3.3.43).

3.3.6 (a) Study Bickley's table (Table 3.3.1) and derive some of the formulas, in particular the expressions for δ and μ in terms of hD, and vice versa.

(b) Show that $h^{-k}\delta^k - D^k$ has an expansion into *even* powers of h when k is even. Find an analogous result for $h^{-k}\mu\delta^k - D^k$ when k is odd.

3.3.7 (a) Compute

$$f'(10)/12, \quad f^{(3)}(10)/720, \quad f^5(10)/30{,}240$$

by means of (3.3.24), given values of $f(x)$ for integer values of x. (This is asked for in applications of Euler–Maclaurin's formula, Sec. 3.4.5.) Do this for $f(x) = x^{-3/2}$. Compare with the correct derivatives. Then do the same for $f(x) = (x^3+1)^{-1/2}$.

(b) Study the backward differentiation formula; see (3.3.23) on a computer. Compute $f'(1)$ for $f(x) = 1/x$, for $h = 0.02$ and $h = 0.03$, and compare with the exact result. Make a semilogarithmic plot of the total error after n terms, $n = 1 : 29$. Study also the sign of the error. For each case, try to find out whether the achievable accuracy is set by the rounding errors or by the semiconvergence of the series.

Hint: A formula mentioned in Problem 3.3.3 (a) can be helpful. Also note that this problem is both similar and very different from the function $\tan x$ that was studied in Example 3.3.6.

(c) Set $x_i = x_0 + ih$, $t = (x - x_2)/h$. Show that

$$y(x) = y_2 + t\Delta y_2 + \frac{t(t-1)}{2}\Delta^2 y_2 + \frac{t(t-1)(t-2)}{6}\Delta^3 y_1$$

equals the interpolation polynomial in $\mathcal{P}_4$ determined by the values $(x_i, y_i), i = 1 : 4$. (Note that $\Delta^3 y_1$ is used instead of $\Delta^3 y_2$ which is located outside the scheme. Is this fine?)

3.3.8 A well-known formula reads

$$P(D)(e^{\alpha t}u(t)) = e^{\alpha t}P(D+\alpha)u(t),$$

where P is an arbitrary polynomial. Prove this, as well as the following analogous formulas:

$$P(E)(a^n u_n) = a^n P(aE)u_n,$$
$$P(\Delta/h)\big((1+\alpha h)^n u_n\big) = (1+\alpha h)^n P((1+\alpha h)\Delta/h + \alpha)u_n.$$

Can you find a more beautiful or more practical variant?

3.3.9 Find the Peano kernel $K(u)$ for the functional $\Delta^2 f(x_0)$. Compute $\int_{\mathbf{R}} K(u)\,du$ both by direct integration of $K(u)$ and by computing $\Delta^2 f(x_0)$ for a suitably chosen function f.

3.3.10 Set $y_j = y(t_j)$, $y'_j = y'(t_j)$. The following relations, due to John Adams,[88] are of great interest in the numerical integration of the differential equations $y' = f(y)$.

(a) **Adams–Moulton's** implicit formula:

$$y_{n+1} - y_n = h\left(a_0 y'_{n+1} + a_1 \nabla y'_{n+1} + a_2 \nabla^2 y'_{n+1} + \cdots\right).$$

Show that $\nabla = -\ln(1-\nabla)\sum a_i \nabla^i$, and find a recurrence relation for the coefficients. The coefficients a_i, $i = 0 : 6$, read as follows (check a few of them):

$$a_i = 1,\quad -\frac{1}{2},\quad -\frac{1}{12},\quad -\frac{1}{24},\quad -\frac{19}{720},\quad -\frac{3}{160},\quad -\frac{863}{60{,}480}.$$

Alternatively, derive the coefficients by means of the matrix representation of a truncated power series.

(b) **Adams–Bashforth's** explicit formula:

$$y_{n+1} - y_n = h\left(b_0 y'_n + b_1 \nabla y'_n + b_2 \nabla^2 y'_n + \cdots\right).$$

Show that $\sum b_i \nabla^i E^{-1} = \sum a_i \nabla^i$, and that $b_n - b_{n-1} = a_n$ $(n \geq 1)$. The coefficients b_i, $i = 0 : 6$, read as follows (check a few of them):

$$b_i = 1,\quad \frac{1}{2},\quad \frac{5}{12},\quad \frac{3}{8},\quad \frac{251}{720},\quad \frac{95}{288},\quad \frac{19{,}087}{60{,}480}.$$

(c) Apply the second order explicit Adams' formula,

$$y_{n+1} - y_n = h\left(y'_n + \tfrac{1}{2}\nabla y'_n\right),$$

to the differential equation $y' = -y^2$, with initial condition $y(0) = 1$ and step size $h = 0.1$. Two initial values are needed for the recurrence: $y_0 = y(0) = 1$, of course, and we choose[89] $y_1 = 0.9090$. Then compute $y'_0 = -y_0^2$, $y'_1 = -y_1^2$. The explicit Adams' formula then yields y_k, $k \geq 2$. Compute a few steps, and compare with the exact solution.[90]

3.3.11 Let $y_j = y_0 + jh$. Find the asymptotic behavior as $h \to 0$ of

$$(5(y_1 - y_0) + (y_2 - y_1))/(2h) - y'_0 - 2y'_1.$$

Comment: This is of interest in the analysis of cubic spline interpolation in Sec. 4.4.2.

[88] John Couch Adams (1819–1892) was an English mathematician. While still an undergraduate he calculated the irregularities of the motion of the planet Uranus, showing the existence of Neptune. He held the position as Professor of Astronomy and Geometry at Cambridge for 32 years.

[89] There are several ways of obtaining $y_1 \approx y(h)$, for example, by one step of Runge's second order method, see Sec. 1.5.3, or by a series expansion, as in Sec. 1.2.4.

[90] For an *implicit* Adams' formula it is necessary, in this example, to solve a quadratic equation in each step.

3.3.12 It sometimes happens that the values of some function $f(x)$ can be computed by some very time-consuming algorithm only, and that one therefore computes it much sparser than is needed for the application of the results. It was common in the pre-computer age to compute sparse tables that needed interpolation by polynomials of a high degree; then one needed a simple procedure for **subtabulation**, i.e., to obtain a denser table for some section of the table. Today a similar situation may occur in connection with the graphical output of the results of (say) a numerical solution of a differential equation.

Define the operators ∇ and ∇_k by the equations

$$\nabla f(x) = f(x) - f(x-h), \qquad \nabla_k f(x) = f(x) - f(x-kh), \quad (k < 1),$$

and set

$$\nabla_k^r = \sum_{s=r}^{\infty} c_{rs}(k) \nabla^s.$$

(a) Suppose that $\Delta^r f(x)$, $r = 0 : m$, has been computed. Suppose that k has been chosen, that the coefficients $c_{rs}(k)$ are known for $r \le m$, $s \le m$, and that $\Delta_k^r f(a)$, $r = 0 : m$, has been computed. Design an algorithm for obtaining $f(x)$, $x = a : kh : a + mkh$, and $\nabla_k^r f(a + mkh)$, $r = 0 : m$. (You can here, e.g., modify the ideas of (3.3.40).) Then you can apply (3.3.40) directly to obtain a tabulation of $f(x)$, $x = a + mkh : kh : b$.

(b) In order to compute the coefficients c_{rs}, $r \le s \le m$, you are advised to use a subroutine for finding the coefficients in the product of two polynomials, truncate the result, and apply the subroutine $m - 1$ times.

(c) Given

f_n	∇f_n	$\nabla^2 f_n$	$\nabla^3 f_n$	$\nabla^4 f_n$
1	0.181269	0.032858	0.005956	0.001080

compute for $k = \frac{1}{2}$, $f_n = f(x_n)$, $\nabla_k^j f_n$ for $j = 1 : 4$. Compute $f(x_n - h)$ and $f(x_n - 2h)$ by means of both $\{\nabla^j f_n\}$ and $\{\nabla_k^j f_n\}$ and compare the results. How big a difference in the results did you expect, and how big a difference do you obtain?

3.3.13 (a) Check Example 3.3.10 and the conclusions about the optimal step length in the text. Investigate how the attainable accuracy varies with u, for these three values of k, if $u = 1.1 \cdot 10^{-16}$.

(b) Study the analogous question for $f''(x_0)$ using the formula

$$f''(x_0) \approx \left(1 - \frac{\delta^2}{12} + \frac{\delta^4}{90} - \frac{\delta^6}{560} + \frac{\delta^8}{3150} - \cdots\right) \frac{\delta^2 f_0}{h^2}.$$

3.3.14 Solve the following difference equations. A solution in complex form should be transformed to real form. As a check, compute (say) y_2 both by recurrence and by your closed form expression.

(a) $y_{n+2} - 2y_{n+1} - 3y_n = 0, \quad y_0 = 0, \quad y_1 = 1$

(b) $y_{n+2} - 4y_{n+1} + 5y_n = 0, \quad y_0 = 0, \quad y_1 = 2$

(c) There exist problems with two-point boundary conditions for difference equations, as for differential equations $y_{n+2} - 2y_{n+1} - 3y_n = 0, \quad y_0 = 0, \quad y_{10} = 1$

(d) $y_{n+2} + 2y_{n+1} + y_n = 0, \quad y_0 = 1, \quad y_1 = 0$

(e) $y_{n+1} - y_n = 2^n, \quad y_0 = 0$

(f) $y_{n+2} - 2y_{n+1} - 3y_n = 1 + \cos\frac{\pi n}{3}, \quad y_0 = y_1 = 0$

Hint: The right-hand side is $\Re(1 + a^n)$, where $a = e^{\pi i/3}$.

(g) $y_{n+1} - y_n = n, \quad y_0 = 0$

(h) $y_{n+1} - 2y_n = n2^n, \quad y_0 = 0$

3.3.15 (a) Prove Lemma 3.3.11.

(b) Consider the difference equation $y_{n+2} - 5y_{n+1} + 6y_n = 2n + 3(-1)^n$. Determine a particular solution of the form $y_n = an + b + c(-1)^n$.

(c) Solve the difference equation $y_{n+2} - 6y_{n+1} + 5y_n = 2n + 3(-1)^n$. Why and how must you change the form of the particular solution?

3.3.16 (a) Show that the difference equation $\sum_{i=0}^{k} b_i \Delta^i y_n = 0$ has the characteristic equation $\sum_{i=0}^{k} b_i (u - 1)^i = 0$.

(b) Solve the difference equation $\Delta^2 y_n - 3\Delta y_n + 2y_n = 0$, with initial condition $\Delta y_0 = 1$.

(c) Find the characteristic equation for the equation $\sum_{i=0}^{k} b_i \nabla^i y_n = 0$.

3.3.17 The influence of wrong boundary slopes for cubic spline interpolation (with equidistant data)—see Sec. 4.4.2—is governed by the difference equation

$$e_{n+1} + 4e_n + e_{n-1} = 0, \quad 0 < n < m,$$

with e_0, e_m given. Show that $e_n \approx u^n e_0 + u^{m-n} e_m$, $u = \sqrt{3} - 2 \approx -0.27$. More precisely,

$$\left| e_n - (u^n e_0 + u^{m-n} e_m) \right| \le \frac{2|u^{3m/2}|}{1 - |u|^m} \max(|e_0|, |e_m|).$$

Generalize the simpler of these results to other difference and differential equations.

3.3.18 The Fibonacci sequence is defined by the recurrence relation

$$y_n = y_{n-1} + y_{n-2}, \quad y_0 = 0, \quad y_1 = 1.$$

(a) Calculate $\lim_{n\to\infty} y_{n+1}/y_n$.

(b) The error of the secant method (see Sec. 6.2.2) satisfies approximately the difference equation $\epsilon_n = C\epsilon_{n-1}\epsilon_{n-2}$. Solve this difference equation. Determine p such that $\epsilon_{n+1}/\epsilon_n^p$ tends to a finite nonzero limit as $n \to \infty$. Calculate this limit.

3.3.19 For several algorithms using the divide and conquer strategy, such as the FFT and some sorting methods, one can find that the work $W(n)$ for the application of them to data of size n satisfies a recurrence relation of the form

$$W(n) = 2W(n/2) + kn,$$

where k is a constant. Find $W(n)$.

3.3.20 When the recursion

$$x_{n+2} = (32x_{n+1} - 20x_n)/3, \quad x_0 = 3, \; x_1 = 2,$$

was solved numerically in low precision (23 bits mantissa), one obtained for x_i, $i = 2 : 12$, the (rounded) values

$$1.33,\ 0.89,\ 0.59,\ 0.40,\ 0.26,\ 0.18,\ 0.11,\ 0.03,\ -0.46,\ -5.05,\ -50.80.$$

Explain the difference from the exact values $x_n = 3(2/3)^n$.

3.3.21 (a) k, N are given integers $0 \le k << \le N$. A "discrete Green's function" $G_{n,k}$, $0 \le n \le N$, for the central difference operator $-\Delta\nabla$ together with the boundary conditions given below, is defined as the solution $u_n = G_{n,k}$ of the difference equation with boundary conditions

$$-\Delta\nabla u_n = \delta_{n,k}, \qquad u_0 = u_N = 0$$

($\delta_{n,k}$ is Kronecker's delta). Derive a fairly simple expression for $G_{n,k}$.

(b) Find (by computer) the inverse of the tridiagonal Toeplitz matrix[91]

$$A = \begin{pmatrix} 2 & -1 & & & \\ -1 & 2 & -1 & & \\ & -1 & 2 & -1 & \\ & & -1 & 2 & -1 \\ & & & -1 & 2 \end{pmatrix}.$$

What is the relation between Problems 3.3.21 (a) and (b)? Find a formula for the elements of A^{-1}. Express the solution of the inhomogeneous difference equation $-\Delta\nabla u_n = b_n$, $u_0 = u_N = 0$, both in terms of the Green function $G_{n,k}$ and in terms of A^{-1} (for general N).

(c) Try to find an analogous formula[92] for the solution of an inhomogeneous boundary value problem for the *differential* equation $-u'' = f(x)$, $u(0) = u(1) = 0$.

3.3.22 (a) Demonstrate the formula

$$\sum_{n=0}^{\infty} \frac{(-x)^n c_n}{n!} = e^{-x} \sum_{n=0}^{\infty} \frac{x^n(-\Delta)^n c_0}{n!}. \tag{3.3.67}$$

Hint: Use the relation $e^{-xE} = e^{-x(1+\Delta)} = e^{-x}e^{-x\Delta}$.

(b) For completely monotonic sequences $\{c_n\}$ and $\{(-\Delta)^n c_0\}$ are typically positive and decreasing sequences. For such sequences, the left-hand side becomes extremely ill-conditioned for large x, (say) $x = 100$, while the graph of the terms on the right-hand side (if exactly computed) is bell-shaped, almost like the normal probability density with mean x and standard deviation $\sqrt{x}$. We have called such a sum a *bell sum*. Such positive sums can be computed with little effort and no trouble with rounding errors, *if their coefficients are accurate*.

Compute the left-hand side of (3.3.67), for $c_n = 1/(n+1)$, $x = 10 : 10 : 100$, and compute the right-hand side, both with numerically computed differences and with

[91] The inverse is a so-called **semiseparable matrix**.

[92] In a *differential* equation, analogous to Problem 3.3.21 (a), the Kronecker delta is to be replaced by the Dirac delta function. Also note that the inverse of the differential operator here can be described as an integral operator with the Green's function as the "kernel."

exact differences; the latter are found in Problem 3.3.3 (a). (In this particular case you can also find the exact sum.)

Suppose that the higher differences $\{(-\Delta)^n c_0\}$ have been computed recursively from rounded values of c_n. Explain why one may fear that the right-hand side of (3.3.67) does not provide much better results than the left-hand side.

(c) Use (3.3.67) to derive the second expansion for $\operatorname{erf}(x)$ in Problem 3.2.8 from the first expansion.

Hint: Use one of the results of Problem 3.3.3 (a).

(d) If $c_n = c_n(a, b)$ is defined as in Problem 3.3.3 (d), then the left-hand side becomes the Maclaurin expansion of the Kummer function $M(a, b, -x)$; see the Handbook [1, Chap. 13]. Show that

$$M(a, b, -x) = e^{-x} M(b - a, b, x)$$

by means of the results of Problems 3.3.23 (a) and 3.3.2 (d).[93]

3.3.23 (a) The difference equation $y_n + 5y_{n-1} = n^{-1}$ was discussed in Example 1.2.1. It can also be written thus: $(6 + \Delta)y_{n-1} = n^{-1}$. The expansion of $(6 + \Delta)^{-1} n^{-1}$ into powers of $\Delta/6$ provides a particular solution of the difference equation.

Compute this numerically for a few values of n. Try to prove the convergence, with or without the expression in Problem 3.3.3 (b). Is this the same as the particular solution $I_n = \int_0^1 x^n (x + 5)^{-1}\, dx$ that was studied in Example 1.2.1?

Hint: What happens as $n \to \infty$? Can more than one solution of this difference equation be bounded as $n \to \infty$?

(b) Make a similar study of the difference equation related to the integral in Problem 1.2.7. Why does the argument suggested by the hint in (a) not work in this case? Try another proof.

3.3.24 (a) Prove Lemma 3.3.15. How is the conclusion to be changed if we do not suppose that $\gamma < \alpha$, even though the coefficients are still positive? Show that a backward recurrence is still to be recommended.

(b) Work out on a computer the numerical details of Example 3.3.16, and compare with the Handbook [1, Example 19.28.1]. (Some deviations are to be expected, since Miller used other rounding rules.) Try to detect the oscillating component by computing the difference scheme of the computed $U(a, 5)$, and estimate roughly the error of the computed values.

3.3.25 (a) For which constant real a does the difference equation

$$y_{n+1} - 2ay_n + y_{n-1} = 0$$

satisfy the root condition? For which values of the real constant a does there exist a solution such that $\lim_{n\to\infty} y_n = 0$? For these values of a, how do you construct a solution $y_n = y_n^*$ by a recurrence and normalization so that this condition as well as the condition $y_0^* + 2\sum_{m=1}^{\infty} y_{2m}^* = 1$ are satisfied? Is y_n^* unique? Give also an explicit expression for y_n^*.

---

[93] This formula is well known in the theory of the confluent hypergeometric functions, where it is usually proved in other ways.

(b) For the other real values of a, show that y_n^* does not exist, but that for any given y_0, y_1 a solution can be accurately constructed by forward recurrence. Give an explicit expression for this solution in terms of Chebyshev polynomials (of the first and the second kind). Is it true that backward recurrence is also stable, though more complicated than forward recurrence?

3.3.26 (a) The Bessel function $J_k(z)$ satisfies the difference equation

$$J_{k+1}(z) - (2k/z)J_k(z) + J_{k-1}(z) = 0, \quad k = 1, 2, 3, \dots,$$

and the identities

$$J_0(z) + 2J_2(z) + 2J_4(z) + 2J_6(z) + \cdots = 1,$$
$$J_0(z) - 2J_2(z) + 2J_4(z) - 2J_6(z) + \cdots = \cos z;$$

see the Handbook [1, Sec. 9.1.27, 9.1.46, and 9.1.47]. Show how one of the identities can be used for normalizing the trial sequence obtained by a backward recurrence. Under what condition does Lemma 3.3.15 give the hint to use the backward recurrence for this difference equation?

(b) Study the section on Bessel functions of integer order in [294]. Apply this technique for $z = 10,\ 1,\ 0.1$ (say). The asymptotic formula (see [1, Sec. 9.3.1])

$$J_k(z) \sim \frac{1}{\sqrt{2\pi k}}\left(\frac{ez}{2k}\right)^k, \quad k \gg 1,\ z \text{ fixed},$$

may be useful in deciding where to start the backward recurrence. Use at least two starting points, and subtract the results (after normalization).

Comment: The above difference equation for $J_k(z)$ is also satisfied by a function denoted $Y_k(z)$:

$$Y_k(z) \sim \frac{-2}{\sqrt{2\pi k}}\left(\frac{ez}{2k}\right)^{-k}, \quad (k \gg 1).$$

How do these two solutions interfere with each other when forward or backward recurrence is used?

3.3.27 *A counterexample* to the technique with frozen coefficients. Consider the difference equation $y_{n+1} - (-1)^n y_n + y_{n-1} = 0$. The technique with frozen coefficients leads to the consideration of the difference equations

$$z_{n+1} - 2az_n + z_{n-1} = 0, \quad a \in [-0.5, 0.5];$$

all of them have only bounded solutions. Find by numerical experiment that, nevertheless, there seems to exist unbounded solutions y_n of the first difference equation.

Comment: A proof of this is found by noting that the mapping $(y_{2n}, y_{2n+1}) \mapsto (y_{2n+2}, y_{2n+3})$ is represented by a matrix that is independent of n and has an eigenvalue that is less than -1.

3.3.28 Let $\{b_n\}_{-\infty}^{\infty}$ be a given sequence, and consider the difference equation

$$y_{n-1} + 4y_n + y_{n+1} = b_n,$$

which can also be written in the form $(6 + \delta^2)y_n = b_n$.

(a) Show that the difference equation has at most one solution that is bounded for $-\infty < n < +\infty$. Find a particular solution in the form of an expansion into powers of the operator $\delta^2/6$. (This is, hopefully, bounded.)

(b) Apply it numerically to the sequence $b_n = (1 + n^2h^2)^{-1}$ for a few values of the step size h, e.g., $h = 0.1, 0.2, 0.5, 1$. Study for $n = 0$ the rate of decrease (?) of the terms in the expansion. Terminate when you estimate that the error is (say) 10^{-6}. Check how well the difference equation is satisfied by the result.

(c) Study theoretically bounds for the terms when $b_n = \exp(i\omega hn)$, $\omega \in \mathbf{R}$. Does the expansion converge? Compare your conclusions with numerical experiments. Extend to the case when $b_n = B(nh)$, where $B(t)$ can be represented by an absolutely convergent Fourier integral,

$$B(t) = \int_{-\infty}^{\infty} e^{i\omega t}\beta(\omega)d\omega.$$

Note that $B(t) = (1 + t^2)^{-1}$ if $\beta(\omega) = \frac{1}{2}e^{-|\omega|}$. Compare the theoretical results with the experimental results in (b).

(d) Put $Q = \delta^2/6$. Show that $\tilde{y}_n \equiv (1 - Q + Q^2 + \cdots \pm Q^{k-1})b_n/6$ satisfies the difference equation $(1 + Q)(\tilde{y}_n - y_n) = Q^k b_n/6$.

Comment: This procedure is worthwhile if the sequence b_n is so smooth that (say) two or three terms give satisfactory accuracy.

3.4 Acceleration of Convergence

3.4.1 Introduction

We have seen that in applied mathematics the solution to many problems can be obtained from a series expansion or a sequence converging to the exact solution. But sometimes the convergence of the series is so slow that the effective use of it is limited.

If a sequence $\{s_n\}_0^\infty$ converges slowly toward a limit s, but has a sort of regular behavior when n is large, it can under certain conditions be transformed into another infinite sequence $\{s'_n\}$, which converges much faster to the same limit. Here s'_n usually depends on the first n elements of the original sequence only. This is called **convergence acceleration**. Such a *sequence* transformation may be iterated to yield a sequence of infinite sequences, $\{s''_n\}$, $\{s'''_n\}$, and so forth, hopefully with improved convergence toward the same limit s. For an *infinite series* convergence acceleration means the convergence acceleration of its sequence of partial sums, because

$$\lim_{n\to\infty} s_n = a \iff a = s_j + \sum_{p=1}^{\infty}(s_{p+j} - s_{p+j-1}).$$

Some algorithms are most easily discussed in terms of sequences, others in terms of series.

Several transformations, linear as well as nonlinear, have been suggested and are successful under various conditions. Some of them, such as Aitken transformation, repeated averages, and Euler's transformation, are most successful on *oscillating sequences* (alternating series or series in a complex variable). Others, such as variants of Aitken acceleration,

Euler–Maclaurin, and Richardson, work primarily on *monotonic sequences* (series with positive terms). Some techniques for convergence acceleration such as continued fractions, Padé approximation, and the ϵ algorithm transform a power series into a sequence of rational functions.

Some of these techniques may even sometimes be successfully applied to *semi-convergent sequences.* Several of them can also use a limited number of coefficients of a power series for the computation of values of an *analytic continuation* of a function, outside the circle of convergence of the series that defined it.

Convergence acceleration cannot be applied to "arbitrary sequences"; somc sort of conditions are necessary that restrict the variation of the future elements of the sequence, i.e., the elements which are not computed numerically. In this section, these conditions are of a rather general type, in terms of *monotonicity, analyticity, or asymptotic behavior* of simple and usual types.

In addition to the "general purpose" techniques to be discussed in this chapter, there are other techniques of convergence acceleration based on the use of more specific knowledge about a problem. For example, the Poisson summation formula

$$\sum_{n=-\infty}^{\infty} f(n) = \sum_{j=-\infty}^{\infty} \hat{f}(j), \quad \hat{f}(\omega) = \int_{-\infty}^{\infty} f(\omega)e^{-2\pi i\omega x}\,dx \tag{3.4.1}$$

($\hat{f}$ is the Fourier transform of f) can be amazingly successful for a certain class of series $\sum a(n)$, namely if $a(x)$ has a rapidly decreasing Fourier transform. The Poisson formula is also an invaluable tool for the design and analysis of numerical methods for several problems; see Theorem 3.4.10.

Irregular errors are very disturbing when these techniques are used. They sometimes set the limit for the reachable accuracy. For the sake of simplicity we therefore use IEEE double precision arithmetic in most examples.

3.4.2 Comparison Series and Aitken Acceleration

Suppose that the terms in the series $\sum_{j=1}^{\infty} a_j$ behave, for large j, like the terms of a series $\sum_{j=1}^{\infty} b_j$, i.e., $\lim_{j\to\infty} a_j/b_j = 1$. Then, if the sum $s = \sum_{j=1}^{\infty} b_j$ is known one can write

$$S = \sum_{j=1}^{\infty} a_j = s + \sum_{j=1}^{\infty} (a_j - b_j),$$

where the series on the right-hand side converges more quickly than the given series. We call this making use of a simple **comparison problem**. The same idea is used in many other contexts—for example, in the computation of integrals where the integrand has a singularity. Usual comparison series are

$$\sum_{j=1}^{\infty} n^{-2} = \pi^2/6, \qquad \sum_{j=1}^{\infty} n^{-4} = \pi^4/90, \quad \text{etc.}$$

A general expression for $\sum_{j=1}^{\infty} n^{-2r}$ is given in (3.4.32). No simple closed form is known for $\sum_{j=1}^{\infty} n^{-3}$.

Example 3.4.1.

The term $a_j = (j^4+1)^{-1/2}$ behaves, for large j, like $b_j = j^{-2}$, whose sum is $\pi^2/6$. Thus

$$\sum_{j=1}^{\infty} a_j = \pi^2/6 + \sum_{j=1}^{\infty}\left((j^4+1)^{-1/2} - j^{-2}\right) = 1.64493 - 0.30119 = 1.3437.$$

Five terms on the right-hand side are sufficient for four-place accuracy in the final result. Using the series on the left-hand side, one would not get four-place accuracy until after 20,000 terms.

This technique is unusually successful in this example. The reader is advised to find out why, and why it is less successful for $a_j = (j^4 + j^3 + 1)^{-1/2}$.

An important comparison sequence is a geometric sequence

$$y_n = s + bk^n$$

for which $\nabla y_n = y_n - y_{n-1} = bk^{n-1}(k-1)$. If this is fitted to the three most recently computed terms of a given sequence, $y_n = s_n$ for (say) $n = j, j-1, j-2$, then $\nabla y_j = \nabla s_j$, $\nabla y_{j-1} = \nabla s_{j-1}$, and

$$k = \nabla s_j / \nabla s_{j-1}, \qquad \nabla s_j = bk^{j-1}(k-1).$$

Hence

$$bk^j = \frac{\nabla s_j}{1 - 1/k} = \frac{\nabla s_j}{1 - \nabla s_{j-1}/\nabla s_j} = \frac{(\nabla s_j)^2}{\nabla^2 s_j}.$$

This yields a comparison sequence for each j. Suppose that $|k| < 1$. Then the comparison sequence has the limit $\lim_{n\to\infty} y_n = s = y_j - bk^j$, i.e.,

$$s \approx s_j' = s_j - \frac{(\nabla s_j)^2}{\nabla^2 s_j}. \tag{3.4.2}$$

This *nonlinear* acceleration method is called **Aitken acceleration**.[94]

Notice that the denominator equals $s_j - 2s_{j-1} + s_{j-2}$, but to minimize rounding errors it should be computed as

$$\nabla s_j - \nabla s_{j-1} = (s_j - s_{j-1}) - (s_{j-1} + s_{j-2})$$

(cf. Lemma 2.3.2). If $\{s_n\}$ is exactly a geometric sequence, i.e., if $s_n - s = k(s_{n-1} - s)$ for all n, then $s_j' = s$ for all j. Otherwise it can be shown (Henrici [193]) that under the assumptions

$$\lim_{j\to\infty} s_j = s, \qquad \lim \frac{s_{j+1} - s_j}{s_j - s_{j-1}} = k^*, \quad |k^*| < 1, \tag{3.4.3}$$

the sequence $\{s_j'\}$ converges faster than the sequence $\{s_j\}$. The above assumptions can often be verified for sequences arising from iterative processes and for many other applications. Note also that Aitken extrapolation is exact for sequences $\{s_n\}$ such that

$$\alpha(s_n - s) + \beta(s_{n+1} - s) = 0 \ \forall\, n,$$

with $\alpha\beta \neq 0$, $\alpha + \beta \neq 0$. This leads to a generalization to be discussed in Sec. 3.5.4.

[94]Named after Alexander Craig Aitken (1895–1967), a Scottish mathematician born in New Zealand.

If you want the sum of slowly convergent *series*, then it may seem strange to compute the sequence of partial sums, and compute the first and second differences of rounded values of this sequence in order to apply Aitken acceleration. The *a-version* of Aitken acceleration works on the terms a_j of an infinite series instead of on its partial sums s_j.

Clearly we have $a_j = \nabla s_j$, $j = 1 : N$. The a-version of Aitken acceleration thus reads

$$s'_j = s_j - a_j^2/\nabla a_j, \quad j = 1 : N. \tag{3.4.4}$$

We want to determine a'_j so that

$$\sum_{k=1}^{j} a'_k = s'_j, \quad j = 1 : N.$$

Then

$$a'_1 = 0, \qquad a'_j = a_j - \nabla(a_j^2/\nabla a_j), \quad j = 2 : N,$$

and $s'_N = s_N - a_N^2/\nabla a_N$ (show this). We may expect that this a-version of Aitken acceleration handles rounding errors better.

The condition $|k^*| < 1$ is a *sufficient* condition only. In practice, Aitken acceleration seems *most efficient if* $k^* = -1$. Indeed, it often converges even if $k^* < -1$; see Problem 3.4.7. It is *much less successful if* $k^* \approx 1$, for example, for slowly convergent series with positive terms.

The Aitken acceleration process can often be *iterated* to yield sequences $\{s''_n\}_0^\infty$, $\{s'''_n\}_0^\infty$, etc., defined by the formulas

$$s''_j = s'_j - \frac{(\nabla s'_j)^2}{\nabla^2 s'_j}, \qquad s'''_j = s''_j - \frac{(\nabla s''_j)^2}{\nabla^2 s''_j} \cdots. \tag{3.4.5}$$

j	s_j	e_j	e'_j	e''_j	e'''_j
6	0.820935	3.5536e−2			
7	0.754268	−3.1130e−2	−1.7783e−4		
8	0.813092	2.7693e−2	1.1979e−4		
9	0.760460	−2.4938e−2	−8.4457e−5	−1.3332e−6	
10	0.808079	2.2681e−2	6.1741e−5	7.5041e−7	
11	0.764601	−2.0797e−2	−4.6484e−5	−4.4772e−7	−1.0289e−8

Example 3.4.2.

By (3.1.13), it follows that for $x = 1$

$$1 - 1/3 + 1/5 - 1/7 + 1/9 - \cdots = \arctan 1 = \pi/4 \approx 0.7853981634.$$

This series converges very slowly. Even after 500 terms there still occur changes in the third decimal. Consider the partial sums $s_j = \sum_{n_0}^{j} (-1)^j (2n+1)^{-1}$, with $n_0 = 5$, and compute the **iterated Aitken** sequences as indicated above.

The (sufficient) theoretical condition mentioned above is not satisfied, since here $\nabla s_n/\nabla s_{n-1} \to -1$ as $n \to \infty$. Nevertheless, we shall see that the Aitken acceleration works well, and that the iterated accelerations converge rapidly. One gains two digits for every pair of terms, in spite of the slow convergence of the original series. The results in the table above were obtained using IEEE double precision arithmetic. The errors of s'_j, $s''_j, \dots$, are denoted by $e'_j, e''_j, \dots$.

Example 3.4.3.

Set $a_n = e^{-\sqrt{n+1}}$, $n \geq 0$. As before, we denote by s_n the partial sums of $\sum a_n$, $s = \lim s_n = 1.67040681796634$, and use the same notations as above. Note that

$$\frac{\nabla s_n}{\nabla s_{n-1}} = \frac{a_n}{a_{n-1}} \approx 1 - \frac{1}{2}n^{-1/2}, \quad (n \gg 1),$$

so this series is slowly convergent. Computations with plain and iterated Aitken in IEEE double precision arithmetic gave the results below.

j	e_{2j}	$e_{2j}^{(j)}$
1	−0.882	−4.10e−1
2	−0.640	−1.08e−1
3	−0.483	−3.32e−2
2	−0.374	−4.41e−3
5	−0.295	−7.97e−4
6	−0.237	−1.29e−4
7	−0.192	−1.06e−5

The sequence $\{e_{2j}^{(j)}\}$ is monotonic until $j = 8$. After this $|e_{2j}^{(j)}|$ is mildly fluctuating around 10^{-5} (at least until $j = 24$), and the differences $\nabla s_{2j}^{(j)} = \nabla e_{2j}^{(j)}$ are sometimes several powers of 10 smaller than the actual errors and are misleading as error estimates. The rounding errors have taken over, and it is almost no use to compute more terms.

It is possible to use more terms for obtaining higher accuracy by applying iterated Aitken acceleration to a **thinned sequence**, for example, $s_4, s_8, s_{12}, \dots$; cf. Problem 3.4.4. Note the thinning is performed on a *sequence* that converges to the limit to be computed, for example, the partial sums of a series. Only in so-called *bell sums* (see Problem 3.4.29) shall we do *a completely different kind of thinning*, namely a thinning of the *terms* of a series.

The convergence ratio of the thinned sequence are much smaller; for the series of the previous example they become approximately

$$\left(1 - \frac{1}{2}n^{-1/2}\right)^4 \approx 1 - 2n^{-1/2}, \quad n \gg 1.$$

The most important point though, is that the rounding errors become more slowly amplified, so that terms far beyond the eighth one of the unthinned sequence can be used in the acceleration, resulting in a much improved final accuracy.

How to realize the thinning depends on the sequence; a different thinning will be used in the next example.

Example 3.4.4.

We shall compute, using IEEE double precision arithmetic,

$$s = \sum_{n=1}^{\infty} n^{-3/2} = 2.612375348685488.$$

If all partial sums are used in Aitken acceleration, it turns out that the error $|e_{2j}^{(j)}|$ is decreasing until $j = 5$, when it is 0.07, and it remains on approximately this level for a long time.

j	0	1	2	3	4	5
E_{2j+1}	−1.61	−0.94	−4.92e−1	−2.49e−1	−1.25e−1	−6.25e−2
$E_{2j+1}^{(j)}$	−1.61	−1.85	−5.06e−2	−2.37e−4	−2.25e−7	2.25e−10

A much better result is obtained by means of thinning, but since the convergence is much slower here than in the previous case, we shall try "geometric" thinning rather than the "arithmetic" thinning used above; i.e., we now set $S_m = s_{2^m}$. Then

$$\nabla S_m = \sum_{1+2^{m-1}}^{2^m} a_n, \qquad S_j = S_0 + \sum_{m=1}^{j} \nabla S_m, \qquad E_j = S_j - s.$$

(If maximal accuracy is wanted, it may be advisable to use the divide and conquer technique for computing these sums (see Problem 2.3.5), but it has not been used here.) By the approximation of the sums by integrals one can show that $\nabla S_m / \nabla S_{m-1} \approx 2^{-1/2}$, $m \gg 1$. The table above shows the errors of the first thinned sequence and the results after iterated Aitken acceleration. The last result has used 1024 terms of the original series, but since

$$s_n - s = -\sum_{j=n}^{\infty} j^{-3/2} \approx -\int_n^{\infty} t^{-3/2}\,dt = -\frac{2}{3} n^{-1/2}, \tag{3.4.6}$$

10^{20} terms would have been needed for obtaining this accuracy without convergence acceleration.

For sequences such that

$$s_n - s = c_0 n^{-p} + c_1 n^{-p-1} + O(n^{-p-2}), \quad p > 0,$$

where s, c_0, c_1 are unknown, the following variant of Aitken acceleration (Bjørstad, Dahlquist, and Grosse [33]) is more successful:

$$s_n' = s_n - \frac{p+1}{p} \frac{\Delta s_n \nabla s_n}{\Delta s_n - \nabla s_n}. \tag{3.4.7}$$

It turns out that s'_n is two powers of n more accurate than s_n, $s'_n - s = O(n^{-p-2})$; see Problem 3.4.12. More generally, suppose that there exists a longer (unknown) asymptotic expansion of the form

$$s_n = s + n^{-p}(c_0 + c_1 n^{-1} + c_2 n^{-2} + \cdots), \quad n \to \infty. \tag{3.4.8}$$

This is a rather common case. Then we can extend this to an *iterative variant*, where p is to be increased by two in each iteration; $i = 0, 1, 2, \ldots$ is a superscript, i.e.,

$$s_n^{i+1} = s_n^i - \frac{p + 2i + 1}{p + 2i} \frac{\Delta s_n^i \nabla s_n^i}{\Delta s_n^i - \nabla s_n^i}. \tag{3.4.9}$$

If p is also unknown, it can be estimated by means of the equation

$$\frac{1}{p+1} = -\Delta \frac{\Delta s_n}{\Delta s_n - \nabla s_n} + O(n^{-2}). \tag{3.4.10}$$

Example 3.4.5.

We consider the same series as in the previous example, i.e., $s = \sum n^{-3/2}$. We use (3.4.9) without thinning. Here $p = -1/2$; see Problem 3.4.13. As usual, the errors are denoted $e_j = s_j - s$, $e_{2j}^i = s_{2j}^i - s$. In the right column of the table below, we show the errors from a computation with 12 terms of the original series.

j	e_{2j}	e_{2j}^j
0	−1.612	−1.612
1	−1.066	−8.217e−3
2	−0.852	−4.617e−5
3	−0.730	+2.528e−7
4	−0.649	−1.122e−9
5	−0.590	−0.634e−11

From this point the errors were around 10^{-10} or a little below. The rounding errors have taken over, and the differences are misleading for error estimation. If needed, higher accuracy can be obtained by arithmetic thinning with more terms.

In this computation only 12 terms were used. In the previous example a less accurate result was obtained by means of 1024 terms of the same series, but we must appreciate that the technique of Example 3.4.4 did not require the existence of an asymptotic expansion for s_n and may therefore have a wider range of application.

There are not yet so many theoretical results that do justice to the practically observed efficiency of iterated Aitken accelerations for oscillating sequences. One reason for this can be that the transformation (3.4.2) which the algorithm is based on is *nonlinear*. For methods of convergence acceleration that are based on *linear* transformations, theoretical estimates of rates of convergence and errors are closer to the practical performance of the methods.

3.4.3 Euler's Transformation

In 1755 Euler gave the first version of what is now called **Euler's transformation**. Euler showed that for an alternating series ($u_j \geq 0$), it holds that

$$S = \sum_{j=0}^{\infty}(-1)^j u_j = \sum_{k=0}^{\infty} \frac{1}{2^k}\Delta^k u_k. \tag{3.4.11}$$

Often it is better to apply Euler's transformation to the tail of a series.

We shall now apply another method of acceleration based on **repeated averaging** of the partial sums. Consider again the same series as in Example 3.4.2, i.e.,

$$\sum_{j=0}^{\infty}(-1)^j(2j+1)^{-1} = 1 - \frac{1}{3} + \frac{1}{5} - \frac{1}{7} + \frac{1}{9} - \cdots = \frac{\pi}{4}. \tag{3.4.12}$$

Let S_N be the sum of the first N terms. The columns to the right of the S_N-column in the scheme given in Table 3.4.1 are formed by building averages.

Each number in a column is the mean of the two numbers which stand to the left and upper left of the number itself. In other words, each number is the mean of its "west" and "northwest" neighbor. The row index of M equals the number of terms used from the original series, while the column index minus one is the number of repeated averaging. Only the digits which are different from those in the previous column are written out.

Table 3.4.1. *Summation by repeated averaging.*

N	S_N	M_2	M_3	M_4	M_5	M_6	M_7
6	0.744012						
7	0.820935	782474					
8	0.754268	787602	5038				
9	0.813092	783680	5641	340			
10	0.760460	786776	5228	434	387		
11	0.808079	784270	5523	376	405	396	
12	0.764601	786340	5305	414	395	400	398

Notice that the values in each column oscillate. In general, for an alternating series, it follows from the next theorem together with (3.3.4) that *if the absolute value of the jth term, considered as a function of j, has a kth derivative which approaches zero monotonically for $j > N_0$, then every other value in column M_{k+1} is larger than the sum, and every other* *is smaller*. This premise is satisfied here, since if $f(j) = (2j+1)^{-1}$, then $f^{(k)}(j) = c_k(2j+1)^{-1-k}$, which approaches zero monotonically.

If roundoff is ignored, it follows from column M_6 that $0.785396 \le \pi/4 \le 0.785400$. To take account of roundoff error, we set $\pi/4 = 0.785398 \pm 3 \cdot 10^{-6}$. The actual error is only $1.6 \cdot 10^{-7}$. In Example 3.4.2 iterated Aitken accelerations gave about one decimal digit more with the same data. It is evident how the above method can be applied to any *alternating series*. The diagonal elements are equivalent to the results from using Euler's transformation.

Euler's transformation and the averaging method can be generalized for the convergence acceleration of a general complex power series

$$S(z) = \sum_{j=1}^{\infty} u_j z^{j-1}. \tag{3.4.13}$$

For $z = -1$ an alternating series is obtained. Other applications include *Fourier series*. They can be brought to this form with $z = e^{i\phi}$, $-\pi \le \phi \le \pi$; see Sec. 4.6.2 and Problem 4.6.7.

The irregular errors of the coefficients play a big role if $|\phi| \ll \pi$, and it is important to reduce their effects by means of a variant of the thinning technique described (for Aitken acceleration) in the previous section. Another interesting application is the *analytic continuation* of the power series outside its circle of convergence; see Example 3.4.7.

Theorem 3.4.1.

The tail of the power series in (3.4.13) *can formally be transformed into the following expansion, where ($z \neq 1$):*

$$S(z) - \sum_{j=1}^{n} u_j z^{j-1} = \sum_{j=n+1}^{\infty} u_j z^{j-1} = \frac{z^n}{1-z} \sum_{s=0}^{\infty} P^s u_{n+1}, \quad P = \frac{z}{1-z}\Delta. \tag{3.4.14}$$

Set $N = n + k - 1$, and set

$$M_{n,1} = \sum_{j=1}^{n} u_j z^{j-1}, \quad M_{N,k} = M_{n,1} + \frac{z^n}{1-z} \sum_{s=0}^{k-2} P^s u_{n+1}, \quad n = N - k + 1. \tag{3.4.15}$$

These quantities can be computed by the following recurrence formula that yields several estimates based on N terms from the original series.[95] *This is called the* **generalized Euler transformation***:*

$$M_{N,k} = \frac{M_{N,k-1} - zM_{N-1,k-1}}{1-z}, \quad k = 2 : N. \tag{3.4.16}$$

For $z = -1$, this is the repeated average algorithm described above, and $P = -\frac{1}{2}\Delta$.

[95] See Algorithm 3.4 for an adaptive choice of a kind of optimal output.

Assume that $|z| \le 1$, that $\sum u_j z^{j-1}$ converges, and that $\Delta^s u_N \to 0$, $s = 0 : k$, as $N \to \infty$. Then $M_{N,k} \to S(z)$ as $N \to \infty$. If, moreover, $\Delta^{k-1} u_j$ has a constant sign for $j \ge N - k + 2$, then the following strict error bounds are obtained:

$$|M_{N,k} - S(z)| \le |z(M_{N,k} - M_{N-1,k-1})| = |M_{N,k} - M_{N,k-1}|, \quad (k \ge 2). \tag{3.4.17}$$

Proof. We first note that as $N \to \infty$, $P^s u_N \to 0$, $s = 0 : k$, and hence, by (3.4.15), $\lim M_{N,k} = \lim M_{N,0} = S(z)$.

Euler's transformation can be formally derived by operators as follows:

$$S(z) - M_{n,1} = z^n \sum_{i=0}^{\infty} (zE)^i u_{n+1} = \frac{z^n}{1 - zE} u_{n+1}$$
$$= \frac{z^n}{1 - z - z\Delta} u_{n+1} = \frac{z^n}{1 - z} \sum_{s=0}^{\infty} P^s u_{n+1}.$$

In order to derive (3.4.16), note that this relation can be written equivalently be written as

$$M_{N,k} - M_{N,k-1} = z(M_{N,k} - M_{N-1,k-1}), \tag{3.4.18}$$

$$M_{N,k-1} - M_{N-1,k-1} = (1 - z)(M_{N,k} - M_{N-1,k-1}). \tag{3.4.19}$$

Remembering that $n = N - k + 1$, we obtain, by (3.4.15),

$$M_{N,k} - M_{N-1,k-1} = \frac{z^{N-k+1}}{1 - z} P^{k-2} u_{N-k+2}, \tag{3.4.20}$$

and it can be shown (Problem 3.4.16) that

$$M_{N,k-1} - M_{N-1,k-1} = z^n P^{k-2} u_{n+1} = z^{N-k+1} P^{k-2} u_{N-k+2}. \tag{3.4.21}$$

By (3.4.20) and (3.4.21), we now obtain (3.4.19) and hence also the equivalent equations (3.4.18) and (3.4.16).

Now substitute j for N into (3.4.21), and add the p equations obtained for $j = N + 1$, $\dots, N + p$. We obtain

$$M_{N+p,k-1} - M_{N,k-1} = \sum_{j=N+1}^{N+p} z^{j-k+1} P^{k-2} u_{j-k+2}.$$

Then substitute $k + 1$ for k, and $N + 1 + i$ for j. Let $p \to \infty$, while k is fixed. It follows that

$$S(z) - M_{N,k} = \sum_{j=N+1}^{\infty} z^{j-k} P^{k-1} u_{j-k+1}$$
$$= \frac{z^{N-k+1} \cdot z^{k-1}}{(1 - z)^{k-1}} \sum_{i=0}^{\infty} z^i \Delta^{k-1} u_{N-k+2+i}; \tag{3.4.22}$$

hence

$$|S(z) - M_{N,k}| \le \left|(z/(1-z))^{k-1} z^{N-k+1}\right| \sum_{i=0}^{\infty} \left|\Delta^{k-1} u_{N-k+2+i}\right|.$$

We now use the assumption that $\Delta^{k-1} u_j$ has constant sign for $j \ge N-k+2$. Since $\sum_{i=0}^{\infty} \Delta^{k-1} u_{N-k+2+i} = -\Delta^{k-2} u_{N-k+2}$, it follows that

$$\begin{aligned} |S(z) - M_{N,k}| &\le \left| z^{N-k+1} \frac{z^{k-1} \Delta^{k-2} u_{N-k+2}}{(1-z)^{k-1}} \right| \\ &= \left| \frac{z \cdot z^{N-k+1}}{1-z} P^{k-2} u_{N-k+2} \right|. \end{aligned}$$

Now, by (3.4.20),

$$|S(z) - M_{N,k}| \le |z| \cdot |M_{N,k} - M_{N-1,k-1}|.$$

This is the first part of (3.4.17). The second part then follows from (3.4.18). □

Remark 3.4.1. Note that the elements $M_{N,k}$ become rational functions of z for fixed N, k. If the term u_n, as a function of n, belongs to $\mathcal{P}_k$, then the classical Euler transformation (for $n = 0$) yields the exact value of $S(z)$ after k terms if $|z| < 1$. This follows from (3.4.14), because $\sum u_j z^j$ is convergent, and $P^s u_{n+1} = 0$ for $s \ge k$. In this particular case, $S(z) = Q(z)(1-z)^{-k}$, where Q is a polynomial; in fact, the Euler transformation gives $S(z)$ correctly for all $z \ne 1$.

The advantage of using the recurrence formula (3.4.16) instead of a more direct use of (3.4.14) is that it provides a whole lower triangular matrix of estimates so that one can, by means of a simple test, decide when to stop. This yields a result with strict error bound, if $\Delta^{k-1} u_j$ has a constant sign (for all j with a given k), and if the effect of rounding errors is evidently smaller than Tol. If these conditions are not satisfied, there is a small risk that the algorithm may terminate if the error estimate is accidentally small, for example, near a sign change of $\Delta^{k-1} u_j$.

The irregular errors of the initial data are propagated to the results. In the long run, they are multiplied by approximately $|z/(1-z)|$ from a column to the next—this is less than one if $\Re z < 1/2$—but in the beginning this growth factor can be as large as $(1+|z|)/|1-z|$. It plays no role for alternating series; its importance when $|1-z|$ is smaller will be commented on in Sec. 4.7.2.

The following algorithm is mainly based on Theorem 3.4.1 with a termination criterion based on (3.4.17). The possibility of the irregular errors becoming dominant has been taken into account (somewhat) in the third alternative of the termination criterion.

The classical Euler transformation would only consider the diagonal elements M_{NN}, $N = 1, 2, \dots$, and the termination would have been based on $|M_{NN} - M_{N-1,N-1}|$. The strategy used in this algorithm is superior for an important class of series.

ALGORITHM 3.4. *Generalized Euler Transformation.*

```
function [sum,errest,N,kk] = euler(z,u,Tol);
% EULER applies the generalized Euler transform to a power
% series with terms u(j)z^j. The elements of M are inspected
% in a certain order, until a pair of neighboring elements
% are found that satisfies a termination criterion.
%
Nmax = length(u);
errest = Inf; olderrest = errest;
N = 1;  kk = 2; M(1,1) = u(1);
while (errest > Tol) & (N < Nmax) & (errest <= olderrest)
      N = N+1;
      M(N,1) = M(N-1,1)+ u(N)*z^(N-1); % New partial sum
      for k = 2:N,
        M(N,k) = (M(N,k-1) - z*M(N-1,k-1))/(1-z);
      temp = abs(M(N,k) - M(N,k-1))/2;
        if temp < errest,
          kk = k; errest = temp;
       end
end
end
sum = (M(N,kk) + M(N,kk-1))/2;
```

An oscillatory behavior of the values $|M_{N,k} - M_{N,k-1}|$ in the same row indicates that the irregular errors have become dominant. The smallest error estimates may then become unreliable.

Remark 3.4.2. If the purpose of the computation is to study the convergence properties of the method rather than to get a numerical result of desired accuracy as quickly as possible, you had better replace the **while** statement by (say) `for N=1:Nmax`, change a few lines in the program, and produce graphical output such as Figure 3.4.1.

The above algorithm gives a strict error bound if, in the notation used in the theorem, $\Delta^{k-1}u_i$ has a constant sign for $i \geq N - k + 2$ (in addition to the other conditions of the theorem). We recall that a sequence for which this condition is satisfied *for every k* is called completely monotonic; see Definition 3.4.2.

It may seem difficult to check if this condition is satisfied. It turns out that many sequences that can be formed from sequences such as $\{n^{-\alpha}\}$, $\{e^{-\alpha n}\}$ by simple operations and combinations belong to this class. The generalized Euler transformation yields a sequence that converges at least as fast as a geometric series. The convergence ratio depends on z; it is less than one in absolute value for any complex z, except for $z > 1$ on the real axis. *Thus, the generalized Euler transformation often provides an analytic continuation of a power series outside its circle of convergence.*

For *alternating series*, with completely monotonic terms, i.e., for $z = -1$, the convergence ratio typically becomes $\frac{1}{3}$. This is in good agreement with Figure 3.4.1. Note that the minimum points for the errors lie almost on a straight line and that the optimal value of $\frac{k}{N}$ is approximately $\frac{2}{3}$, if $N \gg 1$ and if there are no irregular errors.

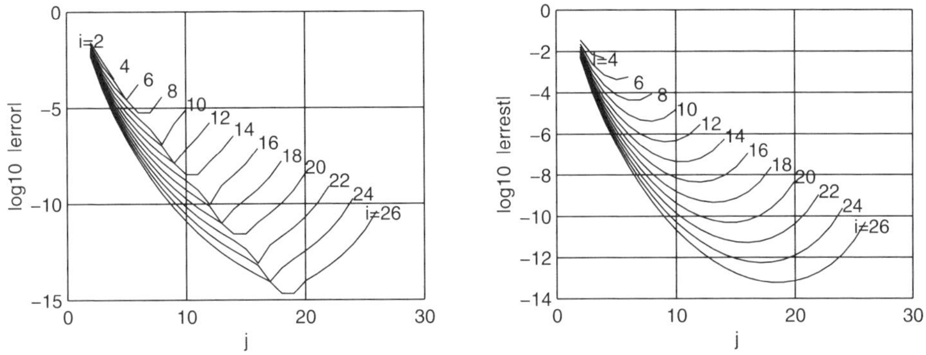

Figure 3.4.1. *Logarithms of the actual errors and the error estimates for $M_{N,k}$ in a more extensive computation for the alternating series in* (3.4.12) *with completely monotonic terms. The tolerance is here set above the level where the irregular errors become important; for a smaller tolerance parts of the lowest curves may become less smooth in some parts.*

Example 3.4.6.

A program, essentially the same as Algorithm 3.4, is applied to the series

$$\sum_{j=1}^{\infty}(-1)^j j^{-1} = 1 - \frac{1}{2} + \frac{1}{3} - \frac{1}{4} + \frac{1}{5} - \cdots = \ln 2 = 0.69314\,71805\,599453$$

with tol $= 10^{-6}$, It stops when $N = 12$, $kk = 9$. The errors $e_k = M_{N,k} - \ln 2$ and the differences $\frac{1}{2}\nabla_k M_{N,k}$ along the last row of M read as shown in the following table.

k	1	2	3	...	10	11	12
e_k	$-3.99\cdot 10^{-2}$	$1.73\cdot 10^{-3}$	$-1.64\cdot 10^{-4}$	...	$5.35\cdot 10^{-7}$	$-9.44\cdot 10^{-7}$	$2.75\cdot 10^{-6}$
$\nabla/2$		$2.03\cdot 10^{-2}$	$-9.47\cdot 10^{-4}$	...	$4.93\cdot 10^{-7}$	$-7.40\cdot 10^{-7}$	$1.85\cdot 10^{-6}$

Note that $|errest| = 4.93 \cdot 10^{-7}$ and $sum - \ln 2 = \frac{1}{2}(e_9 + e_8) = 4.2 \cdot 10^{-8}$. Almost full accuracy is obtained for Tol $= 10^{-16}$, $Nmax = 40$. The results are $N = 32$, $kk = 22$, $errest = 10^{-16}$, $|error| = 2 \cdot 10^{-16}$. Note that $errest < |error|$; this can happen when we ask for such a high accuracy that the rounding errors are not negligible.

Example 3.4.7.

We consider the application to a divergent power series (analytic continuation),

$$S(z) = \sum_{n=1}^{\infty} u_n z^{n-1}, \qquad |z| > 1.$$

As in the previous example we study in detail the case of $u_n = 1/n$. It was mentioned above that in exact arithmetic the generalized Euler transformation converges in the z-plane, cut along the interval $[1, \infty]$. The limit is $-z^{-1}\ln(1-z)$, a single-valued function in this region.

For various z outside the unit circle, we shall see that rounding causes bigger problems here than for Fourier series. The error estimate of Algorithm 3.4, usually underestimated the error, sometimes by a factor of ten. The table below reports some results from experiments without thinning.

z	-2	-4	-10	-100	$2i$	$8i$	$1+i$	$2+i$
|error|	$2\cdot 10^{-12}$	$2\cdot 10^{-8}$	$4\cdot 10^{-5}$	$3\cdot 10^{-3}$	$8\cdot 10^{-11}$	10^{-3}	10^{-7}	$2\cdot 10^{-2}$
N	38	41	43	50	40	39	38	39
kk	32	34	39	50	28	34	22	24

Thinning can be applied in this application, but here not only the argument ϕ is increased (this is good), but also $|z|$ (this is bad). Nevertheless, for $z = 1 + i$, the error becomes 10^{-7}, $3 \cdot 10^{-9}$, 10^{-9}, $4 \cdot 10^{-8}$, for $\tau = 1, 2, 3, 4$, respectively. For $z = 2 + i$, however, thinning improved the error only from 0.02 to 0.01. All this is for IEEE double precision arithmetic.

3.4.4 Complete Monotonicity and Related Concepts

For the class of completely monotonic sequences and some related classes of analytic functions the techniques of convergence acceleration can be put on a relatively solid theoretical basis.

Definition 3.4.2.

A sequence $\{u_n\}$ is **completely monotonic** *(c.m.) for $n \geq a$ if and only if*

$$u_n \geq 0, \quad (-\Delta)^j u_n \geq 0 \quad \forall j \geq 0, \quad n \geq a \text{ (integers)}.$$

Such sequences are also called **totally monotonic**. The abbreviation c.m. will be used, both as an adjective and as a noun, and both in singular and in plural. The abbreviation d.c.m. will similarly be used for *the difference between two completely monotonic sequences*. (These abbreviations are not generally established.)

A c.m. sequence $\{u_n\}_0^\infty$ is **minimal** if and only if it ceases to be a c.m. if u_0 is decreased while all the other elements are unchanged. This distinction is of little importance to us, since we usually deal with a tail of some given c.m. sequence, and it can be shown that *if $\{u_n\}_0^\infty$ is c.m., then $\{u_n\}_1^\infty$ is a minimal c.m. sequence*. Note that, e.g., the sequence $\{1, 0, 0, 0, \dots\}$ is a nonminimal c.m., while $\{0, 0, 0, 0, \dots\}$ is a minimal c.m. Unless it is stated otherwise *we shall only deal with minimal c.m.* without stating this explicitly all the time.

Definition 3.4.3.

A function *$u(s)$ is c.m. for $s \geq a$, $s \in \mathbf{R}$, if and only if*

$$u(s) \geq 0, \quad (-1)^{(j)} u^{(j)}(s) \geq 0, \quad s \geq a \;\forall\; j \geq 0 \text{ (integer)}, \;\forall\, s \geq a \text{ (real)}.$$

$u(s)$ is d.c.m. *if it is a difference of two c.m. on the same interval.*

We also need variants with an open interval. For example, the function $u(s) = 1/s$ is c.m. in the interval $[a, \infty)$ for any positive a, but it is not c.m. in the interval $[0, \infty]$.

The simplest relation of c.m. functions and c.m. sequences reads as follows: if the function $u(s)$ is c.m. for $s \geq s_0$, then the sequence defined by $u_n = u(s_0 + hn)$, $(h > 0)$, $n = 0, 1, 2, \ldots$, is also c.m. since, by (3.3.4), $(-\Delta)^j u_n = (-hD)^j u(\xi) \geq 0$ for some $\xi \geq s_0$.

A function is **absolutely monotonic** in an (open or closed) interval if the function and all its derivatives are nonnegative there.

The main reason why the analysis of a numerical method is convenient for c.m. and d.c.m. sequences is that they are "linear combinations of exponentials," according to the theorem below. The more precise meaning of this requires the important concept of a **Stieltjes integral**.[96]

Definition 3.4.4.

The Stieltjes integral $\int_a^b f(x)\,d\alpha(x)$ *is defined as the limit of sums of the form*

$$\sum_i f(\xi_i)\bigl(\alpha(x_{i+1}) - \alpha(x_i)\bigr), \quad \xi_i \in [x_i, x_{i+1}], \tag{3.4.23}$$

where

$$a = x_0 < x_1 < x_2 < \cdots < x_N = b$$

is a partition of $[a, b]$. *Here* $f(x)$ is bounded and continuous, *and* $\alpha(x)$ *is of* **bounded variation** *in* $[a, b]$, *i.e., the difference between two nondecreasing and nonnegative functions.*

The extension to improper integrals where, for example, $b = \infty$, $\alpha(b) = \infty$, is made in a similar way as for Riemann or Lebesgue integrals. The Stieltjes integral is much used also in probability and mechanics, since it unifies the treatment of continuous and discrete (and mixed) distributions of probability or mass. If $\alpha(x)$ is piecewise differentiable, then $d\alpha(x) = \alpha'(x)\,dx$, and the Stieltjes integral is simply $\int_a^b f(x)\alpha'(x)\,dx$. If $\alpha(x)$ is a *step function*, with jumps (also called point masses) m_i at $x = x_i$, $i = 1 : n$, then $d\alpha(x_i) = \lim_{\epsilon \downarrow 0} \alpha(x_i + \epsilon) - \alpha(x_i - \epsilon) = m_i$,

$$\int_a^b f(x)\,d\alpha(x) = \sum_{i=1}^n m_i f(x_i).$$

(It has been assumed that $f(x)$ is continuous at x_i, $i = 1 : n$.

Integration by parts is as usual; the following example is of interest to us. Suppose that $\alpha(0) = 0$, $\alpha(x) = o(e^{cx})$ as $x \to \infty$, and that $\Re s \geq c$. Then

$$\int_0^\infty e^{-sx}\,d\alpha(x) = s\int_0^\infty \alpha(x)e^{-sx}\,dx. \tag{3.4.24}$$

[96]Thomas Jan Stieltjes (1856–1894) was born in the Netherlands. After working with astronomical calculations at the observatory in Leiden, he accepted a position in differential and integral calculus at the University of Toulouse, France. He did important work on continued fractions and the moment problem, and invented a new concept of the integral.

The integral on the left side is called a **Laplace–Stieltjes transform**, while the integral on the right side is an ordinary Laplace transform. Many properties of power series, though not all, can be generalized to Laplace–Stieltjes integrals—set $z = e^{-s}$. Instead of a disk of convergence, the Laplace–Stieltjes integral has a (right) half-plane of convergence. A difference is that the half-plane of absolute convergence may be different from the half-plane of convergence.

We shall be rather brief and concentrate on the applicability to the study of numerical methods. We refer to Widder [373, 374] for proofs and more precise information concerning Stieltjes integrals, Laplace transforms, and complete monotonicity. Dahlquist [87] gives more details about applications to numerical methods.

The sequence defined by

$$u_n = \int_0^1 t^n \, d\beta(t), \qquad n = 0, 1, 2, \dots, \tag{3.4.25}$$

is called a **moment sequence** if $\beta(t)$ is nondecreasing. We make the *convention that* $t^0 = 1$ *also for* $t = 0$, since the continuity of f is required in the definition of the Stieltjes integral.

Consider the special example where $\beta(0) = 0$, $\beta(t) = 1$ if $t > 0$. This means a unit point mass at $t = 0$, and no more mass for $t > 0$. Then $u_0 = 1$, $u_n = 0$ for $n > 0$. It is then conceivable that making a sequence minimal just means removing a point mass from the origin; thus *minimality means requiring that* $\beta(t)$ *is continuous at* $t = 0$. (For a proof, see [373, Sec. 4.14].)

The following theorem combines parts of several theorems in the books by Widder. It is important that the functions called $\alpha(x)$ and $\beta(t)$ in this theorem *need not to be explicitly known for an individual series* for applications of an error estimate or a convergence rate of a method of convergence acceleration. Some criteria will be given below that can be used for simple proofs that a particular series is (or is not) c.m. or d.c.m.

Theorem 3.4.5.

1. *The sequence* $\{u_n\}_0^\infty$ *is c.m. if and only if it is a moment sequence; it is* minimal *if in addition* $\beta(t)$ *is continuous at* $t = 0$*, i.e., if there is no point mass at the origin. It is a d.c.m. if and only if* (3.4.25) *holds for some* $\beta(t)$ *of bounded variation.*

2. *The function* $u(s)$ *is c.m. for* $s \geq 0$ *if and only if it can be represented as a Laplace–Stieltjes transform,*

$$u(s) = \int_0^\infty e^{-sx} \, d\alpha(x), \qquad s \geq 0, \tag{3.4.26}$$

with a nondecreasing and bounded function $\alpha(x)$*. For the open interval* $s > 0$ *we have the same, except for the boundedness of* $\alpha(x)$*. For a d.c.m. the same is true with* $\alpha(x)$ *of bounded variation (not necessarily bounded as* $x \to \infty$*). The integral representation provides an analytic continuation of* $u(s)$ *from a real interval to a half-plane.*

3. *The sequence* $\{u_n\}_0^\infty$ *is a minimal c.m. if and only if there exists a c.m. function* $u(s)$ *such that* $u_n = u(n)$*,* $n = 0, 1, 2, \dots$.

4. *Suppose that $u(s)$ is c.m. in the interval $s > a$. Then the Laplace–Stieltjes integral converges absolutely and uniformly if $\Re s \geq a'$, for any $a' > a$, and defines an analytic continuation of $u(s)$ that is bounded for $\Re s \geq a'$ and analytic for $\Re s > a$. This is true also if $u(s)$ is a d.c.m.*

Proof. The "only if" parts of these statements are deep results mainly due to Hausdorff[97] and Bernštein,[98] and we omit the rather technical proofs. The relatively simple proofs of the "if" parts of the first three statements will be sketched, since they provide some useful insight.

1. Assume that u_n is a moment sequence, $\beta(0) = 0$, β is continuous at $t = 0$ and nondecreasing for $t > 0$. Note that multiplication by E or Δ outside the integral sign in (3.4.25) corresponds to multiplication by t or $t-1$ inside. Then, for $j, n = 0, 1, 2, \ldots$,

$$(-1)^j \Delta^j u_n = (-1)^j \int_0^1 (t-1)^j t^n \, d\beta(t) = \int_0^1 (1-t)^j t^n \, d\beta(t) \geq 0,$$

and hence u_n is c.m.

2. Assume that $u(s)$ satisfies (3.4.26). It is rather easy to legitimate the differentiation under the integral sign in this equation. Differentiation j times with respect to s yields, for $j = 1, 2, 3, \ldots$,

$$(-1)^j u^{(j)}(s) = (-1)^j \int_0^\infty (-x)^j e^{-sx} \, d\alpha(x) = \int_0^\infty x^j e^{-sx} \, d\alpha(x) \geq 0;$$

and hence $u(s)$ is c.m.

3. Assume that $u_n = u(n) = \int_0^\infty e^{-nx} \, d\alpha(x)$. Define $t = e^{-x}$, $\beta(0) = 0$, $\beta(t) \equiv \beta(e^{-x}) = u(0) - \alpha(x)$, and note that

$$t = 1 \Leftrightarrow x = 0, \qquad t = 0 \Leftrightarrow x = \infty,$$

and that $u(0) = \lim_{x\to\infty} \alpha(x)$. It follows that $\beta(t)$ is nonnegative and nondecreasing, since x decreases as t increases. Note that $\beta(t) \downarrow \beta(0)$ as $t \downarrow 0$. Then

$$u_n = -\int_1^0 t^n \, d\beta(t) = \int_0^1 t^n \, d\beta(t),$$

hence $\{u_n\}$ is a minimal c.m.

4. The distinction is illustrated for $\alpha'(x) = e^{ax}$, $u(s) = (s-a)^{-1}$, for a real a. $u(s)$ is analytic for $\Re s > a$ and bounded only for $\Re s \geq a'$ for any $a' > a$. ☐

[97] Felix Hausdorff (1868–1942), a German mathematician, is mainly known for having created a modern theory of topological and metric spaces.

[98] Sergei Natanovič Bernštein (1880–1968), Russian mathematician. Like his countryman Chebyshev, he made major contributions to polynomial approximation.

The basic formula for the application of complete monotonicity to the summation of power series reads

$$S(z) \equiv \sum_{i=0}^{\infty} u_i z^i = \sum_0^{\infty} \int_0^1 z^i t^i \, d\beta(t) = \int_0^1 \sum_0^{\infty} z^i t^i \, d\beta(t) = \int_0^1 (1 - zt)^{-1} \, d\beta(t). \tag{3.4.27}$$

The inversion of the summation and integration is legitimate when $|z| < 1$. Note that the last integral exists for more general z; a classical principle of complex analysis then yields the following interesting result.

Lemma 3.4.6.

If the sequence $\{u_i\}$ is d.c.m., then the last integral of formula (3.4.27) *provides the unique single-valued* analytic continuation *of $S(z)$ to the whole complex plane, save for a cut along the real axis from* 1 *to* ∞.

Remark 3.4.3. When z is located in the cut, $(1 - zt)^{-1}$ has a nonintegrable singularity at $t = 1/z \in [0, 1]$ unless, e.g., $\beta(t)$ is constant in the neighborhood of this point. If we remove the cut, $S(z)$ will not be single-valued. Check that this makes sense for $\beta(t) = t$.

Next we shall apply the above results to find interesting properties of the (generalized) Euler transformation. For example, we shall see that, for any z outside the cut, there is an optimal strategy for the generalized Euler transformation that provides the unique value of the analytic continuation of $S(z)$. The classical Euler transformation, however, reaches only the half-plane $\Re z < \frac{1}{2}$.

After that we shall see that there are a number of simple criteria for finding out whether a given sequence is c.m., d.c.m., or neither. Many interesting sequences are c.m., for example, $u_n = e^{-kn}$, $u_n = (n+c)^{-k}$, $(k \geq 0, c \geq 0)$, all products of these, and all linear combinations (i.e., sums or integrals) of such sequences with positive coefficients.

The convergence of a c.m. toward zero can be arbitrarily slow, but an alternating series with c.m. terms will, after Euler's transformation, converge as rapidly as a geometric series. More precisely, the following result on the optimal use of a generalized Euler transformation will be shown.

Theorem 3.4.7.

We use the notation of Theorem 3.4.1 *and* (3.4.22). *Suppose that the sequence $\{u_j\}$ is either c.m. or d.c.m. Consider*

$$S(z) = \sum_{j=0}^{\infty} u_j z^j, \quad z \in \mathbf{C},$$

and its analytic continuation (according to the above lemma). Then for the classical Euler transformation the following holds: If $z = -1$, a sequence along a descending diagonal of the scheme M or (equivalently) the matrix $\bar{M}$, i.e., $\{M_{n_0,k}\}_{k=0}^{\infty}$ for a fixed n_0, converges at least as fast as 2^{-k}. More generally, the error behaves like $(z/(1-z))^k$, $(k \gg 1)$. Note that $|z/(1-z)| < 1$ if and only if $\Re z < \frac{1}{2}$. The classical Euler transformation diverges outside this half-plane. If $z = e^{\pm it}$, $\frac{\pi}{3} < t \leq \pi$, it converges as fast as $(2 \sin \frac{t}{2})^{-k}$.

For the generalized *Euler transformation we have the following: If* $z = -1$, *the smallest error in the* i*th row of* $\bar{M}$ *is* $O(3^{-i})$, *as* $i \to \infty$. *More generally, this error is* $O((|z|/(1+|1-z|))^i)$, *hence the smallest error converges exponentially, unless* $z-1$ *is real and positive; i.e., the optimal application of the generalized Euler's transformation provides the analytic continuation, whenever it exists according to Lemma* 3.4.6. *If* $N \gg 1$, *the optimal value*[99] *of* k/N *is* $|1-z|/(1+|1-z|)$. *If* $z = e^{\pm it}$, $0 < t \le \pi$, *the error is* $O((1+2\sin\frac{t}{2})^{-i})$.

Proof. Sketch: The result of the generalized Euler transformation is in Sec. 3.4.3, denoted by $M_{n,k}(z)$. The computation uses $N = n + k$ terms (or partial sums) of the power series for $S(z)$; n terms of the original series—the head—are added, and Euler's transformation is applied to the next k terms—the tail. Set $n/N = \mu$, i.e., $n = \mu N$, $k = (1-\mu)N$, and denote the error of $M_{n,k}$ by $R_{N,\mu}(z)$. Euler's transformation is based on the operator $P = P(z) = \frac{z}{1-z}\Delta$. A multiplication by the operator P corresponds to a multiplication by $\frac{z}{1-z}(t-1)$ inside the integral sign.

First suppose that $|z| < 1$. By the definitions of $S(z)$ and $M_{n,k}(z)$ in Theorem 3.4.1,

$$\begin{aligned} R_{N,\mu}(z) \equiv S - M_{n,k} &= \frac{z^n}{1-z}\sum_{s=k}^{\infty} P^s u_n = \frac{z^n}{1-z}\int_0^1 \sum_{s=k}^{\infty}\left(\frac{z(t-1)}{(1-z)}\right)^s t^n\, d\beta(t) \\ &= \frac{z^n}{1-z}\int_0^1 \left(\frac{z(t-1)}{(1-z)}\right)^k \frac{t^n\, d\beta(t)}{1-z(t-1)/(1-z)} \\ &= (-1)^k \frac{z^N}{(1-z)^k}\int_0^1 (1-t)^k t^n \frac{d\beta(t)}{1-zt}. \end{aligned} \tag{3.4.28}$$

We see that the error oscillates as stated in Sec. 3.4.3. Again, by analytic continuation, this holds for all z except for the real interval $[1, \infty]$. Then

$$|R_{N,\mu}(z)|^{1/N} \le |z/(1-z)^{1-\mu}| \max_{t\in[0,1]}\left((1-t)^{1-\mu}t^{\mu}\right)c^{1/N}, \qquad c = \int_0^1 \frac{|d\beta(t)|}{|1-zt|}.$$

The *first* part of the theorem has $n = 0$, hence $\mu = 0$. We obtain

$$\lim_{N\to\infty} |R_{N,0}|^{1/N} \le |z/(1-z)|$$

as stated. This is less than unity if $|z| < |1-z|$, i.e., if $\Re(z) < \frac{1}{2}$.

Now we consider the *second* part of the theorem. The maximum occurring in the above expression for $|R_{N,\mu}(z)|^{1/N}$ (with N, μ fixed) takes place at $t = \mu$. Hence

$$|R_{N,\mu}(z)|^{1/N} \le |z/(1-z)^{1-\mu}|c^{1/N}(1-\mu)^{1-\mu}\mu^{\mu}.$$

An elementary optimization shows that *the value of* μ *that minimizes this bound for* $|R_{N,\mu}(z)|^{1/N}$ *is* $\mu = 1/(|1-z|+1)$, i.e.,

$$k = (1-\mu)N = \frac{N|1-z|}{|1-z|+1},$$

[99] In practice this is found approximately by the termination criterion of Algorithm 3.4.

and the minimum equals $|z|/(|1-z|+1)$. The details of these two optimizations are left for Problem 3.4.34. This proves the second part of the theorem. □

This minimum turns out to be a rather realistic estimate of the convergence ratio of the *optimal* generalized Euler transformation for power series with d.c.m. coefficients, unless $\beta(t)$ is practically constant in some interval around $t = \mu$; the exception happens, e.g., if $u_n = a^n$, $0 < a < 1$, $a \neq \mu$; see Problem 3.4.33.

Here we shall list a few criteria for higher monotonicity, by which one can often answer the question of whether a *function* is c.m. or d.c.m. or neither. When several c.m. or d.c.m. are involved, the intervals should be reduced to the intersection of the intervals involved. By Theorem 3.4.5, the question is then also settled for the corresponding *sequence*. In simple cases the question can be answered directly by means of the definition or the above theorem, e.g., for $u(s) = e^{-ks}$, s^{-k}, $(k \geq 0)$, for $\Re s \geq 0$ in the first case, for $\Re s > 0$ in the second case.

(A) *If $u(s)$ is c.m., and $a, b \geq 0$, then $g(s) = u(as+b)$ and $(-1)^j u^{(j)}(s)$ are c.m.,* $j = 1, 2, 3, \ldots$. The integral $\int_s^\infty u(t)\,dt$ is also c.m., if it is convergent. (The interval of complete monotonicity may not be the same for g as for f.) Analogous statements hold for sequences.

(B) *The product of two c.m. is c.m. Similarly, the product of two d.c.m. is d.c.m. This can evidently be extended to products of any number of factors, and hence to every positive integral power of a c.m. or d.c.m.* The proof is left for Problem 3.4.34.

(C) *A uniformly convergent positive linear combination of c.m. is itself c.m. The same criterion holds for d.c.m. without the requirement of positivity.* The term "positive linear combination" includes sums with positive coefficients and, more generally, Stieltjes integrals $\int u(s;p)\,d\gamma(p)$, where $\gamma(p)$ is nondecreasing.

(D) *Suppose that $u(s)$ is a d.c.m. for $s \geq a$. $F(u(s))$ is then a d.c.m. for $s > a$, if the radius of convergence of the Taylor expansion for $F(z)$ is greater than* $\max|u(s)|$. *Suppose that $u(s)$ is c.m. for $s \geq a$. We must then add the assumption that the coefficients of the Taylor expansion of $F(z)$ are nonnegative, in order to make sure that $F(u(s))$ is c.m. for $s \geq a$.*

These statements are important particular cases of (C). We also used (B), according to which each term $u(s)^k$ is c.m. (or a d.c.m. in the first statement). Two illustrations: $g(s) = (1-e^{-s})^{-1}$ is c.m. for $s > 0$; $h(s) = (s^2+1)^{-1}$ is a d.c.m. at least for $s > 1$ (choose $z = s^{-2}$). The expansion into powers of s^{-2} also provides an explicit decomposition,

$$h(s) = (s^{-2} + s^{-6} + \cdots) - (s^{-4} + s^{-8} + \cdots) = s^2/(s^4-1) - 1/(s^4-1),$$

where the two components are c.m. for $s > 1$. See also Example 3.4.8.

(E) *If $g'(s)$ is c.m. for $s > a$, and if $u(z)$ is c.m. in the range of $g(s)$ for $s > a$, then $F(s) = u(g(s))$ is c.m. for $s > a$.* (Note that $g(s)$ itself is not c.m.)
For example, we shall show that $1/\ln s$ is c.m. for $s > 1$. Set $g(s) = \ln s$, $u(z) = z^{-1}$, $a = 1$. Then $u(z)$ is completely monotonic for $z > 0$, and $g'(s) = s^{-1}$ is c.m. for $s > 0$, a fortiori for $s > 1$ where $\ln s > 0$. Then the result follows from (E).

The problems of Sec. 3.4 contain many interesting examples that can be treated by means of these criteria. One of the most important is that every rational function that is analytic and bounded in a half-plane is d.c.m. there; see Problem 3.4.35. Sometimes a table of Laplace transforms (see, e.g., the Handbook [1, Chap. 29]) can be useful in combination with the criteria below.

Another set of criteria is related to the *analytic properties of c.m. and d.c.m. functions.* Let $u(s)$ be d.c.m. for $s > a$. According to statement 4 of Theorem 3.4.5, $u(s)$ is analytic and bounded for $s \geq a'$ for any $a' > a$. The converse of this is not unconditionally true. If, however, we add the conditions that

$$\int_{-\infty}^{\infty} |u(\sigma + i\omega)|\, d\omega < \infty, \quad u(s) \to 0, \quad \text{as } |s| \to \infty, \quad \sigma \geq a', \tag{3.4.29}$$

then it can be shown that $u(s)$ is a d.c.m. for $s > a$. This condition is rather restrictive; there are many d.c.m. that do not satisfy it, for example, functions of the form e^{-ks} or $k + b(s-c)^{-\gamma}$ ($k \geq 0, b \geq 0, c > a, 0 < \gamma \leq 1$). The following is a reasonably powerful criterion: *$u(s)$ is a d.c.m. for $s > a$, e.g., if we can make a decomposition of the form*

$$u(s) = f_1(s) + f_2(s) \quad \text{or} \quad u(s) = f_1(s) f_2(s),$$

where $f_1(s)$ is known to be d.c.m. for $s > a$, and $f_2(s)$ satisfies the conditions in (3.4.29).

Theorem 3.4.8.

Suppose that $u(s)$ is c.m. for some s though not for all s. Then a singularity on the real axis, at (say) $s = a$, must be among the rightmost singularities; $u(s)$ is c.m. for $s > a$, hence analytic for $\Re s > a$.

The statement in the theorem is not generally true if $u(s)$ is only d.c.m. Suppose that $u(s)$ is d.c.m. for $s > a$, though not for any $s < a$. Then we cannot even be sure that there exists a singularity s^* such that $\Re s^* = a$.

Example 3.4.8.

This theorem can be used for establishing that a given function is *not* a c.m. For example, $u(s) = 1/(1+s^2)$ is not c.m. since the rightmost singularities are $s = \pm i$, while $s = 0$ is no singularity. $u(s)$ is a d.c.m. for $s > 0$; however, since it is analytic and bounded, and satisfies (3.4.29) for any positive a'. This result also comes from the general statement about rational functions bounded in a half-plane; see Problem 3.4.35.

Another approach: in any text about Laplace transforms you find that, for $s > 0$,

$$\frac{1}{s^2+1} = \int_0^{\infty} e^{-sx} \sin x\, dx = \int_0^{\infty} e^{-sx}(1 + \sin x)\, dx - \int_0^{\infty} e^{-sx}\, dx.$$

Now $\alpha'(x) \geq 0$ in both terms. Hence the formula $(1/s + 1/(s^2+1)) - 1/s$ expresses $1/(s^2+1)$ as the difference of two c.m. sequences for $s > 0$.

The easy application of criterion (D) above gave a smaller interval ($s > 1$), but a faster decrease of the c.m. terms as $s \to \infty$.

Another useful criterion for this kind of negative conclusion is that *a c.m. sequence cannot decrease faster than* every *exponential as $s \to +\infty$, for $s \in \mathbf{R}$*, unless it is identically zero. For there exists a number ξ such that $\alpha(\xi) > 0$, hence

$$u(s) = \int_0^\infty e^{-sx} d\alpha(x) \geq \int_0^\xi e^{-sx} d\alpha(x) \geq e^{-s\xi}\alpha(\xi).$$

For example, e^{-s^2} and $1/\Gamma(s)$ are not c.m. Why does this not contradict the fact that $s^{-1}e^{-s}$ is c.m.?

These ideas can be generalized. Suppose that $\{c_i\}_{i=0}^\infty$ is a given sequence such that the sum $C(t) \equiv \sum_{i=0}^\infty c_i t^i$ is known, and that u_i is c.m. or d.c.m. (c_i and $C(t)$ may depend on a complex parameter z too). Then

$$S_c = \sum_{i=0}^\infty c_i u_i = \sum_{i=0}^\infty c_i \int_0^1 t^i \, d\beta(t) \int_0^1 C(t) \, d\beta(t).$$

It is natural to ask how well S_c is determined if u_i has been computed for $i < N$, if $\{u_n\}_0^\infty$ is constrained to be c.m. A systematic way to obtain *very good* bounds is to find a polynomial $Q \in \mathcal{P}_N$ such that $|C(t) - Q(t)| \leq \epsilon_N$ for all $t \in [0, 1]$. Then

$$|S_c - Q(E)u_0| = \left| \int_0^1 \big(C(t) - Q(t)\big) \, d\beta(t) \right| \leq \epsilon_N \int_0^1 |d\beta(t)|.$$

Note that $Q(E)u_0$ is a linear combination of the computed values u_i, $i < N$, with coefficients independent of $\{u_n\}$. For $C(t; z) = (1 - tz)^{-1}$ the generalized Euler transformation (implicitly) works with a particular array of polynomial approximations, based on Taylor expansion, first at $t = 0$ and then at $t = 1$.

Can we find better polynomial approximations? For $C(t; z) = (1-tz)^{-1}$, **Gustafson's Chebyshev acceleration** (GCA) [177] is in most respects, superior to Euler transformation. Like Euler's transformation this is based on linear transformations of sequences and has the same range of application as the optimal Euler transformation. For GCA

$$\epsilon_N^{1/N} \to 1/(3 + \sqrt{8})$$

if $z = -1$. The number of terms needed for achieving a certain accuracy is thus for GCA about $\ln(3 + \sqrt{8})/\ln 3 \approx 1.6$ times as large as for the optimal Euler transformation.

3.4.5 Euler–Maclaurin's Formula

In the summation of series with essentially positive terms the tail of the sum can be approximated by an integral by means of the trapezoidal rule.

As an example, consider the sum $S = \sum_{j=1}^\infty j^{-2}$. The sum of the first nine terms is, to four decimal places, 1.5398. This suggests that we compare the tail of the series with the integral of x^{-2} from 10 to ∞. We approximate the integral according to the trapezoidal rule (see Sec. 1.1.3),

$$\int_{10}^\infty x^{-2} \, dx = \frac{1}{2}(10^{-2} + 11^{-2}) + \frac{1}{2}(11^{-2} + 12^{-2}) + \cdots = \sum_{j=10}^\infty j^{-2} - \frac{1}{2}10^{-2}.$$

Hence it follows that

$$\sum_{j=1}^{\infty} j^{-2} \approx 1.53977 + [-x^{-1}]_{10}^{\infty} + 0.0050 = 1.53977 + 0.1050 = 1.64477.$$

The correct answer is $\pi^2/6 = 1.64493\,40668\,4823$. We would have needed about 10,000 terms to get the same accuracy by direct addition of the terms!

The above procedure is not a coincidental trick, but a very useful method. A further systematic development of the idea leads to the important Euler–Maclaurin summation formula. We first derive this heuristically by operator techniques and exemplify its use, including a somewhat paradoxical example that shows that a strict treatment with the consideration of the remainder term is necessary for very practical reasons. Since this formula has several other applications, for example, in numerical integration (see Sec. 5.2), we formulate it more generally than needed for the summation of infinite series.

First, consider a rectangle sum on the finite interval $[a, b]$, with n steps of equal length h, $a + nh = b$; with the operator notation introduced in Sec. 3.3.2,

$$h\sum_{i=0}^{n-1} f(a+ih) = h\sum_{i=0}^{n-1} E^i f(a) = h\frac{E^n - 1}{E-1} f(a) = \frac{(E^n-1)}{D}\frac{hD}{e^{hD}-1} f(a).$$

We apply, to the second factor, the expansion derived in Example 3.1.5, with the Bernoulli numbers B_ν (recall that $a + nh = b$, $E^n f(a) = f(b)$):

$$\begin{aligned} h\sum_{i=0}^{n-1} f(a+ih) &= \frac{(E^n-1)}{D}\left(1 + \sum_{\nu=1}^{\infty}\frac{B_\nu (hD)^\nu}{\nu!}\right) f(a) \qquad (3.4.30) \\ &= \int_a^b f(x)\,dx + \sum_{\nu=1}^{k}\frac{h^\nu B_\nu}{\nu!}\left(f^{(\nu-1)}(b) - f^{(\nu-1)}(a)\right) + R_{k+1}. \end{aligned}$$

Here R_{k+1} is a remainder term that will be discussed thoroughly in Theorem 3.4.10. Set $h = 1$, and assume that $f(b), f'(b), \ldots$ tend to zero as $b \to \infty$. Recall that $B_1 = -\frac{1}{2}$, $B_{2j+1} = 0$ for $j > 0$, and set $k = 2r + 1$. This yields **Euler–Maclaurin's summation formula,**[100]

$$\begin{aligned} \sum_{i=0}^{\infty} f(a+i) &= \int_a^{\infty} f(x)\,dx + \frac{f(a)}{2} - \sum_{j=1}^{r}\frac{B_{2j} f^{(2j-1)}(a)}{(2j)!} + R_{2r+2} \qquad (3.4.31) \\ &= \int_a^{\infty} f(x)\,dx + \frac{f(a)}{2} - \frac{f'(a)}{12} + \frac{f^{(3)}(a)}{720} - \cdots, \end{aligned}$$

in a form suitable for the convergence acceleration of series of essentially positive terms. We give in Table 3.4.2 a few coefficients related to the Bernoulli and the Euler numbers.

There are some obscure points in this operator derivation, but we shall consider it as a heuristic calculation only and shall not try to legitimate the various steps of it. With an

[100]Leonhard Euler and the British mathematician Colin Maclaurin apparently discovered the summation formula independently; see Goldstine [159, p. 84]. Euler's publication came in 1738.

Table 3.4.2. *Bernoulli and Euler numbers;* $B_1 = -1/2$, $E_1 = 1$.

$2j$	0	2	4	6	8	10	12
B_{2j}	1	$\frac{1}{6}$	$-\frac{1}{30}$	$\frac{1}{42}$	$-\frac{1}{30}$	$\frac{5}{66}$	$-\frac{691}{2730}$
$\frac{B_{2j}}{(2j)!}$	1	$\frac{1}{12}$	$-\frac{1}{720}$	$\frac{1}{30{,}240}$	$-\frac{1}{1{,}209{,}600}$	$\frac{1}{47{,}900{,}160}$	
$\frac{B_{2j}}{2j(2j-1)}$	1	$\frac{1}{12}$	$-\frac{1}{360}$	$\frac{1}{1260}$	$-\frac{1}{1680}$	$\frac{1}{1188}$	$-\frac{691}{360{,}360}$
E_{2j}	1	-1	5	-61	1385	$-50{,}521$	2,702,765

appropriate interpretation, a more general version of this formula will be proved by other means in Theorem 3.4.10. A general remainder term is obtained there, if you let $b \to \infty$ in (3.4.37). You do not need it often, because the following much simpler error bound is usually applicable—but there are exceptions.

The Euler–Maclaurin expansion (on the right-hand side) is typically semiconvergent only. Nevertheless a few terms of the expansion often give surprisingly high accuracy with simple calculations. For example, if $f(x)$ is c.m., i.e., if

$$(-1)^j f^{(j)}(x) \geq 0, \quad x \geq a, \quad j \geq 0,$$

then the partial sums oscillate strictly around the true result; the first neglected term is then a strict error bound. (This statement also follows from the theorem below.)

Before we prove the theorem we shall exemplify how the summation formula is used in practice.

Example 3.4.9.

We return to the case of computing $S = \sum_{j=1}^{\infty} j^{-2}$ and treat it with more precision and accuracy. With $f(x) = x^{-2}$, $a = 10$, we find $\int_a^{\infty} f(x)\,dx = a^{-1}$, $f'(a) = -2a^{-3}$, $f'''(a) = -24a^{-5}, \ldots$. By (3.4.31), $(r = 2)$,

$$\begin{aligned}\sum_{x=1}^{\infty} x^{-2} &= \sum_{x=1}^{9} x^{-2} + \sum_{i=0}^{\infty} (10+i)^{-2} \\ &= 1.53976\,7731 + 0.1 + 0.005 + 0.00016\,6667 - 0.00000\,0333 + R_6 \\ &= 1.64493\,4065 + R_6.\end{aligned}$$

Since $f(x) = x^{-2}$ is c.m. (see Definition 3.4.2), the first neglected term is a strict error bound; it is less than $720 \cdot 10^{-7}/30{,}240 < 3\cdot 10^{-9}$. (The actual error is approximately $2\cdot 10^{-9}$.)

Although the Euler–Maclaurin expansion in this example seems to converge rapidly, it is in fact only semiconvergent for any $a > 0$, and this is rather typical. We have, namely,

$$f^{(2r-1)}(a) = -(2r)!a^{-2r-1},$$

and, by Example 3.1.5,

$$B_{2r}/(2r)! \approx (-1)^{r+1}2(2\pi)^{-2r}.$$

The ratio of two successive terms is thus $-(2r+2)(2r+1)/(2\pi a)^2$, hence the modulus of terms increases when $2r+1 > 2\pi a$.

The "rule" that one should terminate a semiconvergent expansion at the term of smallest magnitude is, in general, no good for Euler–Maclaurin applications, since the high-order derivatives (on the right-hand side) are typically much more difficult to obtain than a few more terms in the expansion on the left-hand side. Typically, you first choose r, $r \le 3$, depending on how tedious the differentiations are, and then you choose a in order to meet the accuracy requirements.

In this example we were lucky to have access to simple closed expressions for the derivatives and the integral of f. In other cases, one may use the possibilities for the numerical integration on an infinite interval mentioned in Chapter 5. In Problem 3.4.19 you find two formulas that result from the substitution of the formulas (3.3.48) that express higher derivatives in terms of central differences into the Euler–Maclaurin expansion.

An expansion of $f(x)$ into negative powers of x is often useful both for the integral and for the derivatives.

Example 3.4.10.

We consider $f(x) = (x^3+1)^{-1/2}$, for which the expansion

$$f(x) = x^{-3/2}(1+x^{-3})^{-1/2} = x^{-1.5} - \frac{1}{2}x^{-4.5} + \frac{3}{8}x^{-7.5} - \cdots$$

was derived and applied in Example 3.1.6. It was found that

$$\int_{10}^{\infty} f(x)\,dx = 0.632410375,$$

correctly rounded, and that $f'''(10) = -4.13 10^{-4}$ with less than 1% error. The $f'''(10)$-term in the Euler–Maclaurin expansion is thus $-5.73\ 10^{-7}$, with absolute error less than $6\cdot 10^{-9}$. Inserting this into Euler–Maclaurin's summation formula, together with the numerical values of $\sum_{n=0}^{9} f(n)$ and $\frac{1}{2}f(10) - \frac{1}{12}f'(10)$, we obtain $\sum_{n=0}^{\infty} f(n) = 3.7941\,1570 \pm 10^{-8}$. The reader is advised to work out the details as an exercise.

Example 3.4.11.

Let $f(x) = e^{-x^2}$, $a = 0$. Since all derivatives of odd order vanish at $a = 0$, then the expansion (3.4.31) may give the impression that $\sum_{j=0}^{\infty} e^{-j^2} = \int_0^{\infty} e^{-x^2}\,dx + 0.5 = 1.386\,2269$, but the sum (that is easily computed without any convergence acceleration) is actually $1.386\,3186$, hence the remainder R_{2r+2} cannot tend to zero as $r \to \infty$. The infinite Euler–Maclaurin expansion, where all terms but two are zero, *is convergent but is not valid.* Recall the distinction between the convergence and the validity of an infinite expansion made in Sec. 3.1.2.

In this case $f(x)$ is not c.m.; for example, $f''(x)$ changes sign at $x = 1$. With appropriate choice of r, the general error bound (3.4.37) will tell us that the error is very small, but it cannot be used for proving that it is zero—because this is not true.

The mysteries of these examples have hopefully raised the appetite for a more substantial theory, including an error bound for the Euler–Maclaurin formula. We first need some tools that are interesting in their own right.

The **Bernoulli polynomial** $B_n(t)$ is an nth degree polynomial defined by the **symbolic** relation $B_n(t) = (B + t)^n$, where the exponents of B become subscripts after the expansion according to the binomial theorem. The Bernoulli numbers B_j were defined in Example 3.1.5. Their recurrence relation (3.1.19) can be written in the form

$$\sum_{j=0}^{n-1} \binom{n}{j} B_j = 0, \quad n \geq 2,$$

or "symbolically" $(B + 1)^n = B^n = B_n$ (for the computation of B_{n-1}), $n \neq 1$, hence $B_0(t) = 1$, $B_1(t) = t + B_1 = t - 1/2$, and

$$B_n(1) = B_n(0) = B_n, \quad n \geq 2.$$

The **Bernoulli function** $\hat{B}_n(t)$ is a *piecewise polynomial* defined for $t \in \mathbf{R}$ by the equation $\hat{B}_n(t) = B_n(t - \lfloor t \rfloor)$.[101] (Note that $\hat{B}_n(t) = B_n(t)$ if $0 \leq t < 1$.)

Lemma 3.4.9.

(a) $\hat{B}'_{n+1}(t)/(n+1)! = \hat{B}_n(t)/n!$, $(n > 0)$,
$\hat{B}_n(0) = B_n$. *(For $n = 1$ this is the limit from the right.)*

$$\int_0^1 \frac{B_n(t)}{n!} dt = \begin{cases} 1 & \textit{if } n = 0, \\ 0 & \textit{otherwise}. \end{cases}$$

(b) *The piecewise polynomials $\hat{B}_p(t)$ are periodic; $\hat{B}_p(t+1) = \hat{B}_p(t)$. $\hat{B}_1(t)$ is continuous, except when t is an integer. For $n \geq 2$, $\hat{B}_n \in C^{n-2}(-\infty, \infty)$.*

(c) *The Bernoulli functions have the following (modified) Fourier expansions, $(r \geq 1)$,*

$$\frac{\hat{B}_{2r-1}(t)}{(2r-1)!} = (-1)^r 2 \sum_{n=1}^{\infty} \frac{\sin 2n\pi t}{(2n\pi)^{2r-1}}, \qquad \frac{\hat{B}_{2r}(t)}{(2r)!} = (-1)^{r-1} 2 \sum_{n=1}^{\infty} \frac{\cos 2n\pi t}{(2n\pi)^{2r}}.$$

Note that $\hat{B}_n(t)$ is an even (odd) function, when n is even (odd).

(d) $|\hat{B}_{2r}(t)| \leq |B_{2r}|$.

Proof. Statement (a) follows directly from the symbolic binomial expansion of the Bernoulli polynomials.

The demonstration of statement (b) is left for a problem. The reader is advised to draw the graphs of a few low-order Bernoulli functions.

[101]The function $\lfloor t \rfloor$ is the floor function defined as the largest integer $\leq t$, i.e., the integer part of t. In many older and current works the symbol $[t]$ is used instead, but this should be avoided.

The Fourier expansion for $\hat{B}_1(t)$ follows from the Fourier coefficient formulas (3.2.6) (modified for the period 1 instead of 2π). The expansions for $\hat{B}_p(t)$ are then obtained by repeated integrations, term by term, with the use of (a). Statement (d) then follows from the Fourier expansion, because $\hat{B}_{2r}(0) = B_{2r}$. □

Remark 3.4.4. For $t = 0$ we obtain an interesting classical formula, together with a useful asymptotic approximation that was obtained in a different way in Sec. 3.1.2:

$$\sum_{n=1}^{\infty} \frac{1}{n^{2r}} = \frac{|B_{2r}|(2\pi)^{2r}}{2(2r)!}, \qquad \frac{|B_{2r}|}{(2r)!} \sim \frac{2}{(2\pi)^{2r}}. \tag{3.4.32}$$

Also note how the rate of decrease of the Fourier coefficients is related to the type of singularity of the Bernoulli function at the integer points. (It does not help that the functions are smooth in the interval $[0, 1]$.)

The Bernoulli polynomials have a generating function that is elegantly obtained by means of the following "symbolic" calculation:

$$\sum_{0}^{\infty} \frac{B_n(y)x^n}{n!} = \sum_{0}^{\infty} \frac{(B+y)^n x^n}{n!} = e^{(B+y)x} = e^{Bx}e^{yx} = \frac{xe^{yx}}{e^x - 1}. \tag{3.4.33}$$

If the series is interpreted as a power series in the complex variable x, the radius of convergence is 2π.

Theorem 3.4.10 (*The Euler–Maclaurin Formula*).

Set $x_i = a + ih$, $x_n = b$, suppose that $f \in C^{2r+2}(a, b)$, and let $\hat{T}(a : h : b)f$ be the trapezoidal sum

$$\hat{T}(a : h : b)f = \sum_{i=1}^{n} \frac{h}{2}\big(f(x_{i-1}) + f(x_i)\big) = h\left(\sum_{i=0}^{n-1} f(x_i) + \frac{1}{2}(f(b) - f(a))\right). \tag{3.4.34}$$

Then

$$\begin{aligned} \hat{T}(a : h : b)f - \int_a^b f(x)\,dx &= \frac{h^2}{12}\big(f'(b) - f'(a)\big) - \frac{h^4}{720}\big(f'''(b) - f'''(a)\big) \\ &\quad + \cdots + \frac{B_{2r}h^{2r}}{(2r)!}\big(f^{(2r-1)}(b) - f^{(2r-1)}(a)\big) + R_{2r+2}(a, h, b)f. \end{aligned} \tag{3.4.35}$$

The remainder $R_{2r+2}(a, h, b)f$ is $O(h^{2r+2})$. It is represented by an integral with a kernel of constant sign in (3.4.36). An upper bound for the remainder is given in (3.4.37). The estimation of the remainder is very simple in certain important, particular cases:

- *If $f^{(2r+2)}(x)$ does not change sign in the interval $[a, b]$, then $R_{2r+2}(a, h, b)f$ has the same sign as the first neglected term.*[102]

[102] If $r = 0$ all terms of the expansion are "neglected."

- *If $f^{(2r+2)}(x)$ and $f^{(2r)}(x)$ have the same constant sign in $[a, b]$, then the value of the left-hand side of* (3.4.35) *lies between the values of the partial sum of the expansion displayed in* (3.4.35) *and the partial sum with one term less.*[103]

In the limit, as $b \to \infty$, these statements still hold—also for the summation formula (3.4.31)*—provided that the left-hand side of* (3.4.35) *and the derivatives $f^{(\nu)}(b)$ $(\nu = 1 : 2r + 1)$ tend to zero, if it is also assumed that*

$$\int_a^\infty |f^{(2r+2)}(x)|\, dx < \infty.$$

Proof. To begin with we consider a single term of the trapezoidal sum, and set $x = x_{i-1}+ht$, $t \in [0, 1]$, $f(x) = F(t)$. Suppose that $F \in C^p[0, 1]$, where p is an *even* number.

We shall apply *repeated integration by parts*, Lemma 3.2.6, to the integral $\int_0^1 F(t)\, dt = \int_0^1 F(t)B_0(t)\, dt$. Use statement (a) of Lemma 3.4.9 in the equivalent form, $\int B_j(t)/j!\, dt = B_{j+1}(t)/(j + 1)!$

Consider the first line of the expansion in the next equation. Recall that $B_\nu = 0$ if ν is odd and $\nu > 1$. Since $B_{j+1}(1) = B_{j+1}(0) = B_{j+1}$, j will thus be odd in all nonzero terms, except for $j = 0$. Then, with no loss of generality, we assume that p is even.

$$\begin{aligned}
\int_0^1 F(t)\, dt &= \sum_{j=0}^{p-1}(-1)^j F^{(j)}(t)\frac{B_{j+1}(t)}{(j+1)!}\bigg|_{t=0}^{1} + (-1)^p \int_0^1 F^{(p)}(t)\frac{B_p(t)}{p!}\, dt \\
&= \frac{F(1) + F(0)}{2} + \sum_{j=1}^{p-1} \frac{-B_{j+1}}{(j+1)!}\left(F^{(j)}(1) - F^{(j)}(0)\right) + \int_0^1 F^{(p)}(t)\frac{B_p(t)}{p!}\, dt \\
&= \frac{F(1) + F(0)}{2} - \sum_{j=1}^{p-3} \frac{B_{j+1}}{(j+1)!}\left(F^{(j)}(1) - F^{(j)}(0)\right) - \int_0^1 F^{(p)}(t)\frac{B_p - B_p(t)}{p!}\, dt.
\end{aligned}$$

The upper limit of the sum is reduced to $p-3$, since the last term (with $j = p-1$) has been moved under the integral sign, and all values of j are odd. Set $j + 1 = 2k$ and $p = 2r + 2$. Then k is an integer that runs from 1 to r. Hence

$$\sum_{j=1}^{p-3} \frac{B_{j+1}}{(j+1)!}(F^{(j)}(1) - F^{(j)}(0)) = \sum_{k=1}^{r} \frac{B_{2k}}{(2k)!}(F^{(2k-1)}(1) - F^{(2k-1)}(0)).$$

Now set $F(t) = f(x_{i-1} + ht)$, $t \in [0, 1]$. Then $F^{(2k-1)}(t) = h^{2k-1} f^{(2k-1)}(x_{i-1} + ht)$, and make abbreviations such as $f_i = f(x_i)$, $f_i^{(j)} = f^{(j)}(x_i)$.

$$\int_{x_{i-1}}^{x_i} f(x)\, dx = h \int_0^1 F(t)\, dt = \frac{h(f_{i-1} + f_i)}{2} - \sum_{k=1}^{r} \frac{B_{2k}h^{2k}}{(2k)!}(f_i^{(2k-1)} - f_{i-1}^{(2k-1)}) - R,$$

[103]Formally this makes sense for $r \geq 2$ only, but if we interpret $f^{(-1)}$ as "the empty symbol," it makes sense also for $r = 1$. If f is c.m. *the statement holds for every $r \geq 1$*. This is easy to apply, because simple criteria for complete monotonicity are given in Sec. 3.4.4.

where R is the local remainder that is now an integral over $[x_{i-1}, x_i]$. Adding these equations, for $i = 1 : n$, yields a result equivalent to (3.4.35), namely

$$\int_a^b f(x)\,dx = \hat{T}(a:h:b)f - \sum_{k=1}^{r} \frac{B_{2k}h^{2k}}{(2k)!} f^{(2k-1)}(x)\Big|_{x=a}^{b} - R_{2r+2}(a,h,b)f,$$

$$R_{2r+2}(a,h,b)f = h^{2r+2}\int_a^b \Big(B_{2r+2} - \hat{B}_{2r+2}((x-a)/h)\Big)\frac{f^{(2r+2)}(x)}{(2r+2)!}\,dx. \qquad (3.4.36)$$

By Lemma 3.4.9, $|\hat{B}_{2r+2}(t)| \le |B_{2r+2}|$, hence the kernel $B_{2r+2} - \hat{B}_{2r+2}((x-a)/h)$ has the same sign as B_{2r+2}. Suppose that $f^{(2r+2)}(x)$ does not change sign on (a, b). Then

$$\operatorname{sign} f^{(2r+2)}(x) = \operatorname{sign}\big(f^{(2r+1)}(b) - f^{(2r+1)}(a)\big),$$

hence $R_{2r+2}(a,h,b)f$ has the same sign as the first neglected term. The second statement about "simple estimation of the remainder" then follows from Theorem 3.1.4, since the Bernoulli numbers (with even subscripts) have alternating signs.

If $\operatorname{sign} f^{(2r+2)}(x)$ is not constant, then we note instead that

$$|B_{2r+2} - \hat{B}_{2r+2}((x-a)/h)| \le |2B_{2r+2}|,$$

and hence

$$\begin{aligned} |R_{2r+2}(a,h,b)f| &\le h^{2r+2}\frac{|2B_{2r+2}|}{(2r+2)!}\int_a^b |f^{(2r+2)}(x)|\,dx \\ &\approx 2\Big(\frac{h}{2\pi}\Big)^{2r+2}\int_a^b |f^{(2r+2)}(x)|\,dx. \end{aligned} \qquad (3.4.37)$$

If $\int_a^\infty |f^{(2r+2)}(x)|\,dx < \infty$ this holds also in the limit as $b \to \infty$. □

Note that there are (at least) three parameters here that can be involved in *different* natural limit processes. For example, one of the parameters can tend to its limit, while the two others are kept fixed. The remainder formula (3.4.37) contains all you need for settling various questions about convergence.

- $b \to \infty$: natural when Euler–Maclaurin's formula is used as a summation formula, or for deriving an approximation formula valid when b is large.

- $h \to 0$: natural when Euler–Maclaurin's formula is used in connection with numerical integration. You see how the values of derivatives of f at the endpoints a, b can highly improve the estimate of the integral of f, obtained by the trapezoidal rule with constant step size. Euler–Maclaurin's formula is also useful for the design and analysis of other methods for numerical integration; see Romberg's method, Sec. 5.2.2.

- $r \to \infty$: $\lim_{r\to\infty} R_{2r+2}(a,h,b)f = 0$ can be satisfied only if $f(z)$ is an entire function such that $|f^{(n)}(a)| = o((2\pi/h)^n)$ as $n \to \infty$. Fortunately, this type of convergence is rarely needed in practice. With appropriate choice of b and h, the expansion is typically rapidly semiconvergent. Since the derivatives of f are typically more expensive to compute than the values of f, one frequently reduces h (in integration) or increases b (in summation or integration over an infinite interval) and truncates the expansion several terms before one has reached the smallest term that is otherwise the standard procedure with alternating semiconvergent expansion.

Variations of the Euler–Maclaurin summation formula, with *finite differences instead of derivatives* in the expansion, are given in Problem 3.4.19, where you also find *a more general form of the formula*, and two more variations of it.

Euler–Maclaurin's formula can also be used for finding an algebraic expression for a finite sum (see Problem 3.4.31) or, as in the following example, for finding an expansion that determines the asymptotic behavior of a sequence or a function.

Example 3.4.12.

To derive an expansion that generalizes Stirling's formula (3.2.35), we shall use the Euler–Maclaurin formula for $f(x) = \ln x$, $a = m > 0$, $h = 1$, $b = n \geq m$. We obtain

$$\hat{T}(m:1:n)f = \sum_{i=m+1}^{n} \ln i - \frac{1}{2}\ln n + \frac{1}{2}\ln m = \ln(n!) - \frac{1}{2}\ln n - \ln(m!) + \frac{1}{2}\ln m,$$

$$f^{(2k-1)}(x) = (2k-2)!x^{1-2k}, \qquad \int_m^n f(x)\,dx = n\ln n - n - m\ln m + m.$$

Note that $\hat{T}(m:1:n)f$ and $\int_m^n f(x)\,dx$ are unbounded as $n \to \infty$, but their difference is bounded. Putting these expressions into (3.4.35) and separating the terms containing n from the terms containing m gives

$$\ln(n!) - \left(n + \frac{1}{2}\right)\ln n + n - \sum_{k=1}^{r} \frac{B_{2k}}{2k(2k-1)n^{2k-1}} \tag{3.4.38}$$

$$= \ln(m!) - \left(m + \frac{1}{2}\right)\ln m + m - \sum_{k=1}^{r} \frac{B_{2k}}{2k(2k-1)m^{2k-1}} - R_{2r+2}(m:1:n).$$

By (3.4.37),

$$\begin{aligned} |R_{2r+2}(m:1:n)| &\leq \int_m^n \frac{|2B_{2r+2}|}{(2r+2)x^{2r+2}}\,dx \\ &\leq \frac{|2B_{2r+2}|}{(2r+2)(2r+1)|m^{2r+1}|} \approx \frac{(2r)!}{\pi|2\pi m|^{2r+1}}. \end{aligned} \tag{3.4.39}$$

Now let $n \to \infty$ with fixed r, m. First, note that the integral in the error bound converges. Next, in most texts of calculus Stirling's formula is derived in the following form:

$$n! \sim \sqrt{2\pi}\, n^{n+\frac{1}{2}} e^{-n}, \qquad (n \to \infty). \tag{3.4.40}$$

If you take the natural logarithm of this, it follows that the left-hand side of (3.4.38) tends to $\frac{1}{2}\ln(2\pi)$, and hence

$$\ln(m!) = \left(m + \frac{1}{2}\right)\ln m - m + \frac{1}{2}\ln(2\pi) + \sum_{k=1}^{r} \frac{B_{2k}}{2k(2k-1)m^{2k-1}} + R, \qquad (3.4.41)$$

where a bound for R is given by (3.4.39). The numerical values of the coefficients are found in Table 3.4.2.

Remark 3.4.5. You may ask why we refer to (3.4.40). Why not? Well, it is not necessary, because it is easy to prove that the left-hand side of (3.4.38) increases with n and is bounded; it thus tends to some limit C (say). The proof that $C = \ln\sqrt{2\pi}$ *exactly* is harder, without the Wallis product idea (from 1655) or something equally ingenious or exotic. But if you compute the right-hand side of (3.4.38) for $m = 17, r = 5$ (say), and estimate the remainder, you will obtain C to a fabulous guaranteed accuracy, in negligible computer time after a rather short programming time. And you may then replace $\frac{1}{2}\ln 2\pi$ by your own C in (3.4.41), if you like.

Remark 3.4.6. Almost the same derivation works also for $f(x) = \ln(x+z), m = 0$, where z is a complex number not on the negative real axis. A few basic facts about the gamma function are needed; see details in Henrici [197, Sec. 11.11, Example 3].

The result is that *you just replace the integer m by the complex number z in the expansion* (3.4.41). According to the Handbook [1, Sec. 6.1.42] R is to be multiplied by $K(z) = \sup_{u\ge 0}|z^2/(u^2+z^2)|$. For z real and positive, $K(z) = 1$, and since $f'(x) = (z+x)^{-1}$ is c.m., it follows from Theorem 3.4.10 that, *in this case, R is less in absolute value than the first term neglected and has the same sign.*

It is customary to *write* $\ln\Gamma(z+1)$ *instead of* $\ln(z!)$. The gamma function is one of the most important transcendental functions; see, e.g., the Handbook [1, Sec. 6.5] and Lebedev [240].

This formula (with $m = z$) is useful for the practical computation of $\ln\Gamma(z+1)$. Its semiconvergence is best if $\Re z$ is large and positive. If this condition is not satisfied, the situation can easily be improved by means of logarithmic forms of the

- *reflection formula*: $\Gamma(z)\Gamma(1-z) = \pi/\sin\pi z$,
- *recurrence formula*: $\Gamma(z+1) = z\Gamma(z)$.

By simple applications of these formulas the computation of $\ln\Gamma(z+1)$ for an arbitrary $z \in \mathbf{C}$ is reduced to the computation of the function for a number z' such that $|z'| \ge 17$, $\Re z' > \frac{1}{2}$, for which the total error, if $r = 5$, becomes typically less than 10^{-14}. See Problem 3.4.23.

Remark 3.4.7. As you may have noted, we write "the Euler–Maclaurin formula" mainly for (3.4.35), which is used in general theoretical discussions, or if applications other than the summation of an infinite series are the primary issue. The term "the Euler–Maclaurin summation formula" is mainly used in connection with (3.4.31), i.e., when the summation of an infinite series is the issue. "The Euler–Maclaurin expansion" denotes both the right-hand side of (3.4.35), except for the remainder and for the corresponding terms of (3.4.31). These distinctions are convenient for us, but they are neither important nor in general use.

Although, in this section, the main emphasis is on the application of the Euler–Maclaurin formula to the computation of sums and limits, we shall comment a little on its possibilities for other applications.

- It shows that the *global truncation error of the trapezoidal rule* for $\int_a^b f(x)\,dx$ with step size h *has an expansion into powers of* h^2. Note that although the expansion contains derivatives at the boundary points only, the remainder requires that $|f^{(2r+2)}|$ is integrable in the interval $[a, b]$. The Euler–Maclaurin formula is thus the theoretical basis for the application of *repeated Richardson extrapolation* to the results of the trapezoidal rule, known as *Romberg's method*; see Sec. 5.2.2. Note that *the validity depends on the differentiability properties of* f.

- The Euler–Maclaurin formula can be used for highly accurate numerical integration when the values of some derivatives of f are known at $x = a$ and $x = b$. More about this in Sec. 5.2.1.

- Theorem 3.4.10 shows that the trapezoidal rule is second order accurate, unless $f'(a) = f'(b)$, but there exist *interesting exceptions*. Suppose that the function f is infinitely differentiable for $x \in \mathbf{R}$, and that f *has* $[a, b]$ *as an interval of periodicity*, that is $f(x + b - a) = f(x)$ for all $x \in \mathbf{R}$. Then $f^{(k)}(b) = f^{(k)}(a)$, for $k = 0, 1, 2, \ldots$, hence *every term in the Euler–Maclaurin expansion is zero* for the integral over *the whole period* $[a, b]$. One could be led to believe that the trapezoidal rule gives the exact value of the integral, but this is usually not the case; for most periodic functions f, $\lim_{r\to\infty} R_{2r+2}f \neq 0$; *the expansion converges, of course, though not necessarily to the correct result.*

We shall illuminate these amazing properties of the trapezoidal rule from different points of view in several places in this book, for example, in Sec. 5.1.4. See also applications to the so-called bell sums in Problem 3.4.29.

3.4.6 Repeated Richardson Extrapolation

Let $F(h)$ denote the value of a certain quantity obtained with step length h. In many calculations one wants to know the limiting value of $F(h)$ as the step length approaches zero. But the work to compute $F(h)$ often increases sharply as $h \to 0$. In addition, the effects of roundoff errors often set a practical bound for how small h can be chosen.

Often, one has some knowledge of how the truncation error $F(h) - F(0)$ behaves when $h \to 0$. If

$$F(h) = a_0 + a_1 h^p + O(h^r), \quad h \to 0, \quad r > p,$$

where $a_0 = F(0)$ is the quantity we are trying to compute and a_1 is unknown, then a_0 and a_1 can be estimated if we compute F for two step lengths, h and qh, $q > 1$,

$$F(h) = a_0 + a_1 h^p + O(h^r),$$
$$F(qh) = a_0 + a_1 (qh)^p + O(h^r),$$

from which, eliminating a_1, we get

$$F(0) = a_0 = F(h) + \frac{F(h) - F(qh)}{q^p - 1} + O(h^r). \tag{3.4.42}$$

This formula is called **Richardson extrapolation**, or the *deferred approach to the limit.*[104] Examples of this were mentioned in Chapter 1—the application of the above process to the trapezoidal rule for numerical integration (where $p = 2$, $q = 2$), and for differential equations $p = 1$, $q = 2$ for Euler's method, $p = 2$, $q = 2$ for Runge's second order method.

We call the term $(F(h) - F(qh))/(q^p - 1)$ the **Richardson correction**. It is used in (3.4.42) for improving the result. Sometimes it is used only for estimating the error. This can make sense, for example, if the values of F are afflicted by other errors, usually irregular, suspected of being comparable in size to the correction. If the irregular errors are negligible, this error estimate is asymptotically correct. More often, the Richardson correction is used as an error estimate for the improved (or extrapolated) value $F(h)+(F(h)-F(qh))/(q^p-1)$. This is typically a strong overestimate; the error estimate is $O(h^p)$, while the error is $O(h^r)$, $(r > p)$.

Suppose that a more complete expansion of $F(h)$ in powers of h is known to exist,

$$F(h) = a_0 + a_1 h^{p_1} + a_2 h^{p_2} + a_3 h^{p_3} + \cdots, \quad 0 < p_1 < p_2 < p_3 < \cdots, \tag{3.4.43}$$

where the exponents are known while the coefficients are unknown. Then one can *repeat the use of Richardson extrapolation* in a way described below. This process is, in many numerical problems—especially in the numerical treatment of integral and differential equations—one of the simplest ways to get results which have tolerable truncation errors. The application of this process becomes especially simple when the step lengths form a geometric sequence $H, H/q, H/q^2, \ldots$, where $q > 1$ and H is the **basic step length**.

Theorem 3.4.11.

Suppose that there holds an expansion of the form of (3.4.43), *for* $F(h)$, *and set* $F_1(h) = F(h)$,

$$F_{k+1}(h) = \frac{q^{p_k} F_k(h) - F_k(qh)}{q^{p_k} - 1} = F_k(h) + \frac{F_k(h) - F_k(qh)}{q^{p_k} - 1}, \tag{3.4.44}$$

for $k = 1 : (n-1)$, *where* $q > 1$. *Then* $F_n(h)$ *has an expansion of the form*

$$F_n(h) = a_0 + a_n^{(n)} h^{p_n} + a_{n+1}^{(n)} h^{p_{n+1}} + \cdots, \quad a_\nu^{(n)} = \prod_{k=1}^{n-1} \frac{q^{p_k} - q^{p_\nu}}{q^{p_k} - 1} a_\nu. \tag{3.4.45}$$

Note that $a_\nu^{(n)} = 0$ *for all* $\nu < n$.

[104] The idea of a deferred approach to the limit is sometimes used also in the experimental sciences—for example, when some quantity is to be measured in a complete vacuum (difficult or expensive to produce). It can then be more practical to measure the quantity for several different values of the pressure. Expansions analogous to (3.4.43) can sometimes be motivated by the kinetic theory of gases.

Proof. Temporarily set $F_k(h) = a_0 + a_1^{(k)} h^{p_1} + a_2^{(k)} h^{p_2} + \cdots + a_\nu^{(k)} h^{p_\nu} + \cdots$. Put this expansion into the first expression on the right-hand side of (3.4.44) and, substituting $k+1$ for k, put it into the left-hand side. By matching the coefficients for h^{p_ν} we obtain

$$a_\nu^{(k+1)} = a_\nu^{(k)} (q^{p_k} - q^{p_\nu}) / (q^{(p_k)} - 1).$$

By (3.4.43), the expansion holds for $k = 1$, with $a_\nu^{(1)} = a_\nu$. The recursion formula then yields the product formula for $a_\nu^{(n)}$. Note that $a_\nu^{(\nu+1)} = 0$, hence $a_\nu^{(n)} = 0$ for all $\nu < n$. □

The product formula is for theoretical purposes. The recurrence formula is for practical use. If an expansion of the form of (3.4.43) is known to exist, the above theorem gives a way to compute increasingly better estimates of a_0. The leading term of $F_n(h) - a_0$ is $a_n^{(n)} h^{p_n}$; the exponent of h increases with n. A moment's reflection on (3.4.44) will convince the reader that (using the notation of the theorem) $F_{k+1}(h)$ is determined by the $k+1$ values

$$F_1(H), F_1(H/q), \ldots, F_1(H/q^k).$$

With some changes in notation we obtain the following algorithm.

ALGORITHM 3.5. *Repeated Richardson Extrapolation.*

Set

$$T_{m,1} = F(H/q^{m-1}), \quad m = 1 : N, \tag{3.4.46}$$

and for $m = 2 : N$, $k = 1 : m-1$, compute

$$T_{m,k+1} = \frac{q^{p_k} T_{m,k} - T_{m-1,k}}{q^{p_k} - 1} = T_{m,k} + \frac{T_{m,k} - T_{m-1,k}}{q^{p_k} - 1}, \tag{3.4.47}$$

where the second expression is usually preferred.

The computations for repeated Richardson extrapolation can be set up in the following scheme,

$$\begin{array}{cccc} T_{11} & & & \\ T_{21} & T_{22} & & \\ T_{31} & T_{32} & T_{33} & \\ T_{41} & T_{42} & T_{43} & T_{44} \end{array},$$

where an extrapolated value in the scheme is obtained by using the quantity to its left and the correction diagonally above. (In a computer the results are simply stored in a lower triangular matrix.)

According to the argument above, one continues the process until two values *in the same row* agree to the desired accuracy, i.e.,

$$|T_{m,k} - T_{m,k-1}| < \text{Tol} - CU,$$

where Tol is the permissible error and CU is an upper bound of the irregular error (see below). (Tol should, of course, be chosen larger than CU.) If no other error estimate is available, $\min_k |T_{m,k} - T_{m,k-1}| + CU$ is usually chosen as the error estimate, even though it is typically a strong overestimate.

Typically, $k = m$ and T_{mm} is accepted as the numerical result, but this is not always the case. For instance, if H has been chosen so large that the use of the basic asymptotic expansion is doubtful, then the uppermost diagonal of the extrapolation scheme contains nonsense and should be ignored, except for its element in the first column. Such a case is detected by inspection of the difference quotients in a column. If for some k, where $T_{k+2,k}$ has been computed and the modulus of the relative irregular error of $T_{k+2,k} - T_{k+1,k}$ is less than (say) 20%, *and*, most important, the difference quotient $(T_{k+1,k} - T_{k,k})/(T_{k+2,k} - T_{k+1,k})$ is very different from its theoretical value q^{p_k}, then the uppermost diagonal is to be ignored (except for its first element). In such a case, one says that H *is outside the asymptotic regime*.

In this discussion a bound for the inherited irregular error is needed. We shall now derive such a bound. Fortunately, it turns out that the numerical stability of the Richardson scheme is typically very satisfactory (although the total error bound for T_{mk} will never be smaller than the largest irregular error in the first column).

Denote by ϵ_1 the column vector with the irregular errors of the initial data. We neglect the rounding errors committed during the computations.[105] Then the inherited errors satisfy the same linear recursion formula as the $T_{m,k}$, i.e.,

$$\epsilon_{m,k+1} = \frac{q^{p_k}\epsilon_{m,k} - \epsilon_{m-1,k}}{q^{p_k} - 1}.$$

Denote the kth column of errors by ϵ_k, and set $\|\epsilon_k\|_\infty = \max_m |\epsilon_{m,k}|$. Then

$$\|\epsilon_{k+1}\|_\infty \le \frac{q^{p_k}+1}{q^{p_k}-1}\|\epsilon_k\|_\infty.$$

Hence, for every k, $\|\epsilon_{k+1}\|_\infty \le CU$, where $\|\epsilon_1\|_\infty = U$ and C is the infinite product,

$$C = \prod_{k=1}^{\infty} \frac{q^{p_k}+1}{q^{p_k}-1} = \prod_{k=1}^{\infty} \frac{1+q^{-p_k}}{1-q^{-p_k}},$$

that converges as fast as $\sum q^{-p_k}$; the multiplication of ten factors are thus more than enough for obtaining a sufficiently accurate value of C.

The most common special case is an expansion where $p_k = 2k$,

$$F(h) = a_0 + a_1h^2 + a_2h^4 + a_3h^6 + \cdots. \tag{3.4.48}$$

This expansion holds for the error in composite trapezoidal rule and is the basis for *Romberg's method* for numerical integration. The Richardson corrections then become $\Delta/3$, $\Delta/15$, $\Delta/63, \ldots$. In this case *we find that* $C = \frac{5}{3} \cdot \frac{7}{15} \cdots < 2$ (after less than ten factors).

For (systems of) ordinary differential equations there exist some general theorems, according to which the form of the asymptotic expansion (3.4.43) of the global error can be found.

[105] They are usually of less importance for various reasons. One can also *make* them smaller by subtracting a suitable constant from all initial data. This is applicable to all linear methods of convergence acceleration.

- *For Numerov's method* for ordinary differential equations, discussed in Example 3.3.15 and Problem 3.4.27, one can show that we have the same exponents in the expansion for the global error, but $a_1 = 0$ (and the first heading disappears). We thus have the same product as above, except that the first factor disappears, *hence* $C < 2 \cdot \frac{3}{5} = 1.2$.

- For *Euler's method* for ordinary differential equations, presented in Sec. 1.5.1, $p_k = k$; the Richardson corrections are $\Delta/1, \Delta/3, \Delta/7, \Delta/15, \ldots$. Hence $C = 3 \cdot \frac{5}{3} \cdot \frac{9}{7} \cdots = 8.25$.

- For *Runge's second order method*, presented in Sec. 1.5.3, the exponents are the same, but $a_1 = 0$. We thus have the same product as for Euler's method, except that the first factor disappears, and $C = 8.25/3 = 2.75$.

In the special case that $p_j = j \cdot p$, $j = 1, 2, 3, \ldots$ in (3.4.43), i.e., for expansions of the form

$$F(h) = a_0 + a_1 h^p + a_2 h^{2p} + a_3 h^{3p} + \cdots, \tag{3.4.49}$$

it is not necessary that the step sizes form a geometric progression. We can choose any increasing sequence of integers $q_1 = 1, q_2, \ldots, q_k$, set $h_i = H/q_i$, and use an algorithm that looks very similar to repeated Richardson extrapolation. Alternative sequences, that may be suitable in the common case that the cost of evaluating $F(h)$ for small h is high, are the harmonic sequence $1, 2, 3, 4, 5, 6, 7, \ldots$ and the sequence $1, 2, 3, 4, 8, 12, 16, 24, \ldots$, suggested by Bulirsch.

Note that the expansion (3.4.49) is a usual power series in the variable $x = h^p$, which can be approximated by a polynomial in x. Suppose that $k+1$ values $F(H), F(H/q_2), \ldots, F(H/q_k)$ are known. Then by the corollary to Theorem 4.2.1, they are uniquely determined by the interpolation conditions

$$Q(x_i) = F(H/q_i), \quad x_i = (H/q_i)^p, \quad i = 1 : k.$$

Our problem is to find $Q(0)$. **Neville's algorithm** for iterative linear interpolation, which will be derived in Sec. 4.2.4, is particularly convenient in this situation. After a change of notation, Neville's algorithm yields the following recursion: For $m = 1 : N$, set $T_{m,1} = F(H/q_m)$, where $1 = q_1 < q_2 < q_3 \ldots$ is any increasing sequence of integers, and compute, for $m = 2 : N$, $k = 1 : m-1$,

$$T_{m,k+1} = T_{m,k} + \frac{T_{m,k} - T_{m-1,k}}{(q_m/q_{m-k})^p - 1}. \tag{3.4.50}$$

The computations can be set up in a triangle matrix as for repeated Richardson extrapolations.

We remark that Richardson extrapolation does not require an expansion of the form (3.4.43). Let $T_{n,0} = S_n$ be a sequence converging to S and x_n a sequence of parameters converging to zero when $n \to \infty$. Then Richardson extrapolation can be written as

$$T_{n,k+1} = T_{n,k} - \frac{T_{n+1,k} - T_{n,k}}{x_{n+k+1} - x_n}.$$

There are conditions (obtained by P. J. Laurent) such that the columns and the diagonals converge to the same limit S, and conditions for the convergence to be accelerated; see Brezinski [49, Sec. II.3].

Example 3.4.13.

The ancient Greeks computed approximate values of the circumference of the unit circle, 2π, by inscribing a regular polygon and computing its perimeter. Archimedes considered the inscribed 96-sided regular polygon, whose perimeter is $6.28206\ldots = 2\cdot 3.14103\ldots$.

In general, a regular n-sided polygon inscribed (circumscribed) in a circle with radius 1 has perimeter $2a_n$ $(2b_n)$, where

$$a_n = n\sin(\pi/n), \qquad b_n = n\sin(\tan/n).$$

Clearly $a_n < \pi < b_n$, giving lower and upper bound for π. Setting $h = 1/n$, we have

$$a_n = \frac{1}{h}\sin\pi h = \pi - \frac{\pi^3}{3!}h^2 + \frac{\pi^5}{5!}h^4 - \frac{\pi^7}{7!}h^6 + \cdots,$$
$$b_n = \frac{1}{h}\tan\pi h = \pi + \frac{\pi^3}{3}h^2 + \frac{2\pi^5}{15}h^4 - \frac{17\pi^7}{315}h^6 + \cdots.$$

We first derive a recursion formula that leads from a_n and b_n to a_{2n} and b_{2n}. Setting $n_m = n_1\cdot 2^{m-1}$ and

$$s_m = 1/\sin(\pi/n_m), \qquad t_m = 1/\tan(\pi/n_m),$$

we have $a_{n_m} = n_m/s_m$, $b_{n_m} = n_m/t_m$. Using the trigonometric formula $\tan(x/2) = \sin x/(1+\cos x)$, we obtain the recursion

$$t_m = s_{m-1} + t_{m-1}, \qquad s_m = \sqrt{t_m^2 + 1}, \qquad m = 1, 2, \ldots. \tag{3.4.51}$$

Note that no trigonometric functions are used—only the square root, which can be computed by Newton's method.

Taking $n_1 = 6$ gives $a_6 = 6/2 = 3$, and $b_6 = 6/\sqrt{3} = 3.4641\ldots$. The following table gives a_{n_m} for $n_1 = 6$, $m = 1:5$, computed using IEEE double precision arithmetic.

m	n_m	a_{n_m}	b_{n_m}
1	6	3.00000000000000	3.00000000000000
2	12	3.10582854123025	3.21539030917347
3	24	3.13262861328124	3.15965994209750
4	48	3.13935020304687	3.14608621513143
5	96	3.14103195089051	3.14271459964537

From this we can deduce that $3.1410 < \pi < 3.1427$, or the famous, slightly weaker rational lower and upper bounds of Archimedes, $3\frac{10}{71} < \pi < 3\frac{1}{7}$.

The sequences $a(h)$ and $b(h)$ satisfy the assumptions for repeated Richardson extrapolation with $p_k = 2k$. Since the coefficients in the Taylor expansion for $a(h)$ decay faster we use the Richardson scheme with this sequence, giving the results shown in the next table. A correctly rounded value of π to 20 digits reads

$$\pi = 3.14159\,26535\,89793\,23846,$$

and correct digits are shown in boldface.

3.14110472164033			
3.14156197063157	**3.141592**45389765		
3.14159073296874	**3.14159265**045789	**3.1415926535**7789	
3.14159253350506	**3.1415926535**4081	**3.1415926535897**5	**3.14159265358979**

The errors in successive columns decay as $4^{-2k}, 4^{-3k}, 4^{-4k}$, and the final number is correct to all 14 decimals shown. Hence the accuracy used in computing values in the previous table, which could be thought excessive, has been put to good use![106]

Example 3.4.14 (*Application to Numerical Differentiation*).
From Bickley's table (Table 3.3.1) for difference operators in Sec. 3.3.2, we know that

$$\frac{\delta}{h} = \frac{2\sinh(hD/2)}{h} = D + a_2h^2D^3 + a_4h^4D^5 + \cdots,$$
$$\mu = \cosh(hD/2) = 1 + b_2h^2D^2 + b_4h^4D^4 + \cdots,$$

where the values of the coefficients are now unimportant to us. Hence

$$f'(x) - \frac{f(x+h) - f(x-h)}{2h} = Df(x) - \frac{\mu\delta f(x)}{h} \quad \text{and} \quad f''(x) - \frac{\delta^2 f(x)}{h^2}$$

have expansions into *even* powers of h. Repeated Richardson extrapolation can thus be used with step sizes $H, H/2, H/4, \ldots$ and headings $\Delta/3, \Delta/15, \Delta/63, \ldots$. For numerical examples, see the problems for this section.

Richardson extrapolation can be applied in the same way to the computation of higher derivatives. Because of the division by h^k in the difference approximation of $f^{(k)}$, *irregular errors in the values of $f(x)$ are of much greater importance in numerical differentiation than in interpolation and integration.* It is therefore important to use high-order approximations in numerical differentiation, so that larger values of h can be used.

Suppose that the irregular errors of the values of f are bounded in magnitude by u. These errors are propagated to $\mu\delta f(x), \delta^2 f(x), \ldots$ with bounds equal to $u/h, 4u/h^2, \ldots$. As mentioned earlier, the Richardson scheme (in the version used here) is benevolent; it multiplies the latter bounds by a factor less than two.

[106] An extension of this example was used as a test problem for Mulprec, a package for (in principle) arbitrarily high precision floating-point arithmetic in MATLAB. For instance, π was obtained to 203 decimal places with 22 polygons and 21 Richardson extrapolations in less than half a minute. The extrapolations took a small fraction of this time. Nevertheless they increased the number of correct decimals from approximately 15 to 203.

Review Questions

3.4.1 (a) Aitken acceleration is based on fitting three successive terms of a given sequence $\{s_n\}$ to a certain comparison series. Which?

(b) Give sufficient conditions for the accelerated sequence $\{s'_j\}$ to converge faster than $\{s_n\}$.

(c) Aitken acceleration is sometimes applied to a thinned sequence. Why can this give a higher accuracy in the computed limit?

3.4.2 (a) State the original version of Euler's transformation for summation of an alternating series $S = \sum_{j=0}^{\infty}(-1)^j u_j$, $u_j \geq 0$.

(b) State the modified Euler's transformation for this case and discuss suitable termination criteria. What is the main advantage of the modified algorithm over the classical version?

3.4.3 (a) What pieces of information appear in the Euler–Maclaurin formula? Give the generating function for the coefficients. What do you know about the remainder term?

(b) Give at least three important uses of the Euler–Maclaurin formula.

3.4.4 The Bernoulli polynomial $B_n(t)$ have a key role in the proof of the Euler–Maclaurin formula. They are defined by the symbolic relation

$$B_n(t) = (B + t)^n.$$

How is this relation to be interpreted?

3.4.5 (a) Suppose that an expansion of $F(h)$

$$F(h) = a_0 + a_1 h^{p_1} + a_2 h^{p_2} + a_3 h^{p_3} + \cdots, \qquad 0 < p_1 < p_2 < p_3 < \cdots,$$

is known to exist. Describe how $F(0) = a_0$ can be computed by repeated Richardson extrapolation from known values of $F(h)$, $h = H, H/q, H/q^2, \ldots$ for some $q > 1$.

(b) Discuss the choice of q in the procedure in (a). What is the most common case? Give some applications of repeated Richardson extrapolation.

Problems and Computer Exercises

3.4.1 (a) Compute $\sum_{n=1}^{\infty} \frac{1}{(n+1)^3}$ to eight decimal places by using

$$\sum_{n=N}^{\infty} \frac{1}{n(n+1)(n+2)},$$

for a suitable N, as a comparison series. Estimate roughly how many terms you would have to add without and with the comparison series.

Hint: You found the exact sum of this comparison series in Problem 3.3.3.

(b) Compute the sum also by Euler–Maclaurin's formula or one of its variants in Problem 3.4.19.

3.4.2 Study, or write yourself, programs for some of the following methods:[107]

- iterated Aitken acceleration,
- modified iterated Aitken, according to (3.4.9) or an a-version,
- generalized Euler transformation,
- one of the central difference variants of Euler–Maclaurin's formula, given in Problem 3.4.19.

The programs are needed in two slightly different versions.

Version i: For studies of the convergence rate, for a series (sequence) where one knows a sufficiently accurate value exa of the sum (the limit). The risk of drowning in figures becomes smaller if you make graphical output, for example, like Figure 3.4.1.

Version ii: For a run controlled by a tolerance, as in Algorithm 3.4, appropriately modified for the various algorithms. Print also i and, if appropriate, jj. If exa is known, it should be subtracted from the result, because it is of interest to compare $errest$ with the actual error.

Comment: If you do not know exa, find a sufficiently good exa by a couple of runs with very small tolerances, before you study the convergence rates (for larger tolerances).

3.4.3 The formula for Aitken acceleration is sometimes given in the form

$$s_n - \frac{(\Delta s_n)^2}{\Delta^2 s_n} \quad \text{or} \quad s_n - \frac{\Delta s_n \nabla s_n}{\Delta s_n - \nabla s_n}.$$

Show that these are equivalent to s'_{n+2} or s'_{n+1}, respectively, in the notations of (3.4.2). Also note that the second formula is $\lim_{p\to\infty} s'_n$ (not s'_{n+1}) in the notation of (3.4.7).

3.4.4 (a) Try iterated Aitken with thinning for $\sum_1^\infty e^{-\sqrt{n}}$, according to the suggestions after Example 3.4.3.

(b) Study the effect of small random perturbations to the terms.

3.4.5 *Oscillatory series of the form* $\sum_{n=1}^\infty c_n z^n$. Suggested examples:

$$c_n = e^{-\sqrt{n}}, \qquad 1/(1+n^2), \qquad 1/n,\ 1/(2n-1),$$
$$n/(n^2+n+1), \qquad 1/\sqrt{n}, \qquad 1/\ln(n+1),$$

where $z = -1, -0.9, e^{i3\pi/4}, i, e^{i\pi/4}, e^{i\pi/16}$, for the appropriate algorithms mentioned in Problem 3.4.2 above. Apply thinning. Also try classical Euler transformation on some of the cases.

Study how the convergence ratio depends on z, and compare with theoretical results. Compare the various methods with each other.

[107]We have MATLAB in mind, or some other language with complex arithmetic and graphical output.

3.4.6 *Essentially positive series of the form* $\sum_{n=1}^{\infty} c_n z^n$, *where*

$$c_n = e^{-\sqrt{n}}, \quad 1/(1+n^2), \quad 1/(5+2n+n^2), \quad (n \cdot \ln(n+1))^{-2},$$
$$1/\sqrt{n^3+n}, n^{-4/3}, \quad 1/((n+1)(\ln(n+1))^2),$$

$z = 1, 0.99, 0.9, 0.7, e^{i\pi/16}, e^{i\pi/4}, i$. Use appropriate algorithms from Problem 3.4.2. Try also Euler–Maclaurin's summation formula, or one of its variants, if you can handle the integral with good accuracy. Also try to find a good comparison series; it is not always possible.

Study the convergence rate. Try to apply *thinning* to the first two methods.

3.4.7 *Divergent series.* Apply, if possible, Aitken acceleration and the generalized Euler transformation to the following divergent series $\sum_1^\infty c_n z^n$. Compare the numerical results with the results obtained by analytic continuation using the analytic expression for the sum as a function of z.

(a) $c_n = 1, z = -1$; (b) $c_n = n, z = -1$;

(c) c_n is an arbitrary polynomial in n; (d) $c_n = 1, z = i$;

(e) $c_n = 1, z = 2$; (f) $c_n = 1, z = -2$.

3.4.8 Let y_n be the Fibonacci sequence defined in Problem 3.3.18 by the recurrence relation

$$y_n = y_{n-1} + y_{n-2}, \quad y_0 = 0, \quad y_1 = 1.$$

Show that the sequence $\{y_{n+1}/y_n\}_0^\infty$ satisfies the sufficient condition for Aitken acceleration given in the text. Compute a few terms, compute the limit by Aitken acceleration(s), and compare with the exact result.

3.4.9 When the current through a galvanometer changes suddenly, its indicator begins to oscillate with an exponentially damped simple harmonic motion toward a new stationary value s. The relation between the successive turning points $v_0, v_1, v_2, \dots$ is $v_n - s \approx A \cdot (-k)^n$, $0 < k < 1$. Determine, from the following series of measurements, Aitken extrapolated values v_2', v_3', v_4' which are all approximations to s:[108]

$$v_0 = 659, \quad v_1 = 236, \quad v_2 = 463, \quad v_3 = 340, \quad v_4 = 406.$$

3.4.10 (a) Show that the a-version of Aitken acceleration can be *iterated*, for $i = 0 : N-2$,

$$a_{i+1}^{(i+1)} = 0, \quad a_j^{(i+1)} = a_j^{(i)} - \nabla\left((a_j^{(i)})^2/\nabla a_j^{(i)}\right), \quad j = i+2 : N,$$
$$s_N^{(i+1)} = s_N^{(i)} - (a_N^{(i)})^2/\nabla a_N^{(i)}.$$

(Note that $a_j^{(0)} = a_j$, $s_j^{(0)} = s_j$.) We thus obtain N estimates of the sum s. We cannot be sure that the last estimate $s_N^{(N-1)}$ is the best, due to irregular errors in the terms and during the computations. Therefore, accept the average of a few estimates

[108] James Clark Maxwell used Aitken acceleration for this purpose already in 1892 in his "Treatise on Electricity and Magnetism."

that are close to each other, or do you have a better suggestion? This also gives you a (not quite reliable) error estimate.

(b) Although we may expect that the a-version of Aitken acceleration handles rounding errors better than the s-version, the rounding errors may set a limit for the accuracy of the result. It is easy to combine *thinning* with this version. How?

(c) Study or write yourself a program for the a-version, and apply it to one or two problems where you have used the s-version earlier. Also use thinning on a problem, where it is needed. We have here considered N as given. Can you suggest a better termination criterion, or a process for continuing the computation, if the accuracy obtained is disappointing?

3.4.11 A function $g(t)$ has the form

$$g(t) = c - kt + \sum_{n=1}^{\infty} a_n e^{-\lambda_n t},$$

where c, k, a_n, and $0 < \lambda_1 < \lambda_2 < \cdots < \lambda_n$ are unknown constants and $g(t)$ is known numerically for $t_\nu = \nu h$, $\nu = 0, 1, 2, 3, 4$.

Find out how to eliminate c in such a way that a sufficient condition for estimating kh by Aitken acceleration is satisfied. Apply this to the following data, where $h = 0.1$, $g_\nu = g(t_\nu)$:

$$g_0 = 2.14789, \quad g_1 = 1.82207, \quad g_2 = 1.59763, \quad g_3 = 1.40680, \quad g_4 = 1.22784.$$

Then, estimate c.

3.4.12 Suppose that the sequence $\{s_n\}$ satisfies the condition $s_n - s = c_0 n^{-p} + c_1 n^{-p-1} + O(n^{-p-2})$, $p > 0$, $n \to \infty$, and set

$$s_n' = s_n - \frac{p+1}{p} \frac{\Delta s_n \nabla s_n}{\Delta s_n - \nabla s_n}.$$

It was stated without proof in Sec. 3.4.2 that $s_n' - s = O(n^{-p-2})$.

(a) Design an a-version of this modified Aitken acceleration, or look it up in [33].

(b) Since the difference expressions are symmetrical about n one can conjecture that this result would follow from a continuous analogue with derivatives instead of differences. It has been shown [33] that this conjecture is true, but we shall not prove that. Our (easier) problem is just the continuous analogue: suppose that a function $s(t)$ satisfies the condition $s(t) - s = c_0 t^{-p} + c_1 t^{-p-1} + O(t^{-p-2})$, $p > 0$, $t \to \infty$, and set

$$y(t) = s(t) - \frac{p+1}{p} \frac{s'(t)^2}{s''(t)}.$$

Show that $y(t) - s = O(t^{-p-2})$. Formulate and prove the continuous analogue to (3.4.10).

3.4.13 (a) Consider, as in Example 3.4.5, the sum $\sum n^{-3/2}$. Show that the partial sum s_n has an asymptotic expansion of the form needed in that example, with $p = -1/2$.

Hint: Apply Euler–Maclaurin's formula (theoretically).

(b) Suppose that $\sum a_n$ is convergent, and that $a_n = a(n)$ where $a(z)$ is an analytic function at $z = \infty$ (for example a rational function), multiplied by some power of $z - c$. Show that such a function has an expansion such as (3.4.8), and that the same holds for a product of such functions.

3.4.14 Compute and plot

$$F(x) = \sum_{n=0}^{\infty} T_n(x)/(1+n^2), \quad x \in [-1, 1].$$

Find out experimentally or theoretically how $F'(x)$ behaves near $x = 1$ and $x = -1$.

3.4.15 Compute to (say) six decimal places the double sum

$$S = \sum_{m=1}^{\infty}\sum_{n=1}^{\infty} \frac{(-1)^{m+n}}{(m^2+n^2)} = \sum_{n=1}^{\infty}(-1)^m f(m),$$

where

$$f(m) = \sum_{n=1}^{\infty}(-1)^n (m^2+n^2)^{-1}.$$

Compute, to begin with, $f(m)$ for $m = 1:10$ by the generalized Euler transformation. Do you need more values of $f(m)$?

Comment: There exists an explicit formula for $f(m)$ in this case, but you can solve this problem easily without using that.

3.4.16 We use the notation of Sec. 3.4.3 (the generalized Euler transformation). Assume that $N \geq k \geq 1$, and set $n = N - k + 1$. A sum is equal to zero if the upper index is smaller than the lower index.

(a) Prove (3.4.21), which is given without proof in the text; i.e.,

$$M_{N,k-1} - M_{N-1,k-1} = z^n P^{k-2} u_{n+1} \quad (k \geq 2).$$

Hint: By subscript transformations in the definition of $M_{N,k}$, prove that

$$M_{N,k-1} - M_{N-1,k-1} = u_{n+1} z^n + \frac{z^n}{1-z}\sum_{s=0}^{k-3}(zE-1)P^s u_{n+1}.$$

Next, show that $zE - 1 = (1-z)(P-1)$, and use this to simplify the expression.

(b) Derive the formulas

$$M_{k-1,k} = \frac{1}{1-z}\sum_{s=0}^{k-2} P^s u_1, \qquad M_{N,k} = M_{k-1,k} + \sum_{j=0}^{n-1} z^j P^{k-1} u_{j+1}.$$

Comment: The first formula gives the partial sums of the classical Euler transformation. The second formula relates the kth column to the partial sums of the power series with the coefficients $P^{k-1}u_{j+1}$.

3.4.17 (a) If $u_j = a^j$, $z = e^{i\phi}$, $\phi \in [0, \pi]$, for which real values of $a \in [0, 1]$ does the series on the right of (3.4.14) converge faster than the series on the left?

(b) Find how the classical Euler transformation works if applied to the series

$$\sum z^n, \quad |z| = 1, \quad z \neq 1.$$

Compare how it works on $\sum u_n z^n$, for $u_n = a^n$, $z = z_1$, and for $u_n = 1$, $z = az_1$. Consider similar questions for other convergence acceleration methods, which are primarily invented for oscillating sequences.

3.4.18 Compute $\sum_{k=1}^{\infty} k^{1/2}/(k^2 + 1)$ with an error of less than 10^{-6}. Sum the first ten terms directly. Then expand the summand in negative powers of k and use Euler–Maclaurin's summation formula. Or try the central difference variant of Euler–Maclaurin's summation formula given in the next problem; then you do not have to compute derivatives.

3.4.19 *Variations on the Euler–Maclaurin Theme.* Set $x_i = a + ih$, also for noninteger subscripts, and $x_n = b$.

Two variants with central differences instead of derivatives are interesting alternatives, if the derivatives needed in the Euler–Maclaurin formula are hard to compute. Check a few of the coefficients on the right-hand side of the formula

$$\sum_{j=1}^{\infty} \frac{B_{2j}(hD)^{2j-1}}{(2j)!} \approx \frac{\mu\delta}{12} - \frac{11\mu\delta^3}{720} + \frac{191\mu\delta^5}{60{,}480} - \frac{2497\mu\delta^7}{3{,}628{,}800} + \cdots. \tag{3.4.52}$$

Use the expansion for computing the sum given in the previous problem. This formula is given by Fröberg [128, p. 220], who attributes it to Gauss.

Compare the size of its coefficients with the corresponding coefficients of the Euler–Maclaurin formula.

Suppose that $h = 1$, and that the terms of the given series can be evaluated also for noninteger arguments. Then another variant is to compute the central differences for (say) $h = 1/2$ in order to approximate *each* derivative needed more accurately by means of (3.3.48). This leads to the formula[109]

$$\sum_{j=1}^{\infty} \frac{B_{2j}D^{2j-1}}{(2j)!} \sim \frac{\mu\delta}{6} - \frac{7\mu\delta^3}{180} + \frac{71\mu\delta^5}{7560} - \frac{521\mu\delta^7}{226{,}800} + \cdots. \tag{3.4.53}$$

($h = 1/2$ for the central differences; $h = 1$ in the series.) Convince yourself of the reliability of the formula, either by deriving it or by testing it for (say) $f(x) = e^{0.1x}$. Show that the rounding errors of the function values cause almost no trouble in the numerical evaluation of these difference corrections.

[109] The formula is probably very old, but we have not found it in the literature.

3.4.20 (a) Derive formally in a similar way the following formula for an *alternating series*. Set x_i, $h = 1$, $b = \infty$, and assume that $\lim_{x\to\infty} f(x) = 0$.

$$\sum_{i=0}^{\infty}(-1)^i f(a+i) = \frac{1}{2}f(a) - \frac{1}{4}f'(a) + \frac{1}{48}f'''(a) - \cdots$$
$$- \frac{(2^{2r}-1)B_{2r}}{(2r)!} f^{(2r-1)}(a) - \cdots . \qquad (3.4.54)$$

Of course, the integral of f is not needed in this case.[110] Compare it with some of the other methods for alternating series on an example of your own choice.

(b) Derive by using operators (without the remainder R) the following more general form of the Euler–Maclaurin formula (Handbook [1, Sec. 23.1.32]):

$$\sum_{k=0}^{m-1} hf(a+kh+\omega h) = \int_a^b f(t)dt + \sum_{j=1}^{p} \frac{h^j}{j!} B_j(\omega)(f^{(j-1)}(b) - f^{(j-1)}(a))$$
$$- \frac{h^p}{p!}\int_0^1 \hat{B}_p(\omega - t)\sum_{k=0}^{m-1} f^{(p)}(a+kh+th)\,dt.$$

If you use this formula for deriving the midpoint variant in (a) you will find a quite different expression for the coefficients; nevertheless, it is the same formula. See how this is explained in the Handbook [1, Sec. 23.1.10], that is by the "multiplication theorem."[111]

$$B_n(mx) = m^{n-1}\sum_{k=0}^{m-1} B_n(x + k/m), \quad n = 0, 1, 2, \ldots, \quad m = 1, 2, 3, \ldots .$$

3.4.21 Prove statement (b) of Lemma 3.4.9 (concerning the periodicity and the regularity of the Bernoulli functions).

3.4.22 Euler's constant is defined by $\gamma = \lim_{N\to\infty} F(N)$, where

$$F(N) = 1 + \frac{1}{2} + \frac{1}{3} + \cdots + \frac{1}{N-1} + \frac{1}{N} - \ln N.$$

(a) Use the Euler–Maclaurin formula with $f(x) = x^{-1}$, $h = 1$, to show that, for any integer N,

$$\gamma = F(N) + \frac{1}{12}N^{-2} - \frac{6}{720}N^{-4} + \frac{120}{30{,}240}N^{-6} - \cdots,$$

where every other partial sum is larger than γ, and every other is smaller.

[110] Note that the right-hand side yields a finite value if f is a constant or, more generally, if f is a polynomial, although the series on the left-hand side diverges. The same happens to other summation methods.

[111] That formula and the remainder R are derived on p. 21 and p. 30, respectively, in Nörlund [276].

(b) Compute γ to seven decimal places, using $N = 10$, $\sum_{n=1}^{10} n^{-1} = 2.92896825$, $\ln 10 = 2.30258509$.

(c) Show how repeated Richardson extrapolation can be used to compute γ from the following values.

N	1	2	4	8
$F(N)$	0.5	0.55685	0.57204	0.57592

(d) Extend (c) to a computation where a larger number of values of $F(N)$ have been computed as accurately as possible, and so that the final accuracy of γ is limited by the effects of rounding errors. Check the result by looking it up in an accurate table of mathematical constants, for example, in the Handbook [1].

3.4.23 *A digression about the gamma function.*

(a) The Handbook [1, Sec. 6.1.40] gives an expansion for $\ln \Gamma(z)$ that agrees with formula (3.4.41) for $\ln z!$ (if we substitute z for m), except that the Handbook writes $(z - \frac{1}{2}) \ln z$, where we have $(m + \frac{1}{2}) \ln m$. Explain concisely and completely that there is no contradiction here.

(b) An asymptotic expansion for computing $\ln \Gamma(z + 1)$, $z \in \mathbf{C}$, is derived in Example 3.4.12. If r terms are used in the asymptotic expansion, the remainder reads

$$R(z) = K(z)\frac{(2r)!}{\pi|2\pi z|^{2r+1}}, \qquad K(z) = \sup_{u \geq 0} \frac{|z^2|}{|u^2 + z^2|}$$

(see also Remark 3.4.6). Set $z = x + iy$. Show the following more useful bound for $K(z)$, valid for $x > 0$,

$$K(z) \leq \begin{cases} 1 & \text{if } x \geq |y|, \\ \frac{1}{2}(x/|y| + |y|/x) & \text{otherwise.} \end{cases}$$

Find a uniform upper bound for the remainder if $r = 5$, $x \geq \frac{1}{2}$, $|z| \geq 17$.

(c) Write a MATLAB program for the computation of $\ln \Gamma(z+1)$. Use the reflection and recurrence formulas to transform the input value z to another $z = x + iy$ that satisfies $x \geq half$, $|z| \geq 17$, for which this asymptotic expansion is to be used with $r = 5$.

Test the program by computing the following quantities, and compare with their exact values:

$$n!, \qquad \Gamma\left(n + \frac{1}{2}\right)/\sqrt{\pi}, \qquad n = 0, 1, 2, 3, 10, 20;$$

$$\left|\Gamma\left(\frac{1}{2} + iy\right)\right|^2 = \frac{\pi}{\cosh(\pi y)}, \qquad y = \pm 10, \pm 20.$$

If the original input value has a small modulus, there is some cancellation when the output from the asymptotic expansion is transformed to $\ln(1 + z_{input})$, resulting in a loss of (say) one or two decimal digits.

Comment: It is often much better to work with $\ln \Gamma(z)$ than with $\Gamma(z)$. For example, one can avoid exponent overflow in the calculation of a binomial coefficient or a value of the beta function, $B(z, w) = \Gamma(z)\Gamma(w)/\Gamma(z + w)$, where (say) the denominator can become too big, even if the final result is of a normal order of magnitude.

3.4.24 (a) Show that

$$\binom{2n}{n} \sim \frac{2^{2n}}{\sqrt{\pi n}}, \qquad n \to \infty,$$

and give an asymptotic estimate of the relative error of this approximation. Check the approximation as well as the error estimate for $n = 5$ and $n = 10$.

(b) *Random errors in a difference scheme.* We know from Example 3.3.3 that if the items y_j of a difference scheme are afflicted with errors less than ϵ in absolute value, then the inherited error of $\Delta^n y_j$ is at most $2^n \epsilon$ in absolute value. If we consider the errors as independent random variables, uniformly distributed in the interval $[-\epsilon, \epsilon]$, show that the error of $\Delta^n y_j$ has the variance $\binom{2n}{n}\frac{1}{3}\epsilon^2$, hence the standard deviation is approximately

$$2^n \epsilon (9\pi n)^{-1/4}, \qquad n \gg 1.$$

Check the result on a particular case using a Monte Carlo study.

Hint: It is known from probability theory that the variance of $\sum_{j=0}^{n} a_j \epsilon_j$ is equal to $\sigma^2 \sum_{j=0}^{n} a_j^2$, and that a random variable, uniformly distributed in the interval $[-\epsilon, \epsilon]$, has the variance $\sigma^2 = \epsilon^2/3$. Finally, use (3.1.23) with $p = q = n$.

3.4.25 The following table of values of a function $f(x)$ is given.

x	0.6	0.8	0.9	1.0	1.1	1.2	1.4
$f(x)$	1.820365	1.501258	1.327313	1.143957	0.951849	0.752084	0.335920

Compute $f'(1.0)$ and $f''(1.0)$ using repeated Richardson extrapolation.

3.4.26 Compute an approximation to π using Richardson extrapolation with *Neville's algorithm*, based on three simple polygons, with $n = 2$, 3, and 6 sides, not in geometric progression. A 2-sided polygon can be interpreted as a diameter described up and down. Its " perimeter" is thus equal to four. Show that this gives even a little better value than the result (3.14103) obtained for the 96-sided polygon without extrapolations.

3.4.27 *Numerov's method with Richardson extrapolations.*[112]

(a) Show that the formula

$$h^{-2}(y_{n+1} - 2y_n + y_{n-1}) = y_n'' + a(y_{n+1}'' - 2y_n'' + y_{n-1}'')$$

is exact for polynomials of as high degree as possible, if $a = 1/12$. Show that the error has an expansion into *even* powers of h, and determine the first (typically non-vanishing) term of this expansion.

[112] See also Example 3.3.15.

(b) This formula can be applied to the differential equation $y'' = p(x)y$ with given initial values $y(0)$, $y'(0)$. Show that this yields the recurrence relation

$$y_{n+1} = \frac{(2 + \frac{10}{12} p_n h^2) y_n - (1 - \frac{1}{12} p_{n-1} h^2) y_{n-1}}{1 - \frac{1}{12} p_{n+1} h^2}.$$

Comment: If h is small, information about $p(t)$ is lost by outshifting in the factors $1 - \frac{1}{12} p_{n-1} h^2$. It is possible to rewrite the formulas in order to reduce the loss of information, but in the application below this causes no trouble in IEEE double precision.

(c) You proved in (a) that the *local* error has an expansion containing *even powers of h* only. It can be shown that *the same is true for the global error* too. Assume (without proof) that

$$y(x, h) = y(x) + c_1(x)h^4 + c_2(x)h^6 + c_3(x)h^8 + O(h^{10}).$$

Apply this method, together with two Richardson extrapolations in (d), to the problem of computing the solution to the differential equation $y'' = -xy$ with initial values $y(0) = 1$, $y'(0) = 0$, this time over the interval $0 \le x \le 4.8$. Denote the numerical solution by $y(x; h)$, i.e., $y_n = y(x_n; h)$.
Compute the seeds $y_1 = y(h, h)$ by the Taylor expansion given in (1.2.8). The error of $y(0.2, 0, 2)$ should be less than 10^{-10}, since we expect that the (global) errors after two Richardson extrapolations can be of that order of magnitude.
Compute $y(x; h)$, $x = 0 : h : 4.8$, for $h = 0.05$, $h = 0.1$, $h = 0.2$. Store these data in a 100×3 matrix (where you must put zeros into some places). Plot $y(x; 0.05)$ versus x for $x = 0 : 0.05 : 4.8$.

(d) Express with the aid of the Handbook [1, Sec. 10.4] the solution of this initial value problem in terms of Airy functions:[113]

$$y(x) = \frac{\mathrm{Ai}(-x) + \mathrm{Bi}(-x)/\sqrt{3}}{2 \cdot 0.3550280539}.$$

Check a few of your results of the repeated Richardson extrapolation by means of [1, Table 10.11] that, unfortunately, gives only eight decimal places.

3.4.28 (a) Determine the Bernoulli polynomials $B_2(x)$ and $B_3(x)$, and find the values and the derivatives at zero and one. Factorize the polynomial $B_3(x)$. Draw the graphs of a few periods of $\hat{B}_i(x)$, $i = 1, 2, 3$.

(b) In an old textbook, we found a "symbolic" formula, essentially

$$h \sum_{j=0}^{n-1} g'(a + jh) = g(b + hB) - g(a + hB). \tag{3.4.55}$$

The expansion of the right-hand side into powers of hB has been followed by the replacement of the powers of B by Bernoulli numbers; the resulting expansion is

[113] Airy functions are special functions (related to Bessel functions) with many applications to mathematical physics, for example, the theory of diffraction of radio waves along the Earth's surface.

not necessarily convergent, even if the first power series converges for any complex value of hB.

Show that the second expansion is equivalent to the Euler–Maclaurin formula, and that it is to be interpreted according to Theorem 3.4.10.

(c) If g is a polynomial, the expansion is finite. Show the following important formulas, and check them with known results for $k = 1 : 3$.

$$\sum_{j=0}^{n-1} j^{k-1} = \frac{(B+n)^k - B^k}{k} = \frac{B_k(n) - B_k}{k}. \tag{3.4.56}$$

Also find that (3.4.55) makes sense for $g(x) = e^{\alpha x}$, with the "symbolic" interpretation of the power series for e^{Bx}, if you accept the formula $e^{(B+\alpha)x} = e^{Bx}e^{\alpha x}$.

3.4.29 We have called $\sum a_n$ a *bell sum* if a_n as a function of n has a bell-shaped graph, and you must add many terms to get the desired accuracy. Under certain conditions you can get an accurate result by adding (say) every tenth term and multiplying this sum by ten, because both sums can be interpreted as trapezoidal approximations to the same integral, with different step size. Inspired by Euler–Maclaurin's formula, we may hope to be able to obtain high accuracy using an integer step size h, i.e., (say) one quarter of the half-width of "the bell." In other words, we do not have to compute and add more than every hth term. We shall study a class of series

$$S(t) = \sum_{n=0}^{\infty} c_n t^n/n!, \quad t \gg 1, \tag{3.4.57}$$

where $c_n > 0$, $\log c_n$ is rather slowly varying for n large; (say that) $\Delta^p \log c_n = O(n^{-p})$. Let $c(\cdot)$ be a smooth function such that $c(n) = c_n$. We consider $S(t)$ as an approximation to the integral

$$\int_0^{\infty} c(n)t^n/\Gamma(n+1)dn,$$

with a smooth and bell-shaped integrand, almost like the normal frequency function, with standard deviation $\sigma \approx k\sqrt{t}$.

(a) For $p = 1 : 5$, $t = 4^p$, plot $y = \sqrt{2\pi t}e^{-t}t^n/n!$ versus $x = n/t$, $0 \le x \le 3$; include all five curves on the same picture.

(b) For $p = 1 : 5$, $t = 4^p$, plot $y = \ln(e^{-t}t^n/n!)$ versus $x = (n-t)/\sqrt{t}$, $\max(0, t - 8\sqrt{t}) \le n \le t + 8\sqrt{t}$; include all five curves on the same picture. Give bounds for the error committed if you neglect the terms of the series $e^{-t}\sum_0^{\infty} t^n/n!$, which are cut out in your picture.

(c) With the same notation as in (b), use Stirling's asymptotic expansion to show theoretically that, for $t \to \infty$,

$$\frac{e^{-t}t^n}{n!} = \frac{e^{-x^2/2}\big(1 + O(1/\sqrt{t})\big)}{\sqrt{2\pi t}}, \tag{3.4.58}$$

where the $O(1/\sqrt{t})$-term depends on x. Compare this with the plots.

(d) Test these ideas by making numerical experiments with the series

$$e^{-t}\sum_{n\in\mathcal{N}} t^n/n!, \quad \mathcal{N} = \{\text{round}(t - 8\sqrt{t}) : h : \text{round}(t + 8\sqrt{t})\},$$

for some integers h in the neighborhood of suitable fractions of $\sqrt{t}$, inspired by the outcome of the experiments. Do this for t =1000, 500, 200, 100, 50, 30. Compare with the exact result, see how the trapezoidal error depends on h, and try to formulate an error estimate that can be reasonably reliable, in cases where the answer is not known. How large must t be, in order that it should be permissible to choose $h > 1$ if you want (say) six correct decimals?

(e) Compute, with an error estimate, $e^{-t}\sum_{n=1}^{\infty} t^n/(n \cdot n!)$, with six correct decimals for the values of t mentioned in (d). You can also check your result with tables and formulas in the Handbook [1, Chap. 5].

3.4.30 If you have a good program for generating primes, denote the nth prime by p_n and try convergence acceleration to series such as

$$\sum (-1)^n/p_n, \qquad \sum 1/p_n^2.$$

Due to the irregularity of the sequence of primes, you cannot expect the spectacular accuracy of the previous examples. It can be fun to see how these methods work in combination with some comparison series derived from asymptotic results about primes. The simplest one reads $p_n \sim n \ln n$, $(n \to \infty)$, which is equivalent to the classical prime number theorem.

3.4.31 *A summation formula based on the Euler numbers.* The Euler numbers E_n were introduced by (3.1.22). The first values read

$$E_0 = 1, \quad E_2 = -1, \quad E_4 = 5, \quad E_6 = -61.$$

They are all integers (Problem 3.1.7(c)). $E_n = 0$ for odd n, and the sign is alternating for even n. Their generating function reads

$$\frac{1}{\cosh z} = \sum_{j=0}^{\infty} \frac{E_j z^j}{j!}.$$

(a) Show by means of operators the following expansion:

$$\sum_{k=m}^{\infty} (-1)^{k-m} f(k) \approx \sum_{p=0}^{q} \frac{E_{2p} f^{(2p)}(m - \frac{1}{2})}{2^{2p+1}(2p)!}. \tag{3.4.59}$$

No discussion of convergence is needed; the expansion behaves much like the Euler–Maclaurin expansion, and so does the error estimation; see [87].

The coefficient of $f^{(2p)}(m - \frac{1}{2})$ is approximately $2(-1)^p/\pi^{2p+1}$ when $p \gg 1$; e.g., for $p = 3$ the approximation yields $-6.622 \cdot 10^{-4}$, while the exact coefficient is $61/92{,}160 \approx 6.619 \cdot 10^{-4}$.

(b) Apply (3.4.59) to explain the following curious observation, reported by Borwein, Borwein, and Dilcher [43].

$$\sum_{k=1}^{50} \frac{4(-1)^k}{2k-1} = 3.12159465259\ldots,$$
$$(\pi = 3.14159265359\ldots).$$

Note that only three digits disagree. There are several variations on this theme. Reference [43] actually displays the case with 40 decimal places based on 50,000 terms. Make an educated guess concerning how few digits disagreed.

3.4.32 What is $\beta(t)$ (in the notation of (3.4.25)), if $u_n = a^n$, $0 < a < 1$?

3.4.33 Work out the details of the two optimizations in the proof of Theorem 3.4.7.

3.4.34 (a) Show that every rational function $f(s)$ that is analytic and bounded for $\Re s \geq a$ is d.c.m. for $s \geq a$.

(b) Show criterion (B) for higher monotonicity (concerning products).

(c) Which of the coefficient sequences $\{c_n\}$ mentioned in Problems 3.4.5 and 3.4.6 are c.m.? Which are d.c.m.?

(d) Show criterion (E) for higher monotonicity.

3.4.35 Suppose that $u_n = \int_0^1 t^n \, d\beta(t)$, where $\beta(t)$ is of bounded variation in $[0, 1]$. Show that $\lim u_n = 0$ if $\beta(t)$ is continuous at $t = 1$, but that it is not true if $\beta(t)$ has a jump at $t = 1$.

3.5 Continued Fractions and Padé Approximants

3.5.1 Algebraic Continued Fractions

Some functions cannot be well approximated by a power series, but can be well approximated by a quotient of power series. In order to study such approximations we first introduce **continued fractions**, i.e., expressions of the form

$$r = b_0 + \cfrac{a_1}{b_1 + \cfrac{a_2}{b_2 + \cfrac{a_3}{b_3 + \ldots}}} = b_0 + \frac{a_1}{b_1+}\,\frac{a_2}{b_2+}\,\frac{a_3}{b_3+}\cdots. \tag{3.5.1}$$

The second expression is a convenient compact notation. If the number of terms is infinite, r is called an *infinite continued fraction*.

Continued fractions were applied in the seventeenth century to the rational approximation of various algebraic numbers. In such algebraic continued fractions r and the entries a_i, b_i are numbers. Beginning with work by Euler, analytic continued fraction expansions

$$r(z) = b_0 + \frac{a_1 z}{b_1+}\,\frac{a_2 z}{b_2+}\,\frac{a_3 z}{b_3+}\cdots \tag{3.5.2}$$

involving functions of a complex variable $r(z)$ became an important tool in the approximation of special classes of analytic functions of a complex variable.

We first study some algebraic properties of continued fractions. The partial fraction

$$r_n = \frac{p_n}{q_n} = b_0 + \frac{a_1}{b_1+}\ \frac{a_2}{b_2+} \cdots \frac{a_n}{b_n} \tag{3.5.3}$$

is called *the nth approximant* of the continued fraction. There are several essentially different algorithms for evaluating a partial fraction. It can be evaluated *backward* in n divisions using the recurrence

$$y_n = b_n, \qquad y_{i-1} = b_{i-1} + a_i/y_i, \quad i = n : -1 : 1, \tag{3.5.4}$$

for which $r = y_0$. It can happen that in an intermediate step the denominator y_i becomes zero and $y_{i-1} = \infty$. This does no harm if in the next step when you divide by y_{i-1} the result is set equal to zero. If it happens in the last step, the result is ∞.[114]

A drawback of evaluating an infinite continued fraction expansion by the backward recursion (3.5.4) is that you have to decide where to stop in advance. The following theorem shows how *forward* (or top down) evaluation can be achieved.

Theorem 3.5.1.

For the nth convergent $r_n = p_n/q_n$ of the continued fraction (3.5.1), *p_n and q_n, $n \geq 1$, satisfy the recursion formulas*

$$p_n = b_n p_{n-1} + a_n p_{n-2}, \quad p_{-1} = 1, \quad p_0 = b_0, \tag{3.5.5}$$

$$q_n = b_n q_{n-1} + a_n q_{n-2}, \quad q_{-1} = 0, \quad q_0 = 1. \tag{3.5.6}$$

Another useful formula reads

$$p_n q_{n-1} - p_{n-1} q_n = (-1)^{n-1} a_1 a_2 \cdots a_n. \tag{3.5.7}$$

If we substitute $a_n x$ for a_n in (3.5.5)–(3.5.6), *then $p_n(x)$ and $q_n(x)$ become polynomials in x of degree n and $n-1$, respectively.*

Proof. We prove the recursion formulas by induction. First, for $n = 1$, we obtain

$$\frac{p_1}{q_1} = \frac{b_1 p_0 + a_1 p_{-1}}{b_1 q_0 + a_1 q_{-1}} = \frac{b_1 b_0 + a_1}{b_1 + 0} = b_0 + \frac{a_1}{b_1} = r_1.$$

Next, assume that the formulas are valid up to p_{n-1}, q_{n-1} for *every* continued fraction. Note that p_n/q_n can be obtained from p_{n-1}/q_{n-1} by the substitution of $b_{n-1} + a_n/b_n$ for b_{n-1}. Hence

$$\begin{aligned}\frac{p_n}{q_n} &= \frac{(b_{n-1} + a_n/b_n)p_{n-2} + a_{n-1}p_{n-3}}{(b_{n-1} + a_n/b_n)q_{n-2} + a_{n-1}q_{n-3}} = \frac{b_n(b_{n-1}p_{n-2} + a_{n-1}p_{n-3}) + a_n p_{n-2}}{b_n(b_{n-1}q_{n-2} + a_{n-1}q_{n-3}) + a_n q_{n-2}} \\ &= \frac{b_n p_{n-1} + a_n p_{n-2}}{b_n q_{n-1} + a_n q_{n-2}}.\end{aligned}$$

This shows that the formulas are valid also for p_n, q_n. The proof of (3.5.7) is left for Problem 3.5.2. □

[114] Note that this works automatically in IEEE arithmetic, because of the rules of infinite arithmetic; see Sec. 2.2.3.

The evaluation of a continued fraction by forward recursion requires $4n$ multiplications and one division. It is sometimes convenient to write the recursion formulas in matrix form; see Problem 3.5.2. One must also be careful about the numerical stability of these recurrence relations.

In practice the forward recursion for evaluating a continued fraction often generates very large or very small values for the numerators and denominators. There is a risk of *overflow or underflow* with these formulas. Since we are usually not interested in the p_n, q_n themselves, but in the ratios only, we can normalize p_n and q_n by multiplying them by the same factor after they have been computed. If we shall go on and compute p_{n+1}, q_{n+1}, however, *we have to multiply* p_{n-1}, q_{n-1} *by the same factor also*. The formula

$$\frac{a_1}{b_1+}\,\frac{a_2}{b_2+}\,\frac{a_3}{b_3+}\cdots = \frac{k_1a_1}{k_1b_1+}\,\frac{k_1k_2a_2}{k_2b_2+}\,\frac{k_2k_3a_3}{k_3b_3+}\cdots, \tag{3.5.8}$$

where the k_i are any nonzero numbers, is known as an **equivalence transformation**. The proof of (3.5.8) is left for Problem 3.5.6.

Suppose we are given a rational function $R(z) = R_0(z)/R_1(z)$, where $R_0(z)$ and $R_1(z)$ are polynomials. Then by the following division algorithm $R(z)$ can be expressed as a continued fraction that can be evaluated by backward recursion in fewer arithmetic operations; see Cheney [66, p. 151]. The degree of a polynomial $R_j(z)$ is denoted by d_j. By successive divisions (of $R_{j-1}(z)$ by $R_j(z)$) we obtain quotients $Q_j f(z)$ and remainders $R_{j+1}(z)$ as follows.

For $j = 1, 2, \ldots,$ until $d_{j+1} = 0$, set

$$R_{j-1}(z) = R_j(z)Q_j(z) + R_{j+1}(z) \qquad (d_{j+1} < d_j). \tag{3.5.9}$$

Then

$$R(z) = \frac{R_0(z)}{R_1(z)} = Q_1(z) + \frac{1}{R_1(z)/R_2(z)} = \cdots \tag{3.5.10}$$

$$= Q_1(z) + \frac{1}{Q_2(z)+}\,\frac{1}{Q_3(z)+}\cdots\frac{1}{Q_k(z)}. \tag{3.5.11}$$

By means of an equivalence transformation (see (3.5.8)), this fraction can be transformed into a slightly more economic form, where the polynomials in the denominators have leading coefficient unity, while the numerators are in general different from 1.

Example 3.5.1.

In the rational form

$$r(z) = \frac{7z^4 - 101z^3 + 540z^2 - 1204z + 958}{z^4 - 14z^3 + 72z^2 - 151z + 112},$$

the numerator and denominator can be evaluated by Horner's rule. Alternatively, the above algorithm can be used to convert the rational form to the finite continued fraction

$$r(z) = 7 - \frac{3}{z-2-}\,\frac{1}{z-7+}\,\frac{10}{z-2-}\,\frac{2}{z-3}.$$

To evaluate this by backward recursion requires fewer operations than the rational form, but a division by zero occurs at the four points $z = 1, 2, 3, 4$. In IEEE arithmetic the continued fraction evaluates correctly also at these points because of the rules of infinite arithmetic! Indeed, the continued fraction form can be shown to have smaller errors for $z \in [0, 4]$ and to be immune to overflow; see Higham [199, Sec. 27.1].

Every positive number x can be expanded into a regular continued fraction with integer coefficients of the form

$$x = b_0 + \frac{1}{b_1+}\,\frac{1}{b_2+}\,\frac{1}{b_3+}\cdots. \tag{3.5.12}$$

Set $x_0 = x$, $p_{-1} = 1$, $q_{-1} = 0$. For $n = 0, 1, 2, \ldots$ we construct a sequence of numbers,

$$x_n = b_n + \frac{1}{b_{n+1}+}\,\frac{1}{b_{n+2}+}\,\frac{1}{b_{n+3}+}\cdots.$$

Evidently $b_n = \lfloor x_n \rfloor$, the integer part of x_n, and $x_{n+1} = 1/(x_n - b_n)$. Compute p_n, q_n, according to the recursion formulas of Theorem 3.5.1, which can be written in vector form,

$$(p_n, q_n) = (p_{n-2}, q_{n-2}) + b_n(p_{n-1}, q_{n-1})$$

(since $a_n = 1$). Stop when $|x - p_n/q_n| < \text{tol}$ or $n > nmax$. If the number x is rational this expansion is finite. The details are left for Problem 3.5.1. Note that the algorithm is related to the Euclidean algorithm; see Problem 1.2.6.

The above algorithm has been used several times in the previous sections, where some coefficients, known to be rational, have been computed in floating-point. It is also useful for finding near commensurabilities between events with different periods;[115] see Problem 3.5.1 (c).

The German mathematician Felix Klein [228][116] gave the following illuminating description of the sequence $\{(p_n, q_n)\}$ obtained by this algorithm (adapted to our notation):

> Imagine pegs or needles affixed at all the integral points (p_n, q_n), and wrap a tightly drawn string about the sets of pegs to the right and to the left of the ray, $p = xq$. Then the vertices of the two convex string-polygons which bound our two point sets will be precisely the points $(p_n, q_n)\ldots$, the left polygon having the even convergents, the right one the odd.

Klein also points out that "such a ray makes a cut in the set of integral points" and thus makes Dedekind's definition of irrational numbers very concrete. This construction, shown in Figure 3.5.1, illustrates in a concrete way that the successive convergents are closer to x than any numbers with smaller denominators, and that the errors alternate in sign. We omit the details of the proof that this description is correct.

Note that, since $a_j = 1$ for all j, (3.5.7) reads

$$p_n q_{n-1} - p_{n-1} q_n = (-1)^{n-1}.$$

[115] One of the convergents for $\log 2/\log 3$ reads $12/19$. This is, in a way, basic for Western music, where 13 quints make 7 octaves, i.e., $(3/2)^{12} \approx 2^7$.

[116] Felix Christian Klein (1849–1925), a German mathematician, was born 4/25. He delighted in pointing out that the day (5^2), month (2^2), and year (43^2) of his birth was the square of a prime number.

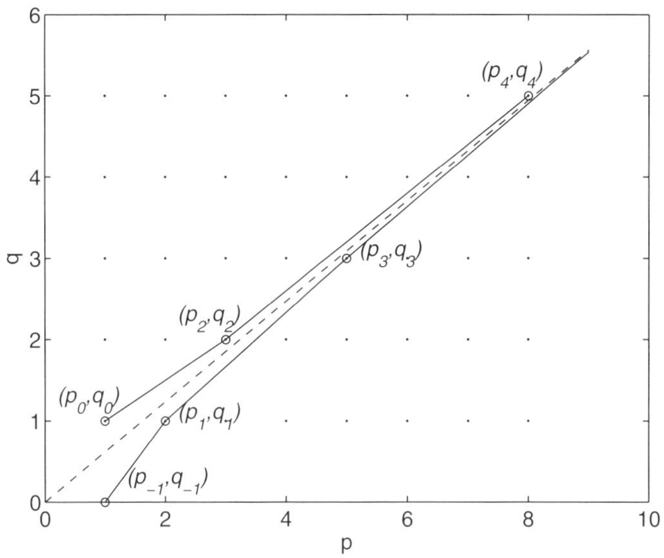

Figure 3.5.1. *Best rational approximations* $\{(p, q)\}$ *to the "golden ratio."*

This implies that the triangle with vertices at the points $(0, 0)$, (q_n, p_n), (q_{n-1}, p_{n-1}) has the smallest possible area among triangles with integer coordinates, and hence there can be no integer points inside or on the sides of this triangle.

Comment: If we know or guess that a result x of a computation is a rational number with a reasonably sized denominator, although it was practical to compute it in floating-point arithmetic (afflicted by errors of various types), we have a good chance of reconstructing the exact result by applying the above algorithm as a postprocessing.

If we just know that the exact x is rational, without any bounds for the number of digits in the denominator and numerator, we must be conservative in claiming that the last fraction that came out of the above algorithm is the exact value of x, even if $|x - p_n/q_n|$ is very small. In fact, the fraction may depend on tol that is to be chosen with respect to the expected order of magnitude of the error of x. If tol has been chosen smaller than the error of x, it may happen that the last fraction obtained at the termination is wrong, while the correct fraction (with smaller numerator and denominator) may have appeared earlier in the sequence (or it may not be there at all).

So a certain judgment is needed in the application of this algorithm. The smaller the denominator and numerator are, the more likely it is that the fraction is correct. In a serious context, it is advisable to check the result(s) by using exact arithmetic. If x is the root of an equation (or a component of the solution of a system of equations), it is typically much easier to check afterward that a suggested result is correct than to perform the whole solution process in exact arithmetic.

The following theorem due to Seidel[117] gives a necessary and sufficient condition for convergence of a continued fraction of the form (3.5.12).

[117] Philipp Ludwig von Seidel (1821–1896), German mathematician and astronomer. In 1846 he submitted his habilitation dissertation entitled "Untersuchungen über die Konvergenz and Divergenz der Kettenbrüche" (Investigations of the Convergence and Divergence of Continued Fractions).

Theorem 3.5.2.

Let all b_n be positive in the continued fraction

$$b_0 + \frac{1}{b_1+} \frac{1}{b_2+} \frac{1}{b_3+} \cdots.$$

Then this converges if and only if the series $\sum b_n$ diverges.

Proof. See Cheney [66, p. 184]. □

Example 3.5.2.

The following are continued fraction expansions of some important irrational numbers:

$$\pi = 3 + \frac{1}{7+} \frac{1}{15+} \frac{1}{1+} \frac{1}{292+} \frac{1}{1+} \frac{1}{1+} \frac{1}{1+} \cdots,$$

$$e = 2 + \frac{1}{1+} \frac{1}{2+} \frac{1}{1+} \frac{1}{1+} \frac{1}{4+} \frac{1}{1+} \frac{1}{1+} \frac{1}{6+} \cdots.$$

For e there is a regular pattern in the expansion, but for π a general formula for the expansion is not known. The partial fractions for π converge rapidly. For example, the error in the third convergent $\pi \approx 355/113$ is $0.266 \cdot 10^{-6}$.

Figure 3.5.1 corresponds to the expansion

$$\frac{\sqrt{5}+1}{2} = 1 + \frac{1}{1+} \frac{1}{1+} \frac{1}{1+} \frac{1}{1+} \frac{1}{1+} \frac{1}{1+} \cdots. \tag{3.5.13}$$

Then, note that $x = 1 + 1/x$, $x > 0$, hence $x = (\sqrt{5}+1)/2$, which is the "golden section ratio" (see also Problem 3.5.3). Note also that, by (3.5.7) with $a_j = 1$,

$$\left| x - \frac{p_n}{q_n} \right| \le \left| \frac{p_{n+1}}{q_{n+1}} - \frac{p_n}{q_n} \right| = \frac{|p_{n+1}q_n - p_n q_{n+1}|}{q_{n+1}q_n} = \frac{1}{q_{n+1}q_n} < \frac{1}{q_n^2}. \tag{3.5.14}$$

3.5.2 Analytic Continued Fractions

Continued fractions have also important applications in analysis. A large number of analytic functions are known to have continued fraction representations. Indeed, some of the best algorithms for the numerical computation of important analytic functions are based on continued fractions. We shall not give complete proofs but refer to classical books of Perron [289], Wall [369], and Henrici [196, 197].

A continued fraction is said to be equivalent to a given series if and only if the sequence of convergents is equal to the sequence of partial sums. There is typically an infinite number of such equivalent fractions. The construction of the continued fraction is particularly simple if we require that the denominators $q_n = 1$ for all $n \ge 1$. For a power series we shall thus have

$$p_n = c_0 + c_1 z + c_2 z^2 + \cdots + c_n z^n, \quad n \ge 1.$$

We must *assume that* $c_j \ne 0$ for all $j \ge 1$.

We shall determine the elements a_n, b_n by means of the recursion formulas of Theorem 3.5.1 (for $n \geq 2$) with initial conditions. We thus obtain the following equations:

$$\begin{aligned} p_n &= b_n p_{n-1} + a_n p_{n-2}, \quad p_0 = b_0, \quad p_1 = b_0 b_1 + a_1, \\ 1 &= b_n + a_n, \qquad b_1 = 1. \end{aligned}$$

The solution reads $b_0 = p_0 = c_0$, $b_1 = 1$, $a_1 = p_1 - p_0 = c_1 z$, and for $n \geq 2$,

$$\begin{aligned} a_n &= (p_n - p_{n-1})/(p_{n-2} - p_{n-1}) = -zc_n/c_{n-1}, \\ b_n &= 1 - a_n = 1 + zc_n/c_{n-1}, \end{aligned}$$

$$c_0 + c_1 z + \cdots + c_n z^n \cdots = c_0 + \frac{zc_1}{1-}\,\frac{zc_2/c_1}{1 + zc_2/c_1 -} \cdots \frac{zc_n/c_{n-1}}{1 + zc_n/c_{n-1} -} \cdots.$$

Of course, an equivalent continued fraction gives by itself *no convergence acceleration, just because it is equivalent.* We shall therefore leave the subject of continued fractions equivalent to a series, after showing two instances of the numerous pretty formulas that can be obtained by this construction. For

$$f(z) = e^z = 1 + z + \frac{z^2}{2!} + \frac{z^3}{3!} + \cdots$$

and

$$f(z) = \frac{\arctan\sqrt{z}}{\sqrt{z}} = 1 - \frac{z}{3} + \frac{z^2}{5} - \frac{z^3}{7} + \cdots,$$

we obtain for $z = -1$ and $z = 1$, respectively, after simple equivalence transformations,

$$e^{-1} = 1 - \frac{1}{1+}\,\frac{1}{1 + y} = \frac{1}{2 + y} \Rightarrow e = 2 + \frac{2}{2+}\,\frac{3}{3+}\,\frac{4}{4+}\,\frac{5}{5+} \cdots,$$

$$\frac{\pi}{4} = \frac{1}{1+}\,\frac{1}{2+}\,\frac{9}{2+}\,\frac{25}{2+}\,\frac{49}{2+} \cdots.$$

There exist, however, other methods to make a correspondence between a power series and a continued fraction. Some of them lead to a considerable convergence acceleration that often makes continued fractions very efficient for the *numerical computation of functions.* We shall return to such methods in Sec. 3.5.3.

Gauss developed a continued fraction for the ratio of two hypergeometric functions (see (3.1.16)),

$$\frac{F(a, b+1, c+1; z)}{F(a, b, c; z)} = \frac{1}{1+}\,\frac{a_1 z}{1+}\,\frac{a_2 z}{1+}\,\frac{a_3 z}{1+} \cdots, \tag{3.5.15}$$

where

$$a_{2n+1} = \frac{(a+n)(c-b+n)}{(c+2n)(c+2n+1)}, \qquad a_{2n} = \frac{(b+n)(c-a+n)}{(c+2n-1)(c+2n)}. \tag{3.5.16}$$

Although the power series converge only in the disk $|z| < 1$, the continued fraction of Gauss converges throughout the complex z-plane cut along the real axis from 1 to $+\infty$. It provides an analytic continuation in the cut plane.

If we set $b = 0$ in (3.5.15), we obtain a continued fraction for $F(a, 1, c+1; z)$. From this, many continued fractions for elementary functions can be derived, for example,

$$\arctan z = \frac{z}{1+}\,\frac{z^2}{3+}\,\frac{2^2z^2}{5+}\,\frac{3^2z^2}{7+}\,\frac{4^2z^2}{9+}\cdots, \tag{3.5.17}$$

$$\tan z = \frac{z}{1-}\,\frac{z^2}{3-}\,\frac{z^2}{5-}\,\frac{z^2}{7-}\cdots. \tag{3.5.18}$$

The expansion for $\tan z$ is valid everywhere, except in the poles. For $\arctan z$ the continued fraction represents a single-valued branch of the analytic function in a plane with cuts along the imaginary axis extending from $+i$ to $+i\infty$ and from $-i$ to $-i\infty$. A continued fraction expansion for arctanh$z$ is obtained by using the relation arctanh$z = -i \arctan iz$. In all these cases the region of convergence as well as the speed of convergence is considerably larger than for the power series expansions. For example, the sixth convergent for $\tan \pi/4$ is almost correct to 11 decimal places.

For the natural logarithm we have

$$\log(1+z) = \frac{z}{1+}\,\frac{z}{2+}\,\frac{z}{3+}\,\frac{2^2z}{4+}\,\frac{2^2z}{5+}\,\frac{3^2z}{6+}\cdots, \tag{3.5.19}$$

$$\frac{1}{2}\log\left(\frac{1+z}{1-z}\right) = z + \frac{z^3}{3} + \frac{z^5}{5} + \frac{z^7}{7} + \cdots \tag{3.5.20}$$

$$= \frac{z}{1-}\,\frac{z^2}{3-}\,\frac{2^2z^2}{5-}\,\frac{3^2z^2}{7-}\,\frac{4^2z^2}{9-}\cdots. \tag{3.5.21}$$

The fraction for the logarithm can be used in the whole complex plane except for the cuts $(-\infty, -1]$ and $[1, \infty)$. The convergence is slow when z is near a cut. For elementary functions such as these, properties of the functions can be used for moving z to a domain where the continued fraction converges rapidly.

Example 3.5.3.

Consider the continued fraction for $\ln(1+z)$ and set $z = 1$. The successive approximations to $\ln 2 = 0.69314\,71806$ are the following.

1/1	2/3	7/10	36/52	208/300	1572/2268	12,876/18,576
1.000000	0.66667	0.700000	0.692308	0.69333	0.693122	0.693152

Note that the fractions give alternatively upper and lower bounds for $\ln 2$. It can be shown that this is the case when the elements of the continued fraction are positive. To get the accuracy of the last approximation above would require as many as 50,000 terms of the series $\ln 2 = \ln(1+1) = 1 - 1/2 + 1/3 - 1/4 + \cdots$.

Continued fraction expansions for the gamma function and the incomplete gamma function are found in the Handbook [1, Sec. 6.5]. For the sake of simplicity we assume that $x > 0$, although the formulas can be used also in an appropriately cut complex plane. The parameter a may be complex in $\Gamma(a, x)$.[118]

$$\Gamma(a, x) = \int_x^\infty t^{a-1} e^{-t}\, dt, \quad \Gamma(a, 0) = \Gamma(a),$$

$$\gamma(a, x) = \Gamma(a) - \Gamma(a, x) = \int_0^x t^{a-1} e^{-t}\, dt, \quad \Re a > 0,$$

$$\Gamma(a, x) = e^{-x} x^a \Big(\frac{1}{x+} \, \frac{1-a}{1+} \, \frac{1}{x+} \, \frac{2-a}{1+} \, \frac{2}{x+} \cdots \Big), \tag{3.5.22}$$

$$\gamma(a, x) = e^{-x} x^a \Gamma(a) \sum_{n=0}^{\infty} \frac{x^n}{\Gamma(a+1+n)}.$$

We mention these functions because they have many applications. Several other important functions can, by simple transformations, be brought to particular cases of this function, for example, the normal probability function, the chi-square probability function, the exponential integral, and the Poisson distribution.

The convergence behavior of continued fraction expansions is much more complicated than for power series. Gautschi [142] exhibits a phenomenon of apparent convergence to the wrong limit for a continued fraction of Perron for ratios of Kummer functions. The sequence of terms initially decreases rapidly, then increases, and finally again decreases to zero at a supergeometric rate.

Continued fractions such as these can often be derived by a theorem of Stieltjes which relates continued fractions to orthogonal polynomials that satisfy a recurrence relation of the same type as the one given above. Another method of derivation is the Padé approximation, studied in the next section, that yields a rational function. Both techniques can be looked upon as a *convergence acceleration of an expansion into powers of* z *or* z^{-1}.

3.5.3 The Padé Table

Toward the end of the nineteenth century Frobenius and Padé developed a more general scheme for expanding a formal power series into rational functions, which we now describe. Let $f(z)$ be a formal power series

$$f(z) = c_0 + c_1 z + c_2 z^2 + \cdots = \sum_{i=0}^{\infty} c_i z^i. \tag{3.5.23}$$

Consider a complex rational form with numerator of degree at most m and denominator of degree at most n such that its power series expansion agrees with that of $f(z)$ as far as possible. Such a rational form is called an (m, n) Padé[119] approximation of $f(z)$.

[118] There are plenty of other notations for this function.

[119] Henri Eugène Padé (1863–1953), French mathematician and student of Charles Hermite, gave a systematic study of Padé forms in his thesis in 1892.

Definition 3.5.3.

The (m, n) Padé approximation of the formal power series $f(z)$ is, if it exists, defined to be a rational function

$$[m, n]_f(z) = \frac{P_{m,n}(z)}{Q_{m,n}(z)} \equiv \frac{\sum_{j=0}^{m} p_j z^j}{\sum_{j=0}^{n} q_j z^j} \tag{3.5.24}$$

that satisfies

$$f(z) - [m, n]_f(z) = Rz^{m+n+1} + O(z^{m+n+2}), \quad z \to 0. \tag{3.5.25}$$

The rational fractions $[m, n]_f$, $m, n \geq 0$ for $f(z)$ can be arranged in a doubly infinite array, called a **Padé table**.

$m \backslash n$	0	1	2	3	$\cdots$
0	$[0, 0]_f$	$[0, 1]_f$	$[0, 2]_f$	$[0, 3]_f$	$\cdots$
1	$[1, 0]_f$	$[1, 1]_f$	$[1, 2]_f$	$[1, 3]_f$	$\cdots$
2	$[2, 0]_f$	$[2, 1]_f$	$[2, 2]_f$	$[2, 3]_f$	$\cdots$
3	$[3, 0]_f$	$[3, 1]_f$	$[3, 2]_f$	$[3, 3]_f$	$\cdots$
$\vdots$	$\vdots$	$\vdots$	$\vdots$	$\vdots$	

The first column in the table contains the partial sums $\sum_{j=0}^{m} c_j z^j$ of $f(z)$.

Example 3.5.4.

The Padé approximants to the exponential function e^z are important because of their relation to methods for solving differential equations. The Padé approximants for $m, n = 0 : 2$ for the exponential function $f(z) = e^z$ are as follows.

$m \backslash n$	0	1	2
0	1	$\dfrac{1}{1 - z}$	$\dfrac{1}{1 - z + \frac{1}{2}z^2}$
1	$1 + z$	$\dfrac{1 + \frac{1}{2}z}{1 - \frac{1}{2}z}$	$\dfrac{1 + \frac{1}{3}z}{1 - \frac{2}{3}z + \frac{1}{6}z^2}$
2	$1 + z + \dfrac{1}{2}z^2$	$\dfrac{1 + \frac{2}{3}z + \frac{1}{6}z^2}{1 - \frac{1}{3}z}$	$\dfrac{1 + \frac{1}{2}z + \frac{1}{12}z^2}{1 - \frac{1}{2}z + \frac{1}{12}z^2}$

There may not exist a rational function that satisfies (3.5.25) for all (m, n). We may have to be content with $k < 1$. However, the closely related problem of finding $Q_{m,n}$ and $P_{m,n}(z)$ such that

$$Q_{m,n} f(z) - P_{m,n}(z) = O(z^{m+n+1}), \quad z \to 0, \tag{3.5.26}$$

always has a solution. The corresponding rational expression is called a Padé form of type (m, n).

Using (3.5.23) and (3.5.24) gives

$$\sum_{k=0}^{\infty} c_k z^k \sum_{j=0}^{n} q_j z^j = \sum_{i=0}^{m} p_i z^i + O(z^{m+n+1}).$$

Matching the coefficients of z^i, $i = 0 : m + n$, gives

$$\sum_{j=0}^{n} c_{i-j} q_j = \begin{cases} p_i & \text{if } i = 0 : m, \\ 0 & \text{if } i = m+1 : m+n, \end{cases} \tag{3.5.27}$$

where $c_i = 0$ for $i < 0$. This is $m + n + 1$ linear equations for the $m + n + 2$ unknowns $p_0, p_1, \dots, p_m, q_0, q_1, \dots, q_n$.

Theorem 3.5.4 (Frobenius).

There always exist Padé forms of type (m, n) for $f(z)$. Each such form is a representation of the same rational function $[m, n]_f$. A reduced representation is possible with $P_{m,n}(z)$ and $Q_{m,n}(z)$ relatively prime, $q_0 = 1$, and $p_0 = c_0$.

We now consider how to determine Padé approximants. With $q_0 = 1$ the last n linear equations in (3.5.27) are

$$\sum_{j=1}^{n} c_{i-j} q_j + c_i = 0, \qquad i = m+1 : m+n, \tag{3.5.28}$$

where $c_i = 0$, $i < 0$. The system matrix of this linear system is

$$C_{m,n} = \begin{pmatrix} c_m & c_{m-1} & \cdots & c_{m-n+1} \\ c_{m+1} & c_m & \cdots & c_{m-n+2} \\ \vdots & \vdots & \cdots & \vdots \\ c_{m+n-1} & c_{m+n-2} & \cdots & c_m \end{pmatrix}. \tag{3.5.29}$$

If $c_{m,n} = \det(C_{m,n}) \neq 0$, then the linear system (3.5.28) has a solution $q_1, \dots, q_n$. The coefficients $p_0, \dots, p_n$ of the numerator are then obtained from

$$p_i = \sum_{j=0}^{\min(i,n)} c_{i-j} q_j, \qquad i = 0 : m. \tag{3.5.30}$$

In the regular case $k = 1$ the error constant R in (3.5.25) is given by

$$R = p_i = \sum_{j=0}^{n} c_{i-j} q_j, \qquad i = m + n + 1.$$

Note that $[m, n]_f$ uses c_l for $l = 0 : m + n$ only; R uses c_{m+n+1} also. Thus, if c_l is given for $l = 0 : r$, then $[m, n]_f$ is defined for $m + n \leq r$, $m \geq 0$, $n \geq 0$.

If n is large, the heavy part of the computation of a Padé approximant

$$[m, n]_f(z) = P_{m,n}(z)/Q_{m,n}(z)$$

of $f(z)$ in (3.5.23) is the solution of the linear system (3.5.28). We see that if m or n is decreased by one, most of the equations of the system will be the same. There are therefore recursive relations between the polynomials $Q_{m,n}(z)$ for adjacent values of m, n, which can be used for computing any sequence of adjacent Padé approximants. These relations have been subject to intensive research that has resulted in several interesting algorithms; see the next section on the epsilon algorithm, as well as the monographs of Brezinski [50, 51] and the literature cited there.

There are situations where the linear system (3.5.28) is singular, i.e.,

$$c_{m,n} = \det(C_{m,n}) = 0.$$

We shall indicate how such singular situations can occur. These matters are discussed more thoroughly in Cheney [66, Chap. 5].

Example 3.5.5.

Let $f(z) = \cos z = 1 - \frac{1}{2}z^2 + \cdots$, set $m = n = 1$, and try to find

$$[1, 1]_f(z) = (p_0 + p_1 z)/(q_0 + q_1 z), \quad q_0 = 1.$$

The coefficient matching according to (3.5.27) yields the equations

$$p_0 = q_0, \qquad p_1 = q_1, \qquad 0 \cdot q_1 = -\frac{1}{2}q_0.$$

The last equation contradicts the condition that $q_0 = 1$. This single contradictory equation is in this case the "system" (3.5.28).

If this equation is ignored, we obtain

$$[1, 1]_f(z) = (1 + q_1 z)/(1 + q_1 z) = 1,$$

with error $\approx \frac{1}{2}z^2$, in spite of the fact that we asked for an error that is $O(z^{m+n+1}) = O(z^3)$. If we instead allow that $q_0 = 0$, then $p_0 = 0$, and we obtain a solution

$$[1, 1]_f(z) = z/z$$

which satisfies (3.5.26) but not (3.5.25). After dividing out the common factor z we get the same result $[1, 1]_f(z) = 1$ as before.

In a sense, this singular case results from a rather stupid request: we ask to approximate the even function $\cos z$ by a rational function where the numerator and the denominator end with odd powers of z. One should, of course, ask for the approximation by a rational function of z^2. What would you do if $f(z)$ is an *odd* function?

It can be shown that these singular cases occur in square blocks of the Padé table, where all the approximants are equal. For example, in Example 3.5.5 we will have $[0, 0]_f =$

$[0,1]_f = [1,0]_f = [1,1]_f = 1$. This property, investigated by Padé, is known as the *block structure of the Padé table*. For a proof of the following theorem, see Gragg [172].

Theorem 3.5.5.
Suppose that a rational function

$$r(z) = \frac{P(z)}{Q(z)},$$

where $P(z)$ and $Q(z)$ are relatively prime polynomials, occurs in the Padé table. Further suppose that the degrees of $P(z)$ and $Q(z)$ are m and n, respectively. Then the set of all places in the Padé table in which $r(z)$ occurs is a square block. If

$$Q(z)f(z) - P(z) = O(z^{m+n+r+1}), \tag{3.5.31}$$

then $r \geq 0$ and the square block consists of $(r+1)^2$ places

$$(m + r_1, n + r_2), \quad r_1, r_2 = 0, 1, \ldots, r.$$

An (m, n) Padé approximant is said to be **normal** if the degrees of $P_{m,n}$ and $Q_{m,n}$ are exactly m and n, respectively, and (3.5.31) holds with $r = 0$. The Padé table is called normal if every entry in the table is normal. In this case all the Padé approximants are different.

Theorem 3.5.6.
An (m, n) Padé approximant $[m,n]_f(z)$ is normal if and only if the determinants

$$\begin{matrix} c_{m,n}, & c_{m1,n+1}, \\ c_{m+1,n}, & c_{m+1,n+1} \end{matrix}$$

are nonzero.
A Padé table is normal if and only if

$$c_{m,n} \neq 0, \quad m, n = 0, 1, 2, \ldots.$$

In particular each Taylor coefficient $c_m, 1 = c_m$, must be nonzero.

Proof. See Gragg [172]. □

Imagine a case where $[m-1, n-1]_f(z)$ happens to be a more accurate approximation to $f(z)$ than usual; say that

$$[m-1, n-1]_f(z) - f(z) = O(z^{m+n+1}).$$

(For instance, let $f(z)$ be the ratio of two polynomials of degree $m-1$ and $n-1$, respectively.) Let b be an arbitrary number, and choose

$$Q_{m,n}(z) = (z+b)Q_{m-1,n-1}(z), \qquad P_{m,n}(z) = (z+b)P_{m-1,n-1}(z). \tag{3.5.32}$$

Then

$$[m,n]_f(z) = P_{m,n}(z)/Q_{m,n}(z)$$
$$= P_{m-1,n-1}(z)/Q_{m-1,n-1}(z) = [m-1,n-1]_f(z),$$

which is an $O(z^{m+n+1})$ accurate approximation to $f(z)$. Hence our request for this accuracy is satisfied by more than one pair of polynomials, $P_{m,n}(z)$, $Q_{m,n}(z)$, since b is arbitrary. This is impossible, unless the system (3.5.28) (that determines $Q_{m,n}$) is singular.

Numerically singular cases can occur in a natural way. Suppose that one wants to approximate $f(z)$ by $[m,n]_f(z)$, although already $[m-1,n-1]_f(z)$ would represent $f(z)$ as well as possible with the limited precision of the computer. In this case we must expect the system (3.5.28) to be very close to a singular system. A reasonable procedure for handling this is to compute the Padé forms for a sequence of increasing values of m, n, to estimate the condition numbers and to stop when it approaches the reciprocal of the machine unit. This illustrates a fact of some generality. *Unnecessary numerical trouble can be avoided by means of a well-designed termination criterion.*

For $f(z) = -\ln(1-z)$, we have $c_i = 1/i$, $i > 0$. When $m = n$ the matrix of the system (3.5.28) turns out to be the notorious Hilbert matrix (with permuted columns), for which the condition number grows exponentially like $0.014 \cdot 10^{1.5n}$; see Example 2.4.7. (The elements of the usual Hilbert matrix are $a_{ij} = 1/(i+j-1)$.)

There is a close connection between continued fractions and Padé approximants. Suppose that in a Padé table the staircase sequence

$$[0,0]_f, [1,0]_f, [1,1]_f, [2,1]_f, [2,2]_f, [3,2]_f, \dots$$

are all normal. Then there exists a regular continued fraction

$$1 + \frac{a_1 z}{1+}\,\frac{a_2 z}{1+}\,\frac{a_3 z}{1+}\dots, \qquad a_n \neq 0, \qquad n = 1,2,3,\dots,$$

with its nth convergent f_n satisfying

$$f_{2m} = [m,m]_f, \qquad f_{2m+1} = [m+1,m]_f, \qquad m = 0,1,2,\dots,$$

and vice versa. For a proof, see [214, Theorem 5.19].

Historically the theory of orthogonal polynomials, to be discussed later in Sec. 4.5.5. originated from certain types of continued fractions.

Theorem 3.5.7.

Let the coefficients of a formal power series (3.5.23) *be the moments*

$$c_n = \int_{-\infty}^{\infty} x^n w(x)\,dx,$$

where $w(t) \geq 0$. *Let* $Q_{m,n}$ *be the denominator polynomial in the corresponding Padé approximation* $[m,n]_f$. *Then the reciprocal polynomials*

$$Q^*_{n,n+1}(z) = z^{n+1} Q_{n,n+1}(1/z), \quad n \geq 0,$$

are the orthogonal polynomials with respect to the inner product

$$(f,g) = \int_{-\infty}^{\infty} f(x)g(x)w(x)\,dx.$$

Example 3.5.6.

The successive convergents of the continued fraction expansion in (3.5.3)

$$\frac{1}{2z}\log\left(\frac{1+z}{1-z}\right) = \frac{1}{1-}\,\frac{z^2}{3-}\,\frac{2^2z^2}{5-}\,\frac{3^2z^2}{7-}\cdots$$

are even functions and staircase Padé approximants. The first few are

$$s_{01} = \frac{3}{3-z^2}, \qquad s_{11} = \frac{15+4z^2}{3(5-3z^2)},$$
$$s_{12} = \frac{105-55z^2}{3(35-30z^2+3z^4)}, \qquad s_{22} = \frac{945-735z^2+64z^4}{15(63-70z^2+15z^4)}.$$

These Padé approximants can be used to evaluate $\ln(1+x)$ by setting $z = x/(2+x)$. The diagonal approximants s_{mm} are of most interest. For example, the approximation s_{22} matches the Taylor series up to the term z^8 and the error is approximately equal to the term $z^{10}/11$.

Chebyshev proved that the denominators in the above Padé approximants are the Legendre polynomials in $1/z$. These polynomials are orthogonal on $[-1,1]$ with respect to the uniform weight distribution $w(x) = 1$; see Sec. 4.5.5.

Explicit expressions for the Padé approximants for e^z were given by Padé (1892) in his thesis. They are

$$P_{m,n}(z) = \sum_{j=0}^{m} \frac{(m+n-j)!\,m!}{(m+n)!\,(m-j)!}\,\frac{z^j}{j!}, \tag{3.5.33}$$

$$Q_{m,n}(z) = \sum_{j=0}^{n} \frac{(m+n-j)!\,n!}{(m+n)!\,(n-j)!}\,\frac{(-z)^j}{j!}, \tag{3.5.34}$$

with the error

$$e^z - \frac{P_{m,n}(z)}{Q_{m,n}(z)} = (-1)^n \frac{m!n!}{(m+n)!(m+n+1)!} z^{m+n+1} + O(z^{m+n+2}). \tag{3.5.35}$$

Note that $P_{m,n}(z) = Q_{n,m}(-z)$, which reflects the property that $e^{-z} = 1/e^z$. Indeed, the numerator and denominator polynomials can be shown to approximate (less accurately) $e^{z/2}$ and $e^{-z/2}$, respectively.

There are several reasons for preferring the diagonal Padé approximants ($m = n$) for which

$$p_j = \frac{(2m-j)!\,m!}{(2m)!\,(m-j)!j!}, \qquad q_j = (-1)^j p_j, \quad j = 0:m. \tag{3.5.36}$$

These coefficients satisfy the recursion

$$p_0 = 1, \quad p_{j+1} = \frac{(m-j)p_j}{(2m-j)(j+1)}, \quad j = 0 : m-1. \tag{3.5.37}$$

For the diagonal Padé approximants the error $R_{m,n}(z)$ satisfies $|R_{m,n}(z)| < 1$, for $\Re z < 0$. This is an important property in applications for solving differential equations.[120] To evaluate a diagonal Padé approximant of even degree we write

$$\begin{aligned} P_{2m,2m}(z) &= p_{2m}z^{2m} + \cdots + p_2 z^2 + p_0 \\ &\quad + z(p_{2m-1}z^{2m-2} + \cdots + p_3 z^2 + p_1) = u(z) + v(z) \end{aligned}$$

and evaluate $u(z)$ and $v(z)$ separately. Then $Q_{2m}(z) = u(z) - v(z)$. A similar splitting can be used for an odd degree.

Recall that in order to compute the exponential function a range reduction should first be performed. If an integer k is determined such that

$$z^* = z - k\ln 2, \quad |z^*| \in [0, \ln 2], \tag{3.5.38}$$

then $\exp(z) = \exp(z^*) \cdot 2^k$. Hence only an approximation of $\exp(z)$ for $|z| \in [0, \ln 2]$ is needed; see Problem 3.5.6.

The problem of convergence of a sequence of Padé approximants when at least one of the degrees tends to infinity is a difficult problem and outside the scope of this book. Padé proved that for the exponential function the poles of the Padé approximants $[m_i, n_i]_f$ tend to infinity when $m_i + n_i$ tends to infinity and

$$\lim_{i\to\infty} [m_i, n_i]_f(z) = e^z$$

uniformly on any compact set of $\mathbf{C}$. For a survey of other results, see [54].

3.5.4 The Epsilon Algorithm

One extension of the Aitken acceleration uses a comparison series with terms of the form

$$c_j = \sum_{\nu=1}^{p} \alpha'_\nu k_\nu^j, \quad j \geq 0, \quad k_\nu \neq 0. \tag{3.5.39}$$

Here α'_ν and k_ν are $2p$ parameters, to be determined, in principle, by means of c_j, $j = 0 : 2p-1$. The parameters may be complex. The power series becomes

$$S(z) = \sum_{j=0}^{\infty} c_j z^j = \sum_{\nu=1}^{p} \alpha'_\nu \sum_{j=0}^{\infty} k_\nu^j z^j = \sum_{\nu=1}^{p} \frac{\alpha'_\nu}{1 - k_\nu z},$$

which is a rational function of z, and thus related to Padé approximation. Note, however, that the poles at k_ν^{-1} should be simple and that $m < n$ for $S(z)$, because $S(z) \to 0$ as $z \to \infty$.

[120] Diagonal Padé approximants are also used for the evaluation of the matrix exponential e^A, $A \in \mathbf{R}^{n\times n}$; see Volume II, Chapter 9.

Recall that the calculations for the Padé approximation determine the coefficients of $S(z)$ *without calculating the* $2n$ *parameters* α'_ν *and* k_ν. It can happen that m becomes larger than n, and if α'_ν and k_ν are afterward determined, by the expansion of $S(z)$ into partial fractions, it can turn out that some of the k_ν are multiple poles. This suggests a generalization of this approach, but how?

If we consider the coefficients q_j, $j = 1 : n$, occurring in (3.5.28) as known quantities, then (3.5.28) can be interpreted as a *linear difference equation*.[121] The general solution of this is given by (3.5.39) if the zeros of the polynomial

$$Q(x) := 1 + \sum_{j=1}^{n} q_j x^j$$

are simple. If multiple roots are allowed, the general solution is, by Theorem 3.3.13 (after some change of notation),

$$c_l = \sum_\nu p_\nu(l) k_\nu^n,$$

where k_ν runs through the different zeros of $Q(x)$ and p_ν is an arbitrary polynomial, the degree of which equals the multiplicity -1 of the zero k_ν. Essentially the same mathematical relations occur in several areas of numerical analysis, such as interpolation and approximation by a sum of exponentials (Prony's method), and in the design of quadrature rules with free nodes (see Sec. 5.3.1).

Shanks [322] considered the sequence transformation

$$e_k(s_n) = \frac{\begin{vmatrix} s_n & s_{n+1} & \cdots & s_{n+k} \\ s_{n+1} & s_{n+2} & \cdots & s_{n+k+1} \\ \vdots & \vdots & \cdots & \vdots \\ s_{n+k} & s_{n+k+1} & \cdots & s_{n+2k} \end{vmatrix}}{\begin{vmatrix} \Delta^2 s_n & \cdots & \Delta^2 s_{n+k-1} \\ \vdots & \cdots & \vdots \\ \Delta^2 s_{n+k-1} & \cdots & \Delta^2 s_{n+2k-2} \end{vmatrix}}, \quad k = 1, 2, 3, \ldots, \tag{3.5.40}$$

and proved that it is exact if and only if the values s_{n+i} satisfy a linear difference equation

$$a_0(s_n - a) + \cdots + a_k(s_{n+k} - a) = 0 \ \forall \ n, \tag{3.5.41}$$

with $a_0 a_k \neq 0, a_0 + \cdots + a_k \neq 0$. For $k = 1$, Shanks' transformation reduces to Aitken's Δ^2 process (the proof is left as Problem 3.5.7). The **Hankel determinants**[122] in the definition of $e_k(s_n)$ satisfy a five-term recurrence relationship, which can be used for implementing the transformation.

[121] This can also be expressed in terms of the z-transform; see Sec. 3.3.5.

[122] A matrix with constant elements in the antidiagonals is called a Hankel matrix, after Hermann Hankel (1839–1873), German mathematician. In his thesis [185] he studied determinants of the class of matrices now named after him.

Here we are primarily interested in the use of Padé approximants as a convergence accelerator in the *numerical* computation of values of $f(z)$ for (say) $z = e^{i\phi}$. A natural question is then whether it is possible to omit the calculation of the coefficients p_j, q_j and find a recurrence relation that gives the function values directly. A very elegant solution to this problem, called the **epsilon algorithm**, was found in 1956 by Wynn [384], after complicated calculations. We shall present the algorithm, but refer to the survey paper by Wynn [386] for proof and more details.

A two-dimensional array of numbers $\epsilon_k^{(n)}$ is computed by the nonlinear recurrence relation,

$$\epsilon_{k+1}^{(p)} = \epsilon_{k-1}^{(p+1)} + \frac{1}{\epsilon_k^{(p+1)} - \epsilon_k^{(p)}}, \qquad p, k = 0, 1, \dots, \tag{3.5.42}$$

which involves four quantities in a rhombus:

$$\begin{array}{ccc} & \epsilon_k^{(p)} & \\ \epsilon_{k-1}^{(p+1)} & & \epsilon_{k+1}^{(p)}. \\ & \epsilon_k^{(p+1)} & \end{array}$$

The sequence transformation of Shanks can be computed by using the boundary conditions $\epsilon_{-1}^{(p)} = 0$, $\epsilon_0^{(p)} = s_p$ in the epsilon algorithm. Then

$$\epsilon_{2k}^{(p)} = e_k(s_p), \qquad \epsilon_{2k+1}^{(p)} = 1/e_k(\Delta s_p), \qquad p = 0, 1, \dots;$$

i.e., the ϵ's with even lower index give the sequence transformation (3.5.40) of Shanks. The ϵ's with odd lower index are auxiliary quantities only.

The epsilon algorithm transforms the partial sums of a series into its Padé quotients or, equivalently, is a process by means of which a series may be transformed into the convergents of its associated and corresponding continued fractions. It is a quite powerful all-purpose acceleration process for slowly converging sequences and usually fully exploits the numerical precision of the data. For an application to numerical quadrature, see Example 5.2.3.

If the boundary conditions

$$\epsilon_{-1}^{(p)} = 0, \qquad \epsilon_0^{(p)} = r_{p,0}(z) = \textstyle\sum_{j=0}^{p} c_j z^j \tag{3.5.43}$$

are used in the epsilon algorithm, this yields for *even* subscripts

$$\epsilon_{2n}^{(p)} = r_{p+n,n}(z) \tag{3.5.44}$$

Thus the epsilon algorithm can be used to compute recursively the lower half of the Padé table. The upper half can be computed by using the boundary conditions

$$\epsilon_{2n}^{(-n)} = r_{0,n}(z) = \frac{1}{\sum_{j=0}^{n} d_j z^j}. \tag{3.5.45}$$

The polynomials $r_{0,n}(z)$ are obtained from the Taylor expansion of $1/f(z)$. Several procedures for obtaining this were given in Sec. 3.1.

It seems easier to program this application of the ϵ-algorithm after a slight change of notation. We introduce an $r \times 2r$ matrix $A = [a_{ij}]$, where

$$a_{ij} = \epsilon_k^{(p)}, \quad k = j - 2, \quad p = i - j + 1.$$

Conversely, $i = k + p + 1$, $j = k + 2$. The ϵ algorithm, together with the boundary conditions, now takes the following form:

for $i = 1 : r$
  $a_{i,1} = 0; \quad a_{i,2} = r_{i-1,0}(z); \quad a_{i,2i} = r_{0,i-1}(z);$
  for $j = 2 : 2(i-1)$
    $a_{i,j+1} = a_{i-1,j-1} + 1/(a_{ij} - a_{i-1,j})$.
  end
end

Results:

$$[m, n]_f(z) = a_{m+n+1,2n+2}, \quad (m, n \ge 0, \quad m + n + 1 \le r).$$

The above program sketch must be improved for practical use. For example, something should be done about the risk for a division by zero.

3.5.5 The qd Algorithm

Let $\{c_n\}$ be a sequence of real or complex numbers and

$$C(z) = c_0 + c_1 z + c_2 z^2 + \cdots \tag{3.5.46}$$

the formal power series formed with these coefficients. The **qd algorithm** of Rutishauser [310][123] forms from this sequence a two-dimensional array, similar to a difference scheme, by alternately taking difference and quotients as follows. Take as initial conditions

$$e_0^{(n)} = 0, \quad n = 1, 2, \ldots, \qquad q_1^{(n)} = \frac{c_{n+1}}{c_n}, \quad n = 0, 1, \ldots, \tag{3.5.47}$$

and form the **quotient-difference scheme**, or qd scheme:

$$\begin{array}{cccccccc}
 & q_1^{(0)} & & & & & \\
0 & & e_1^{(0)} & & & & \\
 & q_1^{(1)} & & q_2^{(0)} & & & \\
0 & & e_1^{(1)} & & e_2^{(0)} & & \\
 & q_1^{(2)} & & q_2^{(1)} & & q_3^{(0)} & , \\
0 & & e_1^{(2)} & & e_2^{(1)} & & \\
 & \vdots & & q_2^{(2)} & & \vdots & \\
 & & \vdots & & \vdots & &
\end{array}$$

[123] Heinz Rutishauser (1912–1970) was a Swiss mathematician, a pioneer in computing, and the originator of many important algorithms. The qd algorithm has had great impact in eigenvalue calculations.

where the quantities are connected by the two **rhombus rules**

$$e_m^{(n)} = q_m^{(n+1)} - q_m^{(n)} + e_{m-1}^{(n+1)}, \tag{3.5.48}$$

$$q_{m+1}^{(n)} = \frac{e_m^{(n+1)}}{e_m^{(n)}} q_m^{(n+1)}, \quad m = 1, 2, \dots, \quad n = 0, 1, \dots. \tag{3.5.49}$$

Each of the rules connects four adjacent elements of the qd scheme. The first rule states that in any rhombus-like configuration of four elements centered in a q-column the sum of the two NE and the two SW elements are equal. Similarly, the second rule states that in any rhombus-like configuration in an e-column the product of the two NE and the two SW elements are equal.

The initial conditions (3.5.47) give the first two columns in the qd scheme. The remaining elements in the qd scheme, if it exists, can then be generated column by column using the rhombus rules. Note the computations break down if one of the denominators in (3.5.49) is zero. If one of the coefficients c_n is zero even the very first q-column fails to exist.

The rhombus rules are based on certain identities between Hankel determinants, which we now describe. These also give conditions for the existence of the qd scheme. The Hankel determinants associated with the formal power series (3.5.46) are, for arbitrary integers n and $k \geq 0$, defined by $H_0^{(n)} = 1$,

$$H_k^{(n)} = \begin{vmatrix} c_n & c_{n+1} & \cdots & c_{n+k-1} \\ c_{n+1} & c_{n+2} & \cdots & c_{n+k} \\ \vdots & \cdots & \cdots & \vdots \\ c_{n+k-1} & c_{n+k-2} & \cdots & c_{n+2k-2} \end{vmatrix}, \quad k > 1, \tag{3.5.50}$$

where $c_k = 0$ for $k < 0$. This definition is valid also for negative values of n. Note, however, that if $n + k \leq 0$, then the entire first row of $H_k^{(n)}$ is zero, i.e.,

$$H_k^{(n)} = 0, \quad k \leq -n. \tag{3.5.51}$$

A formal power series is called **normal** if its associated Hankel determinants $H_m^{(n)} \neq 0$ for all $m, n \geq 0$; it is called k-normal if $H_m^{(n)} \neq 0$ for $m = 0 : k$ and for all $n \geq 0$.

In the following theorem (Henrici [196, Theorem 7.6a]) the elements in the qd scheme can be expressed in terms of Hankel determinants.

Theorem 3.5.8.

Let $H_k^{(n)}$ be the Hankel determinants associated with a formal power series $C = c_0 + c_1 z + c_2 z^2 + \cdots$. If there exists a positive integer k such that the series is k-normal, the columns $q_m^{(n)}$ of the qd scheme associated with C exist for $m = 1 : k$, and

$$q_m^{(n)} = \frac{H_m^{(n+1)} H_{m-1}^{(n)}}{H_m^{(n)} H_{m-1}^{(n+1)}}, \qquad e_m^{(n)} = \frac{H_{m+1}^{(n)} H_{m-1}^{(n+1)}}{H_m^{(n)} H_m^{(n+1)}} \tag{3.5.52}$$

for $m = 1 : k$ and all $n \geq 0$.

The above result is related to Jacobi's identity for Hankel matrices, that for all integers n and $k \geq 1$,

$$(H_k^{(n)})^2 - H_k^{(n-1)} H_k^{(n+1)} + H_{k+1}^{(n-1)} H_{k-1}^{(n+1)} = 0. \tag{3.5.53}$$

This identity can be derived from the following very useful determinant identity.

Theorem 3.5.9 (*Sylvester's Determinant Identity*).
Let $\hat{A} \in \mathbf{C}^{n\times n}$, $n \geq 2$, *be partitioned:*

$$\hat{A} = \begin{pmatrix} \alpha_{11} & a_1^T & \alpha_{12} \\ \hat{a}_{11} & A & \hat{a}_2 \\ \alpha_{21} & a_2^T & \alpha_{22} \end{pmatrix} = \begin{pmatrix} A_{11} & * \\ * & \alpha_{22} \end{pmatrix} = \begin{pmatrix} * & A_{12} \\ \alpha_{21} & * \end{pmatrix}$$
$$= \begin{pmatrix} * & \alpha_{12} \\ A_{21} & * \end{pmatrix} = \begin{pmatrix} \alpha_{11} & * \\ * & A_{21} \end{pmatrix}.$$

Then we have the identity

$$\det(A) \cdot \det(\hat{A}) = \det(A_{11}) \cdot \det(A_{22}) - \det(A_{21}) \cdot \det(A_{12}). \qquad (3.5.54)$$

Proof. If the matrix A is square and nonsingular, then

$$\det(A_{ij}) = \pm \det \begin{pmatrix} A & \hat{a}_j \\ a_i^T & \alpha_{ij} \end{pmatrix} = \pm \det(A) \cdot (\alpha_{ij} - a_i^T A^{-1} \hat{a}_j), \qquad (3.5.55)$$

with negative sign only possible if $i \neq j$. Then, similarly,

$$\det(A) \cdot \det(\hat{A}) = \det(A) \cdot \det \begin{pmatrix} A & \hat{a}_{11} & \hat{a}_2 \\ a_1^T & \alpha_{11} & \alpha_{12} \\ a_2^T & \alpha_{21} & \alpha_{22} \end{pmatrix}$$
$$= (\det(A))^2 \cdot \det \left[\begin{pmatrix} \alpha_{11} & \alpha_{12} \\ \alpha_{21} & \alpha_{22} \end{pmatrix} - \begin{pmatrix} a_1^T \\ a_2^T \end{pmatrix} A^{-1} (\, \hat{a}_1 \quad \hat{a}_2 \,) \right]$$
$$= \det \begin{pmatrix} \beta_{11} & \beta_{12} \\ \beta_{21} & \beta_{22} \end{pmatrix},$$

where $\beta_{ij} = \alpha_{ij} - a_i^T A^{-1} \hat{a}_j$. Using (3.5.55) gives (3.5.54), which holds even when A is singular. □

If the Hankel determinants $H_k^{(n)}$ are arranged in a triangular array,

$$\begin{array}{lllll} 1 & & & & \\ 1 & H_1^{(0)} = c_0 & & & \\ 1 & H_1^{(1)} = c_1 & H_2^{(0)} & & \\ 1 & H_1^{(2)} = c_2 & H_2^{(1)} & H_3^{(0)} & \\ 1 & H_1^{(3)} = c_3 & H_2^{(2)} & H_3^{(1)} & H_4^{(0)} \end{array},$$

then Jacobi's identity links together the entries in a star-like configuration. Since the two first columns are trivial, (3.5.53) may be used to calculate the Hankel determinants recursively from left to right. Further properties of Hankel determinants are given in Henrici [196, Sec. 7.5].

We state without proof an important analytical property of the Hankel determinants that shows how the poles of a meromorphic[124] function can be determined from the coefficients of its Taylor expansion at $z = 0$.

Theorem 3.5.10.

Let $f(z) = c_0 + c_1 z + c_2 z^2 + \cdots$ be the Taylor series of a function meromorphic in the disk $D : |z| < \sigma$ and let the poles $z_i = u_i^{-1}$ of f in D be numbered such that

$$0 < |z_1| \le |z_2| \le \cdots < \sigma.$$

Then for each m such that $|z_m| < |z_{m+1}|$, if n is sufficiently large, $H_m^{(n)} \neq 0$, and

$$\lim_{n\to\infty} H_m^{(n+1)}/H_m^{(n)} = u_1 u_2 \cdots u_m. \tag{3.5.56}$$

In the special case that f is a rational function with a pole of order p at infinity and the sum of orders of all its finite poles is k, then

$$H_k^{(n)} = C_k (u_1 u_2 \cdots u_k)^n, \quad n > p, \tag{3.5.57}$$

where $C_k \neq 0$; furthermore $H_m(n) = 0$, $n > p$, $m > k$.

Proof. The result is a corollary of Theorem 7.5b in Henrici [196]. □

The above results are related to the qd scheme as follows; see Henrici [196, Theorem 7.6b].

Theorem 3.5.11.

Under the hypothesis of Theorem 3.5.10 and assuming that the Taylor series at $z = 0$ is ultimately k-normal for some integer $k > 0$, the qd scheme for f has the following properties:

(a) *For each m such that $0 < m \le k$ and $|z_{m-1}| < |z_m| < |z_{m+1}|$,*

$$\lim_{n\to\infty} q_m^{(n)} = u_m;$$

(b) *For each m such that $0 < m \le k$ and $|z_m| < |z_{m+1}|$,*

$$\lim_{n\to\infty} e_m^{(n)} = 0.$$

From the above results it seems that, under certain restrictions, an algorithm for *simultaneously* computing all the poles of a meromorphic function f directly from its Taylor series at the origin could be constructed, where the qd scheme is computed from left to right. Any q-column corresponding to a simple pole of isolated modulus would tend to the reciprocal value of that pole. The e-columns on both sides would tend to zero. If f is rational, the last e-column would be zero, which could serve as a test of accuracy.

[124] A function which is analytic in a region Ω, except for poles, is said to be meromorphic in Ω.

Unfortunately, as outlined, this algorithm is unstable, i.e., oversensitive to rounding errors, and useless numerically. This fact is related to the occurrence in (3.5.49) of a division of two small quantities, which can have large relative errors. (Recall that e-columns tends to zero.)

A more stable way of constructing the qd scheme is obtained by writing the rhombus rules as

$$q_m^{(n+1)} = \left[e_m^{(n)} - e_{m-1}^{(n+1)}\right] + q_m^{(n)}, \tag{3.5.58}$$

$$e_m^{(n+1)} = \frac{q_{m+1}^{(n)}}{q_m^{(n+1)}} e_m^{(n)}. \tag{3.5.59}$$

Written in this form, the rules can be used to *construct the qd scheme row by row*. The problem now is how to start the algorithm. As seen from the scheme below, to do this it suffices to know *the first two rows of q's and e's*. This, together with the first column of zeros, allows us to proceed along diagonals slanted SW; see scheme below.

$$\begin{array}{cccccccc}
 & q_1^{(0)} & & q_2^{(-1)} & & q_3^{(-2)} & & \cdots \\
0 & & e_1^{(0)} & & e_2^{(-1)} & & e_3^{(-2)} & \cdots \\
 & \times & & \times & & \times & & \\
0 & & \times & & \times & & & \\
 & \times & & \times & & & & \\
0 & & \times & & & & & \\
 & \times & & & & & & \\
0 & & & & & & &
\end{array}$$

This is called the **progressive form** of the qd algorithm. The starting values $q_m^{(n)}$ and $e_m^{(n)}$ for negative values of n can be computed from the relations (3.5.52). In this form the qd algorithm can be used to simultaneously determine the zeros of a polynomial; see Sec. 6.5.4.

The qd algorithm is related to Padé approximants. Consider a continued fraction of the form

$$c(z) = \frac{a_1}{1+}\,\frac{a_2 z}{1+}\,\frac{a_3 z}{1+}\cdots. \tag{3.5.60}$$

The nth approximant

$$w_n(z) = P_n(z)/Q_n(z), \quad n = 1, 2, \dots, \tag{3.5.61}$$

is the finite continued fraction obtained by setting $a_{n+1} = 0$. In the special case that all $a_i > 0$, the continued fraction is called a Stieltjes fraction.[125] The sequence of numerators $\{P_n(z)\}$ and denominators $\{Q_n(z)\}$ in (3.5.61) satisfy the recurrence relations

$$\begin{aligned} P_0 = 0, \quad P_1 = 1, &\qquad P_n = z a_n P_{n-2} + P_{n-1}, \\ Q_0 = Q_1 = 1, &\qquad Q_n = z a_n Q_{n-2} + Q_{n-1}, \quad n \geq 2. \end{aligned}$$

[125] The theory of such fractions was first expounded by Stieltjes in a famous memoir which appeared in 1894, the year of his death.

Hence both P_n and Q_n are polynomials in z of degree $\lfloor (n-1)/2 \rfloor$ and $\lfloor n/2 \rfloor$, respectively. It can be shown that the polynomials P_n and Q_n have no common zero for $n = 1, 2, \ldots$.

From the initial conditions and recurrence relations it follows that $Q_n(0) = 1$, $n = 0, 1, 2, \ldots$. Hence the rational function $w_n(z) = P_n(z)/Q_n(z)$ is analytic at $z = 0$. Hence it can be expanded in a Taylor series

$$\frac{P_n(z)}{Q_n(z)} = c_0^{(n)} + c_1^{(n)} z + c_2^{(n)} z^2 + \cdots \tag{3.5.62}$$

that converges for z sufficiently small. The coefficients $c_k^{(n)}$ in (3.5.62) can be shown to be independent of n for $k < n$. We denote by $c_k := c_k^{(n+1)}$ the ultimate value of $c_k^{(n)}$ for increasing values n and let

$$C(z) = c_0 + c_1 z + c_2 z^2 + \cdots \tag{3.5.63}$$

be the formal power series formed with these coefficients. Then the power series $C(z)$ and the fraction $c(z)$ are said to **correspond** to each other. Note that the formal power series $C(z)$ corresponding to a given fraction $c(z)$ converges for any $z \neq 0$.

The qd algorithm can be used to solve the following problem: Given a (formal) power series $C(z) = c_0 + c_1 z + c_2 z^2 + \cdots$, find a continued fraction $c(z)$ of the form (3.5.60) corresponding to it. Note that we do not require that the formal power series corresponding to the continued fraction converges, merely that the nth approximant w_n of the continued fraction satisfies

$$C(z) - w_n(z) = O(z^n).$$

Theorem 3.5.12 (*Henrici* [197, Theorem 12.4c]).

Given a formal power series $C(z) = c_0 + c_1 z + c_2 z^2 + \cdots$, there exists at most one corresponding continued fraction of the form

$$\frac{a_0}{1-}\,\frac{a_1 z}{1-}\,\frac{a_2 z}{1-}\,\frac{a_3 z}{1-}\,\frac{4z}{1-}\cdots.$$

There exists precisely one such fraction if and only if the Hankel determinants satisfy $H_k{}^{(n)} \neq 0$ for $n = 0, 1$ and $k = 1, 2, \ldots$. If $q_k^{(n)}$ and $e_k^{(n)}$ are the elements of the qd scheme associated with C, then

$$\frac{c_0}{1-}\,\frac{q_1^{(0)} z}{1-}\,\frac{e_1^{(0)} z}{1-}\,\frac{q_2^{(0)} z}{1-}\,\frac{e_2^{(0)} z}{1-}\cdots. \tag{3.5.64}$$

Conversely, this shows that knowing the coefficients of the continued fraction corresponding to f allows us to compute the qd scheme starting from the first diagonal and proceeding in the SW direction. This is called the **progressive** qd algorithm.

Example 3.5.7.

For the power series

$$c(z) = 0! + 1!z + 2!z^2 + 3!z^3 + \cdots,$$

we obtain using the rhombus rules (3.5.48)–(3.5.49) the qd scheme

$$
\begin{array}{ccccccc}
 & 1 & & & & & \\
0 & & 1 & & & & \\
 & 2 & & 2 & & & \\
0 & & 1 & & 2 & & \\
 & 3 & & 3 & & 3 & \\
0 & & 1 & & 2 & & \ddots \\
 & 4 & & 4 & & \ddots & \\
0 & & 1 & & \ddots & & \\
 & 5 & & \ddots & & &
\end{array}
$$

Hence the corresponding continued fraction is

$$
c(z) = \frac{1}{1+}\,\frac{z}{1+}\,\frac{z}{1+}\,\frac{2z}{1+}\,\frac{2z}{1+}\,\frac{3z}{1+}\,\frac{3z}{1+}\cdots.
$$

Review Questions

3.5.1 Define a continued fraction. Show how the convergents can be evaluated either backward or forward.

3.5.2 Show how any positive number can be expanded into a continued fraction with integer elements. In what sense are the convergents the best approximations? How accurate are they?

3.5.3 What is the Padé table? Describe how the Padé approximants can be computed, if they exist. Tell something about singular and almost singular situations that can be encountered, and how to avoid them.

3.5.4 Describe the ϵ algorithm, and tell something about its background.

3.5.5 What are the rhombus rules for the qd algorithm? What is the difference between the standard and the progressive qd algorithm?

3.5.6 Sketch how the qd algorithm, under some restrictions, can be used to compute the zeros of a polynomial. Give necessary conditions for this to work. What governs the rate of convergence?

Problems and Computer Exercises

3.5.1 (a) Write a program for the algorithm of best rational approximations to a real number in Sec. 3.5.1.

Apply it to find a few coefficients of the continued fractions for

$$\tfrac{1}{2}(\sqrt{5}+1), \quad \sqrt{2}, \quad e, \quad \pi, \quad \tfrac{\log 2}{\log 3}, \quad 2^{j/12}$$

for a few integers j, $1 \le j \le 11$.

(b) Check the accuracy of the convergents. What happens when you apply your program to a rational number, e.g., 729/768?

(c) The metonic cycle used for calendrical purposes by the Greeks consists of 235 lunar months, which nearly equal 19 solar years. Show, using the algorithm in Sec. 3.5.1, that $235/19$ is the sixth convergent of the ratio $365.2495/29.53059$ of solar period and the lunar phase (synodic) period.

3.5.2 *A matrix formalism for continued fractions.*

(a) We use the same notations as in Sec. 3.5.1, but set, with no loss of generality, $b_0 = 0$. Set

$$P(n) = \begin{pmatrix} p_{n-1} & p_n \\ q_{n-1} & q_n \end{pmatrix}, \qquad A(n) = \begin{pmatrix} 0 & a_n \\ 1 & b_n \end{pmatrix}.$$

Show that $P(0) = I$,

$$P(n) = P(n-1)A(n), \qquad P(n) = A(1)A(2)\cdots A(n-1)A(n), \qquad n \ge 1.$$

Comment: This does not minimize the number of arithmetic operations but, in a matrix-oriented programming language, it often gives very simple programs.

(b) Write a program for this with some termination criterion and test it on a few cases, such as

$$1 + \frac{1}{1+}\,\frac{1}{1+}\,\frac{1}{1+}\cdots, \qquad 2 + \frac{1}{3+}\,\frac{1}{2+}\,\frac{1}{3+}\,\frac{1}{2+}\,\frac{1}{3+}\cdots, \qquad 2 + \frac{2}{2+}\,\frac{3}{3+}\,\frac{4}{4+}\cdots.$$

As a postprocessing, apply Aitken acceleration in the first two cases in order to obtain a very high accuracy. Does the result look familiar in the last case? See Problem 3.5.3 concerning the exact results in the two other cases.

(c) Write a version of the program with some strategy for scaling $P(n)$ in order to eliminate the risk of overflow and underflow.

Hint: Note that the convergents $x_n = p_n/q_n$ are unchanged if you multiply the $P(n)$ by arbitrary scalars.

(d) Use this matrix form for working out a short proof of (3.5.7).

Hint: What is the determinant of a matrix product?

3.5.3 (a) Explain that $x = 1 + 1/x$ for the continued fraction in (3.5.13).

(b) Compute the periodic continued fraction

$$2 + \frac{1}{3+}\,\frac{1}{2+}\,\frac{1}{3+}\,\frac{1}{2+}\,\frac{1}{3+}\cdots$$

exactly (by paper and pencil). (The convergence is assured by Seidel's theorem (Theorem 3.5.2).)

(c) Suggest a generalization of (a) and (b), where you can always obtain a quadratic equation with a positive root.

(d) Show that

$$\frac{1}{\sqrt{x^2-1}} = \frac{1}{x-}\,\frac{\frac{1}{2}}{x-y}, \qquad \text{where} \quad y = \frac{\frac{1}{4}}{x-}\,\frac{\frac{1}{4}}{x-}\,\frac{\frac{1}{4}}{x-}\cdots.$$

3.5.4 (a) Prove the equivalence transformation (3.5.8). Show that the errors of the convergents have alternating signs if the elements of the continued fraction are positive.

(b) Show how to bring a general continued fraction to the special form of equation (3.5.12).

3.5.5 Show that the (1, 1) Padé approximant of $\sqrt{1+x}$ equals $(4+3x)/(4+x)$. What is the (2, 2) Padé approximant?

3.5.6 Let $P_{m,m}(z)/Q_{m,m}(z)$ be the diagonal Padé approximants of the exponential function.

(a) Show that the coefficients for $P_{m,m}(z)$ satisfy the recursion

$$p_0 = 1, \qquad p_{j+1} = \frac{m-j}{(2m-j)(j+1)} p_j, \quad j = 0 : m-1. \tag{3.5.65}$$

(b) Show that for $m = 6$ we have

$$P_{6,6}(z) = 1 + \frac{1}{2}z + \frac{5}{44}z^2 + \frac{1}{66}z^3 + \frac{1}{792}z^4 + \frac{1}{15{,}840}z^5 + \frac{1}{665{,}280}z^6$$

and $Q_{6,6}(z) = P_{6,6}(-z)$. How many operations are needed to evaluate this approximation for a given z?

(c) Use the error estimate in (3.5.35), neglecting higher-order terms, to compute a bound for the relative error of the approximation in (b) when $|z| \in [0, \ln 2]$. What degree of the diagonal Padé approximant is needed for the relative error to be of the order of the unit roundoff $2^{-53} = 1.11 \cdot 10^{-16}$ in IEEE double precision arithmetic?

3.5.7 For $k = 1$, Shanks' sequence transformation (3.5.40) becomes

$$e_1(s_n) = \begin{vmatrix} s_n & s_{n+1} \\ s_{n+1} & s_{n+2} \end{vmatrix} \Big/ \Delta^2 s_n.$$

Show that this is mathematically equivalent to the result s'_{n+2} from Aitken extrapolation. Why is the direct use of the above expression not safe numerically?

3.5.8 (a) Write a program for computing a Padé approximant and its error term. Apply it (perhaps after a transformation) for various values of m, n to, e.g., e^z, $\arctan z$, $\tan z$. (Note that two of these examples are odd functions.) Use the algorithm of Sec. 3.5.1 for expressing the coefficients as rational numbers. For how large m and n can you use your program (in these examples) without severe trouble with rounding errors?

(b) Let m be an odd number. Try to transform the $(m, m+1)$ Padé approximants of $\arctan z$ and $\tan z$ to continued fractions of the form given in Sec. 3.5.1.

(c) Try to determine for which other functions the Padé table has a similar symmetry as shown in the text for the exponential function e^z.

3.5.9 (a) Show that there is at most one rational function $R(z)$, where the degrees of the numerator and denominator do not exceed, respectively, m and n such that

$$f(z) - R(z) = O(z^{m+n+1}) \quad \text{as} \quad z \to 0,$$

even if the system (3.5.28) is singular. (Note, however, that P_m and Q_n are not uniquely determined if the system is singular; they have common factors.)

(b) Is it true that if $f(z)$ is a rational function of degrees m', n', then

$$[m, n]_f(z) = f(z) \quad \forall\, m \geq m', \quad n \geq n'?$$

3.5.10 Write a program for evaluating the incomplete gamma function. Use the continued fraction (3.5.22) for x greater than about $a + 1$. For x less than about $a + 1$ use the power series for $\gamma(a, x)$.

3.5.11 Compute the infinite sum $1 - \frac{1}{3} + \frac{1}{5} - \frac{1}{7} + \frac{1}{9} - \cdots$ with the epsilon algorithm, and estimate (empirically) the speed of convergence.

3.5.12 Write a program for determining the zeros of a polynomial $p(z)$ of degree n with simple positive zeros. Test it by computing the zeros of some orthogonal polynomials. Discuss how you can shift the zeros so that convergence to a particular zero is enhanced.

Notes and References

Much work on approximations to special functions, for example, the Gauss hypergeometric function and the Kummer function, was done around the end of World War II. A most comprehensive source of information on useful mathematical functions and formulas is the Handbook first published in 1964 by the National Bureau of Standards (renamed National Institute of Standards and Technology (NIST) in 1988), of which more than 150,000 copies have been sold. Tables and formulas in this handbook can be useful in preliminary surveys before turning to computer programs. Methods that are important for the numerical computation of special functions are surveyed in Temme [349].

Although still available and among one of the most cited references, the Handbook is increasingly becoming out of date. A replacement more suited to the needs of today is being developed at NIST. This is planned to be made available both in print and as a free electronic publication on the World Wide Web; see `http://dlmf.nist.gov`. An outline of the features of the new NIST Digital Library of Mathematical Functions is given by D. W. Lozier [249]. The Internet version will come with hyperlinks, interactive graphics, and tools for downloading and searching. The part of the old Handbook devoted to massive tables of values will be superseded. To summarize, data-intensive and operation-preserving methods are replaced by data-conserving and operation-intensive techniques. A complete survey of the available software for computing special functions was given by Lozier and Olver [250]. The latest update of this project appeared in December 2000; see `http://math.nist.gov/mcsd/Reports/2001/nesf/`.

The basic properties of the Gauss hypergeometric function are derived in Lebedev's monograph on special functions [240]. Lebedev's compact book provides a good background to many of the applications of advanced analysis that lack complete proofs in our book. For example, the chapter on the gamma function contains numerous instances of the use of series expansions and analytic continuation that are efficient as well as instructive, important, and beautiful. Codes and other interesting information concerning the evaluation of special functions are also found in [294, Chap. 5 and 6].

A thorough treatment of polynomial interpolation of equidistant data is found in Steffensen [330]; see in particular Sec. 18 about "the calculus of symbols." The history of this topic is presented in Goldstine [159]. Different aspects of automatic differentiation are discussed by Rall [297], Griewank [175], and Corliss et al. [80].

A classic exposition of the theory of infinite series is given in the monograph by Knopp [229]. An exposition of the long and interesting historical development of convergence accelerating methods is given by Brezinski [53]. The use of extrapolation methods in numerical analysis up to 1970 is surveyed in Joyce [215], which contains an extensive bibliography. The book by Brezinski and Redivo-Zaglia [55] covers more recent developments. It also surveys properties of completely monotonic sequences, and how to construct such sequences.

A general extrapolation algorithm that includes almost all known convergence acceleration methods has been given by Håvie [188]. For acceleration of vector sequences several generalizations of scalar sequence transformations have been suggested; see [173]. Used for solving linear equations, these are related to the biconjugate gradient algorithm and projection methods; see the monograph by Brezinski [52]. Some convergence acceleration methods (due to Lindelöf, Plana, and others) transform an infinite series to an integral in the complex plane. With appropriate numerical procedures for computing the integral, these methods can compete with the methods treated in Sec. 3.4. In particular, they are applicable to some difficult *ill-conditioned series*; see Dahlquist [87].

The theory of continued fractions started to develop already in the seventeenth century. The main contributors were Euler, Lambert, and Lagrange; see Brezinski [51]. Algebraic continued fractions and applications to number theory are treated in Riesel [303].

The analytic theory of continued fractions has earlier origins, and contributors include Chebyshev, A. A. Markov, and Stieltjes. Hermite was able to prove the transcendence of e in 1873 using a kind of Padé approximants. His proof was extended in 1892 by Lindemann, who showed that π is a transcendental number, answering a question that had been an open problem for 2000 years. An important survey of theory and applications of continued fractions is given by Jones and Thron [214]; see also Lorenzen and Waadeland [248]. Continued fraction expansions of many special functions are found in Abramowitz and Stegun [1]. Codes and further references are given in Press et al. [294, Chapters 5 and 6].

The basic algorithmic aspects of what we today call Padé approximants were established by Frobenius [127]. Padé [281] gave a systematic study of these approximants and introduced the table named after him. The most complete reference on Padé approximation is Baker and Graves-Morris [14]. An easier to read introduction is Baker [13]. The numerical evaluation of continued fractions is surveyed in Blanche [34]. Gragg [172] gives an excellent survey of the use of the Padé table in numerical analysis.

The theory of the qd algorithm is treated in depth by Henrici [196, Chap. 7]. Rutishauser [311] got the idea for his LR algorithm for computing the eigenvalues of a matrix from the qd algorithm. For recent developments and applications to the matrix eigenvalue problem see Parlett [285].

Chapter 4

Interpolation and Approximation

Far better an approximate answer to the right question,
which is often vague,
than an exact answer to the wrong question,
which can always be made more precise.
—John W. Tukey

4.1 The Interpolation Problem

4.1.1 Introduction

Polynomials are used as the basic means of approximation in nearly all areas of numerical analysis. We have encountered in Sec. 3.3.4 the problem of interpolating the values of a function $f(x)$ in n *equidistant* points by a polynomial $p(x) \in \mathcal{P}_n$.[126] We have also studied application of polynomial approximations to numerical differentiation and integration. It is de facto so, although the polynomials are invisible in the derivations of formulas by operator methods. In the following sections we shall go deeper into the nonequidistant polynomial interpolation problem.

Let $a = x_1 < x_2 < \cdots < x_n = b$ be a grid of distinct points x_i. Find a polynomial $p \in \mathcal{P}_n$ such that

$$p(x_i) = f(x_i), \quad i = 1 : n. \tag{4.1.1}$$

By Theorem 3.3.4, the interpolation polynomial *p is uniquely determined.* This theorem is general, although the rest of Sec. 3.3 dealt with interpolation polynomials in the equidistant case only and their application to numerical differentiation. Note that the formulation and the solution of this problem are *independent of the ordering of the points x_i*.

[126] Recall the definition of $\mathcal{P}_n$ as the space of polynomials in one variable of degree *less than n*; the dimension of the linear space $\mathcal{P}_n$ is n.

4.1.2 Bases for Polynomial Interpolation

A set of polynomials $\{p_1(x), p_2(x), \dots, p_n(x)\}$ such that any polynomial $p \in \mathcal{P}_n$ can be expressed as a linear combination

$$p(x) = \sum_{j=1}^{n} c_j p_j(x)$$

is called a basis in $\mathcal{P}_n$. The column vector $c = (c_1, c_2, \dots, c_n)^T$ can be viewed as a *coordinate vector* of p in the space $\mathcal{P}_n$, with respect to this basis. The interpolation problem (4.1.1) leads to a linear system of equations

$$c_1 p_1(x_i) + c_2 p_2(x_i) + \cdots + c_n p_n(x_i) = f(x_i), \quad i = 1 : n. \tag{4.1.2}$$

If we introduce the matrix

$$P_n = [p_j(x_i)]_{i,j=1}^{n}, \tag{4.1.3}$$

and the column vector $f = (f(x_1), f(x_2), \dots, f(x_n))^T$, then the linear system becomes

$$P_n c = f. \tag{4.1.4}$$

Mathematically, the choice of basis (for a finite-dimensional space) makes no difference. Computationally, working with *rounded values of the coefficients*, the choice of basis can make a great difference. If the purpose is to compute derivatives or integrals of the interpolation polynomial, the power basis or the **shifted power basis**, where $p_j(x) = (x-c)^{j-1}$, i.e.,

$$p(x) = \sum_{j=1}^{n} c_j (x-c)^{j-1},$$

is convenient although not always the best. If a shifted power basis is to be used for polynomial approximation on an interval $[a, b]$, it is often best to choose $c = (a+b)/2$, i.e., equal to the midpoint of the interval.

For the **power basis** $p_j(x) = x^{j-1}$, the coefficients of the interpolation polynomial are given by the solution of the linear system $V_n^T c = f$, where V_n is the Vandermonde matrix

$$V_n = [x_j^{i-1}]_{i,j=1}^{n} = \begin{pmatrix} 1 & 1 & \cdots & 1 \\ x_1 & x_2 & \cdots & x_n \\ \vdots & \vdots & \cdots & \vdots \\ x_1^{n-1} & x_2^{n-1} & \cdots & x_n^{n-1} \end{pmatrix}. \tag{4.1.5}$$

By Theorem 3.3.4 this matrix is nonsingular, since the Vandermonde determinant equals (see (3.3.12))

$$\det(V_n) = \prod_{1 \le i < j \le n} (x_i - x_j).$$

Let $\{p_1(x), p_2(x), \dots, p_n(x)\}$ and $\{q_1(x), q_2(x), \dots, q_n(x)\}$ be two bases for $\mathcal{P}_n$. Then the q_j must be linear combinations of the p_k, $k = 1 : n$. This can be expressed in vector-matrix form:

$$\big(q_1(x), q_2(x), \dots, q_n(x)\big) = \big(p_1(x), p_2(x), \dots, p_n(x)\big) S \quad \forall x, \tag{4.1.6}$$

where S is a constant matrix. *S must be nonsingular*, since if S were singular, then there would exist a nontrivial vector v such that $Sv = 0$, hence

$$(q_1(x), q_2(x), \dots, q_n(x))v = (p_1(x), p_2(x), \dots, p_n(x))Sv = 0 \quad \forall x,$$

and $(q_1(x), q_2(x), \dots, q_n(x))$ would thus not be a basis.

Let $P_n = [p_j(x_i)]_{i,j=1}^n$ and $Q_n = [q_j(x_i)]_{i,j=1}^n$. By putting $x = x_i$, $i = 1 : m$, into (4.1.6), we see that $Q_n = P_n S$, and Q_n is nonsingular for every basis. If we set $p(x) = \sum d_j q_j(x)$, the system (4.1.2) becomes for this basis $Q_n d = f$, and then

$$P_n c = f = Q_n d = P_n S d, \qquad c = P_n^{-1} f = Sd. \tag{4.1.7}$$

The matrix S for the transformation between representations is thus like a coordinate transformation in geometry. Matrix representations of various common bases transformations are given by Gander [130].

The power basis has a bad reputation which is related to the ill-conditioning of the corresponding Vandermonde matrix; see Sec. 4.1.3. There are other bases in $\mathcal{P}_n$ which are often more advantageous to use. By a **triangle family** of polynomials we mean a sequence of polynomials

$$\begin{aligned}
q_1(x) &= s_{11},\\
q_2(x) &= s_{12} + s_{22}x,\\
q_3(x) &= s_{13} + s_{23}x + s_{33}x^2,\\
&\dots\\
q_n(x) &= s_{1n} + s_{2n}x + s_{3n}x^2 + \cdots + s_{nn}x^{n-1},
\end{aligned} \tag{4.1.8}$$

where $s_{jj} \neq 0$ for all j. Note that the coefficients form a lower triangular matrix S.

Conversely, for any j, $p_j(x) = x^{j-1}$ can be expressed recursively and uniquely as linear combinations of $q_1(x), \dots, q_j(x)$. We obtain a triangular scheme also for the inverse transformation:

$$\begin{aligned}
1 &= t_{11}q_1(x),\\
x &= t_{12}q_1 + t_{22}q_2,\\
x^2 &= s_{13}q_1 + t_{23}q_2 + t_{33}q_3,\\
&\dots\\
x^n &= t_{1n}q_1 + t_{2n}q_2 + t_{3n}q_3 + \cdots + t_{nn}q_n,
\end{aligned} \tag{4.1.9}$$

where $t_{jj} \neq 0$ for all j, and the coefficients form a lower triangular matrix $T = S^{-1}$. Thus *every triangle family is a basis for $\mathcal{P}_m$*. (Recall the well-known fact that the inverse of a triangular matrix with nonzero diagonal exists and is triangular.) Among interesting triangle families are the shifted power basis $(x - c)^j$, the Chebyshev polynomials $T_j(x)$, and many other families of orthogonal polynomials.

A triangle family which is often very convenient for solving the interpolation problem is the family of **Newton polynomials**

$$p_1(x) = 1, \qquad p_j(x) = (x - x_1)(x - x_2)\dots(x - x_{j-1}), \quad j = 2 : n, \tag{4.1.10}$$

which has unit leading coefficients. Since $p_j(x_k) = 0$, if $k < j$ we obtain, using the representation

$$p(x) = c_1 p_1 + c_2 p_2(x) + c_3 p_3(x) + \cdots + c_n p_n(x), \tag{4.1.11}$$

lower triangular system $Lc = f$ for the coefficients, where

$$L = \begin{pmatrix} 1 & & & & \\ 1 & (x_2 - x_1) & & & \\ 1 & (x_3 - x_1) & (x_3 - x_1)(x_3 - x_2) & & \\ \vdots & \vdots & \vdots & \ddots & \\ 1 & (x_n - x_1) & (x_n - x_1)(x_n - x_2) & \cdots & \prod_{j=1}^{n-1}(x_n - x_j) \end{pmatrix}. \tag{4.1.12}$$

Hence the coefficients can be computed by forward substitution. In the next section we shall see how this basis leads to Newton's interpolation formula. This is one of the best interpolation formulas with respect to flexibility, computational economy, and numerical stability.

If a polynomial $p(x)$ is given in the form (4.1.11), then it can be evaluated using only n multiplications and $2n$ additions for a given numerical value x using the nested form

$$\begin{aligned} p(x) = (\cdots (c_n(x - x_{n-1}) + c_{n-1})(x - x_{n-2}) \\ + \cdots + c_3)(x - x_2) + c_2)(x - x_1) + c_1. \end{aligned}$$

This can be evaluated by a recursion formula similar to Horner's rule (see Sec. 1.2.2).

Another basis of $\mathcal{P}_n$ that has many advantages is the Lagrange basis of polynomials $\ell_j(x)$. If x_i, $i = 1 : n$, are distinct interpolation points, these are

$$\ell_j(x) = \prod_{\substack{i=1 \\ i \neq j}}^{n} \frac{(x - x_i)}{(x_j - x_i)}, \qquad j = 1 : n. \tag{4.1.13}$$

These bases polynomials of degree $n - 1$ satisfy

$$\ell_j(x_i) = \delta_{ij} = \begin{cases} 1 & \text{if } i = j, \\ 0 & \text{if } i \neq j. \end{cases} \tag{4.1.14}$$

From this property directly follows **Lagrange's interpolation formula**, which directly displays the solution of the interpolation problem for n distinct points.[127]

Theorem 4.1.1.

The unique interpolation polynomial $p \in \mathcal{P}_n$ interpolating the function f at the distinct points x_i, $i = 1 : n$, can be written

$$p(x) = \sum_{j=1}^{n} f(x_j) \prod_{\substack{i=1 \\ i \neq j}}^{n} \frac{(x - x_i)}{(x_j - x_i)}. \tag{4.1.15}$$

[127] Lagrange published this formula in 1794.

It is an easy exercise to show that by l'Hôpital's rule the Lagrange polynomials can be written

$$\ell_j(x) = \frac{\Phi_n(x)}{(x - x_j)\Phi_n'(x_j)}, \qquad \Phi_n(x) = \prod_{i=1}^{n}(x - x_i). \tag{4.1.16}$$

This property characterizes what, in a more general context, is known as a **cardinal basis**.

Lagrange's interpolation formula has been widely regarded as being more suitable for deriving theoretical results than for practical computation. However, in Sec. 4.2.3 two modified forms of Lagrange's interpolation formula will be given, which are also very attractive computationally.

A natural extension of the interpolation problem is to determine a polynomial $p(x) = \sum_{j=1}^{n} c_j p_j(x) \in \mathcal{P}_n$ that, in some sense, best fits to the data $(x_i, f(x_i))$, $i = 1 : m$, where $m > n$. Since the number of data points is larger than the number of parameters, the corresponding linear system $Pc = f$ is overdetermined and can typically be satisfied only approximately. Overdetermination can be used to attain two different types of **smoothing**:

(i) to reduce the effect of random or other irregular errors in the values of the function;

(ii) to give the polynomial a smoother behavior between the grid points.

In least squares approximation one determines the coefficient vector c so that the sum of squared residuals

$$S(c) = \sum_{i=1}^{m}(p(x_i) - f(x_i))^2 \tag{4.1.17}$$

is minimized; see Sec. 1.4. This can in many applications be motivated by statistical arguments; see the Gauss–Markov theorem (Theorem 1.4.1). It also leads to rather simple computations. The conditions for the minimization are

$$\frac{\partial S(c)}{\partial c_k} = 2\sum_{i=1}^{m} p_k(x_i)\big(p(x_i) - f(x_i)\big) = 0, \quad k = 1 : n.$$

A stable method for discrete least squares polynomial approximation, based on using a basis of orthogonal polynomials, will be given in Sec. 4.5.5.

We mention here that a large part of the theory of polynomial interpolation and approximation is valid also for *generalized polynomials*

$$u(x) = \sum_{k=1}^{n} a_k u_k(x),$$

where $\{u_1, u_2, \ldots, u_n\}$ are continuous real-valued functions that form a **Chebyshev system** on a closed finite interval $[a, b]$; see Sec. 4.5.7.

4.1.3 Conditioning of Polynomial Interpolation

Let $p \in \mathcal{P}_n$ be the unique polynomial that interpolates the function values f at the distinct points x_j, $j = 1 : n$. Consider the problem to determine the value of p at the fixed point $\tilde{x}$.

With the terminology of Sec. 2.4.3 the input is the vector $f = (f_1, \dots, f_n)^T \in \mathbf{R}^n$ and the output is the scalar $p(\tilde{x})$.

Using the Lagrange basis (4.1.13) we have

$$p(\tilde{x}) = \sum_{j=1}^{n} f_j \ell_j(\tilde{x}), \qquad \ell_j(\tilde{x}) = \prod_{\substack{i=1\\ i\neq j}}^{n} \frac{(\tilde{x} - x_i)}{(x_j - x_i)}.$$

If the data are perturbed by Δf, where $|\Delta f| \le \epsilon$, then

$$|\Delta p(\tilde{x})| \le \epsilon \sum_{j=1}^{n} |\ell_j(\tilde{x})| \tag{4.1.18}$$

gives an upper bound for the perturbation in $p(\tilde{x})$.

The Lebesgue constant for polynomial interpolation at distinct points $x_i \in [a, b]$, $i = 1 : n$, is defined as

$$\Lambda_n = \max_{a \le x \le b} \sum_{j=1}^{n} |\ell_j(x)|. \tag{4.1.19}$$

From (4.1.18) it follows that Λ_n can be interpreted as the condition number for the interpolation problem with $\tilde{x} \in [a, b]$. We have the inequality

$$\max_{a \le x \le b} |p(x)| \le \Lambda_n \max_{a \le x \le b} |f(x)|,$$

where equality can be obtained for $f(x)$ piecewise linear and such that $f(x_i) = \operatorname{sign} \ell_i(x_i)$.

For equally spaced points Λ_n grows at a rate proportional to $2^n/(n \log n)$; see Cheney and Light [67, Chap. 3]. For the Chebyshev points in $[-1, 1]$, on the other hand,

$$\Lambda_n \le \frac{2}{\pi} \log(n) + 1.$$

The basis consisting of the Lagrange polynomials for Chebyshev nodes is optimally conditioned among all Lagrangian bases. It is indeed optimal among all polynomial bases in the sense of attaining the optimal growth rate $O(\log n)$; see Gautschi [144].

Usually the interpolation problem is solved in two steps. First, a suitable basis $p_j(x)$, $j = 1 : n$, for the space $\mathcal{P}_n$ is chosen and the coefficients c_j in the expansion

$$p(x) = \sum_{j=1}^{n} c_j p_j(x) \tag{4.1.20}$$

are determined. Second, the right-hand side of (4.1.20) is evaluated at some point $x \in [a, b]$. Then we also need to consider the condition number $\kappa(P_n)$ of the matrix P_n which maps the coefficient vector $c = (c_1, \dots, c_n)^T$ to f.

For the power basis $P_n = V_n^T$. Many bounds and asymptotic estimates of $\kappa(V_n)$ are known; see [147, Sec. 1.3.2]; [199, Sec. 22.1]. For equidistant points $x_i = -1 + 2(i - 1)/(n - 1)$ on $[-1, 1]$, it holds that

$$\kappa_\infty(V_n) = \|V_n^{-1}\|_\infty \|V_n\|_\infty \sim \pi^{-1} e^{\pi/4} (3.1)^n; \tag{4.1.21}$$

e.g., $\kappa_\infty(V_{20}) \approx 1.05 \cdot 10^9$. Other point distributions are even worse. For the harmonic points $x_i = 1/i$, $i = 1 : n$, we have $\kappa_\infty(V_n) > n^{n+1}$, which is faster than exponential growth! Surprisingly, some Vandermonde systems, which are so ill-conditioned that Gaussian elimination with pivoting fails to produce a single correct digit, can be solved to full relative accuracy by a fast algorithm given in Sec. 4.2.5.

For the **Chebyshev points** on $[-1, 1]$

$$x_i = \cos\left(\frac{2i-1}{n}\frac{\pi}{2}\right), \quad i = 1 : n, \tag{4.1.22}$$

i.e., the zeros of $T_{n-1}(x)$, the Vandermonde matrix is much better conditioned and

$$\kappa_\infty(V) \sim 0.253^{3/4}(1+\sqrt{2})^n. \tag{4.1.23}$$

In contrast to the power basis the condition of bases of orthogonal polynomials exhibits only polynomial growth in n. For Chebyshev polynomials $p_j = T_j$ on $[-1, 1]$ it holds that

$$\kappa(P_n) \le \sqrt{2}\,n,$$

which is a big improvement on the power basis.

It should be stressed that $\kappa(P_n)$ measures only the sensitivity of the coefficients c_i in the polynomial $p(x) = \sum_{j=1}^n c_j x^{j-1}$ to perturbations in the given data f_i. It is possible that even when these coefficients are inaccurately determined, the interpolation polynomial $p(x)$ does still reproduce the true interpolation polynomial well. For further discussion of these points, see Sec. 4.2.5 and [176].

Review Questions

4.1.1 The interpolation problem in $\mathcal{P}_n$ leads to a linear system $V^T c = f$, where V is a Vandermonde matrix. Write down the expression for the element v_{ij}.

4.1.2 What is meant by the method of undetermined coefficients? Give an example.

4.1.3 What is meant by a triangle family $q_1(x), q_2(x), \dots, q_n(x)$ of polynomials? Are all such families a basis for $\mathcal{P}_n$?

4.1.4 What property characterizes a cardinal basis for $\mathcal{P}_n$?

4.1.5 What good effects can be achieved by using overdetermination in polynomial interpolation?

4.1.6 How is the Lebesgue constant defined and what is its significance for the conditioning of the polynomial interpolation problem?

Problems and Computer Exercises

4.1.1 (a) Study experimentally interpolation in $\mathcal{P}_n$, $n = 2 : 2 : 16$, for $f(x) = (3+x)^{-1}$, $x \in [-1, 1]$. Use the linear system $V^T c = f$ and the power basis. Study both equidistant points and Chebyshev points

$$x_i = -1 + 2\frac{i-1}{n-1}, \qquad x_i = \cos\left(\frac{2i-1}{n}\frac{\pi}{2}\right), \qquad i = 1 : n,$$

respectively. Plot the error curve $y = |f(x) - p(x)|$ in semilogarithmic scale. For the larger values of n, also conduct experiments to illuminate the effects of random perturbations of the function values on the values of $p(x)$.

(b) Also do a few experiments with a random vector f, for $n = 16$ and $n = 8$, in order to compare the grid data and the order of magnitude of $p(x)$ between the grid points.

4.1.2 *A warning for polynomial extrapolation of empirical functions, or ...?*

(a) Write a program $c = polyapp(x, y, n)$ that finds the coefficient vector c for a polynomial in $p \in \mathcal{P}_n$, in a shifted power basis, such that $y_i \approx p(x_i)$, $i = 1 : m$, $m \geq n$, in the least squares sense, or study a program that does almost this.

(b) The following data show the development of the Swedish gross domestic production (GDP), quoted from a table made by a group associated with the Swedish Employer's Confederation. (The data are expressed in prices of 1985 and scaled so that the value for 1950 is 100.)

1950	1955	1960	1965	1970	1975	1980	1985	1990
100.0	117.7	139.3	179.3	219.3	249.1	267.5	291.5	326.4

1952	1957	1962	1967	1972	1977	1982	1987	1992
104.5	124.6	153.5	189.2	226.4	247.7	270.2	307.6	316.6

(c) For the upper pairs of data, compute and plot $p(x), x \in [1950, 2000]$ (say). Mark the given data points. Do this for $m = 9$, and for (say) $n = 9$, and then for $n = 8 : -2 : 2$. Store the results so that comparisons can be made afterward.

(d) Do the same for the lower pairs of data. Organize the plots so that interesting comparisons become convenient. How well are the data points of one of the sets interpolated by the results from the other set?

(e) Make forecasts for 1995 and 2000 with both data sets. Then, use a reduced data set, e.g., for the years 1982 and earlier (so that $m = 7$), and compare the forecasts for 1987 and 1992 with the given data. (Isn't it a reasonable test for every suggested forecast model to study its ability to predict the present from the past?)

(f) See if you obtain better results with the logarithms of the GDP values.

4.2 Interpolation Formulas and Algorithms

4.2.1 Newton's Interpolation Formula

Newton's interpolation formula uses the triangle family of Newton polynomials (4.1.10). Let $p \in \mathcal{P}_n$ be the unique polynomial interpolating a given function $f(x)$ at n distinct real or complex points $x_1, x_2, \dots, x_n$. Suppose that an expansion

$$f(x) = c_1 + c_2(x - x_1) + \cdots + c_n(x - x_1)(x - x_2)\cdots(x - x_{n-1}) \\ + A_n(x)(x - x_1)(x - x_2)\cdots(x - x_n) \tag{4.2.1}$$

holds, where $A_n(x)$ is the coefficient of the remainder term. If $f \in \mathcal{P}_n$, we know from Sec. 4.1.2 that such a formula holds with $A_n(x) \equiv 0$. We shall see that it is correct in general.

For $x = x_1$ we get $c_1 = f(x_1)$. Further,

$$[x_1, x]f = c_2 + c_3(x - x_2) + \cdots + c_n(x - x_2)\cdots(x - x_{n-1}) \\ + A_n(x)(x - x_2)\cdots(x - x_n),$$

where we have set

$$[x_1, x]f = \frac{f(x) - f(x_1)}{x - x_1}.$$

This shows that $c_2 = [x_1, x_2]f$.

We now define **divided differences**[128] for $k > 1$, by the recursion

$$[x_1, \dots, x_{k-1}x_k, x]f = \frac{[x_1, \dots, x_{k-1}, x]f - [x_1, \dots, x_{k-1}, x_k]f}{x - x_k}. \tag{4.2.2}$$

We obtain, for $k = 2$,

$$[x_1, x_2, x]f = c_3 + c_4(x - x_3) + \cdots + c_n(x - x_3)\cdots(x - x_{n-1}) \\ + A_n(x)(x - x_3)\cdots(x - x_n),$$

and $c_3 = [x_1, x_2, x_3]f$. By induction it follows that

$$c_k = [x_1, \dots, x_{k-1}, x_k]f, \quad k = 1 : n; \tag{4.2.3}$$

i.e., c_k in (4.2.1) equals the $(k-1)$th divided difference of f. Further,

$$A_n(x) = [x_1, x_2, \dots, x_n, x]f$$

for the coefficient of the remainder term.

We are now ready to state **Newton's interpolation formula** with exact remainder.

Theorem 4.2.1.

The unique interpolation polynomial $p \in \mathcal{P}_n$ such that $p(x_i) = f(x_i)$, $i = 1 : n$, where the x_i are distinct points, can be written as

$$p(x) = \sum_{k=1}^{n} c_k \Phi_{k-1}(x), \tag{4.2.4}$$

where $c_k = [x_1, x_2, \dots, x_k]f$ is the divided difference and

$$\Phi_0 = 1, \qquad \Phi_k(x) = \Phi_{k-1}(x)(x - x_k), \quad k = 1 : n. \tag{4.2.5}$$

[128]We prefer the modern notation $[\dots]f$ to the older notations $f[\dots]$ or $f(\dots)$, since it emphasizes that $[\dots]$ is an operator. Note that the interpretation $[x]f = f(x)$ is consistent with this.

The formula

$$f(x) = \sum_{k=1}^{n} [x_1, x_2, \dots, x_k]f \,\Phi_{k-1}(x) + [x_1, x_2, \dots, x_n, x]f \,\Phi_n(x) \tag{4.2.6}$$

is an identity, i.e., the exact remainder equals

$$f(x) - p(x) = [x_1, x_2, \dots, x_n, x]f \,\Phi_n(x). \tag{4.2.7}$$

These formulas are valid also for complex x_i and x.

Proof. For $f \in \mathcal{P}_n$ we know that (4.2.1) is correct, hence we can trust the coefficients $c_k = -[x_1, \dots, x_k]f$. Moreover, since $p(x_k) = f(x_k)$, $k = 1 : n$, it follows that $[x_1, x_2, \dots, x_k]p = [x_1, x_2, \dots, x_k]f$, and hence (4.2.4) holds.

We prove the identity (4.2.6) by induction. For $n = 1$ it is true because the right-hand side becomes $f(x_1) + [x_1, x]f \cdot (x - x_1) = f(x)$, which equals the left-hand side. Next suppose that it is true for $n = m$. The difference between the right-hand side for $n = m + 1$ and $n = m$ is

$$\begin{aligned}
&\big([x_1, \dots, x_{m+1}]f - [x_1, \dots, x_m, x]f\big)\,\Phi_m(x) + [x_1, \dots, x_{m+1}, x]f \,\Phi_{m+1}(x) \\
&= \big([x_1, \dots, x_{m+1}]f - [x_1, \dots, x_m, x]f + [x_1, \dots, x_{m+1}, x]f \,(x - x_{m+1})\big)\Phi_m(x) \\
&= \big([x_1, \dots, x_{m+1}, x]f \,(x_{m+1} - x) + [x_1, \dots, x_{m+1}, x]f \,(x - x_{m+1})\big)\Phi_m(x) = 0.
\end{aligned}$$

Hence the conjecture is true for $n = m + 1$. □

Note that to obtain the interpolation polynomial of the next higher degree with Newton's formula, we need only add a term similar to the last term, but involving a new divided difference of one higher order.

In particular, if $f \in \mathcal{P}_n$, then it follows from (4.2.7) that

$$[x_1, x_2, \dots, x_n, x]f = 0 \quad \forall x.$$

For $x = x_{n+1}$, this equation is, by Theorem 3.3.4 the only nontrivial relation of the form $\sum_{j=1}^{n+1} a_j f(x_j) = 0$ that holds for all $f \in \mathcal{P}_n$.

Theorem 4.2.2.

Let x_i, $i = 1 : n$, be pairwise distinct interpolation points. For every n, the divided difference $[x_1, x_2, \dots, x_n]f$ is the unique coefficient of x^{n-1} in the interpolation polynomial $p \in \mathcal{P}_n$.[129] *Further, we have*

$$[x_1, x_2, \dots, x_n]f = \sum_{j=1}^{m} f(x_j) \prod_{\substack{i=1 \\ i \neq j}}^{n} \frac{1}{(x_j - x_i)}. \tag{4.2.8}$$

Proof. The first statement follows from (4.2.3). The right-hand side of (4.2.8) is the coefficients of x^{n-1} obtained by using Lagrange's interpolation formula, Theorem 4.1.1. □

[129] Some authors take this as the *definition* of the divided difference.

It follows from (4.2.8) that the divided differences are symmetric functions of its arguments and continuous functions of its arguments, as long as the points x_i are distinct and $f(x)$ is continuous.

Assume $k > i$. By the definition of divided differences,

$$[x_{i+1}, \dots, x_{k-1}, x_k, x]f = \frac{[x_{i+1}, \dots, x_{k-1}, x]f - [x_{i+1}, \dots, x_{k-1}, x_k]f}{x - x_k}.$$

Now set $x = x_i$ and use the symmetry property (Theorem 4.2.2). We obtain the formula

$$[x_i, x_{i+1}, \dots, x_{k-1}, x_k]f = \frac{[x_i, x_{i+1}, \dots, x_{k-1}]f - [x_{i+1}, \dots, x_{k-1}, x_k]f}{x_i - x_k}. \tag{4.2.9}$$

This formula can be used recursively to compute the divided differences. The computation is conveniently arranged in the table below for $n = 5$ (recall that $[x_i]f = f(x_i)$).

x_1	$[x_1]f$				
		$[x_1, x_2]f$			
x_2	$[x_2]f$		$[x_1, x_2, x_3]f$		
		$[x_2, x_3]f$		$[x_1, x_2, x_3, x_4]f$	
x_3	$[x_3]f$		$[x_2, x_3, x_4]f$		$[x_1, x_2, x_3, x_4, x_5]f$
		$[x_3, x_4]f$		$[x_2, x_3, x_4, x_5]f$	
x_4	$[x_4]f$		$[x_3, x_4, x_5]f$		
		$[x_4, x_5]f$			
x_5	$[x_5]f$				

This table is called a **divided-difference table**. Each entry in the table is computed from the two entries above and below in the previous column. Hence the complete table can be constructed, for example, column by column or diagonal by diagonal.

The exact remainder term in Theorem 4.2.1 is not directly useful since unknown value $f(x)$ occurs in the divided difference $[x_1, \dots, x_n, x]f$. We now derive another expression for the remainder term. In the following, $\operatorname{int}(x, x_1, \dots, x_n)$ denotes the smallest interval that contains the points x and $x_1, \dots, x_n$.

Theorem 4.2.3 (*The Remainder Term for Interpolation*).
Let f be a given real function, with $f^{(n)}$ continuous in $\operatorname{int}(x, x_1, x_2, \dots, x_n)$. Denote by p the polynomial of degree $n-1$ for which $p(x_i) = f(x_i)$, $i = 1 : n$. Then

$$f(x) - p(x) = [x_1, x_2, \dots, x_n, x]f\, \Phi_n(x) = \frac{f^{(n)}(\xi_x)}{n!}\Phi_n(x), \tag{4.2.10}$$

$\Phi_n(x) = \prod_{i=1}^n (x - x_i)$, *for some point* $\xi_x \in \operatorname{int}(x, x_1, x_2, \dots, x_n)$, *and*

$$[x_1, x_2, \dots, x_n, x_{n+1}]f = \frac{f^{(n)}(\xi)}{n!}, \quad \xi \in \operatorname{int}(x_1, \dots, x_{n+1}). \tag{4.2.11}$$

Proof. Following a proof due to Cauchy, we introduce a new variable z and set

$$G(z) = f(z) - p(z) - [x_1, x_2, \dots, x_n, x]f\, \Phi_n(z).$$

We know by Theorem 4.2.1 that

$$f(x) - p(x) = [x_1, x_2, \dots, x_n, x]f\, \Phi_n(x), \tag{4.2.12}$$

hence $G(x) = 0$. Then $G(z) = 0$ for $z = x, x_1, x_2 \dots, x_n$. From repeated use of Rolle's theorem it follows that there exists a point $\xi_x \in \text{int}\,(x, x_1, x_2, \dots, x_n)$ such that $G^{(n)}(\xi_x) = 0$. Since $p^{(n)}(z) = 0$ and $\Phi_n^{(n)}(z) = n!$ for all z, $G^{(n)}(z) = f^{(n)}(z) - [x_1, x_2, \dots, x_n, x]f\, n!$. If we now put $z = \xi_x$, we obtain

$$[x_1, x_2, \dots, x_n, x]f = \frac{f^{(n)}(\xi_x)}{n!}. \tag{4.2.13}$$

Put this into the definition of $G(z)$, and set $z = x$. Since $G(x) = 0$, the first statement follows. The second statement follows from (4.2.13) for $x = x_{n+1}$. □

Notice the similarity to the remainder term in Taylor's formula. Notice also that the right-hand side of (4.2.10) is zero at the grid points—as it should be.

In the proof of this theorem we have assumed that x_i, x, and $f(x)$ are *real*. In Sec. 4.3.1 we shall derive a remainder term for the general case that x_i are complex interpolation points and also consider the case when the points are allowed to coincide.

Theorem 4.2.4.

For equidistant points $x_i = x_1 + (i-1)h$, $f_i = f(x_i)$, it holds that

$$[x_i, x_{i+1}, \dots, x_{i+k}]f = \frac{\Delta^k f_i}{h^k k!}. \tag{4.2.14}$$

Proof. The proof is obtained by induction, with the use of (4.2.9). The details are left to the reader. □

We have noted above that in the notation for the equidistant case $\nabla^k f_n \approx h^k f^{(k)}$, while in the divided difference notation $f[x_n, x_{n-1}, \dots, x_{n-k}] \approx f^{(k)}/k!$. For the basis functions of the interpolation formulas we have, respectively,

$$\binom{x}{k} = O(1), \qquad (x - x_n)(x - x_{n-1}) \cdots (x - x_{n-k+1}) = O(h^k),$$

provided that $x - x_{n-j} = O(h)$, $j = 0 : k-1$.

We are now in a position to give a short proof of the important formula (3.3.4) that we now formulate as a theorem.

Theorem 4.2.5.

Assume that $f \in C^k$. Then

$$\Delta^k f(x) = h^k f^{(k)}(\zeta), \quad \zeta \in [x, x + kh]. \tag{4.2.15}$$

If $f \in \mathcal{P}_k$, then $\Delta^k f(x) = 0$. Analogous results hold, mutatis mutandis, for backward and central differences.

Proof. Combine the result in Theorem 4.2.4 with (4.2.11), after making appropriate substitutions. □

Example 4.2.1.

Suppose we want to compute by linear interpolation the value $f(x)$ at a point $x = x_0 + \theta h$, $h = x_1 - x_0$. Since $\theta(1-\theta)$ takes on its maximum value 1/4 for $\theta = 1/2$, it follows from (4.2.10) that for $0 \le \theta \le 1$ the remainder satisfies

$$|f(x) - p(x)| \le h^2 M_2/8, \quad M_2 = \max_{x \in \text{int}\,[x_0, x_1]} |f''(x)|. \tag{4.2.16}$$

If the values f_0 and f_1 are given to t correct decimal digits, then the roundoff error in evaluating

$$p(x) = (1-\theta)f_0 + \theta f_1, \quad 0 \le \theta \le 1,$$

is bounded by $\frac{1}{2}10^{-t}$. Further, if $\frac{h^2 M_2}{8} \le \frac{1}{2} \cdot 10^{-t}$, then the *total error* in $p(x)$ is bounded by 10^{-t}.

This analysis motivates the rule of thumb that linear interpolation suffices if $|\Delta^2 f_n|/8$ is a tolerable truncation error.

To form the Newton interpolation polynomial we only need one diagonal of the divided-difference table. It is not necessary to store the entire table. The following algorithm replaces (overwrites) the given function values

$$f_i = f(x_i) = [x_1]f, \quad i = 1:n,$$

by the downward diagonal of divided differences of the divided-difference table,

$$[x_1]f, [x_1, x_2]f, \dots, [x_1, x_2, \dots, x_n]f.$$

At step j the jth column of the table is computed as follows:

```
% Compute downward diagonal of the divided-difference
% table. Function values are overwritten
for j = 1:n-1
   for i = n:(-1):j+1
      f(i) = (f(i) - f(j-1))/(x(i) - x(i-j));
   end
end
```

Note that to avoid overwriting data needed later it is necessary to proceed from the bottom of each column. The algorithm uses $n(n-1)/2$ divisions and $n(n-1)$ subtractions.

Newton's interpolation formula has the advantage that if it is not known in advance how many interpolation points are needed to achieve the required accuracy, then one interpolation point can be added at a time. The following *progressive algorithm* computes the divided-difference table one diagonal at a time. In the jth step, $j = 2:n$, the entries

$$[x_{j-1}, x_j]f, \dots, [x_1, x_2, \dots, x_j]f$$

on the upward diagonal of the divided-difference table overwrite the function values $f_{j-1}, \dots, f_1$.

```
% Compute upward diagonal of the divided-difference
% table. Function values are overwritten
for j = 2:n
   for i = j:(-1):2
      f(i-1) = (f(i) - f(i-1))/(x(j) - x(i-1));
   end
end
```

By Theorem 4.2.1 the Newton polynomial has the form

$$p(x) = c_1 + \sum_{j=2}^{n} c_j \prod_{i=1}^{j-1}(x - x_i), \quad c_j = [x_1, \dots, x_j]f. \tag{4.2.17}$$

Substituting z for x we have, by a simple generalization of Horner's rule, that $p(z) = b_1(z)$, where

$$b_n = c_n, \quad b_i(z) = b_{i+1}(z)(z - x_i) + c_i, \quad i = n - 1 : -1 : 1. \tag{4.2.18}$$

This recursion can be used algebraically or to evaluate $p(x)$ for a given numerical value $x = z$. It is straightforward to show that in the latter case the computed result is the exact value corresponding to slightly perturbed divided differences; cf. Problem 2.3.6.

The auxiliary quantities $b_n(z), \dots, b_2(z)$ are of independent interest, since

$$p(x) = b_1 + (x - z)\left(b_2 + \sum_{j=2}^{n-1} b_{j+1}\phi_{j-1}(x)\right). \tag{4.2.19}$$

Hence they give the coefficients of the quotient polynomial in synthetic division of $p(x)$ with $(x - z)$. The proof of this result is left as an exercise. Derivatives of a Newton polynomial can be evaluated by repeated applications of the Horner scheme.

Example 4.2.2.

Compute the interpolation polynomial for the following table:

$$\begin{array}{lcccc}
x_1 = 1 & \mathbf{0} & & & \\
 & & \mathbf{2} & & \\
x_2 = 2 & 2 & & \mathbf{1} & \\
 & & 5 & & \mathbf{0} \\
x_3 = 4 & 12 & & 1 & \\
 & & 8 & & \\
x_4 = 5 & 20 & & &
\end{array}$$

(the entries appearing in the Newton forward interpolation formula are boldface). We get

$$\begin{aligned} p(x) &= 0 + 2(x - 1) + 1(x - 1)(x - 2) + 0(x - 1)(x - 2)(x - 4) \\ &= x^2 - x. \end{aligned}$$

Note that for these particular data the unique interpolation polynomial in $\mathcal{P}_4$ actually belongs to the subspace $\mathcal{P}_3$.

Example 4.2.3.

Set $f(x; z) = 1/(z - x)$; x is the variable, z is a parameter, and both may be complex. The following elementary, though remarkable, expansion can be proved directly by induction.

$$\begin{aligned}\frac{1}{z-x} &= \frac{1}{z-x_1} + \frac{x-x_1}{(z-x_1)(z-x_2)} + \cdots + \frac{(x-x_1)(x-x_2)\cdots(x-x_{n-1})}{(z-x_1)(z-x_2)\cdots(z-x_n)} \\ &\quad + \frac{(x-x_1)\cdots(x-x_n)}{(z-x_1)\cdots(z-x_n)(z-x)} \\ &= \sum_{j=1}^{n} \frac{\Phi_{j-1}(x)}{\Phi_j(z)} + \frac{\Phi_n(x)}{\Phi_n(z)(z-x)}. \end{aligned} \tag{4.2.20}$$

When we match this with Newton's interpolation formula we find that

$$[x_1, x_2, \ldots, x_j]f(x; z) = \frac{1}{\Phi_j(z)}, \tag{4.2.21}$$

$$[x_1, x_2, \ldots, x_j, x]f(x; z) = \frac{1}{\Phi_j(z)(z-x)}. \tag{4.2.22}$$

These formulas can also be proved by induction, by working algebraically with $1/(z-x)$ in the divided-difference table (Problem 4.2.4). See also Problem 3.3.3 (a) for the equidistant case.

This is more than a particular example. Since $1/(z - x)$ is the kernel of Cauchy's integral (and several other integral representations), this expansion is easily generalized to arbitrary analytic functions; see Sec. 4.3.2.

An interesting feature is that these formulas do *not* require that the points x_i be distinct. (They are consistent with the extension to nondistinct points that will be made in Sec. 4.3.1.) Everything is continuous except if $z = x_i$, $i = 1 : n$, or, of course, if $z = x$. If all the x_i coincide, we obtain a geometric series with a remainder.

For given interpolation points the divided differences in Newton's interpolation formula depend on the ordering in which the points x_i are introduced. Mathematically all orderings give the same unique interpolation polynomial. But *the condition number for the coefficients c in the Newton polynomial can depend strongly on the ordering of the interpolation points.* For simple everyday interpolation problems the increasing order $x_1 < x_2 < \cdots < x_n$ will give satisfactory results.

If the point $\tilde{x}$ where the polynomial is to be evaluated is known, then an ordering such that

$$|\tilde{x} - x_1| \leq |\tilde{x} - x_2| \leq \cdots \leq |\tilde{x} - x_n|$$

can be recommended. In the equidistant case this corresponds to using Stirling's interpolation formula (3.3.39). In case convergence is slow and an interpolation polynomial of high order has to be used, another suitable choice is the **Leja ordering**, defined by

$$x_1 = \max_{1\le i\le n} |x_i|, \qquad \prod_{k=1}^{j-1} |x_j - x_k| = \max_{i\ge j} \prod_{k=1}^{j-1} |x_i - x_k|, \quad j = 2 : n-1. \tag{4.2.23}$$

Let $\mathcal{K}$ be a compact set in the complex plane with a connected complement. Any sequence of points $\xi_1, \xi_2, \dots$ which satisfies the conditions

$$|\xi_1| = \max_{\xi\in\mathcal{K}} |\xi|, \qquad \prod_{k=1}^{j-1} |\xi_j - \xi_k| = \max_{\xi\in\mathcal{K}} \prod_{k=1}^{j-1} |\xi - \xi_k|, \quad j = 2, 3, \dots, \tag{4.2.24}$$

are **Leja points** for $\mathcal{K}$. The points may not be uniquely defined by (4.2.24). For a real interval $[a, b]$ the Leja points are distributed similarly to the Chebyshev points. The main advantage of the Leja points is that it is easy to add new Leja points successively to an already computed sequence of Leja points; see Reichel [299].

4.2.2 Inverse Interpolation

It often happens that one has a sequence of pairs $\{(x_i, y_i)\}$ and wants to determine a point where $y(x) = c$. We saw an example in the simulation of the motion of a ball (Sec. 1.5.2), where the landing point was computed by linear inverse interpolation.

In general a natural approach is to reverse the roles of x and y, i.e., to compute the inverse function $x(y)$ for $y = c$ by means of Newton's interpolation formula with the divided differences $[y_i, y_{i+1}, \dots, y_{i+j}]x$. This is called **inverse interpolation**. It is convenient to order the points so that

$$\cdots < y_5 < y_3 < y_1 < c < y_2 < y_4 < \cdots.$$

This approach is successful if the function $x(y)$ is suitable for local approximation by a polynomial.

Sometimes, however, the function $y(x)$ is much better suited for local approximation by a polynomial than the inverse function $x(y)$. Then we can instead, for some m, solve the following equation:

$$y_1 + [x_1, x_2]y \cdot (x - x_1) + \sum_{j=2}^{n-1} [x_1, x_2, \dots, x_{j+1}]y\, \Phi_j(x) = c. \tag{4.2.25}$$

Again it is convenient to order the points so that the root α comes in the middle, for example,

$$\cdots < x_5 < x_3 < x_1 < \alpha < x_2 < x_4 < \cdots.$$

Suppose that $x_i - x_1 = O(h)$, $i > 1$, where h is some small parameter in the context (usually some step size). Then

$$\Phi_j(x) = O(h^j), \qquad \Phi'_j(x) = O(h^{j-1}).$$

The divided differences are $O(1)$, and we assume that $[x_1, x_2]y$ is bounded away from zero. Then the terms of the sum decrease like h^j.

Writing the equation in the form $x = x_1 + F(x)$, where

$$F(x) \equiv \frac{(c - y_1) - \sum_{j=2}^{n-1}[x_1, x_2, \dots, x_{j+1}]y\, \Phi_j(x)}{[x_1, x_2]y}, \tag{4.2.26}$$

we can use the *iteration* $x^{k+1} = x_1 + F(x^k)$ to determine x. We ignore the sum to get the first guess x^0; this means the same as linear inverse interpolation. We stop when x^k and x^{k-1} are sufficiently close. A more careful termination criterion will be suggested in Chapter 6, where the effect on the result of errors like the interpolation error is also discussed. From the discussion of fixed-point iteration in Sec. 1.1.1, we conclude that the iteration converges with linear ratio equal to

$$F'(x) \approx \frac{\Phi_2'(x)[x_1, x_2, x_3]y}{[x_1, x_2]y} = O(h).$$

Thus, if h is small enough, the iterations converge rapidly. If more than two iterations are needed, Aitken acceleration (Sec. 3.4.2) may be practical.

4.2.3 Barycentric Lagrange Interpolation

Lagrange's interpolation formula was introduced in Sec. 4.1.2. For distinct real interpolation points x_i, $i = 1 : n$, this formula uses the cardinal basis of $\mathcal{P}_n$ consisting of the Lagrange polynomials of degree $n - 1$:

$$\ell_j(x) = \prod_{\substack{i=1\\ i\neq j}}^{n} \frac{(x - x_i)}{(x_j - x_i)}, \qquad j = 1 : n, \tag{4.2.27}$$

with the property that

$$\ell_j(x_i) = \delta_{ij} = \begin{cases} 1 & \text{if } i = j, \\ 0 & \text{if } i \neq j. \end{cases}$$

Quite often it is asserted that the Lagrange form is a bad choice for practical computations,[130] since for each new value of x the functions $\ell_j(x)$ have to be recomputed at a cost $\mathcal{O}(n^2)$. Further, adding a new data point x_{n+1}, f_{n+1} will require a new computation from scratch. Therefore, it is concluded that Lagrange's formula is not as efficient as Newton's interpolation formula.

The above assertions are, however, not well-founded. *The Lagrange representation can easily be rewritten in two more attractive forms which are both eminently suitable for computation*; see Rutishauser [312] and Berrut and Trefethen [24]. The key idea is to take out the common factor $\Phi_n(x)$ in (4.1.15) and introduce the **support coefficients**

$$w_j = \frac{1}{\prod_{\substack{i=1\\ i\neq j}}(x_j - x_i)}, \qquad j = 1 : n. \tag{4.2.28}$$

[130]"For actual numerical interpolation Lagrange's formula is, however, as a rule not very suitable." (Steffensen [330, p. 25].)

The Lagrange polynomials can then be written as

$$\ell_j(x) = \Phi_n(x)\frac{w_j}{x - x_j}.$$

Taking out the common factor gives the **modified form** of Lagrange's interpolation formula:

$$p(x) = \Phi_n(x)\sum_{j=1}^{n}\frac{w_j}{x - x_j}f(x_j). \tag{4.2.29}$$

Here w_j depend only on the given points x_j, $j = 1 : n$, and can be computed in $2(n-1)$ flops. This is slightly more than the $\frac{3}{2}n(n-1)$ flops required to compute the divided differences for Newton's interpolation formula. Then, to evaluate $p(x)$ from (4.2.29) for a new value of x we only need to compute $\Phi_n(x)$ and $w_j/(x - x_j)$, $j = 1 : n$, which just requires $\mathcal{O}(n)$ operations.

The product factor $\Phi_n(x)$ in (4.2.29) can be eliminated as follows. Since the interpolation formula is exact for $f(x) \equiv 1$, we have

$$1 = \Phi_n(x)\sum_{j=1}^{n}\frac{w_j}{x - x_j}.$$

Substituting this in (4.2.29),

$$p(x) = \frac{\displaystyle\sum_{j=1}^{n}\frac{w_j}{x - x_j}f(x_j)}{\displaystyle\sum_{j=1}^{n}\frac{w_j}{x - x_j}} \quad \text{if } x \neq x_j, \quad j = 1 : n, \tag{4.2.30}$$

which is the **barycentric form** of Lagrange's interpolation formula. This expresses the value $p(x)$ as a weighted mean of the values f_i. (Note that the coefficients $w_j/(x - x_j)$ need not be positive, so the term "barycentric" is not quite appropriate.)

The barycentric formula (4.2.30) has a beautiful symmetric form and is "eminently suitable for machine computation" (Henrici [195, p. 237]). Unlike Newton's interpolation formula, it does not depend on how the nodes are ordered. The numerical stability of the two modified Lagrange interpolation formulas is, contrary to what is often stated, very good. Note that the interpolation property of $p(x)$ is preserved even if the coefficients w_i are perturbed, but then $p(x)$ is usually no longer a polynomial but a rational function.

There seems to be a stability problem for the formula (4.2.30) when x is very close to one of the interpolation points x_i. In this case $w_i/(x - x_i)$ will be very large and not accurately computed because of the cancellation in the denominator. But this is in fact no problem, since there will be exactly the same error in the denominator. Further, in case $\Delta_i = fl(x - x_i)$ is exactly zero, we can simply put $\Delta_i = u$, where u is the unit roundoff; see Theorem 2.2.2.

The barycentric form of Lagrange's interpolation formula can be as efficiently updated as Newton's formula. Suppose the support coefficients $w_i^{(k-1)}$, $i = 1 : k-1$, for the points $x_1, \dots, x_{k-1}$ are known. Adding the point x_k, the first $k-1$ new support coefficients can be calculated from

$$w_i^{(k)} = w_i^{(k-1)}/(x_i - x_k), \quad i = 1 : k-1,$$

using $(k-1)$ divisions and subtractions. Finally we have $w_k^{(k)} = 1/\prod_{i=1}^{k-1}(x_k - x_i)$. The computation of the support coefficients is summarized in the following program:

```
% Compute support coefficients
w(1) = 1;
for k = 2:n
   w(k) = 1;
   for i = 1:k-1
      w(i) = w(i)/(x(i) - x(k));
      w(k) = w(k)/(x(k) - x(i));
   end
end
```

Note that the support coefficients w_i do not depend on the function to be interpolated. Once they are known interpolating a new function f only requires $\mathcal{O}(n)$ operations. This contrasts favorably with Newton's interpolation formula, which requires the calculation of a new table of divided differences for each new function.

Suppose that we use interpolation points in an interval $[a, b]$ of length $2C$. Then as $n \to \infty$ the scale of the weights will grow or decay exponentially at the rate C^{-n}. If n is large or C is far from 1, the result may underflow or overflow even in IEEE double precision. In such cases there may be a need to rescale the support coefficients.

The computation of the support coefficients can be done in only $n(n-1)$ flops by using the relation (see Problem 4.2.5 and [318, Sec. 3.2.1]) $\sum_{i=1}^{n} w_i = 0$, $n > 1$, to compute $w_n = \sum_{i=1}^{n-1} w_i$. But using this identity destroys the symmetry and can lead to stability problems for large n. Serious cancellation in the sum will occur whenever $\max_i |w_i|$ is much larger than $|w_n|$. Hence the use of this identity is not recommended in general.

Theorem 4.2.6 (*Higham* [200]).

Assume that x_i, f_i, and x are floating-point numbers. Then the computed value $\tilde{p}(x)$ of the interpolation polynomial using the modified Lagrange formula (4.2.29) *satisfies*

$$\tilde{p}(x) = \Phi_n(x) \sum_{i=1}^{n} \frac{w_i}{x - x_i} f(x_i) \prod_{j=1}^{5(n+1)} (1 + \delta_{ij})^{\pm 1}, \tag{4.2.31}$$

where $|\delta_{ij}| \le u$.

Thus the formula (4.2.29) *computes the exact value of an interpolating polynomial corresponding to slightly perturbed function values $f(x_i)$. Hence this formula is* backward stable *in the sense of Definition* 2.4.10.

The barycentric formula is not backward stable. A forward error bound similar to that for the modified formula but containing an extra term proportional to $\sum_{j=1}^{n} |\ell_j(x)|$ can be shown. Hence the barycentric formula can be significantly less accurate than the modified Lagrange formula (4.2.29) only for a poor choice of interpolation points with a large Lebesgue constant Λ_n.

For various important distributions of interpolating points, it is possible to compute the support coefficients w_i analytically.

Example 4.2.4.

For interpolation at the equidistant points x_1, $x_i = x_1 + (i-1)h$, $i = 2 : n$, the support coefficients are

$$\begin{aligned} w_i &= 1\big/ \left((x_i - x_1)\cdots(x_i - x_{i-1})(x_i - x_{i+1})\cdots(x_i - x_n)\right) \\ &= \frac{(-1)^{n-i}}{\left(h^{n-1}(i-1)!\,(n-i)!\right)} \\ &= \frac{(-1)^{n-i}}{h^{n-1}(n-1)!}\binom{n-1}{i}. \end{aligned}$$

In the barycentric formula (4.2.30) a common factor in the coefficients w_i cancels and we may use instead the modified support coefficients

$$w_i^* = (-1)^{i+1}\binom{n-1}{i}. \tag{4.2.32}$$

For a given n these can be evaluated in only $2n$ operations using the recursion

$$w_1^* = n-1, \qquad w_i^* = w_{i-1}^* \frac{n-i}{i}, \qquad i = 2 : n.$$

Example 4.2.5.

Explicit support coefficients are also known for the Chebyshev points of the first and second kind on $[-1, 1]$. For the Chebyshev points of the first kind they are

$$w_i = (-1)^i \sin\frac{(2i-1)}{n}\frac{\pi}{2}, \qquad x_i = \cos\frac{(2i-1)}{n}\frac{\pi}{2}, \qquad i = 1 : n. \tag{4.2.33}$$

For the Chebyshev points of the second kind they are

$$w_i = (-1)^i \delta_j, \qquad x_i = \cos\frac{(i-1)}{(n-1)}\pi, \qquad i = 1 : n, \tag{4.2.34}$$

where $\delta_j = 1/2$ if $i = 1$ or $i = n$, and 1 otherwise. Note that all but two weights are equal. This will be considered from another point of view in Sec. 4.6.

For an interval $[a, b]$ the Chebyshev points can be generated by a linear transformation. The corresponding weight w_i then gets multiplied by $2^n(b-a)^n$. But this factor cancels out in the barycentric formula, and there is no need to include it. Indeed, by *not* doing so the risk of overflow or underflow, when $|b-a|$ is far from 1 and n is large, is avoided.

The two examples above show that with equidistant or Chebyshev points only $\mathcal{O}(n)$ operations are needed to get the weights w_i. For these cases the barycentric formula seems superior to all other interpolation formulas.

Lagrange's interpolation formula can be used to compute the inverse of the Vandermonde matrix V in (4.1.5) in $\mathcal{O}(n^2)$ operations. If we set $V^{-1} = W = (w_{ij})_{i,j=1}^n$, then $WV = I$, the ith row of which can be written

$$\sum_{j=1}^{n} w_{ij} x_k^j = \delta_{ik}, \qquad k = 1 : n.$$

This is an interpolation problem that is solved by the Lagrange basis polynomial

$$\ell_i(x) = \prod_{\substack{k=1\\k\neq i}}^{n} \frac{(x - x_k)}{(x_i - x_k)} = \sum_{j=1}^{n} w_{ij}x^j, \quad j = 1:n. \tag{4.2.35}$$

Clearly V is nonsingular if and only if the points x_i are distinct.

The elements w_{ij} in inverse Vandermonde matrix $W = V^{-1}$ can be computed as follows: First compute the coefficients of the polynomial

$$\Phi_n(x) = (x - x_1)(x - x_2)\cdots(x - x_n) = \sum_{j=1}^{n+1} a_j x^{j-1}.$$

This can be done by the following MATLAB script:

```
% Compute inverse Vandermonde matrix
a(1) = -x(1); a(2) = 1;
for k = 2:n
   a(k+1) = 1;
   for j = k:-1:2
      a(j) = a(j-1) - x(k)*a(j);
   end
   a(1) = -x(k)*a(1);
end
```

Then compute the coefficients of the polynomials

$$q_i(x) = \Phi_n(x)/(x - x_i), \quad i = 1:n,$$

by synthetic division (see Sec. 1.2.2). By (4.2.35) the n^2 elements in W equal the coefficients of the Lagrange polynomials

$$\ell_i(x) = q_i(x)/q_i(x_i), \quad i = 1:n.$$

Here the scalars $q_i(x_i)$ are computed by Horner's rule. The cost of computing the n^2 elements in W by this algorithm is only $6n^2$ flops.

4.2.4 Iterative Linear Interpolation

There are other recursive algorithms for interpolation. Of interest are those based on *successive linear interpolations*. The basic formula is given in the following theorem.

Theorem 4.2.7.

Assume that the two polynomials $p_{n-1}(x)$ and $q_{n-1}(x)$, both in $\mathcal{P}_{n-1}$, interpolate $f(x)$ at the points $x_1, \dots, x_{n-1}$ and $x_2, \dots, x_n$, respectively. If the n points $x_1, x_2, \dots, x_{n-1}, x_n$ are distinct, then

$$p_n(x) = \frac{(x - x_1)q_{n-1}(x) - (x - x_n)p_{n-1}(x)}{x_n - x_1}$$

is the unique polynomial in $\mathcal{P}_n$ that interpolates $f(x)$ at the m points $x_1, x_2, \dots, x_{n-1}, x_n$.

Proof. Since $q_{n-1}(x)$ and $p_{n-1}(x)$ both interpolate $f(x)$ at the points $x_2, \dots, x_{n-1}$ and

$$\frac{(x - x_1) - (x - x_n)}{x_n - x_1} = 1,$$

it follows that $p_n(x)$ also interpolates $f(x)$ at these points. Further, $p_n(x_1) = p_{n-1}(x_1)$ and hence interpolates $f(x)$ at x_1. A similar argument shows that $p_n(x)$ interpolates $f(x)$ at $x = x_n$. Hence $p_n(x)$ is the unique polynomial interpolating $f(x)$ at the distinct points $x_1, x_2, \dots, x_n$. □

A variety of schemes use Theorem 4.2.7 to construct successively higher order interpolation polynomials. Denote by $P_{j,j+1,\dots,k}(x)$, $k > j$, the polynomial interpolating $f(x)$ at the points $x_j, x_{j+1}, \dots, x_k$. The calculations in **Neville's** algorithm can be arranged in a triangular table.

$$\begin{array}{cc|ccccc}
x_1 & f(x_1) & & & & & \\
x_2 & f(x_2) & P_{1,2}(x) & & & & \\
x_3 & f(x_3) & P_{2,3}(x) & P_{1,2,3}(x) & & & \\
x_4 & f(x_4) & P_{3,4}(x) & P_{2,3,4}(x) & P_{1,2,3,4}(x) & & \\
\vdots & \vdots & \vdots & \vdots & \vdots & & \\
x_k & f(x_k) & P_{k-1,k}(x) & P_{k-2,k-1,k}(x) & P_{k-3,k-2,k-1,k}(x) & \dots & P_{1,2,3,\dots,k}
\end{array}$$

Any entry in this table is obtained as a linear combination of the entries to the left and diagonally above in the preceding column.

Note that it is easy to add a new interpolation point in this scheme. To proceed only the last row needs to be retained. This is convenient in applications where the function values are generated sequentially and it is not known in advance how many values are to be generated.

Neville's algorithm is used, for example, in repeated Richardson extrapolation (see Sec. 3.4.6), where polynomial extrapolation to $x = 0$ is to be carried out. Another use of Neville's algorithm is in iterative inverse interpolation; see Isaacson and Keller [208, Chapter 6.2].

If it is known in advance that a fixed number k of points are to be used, then one can instead generate the table column by column. When one column has been evaluated the preceding may be discarded.

Aitken's algorithm uses another sequence of interpolants, as indicated in the table below.

$$\begin{array}{cc|ccccc}
x_1 & f(x_1) & & & & & \\
x_2 & f(x_2) & P_{1,2}(x) & & & & \\
x_3 & f(x_3) & P_{1,3}(x) & P_{1,2,3}(x) & & & \\
x_4 & f(x_4) & P_{1,4}(x) & P_{1,2,4}(x) & P_{1,2,3,4}(x) & & \\
\vdots & \vdots & \vdots & \vdots & & & \\
x_k & f(x_k) & P_{1,k}(x) & P_{1,2,k}(x) & P_{1,2,3,k}(x) & \dots & P_{1,2,3,\dots,k}
\end{array}$$

For a fixed number k of points this table can be generated column by column. To add a new point the upper diagonal $f(x_1), P_{1,2}(x), P_{1,2,3}(x), \dots, P_{1,2,\dots,k}(x)$ needs to be saved. The basic difference between these two procedures is that in Aitken's the interpolants in any row use points with subscripts near 1, whereas Neville's algorithm uses rows with subscripts nearest n.

Neville's and Aitken's algorithms can easily be used in the case of multiple interpolation points also. The modification is similar to that in Newton's interpolation method.

4.2.5 Fast Algorithms for Vandermonde Systems

Given distinct scalars $x_1, x_2, \dots, x_n$, the coefficients for the interpolating polynomial in the power basis $p = \sum_{i=1}^{n} a_i x^{i-1}$ are given by the solution to the linear system

$$V^T a = f, \tag{4.2.36}$$

where V is the Vandermonde matrix

$$V = V(x_1, x_2, \dots, x_n) = \begin{pmatrix} 1 & 1 & \cdots & 1 \\ x_1 & x_2 & \cdots & x_n \\ \vdots & \vdots & \cdots & \vdots \\ x_1^{n-1} & x_2^{n-1} & \cdots & x_n^{n-1} \end{pmatrix}. \tag{4.2.37}$$

Vandermonde systems are appropriate to use for polynomial interpolation when the result is to be expressed in the monomial basis. Often the Newton basis is a better choice. As we shall see, even for solving the Vandermonde system (4.2.36) an algorithm based on Newton's interpolation formula is a much better choice than Gaussian elimination.

Newton's interpolation formula gives the interpolation polynomial in the form

$$p(x) = c_1 p_1(x) + c_2 p_2(x) + \cdots + c_n p_n(x),$$

where the basis polynomials are

$$p_1(x) = 1, \qquad p_k(x) = (x - x_1) \cdots (x - x_{k-1}), \quad k = 2 : n.$$

Here $c_j = [x_1, \dots, x_{j-1}]f$ and the divided differences can be recursively computed as described in Sec. 4.2.1. Then the coefficient vector a of $p(x)$ in the power basis

$$p(x) = a_1 + a_2 x + \cdots + a_n x^{n-1}$$

can be computed by Horner's rule. Note that the matrix V^T is never formed and only storage for a few vectors is needed. It is easily verified that the operation count is $\frac{5}{2}n(n+1)$ flops.

The related **primal Vandermonde system**

$$Vy = b \tag{4.2.38}$$

arises in related problems of determining approximation of linear functionals.

Example 4.2.6.

We shall find a formula for integrals of the form $I(f) = \int_0^1 x^{-1/2} f(x)\,dx$ that is exact for $f \in \mathcal{P}_n$ and uses values $f(x_i)$, $i = 1 : n$. Such integrals need a special treatment because of the singularity at $x = 0$. Set $\mu_j = \int_0^1 x^{-1/2} x^{j-1}\,dx$ and introduce the row vector $\mu^T = (\mu_1, \mu_2, \dots, \mu_n)$. We have that $f(x) \approx p(x) = \sum_{i=1}^{n} c_j x^{j-1}$, where $V^T c = f$, and

$$I(f) \approx \int_0^1 x^{-1/2} p(x)\,dx = \sum_{i=1}^{n} c_j \mu_j = \mu^T V_n^{-T} f, \tag{4.2.39}$$

where V_n is the Vandermonde basis. This can be written $I(f) = a^T f$, where the coefficient vector is $a = V_n^{-1}\mu$, i.e., a is the solution to the primal Vandermonde system $Va = \mu$.

Clearly the approach in Example 4.2.6 can be used for any linear functional. We would also like to have a stable and efficient method for solving the primal system. One possibility would be to use the algorithm given in Sec. 4.2.3 which computes the inverse V^{-1} in about $6n^2$ operations, and then form the product $V^{-1}b = y$.

We shall now derive a more efficient and accurate algorithm for solving primal Vandermonde systems. We start by expressing the solution of the dual problem in terms of a matrix factorization. Using the power basis the unique polynomial satisfying the interpolation conditions $p(x_i) = f_i$, $i = 1 : n$, is

$$p(x) = (1, x, \dots, x^{n-1})a,$$

where the coefficient vector a satisfies the linear system $V^T a = f$.

To derive a corresponding algorithm for solving primal Vandermonde systems the above algorithm can be interpreted as a factorization of the matrix $(V^T)^{-1}$ into a product of diagonal and lower bidiagonal matrices. Let

$$D_k = \operatorname{diag}(1, \dots, 1, (x_{k+1} - x_1), \dots, (x_n - x_{n-k})) \tag{4.2.40}$$

and define the matrices

$$L_k(x) = \begin{pmatrix} I_{k-1} & 0 \\ 0 & B_{n-k+1}(x) \end{pmatrix}, \quad k = 1 : n-1, \tag{4.2.41}$$

where

$$B_p(x) = \begin{pmatrix} 1 & & & \\ -x & 1 & & \\ & \ddots & \ddots & \\ & & -x & 1 \end{pmatrix} \in \mathbf{R}^{p\times p}. \tag{4.2.42}$$

Then the dual Vandermonde algorithm can be written in matrix terms as $c = U^T f$, $a = L^T c$, where

$$U^T = D_{n-1}^{-1} L_{n-1}(1) \cdots D_1^{-1} L_1(1), \tag{4.2.43}$$

$$L^T = L_1^T(x_1) L_2^T(x_2) \cdots L_{n-1}^T(x_{n-1}). \tag{4.2.44}$$

Since $a = V^{-T} f = L^T U^T f$, we have $V^{-T} = L^T U^T$.

ALGORITHM 4.1. *Fast Dual Vandermonde Solver.*

```
function a = dvand(x,f);
  n = length(x);
  a = f;
  % Compute divided differences of f(1:n).
  for k = 1:n-1
     for j = n:(-1):k+1
        a(j) = (a(j) - a(j-1))/(x(j) - x(j-k));
     end
  end
  % Compute coefficients using Horner's scheme.
  for k = n-1:(-1):1
     for j = k:n-1
        a(j) = a(j) - x(k)*a(j+1);
     end
  end
```

We can now obtain a fast algorithm for solving a primal Vandermonde system $Vy = b$ as follows. Transposing the matrix factorization of V^{-T} gives $V^{-1} = UL$. Hence $y = V^{-1}b = U(Lb)$ and the solution to the primal system can be computed from $d = Lb$, $y = Ud$. Transposing (4.2.43)–(4.2.44) gives

$$L = L_{n-1}(x_{n-1}) \cdots L_2(x_2)L_1(x_1),$$

$$U = L_1^T(1)D_1^{-1} \cdots L_{n-1}^T(1)D_{n-1}^{-1}.$$

This leads to an algorithm for solving primal Vandermonde systems. The operation count and storage requirement of this are the same as for Algorithm 4.1.

ALGORITHM 4.2. *Fast Primal Vandermonde Solver.*

```
function y = pvand(x,b);
n = length(x);
  y = b;
  for k = 1:n-1
     for j = n:(-1):k+1
        y(j) = y(j) - x(k)*y(j-1);
     end
  end
  for k = n-1:(-1):1
     for j = k+1:n
        y(j) = y(j)/(x(j) - x(j-k));
      end
    for j = k:n -1
        y(j) = y(j) - y(j+1);
    end
  end
```

The given algorithms can be generalized to confluent Vandermonde matrices (see Example 4.3.1).

The above two algorithms are not only fast. They also give almost full relative accuracy in the solution of some Vandermonde systems which are so ill-conditioned that Gaussian elimination with complete pivoting fails to produce a single correct digit. This was first observed by Björck and Pereyra [31], from which the following example is taken.

Example 4.2.7.

Consider a primal Vandermonde system $V_n y = b$, with

$$x_i = 1/(i+2), \qquad b_i = 1/2^{i-1}, \quad i = 1:n.$$

The exact solution can be shown to be

$$y_i = (-1)^{i-1}\binom{n}{i}(1+i/2)^{n-1}.$$

Let $\bar{y}_i$ be the solution computed by the primal Vandermonde algorithm and take as a measure of the relative error $e_n = \max_{1\le i\le n}|y_i - \bar{y}_i|/|y_i|$. Using a hexadecimal floating-point arithmetic with $u = 16^{-13} = 2.22 \cdot 10^{-16}$, the following results were obtained.

n	5	10	15	20	25
e_n/u	4	5	10	54	81

The computed solution has small componentwise relative error which is remarkable since, for example, $\kappa(V_{10}) = 9 \cdot 10^{13}$.

A forward error analysis given by Higham [198] explains the surprisingly favorable results. If the points are positive and monotonically ordered,

$$0 < x_1 < x_2 < \cdots < x_n, \tag{4.2.45}$$

then the error in the solution $\bar{a}$ of a Vandermonde system $Vy = b$ computed by the primal algorithm can be bounded as

$$|\bar{a} - a| \le 5u|V^{-1}|\,|b| + O(u^2). \tag{4.2.46}$$

If the components of the right-hand side satisfy $(-1)^n b_i \ge 0$, then $|V^{-1}|\,|b| = |V^{-1}b|$, and this bound reduces to

$$|\bar{a} - a| \le 5u|a| + O(u^2); \tag{4.2.47}$$

i.e., the solution is computed with small relative error independent of the conditioning of V. A similar result holds for the dual algorithm. These good results can be shown to be related to the fact that when (4.2.45) holds, the matrix $V(x_1, x_2, \ldots, x_n)$ is **totally positive**, i.e., the determinant of every square's submatrix of V is positive; see [135, 42, 95].

Fast Björck–Pereyra-type algorithms for **Vandermonde-like** matrices of the form

$$P = (p_i(\alpha_j))_{i,j=1}^n,$$

where $p_i(x)$, $i = 1 : n$, are basis polynomials in $\mathcal{P}_n$ that satisfy a three-term recurrence relation, have also been developed; see Higham [199, Sec. 22.2]. Cauchy linear systems $Cx = b$, where the elements of C have the special form

$$c_{ij} = 1/(x_i + y_j), \quad 1 \le i, j \le n,$$

can also be solved with an $O(n^2)$ Björck–Pereyra-type algorithm; see Boros, Kailath, and Olshevsky [42].

4.2.6 The Runge Phenomenon

The remainder term in interpolation is, according to Theorem 4.2.3, equal to

$$\frac{1}{n!}\prod_{i=1}^{n}(x - x_i) f^{(n)}(\xi_x).$$

Here ξ_x depends on x, but one can say that the error curve behaves for the most part like a polynomial curve $y = c\prod_{i=1}^{n}(x - x_i)$. A similar curve is also typical for error curves arising from least squares approximation. This contrasts sharply with the error curve for Taylor approximation, whose behavior is described approximatively by $y = c(x - x_0)^n$.

It is natural to ask what the optimal placing of the interpolation points $x_1, \dots, x_n$ should be in order to minimize the maximum magnitude of $\Phi_n(x) = \prod_{i=1}^{n}(x - x_i)$ in the interval in which the formula is to be used. For the interval $[-1, 1]$ the answer is given directly by the minimax property (Lemma 3.2.4) of the Chebyshev polynomials—choose

$$\Phi_n(x) = T_n(x)/2^{n-1}.$$

Thus the interpolation points should be taken as the zeros of $T_n(x)$. (In the case of an interval $[a, b]$ one makes the linear substitution $x = \frac{1}{2}(a + b) + \frac{1}{2}(b - a)t$.)

Example 4.2.8.

Use the same notations as before. For $f(x) = x^n$ the interpolation error becomes $f(x) - p(x) = \Phi_n(x)$, because $f^{(n)}(x)/n! \equiv 1$. Figure 4.2.1 shows the interpolation error with n equidistant points on $[-1, 1]$ and with n Chebyshev points on the same interval, i.e.,

$$x_i = -1 + 2\frac{i-1}{n-1}, \qquad x_i = \cos\left(\frac{2i-1}{n}\frac{\pi}{2}\right),$$

respectively, for $n = 6$ and $n = 12$. The behavior of the error curves as shown in Figure 4.2.1 is rather typical for functions where $f^{(n)}(x)$ is slowly varying. Also note that the error increases rapidly when x leaves the interval $\text{int}(x_1, x_2, \dots, x_n)$. In the equidistant case, the error is also quite large in the outer parts of the interval.

Equidistant interpolation can give rise to convergence difficulties when the number of interpolation points becomes large. This difficulty is often referred to as **Runge's phenomenon**, and we illustrate it in the following example. A more advanced discussion is given in Sec. 4.3.5, by means of complex analysis.

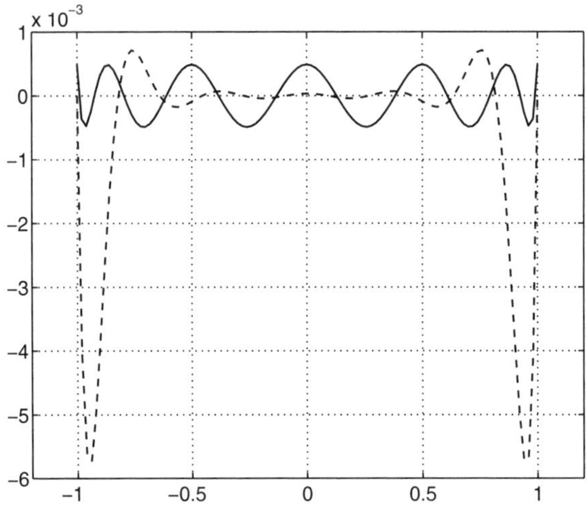

Figure 4.2.1. *Error of interpolation in* $\mathcal{P}_n$ *for* $f(x) = x^n$, *using* $n = 12$: *Chebyshev points (solid line) and equidistant points (dashed line).*

Example 4.2.9.

The graph of the function

$$f = \frac{1}{1 + 25x^2} = \frac{i}{2}\left(\frac{1}{i + 5x} + \frac{1}{i - 5x}\right),$$

where $i = \sqrt{-1}$, is the continuous curve shown in Figure 4.2.2, and is approximated in two different ways by a polynomial of degree 10 in $[-1, 1]$. The dashed curve has been

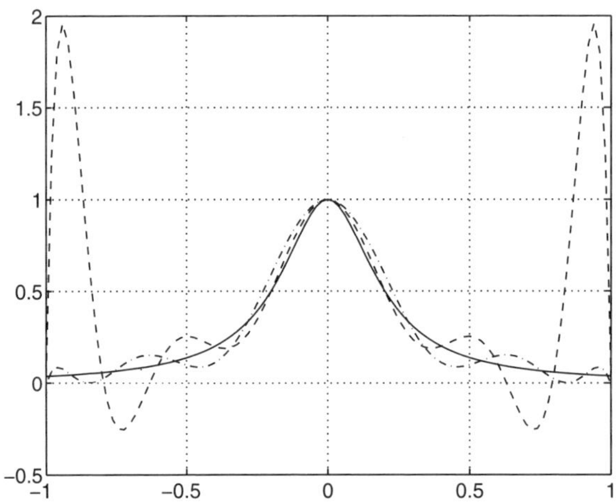

Figure 4.2.2. *Polynomial interpolation of* $1/(1 + 25x^2)$ *in two ways using* 11 *points: equidistant points (dashed curve), Chebyshev abscissae (dashed-dotted curve).*

determined by interpolation on the equidistant grid with $m = 11$ points

$$x_i = -1 + 2(i-1)/(m-1), \quad i = 1 : m. \tag{4.2.48}$$

The dash-dot curve has been determined by interpolation at the Chebyshev points

$$x_i = \cos\left(\frac{2i-1}{m}\frac{\pi}{2}\right), \quad i = 1 : m. \tag{4.2.49}$$

The graph of the polynomial obtained from the equidistant grid has—unlike the graph of f—a disturbing course between the grid points. The agreement with f near the ends of the interval is especially bad, while near the center of the interval $[-\frac{1}{5}, \frac{1}{5}]$ the agreement is fairly good. Such behavior is typical of *equidistant interpolation of fairly high degree*, and can be explained theoretically.

The polynomial obtained from interpolation at Chebyshev points agrees much better with f, but still is not good. The function f is not at all suited for approximation by one polynomial *over the entire interval*. One would get a much better result using approximation with piecewise polynomials; see Sec. 4.4.

Notice that the difference between the values of the two polynomials is much smaller at the grid points of the equidistant grid than in certain points between the grid points, especially in the outer parts of the interval. This intimates that the values which one gets by equidistant interpolation with a polynomial of high degree can be very sensitive to disturbances in the given values of the function. Put another way, *equidistant interpolation using polynomials of high degree is in some cases an ill-conditioned problem, especially in the outer parts of the interval* $[x_1, x_m]$. The effect is even worse if one extrapolates—i.e., if one computes values of the polynomial outside the grid. But equidistant interpolation works well near the center of the interval.

Even with equidistant data one can often get a more well behaved curve by—instead of interpolating—fitting a polynomial of lower degree using the method of least squares. Generally, if one chooses $n < 2\sqrt{m}$, then the polynomial fit is quite well-conditioned, but higher values of n should be avoided.[131]

If one intends to approximate a function in $[-1, 1]$ and one can choose the points at which the function is computed or measured, then one should choose the Chebyshev points. Using these points, interpolation is a fairly *well-conditioned* problem in the entire interval. The risk of disturbing surprises between the grid points is insignificant. One can also conveniently fit a polynomial of lower degree than $n-1$, if one wishes to smooth errors in measurement; see Sec. 4.5.5.

Example 4.2.9 shows *how important it is to study the course of the approximating curve $p^*(x)$ between the points which are used in the calculation before one accepts the approximation*. When one uses procedures for approximation for which one does not have a complete theoretical analysis, one should make an *experimental perturbational calculation*. In the above case such a calculation would very probably reveal that the interpolation polynomial reacts quite strongly if the values of the function are disturbed by small amounts, say $\pm 10^{-3}$. This would give a basis for rejecting the unpleasant dashed curve in the example, even if one knew nothing more about the function than its values at the equidistant grid points.

[131]This fact is related to the shape of the so-called Gram polynomials; see Sec. 4.5.5.

Review Questions

4.2.1 Prove the theorem which says that the interpolation problem for polynomials has a unique solution.

4.2.2 When is linear interpolation sufficient?

4.2.3 Derive Newton's interpolation formula.

4.2.4 Derive Newton's interpolation formula for the equidistant case, starting from Newton's general interpolation formula. How is this formula easily remembered?

4.2.5 Derive the Lagrange interpolation formula. Show how it can be rewritten in barycentric form. When is the latter form more efficient to use?

4.2.6 Neville's and Aitken's interpolation algorithms both perform successive linear interpolation. What is the difference between these?

4.2.7 The fast algorithm in Sec. 4.2.5 for solving primal Vandermonde systems can give surprisingly accurate results provided that the points $x_1, x_2, \ldots, x_n$ satisfy certain conditions. Which?

4.2.8 (a) Why is it important to study the course of the approximating curve $p^*(x)$ between the points which are used in the calculation before one accepts the approximation?

(b) What is a good choice of interpolation points in the interval $[a, b]$, if one wants to get a small error?

Problems and Computer Exercises

4.2.1 (a) Compute $f(3)$ by quadratic interpolation in the following table:

x	1	2	4	5
$f(x)$	0	2	12	21

Use the points 1, 2, and 4, and the points 2, 4, and 5, and compare the results.

(b) Compute $f(3)$ by cubic interpolation.

4.2.2 Compute $f(0)$ using one of the interpolation formulas treated above on the following table:

x	0.1	0.2	0.4	0.8
$f(x)$	0.64987	0.62055	0.56074	0.43609

The interpolation formula is here used for *extrapolation*. Use also Richardson extrapolation and compare the results.

4.2.3 Work out the details of Example 4.2.3 (about divided differences for $1/(z-x)$).

4.2.4 (a) Consider the two polynomials $p(x)$ and $q(x)$, both in $\mathcal{P}_n$, which interpolate $f(x)$ at the points $x_1, \ldots, x_n$, and $x_2, \ldots, x_{n+1}$, respectively. Assume that $\{x_i\}_{i=1}^{n+1}$ is an increasing sequence, and that $f^{(n)}(x)$ has constant sign in the interval $[x_1, x_{n+1}]$.

Show that $f(x)$ is contained between $p(x)$ and $q(x)$ for all $x \in [x_1, x_{n+1}]$.

(b) Suppose that $f(x) = f_1(x) - f_2(x)$, where both $f_1^{(n)}(x)$ and $f_2^{(n)}(x)$ have the same constant sign in $[x_1, x_{n+1}]$. Formulate and prove a kind of generalization of the result in (a).

4.2.5 Using the barycentric formula (4.2.29) the interpolation polynomial can be written as

$$p(x) = \sum_{i=1}^{n} w_i f(x_i) \prod_{\substack{j=1 \\ j \neq i}}^{m} (x - x_j).$$

Show by taking $f(x) \equiv 1$ and equating the coefficients for x^{n-1} on both sides that the support coefficients satisfy $\sum_{i=1}^{n} w_i = 0$.

4.3 Generalizations and Applications

4.3.1 Hermite Interpolation

Newton's interpolation formula can also be used in generalized interpolation problems, where one or more derivatives are matched at the interpolation points. This problem is known as **Hermite interpolation.**[132]

Theorem 4.3.1.

Let $\{z_i\}_{i=1}^{m}$ be m distinct real or complex points. Let $f(z)$ be a given real- or complex-valued function that is defined and has derivatives up to order r_{i-1} ($r_i \geq 1$) at z_i. The **Hermite interpolation problem**, *to find a polynomial $p(z)$ of degree $\leq n-1$, where $n = \sum_{i=1}^{m} r_i$ such that $p(z)$ and its first $r_i - 1$ derivatives agree with those of $f(z)$ at z_i, i.e.,*

$$p^{(j)}(z_i) = f^{(j)}(z_i), \quad j = 0 : r_i - 1, \quad i = 1 : m, \tag{4.3.1}$$

is uniquely solvable. We use here the notation $f^{(0)}(z)$ for $f(z)$.

Proof. Note that (4.3.1) are precisely n conditions on $p(z)$. The conditions can be expressed by a system of n linear equations for the coefficients p, with respect to some basis. This has a unique solution for any right-hand side, unless the corresponding homogeneous problem has a nontrivial solution. Suppose that a polynomial $p \in \mathcal{P}_n$ comes from such a solution of the homogeneous problem, that is,

$$p^{(j)}(z_i) = 0, \quad i = 1 : m, \quad j = 0 : r_i - 1.$$

Then z_i must be a zero of multiplicity r_i of $p(x)$, hence $p(z)$ must have at least $\sum r_i = n$ zeros (counting the multiplicities). But this is impossible, because the degree of p is less than n. This contradiction proves the theorem. □

Hermite interpolation can be viewed as the result of passages to the limit in interpolation at n points, where for $i = 1 : m$, r_i interpolation points coalesce into the point z_i. We

[132]Charles Hermite (1822–1901), a French mathematician, made important contributions to number theory, orthogonal polynomials, and elliptic functions.

say that the point z_i has **multiplicity** r_i. For example, the Taylor polynomial in $\mathcal{P}_n$,

$$p(z) = \sum_{j=0}^{n-1} \frac{f^{(j)}(z_1)}{j!}(z - z_1)^j, \tag{4.3.2}$$

interpolates $f(z)$ at the point z_1 with multiplicity n (or z_1 is repeated n times).

Example 4.3.1.

Consider the problem of finding a polynomial $p(x) \in \mathcal{P}_4$ that interpolates the function f and its first derivative f' at the two points z_1 and z_2, and also its second derivative at z_1. In the notations of Sec. 4.1.1 the linear system for the coefficient vector c becomes $V^T c = f$, where $f = (f(z_1), f'(z_1), f''(z_1), f(z_2), f'(z_2))^T$, and

$$V = \begin{pmatrix} 1 & 0 & 0 & 1 & 0 \\ z_1 & 1 & 0 & z_2 & 1 \\ z_1^2 & 2z_1 & 2 & z_2^2 & 2z_2 \\ z_1^3 & 3z_1^2 & 6z_1 & z_2^3 & 3z_2^2 \\ z_1^4 & 4z_1^3 & 12z_1^2 & z_2^4 & 4z_2^3 \end{pmatrix} \tag{4.3.3}$$

is a **confluent Vandermonde matrix**. Note that the second, third, and fifth column of V is obtained by "differentiating" the previous column. From Theorem 4.3.1 we conclude that such confluent Vandermonde matrices are nonsingular. We remark that the fast algorithms in Sec. 4.2.5 can be generalized to confluent Vandermone systems; see [29].

For the determinant of the general confluent Vandermonde matrix one can show that

$$\det(V) = \prod_{j=1}^{m} \prod_{n=0}^{r_j - 1} n! \prod_{i<j} (z_i - z_j)^{r_i r_j}. \tag{4.3.4}$$

When there are gaps in the sequence of derivatives that are known at a point the interpolation problem is called **Birkhoff interpolation** or **lacunary** interpolation. Such a problem may not have a solution, as is illustrated by the following example.

Example 4.3.2.

Find a polynomial $p \in \mathcal{P}_3$ that interpolates the data $f = (f(-1), f'(0), f(1))^T$. The new feature is that $f(0)$ is missing. If we use the power basis, then we obtain the linear system

$$M_{\mathbf{p}} = \begin{pmatrix} 1 & -1 & 1 \\ 0 & 1 & 0 \\ 1 & 1 & 1 \end{pmatrix}.$$

The determinant is evidently zero, so *there is no solution* for most data. An explanation is that $hf' = \mu\delta f$ for all $f \in \mathcal{P}_3$.

Newton's interpolation formula can be generalized to the confluent case quite easily. We will require some continuity and differentiability properties for divided differences.

These can be obtained through an alternative expression for divided differences that we now give.

Definition 4.3.2.

A set S of points in $\mathbf{C}$ *is called convex if for any* $z, u \in S$, *the straight line* $\{tz+(1-t)u \mid t \in (0,1)\}$ *is also contained in S. The* **convex hull** *of a set S in* $\mathbf{R}^d$ *is the smallest convex subset of* $\mathbf{C}$ *which contains S.*

Let D be the convex hull of the set of points $z, u_1, \dots, u_n$ in $\mathbf{C}$. Assume that f is defined in D and has that its nth derivative exists and is continuous on D. Set $u_0(z) = f(z)$ and consider the functions

$$w_k(z) = \int_0^1 \int_0^{t_1} \cdots \int_0^{t_{k-1}} f^{(k)}\big[u_1 + (u_2 - u_1)t_1 + \cdots + (z - u_k)t_k\big]\, dt_k \cdots dt_1, \quad (4.3.5)$$

$k = 1 : n$. The argument of the integrand lies in the convex hull D, since

$$\begin{aligned}\zeta_k &= u_1 + (z - u_1)t_1 + \cdots + (z - u_k)t_k \\ &= (1 - t_1)u_1 + (t_1 - t_2)u_2 + \cdots + (t_{k-1} - t_k)u_k + t_k z \\ &= \lambda_1 u_1 + \lambda_2 u_2 + \cdots + \lambda_k u_k + \lambda_{k+1} z,\end{aligned}$$

where $1 \le k \le n$. From $1 \ge t_1 \ge t_2 \ge \cdots \ge t_n \ge 0$, it follows that

$$\sum_{i=1}^{k+1} \lambda_i = 1, \quad \lambda_i \ge 0, \quad i = 1 : k+1.$$

If in (4.3.5) we carry out the integration with respect to t_k and express the right-hand side using u_{k-1} we find that the functions $u_k(z)$ can be defined through the recursion

$$w_k(z) = \frac{w_{k-1}(z) - w_{k-1}(x_k)}{z - u_k}, \quad k = 1 : n, \quad (4.3.6)$$

with $w_0(z) = f(z)$. But this is the same recursion (4.3.5) that was used before to define the divided differences and thus

$$[z, u_1, \dots, u_k]f = \int_0^1 \int_0^{t_1} \cdots \int_0^{t_{k-1}} f^{(k)}\big[u_1 + (u_2 - u_1)t_1 + \cdots + (z - u_k)t_k\big]\, dt_k \cdots dt_1, \quad (4.3.7)$$

holds for arbitrary distinct points $z, u_1, \dots, u_n$.

Notice that the integrand on the right-hand side of (4.3.7) is a continuous function of the variables $z, u_1, \dots, u_n$ and hence the right-hand side is a continuous function of these variables. Thus, when the nth derivative of f exists and is continuous on D, (4.3.7) defines the continuous extension of $[z, u_1, \dots, u_k]f$ to the confluent case.

From the continuity of the divided differences it follows that the remainder term for interpolation given in Theorem 4.2.3 remains valid for Hermitian interpolation. Provided the points $x, z_1, \dots, z_m$ are real we have

$$f(x) - p(x) = \frac{f^{(n)}(\xi_x)}{n!}\Phi_n(x), \quad \Phi_n(x) = \prod_{i=1}^{m}(x - z_i)^{r_i}, \quad (4.3.8)$$

with $\xi_x \in \operatorname{int}(x, z_1, \dots, z_m)$. From (4.3.7) follows

$$\big|[z, x_1, \dots, x_k]f\big| \le \max_{z\in D} \big|f^{(n)}(z)\big| \int_0^1 \int_0^{t_1} \cdots \int_0^{t_{n-1}} dt_n \cdots dt_1 \tag{4.3.9}$$

$$= \frac{1}{n!} \max_{z\in D} \big|f^{(n)}(z)\big|, \tag{4.3.10}$$

which can be used to give an upper bound for the remainder in polynomial interpolation for arbitrary interpolation points.

From (4.3.6) it follows that the divided difference $[u_1, \dots, u_{k+s}, x]f$ is equal to the divided difference of $w_k(x)$ at $[u_{k+1}, \dots, u_{k+s}, x]f$, so that by (4.3.7) we can write

$$w_{k+s}(z) = \int_0^1 \cdots \int_0^{t_{s-1}} u_k^{(s)}\big[u_{k+1} + (u_{k+2} - u_{k+1})t_1 + \cdots + (z - u_{k+s})t_k\big]\, dt_s \cdots dt_1.$$

If all points $u_{k+1}, \dots, u_{k+s}$ all tend to the limit z and $w_k(z)$ has a continuous sth derivative at z, then it holds that

$$w_{k+s}(z) = w_k^{(s)}(z) \int_0^1 \cdots \int_0^{t_{s-1}} dt_s \cdots dt_1 = \frac{w_k^{(s)}(z)}{s!}.$$

It can be shown that if $f \in C^k$, the divided differences belong to $C^{k+1-\max r_i}$, and that the interpolation polynomial has this kind of differentiability with respect to the u_i (*nota bene*, if the "groups" do not coalesce further).

Example 4.3.3.

As established above, the usual recurrence formula for divided differences can still be used for the construction of the divided-difference table in case of multiple points. The limit process is just applied to the divided differences, for example,

$$[x_0, x_0]f = \lim_{x_1 \to x_0} \frac{f(x_1) - f(x_0)}{x_1 - x_0} = f'(x_0),$$

$$[x_0, x_0, x_1]f = \frac{[x_0, x_0]f - [x_0, x_1]f}{x_0 - x_1} = \frac{f'(x_0) - [x_0, x_1]f}{x_0 - x_1}.$$

For the interpolation problem considered in Example 4.3.1 we construct the generalized divided-difference table, where $x_1 \ne x_0$.

$$\begin{array}{llllll}
x_0 & f_0 & & & & \\
 & & f_0' & & & \\
x_0 & f_0 & & \frac{1}{2} f_0'' & & \\
 & & f_0' & & [x_0, x_0, x_0, x_1]f & \\
x_0 & f_0 & & [x_0, x_0, x_1]f & & [x_0, x_0, x_0, x_1, x_1]f \\
 & & [x_0, x_1]f & & [x_0, x_0, x_1, x_1]f & \\
x_1 & f_1 & & [x_0, x_1, x_1]f & & \\
 & & f_1' & & & \\
x_1 & f_1 & & & &
\end{array}$$

The interpolating polynomial is

$$p(x) = f_0 + (x - x_0)f_0' + (x - x_0)^2 \frac{1}{2} f_0'' + (x - x_0)^3 [x_0, x_0, x_0, x_1]f,$$
$$+ (x - x_0)^3 (x - x_1)[x_0, x_0, x_0, x_1, x_1]f,$$

and the remainder is

$$\begin{aligned} f(x) - p(x) &= [x_0, x_0, x_0, x_1, x_1, x]f(x - x_0)^3 (x - x_1)^2 \\ &= f^{(5)}(\xi_x)(x - x_0)^3 (x - x_1)^2/5. \end{aligned}$$

An important case of the Hermite interpolation problem is when the given data are $f_i = f(x_i)$, $f_i' = f'(x_i)$, $i = 0, 1$. We can then write the interpolation polynomial as

$$\begin{aligned} p(x) = f_0 &+ (x - x_0)[x_0, x_1]f + (x - x_0)(x - x_1)[x_0, x_0, x_1]f \\ &+ (x - x_0)^2 (x - x_1)[x_0, x_0, x_1, x_1]f. \end{aligned}$$

Set $x_1 = x_0 + h$ and $x = x_0 + \theta h$, and denote the remainder $f(x) - p(x)$ by R_T. Then one can show (Problem 4.3.1) that

$$\begin{aligned} p(x) &= f_0 + \theta \Delta f_0 + \theta(1 - \theta)(hf_0' - \Delta f_0) \\ &\quad - \theta^2 (1 - \theta)\Big[(hf_0' - \Delta f_0) + (hf_1' - \Delta f_0)\Big] \\ &= (1 - \theta) f_0 + \theta f_1 + \theta(1 - \theta)\Big[(1 - \theta)(hf_0' - \Delta f_0) - \theta(hf_1' - \Delta f_0)\Big]. \end{aligned} \tag{4.3.11}$$

For $x \in [x_0, x_1]$ we get the error bound

$$|f(x) - p(x)| \le \frac{1}{384} h^4 \max_{x \in [x_0, x_1]} |f^{(4)}(x)|. \tag{4.3.12}$$

Setting $t = 1/2$, we get the useful approximation formula

$$f\left(\frac{1}{2}(x_0 + x_1)\right) \approx \frac{1}{2}(f_0 + f_1) + \frac{1}{8} h(f_0' - f_1'). \tag{4.3.13}$$

4.3.2 Complex Analysis in Polynomial Interpolation

We shall encounter multivalued functions: the logarithm and the square root. For each of these we choose that branch which is positive for large positive values of the argument z. They will appear in such contexts that we can then keep them nonambiguous by forbidding z to cross the interval $[-1, 1]$. (We can, however, allow z to approach that interval.)

We first consider the general problem of interpolation of an analytic function, at an arbitrary sequence of nodes $u_1, u_2, \dots, u_n$ in $\mathbf{C}$. Multiple nodes are allowed. Set

$$\Phi_n(z) = (z - u_1)(z - u_2) \cdots (z - u_n), \qquad z, u_j \in \mathbf{C}.$$

Let $\mathcal{D}$ be a simply connected open domain in $\mathbf{C}$ that contains the point u and the nodes. The interpolation problem is to find the polynomial $p^* \in \mathcal{P}_n$, that is determined by the conditions $p^*(u_j) = f(u_j)$, $j = 1 : n$, or the appropriate Hermite interpolation problem in the case of multiple nodes. We know that p^* depends linearly on f. In other words, there exists a linear mapping L_n from some appropriate function space so that $p^* = L_n f$.

Assume that f is an analytic function in the closure of $\mathcal{D}$, perhaps except for a finite number of poles p. A pole must not be a node. Recall the elementary identity (4.2.20) in Example 4.2.3,

$$\frac{1}{z-u} = \sum_{j=1}^{n} \frac{\Phi_{j-1}(u)}{\Phi_j(z)} + \frac{\Phi_n(u)}{\Phi_n(z)(z-u)}, \tag{4.3.14}$$

which is valid also for multiple nodes. Introduce the linear operator K_n,

$$(K_n f)(u) = \frac{1}{2\pi i} \int_{\partial\mathcal{D}} \frac{\Phi_n(u) f(z)}{\Phi_n(z)(z-u)}\, dz, \tag{4.3.15}$$

multiply the above identity by $f(z)/(2\pi i)$, and integrate along the boundary of $\mathcal{D}$ to get

$$\frac{1}{2\pi i} \int_{\partial\mathcal{D}} \frac{f(z)}{z-u}\, dz = \sum_{j=1}^{n} \Phi_{j-1}(u) \frac{1}{2\pi i} \int_{\partial\mathcal{D}} \frac{f(z)}{\Phi_j(z)}\, dz + (K_n f)(u). \tag{4.3.16}$$

First, assume that f has no poles in $\mathcal{D}$. By applying the residue theorem to the first two integrals, we note that the equation has the same structure as Newton's unique interpolation formula with exact remainder (4.2.6),

$$f(u) = \sum_{j=1}^{n} \Phi_{j-1}(u)[u_1, \dots, u_j]f + \Phi_n(u)[u_1, u_2, \dots, u_n, u]f.$$

Matching terms in the last two formulas we obtain

$$(f - L_n f)(u) = (K_n f)(u),$$

and, if f is analytic in $\mathcal{D}$, also the formula for the divided-difference,

$$[u_1, u_2, \dots, u_j]f = \frac{1}{2\pi i} \int_{\partial\mathcal{D}} \frac{f(z)\, dz}{(z-u_1)\cdots(z-u_j)}. \tag{4.3.17}$$

From this we obtain an upper bound for the divided-difference

$$\big|[u_1, u_2, \dots, u_j]f\big| = \frac{L}{2\pi} \frac{\max_{z\in\partial\mathcal{D}} |f(z)|}{\min_{z\in\partial\mathcal{D}} |(z-u_1)\cdots(z-u_j)|}, \tag{4.3.18}$$

where $L = \int_{\partial\mathcal{D}} |dz|$.

If there are poles $p \in \mathcal{D}$ with residues $\operatorname{res}_f(p)$, we must add $\sum_p \operatorname{res}_f(p)/(z-p)$ to the left-hand side of (4.3.17) and $\sum_p \operatorname{res}_f(p) \sum_j \Phi_{j-1}(u)/\Phi_j(p)$ to the right-hand side. By (4.3.14) this is, however, equivalent to subtracting

$$\sum_p \operatorname{res}_f(p)\Phi_n(u)/\Phi_n(p)(p-u)$$

from the right-hand side. This yields the following theorem.

Theorem 4.3.3.

Assume that f is analytic in the closure of the open region $\mathcal{D}$, perhaps except for a finite number of poles, $p \in \mathcal{D}$, with residues $\operatorname{res}_f(p)$. *$\mathcal{D}$ also contains the interpolation points $u_1, u_2, \dots, u_n$, as well as the point u. Multiple nodes are allowed. The point u and the poles p must not be nodes, and $p \neq u$.*

Then the interpolation error can be expressed as a complex integral,

$$(f - L_n f)(u) = (K_n f)(u) - \Phi_n(u) \sum_p \frac{\operatorname{res}_f(p)}{\Phi_n(p)(p-u)}, \tag{4.3.19}$$

where K_n is defined by (4.3.15) *and the sum (which may be void) is extended over the poles of f in $\mathcal{D}$.*

This theorem is valid when the interpolation points u_j are in the complex plane, although we shall here mainly apply it to the case when they are located in the interval $[-1, 1]$. An important feature is that this expression for the interpolation error requires no knowledge of $f^{(n)}(z)$.

For the case of *distinct nodes* we have, by the residue theorem,

$$(K_n f)(u) = \sum_{j=1}^{n} \frac{\Phi_n(u)}{(u_j - u)\Phi_n'(u_j)} f(u_j) + f(u), \tag{4.3.20}$$

where the sum, with reversed sign, is Lagrange's form of the interpolation polynomial $L_n f(u)$.

Example 4.3.4 (*Chebyshev Interpolation*).

In Chebyshev interpolation on the interval $[-1, 1]$, the nodes are the zeros of the Chebyshev polynomials $T_n(u)$, and

$$\Phi_n(u) = 2^{1-n} T_n(u).$$

Recall the notation and results in the discussion of Chebyshev expansions in Sec. 3.2.3.

In this example we assume that $f(u)$ is analytic in $\mathcal{D}$. We shall, by means of the integral $K_n f$, show that it yields almost as accurate an approximation as the first n terms of the Chebyshev expansion of the same function.

Let $\mathcal{D} = \mathcal{E}_R$, where $\mathcal{E}_R$ is the ellipse with foci at ± 1, and where R is the sum of the semiaxis. Let $z \in \partial\mathcal{E}_R$ and $u \in [-1, 1]$. Then $|T_n(u)| \le 1$, and it can be shown (Problem 4.3.13) that

$$|T_n(z)| \ge \tfrac{1}{2}(R^n - R^{-n}), \qquad |z - u| \ge a_R - 1, \qquad \int_{\partial\mathcal{E}_R} |dz| \le 2\pi a_R.$$

If we assume that $|f(z)| \le M_R$ for $z \in \partial\mathcal{E}_R$, a straightforward estimation of the line integral (4.3.20) gives

$$\begin{aligned} |f(u) - (L_n f)(u)| = |(K_n f)(u)| &\le \frac{1}{2\pi} \frac{2^{1-n} M_R 2\pi a_R}{2^{1-n}\frac{1}{2}(R^n - R^{-n})(a_R - 1)} \\ &\le \frac{2 M_R a_R R^{-n}}{(1 - R^{-2n})(a_R - 1)}. \end{aligned} \tag{4.3.21}$$

We see that the interpolation error converges at the same exponential rate as the truncation error of the Chebyshev expansion (see Sec. 3.2.3). If $f(z)$ has a singularity arbitrarily close to the interval $[-1, 1]$, then $R \approx 1$, and the exponential convergence rate will be very poor.

This above analysis can be extended to the case of multiple poles by including higher derivatives of f. This yields the general Lagrange formula valid also when there are interpolation points of multiplicity greater than one:

$$p(u) = \sum_{i=1}^{m} \left[\sum_{k=0}^{r_i-1} \frac{1}{k!} \frac{d^k}{du^k} \frac{f(u)}{\prod_{\substack{q=1\\q\neq i}}^{m} (u-u_q)^{r_q}} \Bigg|_{u=u_i} (u-u_i)^k \right] \prod_{\substack{j=1\\j\neq i}}^{m} (u-u_j)^{r_j}. \tag{4.3.22}$$

Clearly, when $r_i = 1$ for all $i = 1 : m$, this formula equals the usual Lagrange interpolation formula. It can be written in the simpler form

$$p(u) = \sum_{i=1}^{n} \sum_{k=0}^{r_i-1} f^{(k)}(u_i) L_{ik}(u), \tag{4.3.23}$$

where $L_{ik}(u)$ are **generalized Lagrange polynomials**. These can be defined starting from the auxiliary polynomials

$$l_{ik}(u) = \frac{(u-u_i)^k}{k!} \prod_{\substack{j=1\\j\neq i}}^{n} \left(\frac{u-u_j}{u_i-u_j} \right)^{r_j}, \qquad i = 1 : m, \quad k = 0 : r_i - 1.$$

Set $L_{i,r_i-1} = l_{i,r_i-1}$, $i = 1 : m$, and form recursively

$$L_{ik}(u) = l_{ik}(u) - \sum_{\nu=k+1}^{r_i-1} l_{ik}^{(\nu)}(u_i) L_{i,\nu}(u), \qquad k = r_i - 2 : -1 : 0.$$

It can be shown by induction that

$$L_{ik}^{(\sigma)}(u_j) = \begin{cases} 1 & \text{if } i = j \text{ and } k = \sigma, \\ 0 & \text{otherwise.} \end{cases}$$

Hence the L_{ik} are indeed the appropriate polynomials.

Example 4.3.5.

When $r_i = 2$, $i = 1 : n$, the Hermite interpolating polynomial is the polynomial of degree less than $n = 2m$, which agrees with $f(u)$ and $f'(u)$ at $u = u_i$, $i = 1 : m$. We have

$$p(u) = \sum_{i=1}^{n} (f(u_i) L_{i0}(u) + f'(u_i) L_{i1}(u)),$$

where $L_{ik}(u)$ can be written in the form

$$L_{i1}(u) = (u-u_i) l_i(u)^2, \qquad L_{i0}(u) = (1 - 2l_i'(u_i)(u-u_i)) l_i(u)^2,$$

and $l_i(u)$, $i = 1 : m$, are the elementary Lagrange polynomials:

$$l_{ik}(u) = \frac{(u-u_i)^k}{k!} \prod_{\substack{j=1\\ j\neq i}}^{n} \left(\frac{u-u_j}{u_i-u_j}\right)^{r_j}, \quad i = 1:n, \quad k = 0:r_i - 1.$$

4.3.3 Rational Interpolation

The rational interpolation problem is to determine a rational function

$$r_{m,n}(z) = \frac{P_m(z)}{Q_n(z)} \equiv \frac{\sum_{j=0}^{m} p_j z^j}{\sum_{j=0}^{n} q_j z^j}, \tag{4.3.24}$$

with numerator of degree m and denominator of degree n so that at distinct points $x_0, x_1, \dots, x_n$ agrees with a function f

$$r_{m,n}(x_i) = f_i, \quad i = 0 : N, \quad N = m + n. \tag{4.3.25}$$

Rational approximation is often superior to polynomial approximation in the neighborhood of a point at which the function has a singularity. Since the coefficients can be determined only up to a common nonzero factor, the number of unknown constants in (4.3.24) equals the number of interpolation points $N + 1 = m + n + 1$.

A *necessary condition* for (4.3.25) to hold clearly is that the linearized condition

$$P_m(x_i) - f_i Q_n(x_i) = 0, \quad i = 0 : N, \tag{4.3.26}$$

is satisfied, i.e., for $i = 0 : m + n$,

$$p_0 x_i + p_1 x_i + \cdots + p_m x_i^m - f_i(q_0 x_i + q_1 x_i + \cdots + q_n x_i^n) = 0. \tag{4.3.27}$$

This is a homogeneous linear system of $(m+n+1)$ equations for the $(m+n+2)$ coefficients in P_m and Q_n. If we introduce the Vandermonde matrices

$$A = \begin{pmatrix} 1 & x_0 & \dots & x_0^m \\ 1 & x_1 & \dots & x_1^m \\ \vdots & \vdots & \ddots & \vdots \\ 1 & x_N & \dots & x_N^m \end{pmatrix}, \qquad B = \begin{pmatrix} 1 & x_0 & \dots & x_0^n \\ 1 & x_1 & \dots & x_1^n \\ \vdots & \vdots & \ddots & \vdots \\ 1 & x_N & \dots & x_N^n \end{pmatrix},$$

this system can be written in matrix form as

$$(A \quad FB)\begin{pmatrix} p \\ q \end{pmatrix} = 0, \qquad F = \operatorname{diag}(f_0, f_1, \dots, f_N), \tag{4.3.28}$$

where $p = (p_0, p_1, \dots, p_m)^T$, $q = (q_0, q_1, \dots, q_n)^T$. This is a homogeneous linear system of $N + 1$ equations in $N + 2$ unknowns. Such a system always has a nontrivial solution. Moreover, since A has full column rank, we must have $q \neq 0$ for such a solution.

We note that the rational interpolant is fully determined by the denominator polynomial. By (4.3.26) $P_m(x)$ is the unique polynomial determined by the interpolation conditions

$$P(x_i) = f_i Q(x_i), \quad i = 1 : m.$$

While for polynomial interpolation there is always a unique solution to the interpolation problem, this cannot be guaranteed for rational interpolation, as shown by the example below. A further drawback is that the denominator polynomial $Q(x)$ may turn out to have zeros in the interval of interpolation.

Example 4.3.6.

Assume that we want to interpolate the four points

x	0	1	2	3
y	2	3/2	4/5	1/2

by a rational function

$$r(x) = \frac{p_0 + p_1 x + p_2 x^2}{q_0 + q_1 x}.$$

Then we must solve the homogeneous linear system

$$\begin{pmatrix} 1 & 0 & 0 \\ 1 & 1 & 1 \\ 1 & 2 & 4 \\ 1 & 3 & 9 \end{pmatrix} \begin{pmatrix} p_0 \\ p_1 \\ p_2 \end{pmatrix} - \begin{pmatrix} 2 & 0 \\ 3/2 & 3/2 \\ 4/5 & 8/5 \\ 1/2 & 3/2 \end{pmatrix} \begin{pmatrix} q_0 \\ q_1 \end{pmatrix} = 0.$$

Setting $p_2 = 1$ we find the solution $p_0 = 8$, $p_1 = -6$, $q_0 = 4$, $q_1 = -2$. The corresponding rational function

$$r(x) = \frac{8 - 6x + x^2}{4 - 2x} = \frac{(4 - x)(2 - x)}{2(2 - x)}$$

has the common factor $(2 - x)$ and is *reducible* to $f_{2,1} = (4 - x)/2$. The original form is indeterminate $0/0$ at $x = 2$, while the reduced form does not take on the prescribed value at $x = 2$.

As shown in the above example, for given data $(x_0, f_0), \dots, (x_n, f_n)$ there can be certain points x_j where the given function value f_j cannot be attained. Such a point x_j is called **unattainable**. This can occur only if x_j is a zero of the denominator $Q_n(x)$. From (4.3.26) it also follows that $P_m(x_j) = 0$. Hence the polynomials $P(x)$ and $Q(x)$ have a common factor $(x - x_j)^d$, where d is chosen maximal. The polynomials pair

$$P^*(x) = \frac{P(x)}{(x - x_j)^d}, \qquad Q^*(x) = \frac{Q(x)}{(x - x_j)^d} \tag{4.3.29}$$

then satisfies (4.3.26) for all points $x_k \neq x_j$. Since d was chosen maximal it holds that $Q^*(x_j) \neq 0$, and the value $P^*(x_j)/Q^*(x_j)$ must be finite. But since when $Q_n(x_j) = 0$, (4.3.26) is satisfied for *any choice of* f_j, one cannot expect the rational function to interpolate a particular value at x_j.

If there are unattainable points, then the polynomials defined in (4.3.29) only solve the linearized equations in the attainable points. It can also happen that the polynomials given by the linearized equations have a common factor $(x - x^*)^d$, $d \geq 1$, with $x^* \neq x_i$, $i = 1 : n$. In this case all polynomials of the form

$$P^*(x) = \frac{P(x)}{(x - x_j)^\nu}, \qquad Q^*(x) = \frac{Q(x)}{(x - x_j)^\nu}, \qquad \nu = 0 : d,$$

satisfy the linearized system and the matrix $(\,A \quad FB\,)$ in (4.3.28) has at most rank $N + 1 - d$. Conversely we have the following result.

Theorem 4.3.4.

If the rank of the matrix $(\,A \quad FB\,)$ *equals* $N + 1 - d$*, then there exists a unique solution* p, q *corresponding to polynomials* P^* *and* Q^* *with degrees at most* $m - d$ *and* $n - d$*. Further, all solutions have the form*

$$r(x) = \frac{s(x)P^*(x)}{s(x)Q^*(x)},$$

where $s(x)$ *is a polynomial of degree at most* d*. A point* x_j *is unattainable if and only if* $Q^*(x_j) = 0$.

Proof. See Schaback and Werner [313, Theorem 12.1.8]. □

Another complication of rational interpolation is that zeros may occur in $Q_n(x)$, which are not common to $P_m(x)$. These zeros correspond to poles in $r(x)$, which if they lie inside the interval $[\min x_k, \max x_k]$ may cause trouble. In general it is not possible to determine *a priori* if the given data (x_k, f_k) will give rise to such poles. Neither do the coefficients in the representation of $r(x)$ give any hint of such occurrences.

An algorithm similar to Newton's algorithm can be used for finding rational interpolants in continued fraction form. Set $v_0(x) = f(x)$, and use a sequence of substitutions:

$$v_k(x) = v_k(x_k) + \frac{x - x_k}{v_{k+1}(x)}, \qquad k = 0, 1, 2, \dots. \tag{4.3.30}$$

The first two substitutions give

$$f(x) = v_0(x) = v_0(x_0) + \frac{x - x_0}{v_1(x)} = v_0(x_0) + \cfrac{x - x_0}{v_1(x_1) + \cfrac{x - x_1}{v_2(x)}}.$$

In general this gives a continued fraction

$$f(x) = a_0 + \cfrac{x - x_0}{a_1 + \cfrac{x - x_1}{a_2 + \cfrac{x - x_2}{a_3 +}}} \dots = a_0 + \frac{x - x_1}{a_1 +}\,\frac{x - x_2}{a_2 +}\,\frac{x - x_3}{a_3 +} \dots, \tag{4.3.31}$$

where $a_k = v_k(x_k)$, and we have used the compact notation introduced in Sec. 3.5.1. This becomes an identity if the expansion is terminated by replacing a_n in the last denominator by $a_n + (x - x_n)/v_{n+1}(x)$. If we set $x = x_k$, $k \le n$, then the fraction terminates before the residual $(x - x_n)/v_{n+1}(x)$ is introduced. This means that setting $1/v_{k+1} = 0$ will give a rational function which agrees with $f(x)$ at the points x_i, $i = 0 : k \le n$, assuming that the constants $a_0, \dots, a_k$ exist. These continued fractions give a sequence of rational approximations $f_{k,k}$, $f_{k+1,k}$, $k = 0, 1, 2, \dots$.

Introducing the notation

$$v_k(x) = [x_0, x_1, \dots, x_{k-1}, x]\phi \tag{4.3.32}$$

we have $a_k = [x_0, x_1, \dots, x_{k-1}, x_k]\phi$. Then by (4.3.30) we have

$$[x]\phi = f(x), \qquad [x_0, x]\phi = \frac{x - x_0}{[x]\phi - [x_0]\phi} = \frac{x - x_0}{f(x) - f(x_0)},$$
$$[x_0, x_1, x]\phi = \frac{x - x_1}{[x_0, x]\phi - [x_0, x_1]\phi},$$

and in general

$$[x_0, x_1, \dots, x_{k-1}, x]\phi = \frac{x - x_{k-1}}{[x_0, \dots, x_{k-2}, x]\phi - [x_0, \dots, x_{k-2}, x_{k-1}]\phi}. \tag{4.3.33}$$

Therefore, we also have

$$a_k = \frac{x_k - x_{k-1}}{[x_0, \dots, x_{k-2}, x_k]\phi - [x_0, \dots, x_{k-2}, x_{k-1}]\phi}. \tag{4.3.34}$$

We call the quantity defined by (4.3.34) the kth **inverse divided difference** of $f(x)$. Note that certain inverse differences can become infinite if the denominator vanishes. They are, in general, *symmetrical only in their last two arguments.*

The inverse divided differences of a function $f(x)$ can conveniently be computed recursively and arranged in a table similar to the divided-difference table.

$$\begin{array}{lllllll}
x_1 & f(x_1) & [x_1]\phi & & & & \\
 & & & [x_1, x_2]\phi & & & \\
x_2 & f(x_2) & [x_2]\phi & & [x_1, x_2, x_3]\phi & & \\
 & & & [x_2, x_3]\phi & & [x_1, x_2, x_3, x_4]\phi & \\
x_3 & f(x_3) & [x_3]\phi & & [x_2, x_3, x_4]\phi & & \\
 & & & [x_3, x_4]\phi & & & \\
x_4 & f(x_4) & [x_4]\phi & & & &
\end{array}$$

Here the upper diagonal elements are the desired coefficients in the expansion (4.3.31).

Example 4.3.7.

Assume that we want to interpolate the four points given in Example 4.3.6 and the additional points (4, 6/17). Forming the inverse differences we get the following table.

x_i	f_i	ϕ_1	ϕ_2	ϕ_3	ϕ_4
0	2				
		−2			
1	3/2		3		
		5/3		0	
2	4/5		∞		−5
		−2		−1/5	
3	1/2		−7		
		−17/7			
4	6/17				

This gives a sequence of rational approximations. If we terminate the expansion

$$f_{2,2} = 2 + \frac{x}{-2+}\,\frac{x-1}{3+}\,\frac{x-2}{0+}\,\frac{x-3}{-5}$$

after a_3 we recover the solution of the previous example. Note that the degeneracy of the approximation is shown by the entry $a_3 = 0$. Adding the last fraction gives the (degenerate) approximation

$$f_{2,2} = \frac{2+x}{1+x^2}.$$

It is verified directly that this rational function interpolates all the given points.

Because the inverse differences are not symmetric in all their arguments the **reciprocal differences** are often preferred. These are recursively defined by

$$[x_i]\rho = f_i, \qquad [x_i, x_{i+1}]\rho = \frac{x_i - x_{i+1}}{f_i - f_{i+1}}, \tag{4.3.35}$$

$$[x_i, x_{i+1}, \dots, x_{i+k}]\rho = \frac{x_i - x_{i+k}}{[x_i, \dots, x_{i+k-1}]\rho - [x_{i+1}, \dots, x_{i+k}]\rho} + [x_{i+1}, \dots, x_{i+k-1}]\rho. \tag{4.3.36}$$

See Hildebrand [201, p. 406]. While this formula is less simple than (4.3.34), the reciprocal differences are *symmetric functions of all their arguments*. The symmetry permits the calculation of the kth reciprocal difference from any two $(k-1)$th reciprocal differences having $k-1$ arguments in common, together with the $(k-2)$th reciprocal difference formed with this argument. Using (4.3.36) a reciprocal difference table may be constructed.

The coefficients in the continued fraction (4.3.31) can then be determined by an interpolation formula due to Thiele [350]:

$$a_0 = f_0, \qquad a_1 = [x_0, x_1]\rho, \qquad a_2 = [x_0, x_1, x_2]\rho - f_0,$$
$$a_3 = [x_0, x_1, x_2, x_3]\rho - [x_0, x_1]\rho, \dots .$$

The formulas using inverse or reciprocal differences are useful if one wants to determine the coefficients of the rational approximation, and use it to compute approximations for several arguments. If one only wants the value of the rational interpolating function for a single argument, then it is more convenient to use an alternative algorithm of Neville type. This is the case in the ρ-algorithm, which is a convergence acceleration procedure using rational interpolation to extrapolate to infinity with the same degree in the numerator and denominator.

If we consider the sequence of rational approximations of degrees (m, n)

$$(0,0), \quad (0,1), \quad (1,1), \quad (1,2), \quad (2,2),$$

the following recursive algorithm results (Stoer and Bulirsch [338, Sec. 2.2]):
For $i = 0, 1, 2, \ldots$, set $T_{i,-1} = 0$, $T_{i,0} = f_i$, and

$$T_{ik} = T_{i,k-1} + \frac{T_{i,k-1} - T_{i-1,k-1}}{\dfrac{x - x_{i-k}}{x - x_i}\left[1 - \dfrac{T_{i,k-1} - T_{i-1,k-1}}{T_{i,k-1} - T_{i-1,k-2}}\right] - 1}, \qquad 1 \le k \le i. \tag{4.3.37}$$

As in Neville interpolation the calculations can be arranged in a table of the following form.

(m,n)	$(0,0)$	$(0,1)$	$(1,1)$	$(1,2)$	$\cdots$
	$f_1 = T_{1,0}$				
0		$T_{2,1}$			
	$f_2 = T_{2,0}$		$T_{3,2}$		
0		$T_{3,1}$		$T_{4,3}$	
	$f_2 = T_{2,0}$		$T_{4,2}$	$\vdots$	$\ddots$
0		$T_{4,1}$	$\vdots$		
	$f_4 = T_{4,0}$	$\vdots$			
$\vdots$	$\vdots$				

Here any entry is determined by a rhombus rule from three entries in the preceding two columns. Note that it is easy to add a new interpolation point in this scheme.

As shown by Berrut and Mittelmann [23], every rational interpolant can be written in barycentric form

$$r(x) = \sum_{k=0}^{N} \frac{u_k}{x - x_k} f_k \Big/ \sum_{k=0}^{N} \frac{u_k}{x - x_k}. \tag{4.3.38}$$

Let $q_k = Q_n(x_k)$, $k = 1 : N$, be the values of the denominator at the nodes. Then the barycentric representation of the denominator $Q_n(x)$ is

$$Q_n(x) = \prod_{i=0}^{N}(x - x_i) \sum_{k=0}^{N} \frac{w_k}{x - x_k} q_k, \qquad w_k = 1 \Big/ \prod_{i=0}^{N}(x_k - x_i).$$

Hence $r(x)$ can be written as in (4.3.38), where $u_k = w_k q_k$ is the weight corresponding to the node x_k. Since $w_k \neq 0$ for all k, it follows that $q_k = 0$ at a node if and only if the corresponding weight equals zero.

The barycentric form has the advantage that the barycentric weights give information about possible unattainable points. However, the determination of the parameter vector $u = (u_0, u_1, \dots, u_N)^T$ is more complicated. An elimination method for the computation of u in $O(n^3)$ operations is given by Berrut and Mittelmann [23].

4.3.4 Multidimensional Interpolation

Polynomial interpolation for functions of several independent variables are generally more difficult than the one-dimensional case. There is in general a lack of uniqueness. In particular, it may not suffice to require that the interpolation points are distinct; see Problem 4.3.9 (b).

As a simple example, consider the problem of interpolating a function given at three distinct points $p_i = (x_i, y_i)$, $i = 1:3$, by a linear function in two variables,

$$p(x, y) = c_1 + c_2 x + c_3 y.$$

This leads to the linear system $Vc = f$, where

$$V = \begin{pmatrix} 1 & x_1 & y_1 \\ 1 & x_2 & y_2 \\ 1 & x_3 & y_3 \end{pmatrix}, \qquad c = \begin{pmatrix} c_1 \\ c_2 \\ c_3 \end{pmatrix}, \qquad f = \begin{pmatrix} f_1 \\ f_2 \\ f_3 \end{pmatrix}.$$

The interpolation problem has exactly one solution if V is nonsingular, i.e., when $\det(V) \neq 0$. But $\frac{1}{2}\det(V)$ is just the area of the triangle with vertices (x_i, y_i), $i = 1:3$. If this area is zero, then the three points lie on a line and the problem has either infinitely many solutions or no solution.

Much of the theory for univariate interpolation can be generalized to multidimensional interpolation problems provided that the function is specified on a Cartesian (tensor) product grid. For simplicity, we first concentrate on functions $f(x, y)$ of two variables, but the extension to more dimensions is not difficult. Assume that we are given function values

$$f_{ij} = f(x_i, y_j), \quad i = 1:n, \quad j = 1:m, \tag{4.3.39}$$

where x_i, $i = 1:n$, and y_j, $j = 1:m$, are distinct points. We seek a polynomial $p(x, y)$ of degree at most $n-1$ in x and at most $m-1$ in y that interpolates these values. Such a polynomial has the form

$$p(x, y) = \sum_{i=0}^{n-1}\sum_{j=0}^{m-1} c_{ij} x^i y^j, \tag{4.3.40}$$

where the mn coefficients c_{ij} are to be determined. Since the number of coefficients equal the number of interpolatory conditions we can expect the polynomial $p(x, y)$ to be uniquely determined. To show this it suffices to show that any polynomial $q(x, y)$ of degree $n-1$ in x and $m-1$ in y that vanishes at the mn distinct points (x_i, y_j), $i = 1:n$, and y_j, $j = 1:m$, must vanish identically.

If we want to compute $p(x, y)$ for given values of x and y, then we can proceed as follows. For each $j = 1:m$, use univariate interpolation to determine the values $p(x, y_j)$, where $p(x, y)$ is the interpolation polynomial in (4.3.40). Next, the m values $p(x, y_j)$,

$j = 1 : m$, are interpolated using univariate interpolation, which determines $p(x, y)$. Note that since the points x_i and y_j are distinct all univariate interpolation polynomials are uniquely determined. It is clear that we will obtain the same result, whether we interpolate first for x and then for y or vice versa. Clearly this approach can also be used in more than two dimensions.

In many cases we are not satisfied with obtaining $p(x, y)$ for specific values of x and y, but want to determine $p(x, y)$ as a polynomial in x and y. We can then use the above procedure *algebraically* to derive a Newton formula for a tensor product interpolation problem in two variables. We set $[x_i; y_j]f = f(x_i, y_j)$, and define bivariate divided differences $[x_1, \dots, x_\nu; y_1, \dots, y_\mu]f$, by recurrences, separately for each variable. We start by forming, for each y_j, $j = 1 : m$, divided differences with respect to the x variable:

$$\begin{array}{llll} [x_1; y_1]f & [x_1, x_2; y_1]f & \cdots & [x_1, x_2, \dots, x_n; y_1]f, \\ [x_1; y_2]f & [x_1, x_2; y_2]f & \cdots & [x_1, x_2, \dots, x_n; y_2]f, \\ \vdots & \vdots & & \vdots \\ [x_1; y_m]f & [x_1, x_2; y_m]f & \cdots & [x_1, x_2, \dots, x_n; y_m]f. \end{array}$$

These are used to form the Newton polynomials

$$p(x, y_j) = \sum_{i=1}^{n} [x_1, \dots, x_i; y_j]f \prod_{\nu=1}^{i-1} (x - x_\nu), \qquad j = 1 : m,$$

which give, for any x, the values of the interpolation polynomial $p(x, y)$ for $y_1, \dots, y_m$. Next we form in each column above the divided differences with respect to y:

$$\begin{array}{llll} [x_1; y_1]f & [x_1, x_2; y_1]f & \cdots & [x_1, \dots, x_n; y_1]f, \\ [x_1; y_1, y_2]f & [x_1, x_2; y_1, y_2]f & \cdots & [x_1, \dots, x_n; y_1, y_2]f, \\ \vdots & \vdots & & \vdots \\ [x_1; y_1, \dots, y_m]f & [x_1, x_2; y_1, \dots, y_m]f & \cdots & [x_1, \dots, x_n; y_1, \dots, y_m]f. \end{array}$$

If these nm divided differences are used for Newton interpolation in the y variable, we obtain Newton's interpolation formula in two variables,

$$p(x, y) = \sum_{i=1}^{n} \sum_{j=1}^{m} [x_1, \dots, x_i; y_1, \dots, y_j]f \prod_{\nu=1}^{i-1} (x - x_\nu) \prod_{\mu=1}^{j-1} (y - y_\mu), \tag{4.3.41}$$

where empty products have the value 1. Note that it is indifferent in which order the divided differences are formed. We could equally well have started to form divided differences with respect to y.

Remainder formulas can be derived from the corresponding one-dimensional error formulas; see Isaacson and Keller [208, Sec. 6.6]. For f sufficiently smooth there exist values ξ, ξ', η, η' such that

$$R(x, y) = \frac{\partial^n f(\xi, y)}{\partial x^n} \frac{\prod_{\nu=1}^{n}(x - x_\nu)}{n!} + \frac{\partial^m f(x, \eta)}{\partial y^m} \frac{\prod_{\mu=1}^{m}(y - y_\mu)}{m!} \tag{4.3.42}$$

$$- \frac{\partial^{n+m} f(\xi', \eta')}{\partial x^n \partial y^m} \frac{\prod_{\nu=1}^{n}(x - x_\nu)}{n!} \frac{\prod_{\mu=1}^{m}(y - y_\mu)}{m!}. \tag{4.3.43}$$

Lagrange's interpolation formula can also be generalized for the tensor product case. We have

$$p(x, y) = \sum_{i=1}^{n} \sum_{j=1}^{m} f(x_i, y_j) \prod_{\substack{\nu=1 \\ \nu \neq i}}^{n} \frac{(x - x_\nu)}{(x_i - x_\nu)} \prod_{\substack{\mu=1 \\ \mu \neq j}}^{n} \frac{(y - y_\mu)}{(y_j - y_\mu)}. \tag{4.3.44}$$

Clearly $p(x, y)$ assumes the values $f(x_i, y_j)$ for $i = 1 : n$, $j = 1 : m$. As the polynomial is of degree $n - 1$ in x and $m - 1$ in y, it must equal the unique interpolating polynomial. Therefore, the remainder must also be the same as for the Newton formula. The Lagrange formula (4.3.44) is easily extended to three and more variables.

The interpolation problem we have treated so far specifies the maximum degree of $p(x, y)$ in x and y separately. Instead we could specify the total degree to be at most $n - 1$. Then the interpolation polynomial must have the form

$$p(x, y) = \sum_{i=0}^{n-1} \sum_{j=0}^{n-i-1} b_{ij} x^i y^j. \tag{4.3.45}$$

There are $\frac{1}{2}n(n + 1)$ coefficients to determine in (4.3.45). We shall show that with the "triangular" array of interpolation points (x_i, y_j), $i + j = 1 : n$, the interpolation polynomial is uniquely determined.

The Newton formula (4.3.41) can be generalized to the case when for each i, $i = 1 : n$, the interpolation given points are (x_i, y_j), $j = 1 : m_i$ with $1 \le m_i \le m$, with a slightly more complicated remainder formula. A particularly interesting case is when $m_i = n - i + 1$, i.e., the interpolation points form a triangle. This gives rise to the interpolating polynomial

$$p(x, y) = \sum_{2 \le i+j \le n+1} [x_1, \dots, x_i; y_1, \dots, y_j] f \prod_{\nu=1}^{i-1}(x - x_\nu) \prod_{\mu=1}^{j-1}(y - y_\mu), \tag{4.3.46}$$

with remainder formula

$$R(x, y) = \sum_{i=1}^{n+1} \frac{\partial^n f(\xi_i, \eta_i)}{\partial x^i \partial y^{n-i}} \frac{\prod_{\nu=1}^{i-1}(x - x_\nu)}{(i - 1)!} \frac{\prod_{\mu=1}^{n-i}(y - y_\mu)}{(n - i)!}. \tag{4.3.47}$$

This formula is due to Biermann [26].

Interpolation formulas for equidistant points $x_i = x_0 + ih$ and $y_j = y_0 + jk$ can readily be obtained from Newton's formula (4.3.41). Using the points (x_i, y_j), $i = 0 : n$, $i = 0 : m$, we get

$$p(x_0 + ph, y + qk) = \sum_{i=0}^{n} \sum_{j=0}^{m} \binom{p}{i} \binom{q}{j} \Delta_x^i \Delta_y^j f(x_0, y_0). \tag{4.3.48}$$

Example 4.3.8.

Formulas for equidistant points can also be obtained by using the operator formulation of Taylor's expansion:

$$\begin{aligned} f(x_0+h, y_0+k) &= \exp\big(hD_x + kD_y\big) f(x_0, y_0) \qquad (4.3.49)\\ &= f_{0,0} + \big(hD_x + kD_y\big) f_{0,0} \\ &\quad + \big(h^2D_x^2 + 2hkD_xD_y + k^2D_y^2\big) f_{0,0} + O(h^3+k^3). \end{aligned}$$

An interpolation formula, exact for all functions $p(x, y)$ in (4.3.40) with $m = n = 3$, can be obtained by replacing in Taylor's formula the derivatives by difference approximations valid for quadratic polynomials,

$$\begin{aligned} f(x_0+ph, y_0+qh) &\approx f_{0,0} + \frac{1}{2}p(f_{1,0} - f_{-1,0}) + \frac{1}{2}q(f_{0,1} - f_{0,-1}) \\ &\quad + \frac{1}{2}p^2(f_{1,0} - 2f_{0,0} + f_{-1,0}) \qquad (4.3.50)\\ &\quad + \frac{1}{4}pq(f_{1,1} - f_{1,-1} - f_{-1,1} + f_{-1,-1}) \\ &\quad + \frac{1}{2}q^2(f_{0,1} - 2f_{0,0} + f_{0,-1}). \end{aligned}$$

This formula uses function values in nine points. (The proof of the expression for approximating the mixed derivative $D_xD_yf_{0,0}$ is left as an exercise; see Problem 4.3.11.)

An important case, to be treated in Sec. 5.4.4, is interpolation formulas in two or more dimensions with function values specified on the vertices and sides of a simplex. These play a fundamental role in the finite element method.

4.3.5 Analysis of a Generalized Runge Phenomenon

In this section we make a more detailed theoretical and experimental study of the Runge phenomenon (see Sec. 4.2.6). We then study interpolation at an infinite equidistant point set from the point of view of complex analysis. This interpolation problem, which was studied by Whittaker and others in the beginning of the twentieth century, became revived and modified in the middle of the same century under the name of the **Shannon sampling theorem**, with important applications to communication theory.

It is well known that the Taylor series of an analytic function converges at an exponential rate inside its circle of convergence, while it diverges at an exponential rate outside. The circle of convergence passes through the nearest singularity. We shall see that similar results hold for certain interpolation processes. In general, the domains of convergence are not disks but bounded by level curves of a logarithmic potential, related to the asymptotic distribution of the interpolation points.

For the sake of simplicity, we now confine the discussion to the case when the points of interpolation are located in the standard interval $[-1, 1]$, but we are still interested in the evaluation of the polynomials in the complex domain. Part of the discussion can, however, be generalized to a case when the interpolation points are on an arc in the complex plane.

We shall study interpolation processes which are *regular* in the following sense. Let the nodes $t_{n,j}$, $j = 1 : n$, $n = 1, 2, 3, \ldots$, be such that there exists an increasing continuously differentiable function $q : [a, b] \mapsto [-1, 1]$ such that

$$q(a + (b-a)(j-1)/n) < t_{n,j} \le q(a + (b-a)j/n),$$

i.e., one node in each of n subintervals of $[-1, 1]$. More precisely, we assume that $q'(\tau) > 0$, $\tau \in (0, 1)$, while $q'(0)$, $q'(1)$ may be zero. Suppose that $z \notin [-1, 1]$, but z may be close to this interval. Then

$$\begin{aligned}\frac{1}{n}\ln \Phi_n(z) &= \frac{1}{n}\sum_{j=1}^{n} \ln (z - t_{n,j}) \\ \to \psi(z) &:= \frac{1}{b-a}\int_a^b \ln (z - q(\tau))\, d\tau, \quad (n \to \infty). \end{aligned} \tag{4.3.51}$$

The crucial factors of the interpolation error are $\Phi_n(u)/\Phi_n(p)$ and $\Phi_n(u)/\Phi_n(z)$. We now obtain a fundamental approximation formula:

$$\begin{aligned}\Phi_n(u)/\Phi_n(z) &= \exp\big(\ln \Phi_n(u) - \ln \Phi_n(z)\big) \\ &= \exp n\big(\psi(u) - \psi(z) + o(1)\big), \quad (n \to \infty).\end{aligned} \tag{4.3.52}$$

If u and z are bounded away from the nodes, the $o(1)$-term is of marginal importance; it is basically $O(1/n)$. For $u \in [-1, 1]$ $\Phi_n(u)$ and $\ln|\Phi_n(u)|$ are oscillatory, and this formula is there to be considered as a one-sided approximate error bound; see Figure 4.3.2 and the more detailed preparatory discussion leading up to Proposition 4.3.6.

We now make a variable transformation in order to be able to utilize results of classical potential theory. Put $q(\tau) = t \in [-1, 1]$, $dt = q'(\tau)d\tau$. Then we can define an asymptotic **node density** for the process,

$$w(t) = \frac{1}{q'_\tau(\tau(t))(b-a)}, \qquad w(t)dt = \frac{d\tau}{b-a}, \tag{4.3.53}$$

where w is the derivative of the inverse function of q divided by $b - a$. Then

$$\psi(z) = \int_{-1}^{1} \ln(z-t)w(t)\,dt, \quad w \in C(-1, 1), \quad w(t) > 0, \quad \int_{-1}^{1} w(t)\,dt = 1, \tag{4.3.54}$$

and $\psi(z)$ is analytic in the whole plane outside the interval $[-\infty, 1]$. Its real part is the **logarithmic potential** of the density w with reversed sign,

$$\frac{1}{n}\ln|\Phi_n(z)| \approx \Re\psi(z) = \int_{-1}^{1} \ln|z-t|w(t)\,dt. \tag{4.3.55}$$

$\Re\psi(z)$ is a harmonic function for all $z = x+iy \notin [-1, 1]$, i.e., it satisfies Laplace's equation $\partial^2\Re\psi/\partial x^2 + \partial^2\Re\psi/\partial y^2 = 0$.

The function $\frac{1}{n}\ln|\Phi_n(z)|$ is itself the logarithmic potential of a discrete distribution of equal weights $1/n$ at the nodes $t_{j,n}$, but it is less pleasant to deal with than $\Re\psi(z)$. For example, it becomes $-\infty$ at the nodes while, according to classical results of potential heory, $\Re\psi(z)$ *is continuous and single-valued everywhere, also on the interval* $[-1, 1]$ (see Figure 4.3.1). The imaginary part, however, is not single-valued. It becomes single-valued

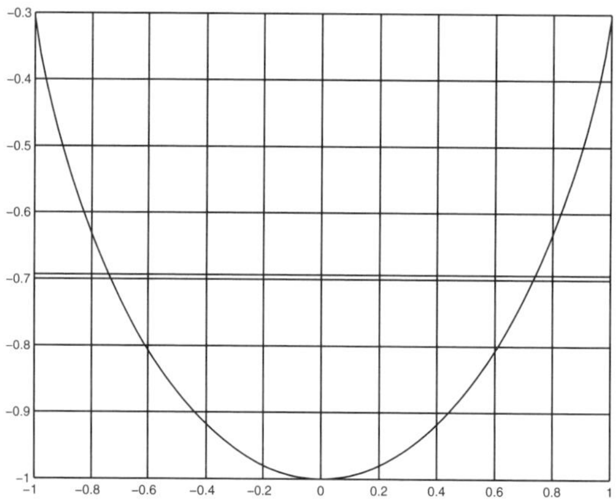

Figure 4.3.1. $\Re\psi(u)$, $u \in [-1,1]$, *for the processes based on equidistant nodes and on Chebyshev nodes. For Chebyshev nodes, the convergence properties are the same over the whole interval* $[-1,1]$, *because* $\Re\psi(u) = -\ln 2 = -0.693$ *for all* $u \in [-1,1]$. *For equidistant nodes, the curve, which is based on* (4.3.61), *partly explains why there are functions* f *such that the interpolation process diverges fast in the outer parts of* $[-1,1]$ *and converges fast in the central parts (often faster than Chebyshev interpolation).*

if we cut the complex plane along the real axis from $-\infty$ to 1, but it tends to $+\pi$ or $-\pi$, depending on whether the cut is approached from above or from below.

Another result from classical potential theory that will be useful later reads

$$\psi'(x-0i) - \psi'(x+0i) = 2\pi i w(x), \quad x \in [-1,1]. \tag{4.3.56}$$

For $|z| \gg 1$, $\psi(z)$ depends only weakly on the node distribution, and the level curves approach circles since, by (4.3.55),

$$\begin{aligned}\psi(z) &= \int_{-1}^{1} \Big(\ln z + \ln(1 - z^{-1}t)\Big) w(t)\,dt \\ &= \ln z - z^{-1}\int_{-1}^{1} t w(t)\,dt - \frac{1}{2}z^{-2}\int_{-1}^{1} t^2 w(t)\,dt \ldots.\end{aligned} \tag{4.3.57}$$

Note that the coefficient of z^{-1} vanishes for a symmetric node distribution.

We have

$$\frac{\partial \Re\psi}{\partial y} = \frac{\Re\partial\psi}{\partial y} = \Re\int_{-1}^{1} \frac{i}{x-t+iy} w(t)\,dt = \int_{-1}^{1} \frac{y}{(x-t)^2+y^2)} w(t)\,dt > 0$$

if $y > 0$. Then $\Re\psi(x+iy)$ is an increasing function of y for $y > 0$.

$$\frac{\partial \Re\psi}{\partial x} = \frac{\Re\partial\psi}{\partial x} = \Re\int_{-1}^{1} \frac{1}{x-t+iy} w(t)\,dt = \int_{-1}^{1} \frac{x-t}{(x-t)^2+y^2)} w(t)\,dt > 0$$

if $x > 1$. Then $\Re\psi(x+iy)$ is an increasing function of x for $x > 1$. Similarly $\Re\psi(x+iy)$ is a decreasing function of y for $y < 0$, and a decreasing function of x for $x < -1$.

We define the region

$$\mathcal{D}(v) = \{z \in \mathbf{C} : \Re\psi(z) < \Re\psi(v)\}. \tag{4.3.58}$$

Set $P^* = \max_{x\in[-1,1]} \Re\psi(x)$, and suppose that $\Re\psi(v) > P^*$. It then follows that $\mathcal{D}(v)$ is a simply connected domain, and the level curve

$$\partial\mathcal{D}(v) = \{z : \Re\psi(z) = \Re\psi(v)\}$$

encloses $[-1, 1]$. Suppose that $a' > a$ and $a' > P^*$. Then the level curve $\{z : \Re\psi(z) = a\}$ is strictly inside the level curve $\{z : \Re\psi(z) = a'\}$. (The proof of these statements essentially utilizes the minimum principle for harmonic functions and the fact that $\Re\psi(z)$ is a regular harmonic function outside $[-1, 1]$ that grows to ∞ with $|z|$.)

Suppose that $f(z)$ is analytic for $|z| = R$, where R is so large that we can set $\psi(z) \approx \ln z$; see (4.3.57). We then obtain, by Theorem 4.3.3, (4.3.20), and (4.3.52),

$$\ln \left|(f - L_n f)(u)\right| \approx n\big(\Re\psi(u) - \ln R + o(1)\big) + \ln M(R). \tag{4.3.59}$$

Example 4.3.9 (*An Entire Function*).

For $f(z) = \exp z^\alpha$, $\alpha > 0$, $\ln M(R) = R^\alpha$. For a fixed n large enough so that the $o(1)$-term can be neglected, the bound in (4.3.59) has a minimum for $R = (n/\alpha)^{1/\alpha}$. Inserting this into (4.3.59) we obtain

$$\ln \left|(f - L_n f)(u)\right| \le n\Big(\Re\psi(u) - \frac{1}{\alpha}(\ln n + 1 + \ln\alpha) + o(1)\Big),$$

which shows that, in this example, the convergence is faster than exponential, and depends rather weakly on the node distribution.

The following estimate comes as a consequence,

$$\left|(f - L_n f)(u)\right| = \Phi_n(u) n^{-n} e^{-n+o(1)n},$$

but this is no surprise. If u is real and $\alpha = 1$, it follows directly from the remainder term (4.2.10) in interpolation

$$(f - L_n f)(u) = \Phi_n(u) e^{\xi_u}/n!,$$

together with Stirling's formula (3.2.36).

The technique used in this example is, however, very general. It can be used for complex u, and for more complicated entire functions.

Example 4.3.10 (*Functions with Poles*).

We choose $\mathcal{D} = \mathcal{D}(v) = \{z \in \mathbf{C} : \Re\psi(z) < \Re\psi(v)\}$ and assume that $f(z)$ has two conjugate poles, p, $\bar{p}$, $\Re\psi(p) < \Re\psi(v)$. (There may be other poles outside $\mathcal{D}(v)$.) Consider the error formula of Theorem 4.3.3. For $n \gg 1$, the ratio of $K_n f$ to the contribution from the poles is $\exp(-n(\Re\psi(v) - \Re\psi(p) + o(1)))$; we can thus neglect $K_n f(u)$. It follows that

$$\begin{aligned}\left|(f - L_n f)(u)\right| &\approx \left|\Phi_n(u) \sum_p \frac{\mathrm{res}_f(p)}{\Phi_n(p)(p-u)}\right| \\ &= \exp n\big(\Re\psi(u) - \Re\psi(p) + o(1)\big) \cdot \left|\frac{2\mathrm{res}_f(p)}{(p-u)}\right|.\end{aligned} \tag{4.3.60}$$

An important conclusion is that *the level curve of the logarithmic potential through the pole p separates the points $u \notin [-1, 1]$, where the interpolation process converges from the points where it diverges.* This **separation statement** holds under more general conditions than we have in this example.

For $u \in [-1, 1]$ there are, however, interpolation processes that may converge in an enumerable point set that is everywhere dense in $[-1, 1]$, even though $\Re\psi(u) > \Re\psi(p)$ in large parts of this interval. It is doubtful if such a process can be regular in the sense defined above, but it can be a subprocess of a regular process.[133] Figure 4.3.3 may give hints how the above separation statement is to be modified in order to make sense also in such a case. The smooth curves of Figure 4.3.3 have been computed by means of (4.3.60) for Runge's example $f(z) = 1/(1 + 25z^2)$ with equidistant nodes; see Example 4.3.11.

We now consider two node distributions.

Example 4.3.11 (*Equidistant Interpolation*).

In this case we may take $q(\tau) = \tau$, $\tau \in [-1, 1]$, $t \in [-1, 1]$, hence $w(t) \equiv 1/2$. For this **equidistant case** we have, if $z \notin [-\infty, 1]$,

$$\psi(z) = \frac{1}{2}\int_{-1}^{1} \ln(z - t)\,dt = \frac{1}{2}\Big((1 - z)\ln(z - 1) + (1 + z)\ln(z + 1)\Big) - 1.$$

The real part $\Re\psi(z)$ is, however, single-valued and continuous everywhere, as mentioned in the comment after (4.3.55). Some level curves are shown in Figure 4.3.2. Note that the tangent of a curve is discontinuous at the intersection with $[-1.1]$. On the imaginary axis,

$$\Re\psi(iy) = \frac{1}{2}\ln(1 + y^2) + y\left(\text{sign}\,(y)\,\frac{\pi}{2} - \arctan y\right) - 1.$$

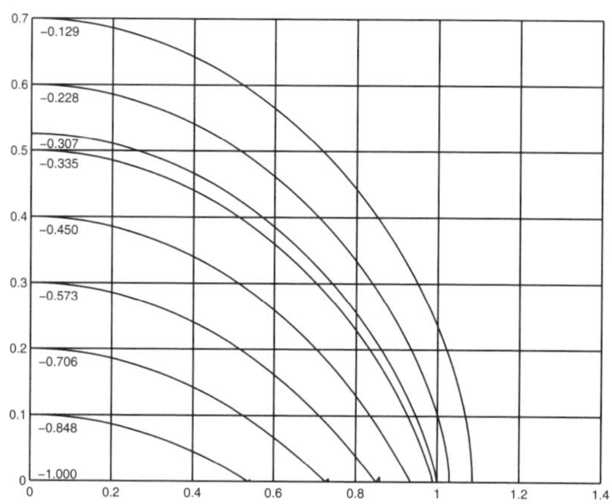

Figure 4.3.2. *Some level curves of the logarithmic potential for $w(t) \equiv \frac{1}{2}$, $t \in [-1, 1]$. Due to the symmetry only one quarter of each curve is shown. The value of $\Re\psi(z)$ on a curve is seen to the left close to the curve. It is explained in Example* 4.3.11 *how the curves have been computed.*

[133]For example, a process generated by successive bisections of the interval $[-1, 1]$ can be a subprocess of one of the equidistant processes studied in next example.

When $z \to x \in [-1, 1]$, from any direction, $\Re\psi(z)$ tends to

$$\Re\psi(x) = \frac{1}{2}\big((1-x)\ln(1-x) + (1+x)\ln(1+x)\big) - 1. \tag{4.3.61}$$

The level curve of $\Re\psi(z)$ that passes through the points ± 1 intersects the imaginary axis at the points $\pm iy$, determined by the equation $\Re\psi(iy) = \Re\psi(1) = \ln 2 - 1$, with the root $y = 0.5255$. Theorem 4.3.5 (below) will tell us that $L_n f(x) \to f(x)$ for all $x \in (-1, 1)$, if $f(z)$ is analytic inside and on this contour.

In the classical example of Runge, $f(z) = 1/(1 + 25z^2)$ has poles inside this contour at $z = \pm 0.2i$. The separation statement in Example 4.3.10 told us that the level curve of $\Re\psi(z)$ which passes through these poles will separate between the points, where the interpolation process converges and diverges. Its intersection with the real axis is determined by the equation $\Re\psi(x) = \Re\psi(0.2i) = -0.70571$. The roots are $x = \pm 0.72668$; see also Figure 4.3.2.

The theory based on the logarithmic potential is of asymptotic nature, and one may ask how relevant it is when the number of nodes is of a reasonable size for practical computation. After all, the behavior of the interpolation error between the nodes is quite different from the smooth logarithmic potential. Figure 4.3.3 indicates, however, that the prediction of this theory can be rather realistic when we ask for local maxima of the modulus of the interpolation error on the real axis.

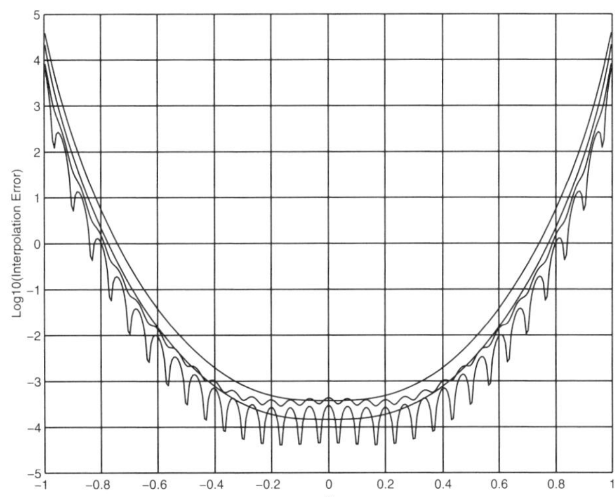

Figure 4.3.3. $\log_{10}|(f - L_n f)(u)|$ *for Runge's classical example* $f(u) = 1/(1 + 25u^2)$ *with* 30 equidistant nodes *in* $[-1, 1]$. *The oscillating curves are the empirical interpolation errors (observed at* 300 *equidistant points), for* $u = x$ *in the lower curve and for* $u = x + 0.02i$ *in the upper curve; in both cases* $x \in [-1, 1]$. *The smooth curves are the estimates of these quantities obtained by the logarithmic potential model; see Examples* 4.3.10 *and* 4.3.11.

Example 4.3.12 (*Chebyshev Interpolation Revisited*).

In this example we have $q(t) = \cos(\pi\tau)$, $\tau \in [-1, 0]$. By (4.3.53)

$$w(t) = \frac{1}{-\pi \sin \pi\tau} = \frac{1}{\pi}(1-t^2)^{-\frac{1}{2}}.$$

Moreover,[134]

$$\Phi_n(z) = 2^{1-n}T_n(z) = 2^{-n}(s^n + s^{-n}),$$

where $z = \frac{1}{2}(s + s^{-1})$, $s = z + \sqrt{z^2-1}$. According to our convention about the choice of branch for the square root $|s| \geq 1$. Hence,

$$\Re\psi(z) = \lim \frac{1}{n} \ln |\Phi_n(z)| - \ln 2 = \ln \frac{|s|}{2} = \ln |z + \sqrt{z^2-1}| - \ln 2.$$

As in the previous example, $\Re\psi(z)$ is single-valued and continuous everywhere, while $\psi(z)$ is unique for $z \notin [-\infty, 1]$ only. Therefore, the family of confocal ellipses $\partial\mathcal{E}_R$ with foci at ± 1 are the level curves of $\Re\psi(z)$. In fact, the interior of $\mathcal{E}_R$ equals $\mathcal{D}(\ln R - \ln 2)$. The family includes, as a limit case ($R \to 1$), the interval $[-1, 1]$, on which $\Re\psi(z) = \ln R - \ln 2 \to -\ln 2 = -0.693$. Note that as $z \to \cos\phi \in [-1, 1]$,

$$|z + \sqrt{(z^2-1)}| = |\cos\phi + i \sin\phi| = 1, \quad \forall\phi.$$

Some level curves are shown in Figure 4.3.4.

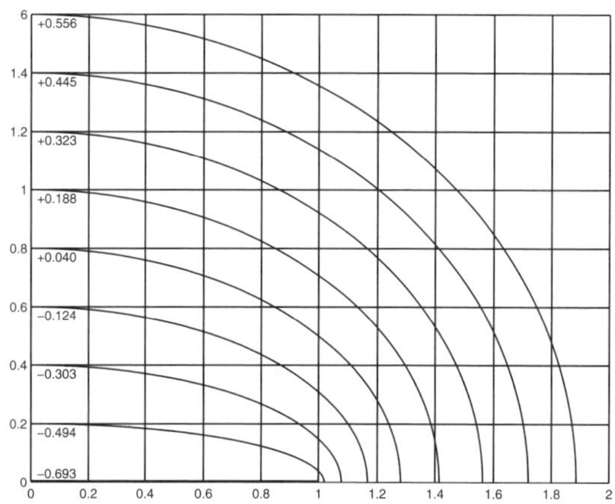

Figure 4.3.4. *Some level curves of the logarithmic potential associated with Chebyshev interpolation. They are ellipses with foci at ± 1. Due to the symmetry only a quarter is shown of each curve. The value of $\Re\psi(z)$ for a curve is seen to the left, close to the curve.*

Note that $\Re\psi(z)$ is smaller in $[-1, 1]$ than anywhere else; this confirms the conclusion in Sec. 4.3.2 that, for any function analytic in a region $\mathcal{D}$ that encloses $[-1, 1]$, Chebyshev interpolation converges at the same exponential rate in (almost) all points u close to $[-1, 1]$.

[134]The complex variable denoted w in Sec. 3.2.3 is now denoted s, in order to avoid a collision with the density $w(t)$.

For the classical Runge case, we have

$$|(f - L_n f)(u)| \approx e^{n(\psi(0)-\psi(0.02i))} = e^{n(-0.693+0.494)} = e^{-0.199n}.$$

This is very different from the equidistant case which diverges if $|u| > 0.72668$ but at the central subinterval has, by Figure 4.3.2,

$$|(f - L_n f)(u)| \approx e^{n(-1+0.706)} = e^{-0.294n},$$

which is better than Chebyshev interpolation.

Also note that if $z \notin [-1, 1]$, we can, by the definition of $\Re\psi(z)$ as a Riemann sum (see (4.3.51)), find a sequence $\{\epsilon_n\}$ that decreases monotonically to zero such that

$$\frac{1}{n}\Big| \ln \Phi_n(z) - \psi(z)\Big| < \epsilon_n, \quad z \notin [-1, 1]. \tag{4.3.62}$$

It is conceivable that the same sequence can be used for all z on a curve that does not touch the interval $[-1, 1]$. (This can be proved, but the proof is omitted.)

We can only claim a *one-sided inequality* if we allow that $u \in [-1, 1]$

$$\frac{1}{n}\Re\big(\ln \Phi_n(u) - \psi(u)\big) < \epsilon_n, \quad u \in \mathbf{C}. \tag{4.3.63}$$

(Recall that $\Re \ln \Phi_n(u) = -\infty$ at the nodes.) We can use the same sequence for z and u. We can also say that $|\Phi_n(u)|$ *behaves like* $\exp((\Re\psi(u) \pm \delta)n)$ *outside the immediate vicinity of the interpolation points.*

Theorem 4.3.5.

Assume that the nodes are in $[-1, 1]$, *and that* $[-1, 1]$ *is strictly inside a simply connected domain* $\mathcal{D} \supseteq \mathcal{D}(v)$.

If $f(\zeta)$ *is analytic in the closure of* $\mathcal{D}$, *then the interpolation error* $(L_n f)(u) - f(u)$ *converges like an exponential to* 0 *for any* $u \in \mathcal{D}(v)$.

Proof. By Theorem 4.3.3, $f(u) - (L_n f)(u) = I_n(u)$, where

$$I_n(u) = \frac{1}{2\pi i}\int_{\partial\mathcal{D}} \frac{\Phi_n(u) f(z)}{\Phi_n(z)(z-u)}\, dz. \tag{4.3.64}$$

Note that $\Re\psi(z) \geq \Re\psi(v)$, because $\mathcal{D} \supseteq \mathcal{D}(v)$. Then, by (4.3.62) and (4.3.63),

$$\Big|\Phi_n(u)/\Phi_n(z)\Big| < \exp n\big(\Re\psi(u) - \Re\psi(v) + 2\epsilon_n\big).$$

Let $|f(z)| \leq M$. For any $u \in \mathcal{D}(v)$, we can choose $\delta > 0$ such that $\Re\psi(u) < \Re\psi(v) - 3\delta$, $|z - u| > \delta$. Next, choose n large enough so that $\epsilon_n < \delta$. Then

$$|f(u) - (L_n f)(u)| < \frac{1}{2\pi} M \exp n(-3\delta + 2\delta) \int_{\partial\mathcal{D}(v)} \frac{|dz|}{\delta} \leq \frac{K\exp(-n\delta)}{\delta},$$

where $K = K(v)$ does not depend on n, δ, and u. Hence, the convergence is exponential. $\square$

We shall now state without proof a complement and a kind of converse to Theorem 4.3.5, for functions $f(z)$ that have simple poles in $\mathcal{D}(v)$.

Proposition 4.3.6.

Assume that $[-1, 1]$ *is strictly inside a domain* $\mathcal{D} \supset \mathcal{D}(v)$, *and that* $f(\zeta)$ *is analytic in the closure of* $\mathcal{D}$, *except for a finite number of simple poles* p *in the interior, all with the same value of* $P(p)$.

Outside the interval $[-1, 1]$, *the curve* $\partial\mathcal{D}(p)$ *then separates the points, where the sequence* $\{(L_n f)(u)\}$ *converges, from the points, where it diverges. The behavior of* $|(L_n f)(u) - f(u)|$, *when* $u \in \mathcal{D}(v)$, $n \gg 1$, *is roughly described by the formula*

$$|(f - L_n f)(u)| \approx K|\Phi_n(u)|e^{(-P(p)\pm\delta)n} / \max_p (1/|p-u|). \tag{4.3.65}$$

There are several interpolation processes with interpolation points in $[-1, 1]$ that converge for all $u \in [-1, 1]$, when the condition of analyticity is replaced by a more modest smoothness assumption, for example, $f \in C^p$. This is the case when the sequence of interpolation points are the zeros of the orthogonal polynomials which belong to a density function that is continuous and strictly positive in $[-1, 1]$. We shall prove the following result.

Proposition 4.3.7.

Consider an interpolation process where the interpolation points have a (perhaps unknown) asymptotic density function $w(x)$, $x \in [-1, 1]$. *Assume that, for some* $k \geq 1$,

$$(L_n f - f)(x) \to 0 \quad \forall x \in [-1, 1], \quad \forall f \in C^k[-1, 1]$$

as $n \to \infty$. *Then the logarithmic potential* $\Re\psi(x)$ *must be constant in* $[-1, 1]$, *and the density function must be the same as for Chebyshev interpolation, i.e.,* $w(x) = \frac{1}{\pi}(1 - x^2)^{-1/2}$.

Proof. Let $f(z)$ be analytic in some neighborhood of $[-1, 1]$, for example, any function with a pole at a point p (arbitrarily) close to this interval. A fortiori, for such a function our interpolation process must converge at all points u in some neighborhood of the interval $[-1, 1]$.

Suppose that $\Re\psi(x)$ is not constant, and let x_1, x_2 be points such that $\Re\psi(x_1) < \Re\psi(x_2)$. We can then choose the pole p so that $\Re\psi(x_1) + \delta < \Re\psi(p) < \Re\psi(x_2) - \delta$. By Proposition 4.3.6, the process would then diverge at some point u arbitrarily close to x_2. This contradiction shows that $\Re\psi(x)$ must be constant in $[-1, 1]$, $\Re\psi(x) = a$ (say).

This gives a Dirichlet problem for the harmonic function $\Re\psi(z)$, $z \notin [-1, 1]$, which has a unique solution, and one can verify that the harmonic function $\Re\psi(z) = a + \Re\ln(z + \sqrt{z^2 - 1})$ satisfies the boundary condition. We must also determine a. This is done by means of the behavior as $z \to \infty$. We find that

$$\begin{aligned}\Re\psi(z) &= a + \Re\ln\left(z + z(1 - z^{-2})^{1/2}\right)\\ &= a + \Re\ln\left(2z - O(z^{-1})\right) = a + \Re\ln z + \ln 2 - O(z^{-2}).\end{aligned}$$

This is to be matched with the result of the discussion of the general logarithmic potential in the beginning of Sec. 4.3.5. In our case, where we have a symmetric distribution and

$\int_1^1 w(x)\,dx = 1$, we obtain $\Re\psi(z) = \Re \ln z + O(z^{-2})$. The matching yields $a = -\ln 2$. Finally, by (4.3.56), we obtain after some calculation $w(x) = (1-x^2)^{-1/2}$. $\square$

Our problem is related to a more conventional application of potential theory, namely the problem of finding the electrical charge distribution of a long insulated charged metallic strip through $[-1, 1]$, perpendicular to the (x, y)-plane. Such a plate will be equipotential. Suppose that such a distribution is uniquely determined by the total charge.[135] The charge density must then be proportional to the density in our example. This density w corresponds to $q(\tau) = \cos \pi t$, i.e., $w(x) = \frac{1}{\pi}(1-x^2)^{-1/2}$. *A fascinating relationship of electricity to approximation*!

The above discussion is related to the derivations and results concerning the asymptotic distribution of the zeros of orthogonal polynomials, given in the standard monograph Szegö [347].

Review Questions

4.3.1 What is meant by Lagrange–Hermite (osculatory) interpolation? Prove the uniqueness result for the Lagrange–Hermite interpolation problem.

4.3.2 (a) Write down the confluent Vandermonde matrix for the Lagrange–Hermite cubic interpolation problem.

(b) Express the divided difference $[x_0, x_0, x_1, x_1]f$ in terms of f_0, f_0', and f_1, f_1'.

4.3.3 What are the *inverse* divided differences of a function $f(x)$ used for? How are they defined? Are they symmetric in all their arguments?

4.3.4 Give the complex integral formula for the interpolation error of a function that is analytic in a domain. Give the assumptions, and explain your notations. Give the interpolation error for the case with poles in the domain also.

4.3.5 How is bilinear interpolation performed? What is the order of accuracy?

4.3.6 What is Chebyshev interpolation, and what can be said about its convergence for analytic functions?

Problems and Computer Exercises

4.3.1 (a) Construct the divided-difference scheme for the simplest Lagrange–Hermite interpolation problem, where the given data are $f(x_i)$, $f'(x_i)$, $i = 0, 1$; $x_1 = x_0 + h$. Prove all the formulas concerning this problem that are stated at the end of Sec. 4.3.1.

(b) For $f(x) = (1+x)^{-1}$, $x_0 = 1$, $x_1 = 1.5$, compute $f(1.25)$ by Lagrange–Hermite interpolation. Compare the error bound and the actual error.

[135] It is unique, but we have not proved this.

(c) Show that for Lagrange–Hermite interpolation

$$|f'(x) - p'(x)| \le \frac{h^3}{72\sqrt{3}}\Big(\max_{x\in[x_0,x_1]} |f^{(iv)}(x)| + O(h|f^{(v)}(x)|)\Big).$$

Hint: $\frac{d}{dx}[x_0, x_0, x_1, x_1, x]f = [x_0, x_0, x_1, x_1, x, x]f \le \cdots$.

4.3.2 Let $p \in \mathcal{P}_6$ be the Lagrange–Hermite interpolation polynomial to the data x_i, $y(x_i)$, $y'(x_i)$, $x_i = x_0 + ih$, $i = 1, 2, 3$.

(a) Find the remainder term, and show that the interpolation error for $x \in [x_1, x_3]$ does not exceed $h^6 \max |f^{(6)}(x)|/4860$ in magnitude.

(b) Write a program that computes $p(x_1 + 2jh/k)$, $j = 0 : k$.

Comment: This is one of several possible procedures for starting a multistep method for an ordinary differential equation $y' = f(x, y)$. Two steps with an accurate one-step method provide values of y, y', and this program then produces starting values (y only) for the multistep method.

4.3.3 Give a short and complete proof of the uniqueness of the interpolation polynomial for distinct points, by the use of the ideas in the proof of Theorem 4.3.1.

4.3.4 Derive an approximate formula for $f'(x_0)$ when the values $f(x_{-1})$, $f(x_0)$, $f(x_1)$ are given at three *nonequidistant* points. Give an approximate remainder term. Check the formula and the error estimate on an example of your own choice.

4.3.5 Consider the problem of finding a polynomial $p \in \mathcal{P}_4$ that interpolates the data $f(1)$, $f(-1)$, $f''(1)$, $f''(-1)$. The new feature is that there are no first derivatives. Show that this problem is uniquely solvable.

Hint: Show that using the power basis one obtains a linear system of the form $Mc = f$, where $f = (f_1, f_{-1}, f''_1, f''_{-1})^T$, and

$$M = \begin{pmatrix} 1 & 1 & 1 & 1 \\ 1 & -1 & 1 & -1 \\ 0 & 0 & 2 & 6 \\ 0 & 0 & 2 & -6 \end{pmatrix}$$

with $\det(M) = 48$.

4.3.6 (a) Given a sequence of function values $f_1, f_2, f_3, \ldots$ at equidistant points $x_j = x_0 + jh$, assume that $\min f_j = f_n$, and let $p(x)$ be the quadratic interpolation polynomial determined by f_{n-1}, f_n, f_{n+1}. Show that

$$\min p(x) = f_n - \frac{(\mu\delta f_n)^2}{2\delta^2 f_n} \quad \text{at} \quad x = x_n - h\frac{\mu\delta f_n}{\delta^2 f_n},$$

and that the error of the minimum value can be bounded by $\max |\Delta^3 f_j|/\sqrt{243}$, where j is in some neighborhood of n. Why and how is the estimate of x less accurate?

(b) Write a handy program that includes the search for all local maxima and minima. Sketch or work out improvements of this algorithm, perhaps with ideas of inverse interpolation and with cubic interpolation, and perhaps for nonequidistant data.

4.3.7 We use the notations and assumptions of Theorem 4.3.3.

Using the representation of the interpolation operator as an integral operator, show that

$$(L_n f)(x) = \frac{1}{2\pi i}\int_{\partial\mathcal{D}} K(x,z)\frac{f(z)}{\Phi(z)}\,dz, \qquad K(x,z) = \frac{\Phi(x)-\Phi(z)}{(x-z)},$$

also if $x \notin \mathcal{D}$. Note that $K(x,z)$ is a polynomial, symmetric in the two variables x, z.

4.3.8 If $f \in \mathcal{P}_n$, then $f - L_n f$ is clearly zero. How would you deduce this obvious fact from the integral $(K_n f)(u)$?

Hint: Let ∂D be the circle $|z| = R$. Make the standard estimation, and let $R \to \infty$.

4.3.9 (a) Show the **bilinear interpolation** formula

$$p(x_0 + ph, y_0 + qk) = (1-p)(1-q)f_{0,0} + p(1-q)f_{1,0} + (1-p)qf_{0,1} + pqf_{1,1} \tag{4.3.66}$$

with error bound

$$\frac{1}{2}\big(p(1-p)h^2|f_{xx}| + q(1-q)k^2|f_{yy}|\big) + O(k^2h^2),$$

where f_{xx} and f_{yy} denote partial derivatives.

(b) Compute by bilinear interpolation $f(0.5, 0.25)$ when

$$f(0,0) = 1, \quad f(1,0) = 2, \quad f(0,1) = 3, \quad f(1,1) = 5.$$

4.3.10 (a) Consider the interpolation problem: Given x_i, y_i, f_i, $i = 1:6$, find $c = (c_1, c_2, c_3, c_4, c_5, c_6)^T$ so that $p(x_i, y_i; c) = f_i$, $i = 1:6$, where

$$p(x, y; c) = c_1 + c_2 x + c_3 y + c_4 x^2 + c_5 xy + c_6 y^2.$$

Choose x_i, y_i, f_i, $i = 1:6$, by 18 independent random numbers, solve the linear system $p(x_i, y_i; c) = f_i$, $i = 1:6$, and look at $\max |c_i|$. Repeat this (say) 25 times. You have a fair chance to avoid singular cases, or cases where $\max |c_i|$ is very large.

(b) Now choose (x_i, y_i) as six distinct points on some circle in $\mathbf{R}^2$, and choose f_i at random. This should theoretically lead to a singular matrix. Explain why, and find experimentally the rank (if your software has convenient commands or routines for that). Find a general geometric characterization of the sextuples of points (x_i, y_i), $i = 1:6$, that lead to singular interpolation problems.

Hint: Brush up on your knowledge of conic sections.

4.3.11 Derive a formula for $f''_{xy}(0,0)$ using f_{ij}, $|i| \le 1$, $|j| \le 1$, which is exact for all quadratic functions.

4.3.12 (Bulirsch and Rutishauser (1968))

(a) The function $\cot x$ has a singularity at $x = 0$. Use values of $\cot x$ for $x = 1°$, $2°, \dots, 5°$, and rational interpolation of order (2,2) to determine an approximate value of $\cot x$ for $x = 2.5°$, and its error.

(b) Use polynomial interpolation for the same problem. Compare the result with that in (a).

4.3.13 Check the omitted details of the derivations in Example 4.3.4. Compare the bounds for Chebyshev interpolation and Chebyshev expansion for $R = 1 + k/n$.

4.3.14 Check the validity of (4.3.56) on the Chebyshev and the equidistant cases. Also show that $\int_{-1}^{1} w(x)\,dx = 1$, and check the statements about the behavior of $P(z)$ for $|z| \gg 1$.

4.3.15 *A generalization of Runge's example.* Let f be an analytic function for which the poles nearest to $[-1, 1]$ are a pair of complex conjugate poles at an arbitrary place on the imaginary axis. Consider interpolation with nodes in $[-1, 1]$.

(a) Suppose that equidistant interpolation converges at $u = 1$. Is it true that Chebyshev interpolation converges faster at $u = 1$?

(b) Is it true that equidistant interpolation converges faster than Chebyshev interpolation in an environment of $u = 0$?

4.3.16 (after Meray (1884) and Cheney [66, p. 65])

(a) Let $L_n f$ be the polynomial of degree less than n which interpolates to the function $f(z) = 1/z$ at the nth roots of unity. Show that $(L_n f)(z) = z^{n-1}$, and that

$$\lim_{n\to\infty} \max_{|u|=1} |(L_n f - f)(u)| > 0.$$

Hint: Solve this directly, without the use of the previous theory.

(b) Modify the theory of Sec. 4.3.2 to the case in (a) with equidistant interpolation points on the unit circle, and make an application to $f(z) = 1/(z - a)$, $a > 0$, $a \neq 1$. Here, $\Phi_n(z) = z^n - 1$. What is $\psi(z)$, $P(z)$? The density function? Check your result by thinking like Faraday. Find out for which values of a, u, ($|u| \neq 1$, $|u| \neq a$), $(L_n f - f)(u) \to 0$, and estimate the speed of convergence (divergence). *Hint:* The integral for $\psi(z)$ is a little tricky, but you may find it in a table. There are, however, simpler alternatives to the integral; see the end of Sec. 4.3.2.

(c) What can be said about the cases excluded above, i.e., $|u| = 1$, $|u| = a$? Also look at the case when $|a| = 1$, $(a \neq 1)$.

(d) Is the equidistant interpolation on the unit circle identical to the Cauchy–FFT method (with $a = 0$, $R = 1$) for the approximate computation of the coefficients in a power series?

4.4 Piecewise Polynomial Interpolation

Interpolating a given function by a single polynomial over its entire range can be an ill-conditioned problem, as illustrated by Runge's phenomenon. On the other hand, polynomials of low degree can give good approximations *locally* in a small interval. In this section we will consider approximation schemes for **piecewise polynomial approximation** with different degrees of global continuity.

With the use of piecewise polynomials, there is no reason to fear equidistant data, as opposed to the situation with higher-degree polynomials. Moreover, if the function to be approximated is badly behaved in a subregion the effect of this can be confined locally, allowing good approximation elsewhere.

In computer graphics and computer aided design (CAD) curves and surfaces have to be represented mathematically, so that they can be manipulated and visualized easily. In 1962 Bézier and de Casteljau, working for the French car companies Renault and Citroën, independently developed Bézier curves for fitting curves and surfaces. Similar work, using bicubic splines, was done in USA at General Motors by Garret Birkhoff and Henry Garabedian [28]. Subsequently, W. J. Gordon of General Motors Research developed the technique of spline blending for fitting smooth surfaces to a rectangular smooth mesh of curves.

Today Bézier curves and spline functions are used extensively in all aircraft and automotive industries. Spline functions can also be used in the numerical treatment of boundary value problems for differential equations. Bézier curves have found use in computer graphics and typography. For example, scalable fonts like PostScript® are stored as piecewise Bézier curves.

4.4.1 Bernštein Polynomials and Bézier Curves

In the following we restrict ourselves to considering polynomial curves, i.e., one-dimensional geometric objects. Parametric curves are often used to find the functional form of a curve given geometrically by a set of points $p_i \in \mathbf{R}^d$, $i = 0 : n$.

Let $c(t) \in \mathbf{R}^d$, $t \in [0, 1]$, be a **parametric curve**. In the simplest case, $n = 1$, we take $c(t)$ to be linear,

$$c(t) = (1 - t)p_0 + tp_1,$$

and connecting the two points p_0 and p_1 so that $p_0 = c(0)$ and $p_1 = c(1)$. For $n > 1$ this will not give a smooth curve and is therefore of limited interest.

We now generalize this approach and take $c(t)$ to be a polynomial of degree n,

$$c(t) = \sum_{i=0}^{n} p_i B_i(t), \qquad t \in [0, 1],$$

where $B_i(t)$, $i = 0 : n$, are the **Bernštein polynomials**[136] defined by

$$B_i^n(t) = \binom{n}{i} t^i (1 - t)^{n-i}, \qquad i = 0 : n. \tag{4.4.1}$$

Using the binomial theorem we have

$$1 = ((1 - t) + t)^n = \sum_{i=0}^{n} \binom{n}{i} t^i (1 - t)^{n-i} = \sum_{i=0}^{n} B_i^n(t).$$

Thus the Bernštein polynomials of degree n are nonnegative on $[0, 1]$ and *give a partition of unity*.

For $n = 3$ the four cubic Bernštein polynomials are

$$B_0^3 = (1 - t)^3, \quad B_1^3 = 3t(1 - t)^2, \quad B_2^3 = 3t^2(1 - t), \quad B_3^3 = t^3. \tag{4.4.2}$$

They are plotted in Figure 4.4.1.

[136] Bernštein introduced the polynomials named after him in 1911.

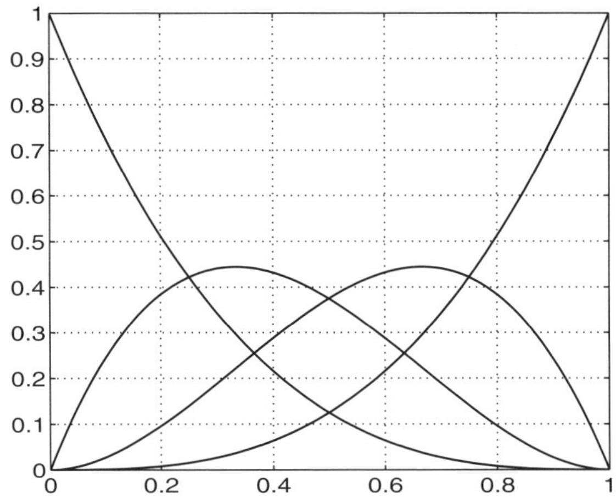

Figure 4.4.1. *The four cubic Bernštein polynomials.*

Some important properties of the Bernštein polynomials are given in the following theorem.

Theorem 4.4.1.

The Bernštein polynomials $B_i^n(t)$ have the following properties:

1. $B_i^n(t) > 0,\ t \in (0, 1)$ *(nonnegativity);*
2. $B_i^n(t) = B_{n-i}^n(1 - t)$ *(symmetry);*
3. $B_i^n(t) = 0$ *has a root* $t = 0$ *of multiplicity* i *and a root* $t = 1$ *of multiplicity* $n - i$*;*
4. *The Bernštein polynomials* $B_i^n(t)$ *have a unique maximum value at* $t = i/n$ *on* $[0, 1]$*;*
5. *The Bernštein polynomials satisfy the following recursion formula:*
$$B_i^n(t) = (1 - t)B_i^{n-1}(t) + tB_{i-1}^{n-1}(t), \quad i = 0 : n; \tag{4.4.3}$$
6. *The Bernštein polynomials of degree* n *form a basis for the space of polynomials of degree* $\leq n$.

Proof. The first four properties follow directly from the definition (4.4.1). The recursion formula is a consequence of the relation

$$\binom{n}{i} = \binom{n-1}{i} + \binom{n-1}{i-1}$$

between the binomial coefficients.

To show the linear independence we observe that if

$$\sum_{i=0}^{n} a_i B_i^n(t) = 0,$$

then according to property 3,

$$\sum_{i=0}^{n} a_i B_i^n(1) = a_n B_i^n(1) = a_n = 0.$$

By repeatedly dividing by $(1-t)$ and using the same argument we find that $a_0 = \cdots = a_{n-1} = a_n = 0$. □

The unique parametric **Bézier curve** corresponding to a given set of $n+1$ **control points** p_i, $i = 0 : n$, equals

$$c(t) = \sum_{i=0}^{n} p_i B_i^n(t), \quad t \in [0, 1], \tag{4.4.4}$$

where $B_i^n(t)$ are Bernštein polynomials of degree n. By property 3 in Theorem 4.4.1 the Bézier curve *interpolates the first and last control points* p_0 and p_n. Often a curve is constructed by smoothly patching together several Bézier curves of low order.

Starting with $B_0^0(t) = 1$ and setting $B_{-1}^n(t) = B_{n+1}^n(t) = 0$, the recursion in Theorem 4.4.1 can be used to evaluate the Bernštein polynomials at a given point t.

It follows directly from the form of (4.4.4) that applying an affine transformation to $c(t)$ can be performed simply by applying the same transformation to the control points. Hence the Bézier curve has the desirable property that it is invariant under translations and rotations.

Example 4.4.1.

A quadratic Bézier curve is given by

$$c(t) = (1-t)^2 p_0 + 2t(1-t)p_1 + t^2 p_2, \quad t \in [0, 1].$$

Clearly $c(0) = p_0$ and $c(1) = p_2$. For $t = 1/2$ we get

$$c\left(\frac{1}{2}\right) = \frac{1}{2}\left(\frac{p_0 + p_2}{2} + p_1\right).$$

Hence we can construct the point $c(1/2)$ geometrically as the intersection between the midpoint of the line between p_0 and p_2 and the point p_1; see Figure 4.4.2.

The **Bézier polygon** is the closed piecewise linear curve connecting the control points p_i and p_{i+1}, $i = 0 : n-1$, and finally p_n and back to p_0. In Figures 4.4.2 and 4.4.3 this is the polygon formed by the dashed lines. This polygon provides a rough idea of the shape of the Bézier curve.

From the definition (4.4.4) of the Bézier curve it follows that for all $t \in [0, 1]$, the curve $c(t)$ is a convex combination of the control points. Therefore, $c(t)$ lies within the convex hull (see Definition 4.3.2) of the control points.

The variation of a function in an interval $[a, b]$ is the least upper bound on the sum of the oscillations in the closed subintervals $[a, x_1], [x_1, x_2], \ldots, [x_n, b]$ for all possible such

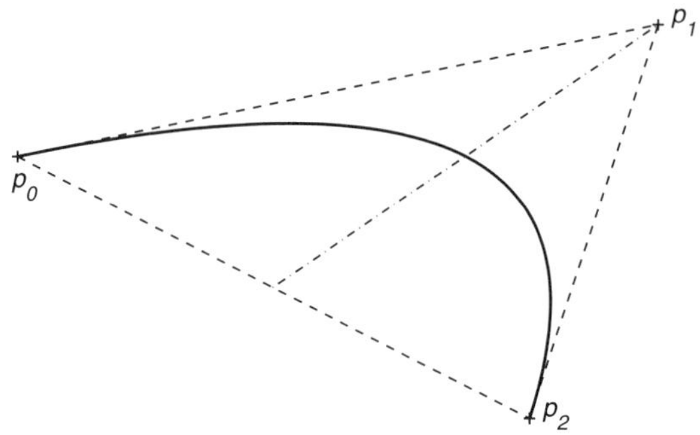

Figure 4.4.2. *Quadratic Bézier curve with control points.*

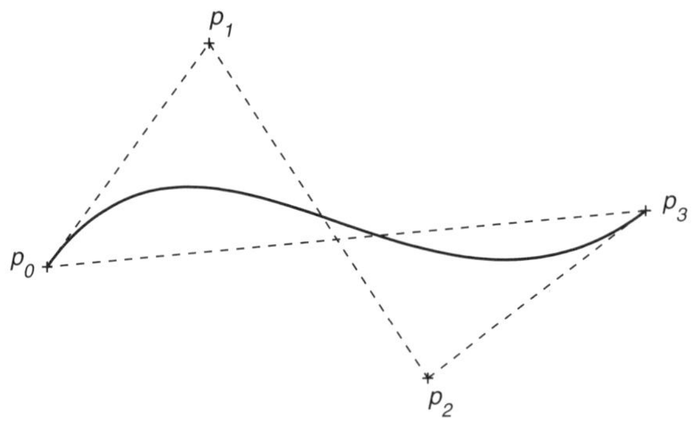

Figure 4.4.3. *Cubic Bézier curve with control points* $p_0, \dots, p_3$.

subdivisions. The Bézier curve is variation diminishing. In particular, if the control points p_i are monotonic, so is $c(t)$.

Usually, not all control points are known in advance. The shape of the curve is controlled by moving the control points until the curve has the desired shape. For example, in the quadratic case moving p_1 has a direct and intuitive effect on the curve $c(t)$. An advantage of the Bernštein basis for representing polynomials is that the coefficients (control points) are closely related to the shape of the curve. This is not the case when using a monomial or Chebyshev basis.

Theorem 4.4.2.

The Bézier curve $c(t)$ is tangent to $p_1 - p_0$ and $p_n - p_{n-1}$ for $t = 0$ and $t = 1$, respectively.

Proof. To show this we compute the derivative of the Bernštein polynomial (4.4.1):

$$\frac{d}{dt} B_i^n(t) = \begin{cases} -nB_0^{n-1}(t) & \text{if } i = 0, \\ n\left(B_{i-1}^{n-1}(t) - B_i^{n-1}(t)\right) & \text{if } 0 < i < n, \\ nB_{n-1}^{n-1}(t) & \text{if } i = n. \end{cases}$$

This follows from

$$\frac{d}{dt}B_i^n(t) = \binom{n}{i}\left(it^{i-1}(1-t)^{n-i} - (n-i)t^i(1-t)^{n-i-1}\right),$$

and using the definition of the Bernštein polynomials. Setting $t = 0$ we find that $\frac{d}{dt}B_i^n(0) = 0$, $i > 1$, and therefore from (4.4.4)

$$\frac{d}{dt}c(t) = n(p_1 - p_0).$$

The result for $t = 1$ follows from symmetry. □

More generally, at a boundary point the kth derivative of the Bézier curve depends only on the k closest control points. This fact is useful for smoothly joining together several pieces of Bézier curves.

De Casteljau's Algorithm

To evaluate the Bézier curve at $t \in [0, 1]$ we use the recursion formula (4.4.3) to obtain

$$\begin{aligned} c(t) &= \sum_{i=0}^{n} p_i B_i^n(t) = (1-t)\sum_{i=0}^{n-1} p_i B_i^{n-1}(t) + t\sum_{i=1}^{n} p_i B_{i-1}^{n-1}(t) \\ &= \sum_{i=0}^{n-1}\left((1-t)p_i + tp_{i+1}\right)B_i^{n-1}(t) = \sum_{i=0}^{n-1} p_i^{(1)}(t)B_i^{n-1}(t). \end{aligned}$$

Here we have introduced the new auxiliary control points

$$p_i^{(1)}(t) = (1-t)p_i + tp_{i+1}, \quad i = 0 : n-1,$$

as convex combinations (depending on t) of the original control points. Using this result we can successively lower the grade of the Bernštein polynomial until we arrive at $B_0^0 = 1$. This gives **de Casteljau's algorithm**, a recursion scheme for the auxiliary control points

$$\begin{aligned} p_i^{(0)}(t) &= p_i, \quad i = 0 : n, \\ p_i^{(r)}(t) &= (1-t)p_i^{(r-1)}(t) + tp_{i+1}^{(r-1)}(t), \quad i = 0 : n-r. \end{aligned} \tag{4.4.5}$$

It follows that

$$c(t) = \sum_{i=0}^{n-r} p_i^{(r)}(t)B_i^{n-r}(t), \quad r = 0 : n, \tag{4.4.6}$$

and in particular $c(t) = p_0^{(n)}$.

De Casteljau's algorithm works by building convex combinations (4.4.5) and is therefore numerically very stable. It can conveniently be arranged in a triangular array.

$$
\begin{array}{ccccc}
p_0 = p_0^{(0)} & & & & \\
 & p_0^{(1)} & & & \\
p_1 = p_1^{(0)} & & p_0^{(2)} & & \\
 & p_1^{(1)} & & & \\
p_2 = p_2^{(0)} & \vdots & p_1^{(2)} & \ddots & \\
\vdots & \vdots & \vdots & \vdots & p_0^{(n)} \\
\vdots & p_{n-2}^{(1)} & \vdots & & \\
p_{n-1} = p_{n-1}^{(0)} & & p_{n-2}^{(2)} & & \\
 & p_{n-1}^{(1)} & & & \\
p_n = p_n^{(0)} & & & &
\end{array}
\tag{4.4.7}
$$

The algorithm uses about n^2 operations and so is less efficient than Horner's algorithm for evaluating a polynomial in the monomial basis.

The kth derivative of $c(t)$ is also available from the de Casteljau scheme. It holds that

$$
\begin{aligned}
c'(t) &= n\left(p_1^{(n-1)} - p_0^{(n-1)}\right), \\
c''(t) &= n(n-1)\left(p_2^{(n-2)} - 2p_1^{(n-2)} + p_0^{(n-2)}\right), \ldots,
\end{aligned}
$$

and in general

$$
c^{(k)}(t) = \frac{n!}{(n-k)!}\Delta^k p_0^{(n-k)}, \quad 0 \le k \le n, \tag{4.4.8}
$$

where the difference operates on the lower index i.

De Casteljau's algorithm is illustrated for the quadratic case in Figure 4.4.4, where the following geometric interpretation can be observed. In the interval $[0, t]$ the Bézier curve is represented by a quadratic spline with control points p_0, $p_0^{(1)}$, $p_0^{(2)}$. In the remaining interval $[t, 1]$ it is represented by a quadratic spline with control points $p_0^{(2)}$, $p_1^{(1)}$, p_2. Note that these

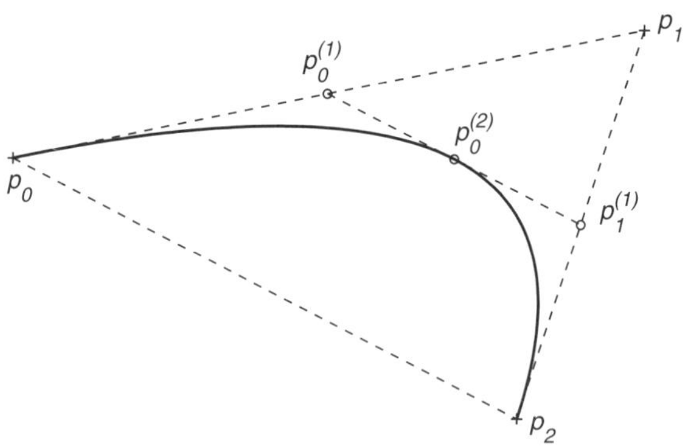

Figure 4.4.4. *De Casteljau's algorithm for* $n = 2$, $t = \frac{1}{2}$.

two sets of control points lie closer to the curve $c(t)$. After a few more subdivisions it will be hard to distinguish the polygon joining the control points from the curve.

De Casteljau's algorithm can also be used to subdivide a Bézier curve into two segments. By repeating this partitioning the Bézier polygons converge fast to the curve. This construction is very well suited to control, for example, a milling machine which can only remove materiel.

4.4.2 Spline Functions

The name **spline** comes from a very old technique in drawing smooth curves in which a thin strip of wood, called a drafting spline, is bent so that it passes through a given set of points; see Figure 4.4.5. The points of interpolation are called **knots** and the spline is secured at the knots by means of lead weights called **ducks**. Before the computer age splines were used in all kinds of geometric design to produce sufficiently smooth profiles. This process was slow and expensive.

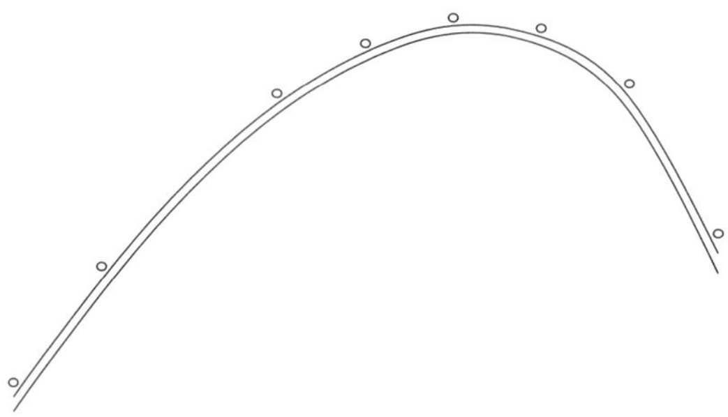

Figure 4.4.5. *A drafting spline.*

The mathematical model of a spline is a special case of the elastic line treated in the theory of elasticity. By Hamilton's principle the spline assumes a shape $y(x)$ which minimizes the strain energy. This is proportional to the line integral of the square of the curvature

$$\kappa(x) = \frac{y''(x)}{(1 + (y'(x))^2)^{3/2}}. \tag{4.4.9}$$

Since $ds = \sqrt{1 + (y'(x))^2}dx$, the energy becomes

$$E(y) = \int_a^b \frac{y''(x)^2}{(1 + (y'(x))^2)^{5/2}}\,dx. \tag{4.4.10}$$

A mathematical model of the **nonlinear spline** was given by Daniel Bernoulli (1742) and Euler (1744).[137] A modern description of the development of Bernoulli and Euler is given by Malcolm in [256], where algorithms for computing discrete approximations of nonlinear splines are also discussed and compared. In general these are very costly.

[137] Euler derived the differential equation satisfied by an elastic line using techniques now known as calculus of variation and Lagrange multipliers. When Euler did this work Lagrange was still a small child!

For slowly varying deflections, i.e., when $(y'(x))^2$ is approximately constant, we have the approximation

$$E(y) \approx \text{const} \cdot \int_a^b y''(x)^2\,dx.$$

Under this assumption, $y(x)$ is built up of *piecewise cubic polynomials* in such a way that $y(x)$ and its two first derivatives are everywhere continuous. Let x_i, $i = 0 : m$, be the points the spline is forced to interpolate. Then the third derivative can have discontinuities at the points x_i.

The mathematical concept of spline functions was introduced in 1946 by Schoenberg in the seminal paper [314]. The importance of the B-spline basis for spline approximation (see Sec. 4.4.3) was also first appreciated by Schoenberg. These were not used in practical calculations for general knot sequences until the early seventies, when a stable recurrence relation was established independently by de Boor [36] and Cox [84].

In the following we restrict ourself to consider curves in the plane. Today B-splines enable the mathematical representation of surfaces far beyond hand techniques. In aircraft design computations they may involve more than 50,000 data points.

Linear and Cubic Splines

We start by formally defining a spline function of order $k \ge 1$.

Definition 4.4.3.

A spline function $s(x)$ of order $k \ge 1$ (degree $k-1 \ge 0$), on a grid

$$\Delta = \{a = x_0 < x_1 < \cdots < x_m = b\}$$

of distinct knots is a real function s with the following properties:

(a) *For $x \in [x_i, x_{i+1}]$, $i = 0 : m-1$, $s(x)$ is a polynomial of degree $< k$.*

(b) *For $k = 1$, $s(x)$ is a piecewise constant function. For $k \ge 2$, $s(x)$ and its first $k-2$ derivatives are continuous on $[a, b]$, i.e., $s(x) \in C^{k-2}[a, b]$.*

The space of all spline functions of order k on Δ is denoted by $S_{\Delta,k}$. From the definition it follows that if $s_1(x)$ and $s_2(x)$ are spline functions of the same degree, so is $c_1 s_1(x) + c_2 s_2(x)$. Clearly $S_{\Delta,k}$ is a *linear vector space*. The space $\mathcal{P}_k$ of polynomials of degree less than k is a linear subspace of $S_{\Delta,k}$. The **truncated power** functions

$$(x - x_j)_+^{k-1} = \max\{(x - x_j)^{k-1}, 0\}, \quad j = 1 : m-1,$$

introduced in Sec. 3.3.3 in connection with the Peano kernel are also elements of $S_{\Delta,k}$.

Theorem 4.4.4.

The monomials and truncated power functions

$$\{1, x, \ldots, x^{k-1}, (x - x_1)_+^{k-1}, \ldots, (x - x_{m-1})_+^{k-1}\} \tag{4.4.11}$$

form a basis for the spline space $S_{\Delta,k}$. In particular, the dimension of this space is $k+m-1$.

Proof. All we need for the first subinterval is a basis of $\mathcal{P}_k$, for example, the power basis $\{1, x, \dots, x^{k-1}\}$. For each additional subinterval $[x_j, x_{j+1})$, $j = 1 : m-1$, we need only add the new basis function $(x - x_j)_+^{k-1}$. It is easy to show that these $k + m - 1$ functions are linearly independent. □

The truncated power basis (4.4.35) has several disadvantages and is not well suited for numerical computations. The basis functions are not local; i.e., they are nonzero on the whole interval $[a, b]$. Further, they (4.4.35) are almost linearly dependent when the knots are close. Therefore, this basis yields dense ill-conditioned linear systems for various tasks. A more suitable basis will be introduced in Sec. 4.4.3.

The simplest case $k = 1$ is the linear spline interpolating given values $y_i = f(x_i)$, $x_i \in [a, b]$, $i = 0 : m$. This is the broken line

$$s(x) = q_i(x) = y_{i-1} + d_i(x - x_{i-1}), \quad x \in [x_{i-1}, x_i), \quad i = 1 : m. \tag{4.4.12}$$

Here

$$d_i = [x_{i-1}, x_i]f(x) = (y_i - y_{i-1})/h_i, \quad h_i = x_i - x_{i-1}, \tag{4.4.13}$$

are the divided differences of f at $[x_{i-1}, x_i]$. By (4.2.16) the error satisfies

$$|f(x) - s(x)| \le \frac{1}{8} \max_i \Big(h_i^2 \max_{x \in [x_{i-1}, x_i]} |f''(x)| \Big), \tag{4.4.14}$$

if $f \in C^2[a, b]$. Hence, we can make the error arbitrarily small by decreasing $\max_i h_i$. An important property of interpolation with a linear spline function is that it preserves monotonicity and convexity of the interpolated function.

The broken line interpolating function has a discontinuous first derivative at the knots which makes it unsuitable for many applications. To get better smoothness piecewise polynomials of higher degree need to be used. Cubic spline functions, which *interpolate* a given function $f(x)$ on the grid Δ and have continuous first and second derivatives, are by far the most important; see Figure 4.4.6.

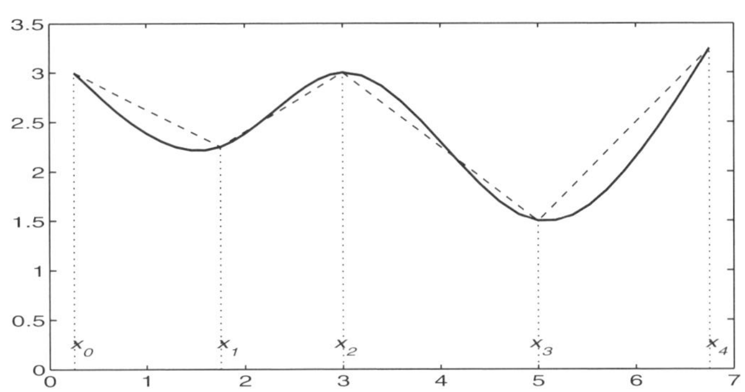

Figure 4.4.6. *Broken line and cubic spline interpolation.*

With piecewise Lagrange–Hermite interpolation we can interpolate given function values and first derivatives of a function on the grid Δ. First recall that in cubic Lagrange–Hermite interpolation (see Theorem 4.3.1 and Problem 4.3.1) a cubic polynomial $q_i(x)$ is fitted to values of a function and its first derivative at the endpoints of the interval $[x_{i-1}, x_i)$. Let

$$\theta = \frac{x - x_{i-1}}{h_i} \in [0, 1), \quad x \in [x_{i-1}, x_i), \tag{4.4.15}$$

be a local variable. Then, by (4.3.11) translated to the notation in (4.4.13), the cubic $q_i(x)$ can be written in the form

$$q_i(x) = \theta y_i + (1-\theta)y_{i-1} + h_i\theta(1-\theta)\left[(k_{i-1} - d_i)(1-\theta) - (k_i - d_i)\theta\right], \tag{4.4.16}$$

where $h_i, d_i, i = 1 : m$, are as defined in (4.4.13), and

$$k_i = q_i'(x_i) \tag{4.4.17}$$

is the derivative at x_i.

The second derivative (and the curvature) of this piecewise polynomial will in general be discontinuous at the grid points. We now show how to choose the parameters $k_i, i = 0 : m$, to also get a continuous *second* derivative. This will yield an interpolating cubic spline. In contrast to piecewise Lagrange–Hermite interpolation, *each piece of the cubic spline will depend on all data points.*

Theorem 4.4.5.

Every cubic spline function, with knots $a = x_0 < x_1 < \cdots < x_m = b$, *which interpolates the function* $y = f(x)$,

$$s(x_i) = f(x_i) = y_i, \quad i = 0 : m,$$

equals for $x \in [x_{i-1}, x_i)$, $i = 1 : m$, *a third degree polynomial of the form* (4.4.16). *The* $m+1$ *parameters* k_i, $i = 0 : m$, *satisfy* $m-1$ *linear equations*

$$h_{i+1}k_{i-1} + 2(h_i + h_{i+1})k_i + h_i k_{i+1} = 3(h_i d_{i+1} + h_{i+1}d_i), \quad i = 1 : m-1, \tag{4.4.18}$$

where $h_i = x_i - x_{i-1}$, $d_i = (y_i - y_{i-1})/h_i$.

Proof. We require the second derivative of the spline $s(x)$ to be continuous at x_i, $i = 1 : m-1$. We have

$$s(x) = \begin{cases} q_i(x), & x \in [x_{i-1}, x_i), \\ q_{i+1}(x), & x \in [x_i, x_{i+1}), \end{cases}$$

where $q_i(x)$ is given by (4.4.22)–(4.4.23). Differentiating $q_i(x)$ twice we get $\frac{1}{2}q_i''(x) = a_{2,i} + 3a_{3,i}(x - x_{i-1})$, and putting $x = x_i$ we obtain

$$\frac{1}{2}q_i''(x_i) = a_{2,i} + 3a_{3,i}h_i = \frac{(k_{i-1} + 2k_i - 3d_i)}{h_i}. \tag{4.4.19}$$

Replacing i by $i+1$ we get $\frac{1}{2}q_{i+1}''(x) = a_{2,i+1} + 3a_{3,i+1}(x - x_i)$, and hence

$$\frac{1}{2}q_{i+1}''(x_i) = a_{2,i+1} = \frac{(3d_{i+1} - 2k_i - k_{i+1})}{h_{i+1}}. \tag{4.4.20}$$

These last two expressions must be equal, which gives the conditions

$$\frac{1}{h_i}(k_{i-1} + 2k_i - 3d_i) = \frac{1}{h_{i+1}}(3d_{i+1} - 2k_i - k_{i+1}), \quad i = 1 : m - 1. \tag{4.4.21}$$

Multiplying both sides by $h_i h_{i+1}$ we get (4.4.18). □

If the cubic spline $s(x)$ is to be evaluated at many points, then it is more efficient to first convert it from the form (4.4.16) to **piecewise polynomial form** (pp-form):

$$q_i(x) = y_{i-1} + a_{1i}(x - x_{i-1}) + a_{2i}(x - x_{i-1})^2 + a_{3i}(x - x_{i-1})^3, \quad i = 1 : m. \tag{4.4.22}$$

From (4.4.16) we obtain after some calculation

$$\begin{aligned} a_{1i} &= q_i'(x_{i-1}) = k_{i-1}, \\ a_{2i} &= \frac{1}{2}q_i''(x_{i-1}) = \frac{(3d_i - 2k_{i-1} - k_i)}{h_i}, \\ a_{3i} &= \frac{1}{6}q_i'''(x_{i-1}) = \frac{(k_{i-1} + k_i - 2d_i)}{h_i^2}. \end{aligned} \tag{4.4.23}$$

Using Horner's scheme $q_i(x)$ can be evaluated from (4.4.22) using only four multiplications. Algorithms for performing conversion to pp-form are given in [37, Chapter X].

The conditions (4.4.18) are $(m - 1)$ linearly independent equations for the $(m + 1)$ unknowns k_i, $i = 0 : m$. Two additional conditions are therefore needed to uniquely determine the interpolating spline. The most important choices are discussed below.

- If the derivatives at the endpoints are known we can take

$$k_0 = f'(a), \qquad k_m = f'(b). \tag{4.4.24}$$

  The corresponding spline function $s(x)$ is called the **complete cubic spline interpolant**.

- If k_0 and k_m are determined by numerical differentiation with a truncation error $O(h^4)$, we call the spline interpolant **almost complete**. For example, k_0 and k_m may be the sum of (at least) four terms of the expansions

$$Df(x_0) = \frac{1}{h}\ln(1 + \Delta)y_0, \qquad Df(x_m) = -\frac{1}{h}\ln(1 - \nabla)y_m,$$

  into powers of the operators Δ and ∇, respectively.[138]

- A physical spline is straight outside the interval $[a, b]$. The corresponding boundary conditions are $q_1''(x_0) = q_m''(x_m) = 0$. From (4.4.20) and (4.4.19) it follows that

$$\frac{1}{2}q_i''(x_{i-1}) = \frac{(3d_i - 2k_{i-1} - k_i)}{h_i}, \qquad \frac{1}{2}q_i''(x_i) = \frac{-(3d_i - k_{i-1} - 2k_i)}{h_i}.$$

[138] Two terms of the central difference expansion in (3.3.46).

Setting $i = 1$ in the first equation and $i = m$ in the second gives the two conditions

$$\begin{aligned} 2k_0 + k_1 &= 3d_1, \\ k_{m-1} + 2k_m &= 3d_m. \end{aligned} \tag{4.4.25}$$

The spline function corresponding to these boundary conditions is called the **natural spline interpolant**. It should be stressed that when a cubic spline is used for the approximation of a smooth function, these boundary conditions are *not natural*! Although the natural spline interpolant in general converges only with order h^2, it has been shown that on any compact subinterval of the open interval (a, b) the order of convergence is $O(h^4)$.

- If the endpoint derivatives are not known, a convenient boundary condition is to require that $s'''(x)$ be continuous across the first and last interior knots x_1 and x_{m-1}. Hence

$$q_1'''(x) = q_2'''(x), \qquad q_{m-1}'''(x) = q_m'''(x).$$

Then x_1 and x_{m-1} are no longer knots, and the corresponding spline interpolant is called the **not-a-knot** spline interpolant.

From (4.4.23) we obtain,

$$\frac{1}{6}q_i'''(x) = a_{3i} = \frac{(k_{i-1} + k_i - 2d_i)}{h_i^2}, \quad x \in [x_{i-1}, x_i), \quad i = 1 : m.$$

Hence the condition $q_1''' = q_2'''$ gives $(k_0 + k_1 - 2d_1)/h_1^2 = (k_1 + k_2 - 2d_2)/h_2^2$, or

$$h_2^2 k_0 + (h_2^2 - h_1^2)k_1 - h_1^2 k_2 = 2(h_2^2 d_1 - h_1^2 d_2).$$

Since this equation would destroy the tridiagonal form of the system, we use (4.4.18), with $i = 1$ to eliminate k_2. This gives the equation

$$h_2 k_0 + (h_2 + h_1)k_1 = 2h_2 d_1 + \frac{h_1(h_2 d_1 + h_1 d_2)}{h_2 + h_1}. \tag{4.4.26}$$

If the right boundary condition is treated similarly we get

$$(h_{m-1} + h_m)k_{m-1} + h_{m-1}k_m = 2h_{m-1}d_m + \frac{h_m(h_{m-1}d_m + h_m d_{m-1})}{h_{m-1} + h_m}. \tag{4.4.27}$$

The error of the not-a-knot spline interpolant is of the same order as that of the complete spline. For functions $f \in C^4[a, b]$ one has an error estimate of the same form as () for $r = 0 : 2$; see Beatson [22]. Indeed, the error analysis below shows that the Lagrange–Hermite interpolation error is, in the whole interval $[a, b]$, asymptotically, the dominant source of error for both complete and not-a-knot spline approximation.

- If the spline is used to represent *a periodic function*, then $y_0 = y_m$ and the boundary conditions

$$s'(a) = s'(b), \qquad s''(a) = s''(b) \tag{4.4.28}$$

suffice to determine the spline uniquely. From the first condition it follows that $k_0 = k_m$, which can be used to eliminate k_0 in (4.4.18) for $k = 1$. Using (4.4.21) the second condition in (4.4.28) gives

$$(k_0 + 2k_1 - 3d_1)/h_1 = -(2k_{m-1} + k_m - 3d_m)/h_m$$

or, after eliminating k_0,

$$2h_m k_1 + 2h_1 k_{m-1} + (h_1 + h_m)k_m = 3(h_m d_1 + h_1 d_m).$$

- The natural spline interpolant has the following best approximation property.

Theorem 4.4.6.

Among all functions g that are twice continuously differentiable on $[a, b]$ and which interpolate f at the points $a = x_0 < x_1 < \cdots < x_m = b$, the natural spline function minimizes

$$\int_a^b \left(s''(t)\right)^2 dt.$$

The same minimum property holds for the complete spline interpolant if the functions g satisfy $g'(a) = f'(a)$ and $g'(b) = f'(b)$.

Proof. See de Boor [37, Chapter 5]. □

Due to this property spline functions yield smooth interpolation curves, except for rather thin oscillatory layers near the boundaries if the "natural" boundary conditions $s''(a) = s''(b) = 0$ are far from being satisfied. For the complete or almost complete cubic spline and for cubic splines determined by the not-a-knot conditions, these oscillations are much smaller as we shall see below. Hence, when a spline is to be used for the approximate representation of a smooth function, the natural spline is *not* a natural choice!

Equations (4.4.18) together with any of these boundary conditions gives a linear system for determining the derivatives k_i. It can be shown that this system is well-conditioned using the following result, which will be proved in Volume II.

Lemma 4.4.7.

Assume that the matrix $A \in \mathbf{R}^{n\times n}$ is strictly row **diagonally dominant**, *i.e.,*

$$\alpha_i := |a_{ii}| - \sum_{j\neq i} |a_{ij}| > 0, \quad i = 1 : n. \tag{4.4.29}$$

Then A is nonsingular, and for row diagonally dominant linear systems Gaussian elimination without pivoting *is stable.*

For the first three boundary conditions the system is **tridiagonal** and, which is easily verified, also strictly row diagonally dominant. In Example 1.3.2, and algorithm was given for solving a tridiagonal linear systems of order m by Gaussian elimination in $\mathcal{O}(m)$ flops.

The linear system resulting from the not-a-knot boundary condition is not diagonally dominant in the first and last row. However, it can be shown that also in this case the system is well-conditioned and can be solved stably by Gaussian elimination without pivoting.

Example 4.4.2.

In the case of spline interpolation with constant step size $h_i = h$, (4.4.18) becomes

$$k_{i-1} + 4k_i + k_{i+1} = 3(d_i + d_{i+1}), \quad i = 1 : m-1. \tag{4.4.30}$$

The not-a-knot boundary conditions (4.4.26)–(4.4.27) become

$$k_0 + 2k_1 = \frac{1}{2}(5d_1 + d_2), \qquad 2k_{m-1} + k_m = \frac{1}{2}(d_{m-1} + 5d_m). \tag{4.4.31}$$

We obtain for $(k_0, k_1, \dots, k_m)$ a tridiagonal system $Tk = g$, where

$$T = \begin{pmatrix} 1 & 2 & & & \\ 1 & 4 & 1 & & \\ & \ddots & \ddots & \ddots & \\ & & 1 & 4 & 1 \\ & & & 2 & 1 \end{pmatrix}, \qquad g = 3\begin{pmatrix} (5d_1 + d_2)/6 \\ d_1 + d_2 \\ \vdots \\ d_{m-1} + d_m \\ (d_{m-1} + 5d_m)/6 \end{pmatrix}.$$

Except for the first and last row, the elements of T are constant along the diagonals. The condition number of T increases very slowly with m; for example, $\kappa(T) < 16$ for $m = 100$.

Consider now the periodic boundary conditions in (4.4.28). Setting $k_m = k_0$ in the last equation we obtain a linear system of equations $Tk = g$ for $k_1, \dots, k_{m-1}$ where

$$T = \left(\begin{array}{ccccc|c} b_1 & c_1 & & & & a_m \\ a_1 & b_2 & c_2 & & & 0 \\ & \ddots & \ddots & \ddots & & \vdots \\ & & a_{m-3} & b_{m-2} & c_{m-2} & 0 \\ & & & a_{m-2} & b_{m-1} & c_{m-1} \\ \hline c_m & 0 & \cdots & 0 & a_{m-1} & b_m \end{array}\right). \tag{4.4.32}$$

Here T is tridiagonal except for its last row and last column, where an extra nonzero element occurs. Such systems, called **arrowhead systems**, can be solved with about twice the work of a tridiagonal system; see [30].

In some applications one wants to smoothly interpolate given points (x_j, y_j), $j = 0 : m$, where a representation of the form $y = f(x)$ is not possible. Then we can use a **parametric spline** representation $x = x(t)$, $y = y(t)$, where the parameter values $0 = t_0 \le t_1 \le \cdots \le t_m$ correspond to the given points. Using the approach described previously, two spline functions $s_x(t)$ and $s_y(t)$ can then be determined that interpolate the points (t_i, x_i) and (t_i, y_i), $i = 0 : m$, respectively. The parametrization is usually chosen as

$$t_i = d_i/d, \quad i = 1 : m,$$

where $d_0 = 0$,

$$d_i = d_{i-1} + \sqrt{(x_i - x_{i-1})^2 + (y_i - y_{i-1})^2}, \quad i = 1 : m,$$

are the cumulative distance, and $d = d_m$.

For boundary conditions we have the same choices as mentioned previously. In particular, using periodic boundary conditions for $s_x(t)$ and $s_y(t)$ allows the representation of *closed curves* (see Problem 4.4.8 (b)). Parametric splines can also be used to approximate curves in higher dimensions.

Error in Cubic Spline Interpolation

The following standard error estimate for the complete cubic spline interpolant is due to Beatson [22].

Theorem 4.4.8.

Let the function f be four times continuously differentiable in $[a, b]$ and let s be the complete cubic spline interpolant on the grid $a = x_0 < x_1 < \cdots < x_m = b$. Then

$$\max_{x\in[a,b]} |f^{(r)}(x) - s^{(r)}(x)| \leq c_r(\beta)h^{4-r} \max_{x\in[a,b]} |f^{(iv)}(x)|, \quad r = 0:3, \tag{4.4.33}$$

where $h = \max_i h_i$, $\beta = h/\min_i h_i$, and

$$c_0 = \frac{5}{384}, \quad c_1 = \frac{1}{216}(9 + \sqrt{3}), \quad c_2(\beta) = \frac{1}{12}(1 + 3\beta), \quad c_3(\beta) = \frac{1}{2}(1 + \beta^2).$$

We comment below on the above result for $r = 0$ and how it is influenced by the choice of other boundary conditions. Let $x \in I_i = [x_{i-1}, x_i]$, $i = 1 : m$, and set

$$t = (x - x_{i-1})/h_i, \qquad y_i = f(x_i), \qquad y_i' = f'(x_i).$$

The error in cubic spline interpolation can be expressed as the sum of two components:

i. The error $E_H(x)$ due to Lagrange–Hermite interpolation with correct values of $f'(x_{i-1})$, $f'(x_i)$. From (4.3.12) we obtain

$$\max_{x\in I_i} |E_H(x)| \leq \frac{1}{384} \max_{x\in I_i} |h_i^4 f^{(iv)}(x)|.$$

ii. The error $E_S(x)$ due to the errors of the slopes $e_i = k_i - y_i'$, $i = 0 : m$. In particular, e_0 and e_m are the errors in the boundary derivatives.

The bound in Theorem 4.4.8 for $r = 0$ is only a factor of five larger than that for piecewise Lagrange–Hermite interpolation with correct derivatives of f at all points x_i, $i = 0 : m$, not just at $x_0 = a$ and $x_m = b$. Indeed, typically the error $E_H(x)$ is the dominant part. By (4.4.16) the second part of the error is

$$E_S(x) = h_i t(1 - t)\big[e_{i-1}(1 - t) - e_i t\big], \quad x = x_{i-1} + th_i, \quad t \in [0, 1).$$

Since $|1 - t| + |t| = 1$, and the maxima of $t(1 - t)$ on $[0, 1]$ equals $1/4$, it follows that

$$|E_S(x)| \leq \frac{1}{4} \max_{1\leq i\leq m} |h_i e_j|, \quad j = i - 1, i, \tag{4.4.34}$$

where $h_i = x_i - x_{i-1}$.

We shall consider the case of constant step size $h_i = h$. For complete splines $e_0 = e_m = 0$ and for almost complete splines $e_0 = O(h^4)$, $e_m = O(h^4)$. It can be shown that then $e_i = O(h^4)$, and its contribution to E_S is $O(h^5)$. Thus if h is sufficiently small, *the Lagrange–Hermite interpolation error is asymptotically the dominant source of error in the whole interval* $[a, b]$.

Finally, we discuss the *effect of the boundary slope errors* for other boundary conditions. Figure 4.4.7 shows (for $m = 20$, $e_0 = e_m = -1$) how rapidly the error component from the boundary dies out. At the midpoint $x = 0.5$ the error is $0.3816 \cdot 10^{-5}$. If $m \gg 1$, $e_0 \neq 0$, and $e_m \neq 0$, the error is negligible outside thin *oscillatory boundary layers* near $x = x_0$ and $x = x_m$. The height and thickness of the layers depend on e_0 and e_m.

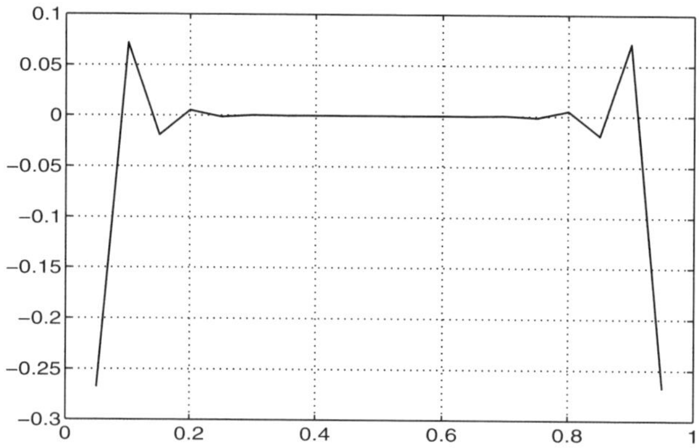

Figure 4.4.7. *Boundary slope errors $e_{B,i}$ for a cubic spline, $e_0 = e_m = -1$; $m = 20$.*

Consider the left boundary; the right one is analogous. For the *natural splines*, there is a peak error near $x = x_0$ of approximately $0.049h^2|y''|$, i.e., 40% of the *linear* interpolation error (instead of cubic). This is often clearly visible in a graph of $s(x)$.

For the not-a-knot splines we obtain approximately $e_0 \sim 0.180h^3y^{(4)}$, and the peak near x_0 becomes $0.031h^4y^{(4)}$, typically very much smaller than we found for natural splines. Still it is about 11.5 times as large as the Lagrange–Hermite interpolation error, but since the oscillations quickly die out, we conclude that *the Lagrange–Hermite interpolation is the dominant error source in cubic not-a-knot spline interpolation in (say) the interval* $[a + 3h, b - 3h]$.

Similar conclusions seem to hold in the case of variable step size also, under the reasonable assumption that $h_{n+1} - h_n = O(h_n^2)$. (The use of variable step size in the context of ordinary differential equations will be treated in a later volume.)

4.4.3 The B-Spline Basis

In the following we will introduce a more convenient **B-spline** basis for the linear space $S_{\Delta,k}$ of spline functions of degree k. The term B-spline was coined by I. J. Schoenberg [314] and is short for basis spline.

It was shown in Sec. 4.4.2 that the set of spline functions of order k, $S_{\Delta,k}$, on the grid

$$\Delta = \{a = x_0 < x_1 < \cdots < x_m = b\},$$

is a linear space of dimension $k + m - 1$. One basis was shown to be the **truncated power basis**:

$$\{1, x, \dots, x^{k-1}\} \cup \{(x - x_1)_+^{k-1}, (x - x_2)_+^{k-1}, \dots, (x - x_{m-1})_+^{k-1}\}. \qquad (4.4.35)$$

In particular, a basis for $S_{\Delta,2}$ consists of continuous piecewise linear functions

$$\{1, x\} \cup \{l_1(x), \dots, l_{m-1}(x)\}, \quad l_i(x) = (x - x_i)_+.$$

Another basis for $S_{\Delta,2}$ is obtained by introducing an artificial exterior knot $x_{-1} \le x_0$. Then it is easy to see that using the functions $l_i(x)$, $i = -1 : m - 1$, every linear spline on $[x_0, x_m]$ can also be written as

$$s(x) = \sum_{i=-1}^{m-1} c_i l_i(x).$$

In anticipation of the fact that it may be desirable to interpolate at other points than the knots, we consider from now on the sequence of knots

$$\Delta = \{\tau_0 \le \tau_1 \le \cdots \le \tau_m\}, \qquad (4.4.36)$$

where $\tau_i < \tau_{i+k}$, $i = 0 : m - k$, i.e., at most k successive knots are allowed to coincide. We start by considering the space $S_{\Delta,1}$. This consists of piecewise constant functions and a basis is

$$N_{i,1}(x) = \begin{cases} 1 & x \in [\tau_i, \tau_{i+1}), \\ 0 & \text{otherwise}, \end{cases} \qquad i = 0 : m - 1. \qquad (4.4.37)$$

The basis functions are arbitrarily chosen to be continuous from the right, i.e., $N_{i,1}(\tau_i) = N_{i,1}(\tau_i + 0)$.

For $k = 2$ we define the **hat functions**[139] by

$$N_{i,2}(x) = \begin{cases} (x - \tau_i)/(\tau_{i+1} - \tau_i), & x \in [\tau_i, \tau_{i+1}], \\ (\tau_{i+2} - x)/(\tau_{i+2} - \tau_{i+1}), & x \in [\tau_{i+1}, \tau_{i+2}), \\ 0, & x \notin (\tau_i, \tau_{i+2}), \end{cases} \quad i = -1 : m - 1. \qquad (4.4.38)$$

We have here introduced two **exterior knots** $\tau_{-1} \le \tau_0$ and $\tau_{m+1} \ge \tau_m$ at the boundaries. (In the following we refer to the knots $\tau_0, \dots, \tau_m$ as **interior knots**.)

At a distinct knot τ_i just one hat function is nonzero, $N_{i+1,2}(\tau_i) = 1$. If all knots are distinct it follows that the spline function of order $k = 2$ interpolating the points (τ_i, y_i), $i = 0 : m$, can be written uniquely as

$$s(x) = \sum_{i=-1}^{m-1} c_i N_{i,2}(x), \qquad (4.4.39)$$

[139]The generalization of hat functions to several dimensions plays a very important role in finite element methods; see Sec. 5.4.4.

where $c_i = y_{i+1}$. This shows that the restriction of the functions $N_{i,2}(x)$, $i = -1 : m - 1$, to the interval $[\tau_0, \tau_m]$ are $(m + 1)$ linearly independent functions in $S_{\Delta,2}$ and form a basis for $S_{\Delta,2}$.

If we allow two interior knots to coalesce, $\tau_i = \tau_{i+1}$, $0 < i < m - 1$, then $N_{i-1,2}(x)$ and $N_{i,2}(x)$ will have a discontinuity at τ_i. This generalizes the concept of a B-spline of order 2 given in Definition 4.4.3 and allows us to model functions with discontinuities at certain knots. Figure 4.4.8 illustrates the formation of a double knot for a linear spline.

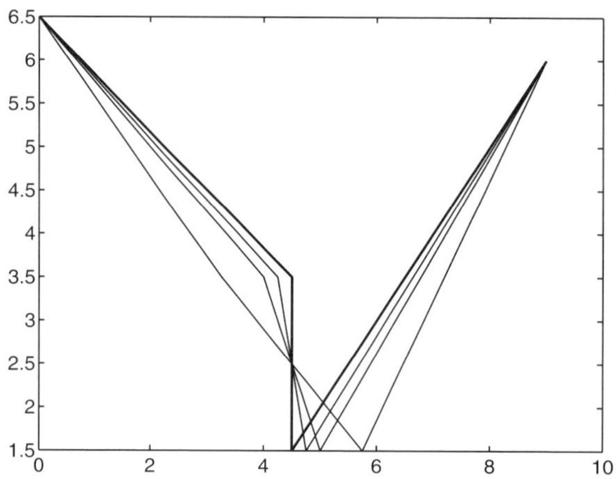

Figure 4.4.8. *Formation of a double knot for a linear spline.*

By definition the basis function $N_{i,2}(x)$ is nonzero only on the interval (τ_i, τ_{i+2}). It follows that for $x \in (\tau_i, \tau_{i+1})$ we have $N_{j,2}(x) = 0$, $j \neq i - 1, i$. *Hence, for any given value of x at most two hat functions will be nonzero* and

$$s(x) = c_{i-1}N_{i-1,2}(x) + c_i N_{i,2}(x), \quad x \in (\tau_i, \tau_{i+1}).$$

The exterior knots are usually taken to coincide with the boundary so that $\tau_{-1} = \tau_0$ and $\tau_{m+1} = \tau_m$. In this case $N_{-1,1}$ and $N_{m-1,1}$ become "half-hats" with a discontinuity at τ_0 and τ_m, respectively.

It is easily verified that the functions $N_{i,2}(x)$ can be written as a linear combination of the basis function $l_i(x) = (x - \tau_i)_+$, $i = 1 : m + 1$. We have

$$\begin{aligned} N_{i,2}(x) &= \big((x - \tau_{i+2})_+ - (x - \tau_{i+1})_+\big)/(\tau_{i+2} - \tau_{i+1}) \\ &\quad -\big((x - \tau_{i+1})_+ - (x - \tau_i)_+\big)/(\tau_{i+1} - \tau_i) \\ &= [\tau_{i+1}, \tau_{i+2}]_t(t - x)_+ - [\tau_i, \tau_{i+1}]_t(t - x)_+ \\ &= (\tau_{i+2} - \tau_i)[\tau_i, \tau_{i+1}, \tau_{i+2}]_t(t - x)_+, \quad i = 1 : m. \end{aligned} \tag{4.4.40}$$

Here $[\tau_i, \tau_{i+1}, \tau_{i+2}]_t$ means the second order divided-difference functional[140] operating on a function of t; i.e., the values τ_i are to be substituted for t, not for x. Recall that divided differences are defined also for *coincident values* of the argument; see Sec. 4.3.1.

The Peano kernel and its basic properties were given in Sec. 3.3.3. The last expression in (4.4.40) tells us that $N_{i,2}$ is the Peano kernel of a second order divided-difference functional

[140]The notation is defined in Sec. 4.2.1.

multiplied by the constant $\tau_{i+2} - \tau_i$. This observation suggests a definition of B-splines of arbitrary order k and a B-spline basis for the space $S_{\Delta,k}$.

Definition 4.4.9.

Let $\Delta = \{\tau_0 \leq \tau_1 \leq \cdots \leq \tau_m\}$ be an arbitrary sequence of knots such that $\tau_i < \tau_{i+k}$, $i = 0 : m - k$. Then a B-spline of order k (apart from a stepsize factor) equals the Peano kernel of a kth order divided-difference functional; more precisely, we define (with the notations used in this chapter)

$$N_{i,k}(x) = (\tau_{i+k} - \tau_i)[\tau_i, \tau_{i+1}, \ldots, \tau_{i+k}]_t (t - x)_+^{k-1}, \tag{4.4.41}$$

where $[\tau_i, \tau_{i+1}, \ldots, \tau_{i+k}]l_x^{k-1}$ denotes the kth divided difference of the function $l_x^{k-1}(\cdot)$ with respect to the set of points $\tau_i, \tau_{i+1}, \ldots, \tau_{i+k}$. This definition remains valid for knots that are not distinct.

It can be shown that $N_{i,k}(x)$ is defined for all x. If the knots are distinct, then by Problem 4.2.7

$$N_{i,k}(x) = (\tau_{i+k} - \tau_i) \sum_{j=i}^{i+k} \frac{(\tau_j - x)_+^{k-1}}{\Phi'_{i,k}(\tau_j)}, \qquad \Phi_{i,k}(x) = \prod_{j=i}^{i+k} (x - \tau_j), \tag{4.4.42}$$

which shows that $N_{i,k}$ is a linear combination of functions $(\tau_j - x)_+^{k-1}$, $j = i : i + k$. Hence it is a spline of order k (as anticipated in the terminology).

For equidistant knots the B-spline is related to the probability density of the sum of k uniformly distributed random variables on $[-\frac{1}{2}, \frac{1}{2}]$. This was known already to Laplace. B-splines of order $k = 1, 2,$ and 3 are shown in Figure 4.4.9.

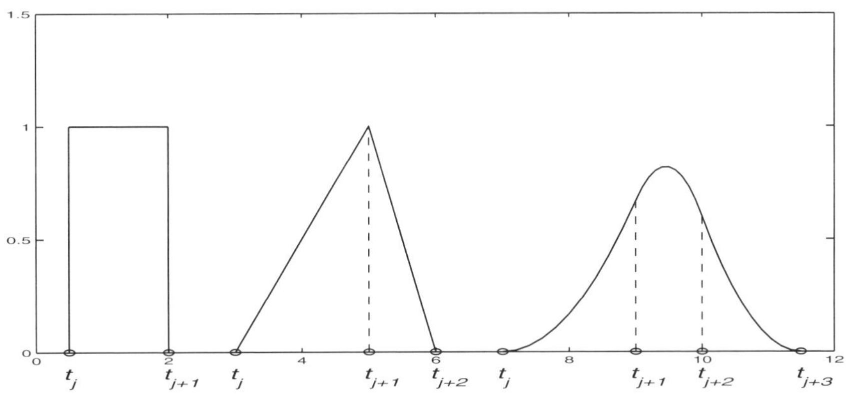

Figure 4.4.9. *B-splines of order $k = 1, 2, 3$.*

Theorem 4.4.10. *The B-splines of order k have the following properties:*

(i) *Positivity:* $N_{i,k}(x) > 0, \quad x \in (\tau_i, \tau_{i+k})$.

(ii) *Compact support:* $N_{i,k}(x) = 0, \quad x \notin [\tau_i, \tau_{i+k}]$.

(iii) *Summation property:* $\sum_i N_{i,k}(x) = 1$ *for all* $x \in [\tau_0, \tau_m]$.

Proof. A proof can be based on the general facts concerning Peano kernels found in Sec. 3.3.3, where also an expression for the B-spline ($k = 3$) is calculated for the equidistant case. (Unfortunately the symbol x means different things here and in Sec. 3.3.3.)

(i) By (4.2.11), $Rf = [\tau_i, \tau_{i+1}, \dots, \tau_{i+k}]f = f^{(k)}(\xi)/k!$, $\xi \in (\tau_i, \tau_{i+k})$, and $Rp = 0$, for $p \in \mathcal{P}_k$. It then follows from the corollary of Peano's remainder theorem that the Peano kernel does not change sign in $[\tau_i, \tau_{i+k}]$. It must then have the same sign as $\int K(u)\,du = R(x-a)^k/k! = 1$. This proves a somewhat weaker statement than (i) ($N_{i,k}(x) \geq 0$ instead of $N_{i,k}(x) > 0$).

(ii) This property follows since a Peano kernel always vanishes outside its interval of support of the functional, in this case $[\tau_i, \tau_{i+k}]$. (A more general result concerning the number of zeros is found, e.g., in Powell [292, Theorem 19.1]. Among other things this theorem implies that the jth derivative of a B-spline, $j \leq k-2$, changes sign exactly j times. This explains the "bell shape" of B-splines.)

(iii) For a sketch of a proof of the summation property,[141] see Problem 4.4.11. □

To get a basis of B-splines for the space $S_{\Delta,k}$, $\Delta = \{\tau_0 \leq \tau_1 \leq \cdots \leq \tau_m\}$, $(m+k-1)$ B-splines of order k are needed. We therefore choose $2(k-1)$ additional knots $\tau_{-k+1} \leq \cdots \leq \tau_{-1} \leq \tau_0$, $\tau_{m+k-1} \geq \cdots \geq \tau_{m+1} \geq \tau_m$, and B-splines $N_{i,k}(x)$, $i = -k+1 : m-1$.

As for $k = 2$ it is convenient to let the exterior knots coincide with the endpoints,

$$\tau_{-k+1} = \cdots = \tau_{-1} = \tau_0, \qquad \tau_m = \tau_{m+1} = \cdots = \tau_{m+k-1}.$$

It can be shown that this choice tends to optimize the conditioning of the B-spline basis. Figure 4.4.10 shows the first four cubic B-splines for $k = 4$ (the four last B-splines are

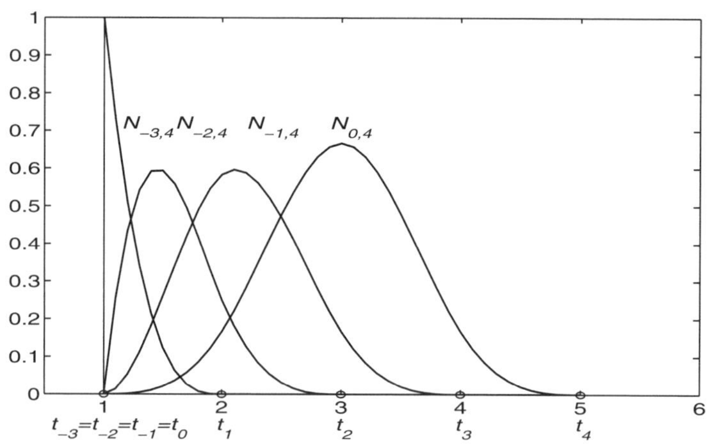

Figure 4.4.10. *The four cubic B-splines nonzero for $x \in (t_0, t_1)$ with coalescing exterior knots $t_{-3} = t_{-2} = t_{-1} = t_0$.*

[141] The B-splines $M_{i,k}$ originally introduced by Schoenberg in 1946 were normalized so that $\int_{-\infty}^{\infty} M_{i,k}\,dx = 1$.

a mirror image of these). We note that $N_{-3,4}$ is discontinuous, $N_{-2,4}$ has a nonzero first derivative, and $N_{-2,4}$ has a nonzero second derivative at the left boundary.

Interior knots of multiplicity $r > 1$ are useful when we want to model a function which has less than $k-2$ continuous derivatives at a particular knot. If $r \le k$ interior knots coalesce, then the spline will only have $k-1-r$ continuous derivatives at this knot.

Lemma 4.4.11.
Let τ_i be a knot of multiplicity $r \le k$, i.e.,

$$\tau_{i-1} < \tau_i = \cdots = \tau_{i+r-1} < \tau_{i+r}.$$

Then $N_{i,k}$ is at least $(k-r-1)$ times continuously differentiable at τ_i. For $r = k$, the B-spline becomes discontinuous.

Proof. The truncated power $(t-\tau_i)_+^{k-1}$ is $(k-2)$ times continuously differentiable and $[\tau_i, \dots, \tau_{i+k}]g$ contains at most the $(r-1)$st derivative of g. Hence the lemma follows. □

Consider the spline function

$$s(x) = \sum_{i=-k+1}^{m-1} c_i N_{i,k}(x). \tag{4.4.43}$$

If $s(x) = 0$, $x \in [\tau_0, \tau_m]$, then $s(\tau_0) = s'(\tau_0) = \cdots = s^{(k-1)}(\tau_0) = 0$, and $s(\tau_i) = 0$, $i = 1 : m-1$. From this it can be determined by induction that in (4.4.43) $c_i = 0$, $i = -k+1 : m-1$. This shows that the $(m+k-1)$ B-splines $N_{i,k}(x)$, $i = -k+1 : m-1$, are linearly independent and form a basis for the space $S_{\Delta,k}$. (A more general result is given in de Boor [37, Theorem IX.1].) Thus any spline function $s(x)$ of order k (degree $k-1$) on Δ can be uniquely written in the form (4.4.43). Note that from the compact support property it follows that for any fixed value of $x \in [\tau_0, \tau_m]$ at most k terms will be nonzero in the sum in (4.4.43), and thus we have

$$s(x) = \sum_{i=j-k+1}^{j} c_i N_{i,k}(x), \quad x \in [\tau_j, \tau_{j+1}). \tag{4.4.44}$$

We will now develop a very important stable recurrence relation for computing B-splines. For this we need the following difference analogue of **Leibniz' formula**.[142]

Theorem 4.4.12 (*Leibniz' Formula*).
Let $f(x) = g(x)h(x)$, and $x_i \le x_{i+1} \le \cdots \le x_{i+k}$. Then

$$[x_i, \dots, x_{i+k}]f = \sum_{r=i}^{i+k} [x_i, \dots, x_r]g \cdot [x_r, \dots, x_{i+k}]h, \tag{4.4.45}$$

[142] Gottfried Wilhelm von Leibniz (1646–1716), a German mathematician who developed his version of calculus at the same time as Newton. Many of the notations he introduced are still used today.

provided that $g(x)$ and $f(x)$ are sufficiently many times differentiable so that the divided differences on the right-hand side are defined for any coinciding points x_j.

Proof. Note that the product polynomial

$$P(x) = \sum_{r=i}^{i+k} (x - x_i) \cdots (x - x_{r-1})[x_i, \dots, x_r]g \cdot \sum_{s=i}^{i+k} (x - x_{s+1}) \cdots (x - x_{i+k})[x_s, \dots, x_{i+k}]h$$

agrees with $f(x)$ at $x_i, \dots, x_{i+k}$ since by Newton's interpolation formula the first factor agrees with $g(x)$ and the second with $h(x)$ there. If we multiply out we can write $P(x)$ as a sum of two polynomials:

$$P(x) = \sum_{r,s=i}^{i+k} \dots = \sum_{r \le s} \dots + \sum_{r>s} \dots = P_1(x) + P_2(x).$$

Since in $P_2(x)$ each term in the sum has $\prod_{j=i}^{i+k}(x - x_j)$ as a factor, it follows that $P_1(x)$ will also interpolate $f(x)$ at $x_i, \dots, x_{i+k}$. The theorem now follows since the leading coefficient of $P_1(x)$, which equals $\sum_{r=i}^{i+k}[x_i, \dots, x_r]g \cdots [x_r, \dots, x_{i+k}]h$, must equal the leading coefficient $[x_i, \dots, x_{i+k}]f$ of the unique interpolation polynomial of degree k. □

Theorem 4.4.13.

The B-splines satisfy the recurrence relation

$$N_{i,k}(x) = \frac{x - \tau_i}{\tau_{i+k-1} - \tau_i} N_{i,k-1}(x) + \frac{\tau_{i+k} - x}{\tau_{i+k} - \tau_{i+1}} N_{i+1,k-1}(x). \tag{4.4.46}$$

Proof. (de Boor [37, pp. 130–131]) The recurrence is derived by applying Leibniz' formula for the kth divided difference to the product

$$(t - x)_+^{k-1} = (t - x)(t - x)_+^{k-2}.$$

This gives

$$\begin{aligned}[\tau_i, \dots, \tau_{i+k}]_t (t - x)_+^{k-1} &= (\tau_i - x)[\tau_i, \dots, \tau_{i+k}]_t (t - x)_+^{k-2} \\ &\quad + 1 \cdot [\tau_{i+1}, \dots, \tau_{i+k}]_t (t - x)_+^{k-2},\end{aligned} \tag{4.4.47}$$

since $[\tau_i]_t(t - x) = (\tau_i - x)$, $[\tau_i, \tau_{i+1}]_t(t - x) = 1$, and $[\tau_i, \dots, \tau_j]_t(t - x) = 0$ for $j > i + 1$. By the definition of a divided difference

$$(\tau_i - x)[\tau_i, \dots, \tau_{i+k}]_t = \frac{\tau_i - x}{\tau_{i+k} - \tau_i} \left([\tau_{i+1}, \dots, \tau_{i+k}]_t - [\tau_i, \dots, \tau_{i+k-1}]_t\right).$$

Substitute this in (4.4.47), simplify, and apply the definition of B-splines. This yields (4.4.46). □

Note that with k multiple knots at the boundaries the denominators in (4.4.46) can become zero. In this case the corresponding numerator is also zero and the term should be set equal to zero.

From Property (ii) in Theorem 4.4.10 we conclude that only k B-splines of order k may be nonzero on a particular interval $[\tau_j, \tau_{j+1}]$. Starting from $N_{i,1}(x) = 1$, $x \in [\tau_i, \tau_{i+1})$ and 0 otherwise (cf. (4.4.37)), these B-splines of order k can be *simultaneously* evaluated using this recurrence by forming successively their values for order $1 : k$ in only about $\frac{3}{2}k^2$ flops. This recurrence is very stable, since it consists of taking a convex combination of two lower-order splines to get the next one.

Suppose that $x \in [\tau_i, \tau_{i+1}]$, and $\tau_i \neq \tau_{i+1}$. Then the B-splines of order $k = 1, 2, 3, \ldots$ nonzero at x can be simultaneously evaluated by computing the following triangular array.

$$
\begin{array}{ccccc}
 & & & 0 & \\
 & & 0 & & \cdots \\
 & 0 & & N_{i-3,4} & \\
0 & & N_{i-2,3} & & \cdots \\
 & N_{i-1,2} & & N_{i-2,4} & \\
N_{i,1} & & N_{i-1,3} & & \cdots \\
 & N_{i,2} & & N_{i-1,4} & \\
0 & & N_{i,3} & & \cdots \\
 & 0 & & N_{i,4} & \\
 & & 0 & & \cdots \\
 & & & 0 &
\end{array}
\tag{4.4.48}
$$

The boundary of zeros in the array is due to the fact that all other B-splines not mentioned explicitly vanish at x. This array can be generated column by column. The first column is known from (4.4.37), and each entry in a subsequent column can be computed as a linear combination with nonnegative coefficients of its two neighbors using (4.4.46). Note that if this is arranged in a suitable order the elements in the new column can overwrite the elements in the old column.

To evaluate $s(x)$, we first determine the index i such that $x \in [\tau_i, \tau_{i+1})$ using, for example, a linear search or bisection (see Sec. 6.1.2). The recurrence above is then used to generate the triangular array (4.4.48) which provides $N_{j,k}(x)$, $j = i - k + 1 : i$, in the sum (4.4.44).

Assume now that $\tau_0 < \tau_1 < \cdots < \tau_m$ are distinct knots. Using the B-spline basis we can formulate a more general interpolation problem, where the $n = m + k - 1$ interpolation points (knots) x_j do not necessarily coincide with the knots τ_i. We consider determining a spline function $s(x) \in S_{\Delta,k}$ such that

$$s(x_j) = f_j, \quad j = 1 : m + k - 1.$$

Since any spline $s(x) \in S_{\Delta,k}$ can be written as a linear combination of B-splines, the interpolation problem can equivalently be written

$$\sum_{i=-k+1}^{m-1} c_i N_{i,k}(x_j) = f_j, \quad j = 1 : m + k - 1. \tag{4.4.49}$$

These equations form a linear system $Ac = f$ for the coefficients, where

$$a_{ij} = N_{i-k,k}(x_j), \quad i, j = 1 : m + k - 1, \tag{4.4.50}$$

and

$$c = (c_{-k+1}, \dots, c_{m-1})^T, \qquad f = (f_1, \dots, f_{m+k-1})^T.$$

The elements $a_{ij} = N_{i-k,k}(x_j)$ of the matrix A can be evaluated by the recurrence (4.4.46). The matrix A will have a banded structure since $a_{ij} = 0$ unless $x_j \in [\tau_i, \tau_{i+k}]$. Hence at most k elements are nonzero in each row of A. (Note that if $x_j = \tau_i$ for some i only $k - 1$ elements will be nonzero. This explains why tridiagonal systems were encountered in cubic spline interpolation in earlier sections.)

Schoenberg and Whitney [315] showed that *the matrix A is nonsingular if and only if its diagonal elements are nonzero*,

$$a_{jj} = N_{j-k,k}(x_j) \neq 0, \quad j = 1 : n,$$

or, equivalently, if the knots x_j satisfy

$$\tau_{j-k} < x_j < \tau_j, \quad j = 1 : n. \tag{4.4.51}$$

Further, the matrix can be shown to be **totally nonnegative**, i.e., the determinant of every submatrix is nonnegative. For such systems, if Gaussian elimination is carried out *without pivoting*, the error bound is particularly favorable; see [40]. This will also preserve the banded structure of A during the elimination.

When the B-spline representation (4.4.43) of the interpolant has been determined it can be evaluated at a given point using the recursion formula (4.4.46). If it has to be evaluated more than two or three times per polynomial piece it is more efficient to first convert the B-spline to pp-form (4.4.22). For hints on how to do that, see Problem 4.4.12 (b) and (c). A detailed discussion is found in de Boor [37, Chapter X].

Unless the Schoenberg–Whitney condition (4.4.51) is well-satisfied the system may become ill-conditioned. For splines of even order k the interior knots

$$\tau_0 = x_0, \quad \tau_{j+1} = x_{j+k/2}, \quad j = 0 : n - k - 1, \quad \tau_m = x_n,$$

is a good choice in this respect. In the important case of cubic splines this means that knots are positioned at each data point except the second- and next-last (cf. the not-a-knot condition in Sec. 4.4.2).

4.4.4 Least Squares Splines Approximation

In some application we are given function values $f_j = f(x_j)$, $j = 1 : n$, that we want to approximate with a spline functions with *much fewer knots* so that $m + k - 1 \leq n$. Then (4.4.49) is an *overdetermined linear system* and the interpolation conditions cannot be satisfied exactly. We therefore consider the linear **least squares spline approximation** problem:

$$\min \sum_{j=1}^{n} \left(\sum_{i=-k+1}^{m-1} c_i N_{i,k}(x_j) - f_j \right)^2. \tag{4.4.52}$$

Using the same notation as above this can be written in matrix form as

$$\min_{c} \|Ac - f\|_2^2. \tag{4.4.53}$$

The matrix A will have full column rank equal to $m + k - 1$ if and only if there is a subset of points τ_j satisfying the Schoenberg–Whitney conditions (4.4.51).

If A has full column rank, then the least squares solution c is uniquely determined by the normal equations $A^T Ac = A^T f$. Since A has at most k nonzero elements in each row the matrix $A^T A$ will have symmetric banded form with at most $2k + 1$ nonzero elements in each row; see Figure 4.4.11.

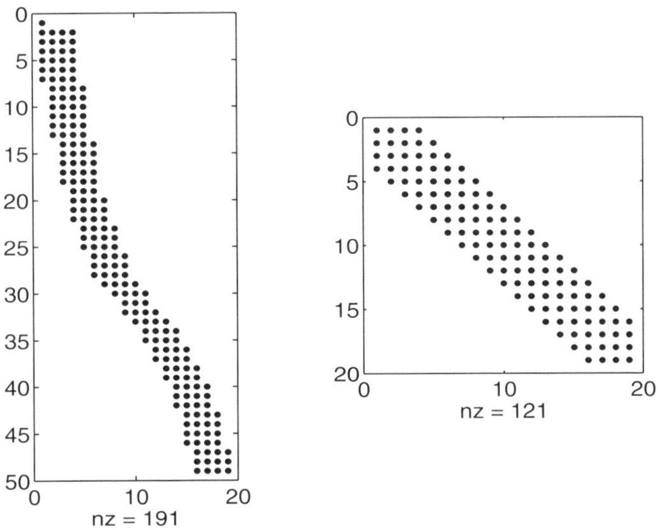

Figure 4.4.11. *Banded structure of the matrices A and $A^T A$ arising in cubic spline approximation with B-splines (nonzero elements shown).*

Example 4.4.3 (*de Boor* [37]).

Consider experimental data describing a property of titanium as a function of temperature. Experimental values for $t_i = 585 + 10i$, $i = 1 : 49$, are given. We want to fit these data using a least squares cubic spline. Figure 4.4.12 shows results from using a least squares fitted cubic spline with 17 knots. The spline with 9 knots shows oscillations near the points where the curve flattens out and the top of the peak is not well matched. Increasing the number of knots to 17 we get a very good fit.

We have in the treatment above assumed that the set of (interior) knots $\{\tau_0 \leq \tau_1 \leq \cdots \leq \tau_m\}$ is given. In many spline approximation problems it is more realistic to consider the location of knots to be free and try to determine a small set of knots such that the given data can be approximated to a some preassigned accuracy. Several schemes have been developed to treat this problem.

One class of algorithms start with only a few knots and iteratively add more knots guided by some measure of the error; see de Boor [36, Chapter XII]. The placement of the

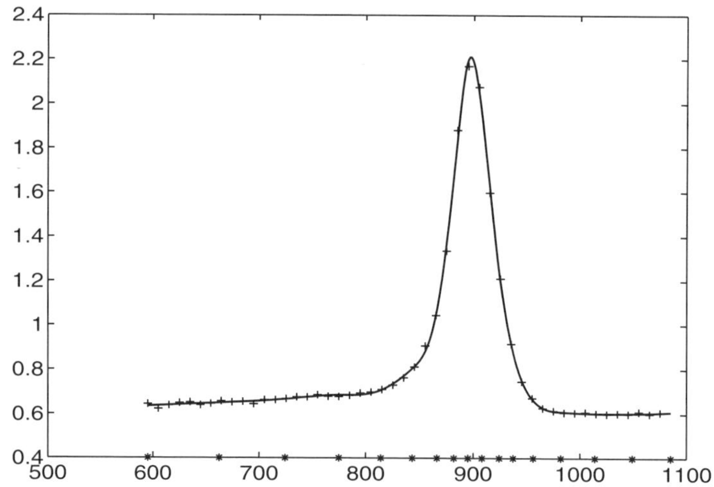

Figure 4.4.12. *Least squares cubic spline approximation of titanium data using* 17 *knots marked on the axes by an "o."*

knots is chosen so that the Schoenberg–Whitney conditions are always satisfied. The iterations are stopped when the approximation is deemed satisfactory. If a knot $\bar{\tau} \in [\tau_j, \tau_{j+1})$ is inserted, then the B-spline series with respect to the enlarged set of knots can cheaply and stably be computed from the old one (see Dierckx [99]).

Other algorithms start with many knots and successively *remove* knots which are not contributing much to the quality of the approximation. In these two classes of algorithms one does not seek an optimal knot placement at each step. This is done in more recent algorithms; see Schwetlick and Schütze [319].

Review Questions

4.4.1 What is meant by a cubic spline function? Give an example where such a function is better suited than a polynomial for approximation over the whole interval.

4.4.2 (a) What is the dimension of the space $S_{\Delta,k}$ of spline functions of order k on a grid $\Delta = \{x_0, x_1, \dots, x_m\}$? Give a basis for this space.

(b) Set up the linear system for cubic spline interpolation in the equidistant case for some common boundary conditions. What do the unknown quantities mean, and what conditions are expressed by the equations? About how many operations are required to interpolate a cubic spline function to $m+1$, $m \gg 1$, given values of a function?

4.4.3 What error sources have influence on the results of cubic spline interpolation? How fast do the boundary errors die out? How do the results in the interior of the interval depend on the step size (asymptotically)? One of the common types of boundary conditions yields much larger error than the others. Which one? Compare it quantitatively with one of the others.

4.4.4 Approximately how many arithmetic operations are required to evaluate the function values of all cubic B-splines that are nonzero at a given point?

4.4.5 Express the restrictions of $f(x) = 1$ and $f(x) = x$ to the interval $[x_0, x_m]$ as linear combinations of the hat functions defined by (4.4.38).

4.4.6 The Schoenberg–Whitney conditions give necessary and sufficient conditions for a certain interpolation problem with B-splines of order k. What is the interpolation problem and what are the conditions?

Problems and Computer Exercises

4.4.1 Consider a cubic Bézier curve defined by the four control points p_0, p_1, p_2, and p_3. Show that at $t = 1/2$

$$c\left(\frac{1}{2}\right) = \frac{1}{4}\frac{p_0 + p_3}{2} + \frac{3}{4}\frac{p_1 + p_2}{2}$$

and interpret this formula geometrically.

4.4.2 (G. Eriksson) Approximate the function $y = \cos x$ on $[-\pi/2, \pi/2]$ by a cubic Bézier curve. Determine the four control points in such a way that it interpolates $\cos x$ and its derivative at $-\pi/2$, 0, and $\pi/2$.

Hint: Use symmetry and the result of Problem 4.4.1 to find the y-coordinate of p_1 and p_2.

4.4.3 Suppose that $f(x)$ and the grid Δ are symmetric around the midpoint of the interval $[a, b]$. You can then considerably reduce the amount of computation needed for the construction of the cubic spline interpolant by replacing the boundary condition at $x = b$ by an adequate condition at the midpoint. Which?

(a) Set up the matrix and right-hand side for this in the case of constant step size h.

(b) Do the same for a general case of variable step size.

4.4.4 (a) Write a program for solving a tridiagonal linear system by Gaussian elimination without pivoting. Assume that the nonzero diagonals are stored in three vectors. Adapt it to cubic spline interpolation with equidistant knots with several types of boundary conditions.

(b) Consider the tridiagonal system resulting from the not-a-knot boundary conditions. Show that after eliminating k_0 between the first two equations and k_m between the last two equations the remaining tridiagonal system for $k_1, \ldots, k_{m-1}$ is diagonally dominant.

(c) Interpolate a cubic spline $s(x)$ through the points $(x_i, f(x_i))$, where

$$f(x) = (1 + 25x^2)^{-1}, \quad x_i = -1 + \frac{2}{10}(i - 1), \quad i = 1 : 11.$$

Compute a natural spline, a complete spline (here $f'(x_1)$ and $f'(x_{11})$ are needed) and a not-a-knot spline. Compute and compare error curves (natural and logarithmic).

(d) Conduct similar runs as in (b), though for $f(x) = 1/x$, $1 \le x \le 2$, with $h = 0.1$

and $h = 0.05$. Compare the "almost complete," as described in the text, with the complete and the natural boundary condition.

4.4.5 If f'' is known at the boundary points, then the boundary conditions can be chosen so that $f'' = s''$ at the boundary points. Show that this leads to the conditions

$$\begin{aligned} 2k_0 + k_1 &= 3d_1 - h_1 f''(x_0), \\ k_{m-1} + 2k_m &= 3d_m + h_m f''(x_m). \end{aligned}$$

4.4.6 Show that the formula

$$\int_{x_0}^{x_m} s(x)dx = \sum_{i=1}^{m} \Big(\frac{1}{2} h_i (y_{i-1} + y_i) + \frac{1}{12} (k_{i-1} - k_i) h_i^2 \Big)$$

is exact for all cubic spline functions $s(x)$. How does the formula simplify if all $h_i = h$?

Hint: Integrate (4.4.16) from x_{i-1} to x_i.

4.4.7 In (4.4.16) the cubic spline $q_i(x)$ on the interval $[x_{i-1}, x_i)$ is expressed in terms of function values y_{i-1}, y_i, and the first derivatives k_{i-1}, k_i.

(a) Show that if $M_i = s''(x_i)$, $i = 0 : m$, are the *second derivatives* (also called **moments**) of the spline function, then

$$k_i - d_i = \frac{h_i}{6}(2M_i + M_{i-1}), \qquad k_{i-1} - d_i = -\frac{h_i}{6}(M_i + 2M_{i-1}).$$

Hence $q_i(x)$ can also be uniquely expressed in terms of y_{i-1}, y_i and M_{i-1}, M_i.

(b) Show that, using the parametrization in (a), the continuity of the *first* derivative of the spline function at an interior point x_i gives the equation

$$h_i M_{i-1} + 2(h_i + h_{i+1})M_i + h_{i+1} M_{i+1} = 6(d_{i+1} - d_i).$$

4.4.8 (a) Develop an algorithm for solving the arrowhead linear system $Tk = g$ (4.4.32), using Gaussian elimination without pivoting. Show that about twice the number of arithmetic operations are needed compared to a tridiagonal system.

(b) At the end of Sec. 4.4.2 parametric spline interpolation to given points (x_i, y_i), $i = 0 : m$, is briefly mentioned. Work out the details on how to use this to represent a closed curve. Try it out on a boomerang, an elephant, or what have you.

4.4.9 (a) Compute and plot a B-spline basis of order $k = 3$ (locally quadratic) and $m = 6$ subintervals of equal length.

Hint: In the equidistant case there is some translation invariance and symmetry, so you do not really need more than essentially three different B-splines. You need one spline with triple knot at x_0 and a single knot at x_1 (very easy to construct), and two more splines.

(b) Set up a scheme to determine a locally quadratic B-spline which interpolates given values at the *midpoints*

$$x_i = (\tau_{i+1} + \tau_i)/2, \quad (\tau_{i+1} \neq \tau_i), \quad i = 0 : m - 1,$$

and the boundary points τ_0, τ_m. Show that the spline is uniquely determined by these interpolation conditions.

4.4.10 Derive the usual formula of Leibniz for the kth derivative from (4.4.45) by a passage to the limit.

4.4.11 Use the recurrence (4.4.46)

$$N_{i,k}(x) = \frac{x - \tau_i}{\tau_{i+k-1} - \tau_i} N_{i,k-1}(x) + \frac{\tau_{i+k} - x}{\tau_{i+k} - \tau_{i+1}} N_{i+1,k-1}(x)$$

to show that

$$\sum_i N_{i,k}(x) = \sum_i N_{i,k-1}(x), \qquad \tau_0 \le x \le \tau_m,$$

where the sum is taken over all nonzero values. Use this to give an induction proof of the summation property in Theorem 4.4.10.

4.4.12 (a) Using the result $\frac{d}{dx}(t - x)_+^{k-1} = -(k-1)(t-k)_+^{k-2}$, $k \ge 1$, show the formula for differentiating a B-spline,

$$\frac{d}{dx} N_{i,k}(x) = (k-1)\left(\frac{N_{i,k-1}(x)}{\tau_{i+k-1} - \tau_i} - \frac{N_{i+1,k-1}(x)}{\tau_{i+k} - \tau_{i+1}} \right).$$

Then use the relation (4.4.46) to show

$$\frac{d}{dx} \sum_{i=r}^{s} c_i N_{i,k}(x) = (k-1) \sum_{i=r}^{s+1} \frac{c_i - c_{i-1}}{\tau_{i+k-1} - \tau_i} N_{i,k-1}(x),$$

where $c_{r-1} := c_{s+1} := 0$.

(b) Given the B-spline representation of a cubic spline function $s(x)$. Show how to find its polynomial representation (4.4.22) by computing the function values and first derivatives $s(\tau_i), s'(\tau_i)$, $i = 0 : m$.

(c) Apply the idea in (a) recursively to show how to compute all derivatives of $s(x)$ up to order $k - 1$. Use this to develop a method for computing the polynomial representation of a spline of arbitrary order k from its B-spline representation.

4.4.13 Three different bases for the space of cubic polynomials of degree ≤ 3 on the interval $[0, 1]$ are the monomial basis $\{1, t, t^2, t^3\}$, the Bernštein basis $\{B_0^3(t), B_1^3(t), B_2^3(t), B_3^3(t)\}$, and the Hermite basis. Determine the matrices for these basis changes.

4.5 Approximation and Function Spaces

Function space concepts have been introduced successively in this book. Recall, for example, the discussion of operators and functionals in Sec. 3.3.2, where the linear space $\mathcal{P}_n$, the n-dimensional space of polynomials of degree less than n, was also introduced. This terminology was used and extended in Sec. 4.1, in the discussion of various bases and coordinate transformations in $\mathcal{P}_n$.

For forthcoming applications of functional analysis to interpolation and approximation it is now time for a digression about

- distances and norms in function spaces;
- a general error bound that we call *the norm and distance formula*;
- inner-product spaces and orthogonal systems.

4.5.1 Distance and Norm

For the study of accuracy and convergence of methods of interpolation and approximation we now introduce the concept of a **metric space**. By this we understand a set of points $\mathcal{S}$ and a real-valued function d, a **distance** defined for pairs of points in $\mathcal{S}$ in such a way that the following axioms are satisfied for all x, y, z in $\mathcal{S}$. (Draw a triangle with vertices at the points x, y, z.)

1. $d(x, x) = 0$ (reflexivity),
2. $d(x, y) > 0$ if $x \neq y$ (positivity),
3. $d(x, y) = d(y, x)$ (symmetry),
4. $d(x, y) \leq d(x, z) + d(z, y)$ (triangle inequality).

The axioms reflect familiar features of distance concepts used in mathematics and everyday life, such as the absolute value of complex numbers, the shortest distance along a geodesic on the surface of the Earth, or the shortest distance along a given road system.[143]

Many other natural and useful relations can be derived from these axioms, such as

$$d(x, y) \geq |d(x, z) - d(y, z)|, \qquad d(x_1, x_n) \leq \sum_{i=1}^{n-1} d(x_i, x_{i+1}), \tag{4.5.1}$$

where $x_1, x_2, \ldots, x_n$ is a sequence of points in $\mathcal{S}$.

Definition 4.5.1.

A sequence of points $\{x_n\}$ in a metric space $\mathcal{S}$ is said to **converge** *to a limit $x^* \in \mathcal{S}$ if $d(x_n, x^*) \to 0$. As $n \to \infty$, we write $x_n \to x^*$ or $\lim_{n\to\infty} x_n = x^*$. A sequence $\{x_n\}$ in $\mathcal{S}$ is called a* **Cauchy sequence** *if for every $\epsilon > 0$ there is an integer $N(\epsilon)$ such that $d(x_m, x_n) < \epsilon$ for all $m, n \geq N(\epsilon)$. Every convergent sequence is a Cauchy sequence, but the converse is not necessarily true. $\mathcal{S}$ is called a* **complete space** *if every Cauchy sequence in $\mathcal{S}$ converges to a limit in $\mathcal{S}$.*

It is well known that **R** satisfies the characterization of a complete space, but the set of *rational* numbers is not complete. For example, the Newton iteration $x_1 = 1$, $x_{n+1} = \frac{1}{2}(x_n + 2/x_n)$, defines a sequence of rational numbers that converges to $\sqrt{2}$, which is not a rational number.

Many important problems in pure and applied mathematics can be formulated as minimization problems. The function space terminology often makes proofs and algorithms less abstract.

---

[143] If $\mathcal{S}$ is a functions space, the points of $\mathcal{S}$ are functions with operands in some other space, for example, in **R** or $\mathbf{R}^n$.

Most spaces that we shall encounter in this book are **linear spaces**. Their elements are called vectors, which is why these spaces are called **vector spaces**. Two operations are defined in these spaces, namely the addition of vectors and the multiplication of a vector by a scalar. They obey the usual rules of algebra.[144] The set of scalars can be either **R** or **C**; the vector space is then called *real* or *complex*, respectively.

Definition 4.5.2.

Let f be in a metric space $\mathcal{B}$ with a distance function $d(x, y)$, and let $\mathcal{S}$ be a subset or linear subspace of $\mathcal{B}$. We define the distance of f to $\mathcal{S}$ by

$$\operatorname{dist}(f, \mathcal{S}) = \inf_{g \in \mathcal{S}} d(f, g). \tag{4.5.2}$$

Much of our discussion also applies to linear spaces of **infinite dimension**, i.e., **function spaces**. The elements (vectors) then are functions of one or several real variables on a compact set, i.e., a closed bounded region. The idea of such a functions space is now illustrated in an example.

Example 4.5.1.

Consider the set of functions representable by a convergent power series on the interval $[-1, 1]$,

$$f(t) = c_0 + c_1 t + c_2 t^2 + \cdots.$$

This is an infinite-dimensional linear space. The functions $1, t, t^2, \ldots$ can be considered as a standard basis of this space. The coordinates of $f(t)$ then are the vector $c_0, c_1, c_2, \ldots$.

We shall be concerned with the problem of **linear approximation**, i.e., a function f is to be approximated using a function f^* that can be expressed as a linear combination of n given linearly independent functions $\phi_1(x), \phi_2(x), \ldots, \phi_n(x)$,

$$f^* = c_1\phi_1(x) + c_2\phi_2(x) + \cdots + c_n\phi_n(x), \tag{4.5.3}$$

where $c_1, c_2, \ldots, c_n$ are parameters to be determined.[145] They may be considered as coordinates of f^* in the functions space spanned by $\phi_1(x), \phi_2(x), \ldots, \phi_n(x)$.

In a vector space the distance of the point f from the origin is called the **norm** of f and denoted by $\|f\|$, typically with some subscript that more or less cryptically indicates the relevant space. The definition of the norm depends on the space. The following axioms must be satisfied.

Definition 4.5.3.

A real-valued function $\|f\|$ is called a norm on a vector space $\mathcal{S}$ if it satisfies conditions 1–3 below for all $f, g \in \mathcal{S}$, and for all scalars λ.

---

[144] See Sec. A.1 in Online Appendix A for a summary about vector spaces. In more detailed texts on linear algebra or functional analysis you find a collection of eight axioms (commutativity, associativity, etc.) required by a linear vector space.

[145] The functions ϕ_j, however, are typically *not* linear. The term "linear interpolation" is, from our present point of view, rather misleading.

1. $\|f\| > 0$, *unless* $f = 0$ *(positivity),*
2. $\|\lambda f\| = |\lambda|\|f\|$ *(homogeneity),*
3. $\|f + g\| \leq \|f\| + \|g\|$ *(triangle inequality).*

A normed vector space is a metric space with the distance

$$d(x, y) = \|x - y\|.$$

If it is also a complete space, it is called a **Banach space**.[146]

The most common norms in spaces of (real and complex) infinite sequences $x = (\xi_1, \xi_2, \xi_3, \ldots)^T$ or spaces of functions on a bounded and closed interval $[a, b]$ are

$$\|x\|_\infty = \max_j |\xi_j|, \qquad \|f\|_\infty = \max_{x\in[a,b]} |f(x)|,$$

$$\|x\|_2 = \left(\sum_{j=1}^{\infty} |\xi_j|^2\right)^{1/2}, \qquad \|f\|_2 = \|f\|_{2,[a,b]} = \left(\int_a^b |f(x)|^2\, dx\right)^{1/2},$$

$$\|x\|_1 = \sum_{j=1}^{\infty} |\xi_j|, \qquad \|f\|_1 = \|f\|_{1,[a,b]} = \int_a^b |f(x)|\, dx.$$

These norms are called

- the ℓ_∞ or **max(imum) norm** (or uniform norm);
- the **Euclidean norm** (or the l_2-norm for coordinate sequences and L_2 norm for integrals);
- the ℓ_1-norm.

It is possible to introduce weight function $w(x)$, which is continuous and strictly positive on the open interval (a, b), into these norms. For example,

$$\|x\|_{2,w} = \left(\sum_{j=1}^{\infty} w_j |\xi_j|^2\right)^{1/2}, \qquad \|f\|_{2,w} = \left(\int_a^b |f(x)|^2 w(x)\, dx\right)^{1/2},$$

is the **weighted Euclidean norm**. We assume that the integrals

$$\int_a^b |x|^k w(x)\, dx$$

exist for all k. Integrable singularities at the endpoints are permitted; an important example is $w(x) = (1 - x^2)^{-1/2}$ on the interval $[-1, 1]$.

[146]Stefan Banach (1892–1945), a Polish mathematician at the University in Lvov. Banach founded modern functional analysis and made major contributions to the theory of topological vector spaces, measure theory, and related topics. In 1939 he was elected President of the Polish Mathematical Society.

The above norms are special cases or limiting cases ($p \to \infty$ gives the max norm) of the l_p- or L_p-norms and weighted variants of these. They are defined for $p \geq 1$ as follows:[147]

$$\|x\|_p = \left(\sum_{j=1}^{\infty} |\xi_j|^p\right)^{1/p}, \qquad \|f\|_p = \left(\int_a^b |f(x)|^p\,dx\right)^{1/p}. \tag{4.5.4}$$

(The sum in the l_p-norm has a finite number of terms, if the space is finite dimensional.)

Convergence in a space, equipped with the max norm, means **uniform convergence**. Therefore, *the completeness of the space $C[a,b]$ follows from a classical theorem of analysis* that tells us that the limit of a uniformly convergent sequence is a continuous function. The generalization of this theorem to several dimensions implies the completeness of the space of continuous functions, equipped with the max norm on a closed bounded region in $\mathbf{R}^n$.

Other classes of functions can be normed with the max norm $\max_{x\in[a,b]} |f(x)|$, for example, $C^1[a,b]$. This space is not complete, but one can often live well with incompleteness.

The notation L_2-norm comes from the function space $L_2[a,b]$, which is the class of functions for which the integral $\int_a^b |f(x)|^2\,dx$ exists, in the sense of Lebesgue. The Lebesgue integral is needed in order to make the space complete and greatly extends the scope of Fourier analysis. No knowledge of Lebesgue integration is needed for the study of this book, but this particular fact can be interesting as background. One can apply this norm also to the (smaller) class of continuous functions on $[a,b]$. In this case the Riemann integral is equivalent. This also yields a normed linear space but *it is not complete*.[148]

Although $C[0,1]$ is an infinite-dimensional space, the restriction of the continuous functions f to the equidistant grid defined by $x_i = ih$, $h = 1/n$, $i = 0:n$, constitutes an $(n+1)$-dimensional space, with the function values on the grid as coordinates. If we choose the norm

$$\|f\|_{2,G_h} = \left(h\sum_{i=0}^{1/h} |f(x_i)|^2\right)^{1/2},$$

then

$$\lim_{h\to 0} \|f\|_{2,G_h} = \left(\int_0^1 |f(x)|^2\,dx\right)^{1/2} = \|f\|_{2,[0,1]}.$$

Limit processes of this type are common in numerical analysis.

Notice that even if $n+1$ functions $\phi_1(x), \phi_1(x), \ldots, \phi_{n+1}(x)$ are linearly independent on the interval $[0,1]$ (say), their restrictions to a grid with n points must be linearly dependent; but if a number of functions are linearly independent on a set $\mathcal{M}$ (a discrete set or continuum), any extensions of these functions to a set $\mathcal{M}' \supset \mathcal{M}$ will also be linearly independent.

[147] The triangle inequality for $\|x\|_p$ is derived from two classical inequalities due to Hölder and Minkowski. Elegant proofs of these are found in Hairer and Wanner [178, p. 327].

[148] A modification of the L_2-norm that also includes higher derivatives of f is used in the Sobolev spaces, which are used as a theoretical framework for the study of the practically very important finite element methods (FEMs), used in particular for the numerical treatment of partial differential equations.

The class of functions, analytic in a simply connected domain $\mathcal{D} \subset \mathbf{C}$, normed with $\|f\|_{\mathcal{D}} = \max_{z\in\partial\mathcal{D}} |f(z)|$, is a Banach space denoted by $\mathbf{Hol}(\mathcal{D})$. (The explanation of this term is that analytic functions are also called **holomorphic**.) By the maximum principle for analytic functions $|f(z)| \le \|f\|_{\mathcal{D}}$ for $z \in \mathcal{D}$.

4.5.2 Operator Norms and the Distance Formula

The concepts of **linear operator** and **linear functional** were introduced in Sec. 3.3.2. Here we extend to a general vector space $\mathcal{B}$ some definitions for a finite-dimensional vector space given in Online Appendix A.

Next we shall generalize the concept **operator norm** that is used for matrices; see Sec. A.4.3 in Online Appendix A. Consider an arbitrary bounded linear operator $A : S_1 \mapsto S_2$ in a normed vector space S:

$$\|A\| = \sup_{\|f\|_{S_1}} \|Af\|_{S_2}. \tag{4.5.5}$$

Note that $\|A\|$ depends on the vector norm in both S_1 and S_2. It follows that $\|Af\| \le \|A\|\|f\|$. Moreover, whenever the ranges of the operators A_1, A_2 are such that $A_1 + A_2$ and A_1A_2 are defined,

$$\|\lambda A\| \le |\lambda|\|A\|, \qquad \|A_1 + A_2\| \le \|A_1\| + \|A_2\|, \qquad \|A_1 \cdot A_2\| \le \|A_1\| \cdot \|A_2\|. \tag{4.5.6}$$

Sums with an arbitrary number of terms and integrals are defined similarly. It follows that $\|A^n\| \le \|A\|^n$, $n = 2, 3, \dots$.

Example 4.5.2.

Let $f \in C[0, 1]$, $\|f\| = \|f\|_\infty$,

$$Af = \sum_{i=1}^{m} a_i f(x_i) \;\Rightarrow\; \|A\| = \sum_{i=1}^{m} |a_i|,$$

$$Af = \int_0^1 e^{-x} f(x)\,dx \;\Rightarrow\; \|A\| = \int_0^1 e^{-x}\,dx = 1 - e^{-1},$$

$$B = \sum_{i=1}^{m} a_i A^i \;\Rightarrow\; \|B\| \le \sum_{i=1}^{m} |a_i|\|A\|^i, \quad (a_i \in \mathbf{C}),$$

$$Kf = \int_0^1 k(x,t) f(t)\,dt \;\Rightarrow\; \|K\|_\infty \le \sup_{x\in[0,1]} \int_0^1 |k(x,t)|\,dt.$$

The proofs of these results are left as a problem. In the last example, approximate the unit square by a uniform grid $(x_i, t_j)_{i,j=1}^m$, $h = 1/m$, and approximate the integrals by Riemann sums. Then approximate $\|K\|_\infty$ by the max norm for the matrix with the elements $k_{i,j} = hk(x_i, t_j)$; see Sec. A.4.3 in Online Appendix A.

Example 4.5.3.

For the forward difference operator Δ we obtain $\|\Delta\|_\infty = 2$, hence $\|\Delta^k\|_\infty \le 2^k$. In fact $\|\Delta^k\|_\infty = 2^k$, because the upper bound 2^k is attained (for every integer k) by the sequence $\{(-1)^n\}_0^\infty$.

Example 4.5.4.

Let $\mathcal{D}$ be a domain in **C**, the interior of which contains the closed interval $[a, b]$. Define the mapping D^k: $\textbf{Hol}\,\mathcal{D} \Rightarrow C[a, b]$ (with max norm), by

$$D^k f(x) = \frac{\partial^k}{\partial x^k} \frac{1}{2\pi i} \int_{\partial\mathcal{D}} f(z) \frac{1}{(z-x)}\,dz = \frac{1}{2\pi i} \int_{\partial\mathcal{D}} \frac{k! f(z)}{(z-x)^{k+1}}\,dz.$$

Then

$$sup_{x\in[a,b]} \|D^k f(x)\| \le \max_{z\in\partial\mathcal{D}} |f(z)| \cdot \sup_{x\in[a,b]} \frac{k!}{2\pi} \int_{\partial\mathcal{D}} \frac{|dz|}{|z-x|^{k+1}} < \infty.$$

Note that D^k is in this setting a *bounded* operator, while if we had considered D^k to be a mapping from $C^k[a, b]$ to $C[a, b]$, where both spaces are normed with the max norm in $[a, b]$, D^k would have been an *unbounded* operator.

Many of the procedures for the approximate computation of linear functionals that we encounter in this book may be characterized as follows. Let A be a linear functional such that Af cannot be easily computed for an arbitrary function f, but it can be approximated by another linear functional $\tilde{A}_k$, more easily computable, such that $\tilde{A}_k f = Af$ for all $f \in \mathcal{P}_k$. A general error bound to such procedures was given in Sec. 3.3.3 by the Peano remainder theorem, in terms of an integral

$$\int f^{(k)}(u) K(u)\,du,$$

where the Peano kernel $K(u)$ is determined by the functional $R = \tilde{A} - A$.

Now we shall give a different type of error bound for more general approximation problems, where other classes of functions and operators may be involved. Furthermore, no estimate of $f^{(k)}(u)$ is required. It is based on the following almost trivial theorem. It yields, however, often less sharp bounds than the Peano formula, in situations when the latter can be applied.

Theorem 4.5.4 (*The Norm and Distance Formula*).

Let A, $\tilde{A}$ be two linear operators bounded in a Banach space $\mathcal{B}$ such that for any vector s in a certain linear subspace S, $\tilde{A}s = As$. Then

$$\|\tilde{A}f - Af\| \le \|\tilde{A} - A\| \operatorname{dist}(f, \mathcal{S}) \quad \forall f \in \mathcal{B}.$$

Proof. Set $R = \tilde{A} - A$. For any positive ϵ, there exists a vector $s_\epsilon \in \mathcal{S}$ such that

$$\|f - s_\epsilon\| = \operatorname{dist}(f, s_\epsilon) < \inf_{s\in\mathcal{S}} \operatorname{dist}(f, s) + \epsilon = \operatorname{dist}(f, \mathcal{S}) + \epsilon.$$

Then $\|Rf\| = \|Rf - Rs_\epsilon\| = \|R(f - s_\epsilon)\| \le \|R\| \|f - s_\epsilon\| < (\operatorname{dist}(f, \mathcal{S}) + \epsilon)\|R\|$. The theorem follows, since this holds for every $\epsilon > 0$. □

The following is a common particular case of the theorem. *If A, A_k are* **linear functionals** *such that $A_k p = Ap$ for all $p \in \mathcal{P}_k$, then*

$$|A_k f - Af| \le (\|A_k\| + \|A\|) \operatorname{dist}(f, \mathcal{P}_k). \qquad (4.5.7)$$

Another important particular case of the theorem concerns projections P from a function space to a finite-dimensional subspace $\mathcal{S}_k$, for example, *interpolation and series truncation operators*, $A = I$, $\tilde{A} = P$; see the beginning of Sec. 4.5.2. Then

$$\|Pf - f\| \leq \|(P - I)f\| \leq (\|P\| + 1)\text{dist}\,(f, \mathcal{S}_k). \tag{4.5.8}$$

The norm and distance formula requires bounds for $\|\tilde{A}\|$, $\|A\|$ and dist $(f, \mathcal{S})$. We have seen examples above of how to obtain bounds for operator norms. Now we shall exemplify how to obtain bounds for the distance of f from some relevant subspace $\mathcal{S}$, in particular spaces of polynomials or trigonometric polynomials restricted to some real interval $[a, b]$. For the efficient estimation of dist $(f, \mathcal{S})$ it may be important to take into account that f is analytic in a larger domain than $[a, b]$.

Theorem 4.5.5 (*Estimation of* $\text{dist}_\infty(f, \mathcal{P}_k)$ *in Terms of* $\|f^{(k)}\|_\infty$).
Let $f \in C^k[a, b] \subset \mathbf{R}$, *and define the norm*

$$\|g\|_{\infty,[a,b]} = \max_{t\in[a,b]} |g(t)| \quad \forall\ g \in C[a, b].$$

Then

$$\text{dist}_{\infty,[a,b]}(f, \mathcal{P}_k) \leq \frac{2}{k!}\left(\frac{b-a}{4}\right)^k \|f^{(k)}\|_{\infty,[a,b]}.$$

Proof. Let $p(t)$ be the polynomial which interpolates $f(t)$ at the points $t_j \in [a, b]$, $j = 1 : k$. By the remainder term (4.2.10) in interpolation,

$$|f(t) - p(t)| \leq \max_{\xi\in[a,b]} \frac{|f^{(k)}(\xi)|}{k!} \prod_{j=1}^{k} |t - t_j|.$$

Set $t = \frac{1}{2}(b+a)+\frac{1}{2}(b-a)x$, and choose $t_j = \frac{1}{2}(b+a)+\frac{1}{2}(b-a)x_j$, where x_j are the zeros of the Chebyshev polynomial $T_k(x)$. Then p is the Chebyshev interpolation polynomial for f on the interval $[a, b]$, and with $M = \max_{t\in[a,b]} |f^{(k)}|/k!$,

$$|f(t) - p(t)| \leq M \prod_{j=1}^{k} \frac{(b-a)|x - x_j|}{2}, \qquad x \in [-1, 1].$$

Since the leading coefficient of $T_k(x)$ is 2^{k-1} and $|T_k(x)| \leq 1$, we have, for $t \in [a, b]$,

$$|f(t) - p(t)| \leq M \left(\frac{b-a}{2}\right)^k \frac{1}{2^{k-1}} |T_k(x)| \leq M \left(\frac{b-a}{2}\right)^k \frac{1}{2^{k-1}}.$$

The bound stated for $\text{dist}_\infty(f, \mathcal{P}_k)$ is thus satisfied. □

Example 4.5.5.

By the above theorem, the function e^t can be approximated on the interval [0, 1] by a polynomial in $\mathcal{P}_6$ with the error bound $2e\cdot 4^{-6}/6! \approx 2\cdot 10^{-6}$. According to the proof this accuracy is attained by Chebyshev interpolation on [0, 1].

If one instead uses the Maclaurin series, truncated to $\mathcal{P}_6$, then the remainder is $e^\theta/(6!) \geq 1.3 \cdot 10^{-3}$. Similarly, with the truncated Taylor series about $t = \frac{1}{2}$ the remainder is $e^\theta/(2^6 6!) \geq 2 \cdot 10^{-5}$, still significantly less accurate than the Chebyshev interpolation. Economization of power series (see Problem 3.2.5), yields approximately the same accuracy as Chebyshev interpolation.

If we do these things on an interval of length h (instead of the interval $[0, 1]$) all the bounds are to multiplied by h^6.

Definition 4.5.6.

Let $f \in C[a, b]$ and $\mathcal{P}_n$ be the space of polynomials of degree less than n. Then we set

$$E_n(f) = \operatorname{dist}_\infty(f, \mathcal{P}_n). \tag{4.5.9}$$

Example 4.5.6.

The use of analyticity in estimates for $\operatorname{dist}_\infty(f, \mathcal{P}_n)$: Denote by $\mathcal{E}_R$ an ellipse in $\mathbf{C}$ with foci at -1 and 1; R is equal to the sum of the semiaxes. Bernštein's approximation theorem, Theorem 3.2.5, gives the following truncation error bound for the Chebyshev expansion for a function $f \in \mathbf{Hol}(\mathcal{E}_R)$, real-valued on $[-1, 1]$ and with $\|f\|_{\mathcal{E}_R} \leq M$:

$$\left| f(x) - \sum_{j=0}^{n-1} c_j T_j(x) \right| \leq \frac{2MR^{-n}}{1 - R^{-1}}, \qquad x \in [-1, 1].$$

This implies that, on the same assumptions concerning f,

$$\operatorname{dist}_{\infty,[-1,1]}(f, \mathcal{P}_n) \leq \frac{2MR^{-n}}{1 - R^{-1}}.$$

By the application of a theorem of Landau concerning bounded power series one can obtain the following bound that can be sharper when $R \approx 1$:

$$\operatorname{dist}_{\infty,[-1,1]}(f, \mathcal{P}_n) \leq 2MR^{-(n+1)}G_n, \tag{4.5.10}$$

where

$$G_n = 2 + \left(\frac{1}{2}\right)^2 + \left(\frac{1 \cdot 3}{2 \cdot 4}\right)^2 + \cdots + \left(\frac{1 \cdot 3 \cdots (2n-1)}{2 \cdot 4 \cdots 2n}\right)^2 \sim \frac{\log n}{\pi}.$$

Suppose that $f \in \mathbf{Hol}(\mathcal{D})$, where $\mathcal{D} \supseteq \frac{1}{2}(b+a) + \frac{1}{2}(b-a)\mathcal{E}_R$. Then transforming from $[-1, 1]$ to a general interval $[a, b] \subset \mathbf{R}$, we have

$$\operatorname{dist}_{\infty,[a,b]}(f, \mathcal{P}_n) \leq 2\|f\|_{\infty,\mathcal{D}} \left(\frac{b-a}{2R}\right)^n \frac{2R}{2R - (b-a)}.$$

Example 4.5.7.

As a first simple example we shall derive an error bound for one step with the trapezoidal rule. Set

$$Af = \int_0^h f(x)\,dx, \qquad A_2 f = \frac{h}{2}\left(f(0) + f(h)\right).$$

We know that $A_2 p = Ap$ if $p \in \mathcal{P}_2$. By Theorem 4.5.5, $\operatorname{dist}_\infty(f, \mathcal{P}_2) \leq \|f''\|_\infty h^2/16$.

Furthermore, $\|A\|_\infty = \int_0^h dx = h$, $\|A_2\|_\infty = h$, hence by (4.5.7) the requested error bound becomes

$$\|A_2 f - Af\|_\infty \le 2h \cdot \|f''\|_\infty h^2/16 = \|f''\|_\infty h^3/8.$$

This general method does not always give the best possible bounds but, typically, it gives no gross overestimate. For the trapezoidal rule we know by Peano's method (Example 3.3.7) that $\|f''\|h^3/12$ is the best possible estimate, and thus we now obtained a 50% overestimate of the error.

The norm and distance formula can also be written in the form

$$\text{dist}\,(f, \mathcal{S}) \ge |Af - \tilde{A}f|/\|A - \tilde{A}\|. \tag{4.5.11}$$

This can be used for finding a simple lower bound for dist $(f, \mathcal{P}_k)$ in terms of an easily computed functional that vanishes on $\mathcal{P}_k$.

Example 4.5.8.

Let $\tilde{A} = 0$. The functional $Af = f(1) - 2f(0) + f(-1)$ vanishes for $f \in \mathcal{P}_2$. If the maximum norm is used on [-1,1], then $\|A\| = 1 + 2 + 1 = 4$. Thus

$$\text{dist}\,(f, \mathcal{P}_2)_{\infty,[-1,1]} \ge \frac{|Af|}{\|A\|} = \frac{1}{4}|f(1) - 2f(0) + f(-1)|.$$

It follows, for example, that the curve $y = e^x$ cannot be approximated by a straight line in $[-1, 1]$ with an error less than $(e - 2 + e^{-1})/4 \approx 0.271$. (This can also be seen without the use of the norm and distance formula.)

It is harder to derive the following generalization without the norm and distance formula. By Example 4.5.3, $\|\Delta^k\| = 2^k$, $\Delta^k p = 0$, if $p \in \mathcal{P}_k$; hence

$$\text{dist}\,(f, \mathcal{P}_k)_{\infty,[x_0,x_k]} \ge 2^{-k}|\Delta^k f(x_0)|. \tag{4.5.12}$$

There is another inequality that is usually sharper but less convenient to use (it follows from the discrete orthogonality property of the Chebyshev polynomials; see Sec. 4.6):

$$\text{dist}\,(f, \mathcal{P}_k)_{\infty,[x_0,x_k]} \ge \frac{1}{k}\left|\sum_{j=0}^{k}(-1)^j a_j f\left(\cos\frac{j\pi}{k}\right)\right|, \tag{4.5.13}$$

where $a_j = \frac{1}{2}$, $j = 0, k$, and $a_j = 1$ otherwise. Inequalities of this type can reveal when one had better use *piecewise polynomial approximation* of a function on an interval instead of using a single polynomial over the whole interval. See also Sec. 4.4.

Denote by $C[a, b]$ the space of continuous functions on a closed bounded interval $[a, b]$. We shall study the approximation of a continuous function $f \in C[a, b]$ by polynomials. We introduce a metric by

$$\|f\| = \max_{x\in[a,b]} |f(x)|.$$

We define the **modulus of continuity** of f by

$$\omega(\delta; f) = \sup_{|x-y|\le\delta} |f(x) - f(y)|. \tag{4.5.14}$$

Then $\omega(\delta; f) \downarrow 0$ as $\delta \downarrow 0$, because a continuous function is uniformly continuous on a closed bounded interval.

One of the fundamental theorems in approximation theory is Weierstrass'[149] approximation theorem from 1885. It is concerned with the uniform approximation of a continuous function f on a closed bounded interval by polynomials. It is of basic importance for many convergence proofs, which we shall see examples of in this book.

Theorem 4.5.7 (*Weierstrass' Approximation Theorem*).

Suppose that f is continuous on a closed, bounded interval $[a, b]$. Let $\mathcal{P}_n$ denote the space of polynomials of degree less than n. Then we can, for any $\epsilon > 0$, find a polynomial $p_n \in \mathcal{P}_n$ such that

$$\max_{x\in[a,b]} |f(x) - p_n(x)| < \epsilon.$$

Two equivalent formulations are

- $\operatorname{dist}(f, \mathcal{P}_n) \downarrow 0$ *as* $n \to \infty$*;*
- *the set of polynomials are dense in* $C[a, b]$.

Proof. Our proof is based on ideas of Lebesgue [241]. For simplicity we assume that the interval is $[0, 1]$. First we interpolate $f(x)$ by a piecewise affine functions on the grid $x_k = k/r$, $k = 0 : r$, i.e., by a linear spline; see Sec. 4.4.2. A basis for the space of linear splines is given by

$$\{1, (x - x_0)_+, (x - x_1)_+, \dots, (x - x_{r-1})_+\};$$

see (4.4.11), where we recall the notation $(x - a)_+ = \max(x - a, 0)$. Hence we can write the interpolating linear spline in the form

$$g(x) = f(0) + \sum_{k=0}^{r-1} c_k (x - x_k)_+. \tag{4.5.15}$$

Set $M_k = \max(f(x_k), f(x_{k+1}))$ and $m_k = \min(f(x_k), f(x_{k+1}))$. For $x \in [x_k, x_{k+1}]$ we have

$$M_k - \omega(1/r) \le f(x) \le m_k + \omega(1/r), \qquad -M_k \le -g(x) \le -m_k,$$

where ω is the modulus of continuity of f. Adding the last two inequalities we obtain

$$-\omega(1/r) \le f(x) - g(x) \le \omega(1/r).$$

[149]Kurt Theodor Wilhelm Weierstrass (1815–1897), German mathematician, whose lectures at Berlin University attracted students from all over the world. He set high standards of rigor in his work and is considered the father of modern analysis. For a more detailed biography and a survey of Weierstrass' work in approximation theory, see Pinkus [291].

This holds for $x \in [x_k, x_{k+1}]$, $k = 0 : r-1$, hence $\|f - g\| \le \omega(1/r)$. We now *choose* r so that $\omega(1/r) < \epsilon/2$ so that $\|f - l\| < \epsilon/2$. Now that r is fixed we see that the coefficients are bounded, (say) $|c_k| \le L$. (In fact we may choose any $L \ge r\epsilon$.)

If we can find any polynomial p such that $\|p - g\| < \epsilon/2$, we are through. In order to achieve this we look at each term in the sum (4.5.15), and set $t = x - x_k$. We shall next find a polynomial P such that

$$|P(t) - t_+| < \frac{\epsilon}{2rL}, \quad t \in [-1, 1]. \tag{4.5.16}$$

Note that

$$t_+ = \frac{1}{2}(t + |t|) = \frac{1}{2}\big(t + \sqrt{1 - (1 - t^2)}\big), \quad t \in [-1, 1].$$

Set $z = 1 - t^2$. We know that the binomial expansion of $\sqrt{1-z}$ is valid for $|z| < 1$. By a classical theorem of Abel, see, e.g., Titchmarsh [351, Sec. 1.22], the sequence of partial sums also converges uniformly to $\sqrt{1-z}$ for $z \in [0, 1]$. After return to the variable t, we thus have a sequence of polynomials that converge uniformly to $|t|$, $t+ \in [-1, 1]$, and this trivially yields a polynomial $p(t)$ that satisfies (4.5.16).

If we now set

$$P(x) = f(0) + \sum_{k=0}^{r-1} c_k p(x - x_k),$$

then

$$|P(x) - g(x)| = \left| \sum_{k=0}^{r-1} l_k (p(x - x_k) - (x - x_k)_+) \right| < 2\frac{\epsilon}{2} = \epsilon.$$

This finishes our proof. □

Several other proofs have been given, e.g., one by S. Bernštein, using Bernštein polynomials; see Davis [92, Sec. 6.2].

The smoother f is, the quicker $\text{dist}(f, P_n)$ decreases, and the narrower the interval is, the less $\text{dist}(f, P_n)$ becomes. In many cases $\text{dist}(f, P_n)$ decreases so slowly toward zero (as n grows) that it is impractical to attempt to approximate f with only one polynomial in the entire interval $[a, b]$.

In infinite-dimensional spaces, certain operators may not be defined everywhere, but only in a set that is everywhere dense in the space. For example, in the space $C[a, b]$ of continuous functions on a bounded interval (with the maximum norm), the operator $A = d/dx$ is not defined everywhere, since there are continuous functions which are not differentiable. By Weierstrass' approximation theorem the set of polynomials is everywhere dense in C, and hence the set of differentiable functions is so too. Moreover, even if Au exists, A^2u may not exist. That A^{-1} may not exist is no novel feature of infinite-dimensional spaces. In $C[a, b]$ the norm of $A = d/dx$ is infinite. This operator is said to be unbounded.

4.5.3 Inner Product Spaces and Orthogonal Systems

An abstract foundation for least squares approximation is furnished by the theory of **inner product spaces** which we now introduce.

Definition 4.5.8.

A normed linear space $\mathcal{S}$ will be called an inner product space if for each two elements f, g in $\mathcal{S}$ there is a scalar designated by (f, g) with the following properties:

1. $(f + g, h) = (f, h) + (g, h)$ *(linearity),*
2. $(f, g) = \overline{(g, f)}$ *(symmetry),*
3. $(f, \alpha g) = \alpha(f, g), \quad \alpha$ *scalar* *(homogeneity),*
4. $(f, f) \geq 0, \qquad (f, f) = 0$ *(positivity).*

The **inner product** (f, g) *is scalar, i.e., real in a real space and complex in a complex space. We set* $\|f\| = (f, f)^{1/2}$. *If* $\|f\| = 0$ *implies that* $f = 0$ *this is a* **norm on** $\mathcal{S}$; *otherwise it is a* **seminorm** *on* $\mathcal{S}$.

We shall show below that the triangle inequality is satisfied. (The other axioms for a norm are obvious.) The standard vector inner products introduced in Sec. A.1.2 in Online Appendix A are particular cases, if we set $(x, y) = y^T x$ in $\mathbf{R}^n$ and $(x, y) = y^H x$ in $\mathbf{C}^n$. A **complete** inner-product space is called a **Hilbert space** and is often denoted $\mathcal{H}$ in this book.

One can make computations using the more general definition of (f, g) given above in the same way that one does with scalar products in linear algebra. Note, however, the *conjugations* necessary in a complex space,

$$(\alpha f, g) = \bar{\alpha}(f, g), \tag{4.5.17}$$

because, by the axioms,

$$(\alpha f, g) = \overline{(g, \alpha f)} = \overline{\alpha(g, f)} = \bar{\alpha}\overline{(g, f)} = \bar{\alpha}(f, g).$$

By the axioms it follows by induction that

$$\left(\phi_k, \sum_{j=0}^{n} c_j \phi_j\right) = \sum_{j=0}^{n} (\phi_k, c_j \phi_j) = \sum_{j=0}^{n} c_j (\phi_k, \phi_j). \tag{4.5.18}$$

Theorem 4.5.9 (*Cauchy–Schwarz inequality*).

The Cauchy–Schwarz inequality in a complex space is

$$|(f, g)| \leq \|f\| \, \|g\|.$$

Proof. Let f, g be two arbitrary elements in an inner-product space. Then,[150] for every *real* number λ,

$$0 \leq (f + \lambda(f, g)g, \, f + \lambda(f, g)g) = (f, f) + 2\lambda|(f, g)|^2 + \lambda^2|(f, g)|^2(g, g).$$

This polynomial in λ with real coefficients cannot have two distinct zeros, hence the discriminant cannot be positive, i.e.,

$$|(f, g)|^4 - (f, f)|(f, g)|^2(g, g) \leq 0.$$

So, even if $(f, g) = 0$, $|(f, g)|^2 \leq (f, f)(g, g)$. □

[150] We found this proof in [304, sec. 83]. The application of the same idea in a *real* space can be made simpler.

By Definition 4.5.8 and the Cauchy–Schwarz inequality,

$$\begin{aligned}\|f+g\|^2 &= (f+g, f+g) = (f,f)+(f,g)+(g,f)+(g,g)\\ &\le \|f\|^2 + \|f\|\,\|g\| + \|g\|\,\|f\| + \|g\|^2 = (\|f\|+\|g\|)^2.\end{aligned}$$

This shows that the **triangle inequality** is satisfied by the norm defined above.

Example 4.5.9.

The set of all complex infinite sequences $\{x_i\}$ for which $\sum_{i=1}^{\infty}|x_i|^2 < \infty$ and which is equipped with the inner product

$$(x, y) = \sum_{i=1}^{\infty} x_i \overline{y}_i$$

constitutes a Hilbert space.

Definition 4.5.10.

Two functions f and g are said to be **orthogonal** *if $(f, g) = 0$. A finite or infinite sequence of functions $\phi_0, \phi_1, \dots, \phi_n$ constitutes an* **orthogonal system** *if*

$$(\phi_i, \phi_j) = 0, \quad i \neq j, \text{ and } \|\phi_i\| \neq 0 \quad \forall i. \tag{4.5.19}$$

If, in addition, $\|\phi_i\| = 1$, for all $i \ge 0$, then the sequence is called an **orthonormal system**.

Theorem 4.5.11 (*Pythagoras' Theorem*).

Let $\{\phi_1, \phi_2, \dots, \phi_n\}$ be an orthogonal system in an inner-product space. Then

$$\left\| \sum_{j=0}^{n} c_j \phi_j \right\|^2 = \sum_{j=0}^{n} |c_j|^2 \|\phi_j\|^2.$$

The elements of an orthogonal system are linearly independent.

Proof. We start as in the proof of the triangle inequality:

$$\|f+g\|^2 = (f,f)+(f,g)+(g,f)+(g,g) = (f,f)+(g,g) = \|f\|^2 + \|g\|^2.$$

Using this result and induction the first statement follows. The second statement then follows because $\sum c_j\phi_j = 0 \Rightarrow |c_j| = 0$ for all j. □

Theorem 4.5.12.

A linear operator P is called **idempotent** *if $P = P^2$. Let $\mathcal{V}$ be the range of P. Then P is a* **projection** *(or projector) onto $\mathcal{V}$ if and only if P is idempotent and $Pv = v$ for each $v \in \mathcal{V}$.*

Proof. If P is a projection, then $v = Px$ for some $x \in \mathcal{B}$, hence $Pv = P^2x = Px = v$. Conversely, if Q is a linear operator such that $Qx \in \mathcal{V}$ for all $x \in \mathcal{B}$, and $v = Qv$ for all $v \in V$, then Q is a projection; in fact $Q = P$. □

Note that $I - P$ is also a projection, because

$$(I - P)(I - P) = I - 2P + P^2 = I - P.$$

Any vector $x \in \mathcal{B}$ can be written *uniquely* in the form

$$x = u + w, \quad u = Px, \quad w = (I - P)x. \tag{4.5.20}$$

Important examples of projections in function spaces are interpolation operators, for example, the mapping of $C[a, b]$ into $\mathcal{P}_k$ by Newton or Lagrange interpolation, because each polynomial is mapped to itself. The two types of interpolation are the same projection, although they use different bases in $\mathcal{P}_k$. Another example is the mapping of a linear space of functions, *analytic* on the unit circle, into $\mathcal{P}_k$ so that each function is mapped to its *Maclaurin expansion truncated to* $\mathcal{P}_k$. There are analogous projections where periodic functions and trigonometric polynomials are involved

In an inner-product space, the **adjoint operator** A^* of a linear operator A is defined by the requirement that

$$(A^*u, v) = (u, Av) \quad \forall\, u, v. \tag{4.5.21}$$

An operator A is called **self-adjoint** if $A = A^*$. In $\mathbf{R}^n$, we define $(u, v) = u^T v$, i.e., the standard scalar product. Then $(A^*u)^T v = u^T Av$, i.e., $u^T((A^*)^T v = u^T Av$, hence $A^* = A^T$. It follows that symmetric matrices are self-adjoint in $\mathbf{R}^n$.

In $\mathbf{C}^n$, with the inner product $(u, v) = u^H v$, it follows that $A^* = A^H$, i.e., Hermitian matrices are self-adjoint. An operator B is **positive definite** if $(u, Bu) > 0$ for all $u \neq 0$.

Example 4.5.10.

An important example of an orthogonal system is the sequence of trigonometric functions $\phi_j(x) = \cos jx$, $j = 0 : N - 1$. These form an orthogonal system, with either of the two inner products

$$(f, g) = \begin{cases} \displaystyle\int_0^{\pi} f(x)g(x)\,dx & \text{(continuous case)}, \\ \displaystyle\sum_{i=0}^{N-1} f(x_i)g(x_i), \quad x_i = \frac{2i+1}{N}\frac{\pi}{2} & \text{(discrete case)}. \end{cases} \tag{4.5.22}$$

Moreover, it holds that

$$\|\phi_j\|^2 = \begin{cases} \frac{1}{2}\pi, \quad j \geq 1, \quad \|\phi_0\|^2 = \pi & \text{(continuous case)}, \\ \frac{1}{2}N, \quad j = 1 : N - 1, \quad \|\phi_0\|^2 = N & \text{(discrete case)}. \end{cases}$$

These results are closely related to the orthogonality of the Chebyshev polynomials; see Theorem 4.5.20. Trigonometric interpolation and Fourier analysis will be treated in Sec. 4.6.

There are many other examples of orthogonal systems. Orthogonal systems of polynomials play an important role in approximation and numerical integration. Orthogonal systems also occur in a natural way in connection with eigenvalue problems for differential equations, which are quite common in mathematical physics.

4.5.4 Solution of the Approximation Problem

Orthogonal systems give rise to extraordinary formal simplifications in many situations. We now consider the least squares approximation problem.

Theorem 4.5.13.

If $\phi_0, \phi_1, \dots, \phi_n$ are linearly independent, then the least squares approximation problem of minimizing the norm of the error function $\|f^* - f\|$ over all functions $f^* = \sum_{j=0}^n c_j\phi_j$ has the unique solution

$$f^* = \sum_{j=0}^{n} c_j^*\phi_j. \tag{4.5.23}$$

The solution is characterized by the orthogonality property that $f^* - f$ is orthogonal to all ϕ_j, $j = 0 : n$. The coefficients c_j^* are called **orthogonal coefficients** *(or* **Fourier coefficients***), and satisfy the linear system of equations*

$$\sum_{j=0}^{n} (\phi_j, \phi_k)c_j^* = (f, \phi_k), \tag{4.5.24}$$

called normal equations. In the important special case when $\phi_0, \phi_1, \dots, \phi_n$ form an orthogonal system, the coefficients are computed more simply by

$$c_j^* = (f, \phi_j)/(\phi_j, \phi_j), \quad j = 0 : n. \tag{4.5.25}$$

Proof. Let $(c_0, \dots, c_n)$ be a sequence of coefficients with $c_j \neq c_j^*$ for at least one j. Then

$$\sum_{j=0}^{n} c_j\phi_j - f = \sum_{j=0}^{n} (c_j - c_j^*)\phi_j + (f^* - f).$$

If $f^* - f$ is orthogonal to all the ϕ_j, then it is also orthogonal to the linear combination $\sum_{j=0}^n (c_j - c_j^*)\phi_j$ and, according to the Pythagorean theorem,

$$\Big\|\sum_{j=0}^{n} c_j\phi_j - f\Big\|^2 = \Big\|\sum_{j=0}^{n} (c_j - c_j^*)\phi_j\Big\|^2 + \|(f^* - f)\|^2 > \|(f^* - f)\|^2.$$

Thus if $f^* - f$ is orthogonal to all ϕ_k, then f^* is a solution to the approximation problem. It remains to show that the orthogonality conditions

$$\left(\sum_{j=0}^{n} c_j^*\phi_j - f, \phi_k\right) = 0, \quad k = 0 : n,$$

can be fulfilled. The above conditions are equivalent to the normal equations in (4.5.24). If $\{\phi_j\}_{j=0}^n$ constitutes an orthogonal system, then the system can be solved immediately, since in each equation all the terms with $j \neq k$ are zero. The formula in (4.5.25) then follows immediately.

Suppose now that we know only that $\{\phi_j\}_{j=0}^n$ are linearly independent. The solution to the normal equations exists and is unique, unless the homogeneous system,

$$\sum_{j=0}^{n}(\phi_j, \phi_k)c_j^* = 0, \quad k = 0 : n,$$

has a solution $c_0, c_1, \dots, c_n$ with at least one $c_i \neq 0$. But this would imply

$$\Big\| \sum_{j=0}^{n} c_j\phi_j \Big\|^2 = \Big(\sum_{j=0}^{n} c_j\phi_j, \sum_{k=0}^{n} c_k\phi_k \Big) = \sum_{k=0}^{n}\sum_{j=0}^{n}(\phi_j, \phi_k)c_jc_k = \sum_{k=0}^{n} 0 \cdot c_k = 0,$$

which contradicts that the ϕ_j were linearly independent. □

In the case where $\{\phi_j\}_{j=0}^n$ form an orthogonal system, *the Fourier coefficients* c_j^* *are independent of* n (see formula (4.5.25)). This has the important advantage that one can increase the total number of parameters without recalculating any previous ones. Orthogonal systems are advantageous not only because they greatly simplify calculations; using them, one can often avoid numerical difficulties with roundoff error which may occur when one solves the normal equations for a nonorthogonal set of basis functions.

With every continuous function f one can associate an infinite series,

$$f \sim \sum_{j=0}^{\infty} c_j^*\phi_j, \quad c_j^* = \frac{(f, \phi_j)}{(\phi_j, \phi_j)}.$$

Such a series is called an **orthogonal expansion**. For certain orthogonal systems this series converges with very mild restrictions on the function f.

Theorem 4.5.14.
If f^* *is defined by formulas* (4.5.23) *and* (4.5.25), *then*

$$\|f^* - f\|^2 = \|f\|^2 - \|f^*\|^2 = \|f\|^2 - \sum_{j=0}^{n}(c_j^*)^2\|\phi_j\|^2.$$

Proof. Since $f^* - f$ is, according to Theorem 4.5.13, orthogonal to all ϕ_j, $0 \le j \le n$, then $f^* - f$ is orthogonal to f^*. The theorem then follows directly from Pythagoras' theorem. □

We obtain as corollary **Bessel's inequality**:

$$\sum_{j=0}^{n}(c_j^*)^2\|\phi_j\|^2 \le \|f\|^2. \tag{4.5.26}$$

The series $\sum_{j=0}^{\infty}(c_j^*)^2\|\phi_j\|^2$ is convergent. If $\|f^* - f\| \to 0$ as $n \to \infty$, then the sum of the latter series is equal to $\|f\|^2$, which is **Parseval's identity**.

Theorem 4.5.15.

If $\{\phi_j\}_{j=0}^m$ *are linearly independent on the grid* $G = \{x_i\}_{i=0}^m$, *then the interpolation problem of determining the coefficients* $\{c_j\}_{j=0}^m$ *such that*

$$\sum_{j=0}^{m} c_j \phi_j(x_i) = f(x_i), \quad i = 0 : m, \tag{4.5.27}$$

has exactly one solution. Interpolation is a special case ($n = m$) *of the method of least squares. If* $\{\phi_j\}_{j=0}^m$ *is an orthogonal system, then the coefficients* c_j *are equal to the orthogonal coefficients in* (4.5.25).

Proof. The system of equations (4.5.27) has a unique solution, since its column vectors are the vectors $\phi_j(G)$, $j = 0 : n$, which are linearly independent. For the solution of the interpolation problem it holds that $\|c_j\phi_j - f\| = 0$; i.e., the error function has the least possible seminorm. The remainder of the theorem follows from Theorem 4.5.13. □

The following collection of important and equivalent properties is named the **fundamental theorem of orthonormal expansions** by Davis [92, Theorem 8.9.1], whom we follow closely at this point.

Theorem 4.5.16.

Let $\phi_0, \phi_1, \phi_2, \cdots$ *be a sequence of orthonormal elements in a complete inner product space* $\mathcal{H}$. *The following six statements are equivalent:*[151]

(A) *The* ϕ_j *is a complete orthonormal system in* $\mathcal{H}$.

(B) *The orthonormal expansion of any element* $y \in \mathcal{H}$ *converges in norm to* y; *i.e.,*

$$\lim_{n\to\infty} \left\| y - \sum_{j=0}^{n} (y, \phi_j)\phi_j \right\|. \tag{4.5.28}$$

(C) *Parseval's identity holds for every* $y \in \mathcal{H}$, *i.e.,*

$$\|y\|^2 = \sum_{j=0}^{\infty} |(y, \phi_j)|^2. \tag{4.5.29}$$

(D) *There is no strictly larger orthonormal system containing* $\phi_1, \phi_2, \ldots$.

(E) *If* $y \in \mathcal{H}$ *and* $(y, \phi_j) = 0$, $j = 0, 2, \ldots$, *then* $y = 0$.

(F) *An element of* $\mathcal{H}$ *is determined uniquely by its Fourier coefficients, i.e., if* $(w, \phi_j) = (y, \phi_j)$, $j = 0, 2, \ldots$, *then* $w = y$.

[151] We assume that $\mathcal{H}$ is not finite-dimensional, in order to simplify the formulations. Only minor changes are needed in order to cover the finite-dimensional case.

Proof. The proof that $A \Leftrightarrow B$ is formulated with respect to the previous statements. By the conjugations necessary in the handling of complex scalars in inner products (see (4.5.17) and (4.5.18)),

$$\left(x - \sum_{j=0}^{\infty}(x, \phi_j)\phi_j,\ y - \sum_{j=0}^{n}(y, \phi_j)\phi_j\right) = (x, y) - \sum_{j=0}^{n}(x, \phi_j)(\phi_j, y).$$

By the Schwarz inequality,

$$\left|(x, y) - \sum_{j=0}^{n}(x, \phi_j)(\phi_j, y)\right| \le \left\|x - \sum_{j=0}^{\infty}(x, \phi_j)\phi_j\right\| \cdot \left\|y - \sum_{j=0}^{n}(y, \phi_j)\phi_j\right\|.$$

For the rest of the proof, see Davis [92, pp. 192 ff.]. □

Theorem 4.5.17.

The converse of statement (F) *holds, i.e., let* $\mathcal{H}$ *be a complete inner product space, and let* a_j *be constants such that* $\sum_{j=0}^{\infty}|a_j|^2 < \infty$. *Then there exists an* $y \in \mathcal{H}$ *such that* $y = \sum_{j=0}^{\infty} a_j\phi_j$ *and* $(y, \phi_j) = a_j$ *for all* j.

Proof. See Davis [92, Sec. 8.9]. □

4.5.5 Mathematical Properties of Orthogonal Polynomials

By a family of **orthogonal polynomials** we mean a triangle family of polynomials (see (4.1.8)), which (in the continuous case) is an orthogonal system with respect to a given inner product. The theory of orthogonal polynomials is also of fundamental importance for many problems which at first sight seem to have little connection with approximation (e.g., numerical integration, continued fractions, and the algebraic eigenvalue problem).

We assume in the following that in the continuous case the inner product is

$$(f, g) = \int_a^b f(x)g(x)w(x)\,dx, \quad w(x) \ge 0, \tag{4.5.30}$$

where $-\infty \le a < b \le \infty$. We assume that the weight function $w(x) \ge 0$ is such that the **moments**

$$\mu_k = (x^k, 1) = \int_a^b x^k w(x)\,dx \tag{4.5.31}$$

are defined for all $k \ge 0$, and $\mu_0 > 0$. In the discrete case, we define the weighted discrete inner product of two real-valued functions f and g on the grid $\{x_i\}_{j=0}^{m}$ of distinct points by

$$(f, g) = \sum_{i=0}^{m} w_i f(x_i)g(x_i), \quad w_i > 0. \tag{4.5.32}$$

Note that both these inner products have the property that

$$(xf, g) = (f, xg). \tag{4.5.33}$$

The continuous and discrete case are both special cases of the more general inner product

$$(f, g) = \int_a^b f(x)g(x)\, d\alpha(x), \tag{4.5.34}$$

where the integral is a Stieltjes integral (see Definition 3.4.4) and $\alpha(x)$ is allowed to be discontinuous. However, in the interest of clarity, we will in the following treat the two cases separately.

The weight function $w(x)$ determines the orthogonal polynomials $\phi_n(x)$ up to a constant factor in each polynomial. The specification of those factors is referred to as **standardization**. These polynomials satisfy a number of relationships of the same general form. In the case of a continuously differentiable weight function $w(x)$ we have an explicit expression

$$\phi_n(x) = \frac{1}{a_n w(x)} \frac{d^n}{dx^n}\{w(x)(g(x))^n\}, \quad n = 0, 1, 2, \dots, \tag{4.5.35}$$

where $g(x)$ is a polynomial in x independent of n. This is **Rodrigues' formula**. The orthogonal polynomials also satisfy a second order differential equation,

$$g_2(x)\phi_n'' + g_1(x)\phi_n' + a_n\phi_n = 0, \tag{4.5.36}$$

where $g_2(x)$ and $g_1(x)$ are independent of n and a_n is a constant only dependent on n.

Let $p_n(x) = k_n x^n + \cdots, n = 0, 1, 2, \dots,$ be a family of real orthogonal polynomials. The symmetric function

$$K_n(x, y) = \sum_{k=0}^{n} p_k(x)p_k(y) \tag{4.5.37}$$

is called the **kernel polynomial** of order n for the orthogonal system. It can be shown that the kernel polynomial has the reproducing property that for every polynomial p of degree at most n

$$(p(x), K_n(x, y)_x) = p(y). \tag{4.5.38}$$

Here the subscript x indicates that the inner product is taken with respect to x. Conversely, if $K(x, y)$ is a polynomial of degree at most n in x and y and if $(p(x), K(x, y)_x) = p(y)$, for all polynomials p of degree at most n, then $K(x, y) = K_n(x, y)$.

An alternative expression, the **Christoffel–Darboux formula**, can be given for the kernel polynomial.

Theorem 4.5.18.

Let $p_n(x) = k_n x^n + \cdots$, $n = 0, 1, 2, \dots,$ be real orthonormal polynomials. Then

$$K_n(x, y) = \frac{k_n}{k_{n+1}} \frac{p_{n+1}(x)p_n(y) - p_n(x)p_{n+1}(y)}{x - y}. \tag{4.5.39}$$

Proof. See Davis [92, Theorem 10.1.6]. □

Given a linearly independent sequence of vectors, an orthogonal system can be derived by a process analogous to Gram–Schmidt orthogonalization.

Theorem 4.5.19.

For every weight function in an inner product space there is a triangle family of orthogonal polynomials $\phi_k(x)$, $k = 0, 1, 2, \dots$, such that $\phi_k(x)$ has exact degree k, and is orthogonal to all polynomials of degree less than k. The family is uniquely determined apart from the fact that the leading coefficients can be given arbitrary positive values.

The monic orthogonal polynomials satisfy the three-term recurrence formula,

$$\phi_{k+1}(x) = (x - \beta_k)\phi_k(x) - \gamma_{k-1}^2\phi_{k-1}(x), \quad k \geq 1, \tag{4.5.40}$$

with initial values $\phi_{-1}(x) = 0$, $\phi_0(x) = 1$. The recurrence coefficients are given by Darboux's formulas

$$\beta_k = \frac{(x\phi_k, \phi_k)}{\|\phi_k\|^2}, \qquad \gamma_{k-1}^2 = \frac{\|\phi_k\|^2}{\|\phi_{k-1}\|^2}. \tag{4.5.41}$$

Proof. The proof is by induction. We have $\phi_{-1} = 0$, $\phi_0 = 1$. Suppose that $\phi_j \neq 0$ have been constructed for $0 \leq j \leq k$, $k \geq 0$. We now seek a polynomial ϕ_{k+1} of degree $k+1$ with leading coefficient equal to 1 which is orthogonal to all polynomials of degree $\leq k$. Since $\{\phi_j\}_{j=0}^k$ is a triangle family, every polynomial of degree k can be expressed as a linear combination of these polynomials. Therefore, we can write

$$\phi_{k+1} = x\phi_k - \sum_{i=0}^{k} c_{k,i}\phi_i, \tag{4.5.42}$$

where ϕ_{k+1} has leading coefficient one. The orthogonality condition is fulfilled if and only if

$$(x\phi_k, \phi_j) - \sum_{i=0}^{k} c_{k,i}(\phi_i, \phi_j) = 0, \quad j = 0 : k.$$

But $(\phi_i, \phi_j) = 0$ for $i \neq j$, and thus $c_{k,j}\|\phi_j\|^2 = (x\phi_k, \phi_j)$. This determines the coefficients uniquely. From the definition of inner product (4.5.30), it follows that

$$(x\phi_k, \phi_j) = (\phi_k, x\phi_j).$$

But $x\phi_j$ is a polynomial of degree $j+1$. Thus if $j < k$, then it is orthogonal to ϕ_k. So $c_{kj} = 0$ for $j < k-1$. From (4.5.42) it then follows that

$$\phi_{k+1} = x\phi_k - c_{k,k}\phi_k - c_{k,k-1}\phi_{k-1}, \tag{4.5.43}$$

with $c_{k,k-1} = 0$ if $k = 0$. This has the same form as the original assertion of the theorem if we set

$$\beta_k = c_{k,k} = \frac{(x\phi_k, \phi_k)}{\|\phi_k\|^2}, \tag{4.5.44}$$

$$\gamma_{k-1}^2 = c_{k,k-1} = \frac{(\phi_k, x\phi_{k-1})}{\|\phi_{k-1}\|^2}, \quad k \geq 1.$$

In the discrete case the division in (4.5.44) can always be performed, as long as $k \leq m$. In the continuous case, no reservation need be made.

The expression for γ_{k-1}^2 can be written in another way. If we take the inner product of (4.5.42) and ϕ_{k+1} we get

$$(\phi_{k+1}, \phi_{k+1}) = (\phi_{k+1}, x\phi_k) - \sum_{i=0}^{k} c_{k,i}(\phi_{k+1}, \phi_i) = (\phi_{k+1}, x\phi_k).$$

Thus $(\phi_{k+1}, x\phi_k) = \|\phi_{k+1}\|^2$, or if we decrease all indices by one, $(\phi_k, x\phi_{k-1}) = \|\phi_k\|^2$. Substituting this in the expression for γ_{k-1}^2 gives the second equation of (4.5.41). $\square$

If the weight distribution $w(x)$ is symmetric about β, i.e., (in the continuous case) $w(\beta - x) = w(x + \beta)$, then $\beta_k = \beta$ for all $k \geq 0$. Further,

$$\phi_k(\beta - x) = (-1)^k \phi_k(x + \beta), \quad k \geq 0; \tag{4.5.45}$$

that is, ϕ_k is symmetric about β for k even and antisymmetric for k odd. The proof is by induction. We have $\phi_0 = 1$ and $\phi_1(x) = x - \beta_0$. Clearly $(\phi_1, \phi_0) = 0$ implies that ϕ_1 is antisymmetric about β and therefore $\beta_0 = \beta$. Thus the hypothesis is true for $k \leq 1$. Now assume that (4.5.45) holds for $k \leq n$. Then

$$\beta_n = \frac{(x\phi_n, \phi_n)}{\|\phi_n\|^2} = \frac{((x - \beta)\phi_n, \phi_n)}{\|\phi_n\|^2} + \beta.$$

Here the first term is zero since it is an integral of an antisymmetric function. It follows that

$$\phi_{n+1}(x) = (x - \beta)\phi_n(x) - \gamma_{n-1}^2 \phi_{n-1}(x),$$

which shows that (4.5.45) holds for $k = n + 1$. An analog result holds for a symmetric discrete inner product.

Often it is more convenient to consider corresponding **orthonormal polynomials** $\hat{\phi}_k(x)$, which satisfy $\|\hat{\phi}_k\| = 1$. We set $\hat{\phi}_0 = 1/\sqrt{\mu_0}$, $\mu_0 = \|\phi_0\|_2^2$, and scale the monic orthogonal polynomials according to

$$\phi_k = (\gamma_1 \cdots \gamma_{k-1})\hat{\phi}_k, \quad k \geq 1; \tag{4.5.46}$$

then we find using (4.5.41) that

$$\frac{\|\hat{\phi}_k\|}{\|\hat{\phi}_{k-1}\|} = \frac{\gamma_1 \cdots \gamma_{k-2}}{\gamma_1 \cdots \gamma_{k-1}} \frac{\|\phi_k\|}{\|\phi_{k-1}\|} = 1.$$

Substituting (4.5.46) in (4.5.40) we obtain the recurrence relation for the orthonormal polynomials

$$\gamma_k \hat{\phi}_{k+1}(x) = (x - \beta_k)\hat{\phi}_k(x) - \gamma_{k-1}\hat{\phi}_{k-1}(x), \quad k \geq 1, \tag{4.5.47}$$

where γ_k is determined by the condition $\|\hat{\phi}_{k+1}\| = 1$.

Perhaps the most important example of a family of orthogonal polynomials is the Chebyshev polynomials $T_n(x) = \cos(n \arccos(x))$ introduced in Sec. 3.2.3. These are orthogonal on $[-1, 1]$ with respect to the weight function $(1 - x^2)^{-1/2}$ and also with respect to a discrete inner product. Their properties can be derived by rather simple methods.

Theorem 4.5.20.

The Chebyshev polynomials have the following two orthogonality properties. Set

$$(f, g) = \int_{-1}^{1} f(x)g(x)(1 - x^2)^{-1/2}\, dx \tag{4.5.48}$$

(the continuous case). Then $(T_0, T_0) = \pi$, *and*

$$(T_j, T_k) = \begin{cases} 0 & \text{if } j \neq k, \\ \pi/2 & \text{if } j = k \neq 0. \end{cases} \tag{4.5.49}$$

Let x_k *be the zeros of* $T_{m+1}(x)$ *and set*

$$(f, g) = \sum_{k=0}^{m} f(x_k)g(x_k), \quad x_k = \cos\left(\frac{2k+1}{m+1}\frac{\pi}{2}\right) \tag{4.5.50}$$

(the discrete case). Then $(T_0, T_0) = m + 1$, *and*

$$(T_j, T_k) = \begin{cases} 0 & \text{if } j \neq k, \\ (m+1)/2 & \text{if } j = k \neq 0. \end{cases} \tag{4.5.51}$$

Proof. In the continuous case, let $j \neq k$, $j \geq 0$, $k \geq 0$. From $x = cos\phi$ it follows that $dx = \sin\phi d\phi = (1 - x^2)^{1/2} d\phi$. Hence

$$\begin{aligned}(T_j, T_k) &= \int_0^{\pi} \cos jx \cos kx\, dx = \int_0^{\pi} \frac{1}{2}\big(\cos(j-k)x + \cos(j+k)x\big)\, dx \\ &= \frac{1}{2}\left(\frac{\sin(j-k)\pi}{j-k} + \frac{\sin(j+k)\pi}{j+k}\right) = 0,\end{aligned}$$

whereby orthogonality is proved.

In the discrete case, set $h = \pi/(m+1)$, $x_\mu = h/2 + \mu h$,

$$(T_j, T_k) = \sum_{\mu=0}^{m} \cos jx_\mu \cos kx_\mu = \frac{1}{2}\sum_{\mu=0}^{m}\big(\cos(j-k)x_\mu + \cos(j+k)x_\mu\big).$$

Using notation from complex numbers ($i = \sqrt{-1}$) we have

$$(T_j, T_k) = \frac{1}{2}\mathrm{Re}\left(\sum_{\mu=0}^{m} e^{i(j-k)h(1/2+\mu)} + \sum_{\mu=0}^{m} e^{i(j+k)h(1/2+\mu)}\right). \tag{4.5.52}$$

The sums in (4.5.52) are geometric series with ratios $e^{i(j-k)h}$ and $e^{i(j+k)h}$, respectively. If $j \neq k$, $0 \leq j \leq m$, $0 \leq k \leq m$, then the ratios are never equal to one, since

$$0 < |\,(j \pm k)h| \leq \frac{2m}{m+1}\pi < \pi.$$

Using the formula for the sum of a geometric series the first sum in (4.5.52) is

$$e^{i(j-k)(h/2)}\frac{e^{i(j-k)(m+1)h}-1}{e^{i(j-k)h}-1} = \frac{e^{i(j-k)\pi}-1}{e^{i(j-k)(h/2)}-e^{-i(j-k)(h/2)}-1}$$
$$= \frac{(-1)^{j-k}-1}{2i\sin(j-k)h/2}.$$

The real part of the last expression is clearly zero. An analogous computation shows that the real part of the other sum in (4.5.52) is also zero. Thus the orthogonality property holds in the discrete case also. It is left to the reader to show that the expressions when $j = k$ given in the theorem are correct. □

For the uniform weight distribution $w(x) = 1$ on $[-1, 1]$ the relevant orthogonal polynomials are the **Legendre polynomials**[152] $P_n(x)$. The Legendre polynomials are defined by Rodrigues' formula

$$P_n(x) = (-1)^n \frac{1}{2^n n!}\frac{d^n}{dx^n}\{(1-x^2)^n\}, \quad n = 0, 1, 2, \dots. \tag{4.5.53}$$

Since $(1 - x^2)^n$ is a polynomial of degree $2n$, $P_n(x)$ is a polynomial of degree n. The Legendre polynomials $P_n = A_n x^n + \cdots$ have leading coefficient and norm

$$A_n = \frac{(2n)!}{2^n(n!)^2}, \qquad \|P_n\| = \frac{2}{2n+1}. \tag{4.5.54}$$

This standardization corresponds to setting $P_n(1) = 1$ for all $n \geq 0$. The extreme values are

$$|P_n(x)| \leq 1, \quad x \in [-1, 1].$$

There seems to be no easy proof for the last result; see [193, p. 219].

Since the weight distribution is symmetric about the origin, these polynomials have the symmetry property

$$P_n(-x) = (-1)^n P_n(x).$$

The Legendre polynomials satisfy the three-term recurrence formula $P_0(x) = 1$, $P_1(x) = x$,

$$P_{n+1}(x) = \frac{2n+1}{n+1} x P_n(x) - \frac{n}{n+1} P_{n-1}(x), \quad n \geq 1. \tag{4.5.55}$$

The first few Legendre polynomials are

$$P_2(x) = \frac{1}{2}(3x^2 - 1), \qquad P_3(x) = \frac{1}{2}(5x^3 - 3x),$$
$$P_4(x) = \frac{1}{8}(35x^4 - 30x^2 + 3), \qquad P_5(x) = \frac{1}{8}(63x^5 - 70x^3 + 15), \dots.$$

A graph of the Legendre polynomial $P_{21}(x)$ is shown in Figure 4.5.1.

[152] Legendre had obtained these polynomials in 1784–1789 in connection with his investigation concerning the attraction of spheroids and the shape of planets.

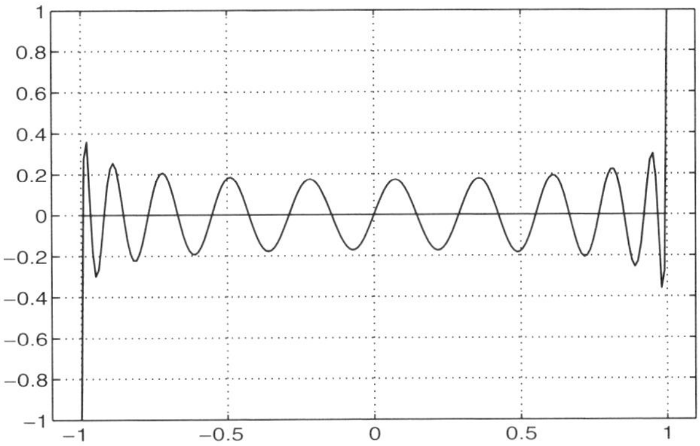

Figure 4.5.1. *The Legendre polynomial* P_{21}.

Often a more convenient standardization is to consider *monic* Legendre polynomials, with leading coefficient equal to one. These satisfy $P_0(x) = 1$, $P_1(x) = x$, and the recurrence formula

$$P_{n+1}(x) = xP_n(x) - \frac{n^2}{4n^2-1}P_{n-1}(x), \quad n \geq 1; \tag{4.5.56}$$

note that we have kept the same notation for the polynomials. It can be shown that

$$P_n = x^n + c_n x^{n-2} + \cdots, \quad c_n = -\frac{n(n-1)}{2(2n-1)}.$$

The **Jacobi polynomials**[153] $J_n(x;\alpha,\beta)$ arise from the weight function

$$w(x) = (1-x)^\alpha (1+x)^\beta, \quad x \in [-1,1], \quad \alpha,\beta > -1.$$

They are special cases of Gauss' hypergeometric function $F(a,b,c:x)$,

$$F(-n, \alpha+1+\beta+n, \alpha+1; x),$$

(see (3.1.16)). The Jacobi polynomials are usually standardized so that the coefficient A_n of x^n in $J_n(x;\alpha,\beta)$ is given by

$$A_n = \frac{1}{2^n n!}\frac{\Gamma(2n+\alpha+\beta+1)}{\Gamma(n+\alpha+\beta+1)}.$$

The Legendre polynomials are obtained as the special case when $\alpha = \beta = 0$. The case $\alpha = \beta = -1/2$, which corresponds to the weight function $w(x) = 1/\sqrt{1-x^2}$, gives the Chebyshev polynomials.

[153]Carl Gustav Jacob Jacobi (1805–1851) was a German mathematician. Jacobi joined the faculty of Berlin University in 1825. Like Euler, he was a proficient calculator who drew a great deal of insight from immense algorithmic work.

The generalized **Laguerre polynomials** $L_n^{(\alpha)}(x)$ are orthogonal with respect to the weight function

$$w(x) = x^\alpha e^{-x}, \quad x \in [0, \infty], \quad \alpha > -1.$$

Setting $\alpha = 0$, we get the Laguerre polynomials $L_n^{(0)}(x) = L_n(x)$. Standardizing these so that $L_n(0) = 1$, they satisfy the three-term recurrence relation

$$(n+1)L_{n+1}(x) = (2n+1-x)L_n(x) - nL_{n-1}(x), \tag{4.5.57}$$

Rodrigues' formula becomes

$$L_n^\alpha(x) = \frac{e^x}{n!x^{(\alpha)}} \frac{d^n}{dx^n}(x^{n-\alpha}e^{-x}).$$

The **Hermite polynomials** are orthogonal with respect to the weight function

$$w(x) = e^{-x^2}, \quad -\infty < x < \infty.$$

With the classic standardization they satisfy the recurrence relation $H_0(x) = 1$, $H_1(x) = 2x$,

$$H_{n+1}(x) = 2xH_n(x) - 2nH_{n-1}(x),$$

and

$$H_n(0) = \begin{cases} (-1)^m(2m)!/m! & \text{if } n = 2m, \\ 0 & \text{if } n = 2m+1. \end{cases}$$

The Hermite polynomials can also be defined by Rodrigues' formula:

$$H_n(x) = (-1)^n e^{x^2} \frac{d^n}{dx^n}\{e^{-x^2}\}.$$

It can be verified that these polynomials are identical to those defined by the recurrence relation.

The properties of some important families of orthogonal polynomials are summarized in Table 4.5.1. Note that here the coefficients in the three-term recurrence relation are given for the *monic orthogonal polynomials*; cf. (4.5.40).

For equidistant data, the **Gram polynomials** $\{P_{n,m}\}_{n=0}^m$ are of interest.[154] These polynomials are orthogonal with respect to the discrete inner product

$$(f, g) = (1/m)\sum_{i=1}^m f(x_i)g(x_i), \quad x_i = -1 + (2i-1)/m.$$

The weight distribution is symmetric around the origin $\alpha_k = 0$. For the *monic* Gram polynomials the recursion formula is (see [16])

$$P_{-1,m}(x) = 0, \qquad P_{0,m} = 1,$$
$$P_{n+1,m}(x) = xP_{n,m}(x) - \beta_{n,m}P_{n-1,m}(x), \quad n = 0 : m-1,$$

[154] Jørgen Pedersen Gram (1850–1916), a Danish mathematician, graduated from Copenhagen University and worked as a director for a life insurance company. He introduced the Gram determinant in connection with his study of linear independence, and his name is also associated with Gram–Schmidt orthogonalization.

Table 4.5.1. *Weight functions and recurrence coefficients for some classical monic orthogonal polynomials.*

$[a, b]$	$w(x)$	Orthog. pol.	μ_0	β_k	γ_k^2
$[-1, 1]$	1	$P_n(x)$ Legendre	2	0	$\dfrac{k^2}{4k^2-1}$
$[-1, 1]$	$(1-x^2)^{-1/2}$	$T_n(x)$ Cheb. 1st	π	0	$\frac{1}{2}$ $(k=1)$ $\frac{1}{4}$ $(k>1)$
$[-1, 1]$	$(1-x^2)^{1/2}$	$U_n(x)$ Cheb. 2nd	$\pi/2$	0	$\frac{1}{4}$
$[-1, 1]$	$(1-x)^{\alpha}(1+x)^{\beta}$	$J_n(x; \alpha, \beta)$ Jacobi			
$[0, \infty]$	$x^{\alpha}e^{-x}, \alpha > -1$	$L_n^{(\alpha)}(x)$ Laguerre	$\Gamma(1+\alpha)$	$2k+\alpha+1$	$k(k+\alpha)$
$[-\infty, \infty]$	e^{-x^2}	$H_n(x)$ Hermite	$\sqrt{\pi}$	0	$\frac{1}{2}k$

where $(n < m)$ and

$$\beta_{n,m} = \frac{n^2}{4n^2-1}\left(1 - \frac{n^2}{m^2}\right).$$

When $n \ll m^{1/2}$, these polynomials are well behaved. But when $n \geq m^{1/2}$, the Gram polynomials have very large oscillations between the grid points, and a large maximum norm in $[-1, 1]$. This fact is related to the recommendation that when fitting a polynomial to *equidistant data*, one should never choose n larger than about $2m^{1/2}$.

Complex Orthogonal Polynomials

So far we have considered the inner products (4.5.30) defined by an integral over the real interval $[a, b]$. Now let Γ be a rectifiable curve (i.e., a curve of finite length) in the complex plane. Consider the linear space of all polynomials with complex coefficients and $z = x+iy$ on Γ. Let $\alpha(s)$ be a function on Γ with infinitely many points of increase and define an inner product by the line integral

$$(p, q) = \int_{\Gamma} p(z)\overline{q(z)}w(s)\, d(s). \tag{4.5.58}$$

The complex monomials $1, z, z^2, \dots$ are independent functions, since if $a_0 + a_1 z + a_2 z^2 + \cdots + a_n z^n \equiv 0$ on Γ it would follow from the fundamental theorem of algebra that $a_0 = a_1 = a_2 = \cdots = a_n = 0$. There is a unique infinite sequence of polynomials $\phi_j = z^j + c_1 z^{j-1} + \cdots + c_j$, $j = 0, 1, 2, \dots$, which are orthonormal with respect to (4.5.58). They can be constructed by Gram–Schmidt orthogonalization as in the real case.

An important case is when Γ is the unit circle in the complex plane. We then write

$$(p,q) = \frac{1}{2\pi}\int_{-\pi}^{\pi} p(z)\overline{q(z)}\,d\alpha(t), \quad z = e^{it}, \tag{4.5.59}$$

where the integral is to be interpreted as a Stieltjes integral. The corresponding orthogonal polynomials are known as **Szegő polynomials** and have applications, e.g., in signal processing.

Properties of Szegő polynomials are discussed in [174]. Together with the reverse polynomials $\tilde{\phi}_j(z) = z^j\phi_j(1/z)$ they satisfy special recurrence relations. A linear combination $\sum_{j=0}^{n} c_j\phi_j(z)$ can be evaluated by an analogue to the Clenshaw algorithm; see [5].

4.5.6 Expansions in Orthogonal Polynomials

Expansions of functions in terms of orthogonal polynomials are very useful. They are easy to manipulate, have good convergence properties in the continuous case, and usually give a well-conditioned representation.

Let $\hat{p}_n$ denote the polynomial of degree less than n for which

$$\|f - \hat{p}_n\|_\infty = E_n(f) = \min_{p\in\mathcal{P}_n} \|f - p\|_\infty,$$

and set

$$p_n = \sum_{j=0}^{n-1} c_j\phi_j,$$

where c_j is the jth Fourier coefficient of f and $\{\phi_j\}$ are the orthogonal polynomials with respect to the inner product (4.5.30). If we use the weighted Euclidean norm, $\hat{p}_n$ is of course not a better approximation than p_n. In fact,

$$\begin{aligned}\|f - p_n\|_w^2 &= \int_a^b |f(x) - p_n(x)|^2 w(x)\,dx \\ &\le \int_a^b |f(x) - \hat{p}_n(x)|^2 w(x)\,dx \le E_n(f)^2 \int_a^b w(x)\,dx.\end{aligned} \tag{4.5.60}$$

This can be interpreted as saying that a kind of weighted mean of $|f(x) - p_n(x)|$ is less than or equal to $E_n(f)$, which is about as good a result as one could demand. The error curve has an oscillatory behavior. In small subintervals, $|f(x) - p_n(x)|$ can be significantly greater than $E_n(f)$. This is usually near the ends of the intervals or in subintervals where $w(x)$ is relatively small. Note that from (4.5.60) and Weierstrass' approximation theorem it follows that

$$\lim_{n\to\infty} \|f - p\|_{2,w}^2 = 0$$

for every continuous function f. From (4.5.60) one gets after some calculations

$$\sum_{j=n}^{\infty} c_j^2\|\phi_j\|^2 = \|f - p_n\|_{2,w}^2 \le E_n(f)^2 \int_a^b w(x)\,dx,$$

which gives one an idea of how quickly the terms in the orthogonal expansion decrease.

Example 4.5.11 (*Chebyshev Interpolation*).

Let $p(x)$ denote the interpolation polynomial in the Chebyshev points $x_k = \cos(\frac{2k+1}{m+1}\frac{\pi}{2}) k = 0 : m$. For many reasons it is practical to write this interpolation polynomial in the form

$$p(x) = \sum_{i=0}^{m} c_i T_i(x). \tag{4.5.61}$$

Then using the discrete orthogonality property (4.5.51)

$$c_i = \frac{(f, T_i)}{\|T_i\|^2} = \frac{1}{\|T_i\|^2} \sum_{k=0}^{m} f(x_k) T_i(x_k) \tag{4.5.62}$$

where

$$\|T_0\|^2 = m + 1, \quad \|T_i\|^2 = \frac{1}{2}(m+1), \quad i > 0.$$

The recursion formula (3.2.20) can be used for calculating the orthogonal coefficients according to (4.5.62).

For computing $p(x)$ with (4.5.61), Clenshaw's algorithm [71] can be used. Clenshaw's algorithm holds for any sum of the form $S = \sum_{k=1}^{n} c_k \phi_k$, where $\{\phi_k\}$ satisfies *a three-term recurrence relation*. It can also be applied to series of Legendre functions, Bessel functions, Coulomb wave functions, etc., because they satisfy recurrence relations of this type, where the α_k, γ_k depend on x; see the Handbook [1] or any text on special functions.

Theorem 4.5.21 (*Clenshaw's Algorithm*).

Suppose that a sequence $\{p_k\}$ *satisfies the three-term recurrence relation*

$$p_{k+1} = \gamma_k p_k - \beta_k p_{k-1}, \quad k = 0 : n-1, \tag{4.5.63}$$

where $p_{-1} = 0$. *Then*

$$S = \sum_{k=0}^{n} c_k p_k = y_0 p_0,$$

where $y_{n+1} = 0$, $y_n = c_n$, *and* y_0 *is obtained by the recursion*

$$y_k = c_k + \gamma_{k-1} y_{k+1} - \beta_k y_{k+2}, \quad k = n-1 : -1 : 0. \tag{4.5.64}$$

Proof. Write the recursion (4.5.63) in matrix form as $L_n p = p_0 e_1$, where

$$L_n = \begin{pmatrix} 1 & & & & \\ -\gamma_0 & 1 & & & \\ \beta_1 & -\gamma_1 & 1 & & \\ & \ddots & \ddots & \ddots & \\ & & \beta_{n-1} & -\gamma_{n-1} & 1 \end{pmatrix}$$

is unit lower triangular and e_1 is the first column of the unit matrix. Then

$$S = c^T p = c^T L_n^{-1} g = g^T (L_n^T)^{-1} c = g^T y,$$

where y is the solution to the upper triangular system $L_n^T y = c$. Solving this by back-substitution we get the recursion (4.5.64). $\square$

It can be proved that Clenshaw's algorithm is componentwise backward stable with respect to the data γ_k and β_k; see Smoktunowicz [329].

Occasionally one is interested in the partial sums of (4.5.61). For example, if the values of $f(x)$ are afflicted with statistically independent errors with standard deviation σ, then the series can be broken off when for the first time

$$\left\| f - \sum_{i=0}^{p} c_i T_i(x) \right\| < \sigma m^{1/2}.$$

The proof of Theorem 4.5.19 is constructive and leads to a unique construction of the sequence of orthogonal polynomials ϕ_k, $n \geq 1$, with leading coefficients equal to one. The coefficients β_k and γ_k and the polynomials ϕ_k can be computed in the order

$$\beta_0, \phi_1(x), \beta_1, \gamma_0, \phi_2(x), \dots .$$

(Recall that $\phi_{-1} = 0$ and $\phi_0 = 1$.) This procedure is called the **Stieltjes procedure**. This assumes that the inner product $(x\phi_k, \phi_k)$ and norm $\|\phi_k\|_2$ can be computed. This clearly can be done for any discrete n-point weight.

For a continuous weight function $w(x)$ this is more difficult. One possible way to proceed is then to approximate the integrals using some appropriate quadrature rule. For a discussion of such methods and examples we refer to Gautschi [148].

There are many computational advantages of using the Stieltjes procedure when computing the discrete least squares approximation

$$f(x) \approx p_m(x) = \sum_{k=0}^{m} c_k \phi_k(x), \tag{4.5.65}$$

where $\phi_0(x), \dots, \phi_m(x)$ are the orthogonal polynomials with respect to the discrete weight function (4.5.32). Recall the family ends with $\phi_m(x)$, since

$$\phi_{m+1}(x) = (x - x_0)(x - x_1) \cdots (x - x_m)$$

is zero at each grid point and $\|\phi_{m+1}\| = 0$. The construction cannot proceed, which is natural since there cannot be more than $m + 1$ orthogonal (or even linearly independent) functions on a grid with $m + 1$ points.

By Theorem 4.5.13 the coefficients are given by

$$c_k = (f, \phi_k)/\|\phi_k\|^2, \quad k = 0 : m. \tag{4.5.66}$$

Approximations of increasing degree can be recursively generated as follows. Suppose that ϕ_j, $j = 0 : k$, and the least squares approximation p_k of degree k, have been computed. In the next step the coefficients β_k, γ_{k-1}, and ϕ_{k+1} are computed and the next approximation is obtained by

$$p_{k+1} = p_k + c_{k+1}\phi_{k+1}, \qquad c_{k+1} = (f, \phi_{k+1})/\|\phi_{k+1}\|^2. \tag{4.5.67}$$

To compute numerical values of $p(x)$ **Clenshaw's algorithm** is used; see Theorem 4.5.21.

For unit weights and a symmetric grid the Stieltjes procedure requires about $4mn$ flops to compute the orthogonal functions ϕ_k at the grid points. If there are differing weights, then about mn additional operations are needed. Similarly, mn additional operations are required if the grid is not symmetric. If the orthogonal coefficients are determined simultaneously for several functions on the same grid, then only about mn additional operations per function are required. (In the above, we assume $m \gg 1, n \gg 1$.) Hence the procedure is much more economical than the general methods based on normal equations, which require $O(mn^2)$ flops.

Since p_k is a linear combination of ϕ_j, $j = 0 : k$, ϕ_{k+1} is orthogonal to p_k. Therefore, an alternative expression for the new coefficient is

$$c_{k+1} = (r_k, \phi_{k+1})/\|\phi_{k+1}\|^2, \qquad r_k = f - p_k, \tag{4.5.68}$$

where r_k is the residual. Mathematically the two formulas (4.5.67) and (4.5.68) for c_{k+1} are equivalent. In finite precision, as higher-degree polynomials p_{k+1} are computed, they will gradually lose orthogonality to previously computed p_j, $j \le k$. In practice there is an advantage in using (4.5.68) since cancellation will then take place mostly in computing the residual $r_k = f - p_k$, and the inner product (r_k, p_{k+1}) is computed more accurately. Theoretically the error $\|p_k - f\|$ must be a nonincreasing function of k. Often the error decreases rapidly with k and then p_k provides a good representation of f already for small values of k. Note that for $n = m$ we obtain the (unique) interpolation polynomial for the given points.

When using the first formula one sometimes finds that *the residual norm increases when the degree of the approximation is increased*! With the modified formula (4.5.68) this is very unlikely to happen; see Problem 4.5.17.[155]

One of the motivations for the method of least squares is that it effectively reduces the influence of random errors in measurements. We now consider some statistical aspects of the method of least squares. First, we need a slightly more general form of Gauss–Markov's theorem.

Corollary 4.5.22.

Assume that in the linear model (1.4.2), ϵ *has zero mean and positive definite covariance matrix* V. *Then the normal equations for estimating* c *are*

$$(A^T V^{-1} A)\hat{c} = A^T V^{-1} y. \tag{4.5.69}$$

In particular, if $V = \operatorname{diag}(\sigma_1^2, \sigma_2^2, \dots, \sigma_n^2)$ *this corresponds to a row scaling of the linear model* $y = Ac + \epsilon$.

[155] The difference between the two variants discussed here is similar to the difference between the so-called classical and modified Gram–Schmidt orthogonalization methods.

Suppose that the values of a function have been measured in the points $x_1, x_2, \dots, x_m$. Let $f(x_p)$ be the measured value, and let $\bar{f}(x_p)$ be the "true" (unknown) function value which is assumed to be the same as the *expected value* of the measured value. Thus *no systematic errors* are assumed to be present. Suppose further that *the errors in measurement at the various points are statistically independent*. Then we have a linear model $f(x_p) = \bar{f}(x_p) + \epsilon$, where

$$E(\epsilon) = 0, \qquad V(\epsilon) = \operatorname{diag}(\sigma_1^2, \dots, \sigma_n^2). \tag{4.5.70}$$

The problem is to use the measured data to estimate the coefficients in the series

$$f(x) = \sum_{j=1}^{n} c_j \phi(x), \quad n \le m,$$

where $\phi_1, \phi_2, \dots, \phi_n$ are known functions. According to the Gauss–Markov theorem and its corollary the estimates c_j^*, which one gets by minimizing the sum

$$\sum_{p=1}^{m} w_p \Bigl(f(x_p) - \sum_{j=1}^{n} c_j \phi_j(x_p) \Bigr)^2, \quad w_p = \sigma_p^{-2},$$

have a smaller variance than the values one gets by any other linear unbiased estimation method. This minimum property holds not only for the estimates of the coefficients c_j, but also for every linear functional of the coefficients, for example, the estimate

$$f_n^*(\alpha) = \sum_{j=1}^{n} c_j^* \phi_j(\alpha)$$

of the value $f(\alpha)$ at an arbitrary point α.

Suppose now that $\sigma_p = \sigma$ for all p and that the functions $\{\phi_j\}_{j=1}^n$ form an *orthonormal system* with respect to the discrete inner product

$$(f, g) = \sum_{p=1}^{m} f(x_p) g(x_p).$$

Then the least squares estimates are $c_j^* = (f, \phi_j)$, $j = 1 : n$. According to (1.4.4) the estimates c_j^* and c_k^* are *uncorrelated if* $j \neq k$ and

$$E\{(c_j^* - \bar{c}_j)(c_k^* - \bar{c}_k)\} = \begin{cases} 0 & \text{if } j \neq k, \\ \sigma^2 & \text{if } j = k. \end{cases}$$

From this it follows that

$$\operatorname{var}\{f_n^*(\alpha)\} = \operatorname{var}\Bigl\{ \sum_{j=1}^{n} c_j^* \phi_j(\alpha) \Bigr\} = \sum_{j=1}^{n} \operatorname{var}\{c_j^*\} |\phi_j(\alpha)|^2 = \sigma^2 \sum_{j=1}^{n} |\phi_j(\alpha)|^2.$$

As an average, *taken over the grid of measurement points*, the variance of the smoothed function values is

$$\frac{1}{m} \sum_{j=1}^{n} \operatorname{var}\{f_n^*(x_i)\} = \frac{\sigma^2}{m} \sum_{j=1}^{n} \sum_{i=1}^{m} |\phi_j(x_i)|^2 = \sigma^2 \frac{n}{m}.$$

Between the grid points, however, the variance can in many cases be significantly larger. For example, when fitting a polynomial to measurements in *equidistant points*, the Gram polynomial $P_{n,m}$ can be much larger between the grid points when $n > 2m^{1/2}$. Set

$$\sigma_I^2 = \sigma^2 \sum_{j=1}^{n} \frac{1}{2} \int_{-1}^{1} |\phi(x)|^2 \, dx.$$

Thus σ_I^2 is an average variance for $f_n^*(x)$ *taken over the entire interval* $[-1, 1]$. The following values for the ratio ρ between σ_I^2 and $\sigma^2(n+1)/(m+1)$ when $m = 42$ were obtained by H. Björk [32]. Related results are found in Reichel [298].

$n+1$	5	10	15	20	25	30	35
ρ	1.0	1.1	1.7	26	$7 \cdot 10^3$	$1.7 \cdot 10^7$	$8 \cdot 10^{11}$

These results are related to the recommendation that one should choose $n < 2m^{1/2}$ when fitting a polynomial to equidistant data. This recommendation seems to contradict the Gauss–Markov theorem, but in fact it just means that one gives up the requirement that the estimates are unbiased. Still, it is remarkable that this can lead to such a drastic reduction of the variance of the estimates f_n^*.

If the measurement points are the Chebyshev abscissae, then no difficulties arise in fitting polynomials to data. The Chebyshev polynomials have a magnitude between grid points not much larger than their magnitude at the grid points. The average variance for f_n^* becomes the same on the interval $[-1, 1]$ as on the net of measurements, $\sigma_2(n+1)/(m+1)$.

The choice of n when m is given is a question of compromising between taking into account the truncation error (which decreases as n increases) and the random errors (which grow when n increases). If f is a sufficiently smooth function, then in the Chebyshev case $|c_j|$ decreases quickly with j. In contrast, the part of c_j which comes from errors in measurements varies randomly with a magnitude of about $\sigma(2/(m+1)^{1/2})$, using (4.5.50) and $\|T_j\|^2 = (m+1)/2$. The expansion should be broken off when the coefficients begin to "behave randomly." An expansion in terms of Chebyshev polynomials can hence be used for *filtering away the "noise" from the signal, even when σ is initially unknown.*

Example 4.5.12.

Fifty-one equidistant values of a certain analytic function were rounded to four decimals. In Figure 4.5.2, a semilog diagram is given which shows how $|c_i|$ varies in an expansion in terms of the Chebyshev polynomials for these data. For $i > 20$ (approximately) the contribution due to noise dominates the contribution due to signal. Thus it is sufficient to break off the series at $n = 20$.

4.5.7 Approximation in the Maximum Norm

In some applications it is more natural to seek an approximation that minimizes the *maximum deviation* instead of the Euclidean norm. Let U be an $(n+1)$-dimensional vector space of continuous (real or complex) functions $u_k(x)$, $k = 0 : n$, defined on an interval $I = [a, b]$.

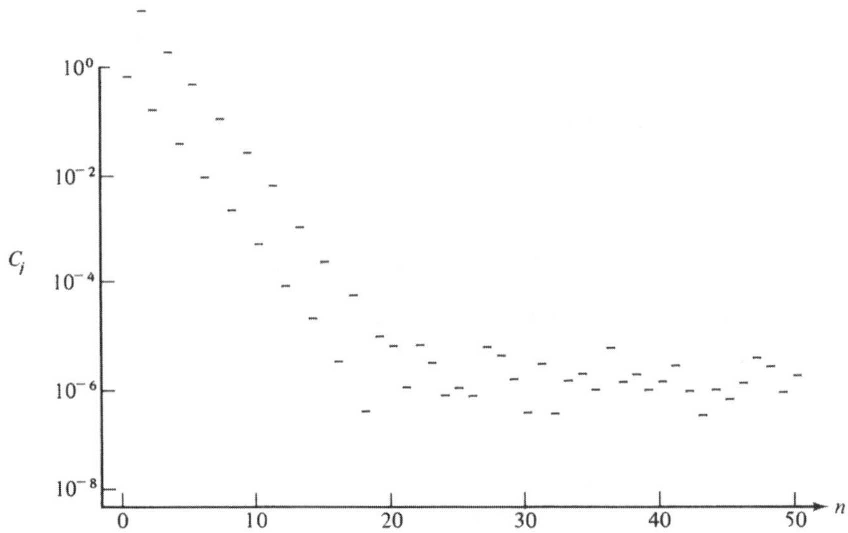

Figure 4.5.2. *Magnitude of coefficients c_i in a Chebyshev expansion of an analytic function contaminated with roundoff noise.*

Given a function $f \in C[a, b]$, we consider the approximation problem of finding a function

$$\tilde{u} = \sum_{k=0}^{n} c_k u_k \in U, \tag{4.5.71}$$

which minimizes the maximum norm of the error,

$$\eta(f, I) := \inf_{u \in U} \|f - u\|_{\infty, I} = \inf_{u \in U} \max_{x \in I} |f(x) - u(x)|.$$

Approximation in the maximum norm is often called Chebyshev approximation. We avoid this terminology, since Chebyshev's name is also naturally associated with two entirely different approximation methods: interpolation in the Chebyshev abscissae, and expansion in a series of Chebyshev polynomials.

The related *discrete approximating problem* is that of minimizing

$$\inf_{u \in U} \|f - u\|_{\infty, X} = \inf_{u \in U} \max_{0 \le i \le m} |f(x_i) - u(x_i)|, \tag{4.5.72}$$

where $X = \{x_0, \dots, x_m\}$ is a finite set of points. We note that the discrete approximation can be posed as a linear programming problem with $2(m + 1)$ linear inequalities:

$$\begin{aligned} &\text{Minimize } \eta \\ &\text{subject to } \; -\eta \le f(x_i) - \sum_{k=0}^{n} c_k u_k(x_i) \le \eta, \quad i = 0 : m. \end{aligned}$$

In principle this problem can be solved by the simplex method. Algorithms for solving this problem are given in [17].

The discrete approximation problem in l_1-*norm* corresponds to the linear programming problem

$$\text{Minimize} \quad \sum_{i=0}^{m} \eta_i$$

$$\text{subject to} \quad -\eta_i \le f(x_i) - \sum_{k=0}^{n} c_k u_k(x_i) \le \eta_i, \quad i = 0 : m.$$

Algorithms for solving this problem are discussed in [18].

As a simple example, consider the problem of best approximation in the maximum norm to the function $f(x) = x^n$ on $[-1, 1]$ by a polynomial of lower degree. From the minimax property of the Chebyshev polynomials it follows that the solution is given by the polynomial

$$f_n^*(x) = x^n - 2^{1-n} T_n(x),$$

which has in fact degree $n - 2$. The error function $-2^{1-n} T_n(x)$ assumes its extrema 2^{1-n} on a sequence of $n + 1$ points, $x_k = \cos(k\pi/n)$. The sign of the error alternates at these points, and thus for $f(x) = x^n$

$$E_{n-1}(f) = E_{n-2}(f) = 2^{1-n}.$$

The above property can be generalized. We shall now give a theorem which is the basis of many algorithms.

Theorem 4.5.23 (*Chebyshev's Equioscillation Theorem*).

Let f be a continuous function on $[a, b]$ and let $\tilde{p}$ be an nth degree polynomial which best approximates f in the maximum norm. Then $\tilde{p}$ is characterized by the fact that there exists at least $n + 2$ points,

$$a \le x_0 < x_1 < \cdots < x_{n+1} \le b,$$

where the error $(f - \tilde{p})$ takes on its maximal magnitude $\eta = \|\tilde{f} - p\|_\infty$ with alternating signs, that is,

$$f(x_k) - \tilde{p}(x_k) = -(f(x_{k+1}) - \tilde{p}(x_{k+1})), \quad k = 0 : n. \tag{4.5.73}$$

This characterization constitutes both a necessary and a sufficient condition. If $f^{(n+1)}$ has constant sign in [a,b], then $x_0 = a$ and $x_{n+1} = b$.

Proof. We prove here the sufficiency of the condition (4.5.73). For the rest of the proof, see, e.g., Cheney [66]. We first derive another characterization of a best approximation. Given a polynomial $\tilde{p}$ of degree n, consider the set of points

$$M = \{x \in [a, b] \mid |f(x) - \tilde{p}(x)| = \|f - \tilde{p}\|_\infty\},$$

where the difference $\tilde{d} = f - \tilde{p}$ takes on the extreme values $\pm\|f - \tilde{p}\|_\infty$. If $\tilde{p}$ is *not* a best approximation, then there is a best approximation that can be written $p^* = \tilde{p} + p$, where $p \ne 0$ is a polynomial of degree at most n. Then for all $x \in M$, we have

$$|f(x) - (\tilde{p}(x) + p(x))| < |f(x) - \tilde{p}(x)|.$$

This can only happen if the sign of $p(x)$ and the sign of $(f(x) - \tilde{p}(x))$ are the same, that is,

$$(f(x) - \tilde{p}(x))p(x) > 0, \quad x \in M.$$

Reversing this implication, it follows that $\tilde{p}$ is a best approximation whenever there is no polynomial satisfying this condition.

Suppose now that $\tilde{p}$ satisfies the conditions in the theorem. Then we claim that the conditions $(f(x_k) - \tilde{p}(x_k))p(x_k) > 0$, $k = 0 : n+2$, cannot hold for any polynomial $p \neq 0$ of degree n. For if they did hold, then p would have at least $n+1$ sign changes in $[a, b]$ and hence also $n+1$ zeros there. But by the fundamental theorem of algebra this is impossible. $\square$

Notice that Theorem 4.5.23 states that the best solution in maximum norm has *at least* $n+2$ points where the error alternates. In general, there can be more points where the maximal deviation is achieved.

Suppose that the function $f \in C[a, b]$ is known at $m > n+1$ distinct points $X = \{\xi_i\}_{i=0}^{m+1}$ in $[a, b]$, and let $\tilde{p}$ be a polynomial of degree n. Then $\tilde{p}$ is a best approximation of f in the discrete maximum norm if there are $n+2$ points $x_k \in X$ such that $(f - \tilde{p})$ assumes its maximal absolute value $d = \|\tilde{f} - p\|_\infty$ at these points with alternating sign. The proof of this is identical to the proof above.

Example 4.5.13.

In certain simple cases the alternating property can be used to construct the best uniform approximation. Suppose we want to find the best linear approximation to a convex function f on $[a, b]$. In this case the maximum error occurs with alternating signs at the three points $a = x_0, x_1, x_2 = b$, where x_1 is chosen so that

$$f'(x_1) = [a, b]f = (f(b) - f(a))/(b - a)$$

gives the solution. The linear polynomial equals the mean value of the tangent of f at x_1 and the linear interpolant through the endpoints of $[a, b]$, i.e.,

$$\tilde{p}(x) = (f(a) + f(b))/2 + (x - (a + x_1)/2)[a, b]f,$$

is illustrated in Figure 4.5.3. Notice that linear interpolation (dotted line) using $f(a)$ and $f(b)$ gives a maximum error which is exactly twice as large.

The only property that was used in the proof of Chebyshev's equioscillation theorem was the fundamental theorem of algebra. This property also holds for other subspaces of $C[a, b]$, which we now define.

Definition 4.5.24.

Let U be an $(n+1)$-dimensional vector space of continuous (real or complex) functions u_i, $i = 0 : n$, defined on a domain $\mathcal{D}$ containing at least $n+1$ distinct points. The set of functions $\{u_0, \dots, u_n\}$ are said to satisfy the **Haar condition**[156] *(or form a* **Chebyshev system***) if one of the three following equivalent conditions is satisfied:*

[156] This concept is named after the Austrian-Hungarian mathematician Alfred Haar (1885–1933) who, together with Friedrich Riesz established a mathematical center in Szeged, Hungary. Many important contributions to modern functional analysis originated in this center.

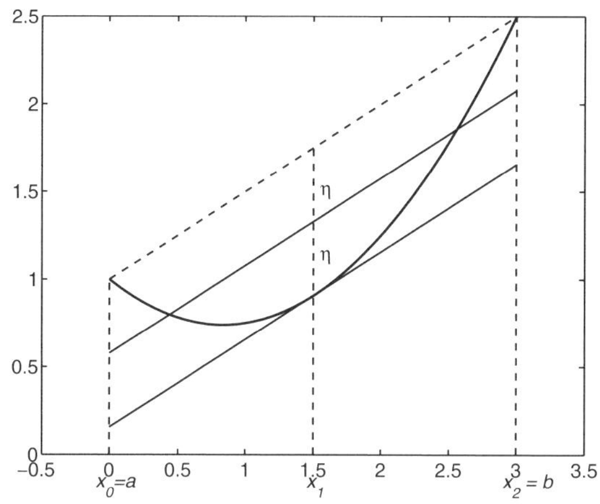

Figure 4.5.3. *Linear uniform approximation.*

1. *For any $n+1$ distinct points x_i, $i = 0 : n$, it holds that*

$$\det\begin{pmatrix} x_0, x_1, \dots, x_n \\ u_0, u_1, \dots, u_n \end{pmatrix} := \det\begin{pmatrix} u_0(x_0) & u_1(x_0) & \cdots & u_n(x_0) \\ u_0(x_1) & u_1(x_1) & \cdots & u_n(x_1) \\ \vdots & \vdots & \vdots & \vdots \\ u_0(x_n) & u_1(x_n) & \cdots & u_n(x_n) \end{pmatrix} \neq 0.$$

2. *The interpolation problem to determine $u \in U$ such that $u(x_i) = f_i$, $i = 0 : n$, is uniquely solvable.*

3. *Each $u \in U$, not identically zero, has at most n distinct zeros on $\mathcal{D}$.*

We remark that continuous functions of more than one variable cannot form a Chebyshev system. A large part of the theory of polynomial interpolation can be generalized to Chebyshev systems. Theorem 4.5.23 holds for any Chebyshev system, since only the Haar condition is needed in the proof.

Example 4.5.14.

The classical example of a Chebyshev system is the set of monomials $u_k(x) = x^k$, $k = 0 : n$. The Vandermonde determinant

$$\det(V) = \prod_{0 \le i, j \le n} (x_i - x_j)$$

is nonzero if the points x_i are pairwise distinct. Some other important examples are as follows:

1. If $\{u_0, u_1, \dots, u_n\}$ is a Chebyshev system on $[a, b]$ and $w(x) \neq 0$, then $w(x)u_0, w(x)u_1, \dots, w(x)u_n$ is a Chebyshev system on $[a, b]$.

2. If $\alpha_0 < \alpha_1 < \cdots \alpha_n$ is a strictly increasing sequence of real numbers, then the power functions $u_k(x) = x^{\alpha_k}$, $k = 0 : n$, form a Chebyshev system on any closed subinterval of $[0, \infty]$.

3. $1, e^x, e^{2x}, \dots, e^{nx}$, $x \in [a, b]$.

4. $1, \sin x, \sin 2x, \dots, \sin mx, \cos x, \cos x, \dots, \cos mx$, $x \in [0, 2\pi]$.

5. Let $s_0, s_1, \dots, s_n$ be distinct values not in $[a, b]$. Then the functions

$$u_i(x) = \frac{1}{x - s_i}, \qquad i = 0 : n,$$

form a Chebyshev system on $[a, b]$.

The last example follows from Cauchy's formula for the determinant

$$\det \begin{pmatrix} x_0, x_1, \dots, x_n \\ u_0, u_1, \dots, u_n \end{pmatrix} = \prod_{i>j}^{n} (x_i - x_j)(s_i - s_j) \Big/ \prod_{i,j=0}^{n} (x_i + s_j);$$

see Davis [92, Lemma 11.3.1].

As shown by Haar, the Haar condition is necessary and sufficient for the best approximation problem in maximum norm to have a unique solution. (Note that because the $\|\cdot\|_\infty$ is not a strict norm for $C[a, b]$, this does not directly follow from general principles.)

Theorem 4.5.25.

Let $U = \operatorname{span}(u_0, u_1, \dots, u_n)$ be a Haar subspace of $C[a, b]$. Then for any $f \in C[a, b]$ there is a unique best uniform approximation $u \in U$.

Even when the alternating property is only approximately satisfied, the following theorem by de la Vallée-Poussin still gives a practical estimate for how good an approximation it is.

Theorem 4.5.26.

Let $f(x)$ and $u_k(x)$, $k = 0 : n$, be real-valued continuous functions defined on the interval $I = [a, b]$ which satisfy the Haar condition on I. Assume that an approximation $\tilde{u}(x) \in U$ has the alternating property

$$(f(x_j) - \tilde{u}(x_j))(-1)^j \sigma = \mu_j > 0 \tag{4.5.74}$$

at $(n + 2)$ distinct points $x_0, \dots, x_{n+1} \in I$, where $\sigma = \pm 1$. Let $\eta(f, D) = \inf_{u \in U} \|f - u\|_{\infty,D}$ denote the error of the best approximation on the domain D. Then it holds that

$$\mu := \min_{0 \le j \le n+1} \mu_j \le \eta(f, X) \le \eta(f, I) \le \|f - \tilde{u}\|_{\infty,I}, \tag{4.5.75}$$

$$\min_{0 \le j \le n+1} \mu_j \le \eta(f, X) \le \max_{0 \le j \le n+1} \mu_j = \|f - \tilde{u}\|_{\infty,X}. \tag{4.5.76}$$

Proof. Since $X \subset I$ we have $\eta(f, X) \le \eta(f, I)$. It then suffices to show that for any $u \in U$, the inequality $\mu \le \|f - u\|_{\infty,X}$ holds. This follows from

$$\begin{aligned}\mu \le \mu_j &= (f(x_j) - \tilde{u}(x_j))(-1)^j\sigma \\ &= (f(x_j) - u(x_j))(-1)^j\sigma + (u(x_j) - \tilde{u}(x_j))(-1)^j\sigma \\ &\le (f(x_j) - u(x_j))(-1)^j\sigma \le \|f - u\|_{\infty,X}\end{aligned}$$

for those $x_j \in X$ for which $(u(x_j) - \tilde{u}(x_j))(-1)^j\sigma \le 0$. Such an x_j must exist since otherwise $u - \tilde{u}$ has at least $n + 1$ zeros and does not vanish. □

It follows that an approximation $\tilde{u}$ with equality in (4.5.76) is the best approximation on X. If X is chosen so that equality also holds in (4.5.75), then $\tilde{u}$ is the best approximation on the whole interval $I = [a, b]$.

The alternating property of the best approximation plays a key role in the solution of the best approximation problem. The idea is to solve a sequence of discrete uniform approximation problems each using $(n + 2)$ reference points:

$$X = \{x_0, \dots, x_{n+1}\}, \quad a \le x_0 \le x_1 \cdots \le x_{n+1} \le b.$$

We therefore now consider how we can determine $\tilde{u} = u^*$ so that equality in (4.5.76) is attained. For a given set of points $X = x_0, \dots, x_{n+1}$ assume that u^* has the alternating property

$$(f(x_j) - u^*(x_j))(-1)^{n+1-j}\eta, \quad j = 0 : n + 1, \tag{4.5.77}$$

where η is to be determined. We extend the function $u_0, \dots, u_n$ with the function $u_{n+1}(x_j) = (-1)^j$ defined on X. Then we obtain a linear system $\tilde{U}_{n+1}c = f$, or

$$\begin{pmatrix} u_0(x_0) & \cdots & u_n(x_0) & 1 \\ \vdots & \ddots & \vdots & \vdots \\ u_0(x_n) & \cdots & u_n(x_n) & (-1)^n \\ u_0(x_{n+1}) & \cdots & u_n(x_{n+1}) & (-1)^{n+1} \end{pmatrix} \begin{pmatrix} c_0 \\ \vdots \\ c_n \\ \eta \end{pmatrix} = \begin{pmatrix} f_0 \\ \vdots \\ f_n \\ f_{n+1} \end{pmatrix}. \tag{4.5.78}$$

The first leading submatrix of order $n + 1$ of $\tilde{U}_{n+1}$ is nonsingular since $u_0, \dots, u_n$ form a Chebyshev system. Hence its first $n + 1$ columns are linearly independent. Further, there cannot exist a linear combination of the first $n + 1$ columns that is parallel to the last column since this would imply that some function $u \in U$ has $n+1$ zeros in $[a, b]$. It follows that the system (4.5.78) is nonsingular and has a unique solution. Note that once d has been determined, then the remaining coefficients c_i, $i = 0 : n$, can be computed by solving an interpolating problem.

In the special case of polynomial approximation the system (4.5.78) can be solved in $O(n^2)$ flops. It can be shown that η can be determined as the quotient between two divided differences

$$\eta = [x_0, x_1, \dots, x_{n+1}]f/[x_0, x_1, \dots, x_{n+1}]s, \tag{4.5.79}$$

where $s(x)$ is a function such that $s(x_i) = (-1)^i$. The Newton form of the solution $p(x)$ is then given by

$$p(x) = \sum_{j=0}^{n} [x_0, \dots, x_j](f - \eta s) \prod_{k=0}^{j-1} (x - x_k). \tag{4.5.80}$$

The **Remez exchange algorithm**, proposed by Remez [300] in 1934, is an iterative method for computing the best uniform approximation on an interval $[a, b]$. (Sometimes the name is transcribed as Remes.) In this method one proceeds as follows:

1. Choose an initial reference set X consisting of $(n + 2)$ points $a \le x_0 < x_1 < \cdots < x_{n+1} \le b$.
2. Compute the (unique) discrete approximation $\tilde{u}_X$ to $f(x)$ on the reference set.
3. Adjust the points in the reference set to be the extreme points of the error curve $f(x) - \tilde{u}_X(x)$.
4. Repeat steps 2 and 3 until convergence.

Usually the initial reference set is chosen as Chebyshev points transformed to the interval $[a, b]$. The Remez algorithm can also be applied to the discrete case where f is only defined on a grid $\xi_0, \xi_1, \dots, \xi_m$, $m \gg n$. In this case the Remez algorithm converges in a finite number of steps.

In the continuous case, step 3, one can use Newton's method on the equation $f(x) - \tilde{u}_X(x) = 0$, with starting values equal to x_i for all interior points. If derivatives are not available, then a combination of inverse parabolic interpolation and golden section search given by Brent [44, Chap. 5] is a good alternative; see Sec. 6.4.3.

A discussion of the exchange algorithm and its convergence can be found in Meinardus [264]. A careful implementation of the Remez algorithm, where the delicate points to consider in an implementation are discussed, is given by Golub and Smith [168]. Assuming that $f(x)$ is sufficiently smooth quadratic, convergence is achieved.

Review Questions

4.5.1 State the axioms that any norm must satisfy. Define the maximum norm and the Euclidean norm for a continuous function f on a closed interval.

4.5.2 Define dist(f, P_n), and state Weierstrass' approximation theorem.

4.5.3 Prove the Pythagorean theorem in an inner product space.

4.5.4 Define and give examples of orthogonal systems of functions.

4.5.5 Formulate and prove Bessel's inequality and Parseval's identity, and interpret them geometrically.

4.5.6 (a) Give some reasons for using orthogonal polynomials in polynomial approximation with the method of least squares.

(b) Give some argument against the assertion that orthogonal polynomials are difficult to work with.

4.5.7 In Chebyshev interpolation we seek the coefficients c_j such that

$$p(x_k) = \sum_{j=0}^{m} c_j T_j(x_k), \quad k = 0 : m,$$

where x_k are the zeros of $T_{m+1}(x)$. How would you compute c_j?

4.5.8 The Gram polynomials are examples of orthogonal polynomials. With respect to what inner product are they orthogonal?

4.5.9 Let $\phi_k(x)$, $k = 1, 2, 3, \dots$, be a triangle family of orthogonal polynomials in an inner product space. What property can be used to generate this sequence of polynomials?

4.5.10 What is the characteristic property of the nth degree polynomial which best approximates a given continuous function in the maximum norm over a closed interval?

4.5.11 Give at least three examples of a Chebyshev system of functions.

Problems and Computer Exercises

4.5.1 Compute $\|f\|_\infty$ and $\|f\|_2$ for the function $f(x) = (1 + x)^{-1}$ on the interval $[0, 1]$.

4.5.2 Determine straight lines which approximate the curve $y = e^x$ such that

(a) the discrete Euclidean norm of the error function on the grid $(-1, -0.5, 0, 0.5, 1)$ is as small as possible;

(b) the Euclidean norm of the error function on the interval $[-1, 1]$ is as small as possible;

(c) the line is tangent to $y = e^x$ at the point $(0, 1)$, i.e., the Taylor approximation at the midpoint of the interval.

4.5.3 Determine, for $f(x) = \pi^2 - x^2$, the "cosine polynomial" $f^* = \sum_{j=0}^{n} c_j \cos jx$ which makes $\|f^* - f\|_2$ on the interval $[0, \pi]$ as small as possible.

4.5.4 (a) Show that on any interval containing the points $-1, -1/3, 1/3, 1$,

$$E_2(f) \ge \frac{1}{8}\left| f(1) - 3f\left(\frac{1}{3}\right) + 3f\left(-\frac{1}{3}\right) - f(-1)\right|.$$

(b) Compute the above bound and the actual value of $E_2(f)$ for $f(x) = x^3$.

4.5.5 (a) Let a scalar product be defined by $(f, g) = \int_a^b f(x)g(x)\,dx$. Calculate the matrix of normal equations, when $\phi_j(x) = x^j$, $j = 0 : n$, when $a = 0$, $b = 1$.

(b) Do the same when $a = -1$, $b = 1$. Show how in this case the normal equations can be easily decomposed into two systems, with approximately $(n+1)/2$ equations in each.

4.5.6 Verify the formulas for $\|\phi_j\|^2$ given in Example 4.5.10.

4.5.7 (a) Show that $\|f - g\| \ge \|f\| - \|g\|$ for all norms. (Use the axioms mentioned in Sec. 4.5.1.)

(b) Show that if $\{c_j\}_0^n$ is a set of real numbers and if $\{f_j\}_0^n$ is a set of vectors, then $\|\sum c_j f_j\| \le \sum |c_j| \|f_j\|$.

4.5.8 Let $G \in \mathbf{R}^{n\times n}$ be a symmetric positive definite matrix. Show that an inner product is defined by the formula $(u, v) = u^T Gv$. Show that $A^* = G^{-1}A^T G$.

4.5.9 In a space of complex-valued twice differentiable functions of t which vanish at $t = 0$ and $t = 1$, let the inner product be

$$(u, v) = \int_0^1 u(t)\bar{v}(t)\, dt.$$

What is the adjoint of the operator $A = d/dt$? Is it true that the operator iA is self-adjoint, and that $-A^2$ is self-adjoint and positive definite?

4.5.10 (a) Show that, in a real inner-product space,

$$4(u, v) = \|u + v\|^2 - \|u - v\|^2.$$

In a complex space this gives only the real part of the inner product. Show that one has to add $\|u - iv\|^2 - \|u + iv\|^2$.

(b) This can be used to reduce many questions concerning inner products to questions concerning norms. For example, in a general inner-product space a **unitary operator** is defined by the requirement that $\|Au\| = \|u\|$ for all u. Show that $(Au, Av) = (u, v)$ for all u, v.

Note, however, that the relation $(u, Au) = (Au, u)$ for all u which, in a real space, holds for every operator A, does *not* imply that $(u, Av) = (Au, v)$ for all u, v. The latter holds only if A is self-adjoint.

4.5.11 Show that $(AB)^* = B^*A^*$. Also show that if C is self-adjoint and positive definite, then A^*CA is so too. (A is not assumed to be self-adjoint.)

4.5.12 Show that

$$(A^{-1})^* = (A^*)^{-1}, \qquad (A^p)^* = (A^*)^p,$$

for all integers p, provided that the operators mentioned exist. Is it true that C^p is self-adjoint and positive definite if C is so?

4.5.13 Show the following **minimum property** of orthogonal polynomials: Among all nth-degree polynomials p_n with leading coefficient 1, the smallest value of

$$\|p_n\|^2 = \int_a^b p_n^2(x)w(x)dx, \quad w(x) \geq 0,$$

is obtained for $p_n = \phi_n/A_n$, where ϕ_n is the orthogonal polynomial with leading coefficient A_n associated with the weight distribution $w(x)$.

Hint: Determine the best approximation to x^n in the above norm or consider the expansion $p_n = \phi_n/A_n + \sum_{j=0}^{n-1} c_j\phi_j$.

4.5.14 Verify the formulas for $\|T_j\|^2$ given in Theorem 4.5.20.

4.5.15 Modify Clenshaw's algorithm to a formula for the derivative of an orthogonal expansion.

4.5.16 (a) Let α_j, $j = 1 : n$, be the zeros of the Chebyshev polynomial $T_n(x)$, $n \geq 1$. (There are, of course, simple trigonometric expressions for them.) Apply Clenshaw's algorithm to compute

$$\sum_{m=0}^{n-1} T_m(\alpha_1)T_m(x), x = \alpha_j, j = 1 : n.$$

It turns out that the results are remarkably simple.

(b) Show that $S = \sum_{k=0}^{n-1} c_k\phi_k$ can be computed by a forward version of Clenshaw's algorithm that reads

$$\begin{aligned}
&y_{-2} = 0; \quad y_{-1} = 0;\\
&\textbf{for } k = 0 : n-1,\\
&\quad y_k = (-y_{k-2} + \alpha_k y_{k-1} + c_k)/\gamma_{k+1};\\
&\textbf{end}\\
&S = c_n\phi_n + \gamma_n y_{n-1}\phi_{n-1} - y_{n-2}\phi_n.
\end{aligned}$$

Add this version as an option to your program, and study [294, Sec. 5.4], from which this formula is quoted (with adaptation to our notation). Make some test example of your own choice.

4.5.17 (a) Write a MATLAB function `c = stieltjes(f,x,w,m,n)` that computes the orthogonal coefficients in a least squares polynomial fit to the data (f_i, x_i) and weights w_i, $i = 0 : m$. Compute the orthogonal polynomials ϕ_k using the Stieltjes procedure. For computing c_k, $k = 0 : n$, use either (4.5.67) or (4.5.68).

(b) Apply the function in (a) to the case $f_i = x_i^7$, $w_i = 1/(f_i)^2$, and $m = 20$. Compute and print the error $\|p_k - f\|$ for $k = 0 : 10$, using the expression (4.5.67) for c_{k+1}. Note that for $k > 7$ the fit should be exact.

(c) Repeat the calculations in (b) using the modified formula (4.5.68). Compare the error with the results in (b).

4.5.18 (a) What first-degree polynomial does one get with Chebyshev interpolation to $f(x) = 1/(3+x)$, $[-1, 1]$? What is the maximum norm of the error?

Answer: $p(x) = (-2x + 6)/17$; 0.0294 attained at $x = -1$.

(b) Determine the first-degree polynomial which best approximates the function $f(x) = 1/(3 + x)$, $[-1, 1]$, in the maximum norm. Determine $E_1(f)$ to at least three decimals.

Hint: The error has maximum magnitude at the two endpoints of the interval and at one inner point.

Answer: $p(x) = (-x + 2\sqrt{2})/8$; $E_1(f) = 0.0214$.

4.5.19 (a) Show that the three conditions in Definition 4.5.24 are equivalent.

(b) Show the formula for eta given in (4.5.79).

4.5.20 Determine, using the Remez algorithm, the second-degree polynomial $p(x) = c_0 + c_1x + c_2x^24$ that best approximates $f(x) = \sin\left(\frac{\pi}{2}x\right)$ on $I = [0, 1]$.

Hint: Since the second derivative f'' has constant sign in I, we can take $x_0 = a$, $x_3 = b$, and only the two interior points x_1 and x_2 are unknown.

4.6 Fourier Methods

Many natural phenomena, for example, acoustical and optical, are of a periodic character. For instance, it was known already by Pythagoras (600 B.C.) that a musical sound is composed of regular oscillations, partly a fundamental tone with a certain frequency f, and partly overtones with frequencies $2f, 3f, 4f, \ldots$. The ratio of the strength of the fundamental tone to that of the overtones is decisive for our impression of the sound. Sounds which are free from overtones occur, for instance, in electronic music, where they are called pure sine tones.

In an electronic oscillator, a current is generated whose strength at time t varies according to the formula $r \sin(\omega t + v)$, where r is called the amplitude of the oscillation; ω is called the angular frequency and is equal to 2π times the frequency; and v is a constant which defines the state at the time $t = 0$. In a loudspeaker, variations of current are converted into variations in air pressure which, under ideal conditions, are described by the same function. In practice, however, there is always a certain distortion; overtones occur. The variations in air pressure which reach the ear can, from this viewpoint, be described as a sum of the form

$$\sum_{k=0}^{\infty} r_k \sin(k\omega t + v_k). \tag{4.6.1}$$

An expansion of this form is called a **Fourier series**.

The separation of a periodic phenomenon into a fundamental tone and overtones permeates not only acoustics, but also many other areas. It is related to an important, purely mathematical theorem first given by Fourier.[157] According to this theorem, every function $f(t)$ with period $2\pi/\omega$ can, under certain very general conditions, be expanded in a Fourier series of the form (4.6.1). (A function has period p if $f(t + p) = f(t)$ for all t.) A more precise formulation will be given later in Theorem 4.6.2.

Fourier series are valuable aids in the study of phenomena which are periodic in time (such as vibrations, sound, light, alternating currents) or in space (waves, crystal structure, etc.). One very important area of application is in digital signal and image processing which is used in interpreting radar and sonar signals. Another is statistical time series, which are used in communications theory, control theory, and the study of turbulence. For the numerical analyst, Fourier analysis is partly a very common computational task and partly an important aid in the analysis of properties of numerical methods.

Modifications of pure Fourier methods are used as a means of analyzing nonperiodic phenomena; see, e.g., Sec. 4.6.3 (periodic continuation of functions) and Sec. 4.6.5 (Fourier transforms). The approximation of Fourier expansions using sampled data and discrete Fourier analysis is treated in Sec. 4.6.2. The fast Fourier transform (FFT) for discrete Fourier analysis and synthesis is treated Sec. 4.7.1. It has caused a complete change of attitude toward what can be done using discrete Fourier methods.

[157] Jean Baptist Joseph Fourier (1768–1830), French mathematician. In 1807 Fourier completed and read to the Paris Institute his important memoir *Théorie Analytique de la Chaleur* [Analytical Theory of Heat], in which he used what is now called Fourier series. It won a prize competition set by the Institute in 1811, but was not published until 1822.

4.6.1 Basic Formulas and Theorems

The basic formulas and theorems derived in this section rely to a great extent on the theory in Sec. 4.5. An expansion of the form of (4.6.1) can be expressed in many equivalent ways. If we set $a_k = r_k \sin v_k$, $b_k = r_k \cos v_k$, then using the addition theorem for the sine function we can write

$$f(t) = \sum_{k=0}^{\infty}(a_k \cos k\omega t + b_k \sin k\omega t), \tag{4.6.2}$$

where a_k, b_k are real constants. Another form, which is often the most convenient, can be found with the help of Euler's formulas,

$$\cos x = \frac{1}{2}(e^{ix} + e^{-ix}), \qquad \sin x = \frac{1}{2i}(e^{ix} - e^{-ix}), \quad (i = \sqrt{-1}).$$

Here and in what follows i denotes the imaginary unit. Then one gets

$$f(t) = \sum_{k=-\infty}^{\infty} c_k e^{ik\omega t}, \tag{4.6.3}$$

where

$$c_0 = a_0, \quad c_k = \frac{1}{2}(a_k - i\,b_k), \qquad c_{-k} = \frac{1}{2}(a_k + i\,b_k), \quad k \ge 1, \dots. \tag{4.6.4}$$

In the rest of this chapter we shall use the term **Fourier series** to denote an expansion of the form of (4.6.1) or (4.6.3). We shall call the partial sums of the form of these series **trigonometric polynomials**. Sometimes the term **spectral analysis** is used to describe the above methods.

We shall study *functions with period* 2π. These are fully defined by their values on the **fundamental interval** $[-\pi, \pi]$. If a function of t has period L, then the substitution $x = 2\pi t/L$ transforms the function to a function of x with period 2π. We assume that the function can have complex values, since the complex exponential function is convenient for manipulations.

In the **continuous case** the inner product of two complex-valued functions f and g of period 2π is defined in the following way (the bar over g indicates complex conjugation):

$$(f, g) = \int_{-\pi}^{\pi} f(x)\bar{g}(x)dx. \tag{4.6.5}$$

(It makes no difference what interval one uses, as long as it has length 2π—the value of the inner product is unchanged.) As usual the norm of the function f is defined by $\|f\| = (f, f)^{1/2}$. Notice that $(g, f) = \overline{(f, g)}$.

Theorem 4.6.1.

The following orthogonality relations hold for the functions

$$\phi_j(x) = e^{ijx}, \quad j = 0, \pm 1, \pm 2, \dots,$$

where

$$(\phi_j, \phi_k) = \begin{cases} 2\pi & \text{if } j = k, \\ 0 & \text{if } j \ne k. \end{cases} \tag{4.6.6}$$

Proof. In the continuous case, if $j \neq k$, it holds that

$$(\phi_j, \phi_k) = \int_{-\pi}^{\pi} e^{ijx} e^{-ikx} dx = \Big|_{-\pi}^{\pi} \frac{e^{i(j-k)x}}{i(j-k)} = \frac{(-1)^{j-k} - (-1)^{j-k}}{i(j-k)} = 0,$$

whereby orthogonality is proved. For $j = k$

$$(\phi_k, \phi_k) = \int_{-\pi}^{\pi} e^{ikx} e^{-ikx} dx = \int_{-\pi}^{\pi} 1\, dx = 2\pi. \qquad \square$$

If one knows that the function $f(x)$ has an expansion of the form

$$f = \sum_{j=-\infty}^{\infty} c_j \phi_j,$$

then from Theorem 4.6.1 it follows formally that

$$(f, \phi_k) = \sum_{j=a}^{b} c_j (\phi_j, \phi_k) = c_k (\phi_k, \phi_k), \quad a \leq k \leq b,$$

since $(\phi_j, \phi_k) = 0$ for $j \neq k$. Thus, changing k to j, we have

$$c_j = \frac{(f, \phi_j)}{(\phi_j, \phi_j)} = \frac{1}{2\pi} \int_{-\pi}^{\pi} f(x) e^{-ijx} dx. \tag{4.6.7}$$

These coefficients are called **Fourier coefficients**; see the more general case in Theorem 4.5.13. In accordance with (4.6.4) set

$$a_j = c_j + c_{-j}, \qquad b_j = i(c_j - c_{-j}).$$

Then

$$\begin{aligned}\sum_{j=-N}^{N} c_j e^{ijx} &= c_0 + \sum_{j=1}^{N} \big(c_j (\cos jx + i \sin jx) + c_{-j}(\cos jx - i \sin jx)\big)\\ &= \frac{1}{2} a_0 + \sum_{j=1}^{N} (a_j \cos jx + b_j \sin jx),\end{aligned}$$

where

$$a_j = \frac{1}{\pi} \int_{-\pi}^{\pi} f(x) \cos jx\, dx, \quad j \geq 0, \qquad b_j = \frac{1}{\pi} \int_{-\pi}^{\pi} f(x) \sin jx\, dx, \quad j \geq 1. \tag{4.6.8}$$

(Notice that the factors preceding the integral are different in the expressions for c_j and for a_j, b_j, respectively.)

From a generalization of Theorem 4.5.13, we also know that the error

$$\Big\| f - \sum_{j=-n}^{n} k_j \phi_j \Big\|, \quad n < \infty,$$

becomes as small as possible if we choose $k_j = c_j$, $-n \le j \le n$. Theorem 4.5.14 and its corollary, Parseval's identity,

$$2\pi \sum_{j=-\infty}^{\infty} |c_j|^2 = \|f\|^2 = \int_{-\pi}^{\pi} |f(x)|^2 \, dx, \tag{4.6.9}$$

are of great importance in many applications of Fourier analysis. The integral in (4.6.9) can be interpreted as the "energy" of the function $f(x)$.

Theorem 4.6.2 (*Fourier Analysis, Continuous Case*).

Assume that the function f is defined at every point in the interval $[-\pi, \pi]$ and that $f(x)$ is finite and piecewise continuous. Associate with f a Fourier series in the following two ways:

$$\frac{1}{2}a_0 + \sum_{j=1}^{\infty}(a_j \cos jx + b_j \sin jx) \quad \textit{and} \quad \sum_{j=-\infty}^{\infty} c_j e^{ijx},$$

where the coefficients a_j, b_j, and c_j are defined by (4.6.8) *in the first case and* (4.6.7) *in the second case. Then the partial sums of the above expansions give the best possible approximations to $f(x)$ by trigonometric polynomials, in the least squares sense.*

If f is of bounded variation and has at most a finite number of discontinuities, then the series is everywhere convergent to $f(x)$. At a point $x = a$ of discontinuity $f(a)$ equals the mean $f(a) = \frac{1}{2}(f(a+) + f(a-))$.

Proof. The proof of the convergence results is outside the scope of this book (see, e.g., Courant and Hilbert [83]). The rest of the assertions follow from previously made calculations in Theorem 4.6.1 and the comments following; see also the proof of Theorem 4.5.13. □

The more regular a function is, the faster its Fourier series converges. The following useful result is relatively easy to prove using (4.6.7) and integrating by parts $k+1$ times (cf. (3.2.8)).

Theorem 4.6.3.

If f and its derivatives up to and including order k are periodic and everywhere *continuous, and if $f^{(k+1)}$ is piecewise continuous, then*

$$|c_j| \le \frac{1}{j^{(k+1)}} \|f^{(k+1)}\|_\infty. \tag{4.6.10}$$

Sometimes it is convenient to separate a function f defined on $[-\pi, \pi]$ into an even and an odd part. We set $f(x) = g(x) + h(x)$, where

$$g(x) = \frac{1}{2}(f(x) + f(-x)), \qquad h(x) = \frac{1}{2}(f(x) - f(-x)) \quad \forall x. \tag{4.6.11}$$

Then $g(x) = g(-x)$ and $h(x) = -h(-x)$. For both $g(x)$ and $h(x)$ it suffices to give the function only on $[0, \pi]$. For the even function $g(x)$ the sine part of the Fourier series drops and we have

$$g(x) = \frac{1}{2}a_0 + \sum_{j=1}^{\infty} a_j \cos jx, \quad a_j = \frac{2}{\pi}\int_0^{\pi} g(x) \cos jx\, dx. \tag{4.6.12}$$

For $h(x)$ the cosine part drops out and the Fourier series becomes a sine series:

$$h(x) = \sum_{j=1}^{\infty} b_j \sin jx, \quad b_j = \frac{2}{\pi}\int_0^{\pi} h(x) \sin jx\, dx. \tag{4.6.13}$$

The proof is left as an exercise to the reader (use the formulas for the coefficients given in (4.6.8)).

Example 4.6.1.

Consider the rectangular wave function obtained by periodic continuation outside the interval $(-\pi, \pi)$ of

$$f(x) = \begin{cases} -1/2, & -\pi < x < 0, \\ 1/2, & 0 < x < \pi; \end{cases}$$

see Figure 4.6.1. The function is odd, so $a_j = 0$ for all j, and

$$b_j = \frac{1}{\pi}\int_0^{\pi} \sin jx\, dx = \frac{1}{j\pi}(1 - \cos j\pi) = \begin{cases} 0 & \text{if } j \text{ even}, \\ 2/(j\pi) & \text{if } j \text{ odd}. \end{cases}$$

Hence

$$f(x) = \frac{2}{\pi}\Big(\sin x + \frac{\sin 3x}{3} + \frac{\sin 5x}{5} + \cdots\Big). \tag{4.6.14}$$

Notice that the coefficients c_j decay as j^{-1} in agreement with Theorem 4.6.3.

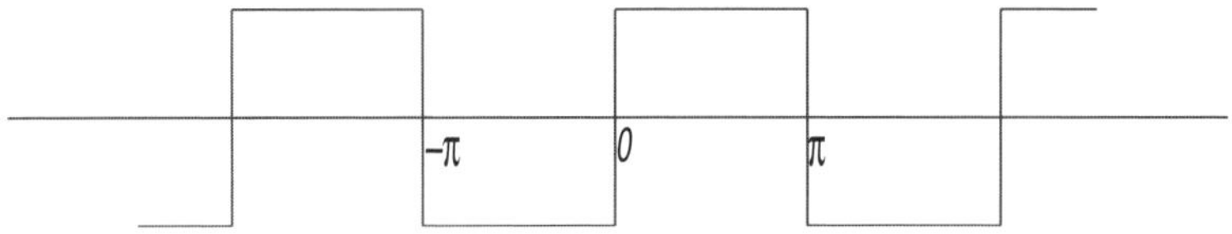

Figure 4.6.1. *A rectangular wave.*

The sum of the series is zero at the points where f has a jump discontinuity; this agrees with the fact that the sum should equal the average of the limiting values to the left and to the right of the discontinuity.

Figure 4.6.2 shows the approximations to the square wave using one, two, five, and ten terms of the series (4.6.14). As can be seen, there is a ringing effect near the discontinuities. The width and energy of this error are reduced when the number of terms in the approximation is increased. However, the height of the overshoot and undershoot near the

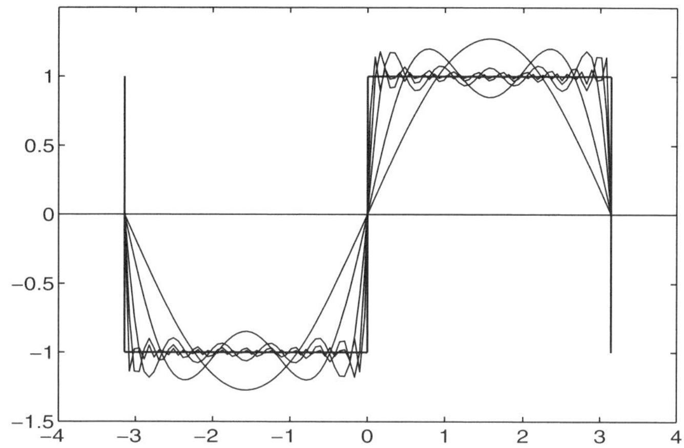

Figure 4.6.2. *Illustration of Gibbs' phenomenon.*

discontinuity converges to a fixed height, which is equal to about 0.179 times the jump in the function value. This artifact is known as **Gibbs' phenomenon.**[158]

The Gibbs' oscillations can be smoothed by multiplying the terms by factors which depend on m, the order of the partial sum. Let us consider the finite expansion

$$f_m(x) = \sum_{k=-(m-1)}^{m-1} c_k e^{ikx}.$$

Then in the smoothed expansion each term in the sum is multiplied by the Lanczos σ-factors,

$$f_m(x) = \sum_{k=-(m-1)}^{m-1} \sigma_k c_k e^{ikx}, \quad \sigma_k = \frac{\sin \pi k/m}{\pi/m} \tag{4.6.15}$$

(see Lanczos [235, Chap. IV, Secs. 6 and 9]). Since the coefficients in the real form of the Fourier series are

$$a_k = c_k + c_{-k}, \qquad b_k = i(c_k - c_{-k}),$$

the same σ-factor applies to them.

The Gibbs' oscillations can also be suppressed by using the epsilon algorithm. For this purpose one adds the conjugate Fourier series, applies the epsilon algorithm, and keeps only the real part of the result.

4.6.2 Discrete Fourier Analysis

Although the data to be treated in Fourier analysis are often continuous in the time or space domain, for computational purposes these data must usually be represented in terms of a

[158] Named after the American physicist J. William Gibbs, who in 1899 proved that this ringing effect will always occur when approximating a discontinuous function with a Fourier series.

finite discrete sequence. For example, a function $f(t)$ of time is recorded at evenly spaced intervals Δt in time. Assume that the function f is known at equidistant arguments in the interval $[0, 2\pi]$,

$$x_k = 2\pi k/N, \quad k = 0 : N-1.$$

Such data can be analyzed by discrete Fourier analysis. Define the inner product

$$(f, g) = \sum_{k=0}^{N-1} f(x_k)\bar{g}(x_k), \quad x_k = 2\pi k/N. \tag{4.6.16}$$

Then, with $\phi_j(x) = e^{ijx}$ we have

$$(\phi_j, \phi_k) = \sum_{k=0}^{N-1} e^{ijx_k} e^{-ikx_k} = \sum_{k=0}^{N-1} e^{i(j-k)hk}.$$

From Lemma 3.2.2 it now follows that

$$(\phi_j, \phi_k) = \begin{cases} N & \text{if } (j-k)/N \text{ is an integer,} \\ 0 & \text{otherwise.} \end{cases} \tag{4.6.17}$$

Theorem 4.6.4 (*Trigonometric Interpolation*).

Every function, defined on the equidistant grid $x_k = 2\pi k/N$, $k = 0 : N-1$, can be interpolated by the trigonometric polynomial

$$f(x) = \begin{cases} \displaystyle\sum_{j=-k}^{k+\theta} c_j e^{ijx}, \\ \displaystyle\frac{1}{2}a_0 + \sum_{j=1}^{k} (a_j \cos jx + b_j \sin jx) + \frac{1}{2}\theta a_{k+1} \cos(k+1)x. \end{cases} \tag{4.6.18}$$

Here

$$\theta = \begin{cases} 1 & \text{if } N \text{ even,} \\ 0 & \text{if } N \text{ odd,} \end{cases} \qquad k = \begin{cases} N/2 - 1 & \text{if } N \text{ even,} \\ (N-1)/2 & \text{if } N \text{ odd,} \end{cases} \tag{4.6.19}$$

and

$$c_j = \frac{1}{N}\sum_{k=0}^{N-1} f(x_k) e^{-ijx_k}, \tag{4.6.20}$$

$$a_j = \frac{2}{N}\sum_{k=0}^{N-1} f(x_k)\cos jx_k, \qquad b_j = \frac{2}{N}\sum_{k=0}^{N-1} f(x_k)\sin jx_k. \tag{4.6.21}$$

If the sums in (4.6.18) *are terminated when $|j| < k + \theta$, then one obtains the trigonometric polynomial which is the best least squares approximation, among all trigonometric polynomials with the same number of terms, to f on the grid.*

Proof. The expression for c_j is justified by (4.6.17). Further, by (4.6.20)–(4.6.21) it follows that

$$a_j = c_j + c_{-j}, \qquad b_j = i(c_j - c_{-j}), \qquad c_{k+1} = \frac{1}{2}a_{k+1}.$$

The two expressions for $f(x)$ are equivalent, because

$$\begin{aligned}\sum_{j=-k}^{k+\theta} c_j e^{ijx} &= c_0 + \sum_{j=1}^{k}\big(c_j(\cos jx + i\sin jx) + c_{-j}(\cos jx - i\sin jx)\big)\\ &\quad + \theta c_{k+1}\cos(k+1)x\\ &= c_0 + \sum_{j=1}^{k}(a_j\cos jx + b_j\sin jx) + \frac{1}{2}\theta a_{k+1}\cos(k+1)x. \qquad \square\end{aligned}$$

The function $f(x)$ coincides *on the grid* $x_0, x_1, \ldots, x_{N-1}$ with the function

$$f^*(x) = \sum_{j=0}^{N-1} c_j e^{ijx} \tag{4.6.22}$$

because

$$e^{-i(N-j)x_k} = e^{ijx_k}, \qquad c_{-j} = c_{N-j}.$$

However, the functions f and f^* are *not* identical between the grid points. If we set $\omega = e^{ix}$ and $\omega_k = e^{ix_k}$, then

$$f^*(x) = P(\omega) = \sum_{j=0}^{N-1} c_j\omega^j,$$

where $P(\omega)$ is a polynomial of degree less than N. It becomes clear that trigonometric interpolation is equivalent to polynomial interpolation at the grid points ω_k. The mapping $\mathbf{C}^N \to \mathbf{C}^N$

$$(f_0, f_1, \ldots, f_{N-1}) \mapsto (c_0, c_1, \ldots, c_{N-1})$$

is called the **discrete Fourier transform** (DFT).

The calculations required to compute the coefficients c_j according to (4.6.20), **Fourier analysis**, are of essentially the same type as the calculations needed to compute $f^*(x)$ at the grid points

$$x_k = 2\pi k/N, \quad k = 0 : N-1,$$

when the expansion in (4.6.22) is known, so-called **Fourier synthesis**. Both calculations can be performed very efficiently using FFT algorithms; see Sec. 4.7.1.

Functions of several variables are treated analogously. Quite simply, one takes one variable at a time. In the discrete case with two variables we set

$$x_k = 2\pi k/N, \qquad y_\ell = 2\pi \ell/N,$$

and assume that $f(x_k, y_\ell)$ is known for $k = 0 : N-1$, $\ell = 0 : N-1$. Set

$$\begin{aligned} c_j(y_\ell) &= \frac{1}{N}\sum_{k=0}^{N-1} f(x_k, y_\ell)e^{-ijx_k},\\ c_{j,k} &= \frac{1}{N}\sum_{\ell=0}^{N-1} c_j(y_\ell)e^{-iky_\ell}.\end{aligned}$$

From Theorem 4.6.4, then (with obvious changes in notations)

$$c_j(y_\ell) = \sum_{k=0}^{N-1} c_{j,k} e^{iky_\ell},$$

$$f(x_k, y_\ell) = \sum_{j=0}^{N-1} c_j(y_\ell) e^{ijx_k} = \sum_{j=0}^{N-1} \sum_{k=0}^{N-1} c_{j,k} e^{(ijx_k + iky_\ell)}.$$

The above expansion is of considerable importance in, e.g., crystallography.

The Fourier coefficients

$$c_j(f) = \frac{1}{2\pi} \int_{-\pi}^{\pi} f(x) e^{ijx}\, dx, \quad j = 0, \pm 1, \pm 2, \dots, \tag{4.6.23}$$

of a function f with period 2π are often difficult to compute. On the other hand the coefficients of the DFT

$$\hat{c}_j(f) = \frac{1}{N} \sum_{k=0}^{N-1} f(x_k) e^{-ijx_k}, \quad x_k = \frac{2\pi k}{N}, \quad j = 0 : N-1, \tag{4.6.24}$$

can be computed very efficiently by the FFT. Now, since $f(x_0) = f(x_N)$, the sum in (4.6.24) can be thought of as a trapezoidal sum

$$\hat{c}_j(f) = \frac{1}{N} \left[\frac{1}{2} f(x_0) + f(x_1) e^{-ijx_1} + \cdots + f(x_{N-1}) e^{-ijx_{N-1}} + \frac{1}{2} f(x_N) \right]$$

approximating the integral (4.6.23). Therefore, one might think of using $\hat{c}_j$ as an approximation to c_j for all $j = 0, \pm 1, \pm 2, \dots$. But $\hat{c}_j(f)$ are periodic in j with period N, whereas by Theorem 4.6.3 the true Fourier coefficients $c_j(f)$ decay as some power $j^{-(k+1)}$ as $j \to \infty$.

We now show a way to remove this deficiency. Let f_k, $k = 0, \pm 1, \pm 2, \dots$, be an infinite N-periodic sequence with $f_k = f(x_k)$. Let $\varphi = Pf$ be a continuous function such that $\varphi(x_k) = f_k$, $k = 0, \pm 1, \pm 2, \dots$, and approximate $c_j(f)$ by $c_j(\varphi)$. Then $c_j(\varphi)$ will decay to zero as $j \to \infty$. It is a remarkable fact that if the approximation scheme P is linear and translation invariant, we have

$$c_j(\varphi) = \tau_j \hat{c}_j(f), \tag{4.6.25}$$

where the **attenuation factors** τ_j *depend only on the approximation scheme* P *and not the function* f. This implies that τ_j can be determined from (4.6.25) by evaluating $c_j(\varphi)$ and $\hat{c}_j(f)$ for some suitable sequence f_k, $k = 0, \pm 1, \pm 2, \dots$. This allows the FFT to be used. For a proof of the above result and a more detailed exposition of attenuation factors in Fourier analysis, we refer to Gautschi [141].

Example 4.6.2.

For a given N-periodic sequence f_k, $k = 0, \pm 1, \pm 2, \dots$, take $\varphi(x) = Pf$ to be defined as the piecewise linear function such that $\varphi(x_k) = f_k$, $k = 0, \pm 1, \pm 2, \dots$. Clearly this approximation scheme is linear and translation invariant. Further, the function φ is continuous and has period 2π.

Consider in particular the sequence $f_k = 1$, $k = 0 \bmod N$, and $f_k = 0$ otherwise. We have

$$\hat{c}_j(f) = \frac{1}{N}\sum_{k=0}^{N-1} f(x_k)e^{-ijx_k} = \frac{1}{N}.$$

Further, setting $h = 2\pi/N$ and using symmetry, we get

$$\begin{aligned} c_j(Pf) &= \frac{1}{2\pi}\int_{-h}^{h}\left(1 - \frac{|x|}{h}\right)e^{ijx}\,dx \\ &= \frac{1}{\pi}\int_0^h \left(1 - \frac{x}{h}\right)\cos jx\,dx = \frac{2}{j^2\pi h}\sin^2\left(\frac{jh}{2}\right). \end{aligned}$$

This gives the attenuation factors

$$\tau_j = \left(\frac{\sin(j\pi/N)}{j\pi/N}\right)^2, \qquad j = 0, \pm 1, \pm 2, \ldots .$$

Note that the coefficients $c_j(Pf)$ decay as j^{-2}, reflecting the fact that the first derivative of Pf is discontinuous. If instead we use a cubic spline approximation, the first and second derivatives are continuous and the attenuation factors decay as j^{-4}.

4.6.3 Periodic Continuation of a Function

Assume now that the function f is from the beginning defined only on the interval $[0, \pi]$. In that case we can extend the function to $[-\pi, 0]$ as either an even or an odd function. Hence the same function can be expanded as either a sine series or a cosine series. Both expansions will converge to the given $f(x)$ in the interval $[0, \pi]$ provided f satisfies the conditions in Theorem 4.6.2. However, the rate of convergence of the two Fourier expansions may be vastly different depending on continuity properties at the points $x = 0$ and $x = \pi$ of the original interval.

If the function f defined in $[0, \pi]$ satisfies $f(0) = f(\pi) = 0$, it can be continued as an odd function (see Figure 4.6.3). Its Fourier expansion then becomes a sine series. In the continuous case,

$$\sum_{j=1}^{\infty} b_j \sin jx, \qquad b_j = \frac{2}{\pi}\int_0^{\pi} f(x)\sin jx\,dx. \tag{4.6.26}$$

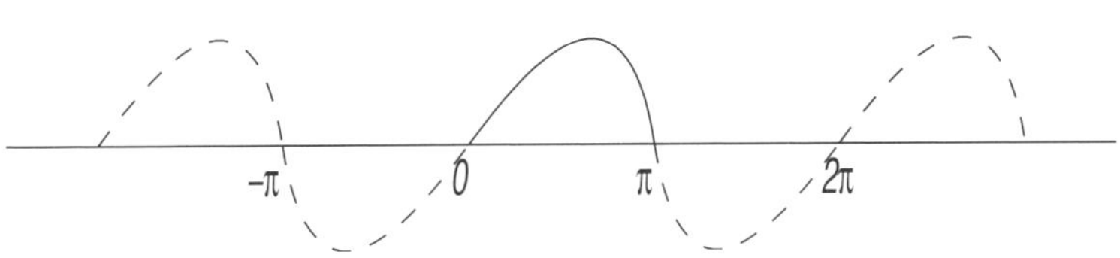

Figure 4.6.3. *Periodic continuation of a function f outside $[0, \pi]$ as an odd function.*

In the discrete case ($x_k = \pi k/N$),

$$\sum_{j=1}^{N-1} b_j \sin jx, \qquad b_j = \frac{2}{N}\sum_{k=1}^{N-1} f(x_k)\sin jx_k. \tag{4.6.27}$$

The reflection as an odd function preserves the continuity in function values and the first derivative in the extended function. According to Theorem 4.6.3 the Fourier coefficients b_j will therefore decrease as j^{-3}.

If $f(0) \neq 0$, one can still use such an expansion but the function will cause a discontinuity at $x = 0$ and $x = \pi$. The discontinuity at $x = 0$ follows from the reflection; the discontinuity at $x = \pi$ follows from the periodicity condition $f(x + 2\pi) = f(x)$ which gives $f(\pi) = f(-\pi)$. This will cause slow convergence (as j^{-1}) of the Fourier coefficients and cause undesirable oscillations due to Gibbs' phenomenon. However, by subtracting a linear function $a + bx$ from f we can always ensure that the condition $f(0) = f(\pi) = 0$ is satisfied.

If $f(0) \neq 0$ and one makes a continuation of f into an *even* function on $[-\pi, \pi]$, the extended function will be continuous at $x = 0$ and $x = \pi$. The Fourier series then becomes a pure cosine series In the continuous case,

$$\frac{1}{2}a_0 + \sum_{j=1}^{\infty} a_j \cos jx, \qquad a_j = \frac{2}{\pi}\int_0^{\pi} f(x)\cos jx\,dx. \tag{4.6.28}$$

In the discrete case ($x_k = \pi k/N$),

$$\frac{1}{2}a_0 + \sum_{j=1}^{N-1} a_j \cos jx, \qquad a_j = \frac{2}{N}\sum_{k=0}^{N-1} f(x_k)\cos jx_k. \tag{4.6.29}$$

The coefficients a_j will decrease as j^{-2}. If $f'(0) = f'(\pi) = 0$, then the first derivative will also be continuous. One can still use such an expansion even if $f'(0) \neq 0$ or $f'(\pi) \neq 0$, but then a discontinuity appears in the first derivative.

4.6.4 Convergence Acceleration of Fourier Series

The generalized Euler transformation described in Sec. 3.4.3 can be used for accelerating the convergence of Fourier series, except in the immediate vicinity of singular points. Consider a complex power series

$$S(z) = \sum_{n=1}^{\infty} u_n z^{n-1}, \qquad z = e^{i\phi}. \tag{4.6.30}$$

A Fourier series that is originally of the form $\sum_{n=-\infty}^{\infty} c_n e^{in\phi}$, or in trigonometric form, can easily be brought to this form; see Problem 4.6.7.

We consider the case

$$S(z) = -\frac{1}{z}\log(1 - z) = \sum_{n=1}^{\infty} \frac{1}{n} z^{n-1}, \tag{4.6.31}$$

which is typical for a power series with completely monotonic terms. (The rates of convergence are the same for almost all series of this class.) Numerical computation, essentially by the above algorithm, gave the following results. The coefficients u_j are computed in IEEE double precision arithmetic. We make the rounding errors during the computations less important by subtracting the first row of partial sums by its last element; it is, of course, added again to the final result.[159] The first table shows, for various ϕ, the most accurate result that can be obtained without thinning. These limits are due to the rounding errors; we can make the pure truncation error arbitrarily small by choosing N large enough.

ϕ	π	$2\pi/3$	$\pi/2$	$\pi/3$	$\pi/4$	$\pi/8$	$\pi/12$	$\pi/180$
$\lvert\text{error}\rvert$	$2\cdot 10^{-16}$	$8\cdot 10^{-16}$	10^{-14}	$6\cdot 10^{-12}$	10^{-9}	$5\cdot 10^{-7}$	$3\cdot 10^{-5}$	$2\cdot 10^{-1}$
N	30	33	36	36	36	40	40	100
kk	21	22	20	21	20	13	10	(3)

τ	80	120	90	15
$\tau\cdot\phi$	π	$2\pi/3$	$\pi/2$	$\pi/12$
$\lvert\text{error}\rvert$	$2\cdot 10^{-14}$	10^{-14}	$3\cdot 10^{-13}$	$3\cdot 10^{-5}$
N	28	31	33	41
kk	20	22	18	10
no. terms	5040	3720	2970	615

Note that a rather good accuracy is also obtained for $\phi = \pi/8$ and $\phi = \pi/12$, where the algorithm is "unstable," since $|\frac{z}{1-z}| > 1$. In this kind of computation "instability" does not mean that the algorithm is hopeless, but it shows the importance of a good termination criterion. The question is to navigate safely between Scylla and Charybdis. For a small value such as $\phi = \pi/180$, the sum is approximately $4.1 + 1.5i$. The smallest error with 100 terms (or less) is 0.02; it is obtained for $k = 3$. Also note that kk/N increases with ϕ.

By the application of thinning the results can often be improved considerably for $\phi \ll \pi$, in particular for $\phi = \pi/180$. Let τ be a positive integer. The thinned form of $S(z)$ reads

$$S(z) = \sum_{p=1}^{\infty} u_p^* z^{\tau\cdot(p-1)}, \qquad u_p^* = \sum_{j=1}^{\tau} u_{j+\tau\cdot(p-1)}\, z^{j-1}.$$

The series (4.6.31) has "essentially positive" terms originally that can become "essentially alternating" by thinning. For example, if $z = e^{i\pi/3}$ and $\tau = 3$, the series becomes an alternating series, perhaps with complex coefficients. It does not matter in the numerical work that u_p^* depends on z.

We present the errors obtained for four values of the parameter τ, with different amounts of work. Compare $|error|$, kk, etc. with appropriate values in the table above. We see that, by thinning, it is possible to calculate the Fourier series very accurately for small values of ϕ also.

[159]Tricks like this can often be applied in linear computations with a slowly varying sequence of numbers. See, for example, the discussion of rounding errors in Richardson extrapolation in Sec. 3.4.6.

Roughly speaking, the optimal rate of convergence of the Euler transformation depends on z in the same way for all power series with completely monotonic coefficients, independently of the rate of convergence of the original series. The above tables from a particular example can therefore—with some safety margin—be used as a guide for the application of the Euler transformation with thinning to any series of this class.

Say that you want the sum of a series $\sum u_n z^n$ for $z = e^{i\phi}$, $\phi = \pi/12$, with relative $|error| < 10^{-10}$. You see in the first table that $|error| = 6\cdot 10^{-12}$ for $\phi = \pi/3 = 4\pi/12$ without thinning. The safety margin is, we hope, large enough. Therefore, try $\tau = 4$. We make two tests with completely monotonic terms: $u_n = n^{-1}$ and $u_n = \exp(-\sqrt{n})$. We hope that tol $= 10^{-10}$ is large enough to make the irregular errors relatively negligible. In both tests the actual magnitude of the error turns out to be $4 \cdot 10^{-11}$, and the total number of terms is $4\cdot 32 = 128$. The values of $errest$ are $6 \cdot 10^{-11}$ and $7\cdot 10^{-11}$; both slightly overestimate the actual errors and are still smaller than tol.

4.6.5 The Fourier Integral Theorem

We have seen how Fourier methods can be used on functions defined on a finite interval usually taken to be $[-\pi, \pi]$. Fourier found that expansion of an arbitrary function in a Fourier series remains possible even if the function is defined on an interval that extends on both sides to infinity. In this case the fundamental frequency converges to zero and the summation process changes into one of integration.

Suppose that the function $f(x)$ is defined on the entire real axis, and that it satisfies the regularity properties which we required in Theorem 4.6.2. Set

$$\varphi(\xi) = f(x), \quad \xi = 2\pi x/L \in [-\pi, \pi],$$

and continue $\varphi(\xi)$ outside $[-\pi, \pi]$ so that it has period 2π. By Theorem 4.6.2, if

$$c_j = \frac{1}{2\pi}\int_{-\pi}^{\pi} \varphi(\xi)e^{-ij\xi}\, d\xi = \frac{1}{L}\int_{-L/2}^{L/2} f(x)e^{-2\pi ixj/L}\, dx, \tag{4.6.32}$$

then $\varphi(\xi) = \sum_{j=-\infty}^{\infty} c_j e^{ij\xi}$, $\xi \in (-\pi, \pi)$, and hence

$$f(x) = \sum_{j=-\infty}^{\infty} c_j e^{2\pi ixj/L}, \quad x \in \left(-\frac{L}{2}, \frac{L}{2}\right).$$

If we set

$$g_L(\omega) = \int_{-L/2}^{L/2} f(x)e^{-2\pi ix\omega}\, dx, \quad \omega = \frac{j}{L}, \tag{4.6.33}$$

then by (4.6.32) we have $c_j = (1/L)g_L(\omega)$, and hence

$$f(x) = \frac{1}{L}\sum_{j=-\infty}^{\infty} g_L(\omega)e^{2\pi ix\omega}, \quad x \in \left(\frac{-L}{2}, \frac{L}{2}\right). \tag{4.6.34}$$

Now by passing to the limit $L \to \infty$, one avoids making an artificial periodic continuation outside a finite interval. The sum in (4.6.34) is a "sum of rectangles" similar to the sum which appears in the definition of a definite integral. But here the argument varies from

$-\infty$ to $+\infty$, and the function $g_L(t)$ depends on L. By a somewhat dubious passage to the limit, then, the pair of formulas (4.6.33) and (4.6.34) becomes the pair

$$g(\omega) = \int_{-\infty}^{\infty} f(x)e^{-2\pi ix\omega}\, dx \iff f(x) = \int_{-\infty}^{\infty} g(\omega)e^{2\pi ix\omega}\, d\omega. \tag{4.6.35}$$

One can, in fact, after a rather complicated analysis, show that the above result is correct; see, e.g., Courant–Hilbert [83]. The proof requires, besides the previously mentioned "local" regularity conditions on f, the "global" assumption that

$$\int_{-\infty}^{\infty} |f(x)|\, dx$$

is convergent. The beautiful, almost symmetric relation of (4.6.35) is called the **Fourier integral theorem**. This theorem, and other versions of it, with varying assumptions under which they are valid, is one of the most important aids in both pure and applied mathematics. The function g is called the **Fourier transform**[160] of f. The Fourier transform is one of the most important tools of applied analysis. It plays a fundamental role in problems relating to input–output relations, e.g., in electrical networks.

Clearly the Fourier transform is a *linear operator*. Another elementary property that can easily be verified is

$$f(ax) \iff \frac{1}{|a|} g\left(\frac{\omega}{a}\right).$$

Whether the function $f(x)$ has even or odd symmetry and is real or purely imaginary leads to relations between $g(\omega)$ and $g(-\omega)$ that can be used to increase computational efficiency. Some of these properties are summarized in Table 4.6.1.

Table 4.6.1. *Useful symmetry properties of the continuous Fourier transform.*

Function	Fourier transform
$f(x)$ real	$g(-\omega) = \overline{g}(\omega)$
$f(x)$ imaginary	$g(-\omega) = -\overline{g}(\omega)$
$f(x)$ even	$g(-\omega) = g(\omega)$
$f(x)$ odd	$g(-\omega) = -g(\omega)$
$f(x)$ real even	$g(\omega)$ real even
$f(x)$ imaginary odd	$g(\omega)$ real odd

Example 4.6.3.

The function $f(x) = e^{-|x|}$ has Fourier transform

$$g(\omega) = \int_{-\infty}^{\infty} e^{-|x|}e^{-2\pi ix\omega}\, dx = \int_{0}^{\infty} \left(e^{-(1+2\pi i\omega)x} + e^{-(1-2\pi i\omega)x}\right) dx$$
$$= \frac{1}{1+2\pi i\omega} + \frac{1}{1-2\pi i\omega} = \frac{2}{1+4\pi^2\omega^2}.$$

[160] The terminology in the literature varies somewhat as to the placement of the factor 2π; it can be taken out of the exponent by a simple change of variable.

Here $f(x)$ is real and an even function. In agreement with Table 4.6.1, the Fourier transform is also real and even.

From (4.6.35) it follows that

$$e^{-|x|} = \int_{-\infty}^{\infty} \frac{2}{1+4\pi^2\omega^2} e^{2\pi i x\omega}\, d\omega = \frac{2}{\pi}\int_0^{\infty} \frac{1}{1+x^2} \cos \pi x\, dx, \qquad (2\pi\omega = x).$$

It is not so easy to prove this formula directly.

Many applications of the Fourier transform involve the use of convolutions.

Definition 4.6.5.

The **convolution** *of f_1 and f_2 is the function*

$$h(\xi) = \operatorname{conv}(f_1, f_2) = \int_{-\infty}^{\infty} f_1(x) f_2(\xi - x)\, dx. \tag{4.6.36}$$

It is not difficult to verify that $\operatorname{conv}(f_1, f_2) = \operatorname{conv}(f_2, f_1)$. The following theorem states that the convolution of f_1 and f_2 can be computed as the inverse Fourier transform of the product $g_1(\omega)g_2(\omega)$. This fact is of great importance in the application of Fourier analysis to differential equations and probability theory.

Theorem 4.6.6.

Let f_1 and f_2 have Fourier transforms g_1 and g_2, respectively. Then the Fourier transform g of the convolution of f_1 and f_2 is the product $g(\omega) = g_1(\omega)g_2(\omega)$.

Proof. By definition the Fourier transform of the convolution is

$$\begin{aligned}
g(\omega) &= \int_{-\infty}^{\infty} e^{-2\pi i\xi\omega}\left(\int_{-\infty}^{\infty} f_1(x) f_2(\xi - x)\, dx\right) d\xi \\
&= \int_{-\infty}^{\infty}\int_{-\infty}^{\infty} e^{-2\pi i(x+\xi-x)\omega} f_1(x) f_2(\xi - x)\, dx\, d\xi \\
&= \int_{-\infty}^{\infty} e^{-2\pi i x\omega} f_1(x)\, dx \int_{-\infty}^{\infty} e^{-2\pi i(\xi-x)\omega} f_2(\xi - x)\, d\xi \\
&= \int_{-\infty}^{\infty} e^{-2\pi i x\omega} f_1(\xi)\, dx \int_{-\infty}^{\infty} e^{-2\pi i x\omega} f_2(x)\, dx = g_1(\omega) g_2(\omega).
\end{aligned}$$

The legitimacy of changing the order of integration is here taken for granted. □

In many physical applications, the following relation, analogous to Parseval's identity (corollary to Theorem 4.5.14), is of great importance. If g is the Fourier transform of f, then

$$\int_{-\infty}^{\infty} |g(\omega)|^2\, d\omega = \int_{-\infty}^{\infty} |f(\xi)|^2\, d\xi. \tag{4.6.37}$$

In signal processing this can be interpreted to mean that the total power in a signal is the same whether computed in the time domain or the frequency domain.

4.6.6 Sampled Data and Aliasing

The ideas of Sec. 4.3.5 can be applied to the derivation of the celebrated **sampling theorem**. This is an interpolation formula that expresses a function that is **band-limited** to the frequency interval $[-W,\ W]$. Such a function has a Fourier representation of the form (see also Strang [339, p. 325])

$$f(z) = \frac{1}{2\pi}\int_{-W}^{W} \hat{f}(k)e^{ikz}\,dk, \quad |\hat{f}(k)| \le M, \tag{4.6.38}$$

in terms of its values at all integer points.

Theorem 4.6.7 (*Shannon's Sampling Theorem*).
Let the function be band-limited to the frequency interval $[-W,\ W]$. *Then*

$$f(z) = \sum_{j=-\infty}^{\infty} f\left(\frac{j\pi}{W}\right)\frac{\sin(Wz - j\pi)}{(Wz - j\pi)}. \tag{4.6.39}$$

Proof. We shall sketch a derivation of this for $W = \pi$. (Strang [339] gives an entirely different derivation, based on Fourier analysis.) We first note that (4.6.38) shows that $f(z)$ is analytic for all z. Then we consider the same Cauchy integral as many times before,

$$I_n(u) = \frac{1}{2\pi i}\int_{\partial\mathcal{D}_n} \frac{\Phi(u)f(z)}{\Phi(z)(z-u)}\,dz, \quad u \in \mathcal{D}_n.$$

Here $\Phi(z) = \sin\pi z$, which vanishes at all integer points, and $\mathcal{D}_n$ is the *open* rectangle with vertices at $\pm(n+1/2) \pm bi$. By the residue theorem, we obtain after a short calculation

$$I_n(u) = f(u) + \sum_{j=-n}^{n} \frac{\Phi(u)f(j)}{\Phi'(j)(j-u)} = f(u) - \sum_{j=-n}^{n} \frac{f(j)\sin\pi(j-u)}{\pi(j-u)}.$$

Set $z = x + iy$. Note that

$$|f(z)| \le \frac{1}{2\pi}\int_{-\pi}^{\pi} Me^{-ky}dk \le \frac{M(e^{|\pi y|} - e^{-|\pi y|})}{|2\pi y|}, \quad |\Phi(z)| \ge e^{|\pi y|}.$$

These inequalities, applied for $y = b$, allow us to let $b \to \infty$ ($2b$ is the height of the symmetric rectangular contour). Then it can be shown that $I_n(u) \to 0$ as $n \to \infty$, which establishes the sampling theorem for $W = \pi$. The general result is then obtained by "regula de tri,"[161] but it is sometimes hard to get it right. □

Note the similarity of (4.6.39) to Lagrange's interpolation formula. Like Lagrange's, it is a so-called *cardinal interpolation formula*. As Wz/π tends to an integer m, all terms except one on the right-hand side become zero; for $j = m$ the term becomes $f(m\pi/W)$.

[161]This rule from Euclid's fifth book of *Elementa* tells us how, given three out of four proportional quantities $a/b = c/d$, one determines the fourth. This name is used in elementary mathematics, e.g., in Germany and Sweden, but does not seem to be known under the same name in English-speaking countries.

Let f be a function which is zero outside the interval $[0, L]$. Its Fourier transform is then given by

$$g(\omega) = \int_0^L f(x)e^{-2\pi i\omega x}\,dx. \tag{4.6.40}$$

We want to approximate $g(\omega)$ using values of $f(x)$ sampled at intervals Δx,

$$f_j = f(j\Delta x), \quad 0 < j < N-1, \quad L = N\Delta x.$$

The integral (4.6.40) can be approximated by

$$g(\omega) \approx \frac{L}{N}\sum_{j=0}^{N-1} f_j e^{-2\pi i\omega j\Delta x}. \tag{4.6.41}$$

Since only N values of f_j are used as input and we want the computed values to be linearly independent, we cannot approximate $g(\omega)$ at more than N points. The wave of lowest frequency associated with the interval $[0, L]$ is $\omega = 1/L = 1/(N\Delta x)$, since then $[0, L]$ corresponds to one full period of the wave. We therefore choose points $\omega_k = k\Delta\omega$, $k = 0 : N$ in the frequency space such that the following **reciprocity relations** hold:

$$LW = N, \qquad \Delta x\Delta\omega = 1/N. \tag{4.6.42}$$

With this choice it holds that

$$W = N\Delta\omega = 1/\Delta x, \qquad L = N\Delta x = 1/\Delta\omega. \tag{4.6.43}$$

Noting that $(j\Delta x)(k\Delta\omega) = jk/N$ we get from the trapezoidal approximation

$$g(\omega_k) \approx \frac{L}{N}\sum_{j=0}^{N-1} f_j e^{-2\pi ikj/N} = Lc_k, \quad k = 0 : N-1,$$

where c_k is the coefficient of the DFT.

The frequency $\omega_c = 1/(2\Delta x) = W/2$ is the so-called **Nyquist's critical frequency**. Sampling the wave $\sin(2\pi\omega_c x)$ with sampling interval Δx will sample exactly *two points per cycle*. It is a remarkable fact that if a function $f(x)$, defined on $[-\infty, \infty]$, is *band-width limited* to frequencies smaller than or equal to ω_c, then $f(x)$ is completely determined by its sample values $j\Delta x$, $\infty \le j \le \infty$; see Shannon's sampling theorem, our Theorem 4.6.7.

If the function is not band-width limited the spectral density outside the critical frequency is moved into that range. This is called **aliasing**. The relationship between the Fourier transform $g(\omega)$ and the DFT of a finite sampled representation can be characterized as follows. Assuming that the reciprocity relations (4.6.42) are satisfied, the DFT of $f_j = f(j\Delta x)$, $0 \le j < N$, will approximate the periodic aliased function

$$\tilde{g}_k = \tilde{g}(k\Delta\omega), \quad 0 \le j < N, \tag{4.6.44}$$

where

$$\tilde{g}(\omega) = g(\omega) + \sum_{k=1}^{\infty}\left(g(\omega + kW) + g(\omega - kW)\right), \quad \omega \in [0, W]. \tag{4.6.45}$$

Since by (4.6.43) $W = 1/\Delta x$, we can *increase* the frequency range $[0, W]$ covered by *decreasing* Δx.

Example 4.6.4.

The function $f(x) = e^{-x}$, $x > 0$, $f(x) = 0$, $x < 0$, has Fourier transform

$$g(\omega) = \frac{1}{1 + 2\pi i \omega} = \frac{1 - i2\pi\omega}{1 + 4\pi^2\omega^2}$$

(cf. Example 4.6.3). Set $f(0) = 1/2$, the average of $f(-0)$ and $f(+0)$, which is the value given by the inverse Fourier transform at a discontinuity.

Set $N = 32$, $T = 8$, and sample the f in the interval $[0, T]$ at equidistant points $j\Delta t$, $j = 0 : N - 1$. Note that T is so large that the aliased function (4.6.44) is nearly equal to f. This sampling rate corresponds to $\Delta t = 8/32 = 1/4$ and $W = 4$.

The effect of aliasing in the frequency domain is evident. The error is significant for frequencies larger than the critical frequency $W/2$. To increase the accuracy W can be increased by decreasing the sampling interval Δx; see Figure 4.6.4.

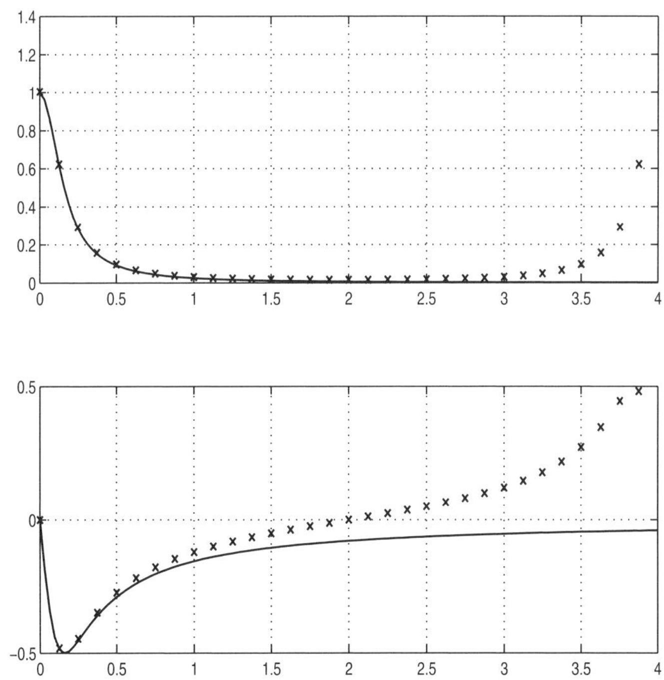

Figure 4.6.4. *The real (top) and imaginary (bottom) parts of the Fourier transform (solid line) of e^{-x} and the corresponding DFT (dots) with $N = 32$, $T = 8$.*

Review Questions

4.6.1 Derive the orthogonality properties and coefficient formulas which are fundamental to Fourier analysis, for both the continuous and the discrete case.

4.6.2 Give sufficient conditions for the Fourier series of the function f to be everywhere convergent to f. To what value will the Fourier series converge at a point $x = a$ of discontinuity of f?

4.6.3 How can the Fourier expansion be generalized to a function $f(x, y)$ of two variables?

4.6.4 (a) Give two ways in which a real function f defined on the interval $[0, \pi]$ can be extended to a periodic function.

(b) What disadvantage (for Fourier analysis) is incurred if the periodic continuation has a discontinuity—for example, in its derivative at the end points of 80, π9?

4.6.5 Describe how the behavior of the coefficients of the discrete Fourier expansion can be modified to improve the approximation of the corresponding continuous Fourier expansion.

4.6.6 Formulate the Fourier integral theorem.

4.6.7 What is meant by aliasing, when approximating the Fourier transform by a discrete transform?

Problems and Computer Exercises

4.6.1 Give a simple characterization of the functions which have a sine expansion containing odd terms only.

4.6.2 Let f be an even function, with period 2π, such that

$$f(x) = \pi - x, \quad 0 \le x \le \pi.$$

(a) Plot the function $y = f(x)$ for $-3\pi \le x \le 3\pi$. Expand f in a Fourier series.

(b) Use this series to show that $1 + 3^{-2} + 5^{-2} + 7^{-2} + \cdots = \pi^2/8$.

(c) Compute the sum $1 + 2^{-2} + 3^{-2} + 4^{-2} + 5^{-2} + \cdots$.

(d) Compute, using (4.6.9), the sum $1 + 3^{-4} + 5^{-4} + 7^{-4} + \cdots$.

(e) Differentiate the Fourier series term by term, and compare with the result for the rectangular wave in Sec. 4.6.1.

4.6.3 (a) Show that the function $G_1(t) = t - 1/2$, $0 < t < 1$, has the expansion

$$G_1(t) = -\sum_{n=1}^{\infty} \frac{\sin 2n\pi t}{n\pi}.$$

(b) Let $G_1(t)$ be as in (a) and consider a sequence of functions such that $G'_{p+1}(t) = G_p(t)$, $p = 1, 2, \ldots$. Derive by termwise integration the expansion for the functions

$G_p(t)$, and show that $c_p - G_p(t)$ has the same sign as c_p. Show also that for p even

$$\sum_{n=1}^{\infty} n^{-p} = \frac{1}{2}|c_p|(2\pi)^p.$$

4.6.4 (a) Using that $\sin x$ is the imaginary part of e^{ix}, prove that

$$\sum_{k=1}^{N-1} \sin \frac{\pi k}{N} = \cot \frac{\pi}{2N}.$$

(b) Determine a sine polynomial $\sum_{j=1}^{n-1} b_j \sin jx$, which takes on the value 1 at the points $x_k = \pi k/n$, $k = 1 : n-1$.

Hint: Use (4.6.26) or recall that the sine polynomial is an odd function.

(c) Compare the limiting value for b_j as $n \to \infty$ with the result in Example 4.6.1.

4.6.5 (a) Prove the inequality in (4.6.10).

(b) Show, under the assumptions on f which hold in (4.6.10), that for $k \geq 1$, f can be approximated by a trigonometric polynomial such that

$$\left\| f - \sum_{j=-n}^{n} c_j e^{ijx} \right\|_\infty < \frac{2}{kn^k} \| f^{(k+1)} \|_\infty.$$

4.6.6 (a) Fourier approximations to the rectangular wave, $f(x) = 1, 0 < x < \pi$, $f(x) = -1$, $-\pi < x < 0$, are shown in Figure 4.6.2 for one, two, five, and ten terms. Plot and compare the corresponding smoothed approximations when the σ-factors in (4.6.15) have been applied.

(b) The delta function $\delta(x)$ is defined as being zero everywhere except between the limits $\pm\epsilon$, where ϵ tends to zero. At $x = 0$ the function goes to infinity in such a way that the area under the function equals one. The formal Fourier expansion of $\delta(x)$ yields

$$f(x) = \frac{1}{\pi}\left(\frac{1}{2} + \cos x + \cos 2x + \cos 3x + \cdots\right),$$

which does not converge anywhere. If the σ-factors in (4.6.15) are applied, we obtain the expansion

$$y_m(x) = \frac{1}{\pi}\left(\frac{1}{2} + \sigma_1 \cos x + \sigma_2 \cos 2x + \cdots + \sigma_{m-1} \cos(m-1)x\right), \quad (4.6.46)$$

which can be considered the trigonometric representation of the delta function. Plot $y_m(x)$, $-6 \leq x \leq 6$, for several values of m.

4.6.7 (a) Consider a *real* function with the Fourier expansion

$$F(\phi) = \sum_{n=-\infty}^{\infty} c_n e^{in\phi}.$$

Show that this rewritten for convergence acceleration with the generalized Euler's method is

$$F(\phi) = c_0 + 2\Re \sum_{n=1}^{\infty} c_n z^n, \quad z = e^{i\phi}.$$

Hint: Show that $c_{-n} = \bar{c}_n$.

(b) Set $c_n = a_n - i b_n$, where a_n, b_n are real. Show that

$$\sum_{n=0}^{\infty} (a_n \cos n\phi + b_n \sin n\phi) = \Re \sum_{n=0}^{\infty} c_n z^n.$$

(c) How would you rewrite the Chebyshev series $\sum_{n=0}^{\infty} T_n(x)/(1+n^2)$?

(d) Consider also how to handle a complex function $F(\phi)$.

In the following problems, we do not require any investigation of whether it is permissible to change the order of summations, integrations, differentiations; it is sufficient to treat the problems in a purely formal way.

4.6.8 The partial differential equation $\frac{\partial u}{\partial t} = \frac{\partial^2 u}{\partial x^2}$ is called the heat equation. Show that the function

$$u(x,t) = \frac{4}{\pi} \sum_{k=0}^{\infty} \frac{\sin(2k+1)x}{2k+1} e^{-(2k+1)^2 t}$$

satisfies the differential equation for $t > 0$, $0 < x < \pi$, with boundary conditions $u(0,t) = u(\pi,t) = 0$ for $t > 0$, and initial condition $u(x,0) = 1$ for $0 < x < \pi$ (see Example 4.6.1).

4.6.9 Show that if $g(t)$ is the Fourier transform of $f(x)$, then

(a) $e^{2\pi k t} g(t)$ is the Fourier transform of $f(x+k)$.

(b) $(2\pi i t)^k g(t)$ is the Fourier transform of $f^{(k)}(x)$, assuming that $f(x)$ and its derivatives up to the kth order tend to zero as $x \to \infty$.

4.6.10 The **correlation** of $f_1(x)$ and $f_2(x)$ is defined by

$$c(\xi) = \int_{-\infty}^{\infty} f_1(x+\xi) f_2(x)\, dx. \tag{4.6.47}$$

Show that if $f_1(x)$ and $f_2(x)$ have Fourier transforms $g_1(t)$ and $g_2(t)$, respectively, then the Fourier transform of $c(\xi)$ is $h(t) = g_1(t) g_2(-t)$.

Hint: Compare Theorem 4.6.6.

4.6.11 (a) Work out the details of the proof of the sampling theorem.

(b) The formulation of the sampling theorem with a general W in Strang [339] does not agree with ours in (4.6.39). Who is right?

4.6.12 Suppose that $f(z)$ satisfies the assumptions of our treatment of the sampling theorem for $W = \pi/h$. Show by integration term by term that

$$\lim_{R\to\infty} \int_{-R}^{R} f(u)\, du = h \lim_{n\to\infty} \sum_{j=-n}^{n} f(jh).$$

Hint: Use the classical formula

$$\int_{-\infty}^{\infty} \frac{\sin x}{x}\,dx = \pi.$$

Full rigor is not necessary.

4.7 The Fast Fourier Transform

4.7.1 The FFT Algorithm

Consider the discrete Fourier transform (DFT)

$$f(x) = \sum_{j=0}^{N-1} c_j e^{ijx}$$

of a function, whose values $f(x_k)$ are known at the grid points $x_k = 2\pi k/N$, $k = 0 : N-1$. According to Theorem 4.6.4 the coefficients are given by

$$c_j = \frac{1}{N}\sum_{k=0}^{N-1} f(x_k)e^{-ijx_k}, \quad j = 0 : N-1. \tag{4.7.1}$$

The evaluation of expressions of the form (4.7.1) occur also in discrete approximations to the Fourier transform.

Setting $\omega_N = e^{-2\pi i/N}$ this becomes

$$c_j = \frac{1}{N}\sum_{k=0}^{N-1} \omega_N^{jk} f(x_k), \quad j = 0 : N-1, \tag{4.7.2}$$

where ω_N is an Nth root of unity, $(\omega_N)^N = 1$. It seems from (4.7.2) that computing the discrete Fourier coefficients would require N^2 complex multiplications and additions. As we shall see, only about $N \log_2 N$ complex multiplications and additions are required using an algorithm called the **fast Fourier transform (FFT)**.

The modern usage of the FFT started in 1965 with the publication of [78] by James W. Cooley of IBM Research and John W. Tukey, Princeton University.[162] In many areas of application (digital signal processing, image processing, time-series analysis, to name a few), the FFT has caused a complete change of attitude toward what can be done using discrete Fourier methods. Without the FFT many modern devices such as cell phones, digital cameras, CT scanners, and DVDs would not be possible. Some applications considered in astronomy require FFTs of several gigapoints.

[162]Tukey came up with the basic algorithm at a meeting of President Kennedy's Science Advisory Committee. One problem discussed at this meeting was that the ratification of a US–Soviet nuclear test ban depended on a fast method to detect nuclear tests by analyzing seismological time-series data.

In the following we will use the common convention *not* to scale the sum in (4.7.2) by $1/N$.

Definition 4.7.1.
The DFT *of the vector* $f \in \mathbf{C}^N$ *is*

$$y = F_N f, \tag{4.7.3}$$

where $F_n \in \mathbf{C}^{N\times N}$ *is the* **DFT matrix** *with elements*

$$(F_N)_{jk} = \omega_N^{jk}, \quad j, k = 0 : N-1, \tag{4.7.4}$$

where $\omega_N = e^{-2\pi i/N}$.[163]

From the definition it follows that the DFT matrix F_N is a complex Vandermonde matrix. Since $\omega_N^{jk} = \omega_N^{kj}$, F_N is symmetric. By Theorem 4.6.4

$$\frac{1}{N} F_N^H F_N = I,$$

where F_N^H is the complex conjugate transpose of F_N. Hence the matrix $\frac{1}{\sqrt{N}} F_N$ is a unitary matrix and the inverse transform can be written as

$$f = \frac{1}{N} F_N^H y.$$

We now describe the central idea of the FFT algorithm, which is based on the divide and conquer strategy (see Sec. 1.2.3). Assume that $N = 2^p$ and set

$$k = \begin{cases} 2k_1 & \text{if } k \text{ is even,} \\ 2k_1 + 1 & \text{if } k \text{ is odd,} \end{cases} \quad 0 \le k_1 \le m-1,$$

where $m = N/2 = 2^{p-1}$. Split the DFT sum into an even and an odd part:

$$y_j = \sum_{k_1=0}^{m-1} (\omega_N^2)^{jk_1} f_{2k_1} + \omega_N^j \sum_{k_1=0}^{m-1} (\omega_N^2)^{jk_1} f_{2k_1+1}, \quad j = 0 : N-1.$$

Let β be the quotient and j_1 the remainder when j is divided by m, i.e., $j = \beta m + j_1$. Then, since $\omega_N^N = 1$,

$$(\omega_N^2)^{jk_1} = (\omega_N^2)^{\beta m k_1} (\omega_N^2)^{j_1 k_1} = (\omega_N^N)^{\beta k_1} (\omega_N^2)^{j_1 k_1} = \omega_m^{j_1 k_1}.$$

Thus if, for $j_1 = 0 : m-1$, we set

$$\phi_{j_1} = \sum_{k_1=0}^{m-1} f_{2k_1} \omega_m^{j_1 k_1}, \qquad \psi_{j_1} = \sum_{k_1=0}^{m-1} f_{2k_1+1} \omega_m^{j_1 k_1}, \tag{4.7.5}$$

[163]Some authors set $\omega_N = e^{2\pi i/N}$. Which convention is used does not much affect the development.

then $y_j = \phi_{j_1} + \omega_N^j \psi_{j_1}$. The two sums on the right are elements of the DFTs of length $N/2$ applied to the parts of f with odd and even subscripts. *The entire DFT of length N is obtained by combining these two DFTs!* Since $\omega_N^m = -1$, we have

$$y_{j_1} = \phi_{j_1} + \omega_N^{j_1} \psi_{j_1}, \tag{4.7.6}$$

$$y_{j_1+N/2} = \phi_{j_1} - \omega_N^{j_1} \psi_{j_1}, \quad j_1 = 0 : N/2 - 1. \tag{4.7.7}$$

These expressions, noted already by Danielson and Lanczos [90], are often called **butterfly relations** because of the data flow pattern. Note that these can be performed in place, i.e., no extra vector storage is needed.

The computation of ϕ_{j_1} and ψ_{j_1} means that one does *two* Fourier transforms with $m = N/2$ terms instead of one with N terms. If $N/2$ is even the same idea can be applied to these two Fourier transforms. One then gets *four* Fourier transforms, each of which has $N/4$ terms. If $N = 2^p$, this reduction can be continued recursively until we get N DFTs with one term. But $F_1 = I$, the identity. A recursive MATLAB implementation of the FFT algorithm is given in Problem 4.7.2.

Example 4.7.1.

For $n = 2^2 = 4$, we have $\omega_4 = e^{-\pi i/2} = -i$, and the DFT matrix is

$$F_4 = \begin{pmatrix} 1 & 1 & 1 & 1 \\ 1 & -i & (-i)^2 & (-i)^3 \\ 1 & (-i)^2 & (-i)^4 & (-i)^6 \\ 1 & (-i)^3 & (-i)^6 & (-i)^9 \end{pmatrix} = \begin{pmatrix} 1 & 1 & 1 & 1 \\ 1 & -i & -1 & i \\ 1 & -1 & 1 & -1 \\ 1 & i & -1 & -i \end{pmatrix}. \tag{4.7.8}$$

It is symmetric and its inverse is

$$F_4^{-1} = \frac{1}{4} \begin{pmatrix} 1 & 1 & 1 & 1 \\ 1 & i & -1 & -i \\ 1 & -1 & 1 & -1 \\ 1 & -i & -1 & i \end{pmatrix}.$$

The number of complex operations (one multiplication and one addition) required to compute $\{y_j\}$ from the butterfly relations when $\{\phi_{j_1}\}$ and $\{\psi_{j_1}\}$ have been computed is 2^p, assuming that the powers of ω are precomputed and stored. Thus, if we denote by q_p the total number of operations needed to compute the DFT when $N = 2^p$, we have

$$q_p \le 2q_{p-1} + 2^p, \quad p \ge 1.$$

Since $q_0 = 0$, it follows by induction that $q_p \le p \cdot 2^p = N \cdot \log_2 N$. *Hence, when N is a power of* 2*, the FFT solves the problem with at most* $N \cdot \log_2 N$ *operations.*

For example, when $N = 2^{20} = 1{,}048{,}576$ the FFT algorithm is theoretically a factor of 84,000 faster than the "conventional" $O(N^2)$ algorithm. On a 3 GHz laptop, a real FFT of this size takes about 0.1 second using MATLAB 6, whereas more than two hours would be required by the conventional algorithm! The FFT not only uses fewer operations to evaluate the DFT, it also is more accurate. Whereas when using the conventional method the roundoff error is proportional to N, for the FFT algorithm it is proportional to $\log_2 N$.

Example 4.7.2.

Let $N = 2^4 = 16$. Then the 16-point DFT (0:1:15) can be split into two 8-point DFTs (0:2:14) and (1:2:15), which can each be split in two 4-point DFTs. Repeating these splittings we finally get 16 one-point DFTs which are the identity $F_1 = 1$. The structure of this FFT is illustrated below.

$$[0\ 1\ 2\ 3\ 4\ 5\ 6\ 7\ 8\ 9\ 10\ 11\ 12\ 13\ 14\ 15]$$

$$\swarrow\searrow$$

$$[0\ 2\ 4\ 6\ 8\ 10\ 12\ 14] \qquad [1\ 3\ 5\ 7\ 9\ 11\ 13]$$

$$\swarrow\searrow \qquad \swarrow\searrow$$

$$[0\ 4\ 8\ 12] \quad [2\ 6\ 10\ 14] \quad [1\ 5\ 9\ 13] \quad [3\ 7\ 11\ 15]$$

$$\swarrow\searrow \quad \swarrow\searrow \quad \swarrow\searrow \quad \swarrow\searrow$$

$$[0\ 8]\ [4\ 12]\ [2\ 10]\ [6\ 14]\ [1\ 9]\ [5\ 13]\ [3\ 11]\ [7\ 15]$$

$$\swarrow\searrow\ \swarrow\searrow\ \swarrow\searrow\ \swarrow\searrow\ \swarrow\searrow\ \swarrow\searrow\ \swarrow\searrow\ \swarrow\searrow$$

$$[0]\ [8]\ [4]\ [12]\ [2]\ [10]\ [6]\ [14]\ [1]\ [9]\ [5]\ [13]\ [3]\ [11]\ [7]\ [15]$$

In most implementations the explicit recursion is avoided. Instead the FFT algorithm is implemented in two stages:

- a reordering stage in which the data vector f is permuted;
- a second stage in which first $N/2$ FFT transforms of length 2 are computed on adjacent elements, followed by $N/4$ transforms of length 4, etc., until the final result is obtained by merging two FFTs of length $N/2$.

We now consider each stage in turn.

Each step of the recursion involves an even–odd permutation. In the first step the points with last binary digit equal to 0 are ordered first and those with last digit equal to 1 are ordered last. In the next step the two resulting subsequences of length $N/2$ are reordered according to the second binary digit, etc. It is not difficult to see that the combined effect of the reordering in stage 1 is a **bit-reversal permutation** of the data points. For $i = 0 : N-1$, let the index i have the binary expansion

$$i = b_0 + b_1 \cdot 2 + \cdots + b_{t-1} \cdot 2^{t-1}$$

and set

$$r(i) = b_{t-1} + \cdots + b_1 \cdot 2^{t-2} + b_0 \cdot 2^{t-1}.$$

That is, $r(i)$ is the index obtained by reversing the order of the binary digits. If $i < r(i)$, then exchange f_i and $f_{r(i)}$. This reordering is illustrated for $N = 16$ below.

Decimal	Binary		Decimal	Binary
0	0000		0	0000
1	0001		8	1000
2	0010		4	0100
3	0011		12	1100
4	0100		2	0010
5	0101		10	1010
6	0110		6	0110
7	0111	$\Longrightarrow$	14	1110
8	1000		1	0001
9	1001		9	1001
10	1010		5	0101
11	1011		13	1101
12	1100		3	0011
13	1101		11	1011
14	1110		7	0111
15	1111		15	1111

We denote the permutation matrix performing the bit-reversal ordering by P_N. Note that if an index is reversed twice we end up with the original index. This means that

$$P_N^{-1} = P_N^T = P_N,$$

i.e., P_N is symmetric. The permutation can be carried out "in place" by a sequence of pairwise interchanges or transpositions of the data points. For example, for $N = 16$ the pairs (1,8), (2,4), (3,12), (5,10), (7,14), and (11,13) are interchanged. The bit-reversal permutation can take a substantial fraction of the total time to do the FFT. Which implementation is best depends strongly on the computer architecture.

We now consider the second stage of the FFT. The key observation to develop a matrix-oriented description of this stage is to note that the Fourier matrices F_N after an odd–even permutation of the columns can be expressed as a 2×2 block matrix, where each block is either $F_{N/2}$ or a diagonal scaling of $F_{N/2}$.

Theorem 4.7.2 (*Van Loan* [366, *Theorem* 1.2.1]).

Let Π_N^T be the permutation matrix which applied to a vector groups the even-indexed components first and the odd-indexed last.[164] *If $N = 2m$, then*

$$F_N \Pi_N = \begin{pmatrix} F_m & \Omega_m F_m \\ F_m & -\Omega_m F_m \end{pmatrix} = \begin{pmatrix} I_m & \Omega_m \\ I_m & -\Omega_m \end{pmatrix} \begin{pmatrix} F_m & 0 \\ 0 & F_m \end{pmatrix},$$

$$\Omega_m = \operatorname{diag}(1, \omega_N, \dots, \omega_N^{m-1}), \quad \omega_N = e^{-2\pi i/N}. \tag{4.7.9}$$

Proof. The proof essentially follows from the derivation of the butterfly relations (4.7.6)–(4.7.7). □

[164]Note that $\Pi_N^T = \Pi_N^{-1}$ is the so-called **perfect shuffle permutation**, which in the permuted vector $\Pi_N^T f$ is obtained by splitting f in half and then "shuffling" the top and bottom halves.

Example 4.7.3.

We illustrate Theorem 4.7.2 for $N = 2^2 = 4$. The DFT matrix F_4 is given in Example 4.7.1. After a permutation of the columns F_4 can be written as a 2×2 block-matrix

$$F_4 \Pi_4^T = \left(\begin{array}{cc|cc} 1 & 1 & 1 & 1 \\ 1 & -1 & -i & i \\ \hline 1 & 1 & -1 & -1 \\ 1 & -1 & i & -i \end{array}\right) = \begin{pmatrix} F_2 & \Omega_2 F_2 \\ F_2 & -\Omega_2 F_2 \end{pmatrix},$$

where

$$F_2 = \begin{pmatrix} 1 & 1 \\ 1 & -1 \end{pmatrix}, \qquad \Omega_2 = \operatorname{diag}(1, -i).$$

When $N = 2^p$ the FFT algorithm can be interpreted as a sparse factorization of the DFT matrix

$$F_N = A_k \cdots A_2 A_1 P_N, \tag{4.7.10}$$

where P_N is the bit-reversal permutation matrix and $A_1, \dots, A_k$ are block-diagonal matrices,

$$A_q = \operatorname{diag}(\underbrace{B_L, \dots, B_L}_{r}), \quad L = 2^q, \quad r = N/L. \tag{4.7.11}$$

Here the matrix $B_k \in \mathbf{C}^{L \times L}$ is the radix-2 butterfly matrix defined by

$$B_L = \begin{pmatrix} I_{L/2} & \Omega_{L/2} \\ I_{L/2} & -\Omega_{L/2} \end{pmatrix}, \tag{4.7.12}$$

$$\Omega_{L/2} = \operatorname{diag}(1, \omega_L, \dots, \omega_L^{L/2-1}), \quad \omega_L = e^{-2\pi i/L}. \tag{4.7.13}$$

The FFT algorithm described above is usually referred to as the Cooley–Tukey FFT algorithm. Using the fact that both the bit-reversal matrix P_N and the DFT matrix F_n are symmetric, we obtain by transposing (4.7.10) the factorization

$$F_N = F_N^T = P_N A_1^T A_2^T \cdots A_k^T. \tag{4.7.14}$$

This gives rise to a "dual" FFT algorithm, referred to as the Gentleman–Sande algorithm [155]. In this the bit-reversal permutation comes *after* the other computations. In many important applications, such as convolution and the solution of discretized Poisson equations (see Sec. 1.1.4), this permits the design of in-place FFT solutions that avoid bit-reversal altogether; see Van Loan [366, Secs. 4.1, 4.5].

In the operation count for the FFT above we assumed that the weights ω_L^j, $j = 1 : L-1$, $\omega_L = e^{-2\pi i/L}$, are precomputed. To do this one could use that

$$\omega_L^j = \cos(j\theta) - i\sin(j\theta), \quad \theta = 2\pi/L,$$

for $L = 2^q$, $q = 2 : k$. This is accurate, but expensive, since it involves $L-1$ trigonometric functions calls. An alternative is to compute $\omega = \cos(\theta) - i\sin(\theta)$ and use repeated multiplication,

$$\omega^j = \omega\omega^{j-1}, \quad j = 2 : L-1.$$

This replaces one sine/cosine call with a single complex multiplication, but has the drawback that accumulation of roundoff errors will give an error in ω_L^j of order ju.

4.7.2 Discrete Convolution by FFT

The most important operation in signal processing is computing the discrete version of the convolution operator. This awkward operation in the time domain becomes very simple in the frequency domain.

Definition 4.7.3.

The convolution of two sequences f_i *and* g_i, $i = 0 : N-1$, *is* $\operatorname{conv}(f, g) = (h_0, h_1, \dots, h_{N-1})^T$, *where*

$$h_k = \sum_{i=0}^{N-1} f_i g_{k-i}, \quad k = 0 : N-1, \tag{4.7.15}$$

where the sequences are extended to have period N by setting $f_i = f_{i+jN}$, $g_i = g_{i+jN}$ *for all integers* i, j.

The discrete convolution can be used to approximate the convolution defined for continuous functions in Definition 4.6.5 in a similar way as the Fourier transform was approximated using sampled values in Sec. 4.6.6.

We can write the sum in (4.7.15) as a matrix-vector multiplication $h = Gf$, where G is a Toeplitz matrix. Writing out components we have

$$\begin{pmatrix} h_0 \\ h_1 \\ h_2 \\ \vdots \\ h_{N-1} \end{pmatrix} = \begin{pmatrix} g_0 & g_{N-1} & g_{N-2} & \cdots & g_1 \\ g_1 & g_0 & g_{N-1} & \cdots & g_2 \\ g_2 & g_1 & g_0 & \cdots & g_3 \\ \vdots & \vdots & \vdots & \cdots & \\ g_{N-1} & g_{N-2} & g_{N-3} & \cdots & g_0 \end{pmatrix} \begin{pmatrix} f_0 \\ f_1 \\ f_2 \\ \vdots \\ f_{N-1} \end{pmatrix}.$$

Note that each column in G is a cyclic down-shifted version of the previous column. Such a matrix is called a **circulant matrix**. We have

$$G = [\, g \quad C_N g \quad \cdots \quad C_N^{N-1} g \,] = g_0 I + g_1 C_N + \cdots + g_{N-1} C_N^{N-1}, \tag{4.7.16}$$

and C_N is the circulant permutation matrix

$$C_N = \begin{pmatrix} 0 & & \cdots & & 1 \\ 1 & & & & \\ & 1 & & & \vdots \\ \vdots & & \ddots & & \\ 0 & & \cdots & 1 & 0 \end{pmatrix}. \tag{4.7.17}$$

The following result is easily verified.

Lemma 4.7.4.

The eigenvalues of the circulant matrix C_N *in* (4.7.17) *are*

$$\omega_j = e^{-2\pi j/N}, \quad j = 0 : N-1,$$

where ω is an Nth root of unity, i.e., $\omega^N = 1$. The columns of the DFT matrix F_N,

$$x_j = (1, \omega_j, \dots, \omega_j^{N-1})^T, \quad j = 0 : N-1,$$

are eigenvectors.

Since the matrix G in (4.7.16) is a polynomial in C_N it has the same set of eigenvectors, and thus G is diagonalized by the DFT matrix F_N,

$$G = F_N \Lambda F_N^{-1}, \quad \Lambda = \operatorname{diag}(\lambda_1, \dots, \lambda_n), \tag{4.7.18}$$

where the eigenvalues of G are

$$\lambda_i = (1, \omega_j, \dots, \omega_j^{N-1}) g, \quad j = 0 : N-1,$$

which is the FFT of the first column in G. Hence $\Lambda = \operatorname{diag}(F_N g)$, where $\operatorname{diag}(x)$ denotes a diagonal matrix with diagonal elements equal to the elements in the vector x.

Theorem 4.7.5.

Let f_i and g_i, $i = 0 : N-1$, be two sequences with DFTs equal to $F_N f$ and $F_N g$. Then the DFT of the **convolution** *of f and g is $F_N f .* F_N g$ ($.*$ denotes the elementwise product).*

Proof. From $G = F_N^{-1} \operatorname{diag}(F_N g) F_N$ it follows that

$$h = Gf = F_N^{-1} \operatorname{diag}(F_N g) F_N f = F_N^{-1}((F_N g) .* (F_N f)). \quad \square \tag{4.7.19}$$

This shows that using the FFT algorithm the discrete convolution can be computed in $O(N \log_2 N)$ operations as follows: First the two FFTs of f and g are computed and multiplied (pointwise) together. Then the inverse DFT of this product is computed. This is one of the most useful properties of the FFT!

Using the Gentleman–Sande algorithm $F_N = P_N A^T$ for the forward DFT and the Cooley–Tukey algorithm $F_N = A P_N$ for the inverse DFT,

$$F_N^{-1} = \frac{1}{N} F_N^H = \frac{1}{N} \bar{A} P_N,$$

we get from (4.7.19)

$$h = \frac{1}{N} \bar{A} P_N((P_N A^T f) .* (P_N A^T g)) = \frac{1}{N} \bar{A}((A^T f) .* (A^T g)). \tag{4.7.20}$$

This shows that h can be computed without the bit-reversal permutation P_N which typically can save 10–30 percent of the overall computation time.

4.7.3 FFTs of Real Data

Frequently the FFT of a real data vector is required. The complex FFT algorithm can still be used, but is inefficient both in terms of storage and operations. By using symmetries in the DFT, which correspond to the symmetries noted in the Fourier transform in Table 4.6.1, better alternatives can be found.

Consider the DFT matrix for $N = 4$ in (4.7.8). Note that the fourth row is the conjugate of the second row. This is not a coincidence; the conjugate transpose of the DFT matrix F_N can be obtained by reversing the order of the last $N-1$ rows. Let T_N be the $N \times N$ permutation matrix obtained by reversing the last $N-1$ columns in the unit matrix I_N. For example,

$$T_4 = \begin{pmatrix} 1 & 0 & 0 & 0 \\ 0 & 0 & 0 & 1 \\ 0 & 0 & 1 & 0 \\ 0 & 1 & 0 & 0 \end{pmatrix}.$$

Then it holds that

$$F_N^H = \overline{F}_N = T_N F_N = F_N T_N. \tag{4.7.21}$$

We first verify that $\overline{F}_N = T_N F_N$, by observing that

$$[T_N F_N]_{jk} = \omega_N^{(N-j)k} = \omega_N^{-jk} = \overline{\omega}_N^{jk} = [\overline{F}_N]_{jk}, \quad 1 \le j \le N-1.$$

Since F_N and T_N are both symmetric, we also have $F_N^H = (T_N F_N)^T = F_N T_N$.

We say that a vector $y \in \mathbf{C}^N$ is **conjugate even** if $\overline{y} = T_N y$, and **conjugate odd** if $\overline{y} = -T_N y$. Suppose now that f is real and $u = F_N f$. Then it follows that

$$\overline{u} = \overline{F}_N f = T_N F_N f = T_N u,$$

i.e., u is conjugate even. If a vector u of even length N is conjugate even, this implies that

$$u_j = \overline{u}_{N-j}, \quad j = 1 : N/2.$$

In particular, u_j is real for $j = 0, N/2$.

For purely imaginary data g and $v = F_N g$, we have

$$\overline{v} = F_N^H \overline{g} = -F_N^H g = -T_N F_N g = -T_N v,$$

i.e., v is conjugate odd. Some other useful symmetry properties are given in Table 4.7.1. We have proved the first two properties; the others are established similarly and we leave the proofs to the reader; see Problem 4.7.4.

Table 4.7.1. *Useful symmetry properties of the DFT.*

Data f	Definition	DFT $F_N f$
real		conjugate even
imaginary		conjugate odd
real even	$f = T_N f$	real
real odd	$f = -T_N f$	imaginary
conjugate even	$\overline{f} = T_N f$	real
conjugate odd	$\overline{f} = -T_N f$	imaginary

We now outline how symmetries can be used to compute the DFTs $u = F_N f$ and $v = F_N g$ of *two* real functions f and g simultaneously. First form the complex function $f + ig$ and compute its DFT

$$w = F_N(f + ig) = u + iv$$

by any complex FFT algorithm. Multiplying by T_N we have

$$T_N w = T_N F_N(f + ig) = T_N(u + iv) = \overline{u} + i\overline{v},$$

where we have used that u and v are conjugate even. Adding and subtracting these two equations we obtain

$$\begin{aligned} w + T_N w &= (u + \overline{u}) + i(v + \overline{v}), \\ w - T_N w &= (u - \overline{u}) + i(v - \overline{v}). \end{aligned}$$

We can now retrieve the two DFTs from

$$u = F_N f = \frac{1}{2}\big[\mathrm{Re}(w + T_N w) + i\,\mathrm{Im}(w - T_N w)\big], \tag{4.7.22}$$

$$v = F_N g = \frac{1}{2}\big[\mathrm{Im}(w + T_N w) - i\,\mathrm{Re}(w - T_N w)\big]. \tag{4.7.23}$$

Note that because of the conjugate even property of u and v there is no need to save the entire transforms.

The above scheme is convenient when, as for convolutions, two real transforms are involved. It can also be used to efficiently compute the DFT of a single real function of length $N = 2^p$. First express this DFT as a combination of the two real FFTs of length $N/2$ corresponding to even and odd numbered data points (as in (4.7.5)). Then apply the procedure above to simultaneously compute these two real FFTs.

4.7.4 Fast Trigonometric Transforms

Two real transforms, the **discrete sine transform (DST)** and **discrete cosine transform (DCT)**, are of interest. There are several variations of these. We define the first two as follows:

- Given real data f_j, $j = 1 : N - 1$, compute

$$y_k = \sum_{j=1}^{N-1} \sin\left(kj\frac{\pi}{N}\right) f_j \qquad \text{(DST-1)}. \tag{4.7.24}$$

- Given real data f_j, $j = 0 : N$, compute

$$y_k = \sum_{j=0}^{N}{}'' \cos\left(kj\frac{\pi}{N}\right) f_j \qquad \text{(DCT-1)}. \tag{4.7.25}$$

(The double prime on the sum means that the first and last term are to be halved.)

The real and imaginary parts of the DFT matrix F_N consist of cosines and sines, $F_N = C_N - iS_N$, where

$$(C_N)_{kj} = \cos\left(kj\frac{2\pi}{N}\right), \qquad (S_N)_{kj} = \sin\left(kj\frac{2\pi}{N}\right), \quad k, j = 0 : N-1. \tag{4.7.26}$$

The DST and DCT transforms can be expressed in matrix form as

$$y = S_{2N}(1 : N-1, 1 : N-1)f, \qquad y = C_{2N}(0 : N, 0 : N)\bar{f},$$

respectively, where $\bar{f} = (\frac{1}{2}f_0, f_1, \dots, f_{N-1}, \frac{1}{2}f_N)^T$.

These two transforms can be computed by applying the FFT algorithm (for real data) to an auxiliary vector $\tilde{f}$ formed by extending the given data vector f either into an odd or even sequence.

For DST-1 the data f_j, $j = 1 : N-1$, are extended to an *odd* sequence of length $2N$ by setting

$$f_0 = f_N = 0, \qquad f_{2N-j} \equiv -f_j, \quad j = 1 : N-1.$$

For example, the data $f = \{f_1, f_2, f_3\}$, $(N = 4)$ are extended to

$$\tilde{f} = \{0, f_1, f_2, f_3, 0, -f_3, -f_2, -f_1\}.$$

The extended vector satisfies $\tilde{f} = -T_{2N}\tilde{f}$, and thus by Table 4.7.1 the DFT of $\tilde{f}$ will be pure imaginary.

For DCT-1 the data f_j, $j = 0 : N$, are extended to an *even* sequence of length $2N$ by setting

$$f_{2N-j} \equiv f_j, \quad j = 1 : N.$$

For example, the data $f = \{f_0, f_1, f_2, f_3, f_4\}$, $(N = 4)$ are extended to

$$\tilde{f} = \{f_0, f_1, f_2, f_3, f_4, f_3, f_2, f_1\}$$

so that $\tilde{f} = T_{2N}\tilde{f}$. By Table 4.7.1 the DFT of $\tilde{f}$ will then be real.

Theorem 4.7.6 (*Van Loan* [366, *Sec.* 4.4]).

Let f_j, $j = 1 : N-1$, form a real data vector f and extend it to a vector $\tilde{f}$ with $\tilde{f}_0 = \tilde{f}_N = 0$ so that $\tilde{f} = -T_{2N}\tilde{f}$. Then $y(1 : N-1)$ is the DST of $\tilde{f}$, where

$$y = y(0 : 2N-1) = \frac{i}{2}F_{2N}\tilde{f}.$$

Let f_j, $j = 0 : N$, form a real data vector f and extend it to a vector $\tilde{f}$ so that $\tilde{f} = T_{2N}\tilde{f}$. Then $y(0 : N)$ is the DCT of f, where

$$y = y(0 : 2N-1) = \frac{1}{2}F_{2N}\tilde{f}.$$

There is an inefficiency factor of two in the above procedure. This can be eliminated by using a different auxiliary vector. For details we refer to [294, pp. 420–421]; [366, Sec. 4.4].

In some application areas variants of the above transforms called DST-2 and DCT-2 turn out to be more useful. We define them as follows:

- Given real data f_j, $j = 1 : N$, compute

$$y_k = \sum_{j=1}^{N} \sin\left(k(2j-1)\frac{\pi}{2N}\right) f_j \qquad \text{(DST-2)}. \tag{4.7.27}$$

- Given real data f_j, $j = 0 : N-1$, compute

$$y_k = \sum_{j=0}^{N-1} \cos\left(k(2j+1)\frac{\pi}{2N}\right) f_j \qquad \text{(DCT-2)}. \tag{4.7.28}$$

The DST-2 and DCT-2 transforms can also be obtained by extending the data vector f.

For DST-2 the data vector f_j, $j = 0 : N-1$, is extended to an *odd* sequence of length $2N$. For example, the data $\{f_1, f_2, f_3, f_4\}$, $(N = 4)$ are extended to

$$\tilde{f} = \{f_1, f_2, f_3, f_4, -f_4, -f_3, -f_2, -f_1\}.$$

The extended vector satisfies $f = -T_{2N} f$, and thus by Table 4.7.1 the DFT of f will be imaginary.

For DCT-2 the data f_j, $j = 0 : N$, are extended to an *even* sequence of length $2N$. For example, the data $\{f_0, f_1, f_2, f_3\}$, $(N = 4)$ are extended to

$$\tilde{f} = \{f_0, f_1, f_2, f_3, f_3, f_2, f_1, f_0\}.$$

so that $f = T_{2N} f$. By Table 4.7.1 the DFT of f will then be real.

We give without proof the following result, which allows the computation of DST-2 and DCT-2 from the FFT.

Theorem 4.7.7.

Let f_j, $j = 1 : N$, form a real data vector f and extend it to a vector $\tilde{f}$ so that $\tilde{f} = -T_{2N}\tilde{f}$. Then $y(1 : N)$ is the DST-2 of f, where

$$y = y(0 : 2N-1) = \frac{i}{2}\Omega_{2N} F_{2N} \tilde{f},$$

where

$$\Omega_{2N} = \operatorname{diag}(1, \omega_{4N}, \ldots, \omega_{4N}^{2N}).$$

Let f_j, $j = 0 : N-1$, form a real data vector f and extend it to a vector $\tilde{f}$ so that $\tilde{f} = T_{2N}\tilde{f}$. Then $y(0 : N-1)$ is the DCT-2 of f, where

$$y = y(0 : 2N-1) = \frac{1}{2}\Omega_{2N} F_{2N} \tilde{f}.$$

The two-dimensional DCT-2 transform has the property that, for a visual image, most of the information is concentrated in the first few coefficients of the DCT. For this reason the DCT-2 transform is often used in image compression algorithms.[165]

[165] This transform is used in the JPEG (Joint Photographic Experts Group) compression algorithm for image processing. Each 8×8 block in the image is transformed by a two-dimensional DCT-2 transform; see Strang [340].

4.7.5 The General Case FFT

It can be argued [294, p. 409] that one should always choose $N = 2^k$ when using the FFT. If necessary the data can be padded with zeros to achieve this. To introduce an odd factor s, let $N = sr$, where r is a power of 2. Then one can combine the power of two algorithm for the r-point subseries with a special algorithm for the s-point subseries. If s is a small number, then one could generate the DFT matrix F_s and use matrix-vector multiplication; (cf. the code given in Problem 4.7.2). But the general case when N is not a power of two is at least of theoretical interest.

Suppose that $N = r_1 r_2 \cdots r_p$. We will describe an FFT algorithm which requires $N(r_1 + r_2 + \cdots + r_p)$ complex operations. Set

$$N_\nu = \prod_{i=\nu+1}^{p} r_i, \quad \nu = 0 : p-1, \quad N_p = 1.$$

Thus

$$N = r_1 r_2 \cdots r_\nu N_\nu, \quad N_0 = N.$$

The algorithm is based on two representations of integers which are generalizations of the position principle (see Sec. 2.2.1).

I. Every integer j, $0 \le j \le N-1$, has a unique representation of the form

$$j = \alpha_1 N_1 + \alpha_2 N_2 + \cdots + \alpha_{p-1} N_{p-1} + \alpha_p, \quad 0 \le \alpha_i \le r_i - 1. \tag{4.7.29}$$

II. For every integer β, $0 \le \beta \le N-1$, β/N has a unique representation of the form

$$\frac{\beta}{N} = \frac{k_1}{N_0} + \frac{k_2}{N_1} + \cdots + \frac{k_p}{N_{p-1}}, \quad 0 \le k_i \le r_i - 1. \tag{4.7.30}$$

Set

$$j_\nu = \sum_{i=\nu+1}^{p} \alpha_i N_i, \qquad \frac{\alpha_\nu}{N_\nu} = \sum_{i=\nu}^{p-1} \frac{k_{i+1}}{N_i}, \quad (j_\nu < N_\nu). \tag{4.7.31}$$

As an exercise, the reader can verify that the coefficients in the above representations can be recursively determined from the following algorithms:[166]

$$j_0 = j, \qquad j_{i-1}/N_i = \alpha_i + j_i/N_i, \quad i = 1 : p,$$

$$\beta_0 = \beta, \qquad \beta_{i-1}/r_i = \beta_i + k_i/r_i, \quad i = 1 : p.$$

From (4.7.29)–(4.7.31), it follows that, since N_i is divisible by N_ν for $i \le \nu$,

$$\frac{j\beta}{N} = \text{integer} + \sum_{\nu=0}^{p-1} \frac{k_{\nu+1}}{N_\nu} \left(\sum_{i=\nu+1}^{p} \alpha_i N_i \right) = \sum_{\nu=0}^{p-1} \frac{k_{\nu+1} j_\nu}{N_\nu} + \text{integer}.$$

166 These algorithms can, in the special case that $r_i = B$ for all i, be used for converting integers or fractions to the number system whose base is B; see Algorithm 2.1.

From this, it follows that

$$\omega^{j\beta} = e^{2\pi i j\beta/N} = \prod_{\nu=0}^{p-1} e^{k_{\nu+1} j_\nu 2\pi i/N_\nu} = \prod_{\nu=0}^{p-1} \omega_\nu^{j_\nu k_{\nu+1}}, \tag{4.7.32}$$

where $\omega_\nu = e^{2\pi i/N_\nu}$, $\omega_0 = \omega$.

A split-radix approach for the general case FFT, which has a lower operation count, is described in Van Loan [366, Sec. 2.1.4].

We now give an illustration of how the factorization in (4.7.32) can be utilized in FFT for the case $p = 3$. Set, in accordance with (4.7.30),

$$f_\beta = c^{(0)}(k_1, k_2, k_3).$$

We have then

$$c_j = \sum_{\beta=0}^{N-1} f_\beta \omega^{j\beta} = \sum_{k_1=0}^{r_1-1}\sum_{k_2=0}^{r_2-1}\sum_{k_3=0}^{r_3-1} c^{(0)}(k_1, k_2, k_3)\omega_2^{j_2 k_3}\omega_1^{j_1 k_2}\omega^{j k_1}.$$

One can thus compute successively (see (4.7.31))

$$c^{(1)}(k_1, k_2, \alpha_3) = \sum_{k_3=0}^{r_3-1} c^{(0)}(k_1, k_2, k_3)\omega_2^{j_2 k_3} \quad (j_2 \text{ depends only on } \alpha_3),$$

$$c^{(2)}(k_1, \alpha_2, \alpha_3) = \sum_{k_2=0}^{r_2-1} c^{(1)}(k_1, k_2, \alpha_3)\omega_1^{j_1 k_2} \quad (j_1 \text{ depends only on } \alpha_2, \alpha_3),$$

$$c_j = c^{(3)}(\alpha_1, \alpha_2, \alpha_3) = \sum_{k_1=0}^{r_1-1} c^{(2)}(k_1, \alpha_2, \alpha_3)\omega^{j k_1} \quad (j \text{ depends on } \alpha_1, \alpha_2, \alpha_3).$$

The quantities $c^{(i)}$ are computed for all $r_1 r_2 r_3 = N$ combinations of the values of the arguments. Thus the total number of operations for the entire Fourier analysis becomes at most $N(r_3 + r_2 + r_1)$. The generalization to arbitrary p is obvious.

Review Questions

4.7.1 Suppose we want to compute the DFT for $N = 2^{10}$. Roughly how much faster is the FFT algorithm compared to the straightforward $O(N^2)$ algorithm?

4.7.2 Show that the matrix $U = \frac{1}{\sqrt{N}} F_N$ is unitary, i.e., $U^H U = I$, where $U^H = (\overline{U})^T$.

4.7.3 Show that the DFT matrix F_4 can be written as a 2×2 block matrix where each block is related to F_2. Give a generalization of this for F_N, $N = 2^m$ that holds for arbitrary m.

4.7.4 Work out, on your own, the bit-reversal permutation of the vector $[0 : N - 1]$ for the case $N = 2^4 = 16$. How many exchanges need to be performed?

Problems and Computer Exercises

4.7.1 The following MATLAB script uses an algorithm due to Cooley et al. to permute the vector $x(1:2^m)$, in bit-reversal order:

```
n = 2^m;
nv2 = n/2; nm1 = n - 1;
j = 1;
for i = 1:nm1
  if i < j
    t = x(j); x(j) = x(i); x(i) = t;
  end
  k = nv2;
  while k < j
     j = j - k; k = k/2;
  end
  j = j + k;
end
```

Plot the time taken by this algorithm on your computer for $m = 5 : 10$. Does the execution time depend linearly on $N = 2^m$?

4.7.2 The following MATLAB program (Moler and Eddins [269]) demonstrates how the FFT idea can be implemented in a simple but efficient recursive MATLAB program. The program uses the fast recursion as long as n is a power of two. When it reaches an odd length it sets up the Fourier matrix and uses matrix-vector multiplication.

ALGORITHM 4.3. *Fast Fourier Transform.*

```
function y = fftx(x);
% FFT computes the Fast Fourier Transform of x(1:n)
x = x(:);
n = length(x);
omega = exp(-2*pi*i/n);
if rem(n,2) == 0
% Recursive divide and conquer
   k = (0:n/2-1)
   w = omega.^k;
   u = fftx(x(1:2:n-1));
   v = w.*fftx(x(2:2:n));
   y = [u+v; u-v];
else
% Generate the Fourier matrix
   j = 0:n-1;
   k = j';
   F = omega.^(k*j);
   y = F*x;
end
```

Apply this program to compute DFT of the rectangular wave in Sec. 4.6.1, sampled at the points $2\pi\alpha/N$, $\alpha = 0 : N-1$. Choose, for instance, $N = 32, 64, 128$.

4.7.3 Write an efficient MATLAB program for computing the DFT of a real data vector of length $N = 2^m$. As outlined in Sec. 4.7.3, first split the data in odd and even data points. Compute the corresponding DFTs using one call of the function `fftx` in Problem 4.7.2 with complex data of length $N/2$.

4.7.4 Verify the last four symmetry properties of DFTs in Table 4.7.1.

Notes and References

A modern treatment of divided differences considered as linear functionals on some vector space of functions is given by de Boor [38]. Our notation $[x_1, \dots, x_k]f$ for the divided difference agrees with that used in de Boor [37]. A general treatment of complex polynomial interpolation on complex nodes is given by Gelfond [151, Sec. 1.4.3]; see also Horn and Johnson [203, Sec. 6.1].

The barycentric form of Lagrange's interpolation formula found in German literature does not seem to have caught on elsewhere, until recently. Berrut and Trefethen [24] argue convincingly that this should be the standard method. The problem of choosing a good ordering of points in Newton and barycentric Lagrange interpolations is discussed in Werner [371]. Newton interpolation using the Leja ordering of points has been analyzed by Reichel [299].

Uniform methods of solving linear problems and computing the inverse of a Vandermonde matrix have been studied by Traub [355]. The algorithm for the inverse Vandermonde matrix given in the text is due to Higham [199, Sec. 22.1]. The $O(n^2)$ Björck–Pereyra algorithm for solving Vandermonde systems appeared in [31].

An introduction to rational interpolation is given in Stoer and Bulirsch [338, Sec. 2.2]. The reciprocal differences of Thiele are also treated by Milne-Thomson [265], with expressions for the truncation error. The ρ-algorithm for convergence acceleration is due to Wynn [385].

A good reference for multidimensional polynomial interpolation is Steffensen [330, Sec. 19]; see also Isaacson and Keller [208, Sec. 6.6]. Methods for multidimensional interpolation of scattered data points, including radial basis functions, are surveyed by Powell in [293]. For a more practical approach the survey by Hayes [189] is still very good. The history of multidimensional interpolation is surveyed in [134].

The survey paper [27] gives a good summary of the pioneering work done by Garret Birkhoff on interpolation and approximation of univariate and bivariate data. The book by de Boor [37] has done much to further the use of B-splines in geometric constructions. It contains a clear outline of the theory and also a library of Fortran programs for computations with spline functions. Several packages are available for computing with splines, e.g., the spline toolbox in MATLAB and FITPACK by Dierckx [97, 98]. A more geometric view of splines is taken in Farin [116].

The computational advantage of the Stieltjes approach for discrete least squares fitting was pointed out by Forsythe [119]. Shampine [321] established the advantage of using the alternative formula involving the residual r_k.

The theory and application of Chebyshev systems in approximation and statistics is surveyed in Karlin and Studden [222]. An application of uniform approximation with polynomials and the Remez algorithm is the determination of nonrecursive digital filters with minimal ripple; see Parks and McClellan [284]. The discrete ℓ_1 approximation problem is related to "sparse approximation" ideas that are currently attracting much attention in the signal processing community.

The series named after Fourier was well established by the work of Bernoulli, Euler, and Lagrange, before the time of Fourier. The Fourier integral is, however, the undisputed discovery of Fourier. A very readable introduction to harmonic analysis is given in Lanczos [235, Chapter IV]. For a more exhaustive treatment of Gibbs' phenomenon, see Tadmor [348].

The rediscovery of the FFT algorithm is surveyed in [76]. Applications are surveyed in [77, 58, 57]. The matrix-oriented framework for the FFT used in this book is due to Van Loan [366]. An early implementation of the FFT is given by Singleton [324, 325]. For a roundoff error analysis of the FFT see [8]. Algorithms for bit-reversal permutations are reviewed in [223].

The FFT algorithm was discovered independently by several mathematicians. Similar ideas were used already by Gauss [137]. The doubling algorithm is contained in the textbook by Runge and König [309] from 1924 and in the paper by Danielson and Lanczos [90].

Some significant developments in approximation theory are not covered in this book, such as wavelets (see Daubechies [91]), radial basis functions (see Buhmann [59]), and box splines (see de Boor, Höllig, and Riemenschneider [39]). The last two constructs are different multidimensional generalizations of univariate spline functions.

Chapter 5

Numerical Integration

Commit your blunders on a small scale and make your profits on a large scale.
—Leo Hendrik Baekeland

5.1 Interpolatory Quadrature Rules

5.1.1 Introduction

In this chapter we study the approximate calculation of a definite integral

$$I[f] = \int_a^b f(x)\,dx, \tag{5.1.1}$$

where $f(x)$ is a given function and $[a, b]$ a finite interval. This problem is often called **numerical quadrature**, since it relates to the ancient problem of the quadrature of the circle, i.e., constructing a square with equal area to that of a circle. The computation of (5.1.1) is equivalent to solving the initial value problem

$$y'(x) = f(x), \quad y(a) = 0, \quad x \in [a, b] \tag{5.1.2}$$

for $y(b) = I[f]$; cf. Sec. 1.5.

As is well known, even many relatively simple integrals cannot be expressed in finite terms of elementary functions, and thus must be evaluated by numerical methods. (For a table of integrals that have closed analytical solutions, see [171].) Even when a closed form analytical solution exists it may be preferable to use a numerical quadrature formula.

Since $I[f]$ is a linear functional, numerical integration is a special case of the problem of approximating a linear functional studied in Sec. 3.3.4. The quadrature rules considered will be of the form

$$I[f] \approx \sum_{i=1}^{n} w_i f(x_i), \tag{5.1.3}$$

where $x_1 < x_2 < \cdots < x_n$ are distinct **nodes** and $w_1, w_2, \ldots, w_n$ the corresponding **weights**. Often (but not always) all nodes lie in $[a, b]$.

The weights w_i are usually determined so that the formula (5.1.3) is exact for polynomials of as high degree as possible. The accuracy therefore depends on how well the integrand $f(x)$ can be approximated by a polynomial in $[a, b]$. If the integrand has a singularity, for example, it becomes infinite at some point in or near the interval of integration, some modification is necessary. Another complication arises when the interval of integration is infinite. In both cases it may be advantageous to consider a weighted quadrature rule:

$$\int_a^b f(x)w(x)\,dx \approx \sum_{i=1}^n w_i f(x_i). \tag{5.1.4}$$

Here $w(x) \geq 0$ is a **weight function** (or density function) that incorporates the singularity so that $f(x)$ can be well approximated by a polynomial. The limits (a, b) of integration are now allowed to be infinite.

To ensure that the integral (5.1.4) is well defined when $f(x)$ is a polynomial, we assume in the following that the integrals

$$\mu_k = \int_a^b x^k w(x)\,dx, \quad k = 1, 2, \ldots, \tag{5.1.5}$$

are defined for all $k \geq 0$, and $\mu_0 > 0$. The quantity μ_k is called the kth **moment** with respect to the weight function $w(x)$. Note that for the formula (5.1.4) to be exact for $f(x) = 1$ it must hold that

$$\mu_0 = \int_a^b 1 \cdot w(x)\,dx = \sum_{i=1}^n w_i. \tag{5.1.6}$$

In the special case that $w(x) = 1$, we have $\mu_0 = b - a$.

Definition 5.1.1.

A quadrature rule (5.1.3) *has* **order of accuracy** *(or degree of exactness) equal to d if it is exact for all polynomials of degree $\leq d$, i.e., for all $p \in \mathcal{P}_{d+1}$.*

In a weighted **interpolatory** quadrature formula the integral is approximated by

$$\int_a^b p(x)w(x)\,dx,$$

where $p(x)$ is the unique polynomial of degree $n-1$ interpolating $f(x)$ at the distinct points $x_1, x_2, \ldots, x_n$. By Lagrange's interpolation formula (Theorem 4.1.1)

$$p(x) = \sum_{i=1}^n f(x_i)\ell_i(x), \qquad \ell_i(x) = \prod_{\substack{j=1\\ j\neq i}}^n \frac{(x - x_j)}{(x_i - x_j)},$$

where $\ell_i(x)$ are the elementary Lagrange polynomials associated with the nodes $x_1, x_2, \ldots, x_n$. It follows that for an interpolatory quadrature formula the weights are given by

$$w_i = \int_a^b \ell_i(x)w(x)\,dx. \tag{5.1.7}$$

In practice, the coefficients are often more easily computed using the method of undetermined coefficients rather than by integrating $\ell_i(x)$.

An expression for the truncation error is obtained by integrating the remainder (see Theorems 4.2.3 and 4.2.4):

$$\begin{aligned} R_n(f) &= \int_a^b [x_1, \dots, x_n, x]f \prod_{i=1}^{n}(x - x_i)w(x)\,dx \\ &= \frac{1}{n!}\int_a^b f^{(n)}(\xi_x)\prod_{i=1}^{n}(x - x_i)w(x)\,dx, \quad \xi_x \in [a, b], \end{aligned} \tag{5.1.8}$$

where the second expression holds if $f^{(n)}$ is continuous in $[a, b]$.

Theorem 5.1.2.

For any given set of nodes $x_1, x_2, \dots, x_n$ an interpolatory quadrature formula with weights given by (5.1.7) has order of exactness equal to at least $n - 1$. Conversely, if the formula has degree of exactness $n - 1$, then the formula must be interpolatory.

Proof. For any $f \in \mathcal{P}_n$ we have $p(x) = f(x)$, and hence (5.1.7) has degree of exactness at least equal to $n - 1$. On the other hand, if the degree of exactness of (5.1.7) is $n - 1$, then putting $f = \ell_i(x)$ shows that the weights w_i satisfy (5.1.7); that is, the formula is interpolatory. □

In general the function values $f(x_i)$ cannot be evaluated exactly. Assume that the error in $f(x_i)$ is e_i, where $|e_i| \le \epsilon$, $i = 1 : n$. Then, if $w_i \ge 0$, the related error in the quadrature formula satisfies

$$\left|\sum_{i=1}^{n} w_i e_i\right| \le \epsilon \sum_{i=1}^{n} |w_i| \le \epsilon\mu_0. \tag{5.1.9}$$

The last upper bound holds only if *all weights in the quadrature rules are positive.*

So far we have assumed that all the nodes x_i of the quadrature formula are fixed. A natural question is whether we can do better by a judicious choice of the nodes. This question is answered positively in the following theorem. Indeed, by a careful choice of nodes the order of accuracy of the quadrature rule can be substantially improved.

Theorem 5.1.3.

Let k be an integer such that $0 \le k \le n$. Consider the integral

$$I[f] = \int_a^b f(x)w(x)\,dx,$$

and an interpolatory quadrature rule

$$I_n(f) = \sum_{i=1}^{n} w_i f(x_i),$$

using n nodes. Let

$$\gamma(x) = \prod_{i=1}^{n}(x - x_i) \tag{5.1.10}$$

be the corresponding **node polynomial**. *Then the quadrature rule $I[f] \approx I_n(f)$ has degree of exactness equal to $d = n + k - 1$ if and only if, for all polynomials $p \in \mathcal{P}_k$, the node polynomial satisfies*

$$\int_a^b p(x)\gamma(x)w(x)\,dx = 0. \tag{5.1.11}$$

Proof. We first prove the *necessity* of the condition (5.1.11). For any $p \in \mathcal{P}_k$ the product $p(x)\gamma(x)$ is in $\mathcal{P}_{n+k}$. Then since $\gamma(x_i) = 0$, $i = 1 : n$,

$$\int_a^b p(x)\gamma(x)w(x)\,dx = \sum_{i=1}^{n} w_i f(x_i)\gamma(x_i) = 0,$$

and thus (5.1.11) holds.

To prove the *sufficiency*, let $p(x)$ be any polynomial of degree $n + k - 1$. Let $q(x)$ and $r(x)$ be the quotient and remainder, respectively, in the division

$$p(x) = q(x)\gamma(x) + r(x).$$

Then $q(x)$ and $r(x)$ are polynomials of degree $k - 1$ and $n - 1$, respectively. It holds that

$$\int_a^b p(x)w(x)\,dx = \int_a^b q(x)\gamma(x)w(x)\,dx + \int_a^b r(x)w(x)\,dx,$$

where the first integral on the right-hand side is zero because of the orthogonality property of $\gamma(x)$. For the second integral we have

$$\int_a^b r(x)w(x)\,dx = \sum_{i=1}^{n} w_i r(x_i),$$

since the weights were chosen such that the formula was interpolatory and therefore exact for all polynomials of degree $n - 1$. Further, since

$$p(x_i) = q(x_i)\gamma(x_i) + r(x_i) = r(x_i), \quad i = 1 : n,$$

it follows that

$$\int_a^b p(x)w(x)\,dx = \int_a^b r(x)w(x)\,dx = \sum_{i=1}^{n} w_i r(x_i) = \sum_{i=1}^{n} w_i p(x_i),$$

which shows that the quadrature rule is exact for $p(x)$. □

How to determine quadrature rules of optimal order will be the topic of Sec. 5.3.

5.1.2 Treating Singularities

When the integrand or some of its low-order derivative is infinite at some point in or near the interval of integration, standard quadrature rules will not work well. It is not uncommon that a single step taken close to such a singular point will give a larger error than all other steps combined. In some cases a singularity can be completely missed by the quadrature rule.

If the singular points are known, then the integral should first be broken up into several pieces so that all the singularities are located at one (or both) ends of the interval $[a, b]$. Many integrals can then be treated by weighted quadrature rules, i.e., the singularity is incorporated into the weight function. Romberg's method can be modified to treat integrals where the integrand has an algebraic endpoint singularity; see Sec. 5.2.2.

It is often profitable to investigate whether one can transform or modify the given problem analytically to make it more suitable for numerical integration. Some difficulties and possibilities in numerical integration are illustrated below in a series of simple examples.

Example 5.1.1.

In the integral

$$I = \int_0^1 \frac{1}{\sqrt{x}} e^x \, dx$$

the integrand is infinite at the origin. By the substitution $x = t^2$ we get

$$I = 2 \int_0^1 e^{t^2} \, dt,$$

which can be treated without difficulty.

Another possibility is to use integration by parts:

$$\begin{aligned} I &= \int_0^1 x^{-1/2} e^x \, dx = 2x^{1/2} e^x \big|_0^1 - 2 \int_0^1 x^{1/2} e^x \, dx \\ &= 2e - 2\frac{2}{3} x^{3/2} e^x \big|_0^1 + \frac{4}{3} \int_0^1 x^{3/2} e^x \, dx = \frac{2}{3} e + \frac{4}{3} \int_0^1 x^{3/2} e^x \, dx. \end{aligned}$$

The last integral has a mild singularity at the origin. If one wants high accuracy, then it is advisable to integrate by parts a few more times before the numerical treatment.

Example 5.1.2.

Sometimes a simple comparison problem can be used. In

$$I = \int_{0.1}^1 x^{-3} e^x \, dx$$

the integrand is infinite near the left endpoint. If we write

$$I = \int_{0.1}^1 x^{-3} \left(1 + x + \frac{x^2}{2}\right) dx + \int_{0.1}^1 x^{-3} \left(e^x - 1 - x - \frac{x^2}{2}\right) dx,$$

the first integral can be computed analytically. The second integrand can be treated numerically. The integrand and its derivatives are of moderate size. Note, however, the cancellation in the evaluation of the integrand.

For integrals over an infinite interval one can try some substitution which maps the interval $(0, \infty)$ to $(0, 1)$, for example, $t = e^{-x}$ of $t = 1/(1 + x)$. But in such cases one must be careful not to introduce an unpleasant singularity into the integrand instead.

Example 5.1.3.

More general integrals of the form

$$\int_0^{2h} x^{-1/2} f(x)\, dx$$

need a special treatment, due to the integrable singularity at $x = 0$. A formula which is exact for any second-degree polynomial $f(x)$ can be found using the method of undetermined coefficients. We set

$$\frac{1}{\sqrt{2h}} \int_0^{2h} x^{-1/2} f(x)\, dx \approx w_0 f(0) + w_1 f(h) + w_2 f(2h),$$

and equate the left- and right-hand sides for $f(x) = 1, x, x^2$. This gives

$$w_0 + w_1 + w_2 = 2, \qquad \frac{1}{2} w_1 + w_2 = \frac{2}{3}, \qquad \frac{1}{4} w_1 + w_2 = \frac{2}{5}.$$

This linear system is easily solved, giving $w_0 = 12/15$, $w_1 = 16/15$, $w_2 = 2/15$.

Example 5.1.4.

Consider the integral

$$I = \int_0^{\infty} (1 + x^2)^{-4/3}\, dx.$$

If one wants five decimal digits in the result, then $\int_R^{\infty}$ is not negligible until $R \approx 10^3$. But one can expand the integrand in powers of x^{-1} and integrate termwise:

$$\begin{aligned} \int_R^{\infty} (1 + x^2)^{-4/3}\, dx &= \int_R^{\infty} x^{-8/3} (1 + x^{-2})^{-4/3}\, dx \\ &= \int_R^{\infty} \left(x^{-8/3} - \frac{4}{3} x^{-14/3} + \frac{14}{9} x^{-20/3} - \cdots \right) \\ &= R^{-5/3} \left(\frac{3}{5} - \frac{4}{11} R^{-2} + \frac{14}{51} R^{-4} - \cdots \right). \end{aligned}$$

If this expansion is used, then one need only apply numerical integration to the interval $[0, 8]$.

With the substitution $t = 1/(1 + x)$ the integral becomes

$$I = \int_0^1 (t^2 + (1 - t)^2)^{-4/3} t^{2/3}\, dt.$$

The integrand now has an infinite derivative at the origin. This can be eliminated by making the substitution $t = u^3$ to get

$$I = \int_0^1 (u^6 + (1-u^3)^2)^{-4/3} 3u^4\, du,$$

which can be computed with, for example, a Newton–Cotes' method.

5.1.3 Some Classical Formulas

Interpolatory quadrature formulas, where the nodes are constrained to be equally spaced, are called **Newton–Cotes**[167] formulas. These are especially suited for integrating a tabulated function, a task that was more common before the computer age. The midpoint, trapezoidal, and Simpson's rules, to be described here, are all special cases of (unweighted) Newton–Cotes' formulas.

The **trapezoidal rule** (cf. Figure 1.1.5) is based on linear interpolation of $f(x)$ at $x_1 = a$ and $x_2 = b$; i.e., $f(x)$ is approximated by

$$p(x) = f(a) + (x-a)[a,b]f = f(a) + (x-a)\frac{f(b)-f(a)}{b-a}.$$

The integral of $p(x)$ equals the area of a trapezoid with base $(b-a)$ times the average height $\frac{1}{2}(f(a)+f(b))$. Hence

$$\int_a^b f(x)\,dx \approx \frac{(b-a)}{2}(f(a)+f(b)).$$

To increase the accuracy we subdivide the interval $[a,b]$ and assume that $f_i = f(x_i)$ is known on a grid of equidistant points

$$x_0 = a, \qquad x_i = x_0 + ih, \qquad x_n = b, \tag{5.1.12}$$

where $h = (b-a)/n$ is the **step length**. The trapezoidal approximation for the ith subinterval is

$$\int_{x_i}^{x_{i+1}} f(x)\,dx = T(h) + R_i, \qquad T(h) = \frac{h}{2}(f_i + f_{i+1}). \tag{5.1.13}$$

Assuming that $f''(x)$ is continuous in $[a,b]$ and using the exact remainder in Newton's interpolation formula (see Theorem 4.2.1) we get

$$R_i = \int_{x_i}^{x_{i+1}} (f(x) - p_2(x))\,dx = \int_{x_i}^{x_{i+1}} (x-x_i)(x-x_{i+1})\,[x_i, x_{i+1}, x]f\,dx. \tag{5.1.14}$$

Since $[x_i, x_{i+1}, x]f$ is a continuous function of x and $(x-x_i)(x-x_{i+1})$ has constant (negative) sign for $x \in [x_i, x_{i+1}]$, the mean value theorem of integral calculus gives

$$R_i = [x_i, x_{i+1}, \xi_i]f \int_{x_i}^{x_{i+1}} (x-x_i)(x-x_{i+1})\,dx, \qquad \xi_i \in [x_i, x_{i+1}].$$

[167] Roger Cotes (1682–1716) was a highly appreciated young colleague of Isaac Newton. He was entrusted with the preparation of the second edition of Newton's *Principia*. He worked out and published the coefficients for Newton's formulas for numerical integration for $n \leq 11$.

Setting $x = x_i + ht$ and using the Theorem 4.2.3, we get

$$R_i = -\frac{1}{2} f''(\zeta_i) \int_0^1 h^2 t(t-1) h \, dt = -\frac{1}{12} h^3 f''(\zeta_i), \quad \zeta_i \in [x_i, x_{i+1}]. \tag{5.1.15}$$

For another proof of this result using the Peano kernel see Example 3.3.16.

Summing the contributions for each subinterval $[x_i, x_{i+1}]$, $i = 0 : n$, gives

$$\int_a^b f(x)\,dx = T(h) + R_T, \quad T(h) = \frac{h}{2}(f_0 + f_n) + h \sum_{i=2}^{n-1} f_i, \tag{5.1.16}$$

which is the **composite trapezoidal rule**. The **global** truncation error is

$$R_T = -\frac{h^3}{12} \sum_{i=0}^{n-1} f''(\zeta_i) = -\frac{1}{12}(b-a) h^2 f''(\xi), \quad \xi \in [a, b]. \tag{5.1.17}$$

(The last equality follows since f'' was assumed to be continuous on the interval $[a, b]$.) This shows that by choosing h small enough we can make the truncation error arbitrarily small. In other words, we have **asymptotic convergence** when $h \to 0$.

In the **midpoint rule** $f(x)$ is approximated on $[x_i, x_{i+1}]$ by its value

$$f_{i+1/2} = f(x_{i+1/2}), \quad x_{i+1/2} = \frac{1}{2}(x_i + x_{i+1}),$$

at the midpoint of the interval. This leads to the approximation

$$\int_{x_i}^{x_{i+1}} f(x)\,dx = M(h) + R_i, \quad M(h) = h f_{i+1/2}. \tag{5.1.18}$$

The midpoint rule approximation can be interpreted as the area of the trapezium defined by the tangent of f at the midpoint $x_{i+1/2}$.

The remainder term in Taylor's formula gives

$$f(x) - (f_{i+1/2} + (x - x_{i+1/2}) f'_{i+1/2}) = \frac{1}{2}(x - x_{i+1/2})^2 f''(\zeta_x), \quad \zeta_x \in [x_i, x_{i+1/2}].$$

By symmetry the integral over $[x_i, x_{i+1}]$ of the linear term vanishes. We can use the mean value theorem to show that

$$R_i = \int_{x_i}^{x_{i+1}} \frac{1}{2} f''(\zeta_x)(x - x_{i+1/2})^2 \, dx = \frac{1}{2} f''(\zeta_i) \int_{-1/2}^{1/2} h^3 t^2 \, dt = \frac{h^3}{24} f''(\zeta_i).$$

Although it uses just *one function value*, the midpoint rule, like the trapezoidal rule, is exact when $f(x)$ is a linear function. Summing the contributions for each subinterval, we obtain the **composite midpoint rule**:

$$\int_a^b f(x)\,dx = M(h) + R_M, \quad M(h) = h \sum_{i=0}^{n-1} f_{i+1/2}. \tag{5.1.19}$$

(Compare the above approximation with the Riemann sum in the *definition* of a definite integral.) For the global error we have

$$R_M = \frac{(b-a)h^2}{24} f''(\zeta), \qquad \zeta \in [a, b]. \tag{5.1.20}$$

The trapezoidal rule is called a **closed rule** because values of f at both endpoints are used. It is not uncommon that f has an integrable singularity at an endpoint. In that case an **open rule**, like the midpoint rule, can still be applied.

If $f''(x)$ has constant sign in each subinterval, then the error in the midpoint rule is approximately half as large as that for the trapezoidal rule and has the opposite sign. But the trapezoidal rule is more economical to use when a sequence of approximations for $h, h/2, h/4, \dots$ is to be computed, since about half of the values needed for $h/2$ were already computed and used for h. Indeed, it is easy to verify the following useful relation between the trapezoidal and midpoint rules:

$$T\left(\frac{h}{2}\right) = \frac{1}{2}(T(h) + M(h)). \tag{5.1.21}$$

If the magnitude of the error in the function values does not exceed $\frac{1}{2}U$, then the magnitude of the propagated error in the approximation for the trapezoidal and midpoint rules is bounded by

$$R_A = \frac{1}{2}(b-a)U, \tag{5.1.22}$$

independent of h. By (5.1.9) this holds *for any quadrature formula of the form* (5.1.3), *provided that all weights* w_i *are positive.*

If the roundoff error is negligible and h sufficiently small, then it follows from (5.1.17) that the error in $T(h/2)$ is about one-quarter of that in $T(h)$. Hence the magnitude of the error in $T(h/2)$ can be estimated by $(1/3)|T(h/2) - T(h)|$, or more conservatively by $|T(h/2) - T(h)|$. (A more systematic use of Richardson extrapolation is made in Romberg's method; see Sec. 5.2.2.)

Example 5.1.5.

Use (5.1.21) to compute the sine integral $\text{Si}(x) = \int_0^x \frac{\sin t}{t}\, dt$ for $x = 0.8$. Midpoint and trapezoidal sums (correct to eight decimals) are given below.

h	$M(h)$	$T(h)$
0.8	0.77883 668	0.75867 805
0.4	0.77376 698	0.76875 736
0.2	0.77251 272	0.77126 217
0.1		0.77188 744

The correct value, to ten decimals, is 0.77209 57855 (see [1, Table 5.2]). We verify that in this example the error is approximately proportional to h^2 for both $M(h)$ and $T(h)$, and we estimate the error in $T(0.1)$ to be $\frac{1}{3}6.26 \cdot 10^{-4} \leq 2.1 \cdot 10^{-4}$.

From the error analysis above we note that the error in the midpoint rule is roughly half the size of the error in the trapezoidal rule and of opposite sign. Hence it seems that the linear combination

$$S(h) = \frac{1}{3}(T(h) + 2M(h)) \tag{5.1.23}$$

should be a better approximation. This is indeed the case and (5.1.23) is equivalent to **Simpson's rule.**[168]

Another way to derive Simpson's rule is to approximate $f(x)$ by a piecewise polynomial of third degree. It is convenient to shift the origin to the midpoint of the interval and consider the integral over the interval $[x_i - h, x_i + h]$. From Taylor's formula we have

$$f(x) = f_i + (x - x_i)f_i' + \frac{(x - x_i)^2}{2} f_i'' + \frac{(x - x_i)^3}{3!} f_i''' + O(h^4),$$

where the remainder is zero for all polynomials of degree three or less. Integrating term by term, the integrals of the second and fourth term vanish by symmetry, giving

$$\int_{x_i-h}^{x_i+h} f(x)\,dx = 2hf_i + 0 + \frac{1}{3}h^3 f_i'' + 0 + O(h^5).$$

Using the difference approximation $h^2 f_i'' = (f_{i-1} - 2f_i + f_{i+1}) + O(h^4)$ (see (4.7.5)) we obtain

$$\begin{aligned} \int_{x_i-h}^{x_i+h} f(x)\,dx &= 2hf_i + \frac{1}{3}h(f_{i-1} - 2f_i + f_{i+1}) + O(h^5) \qquad (5.1.24)\\ &= \frac{1}{3}h(f_{i-1} + 4f_i + f_{i+1}) + O(h^5), \end{aligned}$$

where the remainder term is zero for all third-degree polynomials. We now determine the error term for $f(x) = (x - x_i)^4$, which is

$$R_T = \frac{1}{3}h(h^4 + 0 + h^4) - \int_{x_i-h}^{x_i+h} x^4\,dx = \left(\frac{2}{3} - \frac{2}{5}\right)h^5 = \frac{4}{15}h^5.$$

It follows that an *asymptotic error estimate* for Simpson's rule is

$$R_T = h^5 \frac{4}{15} \frac{f^{(4)}(x_i)}{4!} + O(h^6) = \frac{h^5}{90} f^{(4)}(x_i) + O(h^6). \tag{5.1.25}$$

A strict error estimate for Simpson's rule is more difficult to obtain. As for the midpoint formula, the midpoint x_i can be considered as a **double point** of interpolation; see Problem 5.1.3. The general error formula (5.1.8) then gives

$$R_i(f) = \frac{1}{4!} \int_{x_{i-1}}^{x_{i+1}} f^{(4)}(\xi_x)(x - x_{i-1})(x - x_i)^2(x - x_{i+1})\,dx,$$

[168]Thomas Simpson (1710–1761), English mathematician best remembered for his work on interpolation and numerical methods of integration. He taught mathematics privately in the London coffee houses and from 1737 began to write texts on mathematics.

where $(x - x_{i-1})(x - x_i)^2(x - x_{i+1})$ has constant negative sign on $[x_{i-1}, x_{i+1}]$. Using the mean value theorem gives the error

$$R_T(f) = \frac{1}{90} f^{(4)}(\xi) h^5, \quad \xi \in [x_i - h, x_i + h]. \tag{5.1.26}$$

The remainder can also be obtained from Peano's error representation. It can be shown (see [338, p. 152 ff]) that for Simpson's rule

$$Rf = \int_{\mathbf{R}} f^{(4)}(u) K(u)\, du,$$

where the kernel equals

$$K(u) = -\frac{1}{72}(h - u)^3 (3u + h)^2, \quad 0 \le u \le h,$$

and $K(u) = K(|u|)$ for $u < 0$, $K(u) = 0$ for $|u| > h$. This again gives (5.1.26).

In the **composite Simpson's rule** one divides the interval $[a, b]$ into an *even* number $n = 2m$ steps of length h and uses the formula (5.1.24) on each of m double steps, giving

$$\int_a^b f(x)\, dx = \frac{h}{3}(f_0 + 4S_1 + 2S_2 + f_n) + R_T, \tag{5.1.27}$$

where

$$S_1 = f_1 + f_3 + \cdots + f_{n-1}, \qquad S_2 = f_2 + f_4 + \cdots + f_{n-2}$$

are sums over odd and even indices, respectively. The remainder is

$$R_T(f) = \sum_{i=0}^{m-1} \frac{h^5}{90} f^{(4)}(\xi_i) = \frac{(b-a)}{180} h^4 f^{(4)}(\xi), \quad \xi \in [a, b]. \tag{5.1.28}$$

This shows that we have gained *two orders of accuracy* compared to the trapezoidal rule, without using more function evaluations. This is why Simpson's rule is such a popular general-purpose quadrature rule.

5.1.4 Superconvergence of the Trapezoidal Rule

In general the composite trapezoidal rule integrates exactly polynomials of degree 1 only. It does much better with trigonometric polynomials.

Theorem 5.1.4.
Consider the integral $\int_0^{2\pi} t_m(x)\, dx$, where

$$\begin{aligned} t_m(x) = a_0 + a_1 \cos x + a_2 \cos 2x + \cdots + a_m \cos mx \\ + b_1 \sin x + b_2 \sin 2x + \cdots + b_m \sin mx \end{aligned}$$

is any trigonometric polynomial of degree m. Then the composite trapezoidal rule with step length $h = 2\pi/n$, $n \ge m + 1$, integrates this exactly.

Proof. See Problem 5.1.16. □

Suppose that the function f is infinitely differentiable for $x \in \mathbf{R}$, and that f *has* $[a, b]$ *as an interval of periodicity*, i.e., $f(x + (b - a)) = f(x)$ for all $x \in \mathbf{R}$. Then

$$f^{(k)}(b) = f^{(k)}(a), \quad k = 0, 1, 2, \dots,$$

hence *every term in the Euler–Maclaurin expansion is zero* for the integral over *the whole period* $[a, b]$. One could be led to believe that the trapezoidal rule gives the exact value of the integral, but this is usually not the case. For most periodic functions f, $\lim_{r\to\infty} R_{2r+2} f \neq 0$; *the expansion converges, of course, though not necessarily to the correct result.*

On the other hand, the convergence as $h \to 0$ for a fixed (though arbitrary) r is a different story; the error bound (5.2.10) shows that

$$|R_{2r+2}(a, h, b) f| = O(h^{2r+2}).$$

Since r is arbitrary, this means that *for this class of functions, the trapezoidal error tends to zero faster than any power of h, as $h \to 0$*. We may call this **superconvergence**. The application of the trapezoidal rule to an integral over $[0, \infty)$ of a function $f \in C^\infty(0, \infty)$ often yields similar features, sometimes even more striking.

Suppose that the periodic function $f(z)$, $z = x + iy$, is analytic in a strip, $|y| < c$, around the real axis. It can then be shown that the error of the trapezoidal rule is

$$O(e^{-\eta/h}), \quad h \downarrow 0,$$

where η is related to the width of the strip. A similar result (3.2.19) was obtained in Sec. 3.2.2, for an annulus instead of a strip. There the trapezoidal rule was used in the calculation of the integral of a periodic analytic function over a full period $[0, 2\pi]$ that defined its Taylor coefficients. The error was shown to tend to zero faster than any power of the step length $\Delta\theta$.

As a rule, this discussion does *not* apply to periodic functions which are defined by periodic continuation of a function originally defined on $[a, b]$ (such as the Bernoulli functions). They usually become nonanalytic at a and b, and at all points $a + (b - a)n$, $n = 0, \pm 1, \pm 2, \dots$.

The **Poisson summation formula** is even better than the Euler–Maclaurin formula for the quantitative study of the trapezoidal truncation error on an infinite interval. For convenient reference we now formulate the following surprising result.

Theorem 5.1.5.

Suppose that the trapezoidal rule (or, equivalently, the rectangle rule) is applied with constant *step size h to $\int_{-\infty}^{\infty} f(x)\,dx$. The Fourier transform of f reads*

$$\hat{f}(\omega) = \int_{-\infty}^{\infty} f(x) e^{-i\omega t}\,dt.$$

Then the integration error decreases like $2\hat{f}(2\pi/h)$ as $h \downarrow 0$.

Example 5.1.6.

For the normal probability density, we have

$$f(x) = \frac{1}{\sigma\sqrt{2\pi}} e^{-\frac{1}{2}(\frac{t}{\sigma})^2}, \qquad \hat{f}(\omega) = e^{-\frac{1}{2}(\omega\sigma)^2}.$$

The integration error is thus approximately $2\exp(-2(\pi\sigma/h)^2)$. Roughly speaking, the number of correct digits is doubled if h is divided by $\sqrt{2}$; for example, the error is approximately $5.4\cdot 10^{-9}$ for $h=\sigma$, and $1.4\cdot 10^{-17}$ for $h=\sigma/\sqrt{2}$.

The application of the trapezoidal rule to an integral over $[0,\infty)$ of a function $f\in C^\infty(0,\infty)$ often yields similar features, sometimes even more striking. Suppose that, for $k=1,2,3,\dots,$

$$f^{(2k-1)}(0)=0 \text{ and } f^{(2k-1)}(x)\to 0, \quad x\to\infty,$$

and

$$\int_0^\infty |f^{(2k)}(x)|\,dx<\infty.$$

(Note that for any function $g\in C^\infty(-\infty,\infty)$ the function $f(x)=g(x)+g(-x)$ satisfies such conditions at the origin.) Then all terms of the Euler–Maclaurin expansion are zero, and one can be misled to believe that the trapezoidal sum gives $\int_0^\infty f(x)\,dx$ exactly for any step size h! The explanation is that the remainder $R_{2r+2}(a,h,\infty)$ will typically not tend to zero, as $r\to\infty$ for fixed h. On the other hand, if we consider the behavior of the truncation error as $h\to 0$ for given r, we find that it is $o(h^{2r})$ for any r, just like the case of a periodic function.

For a finite subinterval of $[0,\infty)$, however, the remainder is still typically $O(h^2)$, and for each step the remainder is typically $O(h^3)$. So, there is an *enormous cancellation of the local truncation errors*, when a C^∞-function with vanishing odd-order derivatives at the origin is integrated by the trapezoidal rule over $[0,\infty)$.

Example 5.1.7.

For integrals of the form $\int_{-\infty}^{\infty} f(x)\,dx$, the trapezoidal rule (or the midpoint rule) often gives good accuracy if one integrates over the interval $[-R_1,R_2]$, assuming that $f(x)$ and its lower derivatives are small for $x\le -R_1$ and $x\ge R_2$.

The correct value to six decimal digits of the integral $\int_{-\infty}^{\infty} e^{-x^2}\,dx$ is $\pi^{1/2}=1.772454$. For $x\pm 4$, the integrand is less than $0.5\cdot 10^{-6}$. Using the trapezoidal rule with $h=1/2$ for the integral over $[-4,4]$ we get the estimate 1.772453, an amazingly good result. (The function values have been taken from a six-place table.) The truncation error in the value of the integral is here less than 1/10,000 of the truncation error in the largest term of the trapezoidal sum—a superb example of "cancellation of truncation error." The error committed when we replace ∞ by 4 can be estimated in the following way:

$$|R|=2\int_4^\infty e^{-x^2}\,dx=\int_{16}^\infty e^{-t}\frac{1}{\sqrt{t}}\,dt<\int_{16}^\infty e^{-t}\frac{1}{\sqrt{16}}\,dt=\frac{1}{4}e^{-16}<10^{-7}.$$

5.1.5 Higher-Order Newton–Cotes' Formulas

The classical Newton–Cotes' quadrature rules are interpolatory rules obtained for $w(x)=1$ and equidistant points in $[0,1]$. There are two classes: **closed formulas**, where the endpoints of the interval belong to the nodes, and **open formulas**, where all nodes lie strictly in the interior of the interval (cf. the trapezoidal and midpoint rules).

The closed Newton–Cotes' formulas are usually written as

$$\int_0^{nh} f(x)\,dx = h\sum_{j=0}^{n} w_j f(jh) + R_n(f). \tag{5.1.29}$$

The weights satisfy $w_j = w_{n-j}$ and can, in principle, be determined from (5.1.7). Further, by (5.1.6) it holds that

$$\sum_{j=0}^{n} hw_j = nh. \tag{5.1.30}$$

(Note that here we sum over $n+1$ points in contrast to our previous notation.)

It can be shown that the closed Newton–Cotes' formulas have order of accuracy $d = n$ for n odd and $d = n+1$ for n even. The extra accuracy for n even is, as in Simpson's rule, due to symmetry. For $n \le 7$ all coefficients w_i are positive, but for $n = 8$ and $n \ge 10$ negative coefficients appear. Such formulas may still be useful, but since $\sum_{j=0}^{n} h|w_j| > nh$, they are less robust with respect to errors in the function values f_i.

The closed Newton–Cotes' formulas for $n = 1$ and $n = 2$ are equivalent to the trapezoidal rule and Simpson's rule, respectively. The formula for $n = 3$ is called the $3/8$th rule, for $n = 4$ Milne's rule, and for $n = 6$ Weddle's rule. The weights $w_i = Ac_i$ and error coefficient $c_{n,d}$ of Newton–Cotes' closed formulas are given for $n \le 6$ in Table 5.1.1.

Table 5.1.1. *The coefficients $w_i = Ac_i$ in the n-point closed Newton–Cotes' formulas.*

n	d	A	c_0	c_1	c_2	c_3	c_4	c_5	c_6	$c_{n,d}$
1	1	1/2	1	1						−1/12
2	3	1/3	1	4	1					−1/90
3	3	3/8	1	3	3	1				−3/80
4	5	2/45	7	32	12	32	7			−8/945
5	5	5/288	19	75	50	50	75	19		−275/12,096
6	7	1/140	41	236	27	272	27	236	41	−9/1400

The open Newton–Cotes' formulas are usually written as

$$\int_0^{nh} f(x)\,dx = h\sum_{i=1}^{n-1} w_i f(ih) + R_{n-1,n}(h)$$

and use $n-1$ nodes. The weights satisfy $w_{-j} = w_{n-j}$. The simplest open Newton–Cotes' formula for $n = 2$ is the midpoint rule with step size $2h$. The open formulas have order of accuracy $d = n-1$ for n even and $d = n-2$ for n odd. For the open formulas negative coefficients occur already for $n = 4$ and $n = 6$.

The weights and error coefficients of open formulas for $n \le 5$ are given in Table 5.1.2. We recognize the midpoint rule for $n = 2$. Note that the sign of the error coefficients in the open rules are opposite the sign in the closed rules.

Table 5.1.2. *The coefficients $w_i = Ac_i$ in the n-point open Newton–Cotes' formulas.*

n	d	A	c_1	c_2	c_3	c_4	c_5		$c_{n,d}$
2	1	2	1						1/24
3	1	3/2	1	1					1/4
4	3	4/3	2	−1	2				14/45
5	3	5/24	11	1	1	11			95/144
6	5	3/10	11	−14	26	−14	11		41/140
7	5	7/1440	611	−453	562	562	−453	611	5257/8640

The Peano kernels for both the open and the closed formulas can be shown to have constant sign (Steffensen [330]). Thus the local truncation error can be written as

$$R_n(h) = c_{n,d} h^{d+1} f^{(d)}(\zeta), \qquad \zeta \in [0, nh]. \tag{5.1.31}$$

It is easily shown that the Peano kernels for the corresponding composite formulas also have constant sign.

Higher-order Newton–Cotes' formulas can be found in [1, pp. 886–887]. We now show how they can be derived using the operator methods developed in Sec. 3.3.4. Let m, n be given integers and let h be a positive step size. In order to utilize the symmetry of the problem more easily, we move the origin to the midpoint of the interval of integration. If we set

$$x_j = jh, \qquad f_j = f(jh), \qquad j = -n/2 : 1 : n/2,$$

the Newton–Cotes' formula now reads

$$\int_{-mh/2}^{mh/2} f(x)\,dx = h \sum_{j=-n/2}^{n/2} w_j f_j + R_{m,n}(h), \qquad w_{-j} = w_j. \tag{5.1.32}$$

Note that j, $n/2$, and $m/2$ are not necessarily integers. For a Newton–Cotes' formula, $n/2 - j$ and $m/2 - j$ are evidently integers and hence $(m - n)/2$ is an integer too. There may, however, be other formulas, perhaps almost as good, where this is not the case. The coefficients $w_j = w_{j;m,n}$ are to be determined so that the remainder $R_{m,n}$ vanishes if $f \in \mathcal{P}_q$, with q as large as possible for given m, n.

The left-hand side of (5.1.32), divided by h, reads in operator form

$$(e^{mhD/2} - e^{-mhD/2})(hD)^{-1} f(x_0),$$

which is an even function of hD. By (3.3.42), hD is an odd function of δ. It follows that the left-hand side is an even function of δ; hence we can, for every m, write

$$(e^{hDm/2} - e^{-hDm/2})(hD)^{-1} \mapsto A_m(\delta^2) = a_{1m} + a_{2m}\delta^2 + \cdots + a_{k+1,m}\delta^{2k} \cdots. \tag{5.1.33}$$

We truncate after (say) δ^{2k}; the first neglected term is then $a_{k+2,m}\delta^{2k+2}$. We saw in Sec. 3.3.4 how to bring a truncated δ^2-expansion to $B(E)$-form,

$$b_1 + b_2(E + E^{-1}) + b_3(E^2 + E^{-2}) + \cdots + b_k(E^k + E^{-k}),$$

by matrix multiplication with a matrix M of the form given in (3.3.49). By comparison with (5.1.32), we conclude that $n/2 = k$, that the indices j are integers, and that $w_j = b_{j+1}$ (if $j \geq 0$). *If m is even, this becomes a Newton–Cotes' formula.* If m is odd, it may still be a useful formula, but it does not belong to the Newton–Cotes' family, because $(m - n)/2 = m/2 - k$ is no integer.

If $n = m$, a formula is of the closed type. Its remainder term is the first neglected term of the operator series, truncated after δ^{2k}, $2k = n = m$ (and multiplied by h). Hence the remainder of (5.1.32) can be estimated by $a_{2+m/2}\delta^{m+2} f_0$ or (better)

$$R_{m,m} \sim (a_{m/2+2}/m)H(hD)^{m+2} f_0,$$

where we call $H = mh$ the "bigstep."

If the integral is computed over $[a, b]$ by means of a sequence of bigsteps, each of length H, an estimate of the *global* error has the same form, except that H is replaced by $b - a$ and f_0 is replaced by $\max_{x\in[a,b]} |f(x)|$. The exponent of hD in an error estimate that contains H or $b - a$ is known as the *global order of accuracy* of the method.

If $n < m$, a formula of the open type is obtained. Among the open formulas we shall only consider the case that $n = m - 2$, which gives the open Newton–Cotes' formulas. The operator expansion is truncated after δ^{m-2}, and we obtain

$$R_{m-2,m} \sim (a_{m/2+1}/m)H(hD)^m f_0.$$

Formulas with $n > m$ are rarely mentioned in the literature (except for $m = 1$). We do not understand why; it is rather common that an integrand has a smooth continuation outside the interval of integration.

We next consider the effect of a linear transformation of the independent variable. Suppose that a formula

$$\sum_{j=1}^{N} a_j f(t_j) - \int_0^1 f(t)\,dt \approx c_N f^{(N)}$$

has been derived for the standard interval $[0, 1]$. Setting $x = x_k + th$, $dx = hdt$ we find that the corresponding formula and error constant for the interval $[x_k, x_k + h]$ reads

$$\sum_{j=1}^{N} a_j g(x_k + ht_j) - \int_{x_k}^{x_k+h} g(x)\,dx \approx c_N h^{N+1} g^{(N)}(x_k). \tag{5.1.34}$$

This error estimate is valid asymptotically as $h \to 0$. The **local order of accuracy**, i.e., over one step of length h, is $N + 1$; the **global order of accuracy**, i.e., over $(b - a)/h$ steps of length h, becomes N.

For example, the trapezoidal rule is exact for polynomials of degree 1 and hence $N = 2$. For the interval $[0, 1]$, $L(t^2) = \frac{1}{3}$, $\tilde{L}(t^2) = \frac{1}{2}$, and thus $c_2 = \frac{1}{2}(\frac{1}{2} - \frac{1}{3}) = \frac{1}{12}$. On an interval of length h the asymptotic error becomes $h^3 g''/12$. The local order of accuracy is $N+1=3$; the global order of accuracy is $N=2$.

If the "standard interval" is $[-1, 1]$ instead, the transformation becomes $x = \frac{1}{2}ht$, and h is to be replaced by $\frac{1}{2}h$ everywhere in (5.1.34). Be careful about the exact meaning of a remainder term for a formula of this type provided by a table.

We shall illustrate the use of the Cauchy–FFT method for computing the coefficients a_{im} in the expansion (5.1.33). In this way extensive algebraic calculations are avoided.[169] It can be shown that the exact coefficients are rational numbers, though it is sometimes hard to estimate in advance the order of magnitude of the denominators. The algorithm must be used with judgment. Figure 5.1.1 was obtained for $N=32$, $r=2$; the absolute errors of the coefficients are then less than 10^{-13}. The smoothness of the curves for $j \geq 14$ indicates that the relative accuracy of the values of $a_{m,j}$ are still good there; in fact, other computations show that it is still good when the coefficients are as small as 10^{-20}.

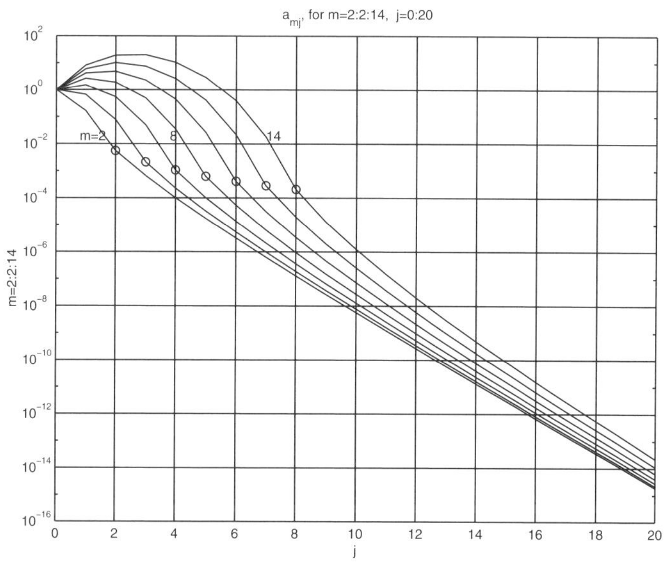

Figure 5.1.1. *The coefficients* $|a_{m,j}|$ *of the* δ^2*-expansion for* $m = 2:2:14$, $j = 0:20$. *The circles are the coefficients for the closed Cotes' formulas, i.e.,* $j = 1 + m/2$.

The coefficients are first obtained in floating-point representation. The transformation to rational form is obtained by a continued fraction algorithm, described in Example 3.5.3. For the case $m = 8$ the result reads

$$A_8(\delta^2) = 8 + \frac{64}{3}\delta^2 + \frac{688}{45}\delta^4 + \frac{736}{189}\delta^6 + \frac{3956}{14,175}\delta^8 - \frac{2368}{467,775}\delta^{10} + \cdots. \tag{5.1.35}$$

[169]These could, however, be carried out using a system such as Maple.

The closed integration formula becomes

$$\int_{-x_4}^{x_4} f(x)dx = \frac{4h}{14{,}175}\Big(-4540 f_0 + 10{,}496(f_1 + f_{-1}) - 928(f_2 + f_{-2})$$

$$+ 5888(f_3 + f_{-3}) + 989(f_4 + f_{-4})\Big) + R, \qquad (5.1.36)$$

$$R \sim \frac{296}{467{,}775} H h^{10} f^{(10)}(x_0). \qquad (5.1.37)$$

It goes without saying that this is not how Newton and Cotes found their methods. Our method may seem complicated, but the MATLAB programs for this are rather short, and to a large extent useful for other purposes. The computation of about 150 Cotes coefficients and 25 remainders ($m = 2 : 14$) took less than two seconds on a PC. This includes the calculation of several alternatives for rational approximations to the floating-point results. For a small number of the 150 coefficients the judicious choice among the alternatives took, however, much more than two (human) seconds; this detail is both science and art.

It was mentioned that *if m is odd*, (5.1.33) does not provide formulas of the Newton–Cotes' family, since $(m - n)/2$ is no integer, nor are the indices j in (5.1.32) integers. Therefore, the operator associated with the right-hand side of (5.1.32) is of the form

$$c_1(E^{1/2} + E^{-1/2}) + c_2(E^{3/2} + E^{-3/2}) + c_3(E^{5/2} + E^{-5/2}) + \cdots.$$

If it is divided algebraically by $\mu = \frac{1}{2}(E^{1/2} + E^{-1/2})$, however, it becomes the $B(E)$-form (say)

$$b_1' + b_2'(E + E^{-1}) + b_3'(E^2 + E^{-2}) + \cdots + b_k(E^k + E^{-k}).$$

If m is odd, we therefore expand

$$(e^{hDm/2} - e^{-hDm/2})(hD)^{-1}/\mu, \quad \mu = \sqrt{1 + \delta^2/4},$$

into a δ^2-series, with coefficients a_j'. Again, this can be done numerically by the Cauchy–FFT method. For each m, two truncated δ^2-series (one for the closed and one for the open case) are then transformed into $B(E)$-expressions numerically by means of the matrix M, as described above. The expressions are then multiplied *algebraically* by $\mu = (\frac{1}{2})(E^{1/2} + E^{-1/2})$. We then have the coefficients of a Newton–Cotes' formula with m odd.

The asymptotic error is

$$a_{m/2+1}' H(hD)^{m+1} \quad \text{and} \quad a_{m/2-1}' H(hD)^{m-1}$$

for the closed type and open type, respectively ($2k = m - 1$). The global orders of accuracy for Newton–Cotes' methods with odd m are thus the same as for the methods where m is one less.

5.1.6 Fejér and Clenshaw–Curtis Rules

Equally spaced interpolation points as used in the Newton–Cotes' formulas are useful for low degrees but can diverge as fast as 2^n as $n \to \infty$. Quadrature rules which use a set of points which cluster near the endpoints of the interval have better properties for large n.

Fejér [117] suggested using the zeros of a Chebyshev polynomial of first or second kind as interpolation points for quadrature rules of the form

$$\int_{-1}^{1} f(x)\,dx = \sum_{k=0}^{n} w_k f(x_k). \tag{5.1.38}$$

Fejér's first rule uses the zeros of the Chebyshev polynomial $T_n(x) = \cos(n \arccos x)$ of the first kind in $(-1, 1)$, which are

$$x_k = \cos\theta_k, \quad \theta_k = \frac{(2k-1)\pi}{2n}, \quad k = 1 : n. \tag{5.1.39}$$

The following explicit formula for the weights is known (see [93]):

$$w_k^{f1} = \frac{2}{n}\left(1 - 2\sum_{j=1}^{\lfloor n/2 \rfloor} \frac{\cos(2j\theta_k)}{4j^2 - 1}\right), \quad k = 1 : n. \tag{5.1.40}$$

Fejér's second rule uses the zeros of the Chebyshev polynomial $U_{n-1}(x)$ of the second kind, which are the extreme points of $T_n(x)$ in $(-1, 1)$ (see Sec. 3.2.3):

$$x_k = \cos\theta_k, \quad \theta_k = \frac{k\pi}{n}, \quad k = 1 : n - 1. \tag{5.1.41}$$

An explicit formula for the weights is (see [93])

$$w_k^{f2} = \frac{4\sin\theta_k}{n} \sum_{j=1}^{\lfloor n/2 \rfloor} \frac{\sin(2j-1)\theta_k}{2j-1}, \quad k = 1 : n - 1. \tag{5.1.42}$$

Both Fejér's rules are open quadrature rules, i.e., they do not use the endpoints of the interval $[-1, 1]$. Fejér's second rule is the more practical, because going from $n+1$ to $2n+1$ points, only n new function values need to be evaluated; cf. the trapezoidal rule.

The quadrature rule of Clenshaw and Curtis [72] is a closed version of Fejér's second rule; i.e., the nodes are the $n+1$ extreme points of $T_n(x)$, in $[-1, 1]$, including the endpoints $x_0 = 1$, $x_n = -1$. An explicit formula for the Clenshaw–Curtis weights is

$$w_k^{cc} = \frac{c_k}{n}\left(1 - \sum_{j=1}^{\lfloor n/2 \rfloor} \frac{b_j}{4j^2 - 1}\cos(2j\theta_k)\right), \quad k = 0 : n, \tag{5.1.43}$$

where

$$b_j = \begin{cases} 1 & \text{if } j = n/2, \\ 2 & \text{if } j < n/2, \end{cases} \qquad c_k = \begin{cases} 1 & \text{if } k = 0, n, \\ 2 & \text{otherwise.} \end{cases} \tag{5.1.44}$$

In particular the weights at the two boundary points are

$$w_0^{cc} = w_n^{cc} = \frac{1}{n^2 - 1 + \operatorname{mod}(n, 2)}. \tag{5.1.45}$$

For both the Fejér and Clenshaw–Curtis rules the weights can be shown to be positive; see Imhof [207]. Therefore the convergence of $I_n(f)$ as $n \to \infty$ for all $f \in C[-1, 1]$ is assured for these rules by the following theorem, which is a consequence of Weierstrass' theorem.

Theorem 5.1.6.

Let x_{nj} and a_{nj}, where $j = 1 : n$, $n = 1, 2, 3, \dots$, be triangular arrays of nodes and weights, respectively, and suppose that $a_{nj} > 0$ for all $n, j \geq 1$. Consider the sequence of quadrature rules

$$I_n f = \sum_{j=1}^{n} a_{nj} f(x_{nj})$$

for the integral $If = \int_a^b f(x)w(x)\,dx$, where $[a, b]$ is a closed, bounded interval, and $w(x)$ is an integrable weight function. Suppose that $I_n p = Ip$ for all $p \in \mathcal{P}_{k_n}$, where $\{k_n\}_{n=1}^{\infty}$ is a strictly increasing sequence. Then

$$I_n f \to If \quad \forall f \in C[a, b].$$

Note that this theorem is not applicable to Cotes' formulas, where some weights are negative.

Convergence will be fast for the Fejér and Clenshaw–Curtis rules provided the integrand is k times continuously differentiable—a property that the user can often check. However, if the integrand is discontinuous, the interval of integration should be partitioned at the discontinuities and the subintervals treated separately.

Despite its advantages these quadrature rules did not receive much use early on, because computing the weights using the explicit formulas given above is costly ($O(n^2)$ flops) and not numerically stable for large values of n. However, it is not necessary to compute the weights explicitly. Gentleman [154, 153] showed how the Clenshaw–Curtis rule can be implemented by means of a discrete cosine transform (DCT; see Sec. 4.7.4). We recall that the FFT is not only fast, but also very resistant to roundoff errors.

The interpolation polynomial $L_n(x)$ can be represented in terms of Chebyshev polynomials

$$L_n(x) = \sum_{k=0}^{n}{}'' c_k T_k(x), \qquad c_k = \frac{2}{n} \sum_{j=0}^{n}{}'' f(x_j) \cos\left(\frac{kj\pi}{n}\right),$$

where $x_j = \cos(j\pi/n)$. This is the real part of an FFT. (The double prime on the sum means that the first and last terms are to be halved.) Then we have

$$I_n(f) = \int_{-1}^{1} L_n(x)\,dx = \sum_{k=0}^{n}{}'' c_k \mu_k, \qquad \mu_k = \int_{-1}^{1} T_k(x)\,dx,$$

where μ_k are the moments of the Chebyshev polynomials. It can be shown (Problem 5.1.7) that

$$\mu_k = \int_{-1}^{1} T_k(x)\,dx = \begin{cases} 0 & \text{if } k \text{ odd}, \\ 2/(1-k^2) & \text{if } k \text{ even}. \end{cases}$$

The following MATLAB program, due to Trefethen [359], is a compact implementation of this version of the Clenshaw–Curtis quadrature.

ALGORITHM 5.1. *Clenshaw–Curtis Quadrature.*

```
function I = clenshaw_curtis(f,n);
% Computes the integral I of f over [-1,1] by the
% Clenshaw-Curtis quadrature rule with n+1 nodes.
x =  cos(pi*(0:n)'/n);
%Chebyshev extreme points
fx = feval(f,x)/(2*n);
%Fast Fourier transform
g = real(fft(fx([1:n+1 n:-1:2])));
%Chebyshev coefficients
a = [g(1); g(2:n)+g(2*n:-1:n+2); g(n+1)];
w = 0*a'; w(1:2:end) = 2./(1-(0:2:n).^2);
I = w*a;
```

A fast and accurate algorithm for computing the weights in the Fejér and Clenshaw–Curtis rules in $O(n \log n)$ flops has been given by Waldvogel [368]. The weights are obtained as the inverse FFT of certain vectors given by explicit rational expressions. On an average laptop this takes just about five seconds for $n = 2^{20} + 1 = 1{,}048{,}577$ nodes!

For Fejér's second rule the weights are the inverse discrete FFT of the vector v with components v_k given by the expressions

$$\begin{aligned} v_k &= \frac{2}{1-4k^2}, \quad k = 0 : \lfloor n/2 \rfloor - 1, \\ v_{\lfloor n/2 \rfloor} &= \frac{n-3}{2\lfloor n/2 \rfloor - 1} - 1, \\ v_{n-k} &= v_k, \quad k = 1 : \lfloor (n-1)/2 \rfloor. \end{aligned} \tag{5.1.46}$$

(Note that this will give zero weights for $k = 0, n$ corresponding to the endpoint nodes $x_0 = -1$ and $x_n = 1$.)

For the Clenshaw–Curtis rule the weights are the inverse FFT of the vector $v + g$, where

$$\begin{aligned} g_k &= -w_0^{cc}, \quad k = 0 : \lfloor n/2 \rfloor - 1, \\ g_{\lfloor n/2 \rfloor} &= w_0 \left[(2 - \text{mod}\,(n, 2))\, n - 1 \right], \\ g_{n-k} &= g_k, \quad k = 1 : \lfloor (n-1)/2 \rfloor, \end{aligned} \tag{5.1.47}$$

and w_0^{cc} is given by (5.1.45). For the weights Fejér's first rule and MATLAB files implementing the algorithm, we refer to [368].

Since the complexity of the inverse FFT is $O(n \log n)$, this approach allows fast and accurate calculation of the weights for rules of high order, in particular when n is a power of 2. For example, using the MATLAB routine `IFFT` the weights for $n = 1024$ only takes a few milliseconds on a PC.

Review Questions

5.1.1 Name three classical quadrature methods and give their order of accuracy.

5.1.2 What is meant by a composite quadrature rule? What is the difference between local and global error?

5.1.3 What is the advantage of including a weight function $w(x) > 0$ in some quadrature rules?

5.1.4 Describe some possibilities for treating integrals where the integrand has a **singularity** or is "almost singular."

5.1.5 For some classes of functions the composite trapezoidal rule exhibits so-called *super-convergence*. What is meant by this term? Give an example of a class of functions for which this is true.

5.1.6 Give an account of the theoretical background of the classical Newton–Cotes' rules.

Problems and Computer Exercises

5.1.1 (a) Derive the closed Newton–Cotes' rule for $m = 3$,

$$I = \frac{3h}{8}(f_0 + 3f_1 + 3f_2 + f_3) + R_T, \quad h = \frac{(b-a)}{3},$$

also known as Simpson's (3/8)-rule.

(b) Derive the open Newton–Cotes' rule for $m = 4$,

$$I = \frac{4h}{3}(2f_1 - f_2 + 2f_3) + R_T, \qquad h = \frac{(b-a)}{4}.$$

(c) Find asymptotic error estimates for the formulas in (a) and (b) by applying them to suitable polynomials.

5.1.2 (a) Show that Simpson's rule is the unique quadrature formula of the form

$$\int_{-h}^{h} f(x)\,dx \approx h(a_{-1}f(-h) + a_0 f(0) + a_1 f(h))$$

that is exact whenever $f \in \mathcal{P}_4$. Try to find several derivations of Simpson's rule, with or without the use of difference operators.

(b) Find the Peano kernel $K_2(u)$ such that $Rf = \int_{\mathbf{R}} f''(u)K_2(u)\,du$, and find the best constants c, p such that

$$|Rf| \le ch^p \max|f''(u)| \quad \forall f \in C^2[-h, h].$$

If you are going to deal with functions that are not in C^3, would you still prefer Simpson's rule to the trapezoidal rule?

5.1.3 The quadrature formula

$$\int_{x_{i-1}}^{x_{i+1}} f(x)\,dx \approx h\big(af(x_{i-1}) + bf(x_i) + cf(x_{i+1})\big) + h^2 df'(x_i)$$

can be interpreted as a Hermite interpolatory formula with a *double point* at x_i. Show that $d = 0$ and that this formula is identical to Simpson's rule. Then show that the error can be written as

$$R(f) = \frac{1}{4!}\int_{x_{i-1}}^{x_{i+1}} f^{(4)}(\xi_x)(x - x_{i-1})(x - x_i)^2(x - x_{i+1})\,dx,$$

where $f^{(4)}(\xi_x)$ is a continuous function of x. Deduce the error formula for Simpson's rule. Setting $x = x_i + ht$, we get

$$R(f) = \frac{h^4}{24} f^{(4)}(\xi_i)\int_{-1}^{1}(t+1)t^2(t-1)h\,dt = \frac{h^5}{90} f^{(4)}(\xi_i).$$

5.1.4 A second kind of Newton–Cotes' open quadrature rule uses the midpoints of the equidistant grid $x_i = ih$, $i = 1 : n$, i.e.,

$$\int_{x_0}^{x_n} f(x)\,dx = \sum_{i=1}^{n} w_i f_{i-1/2}, \qquad x_{i-1/2} = \frac{1}{2}(x_{i-1} + x_i).$$

(a) For $n = 1$ we get the midpoint rule. Determine the weights in this formula for $n = 3$ and $n = 5$. (Use symmetry!)

(b) What is the order of accuracy of these two rules?

5.1.5 (a) Simpson's rule with end corrections is a quadrature formula of the form

$$\int_{-h}^{h} f(x)\,dx \approx h\big(\alpha f(-h) + \beta f(0) + \alpha f(h)\big) + h^2\gamma(f'(-h) - f'(h)),$$

which is exact for polynomials of degree five. Determine the weights α, β, γ by using the test functions $f(x) = 1, x^2, x^4$. Use $f(x) = x^6$ to determine the error term.

(b) Show that in the corresponding composite formula for the interval $[a, b]$ with $b - a = 2nh$, only the endpoint derivatives $f'(a)$ and $f'(b)$ are needed.

5.1.6 (Lyness) Consider the integral

$$I(f, g) = \int_0^{nh} f(x)g'(x)\,dx. \tag{5.1.48}$$

An approximation related to the trapezoidal rule is

$$S_m = \frac{1}{2}\sum_{j=0}^{n-1}\big[f(jh) + f((j+1)h)\big]\big[g((j+1)h) - g(jh)\big],$$

which requires $2(m+1)$ function evaluations. Similarly, an analogue to the midpoint rule is

$$R_m = \frac{1}{2}\sum_{j=0}^{n-1}{}'' f(jh)\big[g((j+1)h) - g((j-1)h)\big],$$

where the double prime on the summation indicates that the extreme values $j = 0$ and $j = m$ are assigned a weighting factor $\frac{1}{2}$. This rule requires $2(m+2)$ function evaluations, two of which lie outside the interval of integration. Show that the difference $S_m - R_m$ is of order $O(h^2)$.

5.1.7 Show the relations

$$\int_{-1}^{x} T_n(t)\,dt = \begin{cases} \dfrac{T_{n+1}(x)}{2(n+1)} - \dfrac{T_{n-1}(x)}{2(n-1)} + \dfrac{(-1)^{n+1}}{n^2-1} & \text{if } n \geq 2, \\ \dfrac{(T_2(x)-1)}{4} & \text{if } n = 1, \\ T_1(x) + 1 & \text{if } n = 0. \end{cases}$$

Then deduce that

$$\int_{-1}^{1} T_n(x)\,dx = \begin{cases} 0 & \text{if } n \text{ odd}, \\ 2/(1-n^2) & \text{if } n \text{ even}. \end{cases}$$

Hint: Make a change of variable in the integral and use the trigonometric identity $2\cos n\phi \sin\phi = \sin(n+1)\phi - \sin(n-1)\phi$.

5.1.8 Compute the integral $\int_0^\infty (1+x^2)^{-4/3}\,dx$ with five correct decimals. Expand the integrand in powers of x^{-1} and integrate termwise over the interval $[R, \infty]$ for a suitable value of R. Then use a Newton–Cotes' rule on the remaining interval $[0, R]$.

5.1.9 Write a program for the derivation of a formula for integrals of the form $I = \int_0^1 x^{-1/2} f(x)\,dx$ that is exact for $f \in \mathcal{P}_n$ and uses the values $f(x_i)$, $i = 1:n$, by means of the power basis.

(a) Compute the coefficients b_i for $n = 6:8$ with equidistant points, $x_i = (i-1)/(n-1)$, $i = 1:n$. Apply the formulas to the integrals

$$\int_0^1 x^{-1/2}e^{-x}\,dx, \qquad \int_0^1 \frac{dx}{\sin\sqrt{x}}, \qquad \int_0^1 (1-t^3)^{-1/2}\,dt.$$

In the first of the integrals compare with the result obtained by series expansion in Problem 3.1.1. A substitution is needed for bringing the last integral to the right form.

(b) Do the same for the case where the step size $x_{i+1} - x_i$ grows proportionally to i; $x_1 = 0$; $x_n = 1$. Is the accuracy significantly different compared to (a), for the same number of points?

(c) Make some very small random perturbations of the x_i, $i = 1:n$ in (a), (say) of the order of 10^{-13}. Of which order of magnitude are the changes in the coefficients b_i, and the changes in the results for the first of the integrals?

5.1.10 Propose a suitable plan (using a computer) for computing the following integrals, for $s = 0.5, 0.6, 0.7, \dots, 3.0$.
(a) $\int_0^\infty (x^3 + sx)^{-1/2}\,dx$; (b) $\int_0^\infty (x^2 + 1)^{-1/2} e^{-sx}\,dx$, error $< 10^{-6}$;
(c) $\int_\pi^\infty (s + x)^{-1/3} \sin x\,dx$.

5.1.11 It is not true that any degree of accuracy can be obtained by using a Newton–Cotes' formula of sufficiently high order. To show this, compute approximations to the integral

$$\int_{-4}^{4} \frac{dx}{1 + x^2} = 2\tan^{-1} 4 \approx 2.6516353\dots$$

using the closed Newton–Cotes' formula with $n = 2, 4, 6, 8$. Which formula gives the smallest error?

5.1.12 For expressing integrals appearing in the solution of certain integral equations, the following modification of the midpoint rule is often used:

$$\int_{x_0}^{x_n} K(x_j, x) y(x)\,dx = \sum_{i=0}^{n-1} m_{ij} y_{i+1/2},$$

where $y_{i+1/2} = y(\frac{1}{2}(x_i + x_{i+1}))$ and m_{ij} is the moment integral

$$m_{ij} = \int_{x_i}^{x_{i+1}} K(x_j, x)\,dx.$$

Derive an error estimate for this formula.

5.1.13 (a) Suppose that you have found a truncated δ^2-expansion, (say) $A(\delta^2) \equiv a_1 + a_2\delta^2 + \cdots + a_{k+1}\delta^{2k}$. Then an equivalent symmetric expression of the form $B(E) \equiv b_1 + b_2(E + E^{-1}) + \cdots + b_{k+1}(E^k + E^{-k})$ can be obtained as $b = M_{k+1}a$, where a, b are column vectors for the coefficients, and M_{k+1} is the $(k+1) \times (k+1)$ submatrix of the matrix M given in (3.3.49).
Use this for deriving (5.1.36) from (5.1.35). How do you obtain the remainder term? If you obtain the coefficients as decimal fractions, multiply them by 14,175/4 in order to check that they agree with (5.1.36).
(b) Use Cauchy–FFT for deriving (5.1.35), and the open formula and the remainder for the same interval.
(c) Set $z_n = \nabla^{-1} y_n - \Delta^{-1} y_0$. We have, in the literature, seen the interpretation that $z_n = \sum_{j=0}^{n} y_j$ if $n \geq 0$. It seems to require some extra conditions to be true. Investigate if the conditions $z_{-1} = y_{-1} = 0$ are necessary and sufficient. Can you suggest better conditions? (The equations $\Delta\Delta^{-1} = \nabla\nabla^{-1} = 1$ mentioned earlier are assumed to be true.)

5.1.14 (a) Write a program for the derivation of quadrature formulas and error estimates using the Cauchy–FFT method in Sec. 5.1.5 for $m = n-1, n, n+1$. Test the formulas and the error estimates for some m, n on some simple (though not too simple) examples. Some of these formulas are listed in the Handbook [1, Sec. 25.4].

In particular, check the closed Newton–Cotes' 9-point formula ($n = 8$).

(b) Sketch a program for the case that $h = 1/(2n+1)$, with the computation of f at $2m$ symmetrical points.

(c) [1, Sec. 25.4] gives several Newton–Cotes' formulas of closed and open types, with remainders. Try to reproduce and extend their tables with techniques related to Sec. 5.3.1.

5.1.15 Compute the integral

$$\frac{1}{2\pi}\int_0^{2\pi} e^{\frac{1}{\sqrt{2}}\sin x}\,dx$$

by the trapezoidal rule, using $h = \pi/2^k$ $k = 0, 1, 2, \ldots$, until the error is on the level of the roundoff errors. Observe how the number of correct digits vary with h. Notice that Romberg is of no use in this problem.

Hint: First estimate how well the function $g(x) = e^{x/\sqrt{2}}$ can be approximated by a polynomial in $\mathcal{P}_8$ for $x \in [-1, 1]$. The estimate found by the truncated Maclaurin expansion is not quite good enough. Theorem 3.1.5 provides a sharper estimate with an appropriate choice of R; remember Scylla and Charybdis.

5.1.16 (a) Show that the trapezoidal rule, with $h = 2\pi/(n+1)$, is exact for all trigonometric polynomials of period 2π, i.e., for functions of the type

$$\sum_{k=-n}^{n} c_k e^{ikt}, \quad i^2 = -1,$$

when it is used for integration over a whole period.

(b) Show that if $f(x)$ can be approximated by a trigonometric polynomial of degree n so that the magnitude of the error is less than ϵ, in the interval $(0, 2\pi)$, then the error with the use of the trapezoidal rule with $h = 2\pi/(n+1)$ on the integral

$$\frac{1}{2\pi}\int_0^{2\pi} f(x)\,dx$$

is less than 2ϵ.

(c) Use the above to explain the sensationally good result in Problem 5.1.15 above, when $h = \pi/4$.

5.2 Integration by Extrapolation

5.2.1 The Euler–Maclaurin Formula

Newton–Cotes' rules have the drawback that they do not provide a convenient way of estimating the error. Also, for high-order rules negative weights appear. In this section we will derive formulas of high order, based on the Euler–Maclaurin formula (see Sec. 3.4.5), which do not share these drawbacks.

Let $x_i = a + ih$, $x_n = b$, and let $T(a : h : b)f$ denote the trapezoidal sum

$$T(a : h : b)f = \sum_{i=1}^{n} \frac{h}{2}\big(f(x_{i-1}) + f(x_i)\big). \tag{5.2.1}$$

According to Theorem 3.4.10, if $f \in C^{2r+2}[a, b]$, then

$$T(a : h : b)f - \int_a^b f(x)\,dx = \frac{h^2}{12}\big(f'(b) - f'(a)\big) - \frac{h^4}{720}\big(f'''(b) - f'''(a)\big)$$
$$+ \cdots + \frac{B_{2r}h^{2r}}{(2r)!}\big(f^{(2r-1)}(b) - f^{(2r-1)}(a)\big) + R_{2r+2}(a, h, b)f.$$

By (3.4.37) the remainder $R_{2r+2}(a, h, b)f$ is $O(h^{2r+2})$ and represented by an integral with a kernel of constant sign in $[a, b]$. The estimation of the remainder is very simple in certain important particular cases. Note that although the expansion contains derivatives at the boundary points only, the remainder requires that $|f^{(2r+2)}|$ is integrable on the whole interval $[a, b]$.

We recall the following simple and useful relation between the trapezoidal sum and the midpoint sum (cf. (5.1.21)):

$$M(a : h : b)f = \sum_{i=1}^{n} hf(x_{i-1/2}) = 2T\left(a : \frac{1}{2}h : b\right)f - T(a : h : b)f. \tag{5.2.2}$$

From this one easily derives the expansion

$$M(a : h : b)f = \int_a^b f(x)\,dx - \frac{h^2}{24}\big(f'(b) - f'(a)\big) + \frac{7h^4}{5760}\big(f'''(b) - f'''(a)\big)$$
$$+ \cdots + \left(\frac{1}{2^{2r-1}} - 1\right)\frac{B_{2r}h^{2r}}{(2r)!}\big(f^{(2r-1)}(b) - f^{(2r-1)}(a)\big) + \cdots,$$

which has the same relation to the midpoint sum as the Euler–Maclaurin formula has to the trapezoidal sum.

The Euler–Maclaurin formula can be used for highly accurate numerical integration when the values of derivatives of f are known at $x = a$ and $x = b$. It is also possible to use difference approximations to estimate the derivatives needed. A variant with uncentered differences is **Gregory's**[170] **quadrature formula**:

$$\int_a^b f(x)\,dx = h\frac{E^n - 1}{hD}f_0 = h\left(\frac{f_n}{-\ln(1-\nabla)} - \frac{f_0}{\ln(1+\Delta)}\right)$$
$$= T(a; h; b) + h\sum_{j=1}^{\infty} a_{j+1}(\nabla^j f_n + (-\Delta)^j f_0),$$

[170] James Gregory (1638–1675), a Scotch mathematician, discovered this formula long before the Euler–Maclaurin formula. It seems to have been used primarily for numerical quadrature. It can be used also for summation, but the variants with central differences are typically more efficient.

where $T(a:h:b)$ is the trapezoidal sum. The operator expansion must be truncated at $\nabla^k f_n$ and $\Delta^l f_0$, where $k \le n$, $l \le n$. (Explain why the coefficients a_{j+1}, $j \ge 1$, occur in the implicit Adams formula too; see Problem 3.3.10 (a).)

5.2.2 Romberg's Method

The Euler–Maclaurin formula is the theoretical basis for the application of repeated Richardson extrapolation (see Sec. 3.4.6) to the results of the trapezoidal rule. This method is known as **Romberg's method.**[171] It is one of the most widely used methods, because it allows a simple strategy for the automatic determination of a suitable step size and order. Romberg's method was made widely known through Stiefel [336]. A thorough analysis of the method was carried out by Bauer, Rutishauser, and Stiefel in [20], which we shall refer to for proof details.

Let $f \in C^{2m+2}[a,b]$ be a real function to be integrated over $[a,b]$ and denote the trapezoidal sum by $T(h) = T(a:h:b)f$. By the Euler–Maclaurin formula it follows that

$$T(h) - \int_a^b f(x)\,dx = c_2h^2 + c_4h^4 + \cdots + c_mh^{2m} + \tau_{m+1}(h)h^{2m+2}, \tag{5.2.3}$$

where $c_k = 0$ if $f \in \mathcal{P}_k$. This suggests the use of repeated Richardson extrapolation applied to the trapezoidal sums computed with step lengths

$$h_1 = \frac{b-a}{n_1}, \quad h_2 = \frac{h_1}{n_2}, \quad \ldots, \quad h_q = \frac{h_1}{n_q}, \tag{5.2.4}$$

where $n_1, n_2, \ldots, n_q$ are strictly increasing positive integers. If we set $T_{m,1} = T(a, h_m, b)f$, $m = 1:q$, then using Neville's interpolation scheme the extrapolated values can be computed from the recursion:

$$T_{m,k+1} = T_{m,k} + \frac{T_{m,k} - T_{m-1,k}}{(h_{m-k}/h_m)^2 - 1}, \qquad 1 \le k < m. \tag{5.2.5}$$

Romberg used step sizes in a geometric progression, $h_m/h_{m-1} = q = 2$. In this case the denominators in (5.2.5) become $2^{2k} - 1$. This choice has the advantage that successive trapezoidal sums can be computed using the relation

$$T\left(\frac{h}{2}\right) = \frac{1}{2}(T(h) + M(h)), \qquad M(h) = \sum_{i=1}^{n} hf(x_{i-1/2}), \tag{5.2.6}$$

where $M(h)$ is the midpoint sum. This makes it possible to reuse the function values that have been computed earlier.

We remark that, usually, a composite form of Romberg's method is used; the method is applied to a sequence of interval $[a + iH, a + (i+1)H]$ for some bigstep H. The

[171] Werner Romberg (1909–2003) was a German mathematician. For political reasons he fled Germany in 1937, first to Ukraine and then to Norway, where in 1938 he joined the University of Oslo. He spent the war years in Sweden and then returned to Norway. In 1949 he joined the Norwegian Institute of Technology in Trondheim. He was called back to Germany in 1968 to take up a position at the University of Heidelberg.

applications of repeated Richardson extrapolation and the Neville algorithms to differential equations belong to the most important.

Rational extrapolation can also be used. This gives rise to a recursion of a form similar to (5.2.5):

$$T_{m,k+1} = T_{m,k} + \frac{T_{m,k} - T_{m-1,k}}{\left(\dfrac{h_{m-k}}{h_m}\right)^2 \left[1 - \dfrac{T_{m,k} - T_{m-1,k}}{T_{m,k} - T_{m-1,k-1}}\right] - 1}, \quad 1 \le k \le m; \tag{5.2.7}$$

see Sec. 4.3.3.

For practical numerical calculations the values of the coefficients c_k in (5.2.3) are not needed, but they are used, for example, in the derivation of an error bound; see Theorem 5.2.1. It is also important to remember that the coefficients depend on derivatives of increasing order; the success of repeated Richardson extrapolations is thus related to the behavior in $[a, b]$ of the higher derivatives of the integrand.

Theorem 5.2.1 (*Error Bound for Romberg's Method*).

The items $T_{m,k}$ in Romberg's method are estimates of the integral $\int_a^b f(x)\,dx$ that can be expressed as a linear functional,

$$T_{m,k} = (b-a)\sum_{j=0}^{n} \alpha_{m,j}^{(k)} f(a+jh), \tag{5.2.8}$$

where $n = 2^{m-1}$, $h = (b-a)/n$, and

$$\sum_{j=0}^{n} \alpha_{m,j}^{(k)} = 1, \quad \alpha_{m,j}^{(k)} > 0. \tag{5.2.9}$$

The remainder functional for $T_{m,k}$ is zero for $f \in \mathcal{P}_{2k}$, and its Peano kernel is positive in the interval (a, b). The truncation error of $T_{m,k}$ reads

$$\begin{aligned} T_{m,k} - \int_a^b f(x)dx &= r_k h^{2k}(b-a) f^{(2k)}\left(\frac{1}{2}(a+b)\right) + O(h^{2k+2}(b-a)f^{(2k+2)}) \\ &= r_k h^{2k}(b-a) f^{(2k)}(\xi), \quad \xi \in (a, b), \end{aligned} \tag{5.2.10}$$

where

$$r_k = 2^{k(k-1)}|B_{2k}|/(2k)!, \quad h = 2^{1-m}(b-a).$$

Proof. Sketch: Equation (5.2.8) follows directly from the construction of the Romberg scheme. (It is for theoretical use only; the recursion formulas are better for practical use.) The first formula in (5.2.9) holds because $T_{m,k}$ is exact if $f = 1$. The second formula is easily proved for low values of k. The general proof is more complicated; see [20, Theorem 4].

The Peano kernel for $m = k = 1$ (the trapezoidal rule) was constructed in Example 3.3.7. For $m = k = 2$ (Simpson's rule), see Sec. 5.1.3. The general case is more complicated. Recall that, by Corollary 3.3.9 of Peano's remainder theorem, a remainder formula with a mean value $\xi \in (a, b)$ exists if and only if the Peano kernel does not change sign.

Bauer, Rutishauser, and Stiefel [20, pp. 207–210] constructed a recursion formula for the kernels, and succeeded in proving that they are all positive, by an ingenious use of the recursion. The expression for r_k is also derived there, although with a different notation. □

From (5.2.9) it follows that if the magnitude of the irregular error in $f(a + jh)$ is at most ϵ, then the magnitude of the inherited irregular error in $T_{m,k}$ is at most $\epsilon(b - a)$. There is another way of finding r_k. Note that for each value of k, the error of $T_{k,k}$ for $f(x) = x^{2k}$ can be determined numerically. Then r_k can be obtained from (5.2.10). $T_{m,k}$ is the same formula as $T_{k,k}$, although with a different h.

According to the discussion of repeated Richardson extrapolation in Sec. 3.4.6, one continues the process until two values *in the same row* agree to the desired accuracy. If no other error estimate is available, $\min_k |T_{m,k} - T_{m,k-1}|$ is usually chosen as an estimate of the truncation error, even though it is usually a strong overestimate. A feature of the Romberg algorithm is that it also contains exits with lower accuracy at a lower cost.

Example 5.2.1 (*A Numerical Illustration to Romberg's Method*).

Use Romberg's method to compute the integral (cf. Example 5.1.5)

$$\int_0^{0.8} \frac{\sin x}{x}\, dx.$$

The midpoint and trapezoidal sums are with ten correct decimals equal to

h	$M(h)f$	$T(h)f$
0.8	0.77883 66846	0.75867 80454
0.4	0.77376 69771	0.76875 73650
0.2	0.77251 27161	0.77126 21711
0.1		0.77188 74436

It can be verified that in this example the error is approximately proportional to h^2 for both $M(h)$ and $T(h)$. We estimate the error in $T(0.1)$ to be $\frac{1}{3}6.26 \cdot 10^{-4} \le 2.1 \cdot 10^{-4}$.

The trapezoidal sums are then copied to the first column of the Romberg scheme. Repeated Richardson extrapolation is performed giving the following table.

m	T_{m1}	T_{m2}	T_{m3}	T_{m4}
1	0.75867 80454			
2	0.76875 73650	0.77211 71382		
3	0.77126 21711	0.77209 71065	0.77209 57710	
4	0.77188 74437	0.77209 58678	0.77209 57853	0.77209 57855
5	0.77204 37039	0.77209 57906	0.77209 57855	0.77209 57855

We find that $|T_{44} - T_{43}| = 2 \cdot 10^{-10}$, and the irregular errors are less than 10^{-10}. Indeed, all ten digits in T_{44} are correct, and $I = 0.77209\,57854\,82\ldots$. Note that the rate of convergence in successive columns is $h^2, h^4, h^6, h8, \ldots$.

The following MATLAB program implements Romberg's method. In each major step a new row in the Romberg table is computed.

ALGORITHM 5.2. *Romberg's Method.*

```
function [I, md, T] = romberg(f,a,b,tol,q);
% Romberg's method for computing the integral of f over [a,b]
% using at most q extrapolations. Stop when two adjacent values
% in the same column differ by less than tol or when q
% extrapolations have been performed. Output is an estimate
% I of the integral with error bound md and the active part
% of the Romberg table.
%
T = zeros(q+2,q+1);
h = b - a;   m = 1; P = 1;
T(1,1) = h*(feval(f,a) + feval(f,b))/2;
for m = 2:q+1
  h = h/2; m = 2*m;
  M = 0;                 % Compute midpoint sum
  for k = 1:2:m
    M = M + feval(f, a+k*h);
  end
  T(m,1) = T(m-1,1)/2 + h*M;
  kmax = min(m-1,q);
  for k = 1:kmax       % Repeated Richardson extrapolation
    T(m,k+1) = T(m,k) + (T(m,k) - T(m-1,k))/(2^(2*k) - 1);
  end
  [md, kb] = min(abs(T(m,1:kmax) - T(m-1,1:kmax)));
  I = T(m,kb);
  if md <= tol             % Check accuracy
    T = T(1:m,1:kmax+1); % Active part of T
  return
  end
end
```

In the above algorithm the value $T_{m,k}$ is accepted when $|T_{m,k} - T_{m-1,k}| \leq \text{tol}$, where tol is the permissible error. Thus one extrapolates until two values *in the same column* agree to the desired accuracy. In most situations, this gives, if h is sufficiently small, with a large margin a bound for the truncation error in the lower of the two values. Often instead the subdiagonal error criterion $|T_{m,m-1} - T_{m,m}| < \delta$ is used, and T_{mm} taken as the numerical result.

If the use of the basic asymptotic expansion is doubtful, then the uppermost diagonal of the extrapolation scheme should be ignored. Such a case can be detected by inspection of the difference quotients in a column. If for some k, where $T_{k+2,k}$ has been computed and the modulus of the relative irregular error of $T_{k+2,k} - T_{k+1,k}$ is less than (say) 20%, *and*, most important, the difference quotient

$$(T_{k+1,k} - T_{k,k})/(T_{k+2,k} - T_{k+1,k})$$

is very different from its theoretical value q^{p_k}, then the uppermost diagonal is to be ignored (except for its first element).

Sometimes several of the uppermost diagonals are to be ignored. For the integration of a class of periodic functions the trapezoidal rule is superconvergent; see Sec. 5.1.4. In this case all the difference quotients in the first column are much larger than $q^{p_1} = q^2$. According to the rule just formulated, every element of the Romberg scheme outside the first column should be ignored. This is correct; *in superconvergent cases Romberg's method is of no use*; it destroys the excellent results that the trapezoidal rule has produced.

Example 5.2.2.

The remainder for $T_{k,k}$ in Romberg's method reads

$$T_{k,k} - \int_a^b f(x)\,dx = r_k h^{2^k}(b-a)f^{2^k}(\xi).$$

For $k = 1$, T_{11} is the trapezoidal rule with remainder $r_1 h^2(b-a)f^{(2)}(\xi)$. By working algebraically in the Romberg scheme, we see that T_{22} is the same as Simpson's rule. It can also be shown that T_{33} is the same as Milne's formula, i.e., the five-point closed Newton–Cotes' formula. It follows that for $k = \{1, 2, 3\}$ both methods give, with $k' = \{2, 3, 5\}$ function values, exact results for $f \in \mathcal{P}_{k'}$.

This equivalence can also be proved by the following argument. By Corollary 3.3.5, there is only one linear combination of the values of the function f at $n+1$ given points that can yield $\int_a^b f(x)\,dx$ exactly for all polynomials $f \in \mathcal{P}_{n+1}$. It follows that the methods of Cotes and Romberg for $T_{k,k}$ are identical for $k = 1, 2, 3$.

For $k > 3$ the methods are not identical. For $k = 4$ (9 function values), Cotes is exact in $\mathcal{P}_{10}$, while T_{44} is exact in $\mathcal{P}_8$. For $k = 5$ (17 function values), Cotes is exact in $\mathcal{P}_{18}$, while T_{55} is exact in $\mathcal{P}_{10}$. This sounds like an advantage for Cotes, but one has to be sceptical about formulas that use equidistant points in polynomial approximation of very high degree; see the discussion of Runge's phenomena in Chapter 4.

Note that the remainder of T_{44} is

$$r_4 h^8(b-a)f^{(8)}(\xi) \approx r_4(b-a)\Delta^8 f(a), \qquad r_4 = 16/4725,$$

where $\Delta^8 f(a)$ uses the same function values as T_{44} and C_8. So we can use $r_4(b-a)\Delta^8 f(a)$ as an asymptotically correct error estimate for T_{44}.

We have assumed so far that the integrand is a real function $f \in C^{2m+2}[a,b]$. For example, if the integrand $f(x)$ has an algebraic endpoint singularity,

$$f(x) = x^\beta h(x), \qquad -1 < \beta \le 0,$$

where $h(x) \in C^{p+1}[a,b]$, this assumption is not valid. In this case an asymptotic error expansion of the form

$$T(h) - I = \sum_{q=1}^{n} a_q h^{q+\beta} + \sum_{q=2}^{q} b_q h^q + O(h^{q+1}) \tag{5.2.11}$$

can be shown to hold for a trapezoidal sum. Similar but more complicated expansions can be obtained for other classes of singularities. If $p = -1/2$, then $T(h)$ has an error expansion in $h^{1/2}$:

$$T(h) - I = a_1 h^{3/2} + b_2 h^2 + a_2 h^{5/2} + b_3 h^3 + a_3 h^{5/2} + \cdots.$$

Richardson extrapolation can then be used with the denominators

$$2^{p_j} - 1, \quad p_j = 1.5, 2, 2.5, 3, \dots.$$

Clearly the convergence acceleration will be much less effective than in the standard Romberg case.

In Richardson extrapolation schemes the exponents in the asymptotic error expansions have to be known *explicitly*. In cases when *the exponents are unknown* a nonlinear extrapolation scheme like the ϵ algorithm should be used. In this a two-dimensional array of numbers $\epsilon_k^{(p)}$, initialized with the trapezoidal approximations $T_m = T(h_m)$, $h_m = (b-a)/2^m$, is computed by the recurrence relation

$$\begin{aligned}
&\epsilon_{-1}^{(m)} = 0, \quad m = 1 : n-1, \dots, \\
&\epsilon_0^{(m)} = T_m, \quad m = 0 : n, \\
&\epsilon_{k+1}^{(m)} = \epsilon_{k-1}^{(m+1)} + \frac{1}{\epsilon_k^{(m+1)} - \epsilon_k^{(m)}}, \qquad k = 0 : n-2, \quad m = 0 : n-k-1.
\end{aligned}$$

Example 5.2.3.

Accelerating the sequence of trapezoidal sums using the epsilon algorithm may work when Romberg's method fails. In the integral

$$\int_0^1 \sqrt{x}\,dx = 2/3,$$

the integrand has a singularity at the left endpoint.

Using the trapezoidal rule with $2^k + 1$ points, $k = 0 : 9$, the error is divided roughly by $2\sqrt{2} \approx 2.828$ when the step size is halved. For $k = 9$ we get he approximation $I \approx 0.66664\,88815$ with an error $0.18 \cdot 10^{-4}$.

Applying the ϵ algorithm to these trapezoidal sums, we obtained the accelerated values displayed in the table below. (Recall that the quantities in odd-numbered columns are only intermediate quantities.) The magnitude of the error in $\epsilon_8^{(1)}$ is close to full IEEE double precision. Note that we did not use any a priori knowledge of the error expansion.

k	$\epsilon_{2k}^{(9-2k)}$	Error
0	0.66664 88815 4995	$-0.1779 \cdot 10^{-4}$
1	0.66666 67335 1817	$0.6685 \cdot 10^{-7}$
2	0.66666 66666 6037	$-0.6292 \cdot 10^{-11}$
3	0.66666 66666 6669	$0.268 \cdot 10^{-13}$
4	0.66666 66666 6666	$-0.044 \cdot 10^{-13}$

An application of the epsilon algorithm to computing the integral of an oscillating integrand to high precision is given in Example 5.2.5.

5.2.3 Oscillating Integrands

Highly oscillating integrals of the form

$$I[f] = \int_a^b f(x)e^{i\omega g(x)}\,dx, \tag{5.2.12}$$

where $f(x)$ is a slowly varying function and $e^{i\omega g(x)}$ is oscillating, frequently occur in applications from electromagnetics, chemistry, fluid mechanics, etc. Such integrals are allegedly difficult to compute. When a standard numerical quadrature rule is used to compute (5.2.12), using a step size h such that $\omega h \ll 1$ is required. For large values of ω this means an exceedingly small step size and a large number of function evaluations.

Some previously mentioned techniques such as using a simple comparison problem, or a special integration formula, can be effective also for an oscillating integrand. Consider the case of a Fourier integral, where $g(x) = x$, in (5.2.12). The trapezoidal rule gives the approximation

$$I[f] \approx \frac{1}{2}h\big(f_0 e^{i\omega a} + f_N e^{i\omega b}\big) + h\sum_{j=1}^{N-1} f_j e^{i\omega x_j}, \tag{5.2.13}$$

where $h = (b-a)/N$, $x_j = a + jh$, $f_j = f(x_j)$. This formula cannot be used unless $\omega h \ll 1$, since its validity is based on the assumption that the whole integrand varies linearly over an interval of length h.

A better method is obtained by approximating just $f(x)$ by a piecewise linear function,

$$p_j(x) = f_j + \frac{x - x_j}{h}(f_{j+1} - f_j), \quad x \in [x_j, x_{j+1}], \quad j = 0 : N-1.$$

The integral over $[x_j, x_{j+1}]$ can then be approximated by

$$\int_{x_j}^{xj+1} p_j(x)e^{i\omega x}\,dx = he^{i\omega x_j}\left(f_j\int_0^1 e^{i\omega ht}\,dt + (f_{j+1} - f_j)\int_0^1 te^{i\omega ht}\,dt\right),$$

where we have made the change of variables $x - x_j = th$. Let $\theta = h\omega$ be the **characteristic frequency**. Then

$$\int_0^1 e^{i\theta t}\,dt = \frac{1}{i\theta}(e^{i\theta} - 1) = \alpha,$$

and using integration by parts

$$\int_0^1 te^{i\theta t}\,dt = \frac{1}{i\theta}te^{i\theta t}\Big|_0^1 - \frac{1}{i\theta}\int_0^1 e^{i\theta t}\,dt = \frac{1}{i\theta}e^{i\theta} + \frac{1}{\theta^2}(e^{i\theta} - 1) = \beta.$$

Here α and β depend on θ but not on j. Summing the contributions from all intervals we obtain

$$\begin{aligned} &h(\alpha - \beta)\sum_{j=0}^{N-1} f_j e^{i\omega x_j} + h\beta\sum_{j=0}^{N-1} f_{j+1}e^{i\omega x_j} \\ &= h(\alpha - \beta)\sum_{j=0}^{N-1} f_j e^{i\omega x_j} + h\beta e^{-i\theta}\sum_{j=1}^{N} f_j e^{i\omega x_j}. \end{aligned}$$

The resulting quadrature formula has the same form as (5.2.13),

$$I[f] \approx hw(\theta) \sum_{j=0}^{N-1} f_j e^{i\omega x_j} + hw_N(\theta)\big(f_N e^{i\omega x_N} - f_0 e^{i\omega x_0}\big), \tag{5.2.14}$$

with the weights $w_0(\theta) = \alpha - \beta$, $w_N(\theta) = \beta e^{-i\theta}$, and $w(\theta) = w_0 + w_N$. Then

$$w_0(\theta) = w_N(-\theta) = \frac{1 - i\theta - e^{-i\theta}}{\theta^2}, \qquad w(\theta) = \frac{(\sin \frac{1}{2}\theta)^2}{(\frac{1}{2}\theta)^2}. \tag{5.2.15}$$

Note that the same trigonometric sum is involved, now multiplied with the real factor $w(\theta)$. The sum in (5.2.14) can be computed using the FFT; see Sec. 4.7.3.

The weights tend to the trapezoidal weights when $\omega h \to 0$ (check this!). For small values of $|\theta|$ there will be cancellation in these expressions for the coefficients and the Taylor expansions should be used instead; see Problem 5.2.8.

A similar approach for computing trigonometric integrals of one of the forms

$$\int_a^b f(x)\cos(\omega x)\,dx, \qquad \int_a^b f(x)\sin(\omega x)\,dx \tag{5.2.16}$$

was advanced by Filon [118] already in 1928. In this the interval $[a, b]$ is divided into an even number of $2N$ subintervals of equal length $h = (b - a)/(2N)$. The function $f(x)$ is approximated over each double interval $[x_{2i}, x_{2(i+1)}]$ by the quadratic polynomial $p_i(x)$ interpolating $f(x)$ at x_{2i}, x_{2i+1}, and $x_{2(i+1)}$. Filon's formula is thus related to Simpson's rule. (The formula (5.2.14) is often called the Filon-trapezoidal rule.) For $\omega = 0$, Filon's formula reduces to the composite Simpson's formula, but it is not exact for cubic functions $f(x)$ when $\omega \neq 0$.

The integrals

$$\int_{x_{2i}}^{x_{2(i+1)}} p_i(x)\cos(\omega x)\,dx, \qquad \int_{x_{2i}}^{x_{2(i+1)}} p_i(x)\sin(\omega x)\,dx$$

can be computed analytically using integration by parts. This leads to Filon's integration formula; see the Handbook [1, Sec. 25.4.47].

Similar formulas can be developed by using different polynomial approximations of $f(x)$. Einarsson [105] uses a cubic spline approximation of $f(x)$ and assumes that the first and second derivatives of f at the boundary are available. The resulting quadrature formula has an error which usually is about four times smaller than that for Filon's rule.

Using the Euler–Maclaurin formula on the function it can be shown (see Einarsson [106, 107]) that the expansion of the error for the Filon-trapezoidal rule, the Filon–Simpson method, and the cubic spline method contain only even powers of h. Thus the accuracy can be improved by repeated Richardson extrapolation. For example, if the Filon-trapezoidal rule is used with a sequence of step sizes h, $h/2$, $h/4$, $\dots$, then one can proceed as in Romberg's method. Note that the result after one extrapolation is not exactly equal to the Filon–Simpson rule, but gives a marginally better result when $\omega h = O(1)$.

Example 5.2.4 (*Einarsson* [106]).

Using the standard trapezoidal rule to compute the Fourier integral

$$I = \int_0^\infty e^{-x} \cos \omega x \, dx = \frac{1}{1+\omega^2}$$

gives the result

$$I_T = h\left(\frac{1}{2} + \Re \sum_{j=1}^{\infty} e^{-jh} e^{ih\omega j}\right) = \frac{h}{2} \frac{\sinh h}{\cosh h - \cosh h\omega},$$

where h is the step length. Assuming that $h\omega$ is sufficiently small we can expand the right-hand side in powers of h, obtaining

$$I_T = I\left(1 + \frac{h^2}{12}(1+\omega^2) + \frac{h^4}{720}(1+\omega^2)(3\omega^2 - 1) + O(h^6)\right).$$

For the Filon-trapezoidal rule the corresponding result is

$$I_{FT} = \left(\frac{\sin \frac{1}{2}\omega h}{\frac{1}{2}\omega h}\right)^2 I_T = I\left(1 + \frac{h^2}{12} - \frac{h^4}{720}(3\omega^2 + 1) + O(h^6)\right).$$

For small values of ω the two formulas are seen to be equivalent. However, for larger values of ω, the error in the standard trapezoidal rule increases rapidly.

The expansions only have even powers of h. After one step of extrapolation the Filon-trapezoidal rule gives a relative error equal to $h^4(3\omega^2 + 1)/180$, which can be shown to be slightly better than for the Filon–Simpson rule.

More general Filon-type methods can be developed as follows. Suppose we wish to approximate the integral

$$I[f] = \int_0^h f(x) e^{i\omega x} \, dx = h \int_0^1 f(ht) e^{ih\omega t} \, dt, \tag{5.2.17}$$

where f is itself sufficiently smooth. We choose distinct nodes $0 \le c_1 < c_2 < \cdots < c_\nu \le 1$ and consider the quadrature formula interpolatory weights $b_1, b_2, \ldots, b_\nu$. Let s be the largest integer j so that

$$\int_0^1 t^{j-1} \gamma(t) \, dt = 0, \quad \gamma(t) = \prod_{i=1}^{\nu} (t - c_i). \tag{5.2.18}$$

Then by Theorem 5.1.3, $s \le \nu$, and the order of the corresponding quadrature formula is $p = \nu + s$. A Filon-type quadrature rule is now obtained by interpolating f by the polynomial

$$p(x) = \sum_{k=1}^{\nu} \ell_k(x/h) f(c_k h),$$

where ℓ_k is the kth cardinal polynomial of Lagrange interpolation. Replacing f by p in (5.2.17), we obtain

$$Q_h[f] = \sum_{k=1}^{\nu} \beta_k(\theta) f(c_k h), \quad \beta_k(\theta) = \int_0^1 \ell_k(t) e^{ih\omega t}\, dt. \tag{5.2.19}$$

The coefficients $\beta_k(\theta)$ can be computed also from the moments

$$\mu_k(\theta) = \int_0^1 t^k e^{i\theta t}\, dt, \quad k = 0 : \nu - 1,$$

by solving the Vandermonde system

$$\sum_{j=1}^{\nu} \beta_j(\theta) c_j^k = \mu_k(\theta), \quad k = 0 : \nu - 1.$$

The derivation of the Filon-type quadrature rule is analogous to considering $e^{i\theta t}$ as a complex-valued weight function. However, any attempt to choose the nodes c_j so that the order of the integration rule is increased over ν is likely to lead to complex nodes and useless formulas.

The general behavior of Filon-type quadrature rules is that for $0 < \theta \ll 1$ they show similar accuracy to the corresponding standard interpolatory rule. For $\theta = O(1)$ they are also very effective, although having order $\nu \le p$. The common wisdom is that if used in the region where θ is large they can give large errors. However, Einarsson [106] observed that the cubic spline method gives surprisingly good results also for large values of θ, seemingly in contradiction to the condition in the sampling theorem that at least two nodes per full period are needed.

Iserles [209] shows that once appropriate Filon-type methods are used the problem of highly oscillatory quadrature becomes relatively simple. Indeed, *the precision of the calculation actually increases as the oscillation grows.* This is quantified in the following theorem.

Theorem 5.2.2 (*Iserles* [209, Theorem 2]).

Let $\theta = h\omega$ be the characteristic frequency. Then the error $E_h[f]$ in the Filon-type quadrature formula (5.2.19) *is*

$$E_h[f] \sim \mathcal{O}(h^{\nu+1}\theta^{-p}), \tag{5.2.20}$$

where $p = 2$ if $c_1 = 0$ and $c_\nu = 0$; $p = 1$ otherwise.

To get the best error decay the quadrature formula should include the points $c_1 = 0$ and $c_\nu = 1$. This is the case both for the Filon-trapezoidal method and the Filon–Simpson rule. Figure 5.2.1 shows the absolute value of the integral

$$I = \int_0^h e^x e^{i\omega x}\, dx = (e^{(1+i\omega)h} - 1)/(1 + i\omega),$$

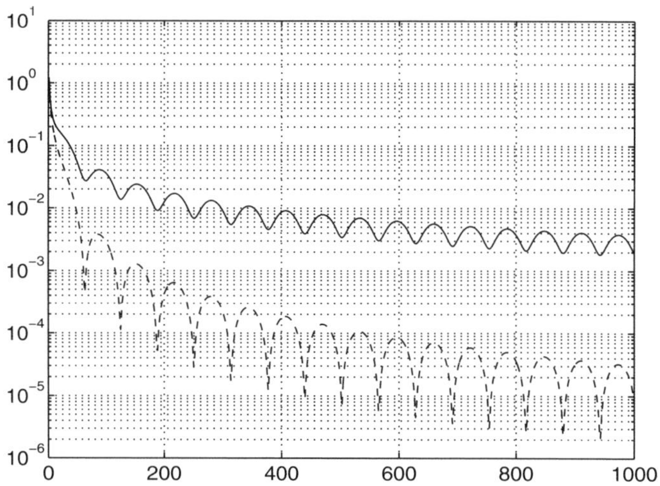

Figure 5.2.1. *The Filon-trapezoidal rule applied to the Fourier integral with $f(x) = e^x$, for $h = 1/10$, and $\omega = 1 : 1000$; solid line: exact integral; dashed line: absolute value of the error.*

and the absolute value of the error in the Filon-trapezoidal approximation for $h = 0.1$ and $\omega = 1 : 1000$. Clearly the error is small and becomes smaller as the characteristic frequency grows!

Sometimes convergence acceleration of a related series can be successfully employed for the evaluation of an integral with an oscillating integrand. Assume that the integral has the form

$$I[f] = \int_0^\infty f(x)\sin(g(x))\,dx,$$

where $g(x)$ is an increasing function and both $f(x)$ and $g(x)$ can be approximated by a polynomial. Set

$$I[f] = \sum_{n=0}^{\infty}(-1)^N u_n, \quad u_n = \int_{x_n}^{x_{n+1}} f(x)\,|\sin(g(x))|\,dx,$$

where $x_0, x_1, x_2, \ldots$ are the successive zeros of $\sin(g(x))$. The convergence of this alternating series can then be improved with the help of repeated averaging; see Sec. 3.4.3. Alternatively a sequence of partial sums can be computed, which then is accelerated by the epsilon algorithm. Sidi [323] has developed a useful extrapolation method for oscillatory integrals over an infinite interval.

Example 5.2.5 (*Gautschi* [149]).

The first problem in "The 100-digit Challenge"[172] is to compute the integral

$$I = \lim_{\epsilon\to 0}\int_\epsilon^1 t^{-1}\cos(t^{-1}\ln t)\,dt \tag{5.2.21}$$

[172] See [41] and `www.siam.org/books/100digitchallenge`.

to ten decimal places. Since the integrand is densely oscillating as $t \downarrow 0$ and at the same time the oscillations tend to infinity (see Figure 5.2.2), this is a challenging integral to compute numerically. (Even so the problem has been solved to an accuracy of 10,000 digits!)

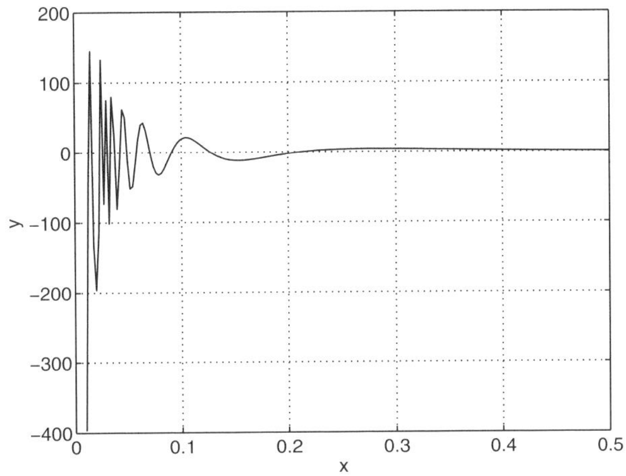

Figure 5.2.2. *The oscillating function* $x^{-1}\cos(x^{-1}\ln x)$.

With the change of variables $u = t^{-1}$, $du = -t^{-2}dt$, we get

$$I = \int_1^\infty u^{-1}\cos(u\ln u)\,du. \tag{5.2.22}$$

Making the further change of variables $x(u) = u\ln u$, we have $dx = (1+\ln u)du = (u+x)u^{-1}du$, and the integral becomes

$$I = \int_0^\infty \frac{\cos x}{x+u(x)}\,dx. \tag{5.2.23}$$

The inverse function $u(x)$ is smooth and relatively slowly varying, with $u(0) = 1$, $u'(0) = 1$. For $x > 0$, $u'(x)$ is positive and decreasing, while $u''(x)$ is negative and decreasing in absolute value. The function $u(x)$ is related to Lambert's W-function, which is the inverse of the function $x = we^w$ (see Problem 3.1.12). Clearly $u(x) = e^{w(x)}$.

The zeros of the integrand in (5.2.23) are at odd multiples of $\pi/2$. We split the interval of integration into intervals of constant sign for the integrand

$$I = \int_0^{\pi/2} \frac{\cos x}{x+u(x)}\,dx + \sum_{k=1}^{\infty} I_k, \quad I_k = \int_{(2k-1)\pi/2}^{(2k+1)\pi/2} \frac{\cos x}{x+u(x)}\,dx.$$

Changing variables $x = t + k\pi$ in the integrals I_k,

$$I_k = (-1)^k \int_{-\pi/2}^{\pi/2} \frac{\cos t}{t + k\pi + u(t+k\pi)}\,dt. \tag{5.2.24}$$

The terms form an alternating series with terms decreasing in absolute values. It is, however, slowly converging and for an error bound of $\frac{1}{2}10^{-5}$ about 116,000 terms would be needed. Accelerating the convergence using the epsilon algorithm, Gautschi found that using only 21 terms in the series suffices to give an accuracy of about 15 decimal digits:

$$I = 0.32336\,74316\,77779.$$

The integrand in the integrals (5.2.24) is regular and smooth. For computing these, for example, a Clenshaw–Curtis quadrature rule can be used after shifting the interval of integration to $[-1, 1]$; see also Problem 5.3.11.

5.2.4 Adaptive Quadrature

Suppose the integrand $f(x)$ (or some of its low-order derivatives) has strongly varying orders of magnitude in different parts of the interval of integration $[a, b]$. Clearly, one should then use *different step sizes in different parts of the integration interval.* If we write

$$\int_a^b = \int_a^{c_1} + \int_{c_1}^{c_2} + \cdots + \int_{c_1}^{b},$$

then the integrals on the right-hand side can be treated as independent subproblems. In **adaptive quadrature methods** step sizes are automatically adjusted so that the approximation satisfies a prescribed error tolerance:

$$\left| I - \int_a^b f(x)\,dx \right| \leq \epsilon. \tag{5.2.25}$$

A common difficulty is when the integrand exhibits one or several sharp peaks as exemplified in Figure 5.2.3. It should be realized that without further information about the location of the peaks all quadrature algorithms can fail if the peaks are sharp enough.

We consider first a fixed-order adaptive method based on Simpson's rule. For a subinterval $[a, b]$, set $h = (b - a)$ and compute the trapezoidal approximations

$$T_{00} = T(h), \qquad T_{10} = T(h/2), \qquad T_{20} = T(h/4).$$

The extrapolated values

$$T_{11} = (4T_{10} - T_{00})/3, \qquad T_{21} = (4T_{20} - T_{10})/3$$

are equivalent to (the composite) Simpson's rule with step length $h/2$ and $h/4$, respectively. We can also calculate

$$T_{22} = (16T_{21} - T_{11})/15,$$

which is Milne's method with step length $h/4$ with remainder equal to

$$(2/945)(h/4)^6(b - a)f^{(6)}(\xi).$$

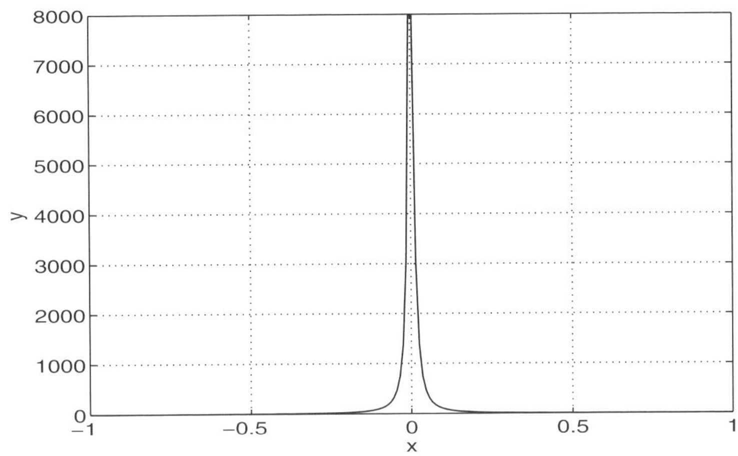

Figure 5.2.3. *A needle-shaped function.*

For T_{22} we can estimate the truncation error by $|T_{22} - T_{21}|$, which is usually a strong overestimate. We *accept* the approximation if

$$|T_{22} - T_{21}| < \frac{h_j \epsilon}{b-a}, \tag{5.2.26}$$

i.e., we require the error to be *less than* $\epsilon/(b-a)$ *per unit step*. Otherwise we *reject* the approximation, and subdivide the interval in two intervals, $[a_j, \frac{1}{2}(a_j+b_j)]$, $[\frac{1}{2}(a_j+b_j), b_j]$. The same rule is now applied to these two subintervals.

Note that if the function values computed previously are saved, these can be reused for the new intervals. We start with one interval $[a, b]$ and carry on subdivisions until the error criterion in (5.2.26) is satisfied for all intervals. Since the total error is the sum of errors for all subintervals, we then have the required error estimate:

$$R_T < \sum_j \frac{h_j \epsilon}{b-a} = \epsilon.$$

The possibility that a user might try to integrate a nonintegrable function (e.g., $f(x) = x^{-1}$ on $[0, 1]$) cannot be neglected. In principle it is not possible to decide whether a function $f(x)$ is integrable on the basis of a finite sample $f(x_1), \dots, f(x_n)$ of function values. Therefore, it is necessary to impose

1. an upper limit on the number of function evaluation,

2. a lower limit on the size of the subregions.

This means that premature termination may occur even when the function is only close to being nonintegrable, for example, $f(x) = x^{-0.99}$.

Many different adaptive quadrature schemes exist. Here we shall illustrate one simple scheme based on a five-point closed Newton–Cotes' rule, which applies bisection in a locally adaptive strategy. All function evaluations contribute to the final estimate. In many situations it might be preferable to specify a *relative error tolerance*:

$$tol = \eta \left| \int_a^b f(x)\,dx \right|.$$

A more complete discussion of the choice of termination criteria in adaptive algorithms is found in Gander and Gautschi [131].

Algorithm 5.3. *Adaptive Simpson.*

Let f be a given function to be integrated over $[a, b]$. The function `adaptsimp` uses a recursive algorithm to compute an approximation with an error less than a specified tolerance $\tau > 0$.

```
function [I,nf] = adaptsimp(f,a,b,tol);
% ADAPTSIMP calls the recursive function ADAPTREC to compute
% the integral of the vector-valued function f over [a,b];
% tol is the desired absolute accuracy; nf is the number of
% function evaluations.
%
ff = feval(f,[a, (a+b)/2, b]);
nf = 3; % Initial Simpson approximation
I1 = (b - a)*[1, 4, 1]*ff'/6;
% Recursive computation
[I,nf] =  adaptrec(f,a,b,ff,I1,tol,nf);

function [I,nf] = adaptrec(f,a,b,ff,I1,tol,nf);
h = (b - a)/2;
fm = feval(f, [a + h/2, b - h/2]);
nf = nf + 2;
% Simpson approximations for left and right subinterval
fR = [ff(2); fm(2); ff(3)];
fL = [ff(1); fm(1); ff(2)];
IL = h*[1, 4, 1]*fL/6;
IR = h*[1, 4, 1]*fR/6;
I2 = IL + IR;
I  = I2 + (I2 - I1)/15;    % Extrapolated approximation
if  abs(I - I2) > tol      % Refine both subintervals
   [IL,nf] = adaptrec(f,a,a+h,fL,IL,tol/2,nf);
   [IR,nf] = adaptrec(f,b-h,b,fR,IR,tol/2,nf);
   I   = IL + IR;
 end
```

Note that in a **locally adaptive** algorithm using a recursive partitioning scheme, the subintervals are processed from left to right until the integral over each subinterval satisfies some error requirement. This means that an a priori initial estimate of the whole integral, needed for use in a relative local error estimate cannot be updated until all subintervals are processed and the computation is finished. Hence, if a relative tolerance is specified, then an estimate of the integral is needed before the recursion starts. This is complicated by the fact that the initial estimate might be zero, for example, if a periodic integrand is sampled at equidistant intervals. Hence a combination of relative and absolute criteria might be preferable.

Example 5.2.6.

This algorithm was used to compute the integral

$$\int_{-4}^{4} \frac{dx}{1+x^2} = 2.65163532733607$$

with an absolute tolerance 10^{-p}, $p = 4, 5, 6$. The following approximations were obtained.

I	tol	n	Error
2.65162 50211	10^{-4}	41	$1.0\,10^{-5}$
2.65163 52064	10^{-5}	81	$1.2\,10^{-7}$
2.65163 5327353	10^{-6}	153	$-1.7\,10^{-11}$

Note that the actual error is much smaller than the required tolerance.

So far we have considered adaptive routines, which use fixed quadrature rules on each subinterval but where the partition of the interval depends on the integrand. Such an algorithm is said to be **partition adaptive**. We can also consider **doubly adaptive** integration algorithms. These can choose from a sequence of increasingly higher-order rules to be applied to the current subinterval. Such algorithms use a selection criterion to decide at each stage whether to subdivide the current subinterval or to apply a higher-order rule. Doubly adaptive routines cope more efficiently with smooth integrands.

Many variations on the simple scheme outlined above are possible. For example, we could base the method on a higher-order Romberg scheme, or even try to choose an optimal order for each subinterval. Adaptive methods work even when the integrand $f(x)$ is badly behaved. But if f has singularities or unbounded derivatives, the error criterion may never be satisfied. To guard against such cases it is necessary to include some bound of the number of recursion levels that are allowed. It should be kept in mind that although adaptive quadrature algorithms are convenient to use they are in general less efficient than methods which have been specially adapted for a particular problem.

We finally warn the reader that *no automatic quadrature routine can always be guaranteed to work*. Indeed, any estimate of $\int_a^b f(x)\,dx$ based solely on the value of $f(x)$ on finitely many points can fail. The integrand $f(x)$ may, for example, be nonzero only on a small subset of $[a, b]$. An adaptive quadrature rule based only on samples $f(x)$ in a finite number of points theoretically may return the value zero in such a case!

We recall the remark that evaluation of the integral $\int_a^b f(x)\,dx$ is equivalent to solving an initial value problem $y' = f(x)$, $y(a) = 0$, for an ordinary differential equation. For such problems sophisticated techniques for adaptively choosing step size and order in the integration have been developed. These may be a good alternative choice for handling difficult cases.

Review Questions

5.2.1 (a) Give an account of the theoretical background of Romberg's method.

(b) For which values of k are the elements T_{kk} in the Romberg scheme identical to closed Newton–Cotes' formulas?

5.2.2 Romberg's method uses extrapolation of a sequence of trapezoidal approximations computed for a sequence of step sizes $h_0, h_1, h_2, \ldots$. What sequences have been suggested and what are their relative merits?

5.2.3 When the integrand has a singularity at one of the endpoints, many quadrature methods converge very slowly. Name a few possible ways to resolve this problem.

5.2.4 Romberg's method works only when the error of the trapezoidal rule has an expansion in even powers of h. If this is not the case, what other extrapolations methods should be tried?

5.2.5 Describe at least two methods for treating an integral with an oscillating integrand.

5.2.6 In partition adaptive quadrature methods the step sizes are locally adopted. Discuss how the division into subintervals can be controlled.

Problems and Computer Exercises

5.2.1 Is it true that (the short version of) Simpson's formula is a particular case of Gregory's formula?

5.2.2 Use Romberg's method to compute the integral $\int_0^4 f(x)\,dx$, using the following (correctly rounded) values of $f(x)$. Need all the values be used?

x	0.0	0.5	1.0	1.5	2.0	2.5	3.0	3.5	4.0
$f(x)$	−4271	−2522	−499	1795	4358	7187	10,279	13,633	17,247

5.2.3 (a) Suppose that the form of the error of Romberg's method is known, but not the error constant r_k. Determine r_k numerically for $k = 3$ and $k = 4$, by computing the Romberg scheme for $f(x) = x^{2k}$.

(b) Prove the formula for the error constant of Romberg's method.

5.2.4 Compute by the Euler–Maclaurin formula, or rather the composite trapezoidal rule,

$$\text{(a)} \int_0^\infty e^{-x^2/2}dx, \qquad \text{(b)} \int_0^\infty \frac{dx}{\cosh(\pi x)}$$

as accurately as you can with the normal precision of your computer (or software). Then find out empirically how the error depends on h. Make semilogarithmic plots on the same screen. How long a range of integration do you need?

5.2.5 (a) Use Romberg's method and Aitken acceleration to compute the integral

$$I[f] = \int_1^\infty \frac{1}{1+x^2}\,dx = \int_1^2 + \int_2^4 + \int_4^8 + \cdots .$$

Determine where to terminate the expansion, and then use Aitken acceleration to find $I[f]$. Compare with the exact result. Think of an error estimate that can be used if the exact result is not known.

(b) Treat in the same way

$$\int_1^\infty \frac{1}{\sqrt{x+x^3}}.$$

Compare the computational effort for the computation of the tail $\int_R^\infty$ by acceleration and by series expansion with the same accuracy.

5.2.6 Modify the MATLAB function `romberg` so that it uses rational extrapolation according to the recursion (5.2.7) instead of polynomial extrapolation. Use the modified program to compute the integral in Example 5.2.2. Compare the results for the two different extrapolation methods.

5.2.7 Apply the MATLAB program `romberg` in Sec. 5.2.2 and repeated averages on the integral

$$\int_0^{1000} x\cos(x^3)\,dx.$$

Try to obtain the results with 10 decimal places.

5.2.8 (a) Show the following series expansions for the coefficients in the Filon-trapezoidal formula:

$$w_0(\theta) = w_N(-\theta) = \frac{1}{2} - \frac{\theta^2}{24} + \frac{\theta^4}{720} - \cdots + i\left(\frac{\theta}{6} - \frac{\theta^3}{120} + \frac{\theta^5}{5040} - \cdots\right),$$

$$w(\theta) = w_0(\theta) + w_N(-\theta) = 1 - \frac{\theta^2}{12} + \frac{\theta^4}{360} - \cdots.$$

(b) For what value of θ should you switch to using the series expansions above, if you want to minimize an upper bound for the error in the coefficients?

5.3 Quadrature Rules with Free Nodes

5.3.1 Method of Undetermined Coefficients

We have previously seen how to derive quadrature rules using Lagrange interpolation or operator series. We now outline another general technique, the method of undetermined coefficients, for determining quadrature formulas of maximum order with both free and prescribed nodes.

Let L be a linear functional and consider approximation formulas of the form

$$Lf \approx \tilde{L}f = \sum_{i=1}^{p} a_i f(x_i) + \sum_{j=1}^{q} b_j f(z_j), \tag{5.3.1}$$

where the x_i are p given nodes, while the z_j are q **free nodes**. The latter are to be determined together with the weight factors a_i, b_j. The altogether $p + 2q$ parameters in the formula are to be determined, if possible, so that the formula becomes exact for all polynomials of degree less than $N = p + 2q$. We introduce the two node polynomials

$$r(x) = (x - x_1)\cdots(x - x_p), \qquad s(x) = (x - z_1)\cdots(x - z_q) \tag{5.3.2}$$

of degree p and q, respectively.

Let $\phi_1, \phi_2, \dots, \phi_N$ be a basis of the space of polynomials of degree less than N. We assume that the quantities $L\phi_k$, $k = 1 : p + 2q$ are known. Then we obtain the *nonlinear* system

$$\sum_{i=1}^{p} \phi_k(x_i)a_i + \sum_{j=1}^{q} \phi_k(z_j)b_j = L\phi_k, \quad k = 1, 2, \dots, p + 2q, \tag{5.3.3}$$

for the $p + 2q$ parameters. This system is nonlinear in z_j, but of a very special type. Note that the free nodes z_j appear in a symmetric fashion; the system (5.3.3) is invariant with respect to permutations of the free nodes together with their weights. We therefore first ask for their **elementary symmetric functions**, i.e., for the coefficients g_j of the node polynomial

$$s(x) = \phi_{q+1}(x) - \sum_{j=1}^{q} g_j\phi_j(x) \tag{5.3.4}$$

that has the free nodes $z_1, \dots, z_q$ as zeros. We change the basis to the set

$$\phi_1(x), \dots, \phi_q(x), s(x)\phi_1(x), \dots, s(x)\phi_{p+q}(x).$$

In the system (5.3.3), the equations for $k = 1 : q$ will not be changed, but the equations for $k = 1 + q : p + 2q$ become

$$\sum_{i=1}^{p} \phi_{k'}(x_i)s(x_i)a_i + \sum_{j=1}^{q} \phi_{k'}(z_j)s(z_j)b_j = L(s\phi_{k'}), \quad 1 \le k' \le p + q. \tag{5.3.5}$$

Here the second sum disappears since $s(z_j) = 0$ for all j. (This is the nice feature of this treatment.) Further, by (5.3.4),

$$L(s\phi_{k'}) = L(\phi_{k'}\phi_{q+1}) - \sum_{j=1}^{q} L(\phi_{k'}\phi_j)g_j, \quad 1 \le k' \le p + q. \tag{5.3.6}$$

We thus obtain the following *linear* system for the computation of the $p + q$ quantities, g_j, and $A_i = s(x_i)a_i$:

$$\sum_{j=1}^{q} L(\phi_{k'}\phi_j)g_j + \sum_{i=1}^{p} \phi_{k'}(x_i)A_i = L(\phi_{k'}\phi_{q+1}), \quad k' = 1 : p + q. \tag{5.3.7}$$

The weights of the fixed nodes are $a_i = A_i/s(x_i)$. The free nodes z_j are then determined by finding the q roots of the polynomial $s(x)$. Methods for computing roots of a polynomial are given in Sec. 6.5. Finally, with a_i and z_j known, the weights b_j are obtained by the solution of the first q equations of the system (5.3.3) which are linear in b_j.

The remainder term $Rf = (Lf - \tilde{L}f)$ of the method, exact for all polynomials of degree less than $N = p + 2q$, is of the form

$$Rf = R(f - P_N) \approx c_N f^{(N)}(\xi), \quad c_N = R(x^N)/N!,$$

where c_N is called the error constant. Note that $R(x^N) = R(\phi_{N+1})$, where $\phi_{(N+1)}$ is any monic polynomial of degree N, since $x^N - \phi_{(N+1)}$ is a polynomial of degree less than N. Hence, for the determination of the error constant we compute the difference between the right-hand and the left-hand sides of

$$\sum_{i=1}^{p} \phi_k(x_i)\, a_i + \sum_{j=1}^{q} \phi_k(z_j)\, b_j + N!c_N = L\phi_{N+1}, \qquad N = p + 2q, \tag{5.3.8}$$

and divide by $(N)!$. If, for example, a certain kind of symmetry is present, then it can happen that $c_{p+2q} = 0$. The formula is then more accurate than expected, and we take $N = p + 2q + 1$ instead. The case that also $c_{p+2q+1} = 0$ may usually be ignored. It *can* occur if several of the given nodes are located, where free nodes would have been placed.

From a pure mathematical point of view all bases are equivalent, but equation (5.3.3) may be better conditioned with some bases than with others, and this turns out to be an important issue when $p + 2q$ is large. We mention three different situations.

(i) The most straightforward choice is to set $[a, b] = [0, 1]$ and use the monomial basis $\phi_k(x) = x^{k-1}$, $x \in (0, b)$ (b may be infinite). For this choice the condition number of (5.3.3) increases exponentially with $p + 2q$. Then the free nodes and corresponding weights may become rather inaccurate when $p + 2q$ is large. It is usually found, however, that unless the condition number is so big that the solution breaks down completely, the computed solution will satisfy equation (5.3.3) with a small residual. This is what really matters for the application of formula (5.3.1).

(ii) Take $[a, b] = [-1, 1]$, and assume that the weight function $w(x)$ and the given nodes x_i are symmetrical with respect to the origin. Then the weights a_i and b_i, and the free nodes z_j will also be symmetrically located, and with the monomial basis it holds that $L(\phi_k(x)) = 0$, when k is even. If $p = 2p'$ is even, the number of parameters will be reduced to $p' + q$ by the transformation $x = \sqrt{\xi}$, $\xi \in [0, b^2]$. Note that $w(x)$ will be replaced by $w(\sqrt{\xi})/\sqrt{\xi}$. If p is odd, one node is at the origin, and one can proceed in an analogous way. This should also reduce the condition number approximately to its square root, and it is possible to derive in a numerically stable way formulas with about twice as high an order of accuracy as in the unsymmetric case.

(iii) Taking ϕ_k to be the orthogonal polynomials for the given weight function will give a much better conditioned system for determining the weights. This case will be considered in detail in Sec. 5.3.5.

Example 5.3.1.

Consider the linear functional $L(f) = \int_0^1 f(x)\, dx$. Set $p = 0$, $q = 3$ and choose the monomial basis $\phi_i(x) = x^{i-1}$. Introducing the node polynomial

$$s(x) = (x - z_1)(x - z_2)(x - z_3) = x^3 - s_3 x^2 - s_2 x - s_1,$$

the linear system (5.3.6) becomes

$$\begin{pmatrix} 1 & 1/2 & 1/3 \\ 1/2 & 1/3 & 1/4 \\ 1/3 & 1/4 & 1/5 \end{pmatrix} \begin{pmatrix} s_1 \\ s_2 \\ s_3 \end{pmatrix} = \begin{pmatrix} 1/4 \\ 1/5 \\ 1/6 \end{pmatrix}.$$

The exact solution is $s_1 = 1/20$, $s_2 = -3/5$, and $s_3 = 3/2$. The free nodes thus are the zeros of $s(x) = x^3 - 3x^2/2 + 3x/5 - 1/20$, which are $z_2 = 1/2$ and $z_{1,3} = 1/2 \pm \sqrt{3/20}$. The weights b_1, b_2, b_3 are then found by solving (5.3.3) for $k = 1 : 3$.

The matrix of the above system is a Hankel matrix. The reader should verify that when $p > 0$ the matrix becomes a kind of combination of a Hankel matrix and a Vandermonde matrix.

5.3.2 Gauss–Christoffel Quadrature Rules

Assume that the n nodes in a quadrature formula are chosen so that

$$(f, s) = \int_a^b p(x)s(x)w(x)\,dx = 0 \quad \forall\ p(x) \in \mathcal{P}_n, \tag{5.3.9}$$

where $s(x) = (x - x_1)(x - x_2)\cdots(x - x_n)$ is the node polynomial. Then, by Theorem 5.1.3, the corresponding interpolatory quadrature rule will have the maximum possible order $2n - 1$.

We define an inner product with respect to a weight function $w(x) \geq 0$ by

$$(f, g) = \int_a^b f(x)g(x)w(x)\,dx, \tag{5.3.10}$$

and assume that the moments

$$\mu_k = (x^k, 1) = \int_a^b x^k w(x)\,dx \tag{5.3.11}$$

are defined for all $k \geq 0$, and $\mu_0 > 0$. This inner product has the important property that $(xf, g) = (f, xg)$. The condition (5.3.9) on the node polynomial can then be interpreted to mean that $s(x)$ is orthogonal to all polynomials in $\mathcal{P}_n$.

For the weight function $w(x) \equiv 1$ the corresponding quadrature rules were derived in 1814 by Gauss [136]. Formulas for more general weight functions were given by Christoffel [69] in 1858,[173] which is why these are referred to as **Gauss–Christoffel quadrature rules**.

The construction of Gauss–Christoffel quadrature rules is closely related to the theory of orthogonal polynomials. In Sec. 4.5.5 we showed how the orthogonal polynomials corresponding to the inner product (5.3.10) could be generated by a three-term recurrence formula. The zeros of these polynomials are the nodes in a Gauss–Christoffel quadrature formula. As for all interpolatory quadrature rules the weights can be determined by

[173] Elwin Bruno Christoffel (1829–1900) worked mostly in Strasbourg. He is best known for his work in geometry and tensor analysis, which Einstein later used in his theory of relativity.

integrating the elementary Lagrange polynomials (5.1.7)

$$w_i = \int_a^b \ell_i(x)w(x)\,dx, \quad \ell_i(x) = \prod_{\substack{j=1\\ j\neq i}}^{n} \frac{(x - x_j)}{(x_i - x_j)}, \qquad i = 1:n.$$

In Sec. 5.3.5 we will outline a more stable algorithm that determines the nodes and weights by solving the eigenvalue problem for a symmetric tridiagonal matrix defined by the coefficients in the recurrence relation.

We shall now prove some important properties of Gauss–Christoffel quadrature rules using the general theory of orthogonal polynomials.

Theorem 5.3.1.

The zeros x_i, $i = 1 : n$, of the orthogonal polynomial polynomial $\varphi_{n+1}(x)$ of degree n, associated with the weight function $w(x) \geq 0$ on $[a, b]$, are real, distinct, and contained in the open interval (a, b).

Proof. Let $a < x_1 < x_2 < \cdots < x_m < b$ be the roots of $\varphi_{n+1}(x)$ of odd multiplicity, which lie in (a, b). At these roots $\varphi_{n+1}(x)$ changes sign and therefore the polynomial $q(x)\varphi_{n+1}(x)$, where

$$q(x) = (x - x_1)(x - x_2)\cdots(x - x_m),$$

has constant sign in $[a, b]$. Hence,

$$\int_a^b \varphi_{n+1} q(x)w(x)\,dx > 0.$$

But this is possible only if the degree of $q(x)$ is equal to n. Thus $m = n$ and the theorem follows. □

Corollary 5.3.2.

If $x_1, x_2, \ldots, x_n$ are chosen as the n distinct zeros of the orthogonal polynomial φ_{n+1} of degree n in the family of orthogonal polynomials associated with $w(x)$, then the formula

$$\int_a^b f(x)w(x)\,dx \approx \sum_{i=1}^{n} w_i f(x_i), \quad w_i = \int_a^b \ell_i(x)w(x)\,dx, \qquad (5.3.12)$$

is exact for polynomials of degree $2n - 1$.

Apart from having optimal degree of exactness equal to $2n - 1$, Gaussian quadrature rules have several important properties, which we now outline.

Theorem 5.3.3.

All weights in a Gaussian quadrature rule are real, distinct, and positive.

Proof. Let

$$\ell_i(x) = \prod_{\substack{j=1\\ j\neq i}}^{n} \frac{(x - x_j)}{(x_i - x_j)}, \qquad i = 1:n,$$

be the Lagrange polynomials. Then the quadrature formula (5.3.12) is exact for $p(x) = (\ell_i(x))^2$, which is of degree $2(n-1)$. Further, $\ell_i(x_j) = 0$, $j \neq i$, and therefore

$$\int_a^b (\ell_i(x))^2 w(x)\,dx = w_i(\ell_i(x_i))^2 = w_i.$$

Since $w(x) > 0$ it follows that $w_i > 0$. □

Gaussian quadrature formulas can also be derived by Hermite interpolation on the nodes x_k, *each counted as a double node* and requiring that coefficients of the derivative terms should be zero. This interpretation gives a convenient expression for the error term in Gaussian quadrature.

Theorem 5.3.4.

The remainder term in Gauss' quadrature rule (5.3.12) *with n nodes is given by the formula*

$$I[f] - I_n(f) = \frac{f^{(2n)}(\xi)}{(2n)!}\int_a^b \left[\prod_{i=1}^n (x - x_i)\right]^2 w(x)\,dx = c_n f^{(2n)}(\xi), \quad a < \xi < b. \tag{5.3.13}$$

The constant c_n can be determined by applying the formula to some polynomial of degree $2n$.

Proof. Denote by $q(x)$ the polynomial of degree $2n-1$ which solves the Hermite interpolation problem (see Sec. 4.3.1)

$$q(x_i) = f(x_i), \qquad q'(x_i) = f'(x_i), \quad i = 1:n.$$

The Gauss quadrature formula is exact for $q(x)$, and hence

$$\int_a^b q(x)w(x)\,dx = \sum_{i=1}^n w_i q(x_i) = \sum_{i=1}^n w_i f(x_i).$$

Thus

$$\sum_{i=1}^n w_i f(x_i) - \int_a^b f(x)w(x)\,dx = \int_a^b (q(x) - f(x))w(x)\,dx.$$

Using the remainder term (4.3.4) in Hermite interpolation gives

$$f(x) - q(x) = \frac{f^{(2n)}(\xi)}{(2n)!}(\varphi_n(x))^2, \quad \varphi_n(x) = \prod_{i=1}^n (x - x_i),$$

and the theorem now follows. □

Using Bernštein's approximation theorem (Theorem 3.2.5) we get the following corollary.

Corollary 5.3.5.

Let f real-valued for $z \in [-1, 1]$, and analytic and single-valued $|f(z)| \le M$ in the region $z \in \mathcal{E}_R$, $R > 1$, where

$$\mathcal{E}_R = \{z : |z-1| + |z+1| \le R + R^{-1}\},$$

be an ellipse with foci at 1 and −1. Then the remainder term in a Gauss quadrature rule with n nodes for the interval $[-1, 1]$ satisfies

$$|I[f] - I_n(f)| \le \frac{2M\mu_0}{1 - 1/R} R^{-2n}. \tag{5.3.14}$$

This shows the rapid convergence of Gauss' quadrature rules for functions analytic in a region $\mathcal{E}_R$, with $R \gg 1$.

We now mention some classical Gauss–Christoffel quadrature rules, which are related to the orthogonal polynomials surveyed in Sec. 4.5.5. For an integral $\int_{-1}^{1} f(x)\,dx$, with uniform weight distribution $w(x) = 1$, the relevant orthogonal polynomials are the Legendre polynomials $P_n(x)$.

As a historical aside, Gauss derived his quadrature formula by considering the continued fraction

$$\frac{1}{2}\int_{-1}^{1} \frac{dx}{z-x} = \frac{1}{2}\ln\left(\frac{z+1}{z-1}\right) = \frac{1}{z-}\,\frac{1/3}{z-}\,\frac{4/(3\cdot 5)}{z-}\,\frac{9/(5\cdot 7)}{z-}\cdots, \tag{5.3.15}$$

which he had derived in an earlier paper. The nth convergent of this continued fraction is a rational function with a numerator of degree $n-1$ in z and denominator of degree n which is the $(n-1, n)$ Padé approximant to the function. Decomposing this fraction in partial fractions the residues and the poles can be taken as nodes of a quadrature formula. Using the accuracy properties of the Padé approximants Gauss showed that the quadrature formula will have order $2n-1$.

The reciprocal of the denominators' polynomials $P_n(z) = z^n Q_n(1/z)$ are precisely the Legendre polynomials; see Example 3.5.6. Recall that the monic Legendre polynomials satisfy the recurrence formula $P_0 = 1$, $P_1 = x$,

$$P_{n+1}(x) = xP_n(x) - \frac{n^2}{4n^2-1}P_{n-1}(x), \quad n \ge 1.$$

The first few monic Legendre polynomials are

$$P_2(x) = \frac{1}{3}(3x^2 - 1), \qquad P_3(x) = \frac{1}{5}(5x^3 - 3x),$$

$$P_4(x) = \frac{1}{35}(35x^4 - 30x^2 + 3), \qquad P_5(x) = \frac{1}{63}(63x^5 - 70x^3 + 15x), \ldots.$$

Example 5.3.2.

For a two-point Gauss–Legendre quadrature rule the two abscissae are the zeros of $P_2(x) = \frac{1}{3}(3x^2 - 1)$, i.e., $\pm 3^{-1/2}$. Note that they are symmetric with respect to the origin.

The weights can be determined by application of the formula to $f(x) = 1$ and $f(x) = x$, respectively. This gives

$$w_0 + w_1 = 2, \qquad -3^{-1/2}w_0 + 3^{-1/2}w_1 = 0,$$

with solution $w_0 = w_1 = 1$. Hence the formula

$$\int_{-1}^{1} f(x)\,dx \approx f(-3^{-1/2}) + f(3^{-1/2})$$

is exact for polynomials of degree ≤ 3. For a three-point Gauss formula, see Problem 5.3.1.

Abscissae and weights for Gauss formulas using $n = m + 1$ points, for $n = 2 : 10$, with 15 decimal digits and $n = 12, 16, 20, 24, 32, 40, 48, 64, 80$, and 96 with 20 digits are tabulated in [1, Table 25.4]; see Table 5.3.1 for a sample. Instead of storing these constants, it might be preferable to use a program that generates abscissae and weights as needed.

Table 5.3.1. *Abscissae and weight factors for some Gauss–Legendre quadrature from* [1, Table 25.4].

x_i		w_i
	$n = 3$	
0.00000 00000 00000		0.88888 88888 88889
±0.77459 66692 41483		0.55555 55555 55556
	$n = 4$	
±0.33998 10435 84856		0.65214 51548 62546
±0.86113 63115 94053		0.34785 48451 37454
	$n = 5$	
0.00000 00000 00000		0.56888 88888 88889
±0.53846 93101 05683		0.47862 86704 99366
±0.90617 98459 38664		0.23692 68850 56189

For the weight function

$$w(x) = (1-x)^{\alpha}(1+x)^{\beta}, \quad x \in [-1, 1], \quad \alpha, \beta > -1,$$

the nodes are obtained from the zeros of the Jacobi polynomials $J_n(x; \alpha, \beta)$. In the special case when $\alpha = \beta = 0$ these equal the Legendre polynomials. The case $\alpha = \beta = -1/2$, which corresponds to the weight function $w(x) = 1/\sqrt{1-x^2}$, gives the Chebyshev polynomials $T_n(x)$ of the first kind. Similarly, $\alpha = \beta = 1/2$ gives the Chebyshev polynomials $U_n(x)$ of the second kind.

If a quadrature rule is given for the standard interval $[-1, 1]$, the corresponding formula for an integral over the interval $[a, b]$ is obtained by the change of variable $t = \frac{1}{2}((b-a)x + (a+b))$, which maps the interval $[a, b]$ onto the standard interval $[-1, 1]$:

$$\int_a^b f(t)dt = \frac{b-a}{2}\int_{-1}^{1} g(x)\,dx, \qquad g(x) = f\left(\frac{1}{2}((b-a)x + (a+b))\right).$$

If $f(t)$ is a polynomial, then $g(x)$ will be a polynomial of the same degree, since the transformation is linear. Hence the order of accuracy of the formula is not affected.

Two other important cases of Gauss quadrature rules deal with infinite intervals of integration. The generalized Laguerre polynomials $L_n^{(\alpha)}(x)$ are orthogonal with respect to the weight function

$$w(x) = x^\alpha e^{-x}, \quad x \in [0, \infty], \quad \alpha > -1.$$

Setting $\alpha = 0$, we get the Laguerre polynomials $L_n^{(0)}(x) = L_n(x)$.

The Hermite polynomials are orthogonal with respect to the weight function

$$w(x) = e^{-x^2}, \quad -\infty < x < \infty.$$

Recall that weight functions and recurrence coefficients for the above monic orthogonal polynomials are given in Table 4.5.1.

Rather little is found in the literature on numerical analysis about densities on infinite intervals, except the classical cases above. It follows from two classical theorems of Hamburger in 1919 and M. Riesz in 1923 that the system of orthogonal polynomials for the density w over the infinite interval $[-\infty, \infty]$ is complete if, for some $\beta > 0$,

$$\int_{-\infty}^{\infty} e^{\beta|x|} w(x)\, dx < \infty;$$

see Freud [126, Sec. II.4–5]. For densities on $[0, \infty]$, x is to be replaced by $\sqrt{x}$ in the above result. (Note that a density function on the positive real x-axis can be mapped into an even density function on the whole real t-axis by the substitution $x = t^2$.

5.3.3 Gauss Quadrature with Preassigned Nodes

In many applications it is desirable to use Gauss-type quadrature where some nodes are preassigned and the rest chosen to maximize the order of accuracy. In the most common cases the preassigned nodes are at the endpoints of the interval. Consider a quadrature rule of the form

$$\int_a^b f(x)w(x)\, dx = \sum_{i=1}^{n} w_i f(x_i) + \sum_{j=1}^{m} b_j f(z_j) + R(f), \tag{5.3.16}$$

where z_j, $j = 1 : m$, are fixed nodes in $[a, b]$ and the x_i are determined so that the interpolatory rule is exact for polynomials of order $2n + m - 1$. By a generalization of Theorem 5.3.4 the remainder term is given by the formula

$$R(f) = \frac{f^{(2n+m)}(\xi)}{(2n)!} \int_a^b \prod_{i=1}^{m}(x - z_i)\Big[\prod_{i=1}^{n}(x - x_i)\Big]^2 w(x)\, dx, \quad a < \xi < b. \tag{5.3.17}$$

In **Gauss–Lobatto** quadrature both endpoints are used as abscissae, $z_1 = a$, $z_2 = b$, and $m = 2$. For the standard interval $[a, b] = [-1, 1]$ and the weight function $w(x) = 1$, the quadrature formula has the form

$$\int_{-1}^{1} f(x)\, dx = w_0 f(-1) + w_{n+1} f(1) + \sum_{i=1}^{n} w_i f(x_i) + E_L. \tag{5.3.18}$$

The abscissae $a < x_i < b$ are the zeros of the orthogonal polynomial ϕ_n corresponding to the weight function $\tilde{w}(x) = (1 - x^2)$, i.e., up to a constant factor equal to the Jacobi polynomial $J_n(x, 1, 1) = P'_{n+1}(x)$. The nodes lie symmetric with respect to the origin. The corresponding weights satisfy $w_i = w_{n+1-i}$, and are given by

$$w_0 = w_{n+1} = \frac{2}{(n+2)(n+1)}, \qquad w_i = \frac{w_0}{(P_{n+1}(x_i))^2}, \qquad i = 1 : n. \tag{5.3.19}$$

The Lobatto rule (5.3.18) is exact for polynomials of order $2n+1$, and for $f(x) \in C^{2m}[-1, 1]$ the error term is given by

$$R(f) = -\frac{(n+2)(n+1)^3 2^{2n+3}(n!)^4}{(2n+3)[(2n+2)!]^3} f^{(2n+2)}(\xi), \quad \xi \in (-1, 1). \tag{5.3.20}$$

Nodes and weights for Lobatto quadrature are found in [1, Table 25.6].

In **Gauss–Radau** quadrature rules $m = 1$ and one of the endpoints is taken as the abscissa, $z_1 = a$ or $z_1 = b$. The remainder term (5.3.17) becomes

$$R(f) = \frac{f^{(2n+1)}(\xi)}{(2n)!} \int_a^b (x - z_1) \left[\prod_{i=1}^{n} (x - x_i) \right]^2 w(x)\, dx, \quad a < \xi < b. \tag{5.3.21}$$

Therefore, if the derivative $f^{(n+1)}(x)$ has constant sign in $[a, b]$, then the error in the Gauss–Radau rule with $z_1 = b$ will have opposite sign to the Gauss–Radau rule with $z_1 = a$. *Thus, by evaluating both rules we obtain lower and upper bounds for the true integral.* This has many applications; see Golub [163].

For the standard interval $[-1, 1]$ the Gauss–Radau quadrature formula with $z_1 = 1$ has the form

$$\int_{-1}^{1} f(x)\, dx = w_0 f(-1) + \sum_{i=1}^{n} w_i f(x_i) + E_{R1}. \tag{5.3.22}$$

The n free abscissae are the zeros of

$$\frac{P_n(x) + P_{n+1}(x)}{x - 1},$$

where $P_m(x)$ are the Legendre polynomials. The corresponding weights are given by

$$w_0 = \frac{2}{(n+1)^2}, \qquad w_i = \frac{1}{(n+1)^2} \frac{1 - x_i}{(P_n(x_i))^2}, \qquad i = 1 : n. \tag{5.3.23}$$

The Gauss–Radau quadrature rule is exact for polynomials of order $2n$. Assuming that $f(x) \in C^{2m-1}[-1, 1]$, then the error term is given by

$$E_{R1}(f) = \frac{(n+1) 2^{2n+1}}{[(2n+1)!]^3} (n!)^4 f^{(2n+1)}(\xi_1), \quad \xi_1 \in (-1, 1). \tag{5.3.24}$$

A similar formula can be obtained with the fixed point $+1$ by making the substitution $t = -x$.

By modifying the proof of Theorem 5.3.3 it can be shown that the weights in Gauss–Radau and Gauss–Lobatto quadrature rules are positive if the weight function $w(x)$ is nonnegative.

Example 5.3.3.

The simplest Gauss–Lobatto rule is Simpson's rule with $n = 1$ interior node. Taking $n = 2$ the interior nodes are the zeros of $\phi_2(x)$, where

$$\int_{-1}^{1} (1 - x^2)\phi_2(x)p(x)\,dx = 0 \quad \forall p \in P_2.$$

Thus, ϕ_2 is, up to a constant factor, the Jacobi polynomial $J_2(x, 1, 1) = (x^2 - 1/5)$. Hence the interior nodes are $\pm 1/\sqrt{5}$ and by symmetry the quadrature formula is

$$\int_{-1}^{1} f(x)\,dx = w_0(f(-1) + f(1)) + w_1(f(-1/\sqrt{5}) + f(1/\sqrt{5})) + R(f), \qquad (5.3.25)$$

where $R(f) = 0$ for $f \in P_6$. The weights are determined by exactness for $f(x) = 1$ and $f(x) = x^2$. This gives $2w_0 + 2w_1 = 2$, $2w_0 + (2/5)w_1 = \frac{2}{3}$, i.e., $w_0 = \frac{1}{6}$, $w_1 = \frac{5}{6}$.

A serious drawback with Gaussian rules is that as we increase the order of the formula, *all interior abscissae change*, except that at the origin. Thus function values computed for the lower-order formula are not used in the new formula. This is in contrast to Romberg's method and Clenshaw–Curtis quadrature rules, where *all old function values* are used also in the new rule when the number of points is doubled.

Let G_n be an n-point Gaussian quadrature rule

$$\int_a^b f(x)w(x)\,dx \approx \sum_{i=0}^{n-1} a_i f(x_i),$$

where x_i, $i = 0 : n - 1$, are the zeros of the nth degree orthogonal polynomial $\pi_n(x)$. Kronrod [232, 233] considered extending G_n by finding a new quadrature rule

$$K_{2n+1} = \sum_{i=0}^{n-1} a_i f(x_i) + \sum_{i=0}^{n} b_i f(y_i), \qquad (5.3.26)$$

where the new $n + 1$ abscissae y_i are chosen such that the degree of the rule K_{2n+1} is equal to $3n + 1$. The new nodes y_i should then be selected as the zeros of a polynomial $p_{n+1}(x)$ of degree $n + 1$, satisfying the orthogonality conditions

$$\int_a^b \pi_n(x)p_{n+1}(x)w(x)\,dx = 0. \qquad (5.3.27)$$

If the zeros are real and contained in the closed interval of integration $[a, b]$ such a rule is called a **Kronrod extension** of the Gaussian rule. The two rules (G_n, K_{2n+1}) are called a **Gauss–Kronrod** pair. Note that the number of new function evaluations are the same as for the Gauss rule G_{n+1}.

It has been proved that a Kronrod extension exists for the weight function $w(x) = (1 - x^2)^{\lambda - 1/2}$, $\lambda \in [0, 2]$, and $[a, b] = [-1, 1]$. For this weight function the new nodes interlace the original Gaussian nodes, i.e.,

$$-1 \le y_0 < x_0 < y_1 < x_1 < y_2 < \cdots < x_{n-1} < y_n < 1.$$

This interlacing property can be shown to imply that all weights are positive. Kronrod considered extensions of Gauss–Legendre rules, i.e., $w(x) = 1$, and gives nodes and weights in [233] for $n \leq 40$.

It is not always the case that all weights are positive. For example, it has been shown that Kronrod extensions of Gauss–Laguerre and Gauss–Hermite quadrature rules with positive weights do not exist when $n > 0$ in the Laguerre case and $n = 3$ and $n > 4$ in the Hermite case. On the other hand, the Kronrod extensions of Gauss–Legendre rules can be shown to exist and have positive weights.

Gauss–Kronrod rules are one of most effective methods for calculating integrals. Often one takes $n = 7$ and uses the Gauss–Kronrod pair (G_7, K_{15}), together with the realistic but still conservative error estimate $(200|G_n - K_{2n+1}|)^{1.5}$; see Kahaner, Moler, and Nash [220, Sec. 5.5].

Kronrod extension of Gauss–Radau and Gauss–Lobatto rules can also be constructed. Kronrod extension of the Lobatto rule (5.3.25) is given by Gander and Gautschi [131] and used in an adaptive Lobatto quadrature algorithm. The simplest extension is the four-point Lobatto–Kronrod rule

$$\int_{-1}^{1} f(x)\,dx = \frac{11}{210}(f(-1) + f(1)) + \frac{72}{245}\left(f\left(-\sqrt{\frac{2}{3}}\right) + f\left(\sqrt{\frac{2}{3}}\right)\right) + \frac{125}{294}\left(f\left(-\frac{1}{\sqrt{5}}\right) + f\left(\frac{1}{\sqrt{5}}\right)\right) + \frac{16}{35}f(0) + R(f). \qquad (5.3.28)$$

This rule is exact for all $f \in \mathcal{P}_{10}$. Note that the Kronrod points $\pm\sqrt{2/3}$ and 0 interlace the previous nodes.

5.3.4 Matrices, Moments, and Gauss Quadrature

We first collect some classical results of Gauss, Christoffel, Chebyshev, Stieltjes, and others, with a few modern aspects and notations appropriate for our purpose.

Let $\{p_1, p_2, \dots, p_n\}$, where p_j is of exact degree $j - 1$, be a basis for the space $\mathcal{P}_n$ of polynomials of degree $n - 1$. We introduce the row vector

$$\pi(x) = [p_1(x),\ p_1(x), \dots, p_n(x)] \qquad (5.3.29)$$

containing these basis functions. The **modified moments** with respect to the basis $\pi(x)$ are

$$\nu_k = (p_k, 1) = \int_a^b p_k(x)w(x)\,dx, \qquad k = 1 : n. \qquad (5.3.30)$$

We define the two symmetric matrices

$$G = \int \pi(x)^T \pi(x) w(x)\,dx, \qquad \hat{G} = \int x\pi(x)^T \pi(x) w(x)\,dx \qquad (5.3.31)$$

associated with the basis defined by π. These have elements

$$g_{ij} = (p_i, p_j) = (p_j, p_i), \qquad \hat{g}_{ij} = (xp_i, p_j) = (xp_j, p_i),$$

respectively. Here

$$G = \begin{pmatrix} (p_1, p_1) & (p_1, p_2) & \dots & (p_1, p_n) \\ (p_2, p_1) & (p_2, p_2) & \dots & (p_2, p_n) \\ \vdots & \vdots & \ddots & \vdots \\ (p_n, p_1) & (p_n, p_2) & \dots & (p_n, p_n) \end{pmatrix} \tag{5.3.32}$$

is called the **Gram matrix**.

In particular, for the power basis

$$\theta(x)(1, x, x^2, \dots, x^{n-1}) \tag{5.3.33}$$

we have $g_{ij} = (x^{i-1}, x^{j-1}) = \mu_{i+j-2}$, where $\mu_k = (x^k, 1) = \int_a^b x^k w(x)\,dx$ are the ordinary moments. In this case the matrices G and $\hat{G}$ are the Hankel matrices,

$$G = \begin{pmatrix} \mu_0 & \mu_1 & \cdots & \mu_{n-1} \\ \mu_1 & \mu_2 & \cdots & \mu_n \\ \vdots & \vdots & \dots & \vdots \\ \mu_{n-1} & \mu_n & \cdots & \mu_{2n-2} \end{pmatrix}, \qquad \hat{G} = \begin{pmatrix} \mu_1 & \mu_2 & \cdots & \mu_n \\ \mu_2 & \mu_3 & \cdots & \mu_{n+1} \\ \vdots & \vdots & \dots & \vdots \\ \mu_n & \mu_{n+1} & \cdots & \mu_{2n-1} \end{pmatrix}.$$

In particular, for $w(x) \equiv 1$ and $[a, b] = [0, 1]$ we have $\mu_k = \int_0^1 x^{k-1}\,dx = 1/k$, and G is the notoriously ill-conditioned Hilbert matrix.

Let u and v be two polynomials in $\mathcal{P}_n$ and set

$$u(x) = \pi(x)u_\pi, \qquad v(x) = \pi(x)v_\pi,$$

where u_π, v_π are column vectors with the coefficients in the representation of u and v with respect to the basis defined by $\pi(x)$. Then

$$(u, v) = \int_a^b u_\pi^T \pi(x)^T \pi(x) v_\pi w(x)\,dx = u_\pi^T G v_\pi.$$

For $u = v \neq 0$ we find that $u_\pi^T G u_\pi = (u, u) > 0$, i.e., the Gram matrix G is positive definite. (The matrix $\hat{G}$ is, however, usually indefinite.)

A polynomial of degree n that is orthogonal to all polynomials of degree less than n can be written in the form

$$\phi_{n+1}(x) = xp_n(x) - \pi(x)c_n, \quad c_n \in \mathbf{R}^n. \tag{5.3.34}$$

Here c_n is determined by the linear equations

$$0 = (\pi(x)^T, \phi_{n+1}(x)) = (\pi(x)^T, xp_n(x)) - (\pi(x)^T, \pi(x))c_n,$$

or in matrix form

$$Gc_n = \hat{g}_n, \tag{5.3.35}$$

where $\hat{g}_n = \hat{G}e_n$ is the last column of the matrix $\hat{G}$. Further, there are coefficients $c_{k,j}$ depending on the basis only such that

$$xp_j(x) = \sum_{k=1}^{j+1} c_{k,j} p_k(x), \quad j = 1 : n-1.$$

Together with (5.3.34) this can be summarized in the (row) vector equation

$$x\pi(x) = \pi(x)\overline{C} + \phi_{n+1}(x)e_n^T, \quad \overline{C} = (C, c_n). \tag{5.3.36}$$

Here $e_n^T = (0, 0, \dots, 1)$ and $C = (c_{k,j}) \in \mathbf{R}^{n\times(n-1]}$ is an upper Hessenberg matrix. Note that C depends on the basis only, while c_n also depends on the weight function.

For the power basis $p_j(x) = x^{j-1}$, the matrix C is a **shift matrix**; the only nonzero elements are ones in the first main subdiagonal. If the basis is some family of orthogonal polynomials (possibly with respect to weight function other than w) C is a tridiagonal matrix, obtained by means of the three-term recurrence relation for this family.

After multiplication of (5.3.36) by $\pi(x)^T w(x)$ and integration we obtain by (5.3.31)

$$G\overline{C} = \hat{G}, \tag{5.3.37}$$

where the last column of this matrix equation is the same as (5.3.35). Let G^*, C^* be defined like G, C, with n increased by one. Note that G and C are principal submatrices of G^* and C^*. Then $\hat{G}$ equals the n first rows of the product G^*C^*. Thus, no integrations are needed for g_n, except for the Gram matrix G.

Theorem 5.3.6.

Denote by R the matrix of coefficients of the expansions of the general basis functions $\pi(x) = [p_1(x), p_1(x), \dots, p_n(x)]$ into the orthonormal basis polynomials with respect to the weight function w, i.e.,

$$\pi(x) = \varphi(x)R, \quad \varphi(x) = (\phi_1(x), \phi_2(x), \dots, \phi_n(x)). \tag{5.3.38}$$

(Conversely, the coefficients of the expansions of the orthogonal polynomials into the original basis functions are found in the columns of R^{-1}.) Then $G = R^TR$, i.e., R is the upper triangular Cholesky factor of the Gram matrix G. Note that up to the mth row this factorization is the same for all $n \geq m$. Further, $\hat{G} = R^T JR$, where J is a symmetric tridiagonal matrix.

Proof. R is evidently an upper triangular matrix. Further, we have

$$G = \int \pi(x)^T\pi(x)w(x)\,dx = \int R^T\varphi(x)^T\varphi(x)Rw(x)\,dx$$
$$= R^T IR = R^TR,$$

since the elements of $\varphi(x)$ are an orthonormal system. This shows that R is the Cholesky factor of G. We similarly find that

$$\hat{G} = R^T JR, \quad J = \int x\varphi(x)^T\varphi(x)w(x)\,dx,$$

and thus J clearly is a symmetrical matrix. J is a particular case of $\hat{G}$ and from (5.3.37) and $G = I$ it follows that $J = \overline{C}$, a Hessenberg matrix. Hence J is a symmetric tridiagonal matrix. □

From (5.3.37) and Theorem 5.3.6 it follows that

$$\hat{G} = R^T J R = G\overline{C} = R^T R \overline{C}.$$

Since R is nonsingular we have $R\overline{C} = JR$, or

$$J = R\overline{C}R^{-1}. \tag{5.3.39}$$

This shows that the spectrum of $\overline{C}$ equals the spectrum of J, for every choice of basis. We shall see that it is equal to the set of zeros of the orthogonal polynomial ϕ_{n+1}. For the power basis $p_j(x) = x^{j-1}$ (5.3.34) reads

$$\phi_{n+1}(x) = x^n - \sum_{k=1}^{n} c_{n,k} x^{k-1},$$

and hence

$$\overline{C} = \begin{pmatrix} 0 & & & & c_{n,1} \\ 1 & 0 & & & c_{n,2} \\ & 1 & \ddots & & \vdots \\ & & \ddots & 0 & c_{n,n-1} \\ & & & 1 & c_{n,n} \end{pmatrix} \in \mathbf{R}^{n\times n}.$$

This is the companion matrix of $\phi_{n+1}(x)$, and it can be shown that (see Sec. 6.5.2)

$$\det(zI - \overline{C}) = \phi_{n+1}(x). \tag{5.3.40}$$

Thus the eigenvalues λ_j, $j = 1:n$, *of* $\overline{C}$ *are the zeros of* $\phi_{n+1}(x)$, and hence the nodes for the Gauss–Christoffel quadrature formula.

It can be verified that the row eigenvector of $\overline{G}$ corresponding to λ_j is

$$\theta(\lambda_j) = (1, \lambda_j, \lambda_j^2, \dots, \lambda_j^{n-1}); \tag{5.3.41}$$

i.e., it holds that

$$\theta(\lambda_j)\overline{C} = \lambda_j \theta(\lambda_j), \quad j = 1:n. \tag{5.3.42}$$

This yields a diagonalization of $\overline{C}$, since, by the general theory of orthogonal polynomials (see Theorem 5.3.3), the roots are simple roots, located in the interior of the smallest interval that contains the weight distribution.

To summarize, we have shown that if C and the Gram matrix G are known, then c_n can be computed by performing the Cholesky decomposition $G = R^T R$ and then solving $R^T R c_n = \hat{g}_n$ for c_n. The zeros of $\phi_{n+1}(x)$ are then equal to the eigenvalues of $\overline{C} = (C, c_n)$ or, equivalently, the eigenvalues of the symmetric tridiagonal matrix $J = R\overline{C}R^{-1}$. *This is true for any basis* $\pi(x)$. Note that J can be computed by solving the matrix equation $JR = R\overline{C}$ or

$$R^T J = (R\overline{C})^T. \tag{5.3.43}$$

Here R^T is a lower triangular matrix and the right-hand side a lower Hessenberg matrix. This and the tridiagonal structure of J considerably simplifies the calculation of J. In the next section we show how the theory developed here leads to a stable and efficient algorithm for computing Gauss quadrature rules.

5.3.5 Jacobi Matrices and Gauss Quadrature

The computations are most straightforward for the power basis, $\theta(x)$, using the moments of the weight function as the initial data. But the condition number of the Gram matrix G, which in this case is a Hankel matrix, increases rapidly with n. This is related to the by now familiar fact that, when n is large, x^n can be accurately approximated by a polynomial of lower degree. Thus the moments for the power basis are not generally a good starting point for the numerical computation of the matrix J.

For the orthonormal basis $\varphi(x)$, we have $G = I$, and

$$\overline{C} = \hat{G} = J = \begin{pmatrix} \beta_1 & \gamma_1 & & & 0 \\ \gamma_1 & \beta_2 & \gamma_2 & & \\ & \gamma_2 & & \ddots & \\ & & \ddots & \ddots & \gamma_{n-1} \\ 0 & & & \gamma_{n-1} & \beta_n \end{pmatrix} \tag{5.3.44}$$

is a symmetric tridiagonal matrix with nonzero off-diagonal elements. Such a tridiagonal matrix is called a **Jacobi matrix** and has n real distinct eigenvalues λ_j. The row eigenvectors $\varphi(\lambda_j)$ satisfy

$$\varphi(\lambda_j)J = \lambda_j\varphi(\lambda_j), \quad j = 1:n, \tag{5.3.45}$$

and are mutually orthogonal. Setting

$$\Phi = (\varphi(\lambda_1)^T, \dots, \varphi(\lambda_n)^T), \qquad \Lambda = \operatorname{diag}(\lambda_1, \dots, \lambda_n),$$

we obtain by (5.3.45) and the symmetry of J the important matrix formula

$$J\Phi = \Phi\Lambda. \tag{5.3.46}$$

It also follows from (5.3.36) that for all x

$$x\varphi(x) = J\varphi(x) + \gamma_n\phi_{n+1}(x)e_n^T, \tag{5.3.47}$$

where γ_n is to be chosen so that $\|\phi_{n+1}\| = 1$. The last column of this equation gives

$$(x - \beta_n)\phi_n(x) = \gamma_{n-1}\phi_{n-1}(x) + \gamma_n\phi_{n+1}(x), \tag{5.3.48}$$

which is the three-term recurrence relation (4.5.36) for orthogonal polynomials.

Let V be an orthogonal matrix that diagonalizes J, i.e.,

$$JV = V\Lambda, \quad V^TV = VV^T = I,$$

where Λ is the diagonal in (5.3.46). It follows that $V = \Phi D$ for some diagonal matrix $D = \operatorname{diag}(d_i)$, and

$$VV^T = \Phi D^2 \Phi^T = I,$$

i.e.,

$$\sum_{k=1}^{n} d_k^2 \phi_i(\lambda_k)\phi_j(\lambda_k) = \delta_{ij} = (\phi_i, \phi_j), \quad i, j = 1:n.$$

This equality holds also for $i = n+1$, because $\phi_{n+1}(\lambda_k) = 0$, for all k, and $(\phi_{n+1}, \phi_j) = 0$, $j = 1 : k$.

Since every polynomial p of degree less than $2n$ can be expressed as a linear combination of polynomials of the form $\phi_i \phi_j$ (in infinitely many ways) it follows that

$$\sum_{k=1}^{n} d_k^2 p(\lambda_k) = \int p(x) w(x)\, dx, \tag{5.3.49}$$

for any polynomial p of degree less than $2n$. This yields the **Gauss–Christoffel quadrature rule**:

$$\int f(x) w(x)\, dx = \sum_{k=1}^{n} d_k^2 f(\lambda_k) + R, \tag{5.3.50}$$

where

$$R = \int (f(x) - p(x)) w(x)\, dx,$$

for any polynomial p *of degree less than* $2n$ *such that* $p(\lambda_k) = f(\lambda_k)$, $k = 1 : n$.

The familiar form for the remainder term

$$R = k_n f^{(2n)}(\xi)/(2n)! \tag{5.3.51}$$

is obtained by choosing a Hermite interpolation polynomial for p and then applying the mean value theorem. The constant k_n is independent of f. The choice $f(x) = A_n^2 x^{2n} + \cdots$ gives $k_n = A_n^{-2}$. A recurrence relation for the leading coefficient A_j is obtained by (5.3.48). We obtain

$$A_0 = \mu_0^{-1/2}, \qquad A_{k+1} = A_k/\gamma_k. \tag{5.3.52}$$

The mean value form for R may be inappropriate when the interval is infinite. Some other estimate of the above integral for R may then be more adequate, for example, in terms of the best approximation of f by a polynomial in some weighted L_p-norm.

A simple formula for the weights d_k^2, due to Golub and Welsch, is obtained by matching the first rows of the equality $V = \Phi D$. Since the elements in the first row of Φ are all equal to the constant $\phi_1 = \mu_0^{-1/2}$, we obtain

$$e_1^T V = \mu_0^{-1/2} d^T, \qquad d_k^2 = \mu_0 v_{1,k}^2, \quad k = 1 : n. \tag{5.3.53}$$

The well-known fact that the weights are positive and their sum equals μ_0 follows immediately from this simple formula for the weights. We summarize these results in the following theorem.

When the three-term recurrence relation for the orthonormal polynomials associated with the weight function $w(x)$ is known, or can be computed by the Stieltjes procedure in Sec. 4.5.5, the Gauss–Christoffel rule can be obtained elegantly as follows. The nodes of the Gauss–Christoffel rule are the eigenvalues of the tridiagonal matrix J, and by (5.3.53) the weights equal the square of the first components of the corresponding eigenvectors. These quantities can be computed in a stable and efficient way by the QR algorithm; see Volume II. In [162] this scheme is extended to the computation of nodes and weights for Gauss–Radau and Gauss–Lobatto quadrature rules.

When the coefficients in the three-term relation cannot be obtained by theoretical analysis or numerical computation, we consider the matrices $\overline{C}$ and $G = R^T R$ as given data about the basis and weight function. Then J can be computed by means of (5.3.39) and the nodes and weights are computed according to the previous case. Note that R and J can be determined simultaneously for all $k \le n$; just take the submatrices of the largest ones.

The following concise and applicable result was found independently by Golub and Meurant (see [165, Theorem 3.4]) and the first-named author (see [88, Theorem 2.2]).

Theorem 5.3.7.

Let J be the symmetric tridiagonal $n \times n$ matrix that contains the coefficients in the three-term recurrence relation for the orthogonal polynomials associated with a positive weight function $w(x)$ (with any sequence of leading coefficients). Let $e_1 = (1, 0, 0, \dots, 0)^T$ and f be an analytic function in a domain that contains the spectrum of J.

Then the formula

$$\frac{1}{\mu_0}\int f(x)w(x)\,dx \approx e_1^T f(J)e_1 \tag{5.3.54}$$

is exact when f is a polynomial of degree less than $2n$.

Proof. If $J = V\Lambda V^T$ is the spectral decomposition of J, then we have

$$f(J) = V^T \operatorname{diag}(f(\lambda_1, \dots, f(\lambda_n)))V.$$

Let p be a polynomial of degree less than $2n$. We obtain using (5.3.53)

$$e_1^T V\Lambda V^T e_1^T = \mu_0^{-1/2} d^T p(\Lambda)\mu_0^{-1/2} d = \mu_0^{-1}\sum_{j=1}^{n} p(\lambda_j)d_j^2 = \mu_0^{-1}\int p(x)w(x)\,dx,$$

since Gauss–Christoffel quadrature is exact for p. ☐

If $f(J)$ is evaluated by means of the diagonalization of J, (5.3.54) becomes exactly the Gauss–Christoffel rule, but it is noteworthy that $e_1^T V^T f(\Lambda)Ve_1$ can sometimes be evaluated without a diagonalization of J. The accuracy of the estimate of the integral still depends on how well $f(z)$ can be approximated by a polynomial of degree less than twice the size of J in the weighted L_1-norm with weight function $w(x)$.

In many important cases the weight function $w(x)$ is *symmetric about the origin.* Then the moments of odd order are zero, and the orthogonal polynomials of odd (even) degree are odd (even) functions. By Theorem 4.5.19 the coefficients $\beta_k = 0$ for all k, i.e., *the matrix J will have a zero diagonal.* The eigenvalues of J will then appear in pairs, $\pm\lambda_k$. If n is odd, there is also a simple zero eigenvalue. The weights are symmetric so that the weights corresponding to the two eigenvalues $\pm\lambda_i$ are the same.

We shall see that in the symmetric case the eigenvalue problem for the tridiagonal matrix $J \in \mathbf{R}^{n\times n}$ can be reduced to a singular value problem for a smaller bidiagonal matrix B, where

$$B \in \begin{cases} \mathbf{R}^{n/2\times n/2} & \text{if } n \text{ even},\\ \mathbf{R}^{(n+1)/2\times(n-1)/2} & \text{if } n \text{ odd}.\end{cases}$$

We permute rows and columns in J, by an odd-even permutation; for example, if $n = 7$, then $(1, 2, 3, 4, 5, 6, 7) \mapsto (1, 3, 5, 7, 2, 4, 6)$, and

$$\tilde{J} = T^{-1}JT = \begin{pmatrix} 0 & B \\ B^T & 0 \end{pmatrix}, \quad B = \begin{pmatrix} \gamma_1 & 0 & 0 \\ \gamma_2 & \gamma_3 & 0 \\ 0 & \gamma_4 & \gamma_5 \\ 0 & 0 & \gamma_6 \end{pmatrix},$$

where T is the permutation matrix effecting the permutation. Then, J and $\tilde{J}$ have the same eigenvalues. If the orthogonal matrix V diagonalizes J, i.e., $J = V\Lambda V^T$, then $\tilde{V} = T^{-1}V$ diagonalizes $\tilde{J} = T^TJT$, i.e., $\tilde{J} = T^{-1}JT = T^{-1}V\lambda V^TT$. Note that the first row of V is just a permutation of $\tilde{V}$. We can therefore substitute $\tilde{V}$ for V in equation (5.3.53), which gives the weights in the Gauss–Christoffel formula.

The following relationship between the singular value decomposition (SVD) and a Hermitian eigenvalue problem, exploited by Lanczos [236, Chap. 3], can easily be verified.

Theorem 5.3.8.

Let the SVD of $B \in \mathbf{R}^{m\times n}$ $(m \geq n)$ be $B = P\Sigma Q^T$, where

$$\Sigma = \operatorname{diag}(\Sigma_1,\ 0), \quad \Sigma_1 = \operatorname{diag}(\sigma_1, \sigma_2, \dots, \sigma_n),$$

and

$$P = (P_1,\ P_2) \in \mathbf{C}^{m\times m}, \quad P_1 \in \mathbf{C}^{m\times n}, \quad Q \in \mathbf{C}^{n\times n}.$$

Then the symmetric matrix $C \in \mathbf{R}^{(m+n)\times(m+n)}$ has the eigendecomposition

$$C = \begin{pmatrix} 0 & B \\ B^T & 0 \end{pmatrix} = V \begin{pmatrix} \Sigma_1 & 0 & 0 \\ 0 & 0 & 0 \\ 0 & 0 & -\Sigma_1 \end{pmatrix} V^T, \tag{5.3.55}$$

where $V \in$ is orthogonal:

$$V = \frac{1}{\sqrt{2}} \begin{pmatrix} P_1 & \sqrt{2}\,P_2 & P_1 \\ Q & 0 & -Q \end{pmatrix}^T. \tag{5.3.56}$$

Hence the eigenvalues of C are $\pm\sigma_1, \pm\sigma_2, \dots, \pm\sigma_r$, and zero repeated $(m-n)$ times.

The QR algorithm for symmetric tridiagonal matrices can be adopted to compute the singular values σ_i and the first components of the matrix $P = (P_1, P_2)$ of left singular vectors of the bidiagonal matrix B; see Volume II.

Example 5.3.4.

The monic Legendre polynomials are symmetric around the origin, and thus $\beta_n = 0$ for all n and $\mu_0 = 2$. According to (4.5.56) we have

$$\gamma_n = \frac{n}{\sqrt{4n^2 - 1}} = \frac{1}{\sqrt{4 - n^{-2}}}.$$

ALGORITHM 5.4. *Gauss–Legendre Quadrature.*

The following MATLAB function computes the nodes and weights of the Gauss–Legendre rule with n points by generating the bidiagonal matrix B and its SVD.

```
function [x,w] = legendre(n);
% LEGENDRE(n) computes the nodes and weights in the
% Gauss-Legendre quadrature rule with n+1 nodes (n > 1).
%
gamma = 1./sqrt(4 - [1:n].^(-2));
gamma(n+1) = 0;
b0(1) = gamma(1:2:n+1);
b1(k) = gamma(2:2:n);
B = diag(b0,0) + diag(b1,1);
[P,S,Q] = svd(B);
x = diag(S); [x,i] = sort(x);
w = P(1,i).^2;
if rem(n,2) == 0 w(1) = 2*w(1); end
```

For $n = 6$ the upper bidiagonal matrix becomes

$$B = \begin{pmatrix} 1/\sqrt{3} & 2/\sqrt{15} & & \\ & 3/\sqrt{35} & 4/\sqrt{63} & \\ & & 5/\sqrt{99} & 6/\sqrt{143} \\ & & & 0 \end{pmatrix} \in \mathbf{R}^{4\times 4},$$

and we obtain the nonnegative nodes (cf. Table 5.3.1) $x_1 = 0$,

$$x_2 = 0.40584515137740, \qquad x_3 = 0.74153118559939, \qquad x_4 = 0.94910791234276.$$

The first row of $P = (\, P_1 \quad P_2 \,)$ is

$$-0.45714285714286,\ -0.61792398440675,\ 0.52887181007242,\ -0.35984019532130,$$

where the first entry corresponds to node $x_1 = 0$. Dividing the last three components by $\sqrt{2}$, squaring, and multiplying with $\mu_0 = 2$, gives the weights

$$w_1 = 0.41795918367347,\ w_2 = 0.38183005050512,\ w_3 = 0.27970539148928,$$
$$w_4 = 0.12948496616887.$$

We remark that the given program is inefficient in that the full matrices of left and right singular vectors are computed. Unless n is very large the execution time is negligible anyway.

In the computation of harmonic transforms used in spectral weather analysis, Gauss–Legendre quadrature rules with values of n in excess of 1000 are required. Methods for computing points and weights accurate to double precision for such high values of n are discussed by Swarztrauber in [345].

Review Questions

5.3.1 What increase in order of accuracy can normally be achieved by a judicious choice of the nodes in a quadrature formula?

5.3.2 What are orthogonal polynomials? Give a few examples of families of orthogonal polynomials together with the three-term recursion formula, which its members satisfy.

5.3.3 Formulate and prove a theorem concerning the location of zeros of orthogonal polynomials.

5.3.4 Give an account of Gauss quadrature formulas, including accuracy and how the nodes and weights are determined. What important properties are satisfied by the weights?

5.3.5 What is the orthogonality property of the Legendre polynomials?

Problems and Computer Exercises

5.3.1 Prove that the three-point quadrature formula

$$\int_{-1}^{1} f(x)\,dx \approx \frac{1}{9}\left(5f\left(-\sqrt{\frac{3}{5}}\right) + 8f(0) + 5f\left(\sqrt{\frac{3}{5}}\right)\right)$$

is exact for polynomials of degree five. Apply it to the computation of

$$\int_{0}^{1} \frac{\sin x}{1+x}\,dx,$$

and estimate the error in the result.

5.3.2 (a) Calculate the Hermite polynomials H_n for $n \le 4$ using the recurrence relation.

(b) Express, conversely, $1, x, x^2, x^3, x^4$ in terms of the Hermite polynomials.

5.3.3 (a) Determine the orthogonal polynomials $\phi_n(x)$, $n = 1, 2, 3$, with leading coefficient 1, for the weight function $w(x) = 1 + x^2$, $x \in [-1, 1]$.

(b) Give a two-point Gaussian quadrature formula for integrals of the form

$$\int_{-1}^{1} f(x)(1+x^2)\,dx$$

which is exact when $f(x)$ is a polynomial of degree three.

Hint: Use either the method of undetermined coefficients taking advantage of symmetry, or the three-term recurrence relation in Theorem 5.3.1.

5.3.4 (Gautschi)

(a) Construct the quadratic polynomial ϕ_2 orthogonal on $[0, \infty]$ with respect to the weight function $w(x) = e^{-x}$. *Hint:* Use $\int_0^\infty t^m e^{-t}\,dt = m!$.

(b) Obtain the two-point Gauss–Laguerre quadrature formula

$$\int_0^\infty f(x)e^{-x}\,dx = w_1 f(x_1) + w_2 f(x_2) + E_2(f),$$

including a representation for the remainder $E_2(f)$.

(c) Apply the formula in (b) to approximate

$$I = \int_0^\infty (x+1)^{-1}e^{-x}\,dx.$$

Use the remainder term to estimate the error, and compare your estimate with the true error ($I = 0.596347361\ldots$).

5.3.5 Show that the formula

$$\int_{-1}^1 f(x)(1-x^2)^{-1/2}\,dx = \frac{\pi}{n}\sum_{k=1}^n f\left(\cos\frac{2k-1}{2n}\pi\right)$$

is exact when $f(x)$ is a polynomial of degree at most $2n-1$.

5.3.6 (a) Use the MATLAB program in Example 5.3.4 to compute nodes and weights for the Gauss–Hermite quadrature rule. Use it to compute a 10-point rule; check the result using a table.

(b) Write a program for computing nodes and weights for Gauss quadrature rules when $w(x)$ is not symmetric. In MATLAB use the function `[v,d] = eig(J)` to solve the eigenvalue problems. Use the program to compute some Gauss–Laguerre quadrature rules.

5.3.7 Derive the Gauss–Lobatto quadrature rule in Example 5.3.3, with two interior points by using the Ansatz

$$\int_{-1}^1 f(x)\,dx = w_1(f(-1)+f(1)) + w_2(f(-x_1)+f(x_1)),$$

and requiring that it be exact for $f(x) = 1, x^2, x^4$.

5.3.8 Compute an approximate value of

$$\int_{-1}^1 x^4 \sin^2 \pi x\,dx = 2\int_0^1 x^4 \sin^2 \pi x\,dx,$$

using a five-point Gauss–Legendre quadrature rule on [0, 1] for the weight function $w(x) = 1$. For nodes and weights see Table 5.3.1 or use the MATLAB function `legendre(n)` given in Example 5.3.4. (The true value of the integral is 0.11407 77897 39689.)

5.3.9 (a) Determine exactly the Lobatto formulas with given nodes at -1 and 1 (and the remaining nodes free), for the weight functions

$$w(x) = (1-x^2)^{-1/2}, \quad x \in [-1, 1].$$

Determine for this weight function also the nodes and weights for the Gauss quadrature formula (i.e., when all nodes are free).

Hint: Set $x = \cos\phi$, and formulate equivalent problems on the unit circle. Note that you obtain (at least) two different discrete orthogonality properties of the Chebyshev polynomials this way.

(b) Lobatto–Kronrod pairs are useful when a long interval has been divided into several shorter intervals (cf. Example 5.3.3). Determine Lobatto–Kronrod pairs (exactly) for $w(x) = (1 - x^2)^{-1/2}$.

5.3.10 Apply the formulas in Problem 5.3.9 to the case $w(x) = 1$, $x \in [-1, 1]$, and some of the following functions.

(a) $f(x) = e^{kx}$, $k = 1, 2, 4, 8, \dots$; (b) $f(x) = 1/(k + x)$, $k = 1, 2, 1.1, 1.01$;

(c) $f(x) = k/(1 + k^2x^2)$, $k = 1, 4, 16, 64$.

Compare the actual errors with the error estimates.

5.3.11 For $k = 1$ the integral (5.2.24) in Example 5.2.5 is

$$\int_{-\pi/2}^{\pi/2} \frac{\cos t}{t + \pi + u(t + \pi)}\, dt.$$

Compute this integral with at least ten digits of accuracy, using a Gauss–Legendre rule of sufficiently high order. Use the MATLAB function `legendre(n)` given in Example 5.3.4 to generate the nodes and weights.

5.3.12 Write a MATLAB function for the evaluation of the Sievert[174] integral,

$$S(x, \theta) = \int_0^\theta e^{-x/\cos\phi}\, d\phi,$$

for any $x \geq 0$, $x \leq \theta \leq 90°$, with at least six decimals relative accuracy. There may be useful hints in [1, Sec. 27.4].

5.4 Multidimensional Integration

Numerical integration formulas in several dimensions, sometimes called **numerical cubature**, are required in many applications. Several new difficulties are encountered in deriving and applying such rules.

In one dimension any finite interval of integration $[a, b]$ can be mapped by an affine transformation onto $[-1, 1]$ (say). Quadrature rules need therefore only be derived for this standard interval. The order of accuracy of the rule is preserved since affine transformations preserve the degree of the polynomial. In d dimensions the boundary of the region of integration has dimension $d - 1$, and can be complicated manifold. For any dimension $d \geq 2$ there are infinitely many connected regions in $\mathbf{R}^d$ which cannot be mapped onto each other using affine transformations. Quadrature rules with a certain polynomial accuracy designed for any of these regions are fundamentally different than for any other region.

[174]Sievert was a Swedish radiophysicist who was so great that doses of radiation are measured in millisieverts, or even microsieverts, all over the world.

The number of function values needed to obtain an acceptable approximation tends to increase exponentially in the number of dimensions d. That is, if n points are required for an integral in one dimension, then n^d points are required in d dimensions. Thus, even for a modest number of dimensions, achieving an adequate accuracy may be an intractable problem. This is often referred to as **the curse of dimensionality**, a phrase coined by Richard Bellman.[175]

5.4.1 Analytic Techniques

It is advisable to try, if possible, to reduce the number of dimensions by applying analytic techniques to parts of the task.

Example 5.4.1.

The following triple integral can be reduced to a single integral:

$$\int_0^\infty \int_0^\infty \int_0^\infty e^{-(x+y+z)} \sin(xz) \sin(yx)\, dx dy dz$$
$$= \int_0^\infty e^{-x}\, dx \int_0^\infty e^{-y} \sin(yx) dy \int_0^\infty e^{-z} \sin(zx)\, dz = \int_0^\infty \left(\frac{x}{1+x^2}\right)^2 e^{-x}\, dx.$$

This follows because

$$\int_0^\infty e^{-z} \sin(zx) dz = \int_0^\infty e^{-y} \sin(yx) dz = \frac{x}{1+x^2}.$$

The remaining single integral is simply evaluated by the techniques previously studied.

Often a transformation of variable is needed for such a reduction. Given a region D in the (x, y)-plane, this is mapped onto a region D' in the (u, v)-plane by the variable transformation

$$x = \phi(u, v), \qquad y = \psi(u, v). \tag{5.4.1}$$

If ϕ and ψ have continuous partial derivatives and the Jacobian

$$J(u, v) = \begin{vmatrix} \partial\phi/\partial u & \partial\phi\partial u \\ \partial\psi\partial u & \partial\psi\partial u \end{vmatrix} \tag{5.4.2}$$

does not vanish in D', then

$$\int\int_D f(x, y)\, dx\, dy = \int\int_{D'} f(\phi(x, y), \psi(x, y)) |J(u, v)|\, du\, dv. \tag{5.4.3}$$

It is important to take into account any symmetries that the integrand can have. For example, the integration of a spherically symmetric function over a spherical region reduces in polar coordinates to a one-dimensional integral.

[175]Richard Ernest Bellman (1920–1984) was an American mathematician. From 1949 to 1965 he worked at the Rand Corporation and made important contributions to operations research and dynamic programming.

Example 5.4.2.

To evaluate the integral

$$I = \int\int_D \frac{y \sin(ky)}{x^2 + y^2}\, dx\, dy,$$

where D is the unit disk $x^2 + y^2 \leq 1$, we introduce polar coordinates (r, φ), $x = r \cos \varphi$, $y = r \sin \varphi$, $dx\, dy = r\, dr\, d\varphi$. Then, after integrating in the r variable, this integral is reduced to the single integral

$$I = \frac{1}{k} \int_0^{2\pi} [1 - \cos (k \sin \varphi)]\, d\varphi.$$

This integral is not expressible in finite terms of elementary functions. Its value is in fact $(1 - J_0(k))2\pi/k$, where J_0 is a Bessel function. Note that the integrand is a periodic function of φ, in which the trapezoidal rule is very efficient (see Sec. 5.1.4). This is a useful device for Bessel functions and many other transcendental functions which have integral representations.

If the integral cannot be reduced, then several approaches are possible:

(a) Tensor products of one-dimensional quadrature rules can be used. These are particularly suitable if the boundary of the region is composed of straight lines. Otherwise numerical integration in one direction at a time can be used; see Sec. 5.4.3.

(b) For more general boundaries an irregular triangular grid can be used; see Sec. 5.4.4.

(c) Monte Carlo or quasi–Monte Carlo methods can be used, mainly for problems with complicated boundaries and/or a large number of dimensions; see Sec. 5.4.5.

5.4.2 Repeated One-Dimensional Integration

Consider a double integral

$$I = \int\int_D u(x, y)\, dxdy, \tag{5.4.4}$$

over a region D in the (x, y)-plane such that lines parallel with the x-axis have at most one segment in common with D (see Figure 5.4.1). Then I can be written in the form

$$I = \int_a^b \left(\int_{c(x)}^{d(x)} f(x, y) dy \right) dx,$$

or

$$I = \int_a^b \varphi(x)\, dx, \qquad \varphi(x) = \int_{c(x)}^{d(x)} f(x, y) dy. \tag{5.4.5}$$

The one-dimensional integral $\varphi(x)$ can evaluated for the sequence of abscissae x_i, $i = 1, \ldots, n$, used in another one-dimensional quadrature rule for J. Note that if D is a more

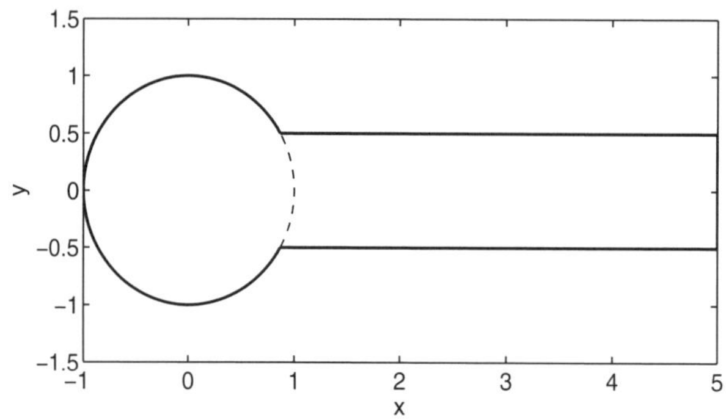

Figure 5.4.1. *Region D of integration.*

general domain, it might be possible to decompose D into the union of simpler domains on which these methods can be used.

Example 5.4.3.

Compute

$$I = \int\int_D \sin^2 y \sin^2 x (1 + x^2 + y^2)^{-1/2}\, dx\, dy,$$

where

$$D = \{(x, y) \mid x^2 + y^2 \le 1\} \cup \{(x, y) \mid 0 \le x \le 3, |y| \le 0.5\}$$

is a composite region (see Figure 5.4.1). Then

$$I = \int_{-1}^{3} \sin^2 x\, \varphi(x)\, dx, \qquad (5.4.6)$$

$$\varphi(x) = \int_{-c(x)}^{c(x)} \sin^2 y\, (1 + x^2 + y^2)^{-1/2} dy, \qquad (5.4.7)$$

where

$$c(x) = \begin{cases} (1 - x^2)^{1/2}, & x < \frac{1}{2}\sqrt{3}, \\ \frac{1}{2}, & x \ge \frac{1}{2}\sqrt{3}. \end{cases}$$

Values of $\varphi(x)$ were obtained by the application of Romberg's method to (5.4.7) and numerical integration applied to the integral (5.4.6) yielded the value of $I = 0.13202 \pm 10^{-5}$. Ninety-six values of x were needed, and for each value of x, 20 function evaluations used, on the average. The grid is chosen so that $x = \frac{1}{2}\sqrt{3}$, where $\varphi'(x)$ is discontinuous, is a grid point.

5.4.3 Product Rules

In $d = 2$ dimensions, common boundaries are a rectangle, circle, or triangle, or a combination of these. Consider a **double integral** over a rectangular region $D = \{(x, y) \mid a \le$

$x \le b,\ c \le y \le d\}$. Introduce an equidistant **rectangular grid** in the (x, y)-plane, with grid spacings h and k in the x and y directions,

$$x_i = a + ih, \quad y_j = c + jk, \quad h = (b-a)/n, \quad k = (d-c)/m,$$

and set $u_{ij} = u(x_i, y_j)$. Then the following **product rule** for the double integral generalizes the compound midpoint rule:

$$I \approx hk \sum_{i=1}^{m} \sum_{j=1}^{n} u_{i-1/2,j-1/2}. \tag{5.4.8}$$

The product trapezoidal rule is

$$\begin{aligned} I &\approx hk \sum_{i=1}^{m} \sum_{j=1}^{n} \frac{1}{4}(u_{i-1,j-1} + u_{i-1,j} + u_{i,j-1} + u_{i,j}) \\ &= hk \sum_{i=0}^{m} \sum_{j=0}^{n} w_{ij} u_{ij}. \end{aligned} \tag{5.4.9}$$

Here $w_{ij} = 1$ for the interior grid points, i.e., when $0 < i < m$, and $0 < j < n$. For the trapezoidal rule $w_{ij} = \frac{1}{4}$ for the four corner points, while $w_{ij} = \frac{1}{2}$ for the other boundary points. Both formulas are exact for all **bilinear functions** $x^i y^j$, $0 \le i, j \le 1$. The error can be expanded in even powers of h and k so that Romberg's method can be used to get more accurate results. The generalization to integrals over the hypercube $[0, 1]^d$ is straightforward.

It is not necessary to use the same quadrature rule in both dimensions. Suppose we have the two one-dimensional quadrature rules

$$\int_a^b f(x)\,dx \approx \sum_{i=1}^{n} w_i f(x_i) + (b-a)R_1, \qquad \int_c^d g(y)\,dy \approx \sum_{j=1}^{m} v_j g(y_j) + (d-c)R_2. \tag{5.4.10}$$

Combining these two rules over the rectangular region D gives the product rule

$$\begin{aligned} \int_a^b \int_c^d u(x, y)\,dx\,dy &\approx \int_a^b \left(\sum_{j=1}^{m} v_j u(x, y_j) + (d-c)R_2 \right) dx \\ &= \sum_{j=1}^{m} v_j \int_a^b u(x, y_j)\,dx + \int_a^b (d-c)R_2\,dx \approx \sum_{i=1}^{n} \sum_{j=1}^{m} w_i v_j u(x_i, y_j) + R, \end{aligned}$$

where

$$R = (d-c) \int_a^b R_2\,dx + (b-a) \sum_{j=1}^{m} v_j R_1 \approx (b-a)(d-c)(R_1 + R_2).$$

The following property of product rules follows easily.

Theorem 5.4.1.

If the two one-dimensional rules (5.4.10) *integrate* $f(x)$ *exactly over* $[a, b]$ *and* $g(y)$ *exactly over* $[c, d]$, *then the product rule* (5.4.11) *integrates* $u(x, y) = f(x)g(y)$ *exactly over the region* $[a, b] \times [c, d]$.

If the one-dimensional rules are exact for polynomials of degree d_1 *and* d_2, *respectively, then the product rule will be exact for all bivariate polynomials* $x^p y^q$, *where* $p \le d_1$ *and* $q \le d_2$.

Example 5.4.4.

The product Simpson's rule for the square $|x| \le h$, $|y| \le h$ has the form

$$\int_{-h}^{h}\int_{-h}^{h} u(x, y)\,dx dy = 4h^2 \sum_{j=-1}^{1}\sum_{i=-1}^{1} w_{i,j} u(x_i, y_j).$$

It uses $3^2 = 9$ function values, with abscissae and weights given by

(x_i, y_j)	(0,0)	$(\pm h, \pm h)$	$(\pm h, 0)$	$(0, \pm h)$
$w_{i,j}$	4/9	1/36	1/9	1/9

Of similar accuracy is the product rule obtained from a two-point Gauss–Legendre rule, which uses the four points

$$(x_i, y_i) = \left(\pm\frac{h}{\sqrt{3}}, \pm\frac{h}{\sqrt{3}}\right) \quad w_i = \frac{1}{4}.$$

For both rules the error is $O(h^4)$. Note that for the corresponding composite rules, the functions values at corner points and midpoints in the product Simpson's rule are shared with other subsquares. Effectively this rule also uses four function values per subsquare.

Higher-accuracy formulas can also be derived by **operator** techniques, based on an operator formulation of Taylor's expansion (see Example 4.3.8),

$$u(x_0 + h, y_0 + k) = e^{(hD_x + kD_y)} u(x_0, y_0). \tag{5.4.11}$$

For regions D, such as a square, cube, cylinder, etc., which are the Cartesian product of lower-dimensional regions, product integration rules can be developed by multiplying together the lower-dimensional rules. Product rules can be used on nonrectangular regions, if these can be mapped into a rectangle. This can be done, for example, for a triangle, but product rules derived in this way are often not very efficient and are seldom used.

For nonrectangular regions, the rectangular grid may also be bordered by triangles or "triangles" with one curved side, which may be treated with the techniques outlined in the next section.

So far we have restricted ourselves to the two-dimensional case. But the ideas are more general. Let $(x_1, \dots, x_r) \in C$, where C is a region in $\mathbf{R}^r$ and $(y_1, \dots, y_s) \in D$, where D is a region in $\mathbf{R}^r$. Let $C \times D$ denote the Cartesian product of C and D, i.e., the region in $\mathbf{R}^{r+s}$ consisting of points (using vector notations) $(\mathbf{x}, \mathbf{y})$ such that $\mathbf{x} \in C$ and $\mathbf{y} \in D$.

Suppose we have two quadrature rules for the regions C and D

$$\int_C f(\mathbf{x})\,d\mathbf{x} \approx \sum_{i=1}^{n} w_i f(\mathbf{x}_i), \qquad \int_D g(\mathbf{y})\,d\mathbf{y} \approx \sum_{j=1}^{m} v_i g(\mathbf{y}_i). \tag{5.4.12}$$

We can combine these two rules to give a product rule for the region $C \times D$:

$$\int_{C\times D} u(\mathbf{x},\mathbf{y})\,d\mathbf{x}\,d\mathbf{y} \approx \sum_{i=1}^{n}\sum_{i=1}^{n} w_i v_j u(\mathbf{x}_i,\mathbf{y}_j). \tag{5.4.13}$$

Product rules are not necessarily the most economical rules. More efficient quadrature rules exist, which are not the result of applying one-dimensional rules to several dimensions. We could try to determine such rules by selecting n nodes and weights so that the rule integrates bivariate polynomials of as high a degree as possible. This is much more difficult in several dimensions than in one dimension, where this approach led to Gaussian rules. The solution is in general not unique; there may be several rules with different nodes and weights. For most regions it is not known what the best rules are. Some progress has been made in developing nonproduct quadrature rules of optimal order for triangles.

Some simple quadrature rules for circles, triangles, hexagons, spheres, and cubes are given in [1, pp. 891–895], for example, the following quadrature rule for a double integral over a disk $C = \{(x, y) \in C \mid x^2 + y^2 \le h^2\}$:

$$\int\int_C f(x, y)\,dxdy = \pi h^2 \sum_{i=1}^{4} w_i f(x_i, y_i) + O(h^4),$$

where

$$(x_i, y_i) = (\pm h/2, \pm h/2), \quad w_i = 1/4, \quad i = 1:4.$$

This four-point rule has the same order of accuracy as the four-point Gaussian product for the square given in Example 5.4.4. A seven-point $O(h^6)$ rule uses the points (see Figure 5.4.2)

$$\begin{aligned} (x_1, y_1) = (0, 0), \quad (x_i, y_i) &= (\pm h\sqrt{2/3}, 0), \quad i = 2, 3 \\ (x_i, y_i) &= (\pm h/\sqrt{6}, \pm h/\sqrt{2}), \quad i = 4:7, \end{aligned}$$

with weights $w_1 = 1/4$, and $w_i = 1/8$, $i = 2:7$.

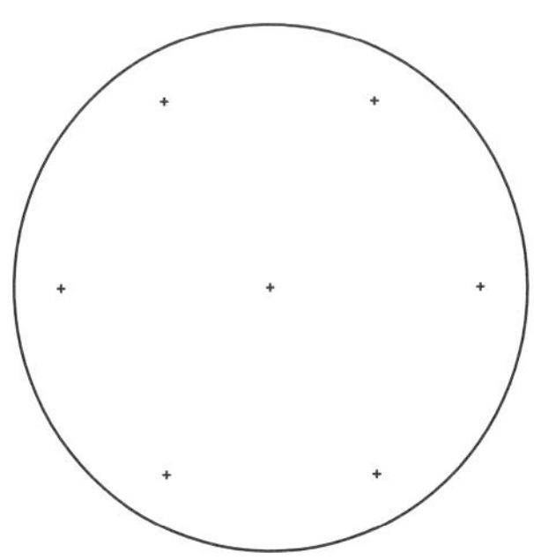

Figure 5.4.2. *A seven-point $O(h^6)$ rule for a circle.*

Example 5.4.5.

We seek a quadrature rule

$$I = \int\int_T f(x, y)\,dxdy = A\sum_{i=1}^{n} w_i f(x_i, y_i) + R, \tag{5.4.14}$$

where T is an equilateral triangle with sides of length h and area $A = h^2\sqrt{3}/4$. We use function values at the "center of mass" $(x_1, y_1) = (0, h/(2\sqrt{3}))$ of the triangle and at the corner nodes

$$(x_i, y_i) = (\pm h/2, 0), \quad i = 2, 3, \quad \text{and} \quad (x_4, y_4) = (0, h\sqrt{3}/2).$$

Then, taking $w_1 = 3/4$, and $w_i = 1/12$, $i = 2 : 4$, we get a four-point rule with error $R = O(h^3)$.

Adding nodes at the midpoint of the sides

$$(x_5, y_5) = (0, 0) \quad \text{and} \quad (x_i, y_i) = (\pm h/4, h\sqrt{3}/4), \quad i = 6, 7,$$

and using weights $w_1 = 9/20$, and $w_i = 1/20$, $i = 2 : 4$, $w_i = 2/15$, $i = 5 : 7$, gives a seven-point rule for which $R = O(h^4)$ in (5.4.14).

5.4.4 Irregular Triangular Grids

A grid of triangles of arbitrary form is a convenient means for approximating a complicated plane region. It is fairly easy to program a computer to refine a coarse triangular grid automatically; see Figure 5.4.3. It is also easy to adapt the density of points to the behavior of the function.

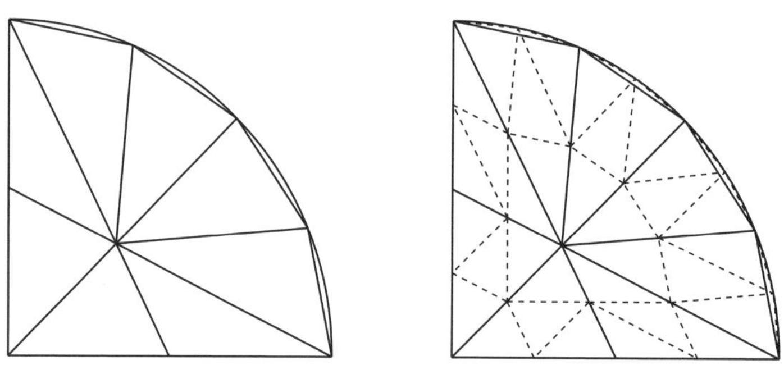

Figure 5.4.3. *Refinement of a triangular grid.*

Triangular grids are thus more flexible than rectangular ones. On the other hand, the administration of a rectangular grid requires less storage and a simpler program. Sometimes the approximation formulas are also a little simpler. Triangular grids are used, for example, in the **finite element method** (FEM) for problems in continuum mechanics and other applications of partial differential equations; see [113].

Let the points P_j, $j = 1, 2, 3$, with coordinates $p_j = (x_j, y_j)$, be the vertices of a triangle T with area $Y > 0$. Then any point $p = (x, y)$ in the plane can be uniquely expressed by the vector equation

$$p = \theta_1 p_1 + \theta_2 p_2 + \theta_3 p_3, \quad \theta_1 + \theta_2 + \theta_3 = 1. \tag{5.4.15}$$

The θ_i, which are called homogeneous **barycentric coordinates** of P, are determined from the following nonsingular set of equations:

$$\begin{aligned} \theta_1 x_1 + \theta_2 x_2 + \theta_3 x_3 &= x, \\ \theta_1 y_1 + \theta_2 y_2 + \theta_3 y_3 &= y, \\ \theta_1 + \theta_2 + \theta_3 &= 1. \end{aligned} \tag{5.4.16}$$

Barycentric coordinates were discovered by Möbius[176] in 1827; see Coxeter [85, Sec. 13.7]. In engineering literature the barycentric coordinates for a triangle are often called **area coordinates** since they are proportional to the area of the three subtriangles induced by P; see Figure 5.4.4.

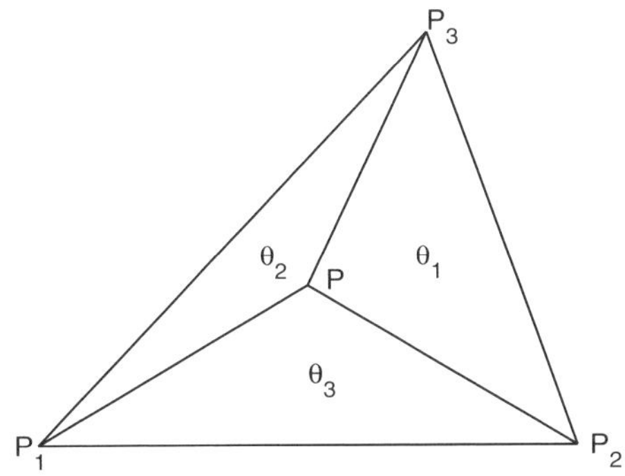

Figure 5.4.4. *Barycentric coordinates of a triangle.*

The interior of the triangle is characterized by the inequalities $\theta_i > 0$, $i = 1, 2, 3$. In this case P is the center of mass (centroid) of the three masses $\theta_1, \theta_2, \theta_3$ located at the vertices of the triangle. This explains the term "barycentric coordinates." The equation for the side $P_2 P_3$ is $\theta_1 = 0$; similarly $\theta_2 = 0$ and $\theta_3 = 0$ describe the other two sides. Note that if θ and θ' $(i = 1, 2, 3)$ are the barycentric coordinates of the points P_i and P_j, respectively, then the barycentric coordinates of $\alpha P + (1 - \alpha) P'$ are $\alpha\theta + (1 - \alpha)\theta'$.

Barycentric coordinates are useful also for $d > 2$ dimensions. By a **simplex** in $\mathbf{R}^d$ we mean the convex hull of $(d + 1)$ points $p_j = (p_{1j}, p_{2j}, \dots, p_{dj})^T \in \mathbf{R}^d$, which are called the vertices of the simplex. We assume that the vertices are not contained in a hyper-plane. This is the case if and only if the $(d + 1) \times (d + 1)$ matrix

$$A = \begin{pmatrix} p_1 & p_2 & \cdots & p_{d+1} \\ 1 & 1 & \cdots & 1 \end{pmatrix} \tag{5.4.17}$$

is nonsingular. For $d = 2$ the simplex is a triangle and for $d = 3$ a tetrahedron.

[176] August Ferdinand Möbius (1790–1868) was a German astronomer and mathematician, and a professor at the University of Leipzig. His 1827 work *Barycentric Calculus* became a classic and played an important role in the development of projective geometry.

The barycentric coordinates of a point p are the components of the unique vector $\theta \in \mathbf{R}^{d+1}$ such that

$$(p_1, \dots, p_{d+1})\theta = p, \quad e^T\theta = 1 \tag{5.4.18}$$

or, equivalently, $\theta = A^{-1}\binom{p}{1}$. The center of gravity of the simplex is the point with coordinates $\theta_i = 1/(d+1)$, $i = 1 : d+1$.

If u is a *nonhomogeneous linear function* of p, i.e., if

$$u(p) = a^T p + b = (a^T, b)\begin{pmatrix} p \\ 1 \end{pmatrix},$$

then the reader can verify that

$$u(p) = \sum_{j=1}^{d+1} \theta_j u(p_j), \quad u(p_j) = a^T p_j + b. \tag{5.4.19}$$

This is a form of *linear interpolation* and shows that a linear function is uniquely determined by its values at the vertices.

Using also the midpoints of the edges $p_{ij} = \frac{1}{2}(p_i + p_j)$ a *quadratic interpolation* formula can be obtained.

Theorem 5.4.2.
Define

$$\Delta''_{ij} = u(p_i) + u(p_j) - 2u\left(\frac{1}{2}(p_i + p_j)\right), \quad i < j. \tag{5.4.20}$$

Then the interpolation formula

$$u(p) = \sum_j \theta_j u(p_j) - 2\sum_{i<j} \theta_i \theta_j \Delta''_{ij}, \tag{5.4.21}$$

where the summation indices i, j are assumed to take all values $1 : d+1$ unless otherwise specified, is exact for all quadratic functions.

Proof. The right-hand is a quadratic function of p, since it follows from (5.4.16) that the θ_i are (nonhomogeneous) linear functions of the coordinates of p. It remains to show that the right-hand side is equal to $u(p)$ for $p = p_j$, and $p = (p_i + p_j)/2$, $i, j = 1 : d+1$.

For $p = p_j$, $\theta_j = 1$, $\theta_i = 0$, $i \neq j$, hence the right-hand side equals u_i. For $p = (p_i + p_j)/2$, $\theta_i = \theta_j = 1/2$, $\theta_k = 0$, $k \neq i, j$, and hence the right-hand side becomes

$$\frac{1}{2}(u_i + u_j) - 2 \cdot \frac{1}{2}\left(u_i + u_j - 2u\left(\frac{1}{2}(p_i + p_j)\right)\right) = u\left(\frac{1}{2}(p_i + p_j)\right). \qquad \square$$

The following theorem for triangles ($d = 2$) is equivalent to a rule which has been used in mechanics for the computation of moments of inertia since the nineteenth century.

Theorem 5.4.3.

Let T be a triangle with vertices p_1, p_2, p_3 and area Y. Then the integration formula

$$\int_T u(x,y)\,dxdy = \frac{Y}{3}\left(u\left(\frac{1}{2}(p_1+p_2)\right) + u\left(\frac{1}{2}(p_1+p_3)\right) + u\left(\frac{1}{2}(p_2+p_3)\right)\right) \tag{5.4.22}$$

is exact for all quadratic functions.

Proof. Using the interpolation formula (5.4.21), the integral equals

$$\int_T u(x,y)\,dxdy = \sum_j u(p_j)\int_T \theta_j dxdy - 2\sum_{i<j}\Delta''_{ij}\int_T \theta_i\theta_j dxdy.$$

By symmetry, $\int_T \theta_i\,dxdy$ is the same for $i = 1, 2, 3$. Similarly, $\int_T \theta_i\theta_j\,dxdy$ is the same for all $i < j$. Hence using (5.4.20)

$$\begin{aligned}\int_T u(x,y)\,dxdy &= a(u_1+u_2+u_3) - 2b(\Delta''_{23}+\Delta''_{13}+\Delta''_{12})\\ &= (a-4b)(u_1+u_2+u_3)\\ &\quad + 4b\left(u\left(\frac{1}{2}(p_1+p_2)\right) + u\left(\frac{1}{2}(p_2+p_3)\right) + u\left(\frac{1}{2}(p_3+p_1)\right)\right),\end{aligned} \tag{5.4.23}$$

where

$$a = \int_T \theta_1\,dxdy, \qquad b = \int_T \theta_1\theta_2\,dxdy.$$

Using θ_1, θ_2 as new variables of integration, we get by (5.4.16) and the relation $\theta_3 = 1 - \theta_1 - \theta_2$

$$\begin{aligned}x &= \theta_1(x_1-x_3) + \theta_1(x_1-x_3) + x_3,\\ y &= \theta_1(y_1-y_3) + \theta_1(y_1-y_3) + y_3.\end{aligned}$$

The functional determinant is equal to

$$\begin{vmatrix} x_1-x_3 & x_2-x_3 \\ y_1-y_3 & y_2-y_3 \end{vmatrix} = 2Y,$$

and (check the limits of integration!)

$$a = \int_{\theta_1=0}^{1}\int_{\theta_2=0}^{1-\theta_1} 2\theta_1 d\theta_1 d\theta_2 = 2Y\int_0^1 \theta_1(1-\theta_1)d\theta_1 = \frac{Y}{3},$$

$$b = \int_{\theta_1=0}^{1}\int_{\theta_2=0}^{1-\theta_1} 2\theta_1\theta_2 d\theta_1 d\theta_2 = 2A\int_0^1 \theta_1\frac{(1-\theta_1)^2}{2}d\theta_1 = \frac{Y}{12}.$$

The results now follow by insertion of this into (5.4.23). □

A numerical method for a two-dimensional region can be based on Theorem 5.4.2, by covering the domain D by triangles. For each curved boundary segment (Figure 5.4.5) the correction

$$\frac{4}{3}f(S)A(PRQ) \tag{5.4.24}$$

is to be added, where $A(PRQ)$ is the area of the triangle with vertices P, R, Q. The error of the correction can be shown to be $O(\|Q-P\|^5)$ for each segment, if R is close to the midpoint of the arc PQ. If the boundary is given in parametric form, $x = x(x)$, $y = y(x)$, where x and y are twice differentiable on the arc PQ, then one should choose $t_R = \frac{1}{2}(t_P + t_Q)$. Richardson extrapolation can be used to increase the accuracy; see the examples below.

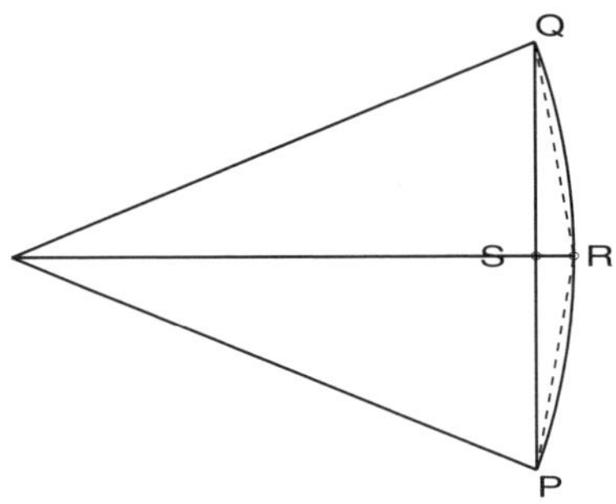

Figure 5.4.5. *Correction for curved boundary segment.*

Example 5.4.6.
Consider the integral

$$I = \int\int_D (x^2+y^2)^k \,dxdy,$$

where the region D and the grids for I_4 and I_{16} are shown in Figure 5.4.6 and I_n denotes the result obtained with n triangles. Because of symmetry the error has an expansion in even powers of h. Therefore, we can use repeated Richardson extrapolation and put

$$R'_n = I_{4n} + \frac{1}{15}(I_{4n} - I_n), \qquad R''_n = R'_{4n} + \frac{1}{63}(R'_{4n} - R'_n).$$

The results are shown in the table below. In this case the work could be reduced by a factor of four, because of symmetry.

k	I_4	I_{16}	I_{64}	R'_4	R'_{16}	R''_4	Correct
2	0.250000	0.307291	0.310872	0.311111	0.311111	0.311111	28/90
3	0.104167	0.161784	0.170741	0.165625	0.171338	0.171429	0.171429
4	0.046875	0.090678	0.104094	0.093598	0.104988	0.105169	0.105397

It is seen that R'-values have full accuracy for $k = 2$ and that the R''-values have high accuracy even for $k = 4$. In fact, it can be shown that R'-values are exact for any

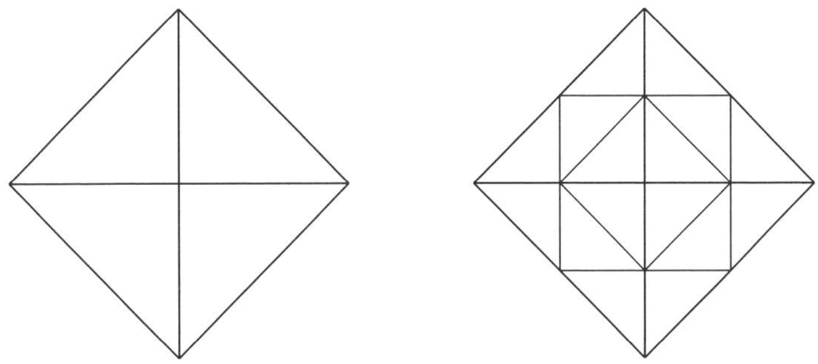

Figure 5.4.6. *The grids for I_4 and I_{16}.*

fourth-degree polynomial and R''-values are exact for any sixth-degree polynomial, when the region is covered exactly by the triangles.

Example 5.4.7.

The integral

$$a \int\int (a^2 - y^2)^{-1/2}\, dxdy,$$

over a quarter of the unit circle $x^2+y^2 \leq 1$, is computed with the grids shown in Figure 5.4.6, and with boundary corrections according to (5.4.15). The following results, using the notation of the previous example, were obtained and compared with the exact values.

a	I_8	I_{32}	R'_8	Correct
2	0.351995	0.352077	0.352082	0.352082
4	0.337492	0.337608	0.337615	0.337616
6	0.335084	0.335200	0.335207	0.335208
8	0.334259	0.334374	0.334382	0.334382

Note, however, that Richardson extrapolation may not always give improvement, for example, when the rate of convergence of the basic method is *more rapid* than usual.

5.4.5 Monte Carlo Methods

Multidimensional integrals arise frequently in physics, chemistry, computational economics,[177] and other branches of science. If a product rule is used to evaluate a multidimensional integral in d dimensions the work will increase exponentially with the number of dimensions d. For example, the product rule of an 8-point one-dimensional rule will require $(8)^8 = 2^{24} \approx 1.6 \cdot 10^7$ function evaluations in eight dimensions. This means that the problem may quickly become intractable when d increases.

One important application of the Monte Carlo method described in Section 1.4.2 is the numerical calculation of integrals of high dimension. For the Monte Carlo method the

[177]The valuation of financial derivatives can require computation of integrals in 360 dimensions!

accuracy achieved is always proportional to $1/\sqrt{n}$, where n is the number of function evaluations *independent of the dimension* d. Thus, if approached randomly multidimensional integration becomes tractable! The Monte Carlo method can be said to break "the curse of dimension" inherent in other methods. (For smooth integrands the Monte Carlo method is, however, not of optimal complexity.)

We shall briefly describe some ideas used in integration by the Monte Carlo method. For simplicity, we first consider integrals in *one* dimension, even though the Monte Carlo method cannot really compete with traditional numerical methods for this problem.

Let R_i, $i = 1 : N$, be a sequence of random numbers rectangularly distributed on $[0, 1]$, and set

$$I = \int_0^1 f(x)\,dx \approx I_1, \quad I_1 = \frac{1}{N}\sum_{i=1}^{N} f(R_i).$$

Then the expectation of the variable I_1 is I and its standard deviation decreases as $N^{-1/2}$. I_1 can be interpreted as a stochastic estimate of the mean value of $f(x)$ in the interval $[0, 1]$. This generalizes directly to multidimensional integrals over the unit hypercube. Let $R_i \in \mathbf{R}^d$, $i = 1 : N$, be a sequence of random points uniformly distributed on $[0, 1]^d$. Then

$$I = \int_{[0,1]^d} f(x)\,dx \approx I_1, \quad I_1 = \frac{1}{N}\sum_{i=1}^{N} f(R_i). \tag{5.4.25}$$

If the integral is to be taken over a subregion $\mathcal{D} \subset [0, 1]^d$, we can simply set $f(x) = 0$, $x \notin \mathcal{D}$. In contrast to interpolatory quadrature methods smooth functions are not integrated more efficiently than discontinuous functions. According to the law of large numbers, the convergence

$$I_N(f) \to \text{vol}(\mathcal{D})\mu(f) \quad \text{as} \quad N \to \infty,$$

where $\mu(f)$ is the mean value of $f(X)$, where X is a continuous random variable uniformly distributed in $\mathcal{D} \subset [0, 1]^d$.

A probabilistic error estimate can be obtained by estimating the standard deviation of $\mu(f)$ by the empirical standard deviation $s_N(f)$, where

$$s_N(f)^2 = \frac{1}{N-1}\sum_{i=1}^{N} (f(R_i) - I_N(f))^2. \tag{5.4.26}$$

If the integral is over a subregion $\mathcal{D} \subset [0, 1]^d$, we should use the mean value over $\mathcal{D}$, that is, neglect all points $R_i \notin \mathcal{D}$.

The standard deviation of the Monte Carlo estimate in (5.4.25) decreases as $N^{-1/2}$. This is very slow even compared to the trapezoidal rule, for which the error decreases as N^{-2}. To get one extra decimal place of accuracy we must increase the number of points by a factor of 100. To get three-digit accuracy the order of one million points may be required!

But if we consider, for example, a six-dimensional integral this is not exorbitant. Using a product rule with ten subdivisions in each dimension would also require 10^6 points.

The above Monte Carlo estimate is a special case of a more general one. Suppose X_i, $i = 1 : N$, has density function $g(x)$. Then

$$I_2 = \frac{1}{N} \sum_{i=1}^{N} \frac{f(X_i)}{g(X_i)}$$

has expected value I, since

$$E\left(\frac{f(X_i)}{g(X_i)}\right) = \int_0^1 \frac{f(x)}{g(x)} f(x)\, dx = \int_0^1 f(x)\, dx = I.$$

If one can find a frequency function $g(x)$ such that $f(x)/g(x)$ fluctuates less than $f(x)$, then I_2 will have smaller variance than I_1. This procedure is called **importance sampling**; it has proved very useful in particle physics problems, where important phenomena (for example, dangerous radiation which penetrates a shield) are associated with certain events of low probability.

We have previously mentioned the method of using a simple comparison problem. The Monte Carlo variant of this method is called the **control variate method**. Suppose that $\varphi(x)$ is a function whose integral has a known value K, and suppose that $f(x) - \varphi(x)$ fluctuates much less than $f(x)$. Then

$$I = K + \int_0^1 (f(x) - \varphi(x))\, dx,$$

where the integral to the right can be estimated by

$$I_3 = \frac{1}{N} \sum_{i=1}^{N} (f(R_i) - \varphi(R_i)),$$

which has less variance than I_1.

5.4.6 Quasi–Monte Carlo and Lattice Methods

In Monte Carlo methods the integrand is evaluated at a sequence of points which are supposed to be a sample of independent random variables. In **quasi–Monte Carlo** methods the accuracy is enhanced by using specially chosen deterministic points not necessarily satisfying the statistical tests discussed in Sec. 1.6.2. These points are constructed to be approximately equidistributed over the region of integration.

If the region of integration D is a subset of the d-dimensional unit cube $C_n = [0, 1]^d$ we set $f(x) \equiv 0$, for $x \notin D$. We can then always formulate the problem as the approximation of an integral over the d-dimensional unit cube $C_n = [0, 1]^d$:

$$I[f] = \int_{C_n} f(x_1, \ldots, x_d)\, dx_1 \cdots dx_d. \tag{5.4.27}$$

An infinite sequence of vectors $x_1, x_2, x_3, \dots$ in $\mathbf{R}^d$ is said to be **equidistributed** in the cube $[0, 1]^d$ if

$$I[f] = \lim_{N\to\infty} Q_N(f), \quad Q_N(f) = \frac{1}{N}\sum_{i=1}^{N} f(x_i), \tag{5.4.28}$$

for all Riemann integrable functions $f(x)$. The quadrature rules Q_N are similar to those used in Monte Carlo methods; this explains the name quasi–Monte Carlo methods.

In the **average case setting** the requirement that the worst case error is smaller than ϵ is replaced by the weaker guarantee that the expected error is at most ϵ. This means that we make some assumptions about the distribution of the functions to be integrated. In this setting the complexity of multidimensional integration has been shown to be proportional to $1/\epsilon$, compared to $(1/\epsilon)^2$ for the Monte Carlo method. Hence the Monte Carlo method is not optimal.

The convergence of the quadrature rules Q_N in (5.4.28) depends on the variation of f and the distribution of the sequence of points $x_1, \dots, x_N$. The **discrepancy** of a finite sequence of points $x_1, x_2, \dots, x_N$ is a measure of how much the distribution of the sequence deviates from an equidistributed sequence. The deterministic set of points used in quasi–Monte Carlo methods are constructed from **low discrepancy sequences**, which are, roughly speaking, uniformly spread as $N \to \infty$; see Niederreiter [275].

Let $0 < a_i \le 1$, $i = 1 : d$, and restrict $f(x)$, $x \in \mathbf{R}^n$, to the class of functions

$$f(x) = \begin{cases} 1 & \text{if } 0 \le x_i \le a_i, \\ 0 & \text{otherwise.} \end{cases}$$

We require the points to be such that every $Q_N(f)$ gives a good approximation of the integral of $f(x)$ over the hypercube $[0, 1]^d$ for all functions in this class.

Low discrepancy sequences are usually generated by algorithms from number theory, a branch of mathematics seemingly far removed from analysis. Recall from Sec. 2.2.1 that each integer i has a unique representation $d_k \cdots d_2 d_1 d_0$ with respect to a integer basis $b \ge 2$. The **radical inverse function** φ_b maps an integer i onto the real number

$$\varphi_b(i) = 0.d_0 d_1 d_2 \cdots d_k \in [0, 1).$$

The **Van der Corput sequence** (see [365]) with respect to the base b is the infinite sequence defined by

$$x_i = \varphi_b(i), \quad i = 1, 2, 3, \dots.$$

These sequences can be shown to have an asymptotic optimal discrepancy. The first few elements in the sequence for $b = 2$ are shown in the table below.

i			$\varphi_2(i)$
1	1	.1	0.5
2	10	.01	0.25
3	11	.11	0.75
4	100	.001	0.125
5	101	.101	0.625
6	110	.011	0.375
7	111	.111	0.875

Halton sequences (see [180]) are multidimensional extensions of Van der Corput sequences.

Definition 5.4.4.

Let the bases $b_1, b_2, \dots, b_d$ *be pairwise relative prime. Then the Halton sequence* $x_i \in [0, 1]^d$ *with respect to these bases is defined by*

$$x_i = (\varphi_{b_1}(i), \varphi_{b_2}(i), \dots, \varphi_{b_d}(i))^T, \quad i = 0, 1, 2, \dots, \tag{5.4.29}$$

where $\varphi_{b_k}(i)$ *is the radical inverse function with respect to the basis* b_k, $k = 1 : d$.

The **Hammersley sequences** [183] are similar to Halton sequences but are finite and differ in the definition of the first component. The N-point Hammersley sequence with respect to the bases $b_1, b_2, \dots, b_{d-1}$ is the sequence of points $x_i \in [0, 1]^d$ defined by

$$x_i = \Big(i/N, \varphi_{b_1}(i), \varphi_{b_2}(i), \dots, \varphi_{b_{d-1}}(i)\Big)^T, \quad i = 0 : N - 1. \tag{5.4.30}$$

The Hammersley points in $[0, 1]^2$ for $N = 512$ and $b_1 = 2$ are shown in Figure 5.4.7.

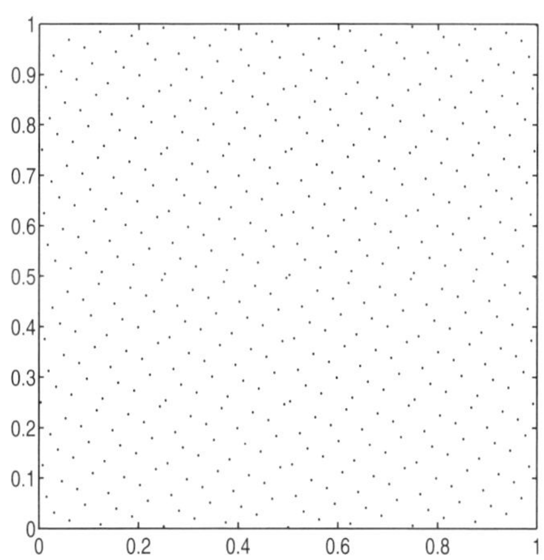

Figure 5.4.7. *Hammersley points in* $[0, 1]^2$.

Wozniakowski [383] showed that the Hammersley points are optimal sampling points for multidimensional integration. With n function evaluations a worst case error of quasi–Monte Carlo methods is bounded by a multiple of $(\log n)^d/n$, which can be compared to $n^{-1/2}$ for Monte Carlo methods.

ALGORITHM 5.5. *Hammersley Points.*

The following MATLAB program generates N Hammersley points in the two-dimensional square $[0, 1]^2$ for $b_1 = 2$.

```
[x,y] = Hammersley(N);
n = ceil(log2(N));
for i = 1:N
  x(i) = (i-1)/N;
  j = i-1;
  for p = 1:n
    j = j/2; d(p) = 0;
    if j > floor(j)
      d(p) = 1; j = floor(j);
    end;
  end;
  phi = d(n)/2;
  for p = n-1:-1:1
    phi = (phi + d(p))/2;
  end;
  y(i) = phi;
end;
```

Using harmonic analysis it was shown in Sec. 5.1.4 that the composite trapezoidal rule can be very accurate for periodic integrands. These results can be extended to multidimensional integration of periodic integrands. A **lattice rule** for the numerical integration over the d-dimensional unit cube $C_n = [0, 1]^d$ is an equal weight rule,

$$Q_N(f) = \frac{1}{N}\sum_{j=0}^{N-1} f(x_i), \tag{5.4.31}$$

where the sampling points x_i, $i = 0 : N - 1$, are points of an integration lattice in the cube $[0, 1]^d$. A multidimensional extension of the compound trapezoidal rule is obtained by taking

$$x_i = \text{fraction}\left(\frac{i}{N}p\right),$$

where fraction(x) returns the fractional part of x. Lattice rules can be studied by expanding the integrand in a Fourier series. The "curse of dimension" can be lifted by using a class of randomly shifted lattice rules introduced by Ian H. Sloane.

Review Questions

5.4.1 What is meant by a product integration rule for computing a multidimensional integral? What is the drawback with such rules for high dimensions?

5.4.2 Give the generalization of the composite trapezoidal and midpoint rules for a two-dimensional rectangular grid.

5.4.3 Define barycentric coordinates in two dimensions. Give a formula for linear interpolation on a triangular grid.

5.4.4 For high-dimensional integrals and difficult boundaries Monte Carlo methods are often used. How does the accuracy of such methods depend on the number n of evaluations of the integrand?

5.4.5 How do quasi–Monte Carlo methods differ from Monte Carlo methods?

Problems and Computer Exercises

5.4.1 Let E be the ellipse $\{(x, y) \mid (x/a)^2 + (y/b)^2 \leq 1\}$. Transform

$$I = \int\int_E f(x, y)\, dxdy$$

into an integral over a rectangle in the (r, t)-plane with the transformation $x = a\, r \cos t$, $y = b\, r \sin t$.

5.4.2 Consider the integral I of $u(x, y, z)$ over the cube $|x| \leq h$, $|y| \leq h$, $|z| \leq h$. Show that the rule

$$I \approx \frac{4}{3} h^3 \sum_{i=1}^{6} f(x_i, y_i, z_i),$$

where $(x_i, y_i, z_i) = (\pm h, 0, 0)$, $(0, \pm h, 0)$, $(0, 0, \pm h)$, is exact for all monomials $x^i y^j$, $0 \leq i, j \leq 1$.

5.4.3 (a) In one dimension Simpson's rule can be obtained by taking the linear combination $S(h) = (T(h) + 2M(h))/3$ of the trapezoidal and midpoint rule. Derive a quadrature rule

$$\int_{-h}^{h}\int_{-h}^{h} f(x, y)\, dxdy = \frac{4h^2}{6}(f_{1,0} + f_{0,1} + f_{-1,0} + f_{0,-1} + 2f_{0,0})$$

for the square $[-h, h]^2$ by taking the same linear combination of the *product* trapezoidal and midpoint rules. Note that this rule is not equivalent to the product Simpson's rule.

(b) Show that the rule in (a) is exact for all cubic polynomials. Compare its error term with that of the product Simpson's rule.

(c) Generalize the midpoint and trapezoidal rules to the cube $[-h, h]^3$. Then derive a higher-order quadrature rule using the idea in (a).

5.4.4 Is a quadratic polynomial uniquely determined, given six function values at the vertices and midpoints of the sides of a triangle?

5.4.5 Show that the boundary correction of (5.4.15) is exact if $f \equiv 1$, and if the arc is a parabola where the tangent at R is parallel to PQ.

5.4.6 Formulate generalizations to several dimensions of the integral formula of Theorem 5.4.2, and convince yourself of their validity.

Hint: The formula is most simply expressed in terms of the values in the vertices and in the centroid of a simplex.

5.4.7 (a) Write a program which uses the Monte Carlo method to compute $\int_0^1 e^x\,dx$. Take 25, 100, 225, 400, and 635 points. Plot the error on a log-log scale. How does the error depend (approximately) on the number of points?

(b) Compute the integral in (a) using the control variate method. Take $\varphi(x) = 1 + x + x^2/2$. Use the same number of points as in (a).

5.4.8 Use the Monte Carlo method to estimate the multiple integral

$$I = \int_{[0,1]^n} \prod |x_k - 1/3|^{1/2} \approx 0.49^n,$$

for $n = 6$. What accuracy is attained using $N = 10^3$ random points uniformly distributed in six dimensions?

5.4.9 Write a program to generate Halton points in two dimensions for $b_1 = 2$ and $b_2 = 5$. Then plot the first 200 points in the unit square.

Hint: For extracting the digits to form x_i, see Algorithm 2.1. You can also use TOMS Algorithm 247; see [181].

Notes and References

Numerical integration is a mature and well-understood subject. There are several comprehensive monographs devoted to this area; see in particular Davis and Rabinowitz [93] and Engels [112]. Examples of integrals arising in practice and their solution are found in [93, Appendix 1]. Newton–Cotes' and other quadrature rules can also be derived using computer algebra systems; see [132]. A collection of numerical quadrature rules is given in the Handbook [1, Sec. 25].

The idea of adaptive Simpson quadrature is old and treated fully by Lyness [251]. Further schemes, computer programs, and examples are given in [93]. A recent discussion of error estimates and reliability of different codes is given by Espelid [115].

The literature on Gauss–Christoffel quadrature and its computational aspects is extensive. Gauss–Legendre quadrature was derived by Gauss in 1814 using a continued fraction expansion. In 1826 Jacobi showed that the nodes were the zeros of the Legendre polynomials and that they were real, simple, and in $[-1, 1]$. The convergence of Gaussian quadrature methods was first studied by Stieltjes in 1884. More on the history can be found in Gautschi [143]. Recent results by Trefethen [359] suggest that the Clenshaw–Curtis rule may be as accurate as Gauss–Legendre quadrature with an equal number of nodes.

The importance of the eigenvalues and eigenvectors of the Jacobi matrices for computing Gauss' quadrature rules was first elaborated by Golub and Welsch [170]. The generalization to Radau and Lobatto quadrature was outlined in Golub [162] and further generalized by Golub and Kautsky [164].

The presentation in Sec. 5.3.4 was developed in Dahlquist [88]. Related ideas can be found in Gautschi [140, 146] and Mysovskih [273]. The encyclopedic book by Gautschi [148] describes the current state-of-the-art of orthogonal polynomials and Gauss–Christoffel quadrature computation; see also the survey by Laurie [238].

There is an abundance of tables giving abscissae and weights for various quadrature rules. Abramowitz and Stegun [1, Sec. 25] give tables for Newton–Cotes' and several Gauss–Christoffel rules. Many Gaussian quadrature formulas with various weight functions are tabulated in Stroud and Secrest [343]. A survey of other tables is given in [93, Appendix 4].

Many algorithms and codes for generating integration rules have appeared in the public domain. In [93, Appendices 2, 3] several useful Fortran programs are listed and a bibliography of Algol and Fortran programs published before 1984 is given. Kautsky and Elhay [224] have developed algorithms and a collection of Fortran subroutines called IQPACK [110] for computing weights of interpolatory quadratures. QUADPACK is a collection of Fortran 77 and 90 subroutines for integration of functions available at `www.netlib.org`. It is described in the book by R. Piessens et al. [290].

A software package in the public domain by Gautschi [145] includes routines for generating Gauss–Christoffel rules with arbitrary weight functions. A package QPQ consisting of MATLAB programs for generating orthogonal polynomials as well as dealing with applications is available at `www.cs.purdue.edu/archives/2002/wxg/codes`. Part of these programs are described in Gautschi [150]. Maple programs for Gauss quadrature rules are given by von Matt [262]. An overview of results related to Gauss–Kronrod rules is given by Monegato [270]. The calculation of Gauss–Kronrod rules is dealt with in [237, 62].

Gaussian product rules for integration over the n-dimensional cube, sphere, surface of a sphere, and tetrahedron are derived in Stroud and Secrest [343, Ch. 3]. Some simple formulas of various accuracy are tabulated in [1, Sec. 25]. The derivation of such formulas are treated by Engels [112]. Nonproduct rules for multidimensional integration are found in Stroud [342].

A good introduction to multidimensional integration formulas and Monte Carlo methods is given by Ueberhuber [364, Chap. 12]. Construction of fully symmetric numerical multidimensional integration formulas over the hypercube $[-h, h]^d$ using a rectangular grid is treated by McNamee and Stenger [263]. For a survey of recent results on sparse grids and breaking "the curse of dimensionality," see Bungartz and Griebel [60]. The efficiency of quasi–Monte Carlo methods are discussed in [328]. Lattice rules are treated in the monograph by Sloan and Joe [327]. For an introduction see also [364, Sec. 12.4.5].

Chapter 6

Solving Scalar Nonlinear Equations

If we start with an approximation to a zero that is appreciably more correct than the limiting accuracy, a single (Newton) iteration will usually spoil this very good approximation and produce one with an error which is typically of the limiting accuracy.
—J. H. Wilkinson

6.1 Some Basic Concepts and Methods

6.1.1 Introduction

In this chapter we study numerical methods for computing accurate approximations to the roots of a scalar nonlinear equation

$$f(x) = 0, \tag{6.1.1}$$

where $f(x)$ is a real-valued function of one variable. This problem has occupied mathematicians for many centuries and several of the basic methods date back a long time. In general the roots of (6.1.1) cannot be expressed in closed form. Even when an explicit solution is available (as, for example, for the reduced cubic equation), this is often so complicated that using Newton's method is much more practical; see Problem 2.3.8.

Numerical methods are iterative in nature. Starting from one or more initial approximations, they produce a sequence of approximations, which presumably converges to the desired root. Note that it suffices that the function $f(x)$, and preferably some of its derivatives, can be evaluated for given numerical values of x for numerical methods to be applicable.

It is not uncommon in applications that each function value is obtained by a complicated computation, for example, by the numerical solution of a differential equation. The object is then to use as few function evaluations as possible in order to approximate the root with a prescribed accuracy.

Iterative methods have to be truncated after a finite number of steps and therefore can yield only approximations to the desired roots. Further, the roundoff errors that occur in the evaluation of $f(x)$ will limit the accuracy limiting by *any numerical method.* The effect of such rounding errors depends on the conditioning of the roots and is discussed in Sec. 6.1.3.

With certain methods it is sufficient for convergence to know an initial interval $[a, b]$, which contains the desired root (and no other root). An important example is the bisection method described in Sec. 6.1.2. It is often suitable to use a hybrid method in which the bisection method is used to roughly locate the root. A more rapidly convergent method is then used to refine this approximation. These latter methods make use of more regularity assumptions about $f(x)$, and usually also require an initial approximation close to the desired root.

The theory of fixed-point iteration methods is treated in Sec. 6.1.4 and the concepts of convergence order and efficiency introduced in Sec. 6.1.5. The secant method and other methods based on interpolation are described in Sec. 6.2. In Sec. 6.4 we briefly consider methods for solving the related problem of finding the minimum or maximum of a real-valued function $g(x)$. Newton's method and other methods of higher order are analyzed in Sec. 6.3. A classical problem is that of determining all real or complex roots of an algebraic equation. Special features and methods for this problem are taken up in Sec. 6.5.

Many of the methods for a single equation, such as Newton's method, are easily generalized for *systems of nonlinear equations.* But unless good approximations to the roots are known, several modifications of the basic methods are required; see Volume II, Chapter 11.

6.1.2 The Bisection Method

It is often advisable to start by collecting some qualitative information about the roots to be computed. One should try to determine how many roots there are and their approximate location. Such information can often be obtained by graphing the function $f(x)$. This can be a useful tool for determining the number of roots and intervals containing each root.

Example 6.1.1.

Consider the equation

$$f(x) = (x/2)^2 - \sin x = 0.$$

In Figure 6.1.1 the graphs of $y = (x/2)^2$ and $y = \sin x$ are shown. Observing the intersection of these we find that the unique positive root lies in the interval $(1.8, 2)$, probably close to $\alpha \approx x_0 = 1.9$.

The following **intermediate value theorem** can be used to infer that an interval $[a, b]$ contains *at least one root* of $f(x) = 0$.

Theorem 6.1.1 (*Intermediate Value Theorem*).

Assume that the function $f(x)$ is continuous for $x \in [a, b]$, $f(a) \neq f(b)$, and k is between $f(a)$ and $f(b)$. Then there is a point $\xi \in (a, b)$ such that $f(\xi) = k$. In particular, if $f(a)f(b) < 0$, then the equation $f(x) = 0$ has at least one root in the interval (a, b).

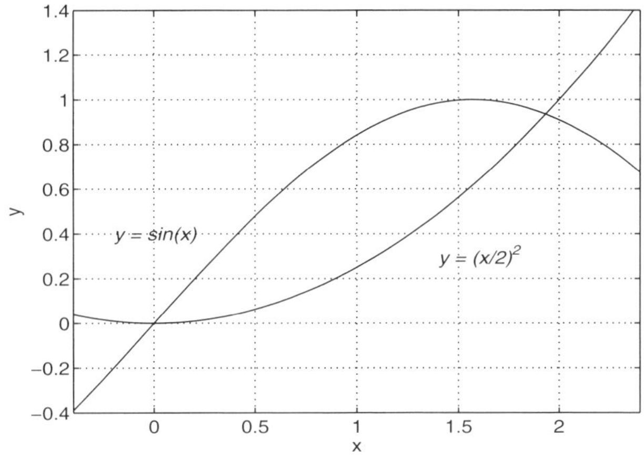

Figure 6.1.1. *Graph of curves* $y = (x/2)^2$ *and* $\sin x$.

A systematic use of the intermediate value theorem is made in the **bisection method**. Assume that $f(x)$ is continuous in the interval (a_0, b_0) and that $f(a_0)f(b_0) < 0$. We shall determine a nested sequence of intervals $I_k = (a_k, b_k)$, $k = 1, 2, 3, \ldots$, such that

$$(a_0, b_0) \supset (a_1, b_1) \supset (a_2, b_2) \supset \cdots$$

and which all contain a root of the equation $f(x) = 0$. The intervals are determined recursively as follows. Given $I_k = (a_k, b_k)$ compute the midpoint

$$m_k = \frac{1}{2}(a_k + b_k) = a_k + \frac{1}{2}(b_k - a_k) \ldots \tag{6.1.2}$$

and $f(m_k)$. The latter expression has the advantage that using this to compute the midpoint, no rounding error occurs in the subtraction (see Theorem 2.2.2 and Example 6.1.3).

We can assume that $f(m_k) \neq 0$, since otherwise we have found a root. The new interval is then determined by

$$I_{k+1} = (a_{k+1}, b_{k+1}) = \begin{cases} (m_k, b_k) & \text{if } f(m_k)f(a_k) > 0, \\ (a_k, m_k) & \text{if } f(m_k)f(a_k) < 0. \end{cases} \tag{6.1.3}$$

From the construction it follows immediately that $f(a_{k+1})f(b_{k+1}) < 0$ (see also Figure 6.1.1) and therefore the interval I_{k+1} also contains a root of $f(x) = 0$.

If the sign of $f(m_k)$ has been correctly evaluated for $k = 1 : n$, then after n bisection steps we have contained a root in the interval (a_n, b_n) of length $2^{-n}(b_0 - a_0)$. If we take m_n as an estimate of the root α, we have the error estimate

$$|\alpha - m_n| < 2^{-(n+1)}(b_0 - a_0). \tag{6.1.4}$$

At each step we gain one binary digit in accuracy or, since $10^{-1} \approx 2^{-3.3}$, on average one decimal digit per 3.3 steps. To find an interval of length δ which includes a root will require about $\log_2((b-a)/\delta)$ evaluations of f. Note that the bisection algorithm makes no quantitative use of the magnitude of computed function values.

Example 6.1.2.

The bisection method applied to the equation $(x/2)^2 - \sin x = 0$, with $I_0 = (1.8, 2)$ gives the sequence of intervals $[a_n, b_n]$, where

k	a_k	b_k	m_k	$f(m_k)$
1	1.8	2	1.9	< 0
2	1.9	2	1.95	> 0
3	1.9	1.95	1.925	< 0
4	1.925	1.95	1.9375	> 0
5	1.925	1.9375	1.93125	< 0
6	1.93125	1.9375	1.934375	> 0

Here after six function evaluations we have $\alpha \in (1.93125, 1.934375)$, an interval of length $0.2 \cdot 2^{-6} = 0.003125$ (see Figure 6.1.2).

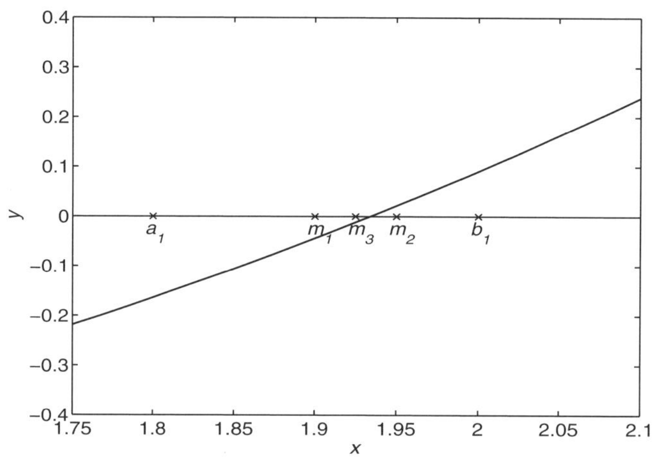

Figure 6.1.2. *The bisection method.*

Example 6.1.3.

The inequalities $a \le fl(\frac{1}{2}(a+b)) \le b$, where a and b are floating-point numbers with $a \le b$, can be violated in floating-point arithmetic. For example, assume that base 10 arithmetic with six digits is used. Taking $a = 0.742531$ and $b = 0.742533$ we obtain $fl(a+b) = 1.48506$ (rounded) and $\frac{1}{2}(a+b) = 0.742530 < a$.

On the other hand the inequalities $a \le fl(a + \frac{1}{2}(b-a)) \le b$ hold in floating-point arithmetic, for any base β. (Recall that by Lemma 2.3.2 $fl(b-a)$ is evaluated *exactly* in binary arithmetic if $b/2 \le a \le 2b$.) With a and b as given, we get the correct value 0.742532.

An algorithmic description of the bisection method is given below. Let f be a given function and $I = [a, b]$ an interval such that $b > a$ and $f(a)f(b) \leq 0$. The bisection algorithm attempts to compute an approximation to a root $m \in I$ of $f(x) = 0$, with an error less than a specified tolerance $\tau > 0$. Note that the given tolerance τ is increased by the amount $u \max(|a|, |b|)$, where u is the machine precision. This is to guard against the possibility that δ has been chosen too small (e.g., smaller than the spacing between the floating-point numbers between a and b).

Algorithm 6.1. *Bisection.*

The function `bisect` computes an approximation r to a local root of a given function in the interval $I = (a, b)$ with an error less than a specified tolerance $\tau + u \max(|a|, |a|)$, where u is the unit roundoff.

```
function root = bisect(fname,a,b,tau);
%
% [a,b] is an initial interval such that
% f(a)*f(b) < 0.
fa = feval(fname,a);
fb = feval(fname,b);
while abs(b-a) > tau + eps*max(abs(a),abs(b))
   mid = a + (b - a)/2;
   fmid = feval(fname,mid);
   if fa*fmid <= 0
   % Keep left endpoint a
      b = mid;
   else
   % Keep right endpoint b
      a = mid; fa = fmid;
end;
root = a  + (b - a)/2;
end;
```

The time required by the bisection algorithm is typically proportional to the number of function evaluations, other arithmetic operations being insignificant. The correct subinterval will be chosen in the algorithm as long as the sign of the computed function value $f(m)$ is correctly determined. If the tolerance τ is taken too "small" or the root is ill-conditioned, this may fail to be true in the later steps. Even then the computed midpoints will stay within a certain domain of uncertainty. Due to rounding errors there is a limiting accuracy with which a root can be determined from approximate function values; see Sec. 6.1.3.

The bisection method is optimal for the class of functions that changes sign on $[a, b]$ in the sense that it minimizes the maximum number of steps over all such functions. The convergence is rather slow, but *independent of the regularity of* $f(x)$. For other classes of functions, for example, functions that are continuously differentiable on $[a, b]$, methods like Newton's method which assume some regularity of $f(x)$ can achieve significantly faster convergence.

If $f(a)f(b) < 0$, then by the intermediate value theorem the interval (a, b) contains *at least one root* of $f(x) = 0$. If the interval (a, b) contains several roots of $f(x) = 0$, then the bisection method will converge to just one of these. (Note that there may be one or several roots in (a, b), and in the case $f(a)f(b) > 0$.)

If we only know (say) a lower bound $a < \alpha$ for the root to be determined we can proceed as follows. We choose an initial step length h and in the first **hunting phase** compute successively function values $f(a+h)$, $f(a+2h)$, $f(a+4h), \dots$, i.e., we double the step, until a function value is found such that $f(a)f(a + 2^k h) < 0$. At this point we have bracketed a root and can initiate the bisection algorithm.

In the bisection method the interval of interest is in each step split into *two* subintervals. An obvious generalization is to partition instead into k subintervals, for $p \geq 2$. In such a **multisection method** of order p the interval $I = [a, b]$ is divided into k subintervals $I_i = [x_i, x_{i+1}]$, where

$$x_i = a + i\,[(b-a)/p], \quad i = 0 : p.$$

If there exists only one root in the interval I and we wish to compute it with an absolute error ϵ, then it is necessary to perform

$$n_k = \log_2\left(\frac{b-a}{2\epsilon}\right) \Big/ \log_2(p)$$

multisections of order p. Thus, the efficiency of multisection of order p compared to bisection ($p = 2$) is

$$n_2/(pn_p) = \log_2(p)/p.$$

Hence if there is a single root in the interval bisection is always preferable. If there are several roots in the interval multisection may perform better if the subintervals can be processed in parallel.

There are several other applications of the bisection algorithm. For example, in Sec. 4.4.3 we considered evaluating the nonzero B-splines for a given argument x. Then we first have to search an ordered sequence of knots $\tau_0, \dots, \tau_m$ to find the interval such that $\tau_j \leq x < \tau_{j+1}$. This can be achieved by a slight modification of the bisection method.

A similar problem, important in computer science, is *searching in an ordered register*, for example, a register of employees ordered according to increasing Social Security number. If the nth number in the register is denoted by $f(n)$, then searching for a certain number a means that an equation $f(n) = a$ is to be solved (here f is an increasing, discontinuous function). The bisection method can also be used in searching an *alphabetically* ordered register.

In later sections we will study methods for solving a nonlinear equation which make more efficient use of computed function values than the bisection method and possibly also use values of derivatives of $f(x)$. If $f(x)$ is sufficiently regular such methods can achieve significantly faster convergence.

6.1.3 Limiting Accuracy and Termination Criteria

In the following we denote by $\overline{f}(x)$ the limited-precision approximation obtained when $f(x)$ is evaluated in floating-point arithmetic. When a monotone function $f(x)$ is evaluated in floating-point arithmetic the resulting approximation $\overline{f}(x)$ is not, in general, monotone. The effect of rounding errors, in evaluating a certain polynomial of fifth degree with a simple zero at $x = 1$, is illustrated in Figure 6.1.3. Note the loss of monotonicity caused by rounding errors. This figure also shows that even if $\overline{f}(a)\overline{f}(b) < 0$, the true equation $f(x) = 0$ may not have a zero in $[a, b]$!

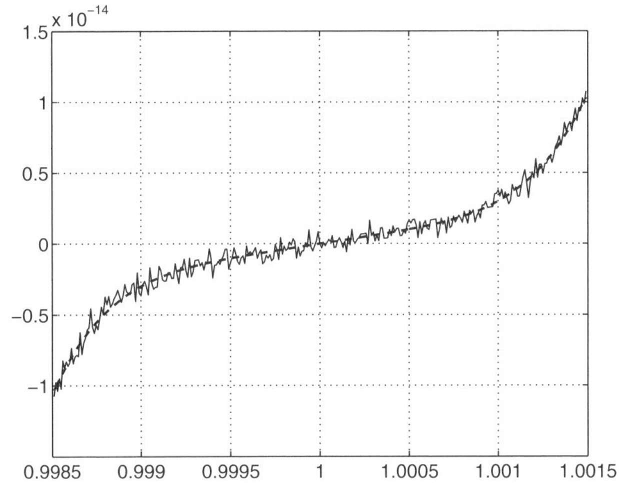

Figure 6.1.3. *Limited-precision approximation of a continuous function.*

Even if the true function value $|f(x_n)|$ is "small" one cannot deduce that x_n is close to a zero of $f(x)$ without some assumption about the size of the derivative of f. We recall some basic results from analysis; for proofs see, for example, Ostrowski [279, Chapter 2].

Theorem 6.1.2.

Let $f(x)$ be continuous and differentiable in the interval $J = [x_n - \eta, x_n + \eta]$ for some $\eta > 0$. If $|f'(x)| \geq m_1$ for all $x \in J$ and $|f(x_n)| \leq \eta m_1$, then $f(x)$ has exactly one zero in J.

A root α of $f(x) = 0$ is said to be a **simple** root if $f'(\alpha) \neq 0$. We now derive an error estimate for a simple root α of $f(x)$, which takes into account errors in the computed values of $f(x)$. Assume that

$$\overline{f}(x) = f(x) + \delta(x), \quad |\delta(x)| \leq \delta, \quad x \in J, \tag{6.1.5}$$

where δ is an upper bound for rounding and other errors in computed function values of $f(x)$. Using Theorem 6.1.2 we obtain

$$|x_n - \alpha| \leq \eta, \quad \eta = (|\overline{f}(x_n)| + \delta)/m_1, \quad |f'(x)| \geq m_1, \quad x \in J. \tag{6.1.6}$$

Obviously the best we can hope for is to find an approximation x_n such that the computed function value $\overline{f}(x_n) = 0$. It follows that *for any numerical method*, δ/m_1 is an approximate limit for the accuracy with which a simple zero α can be determined. If $f'(x)$ does not vary much near $x_n = \alpha$, then we have the approximate error bound

$$|x_n - \alpha| \le \delta/m_1 \approx \epsilon_\alpha, \qquad \epsilon_\alpha = \delta/|f'(\alpha)|. \tag{6.1.7}$$

Since this is the best error bound for *any* method, we call ϵ_α the **limiting accuracy** for the simple root α, and the interval $[\alpha - \epsilon_\alpha, \alpha + \epsilon_\alpha]$ the **domain of uncertainty** for the root α. If $|f'(\alpha)|$ is small, then ϵ_α is large and the problem of computing the root α is ill-conditioned (see again Figure 6.1.3).

Example 6.1.4.

Suppose we have computed the approximation $x = 1.93375$ to the positive root to the equation $f(x) = \sin x - (x/2)^2$. Now $f'(x) = \cos x - x/2$ and it is easily verified that

$$|f'(x)| > 1.31 = m_1, \qquad x \in [1.93, 1.94].$$

Further, using six decimals we get $\sin 1.93375 = 0.934852 \pm 0.5 \cdot 10^{-6}$, and $(x/2)^2 = 0.966875^2 = 0.934847 \pm 0.5 \cdot 10^{-6}$. Then (6.1.6) gives the strict error estimate

$$|x - \alpha| < 6 \cdot 10^{-6}/1.31 < 5.6 \cdot 10^{-6}.$$

Using the following theorem, an analogous result can be shown for zeros of a complex function $f(z)$ of a complex variable z.

Theorem 6.1.3.

Let $f(z)$ be analytic in the disk $K = \{z \mid |z - z_0| \le \eta\}$ for some $\eta > 0$. If $|f'(z)| \ge m_1$ in K and $|f(z_0)| \le \eta m_1$, then $f(z)$ has a zero inside K.

The **multiplicity** of a root is defined as follows.

Definition 6.1.4.

Suppose that $f(x)$ is q times continuously differentiable in a neighborhood of a root α to the equation $f(x) = 0$. Then α is said to have multiplicity q if

$$0 \ne \lim_{x \to \alpha} \left| f(x)/(x - \alpha)^q \right| < \infty. \tag{6.1.8}$$

If a root α has multiplicity q, then by (6.1.8) $f^{(j)}(\alpha) = 0$, $j < q$, and from Taylor's formula

$$f(x) = \frac{1}{q!}(x - \alpha)^q f^{(q)}(\xi), \qquad \xi \in \operatorname{int}(x, \alpha). \tag{6.1.9}$$

Assuming that $|f^{(q)}(x)| \ge m_q$, $x \in J$, and proceeding as before, we find that the limiting accuracy for a root of multiplicity q is given by

$$|x_n - \alpha| \le (q!\,\delta/m_q)^{1/q} \approx \epsilon_\alpha, \qquad \epsilon_\alpha = (q!\,\delta/|f^{(q)}(\alpha)|)^{1/q}. \tag{6.1.10}$$

Comparing this with (6.1.7), we see that because of the exponent $1/q$ multiple roots are in general very ill-conditioned. A similar behavior can also be expected when there are several distinct but "close" roots. An instructive example is the Wilkinson polynomial, studied in Sec. 6.5.2.

Example 6.1.5.

The equation $f(x) = (x - 2)x + 1 = 0$ has a double root $x = 1$. The (exact) value of the function at $x = 1 + \epsilon$ is

$$f(1 + \epsilon) = (\epsilon - 1)(1 + \epsilon) + 1 = -(1 - \epsilon^2) + 1 = \epsilon^2.$$

Now, suppose that we use a floating-point arithmetic with eight decimal digits in the mantissa. Then

$$fl(1 - \epsilon^2) = 1, \quad |\epsilon| < \frac{1}{2}\sqrt{2}\cdot 10^{-4},$$

and for $0.99992929 \le x \le 1.0000707$, the computed value of $f(x)$ will be zero when $f(x)$ is evaluated using Horner's rule. Hence the root can only be computed with about four correct digits, i.e., with a relative error equal to the *square root of the machine precision.*

An important practical question concerning iterative methods is how to stop them. There are a few different **termination criteria** in practical use. The simplest is to stop after a preset number of iterations. This is, in general, too crude, but it is advisable always to specify the maximum number of iterations allowed, to guard against unforeseen problems. For example, programming errors could otherwise lead to an infinite loop being executed.

If the iteration method produces a sequence of bracketing intervals $[a_k, b_k]$ the iterations can be terminated when

$$|b_k - a_k| \le \text{tol}, \quad \text{tol} = \tau + 2u|x_n|, \tag{6.1.11}$$

where $\tau > 0$ is a user specified absolute tolerance and u is the rounding unit (see Sec. 2.2.2). The second term assures that the iterations are terminated when the roundoff level of the floating-point arithmetic is reached.

If for a simple root a lower bound for $|f'(\alpha)|$ is known, the iterations can be terminated on the basis of the error estimate (6.1.7). But it is usually more effective to iterate a few extra times, rather than make the effort to use a special formula for error estimation.

The termination criteria must be able to deal with the possibility that the user specified tolerance is too small (or the root too ill-conditioned) and cannot be attained. In this case from some step onward rounding errors will dominate in the evaluation of $f(x_n)$ and the computed values of $f(x)$ may vary quasi-randomly in the interval $(-\delta, \delta)$ of limiting accuracy. If we are using a bracketing method like the bisection method, the iterations will continue until the criterion (6.1.11) is satisfied. This, of course, does *not* ensure that the root actually has been determined to this precision.

For iterative methods with fast convergence (like Newton's method) the following termination criterion can often be used: Stop when for the first time the approximation x_n satisfies the two conditions

$$|x_{n+1} - x_n| \ge |x_n - x_{n-1}|, \qquad |x_n - x_{n-1}| \le tol. \tag{6.1.12}$$

Here tol is a coarse tolerance, used only to prevent the iterations from being terminated before x_n even has come close to α. When (6.1.12) is satisfied the limiting accuracy has been reached and the quantity $|x_{n+1} - x_n|$ usually is a good estimate of the error $|x_n - \alpha|$. Using this criterion the risk of never terminating the iterations for an ill-conditioned root is quite small. Note also that iteration methods of superlinear convergence ultimately converge so fast that the cost of always iterating until the limiting accuracy is obtained may be small, even if the user specified tolerance is much larger than ϵ_α.

6.1.4 Fixed-Point Iteration

We now introduce a very general class of iteration methods, which includes many important root finding methods as special cases.

Let ϕ be a continuous function and $\{x_n\}$ the sequence generated by

$$x_{n+1} = \phi(x_n), \quad n = 0, 1, 2, \dots \tag{6.1.13}$$

for some initial value x_0. Assuming that $\lim_{n\to\infty} x_n = \alpha$, it follows that

$$\alpha = \lim_{n\to\infty} x_n = \lim_{n\to\infty} \phi(x_n) = \phi(\alpha), \tag{6.1.14}$$

i.e., the limiting value α is a root of the equation $x = \phi(x)$. We call α a **fixed point** of the mapping $x \to \phi(x)$ and the iteration (6.1.13) a **fixed-point iteration**.

An iterative method for solving an equation $f(x) = 0$ can be constructed by rewriting it in the equivalent form $x = \phi(x)$, which then defines a fixed-point iteration (6.1.13). Clearly this can be done in many ways. For example, let $g(x)$ be any function such that $g(\alpha) \neq 0$ and set

$$\phi(x) = x - f(x)g(x). \tag{6.1.15}$$

Then α is a solution to $f(x) = 0$ if and only if α is a fixed point of ϕ.

Example 6.1.6.

The equation $x + \ln x = 0$ can, for example, be written as

$$\text{(i) } x = -\ln x; \qquad \text{(ii) } x = e^{-x}; \qquad \text{(iii) } x = (x + e^{-x})/2.$$

Each of these give rise to a different fixed-point iteration. Results from the first eight iterations,

$$x_{n+1} = e^{-x_n}, \quad x_0 = 0.3,$$

are pictured in Figure 6.1.4. The convergence is slow and we get $x_9 = 0.5641$ (correct value 0.567143).

As was shown already in Sec. 1.1.1 the iteration (6.1.13) may not converge even if the initial value x_0 is chosen arbitrarily close to a root. If $\lim_{n\to\infty} x_n = \alpha$ for all x_0 in a sufficiently close neighborhood of α, then α is called a **point of attraction**; otherwise α is a **point of repulsion**. The case $|\phi'(\alpha)| = 1$ is indeterminate; the fixed-point iteration $x_{n+1} = \phi(x_n)$ can either converge or diverge (see Problem 6.1.10).

We shall see that under certain conditions the fixed-point problem has a unique solution and that the iteration defined by (6.1.13) converges to this solution. The following important theorem not only provides a solid basis for iterative numerical techniques, but also is an

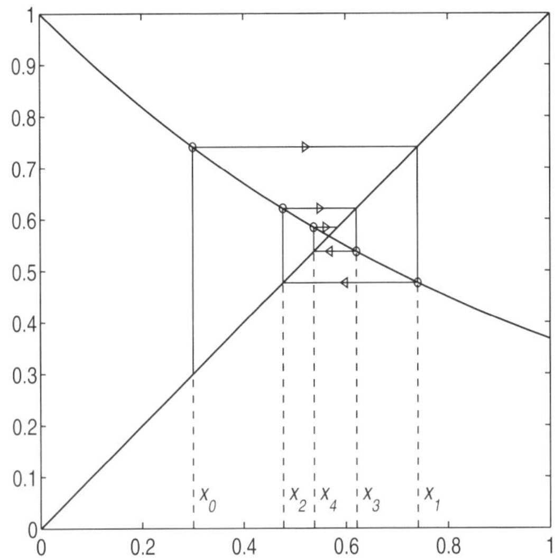

Figure 6.1.4. *The fixed-point iteration* $x_{k+1} = e^{-x_k}$, $x_0 = 0.3$.

important tool in theoretical analysis. Note that *the existence of a fixed point is not assumed a priori*.

Theorem 6.1.5 (Contraction Mapping Theorem).

Let x_0 *be a starting point, and consider the iteration* $x_{n+1} = \phi(x_n)$, $n = 1, 2, \ldots$. *Let*

$$J_\rho = \{x \mid |x - x_0| < \rho\}$$

be an open interval of length ρ *around* x_0 *and assume that* $x \to \phi(x)$ *is a* **contraction mapping**, *i.e., for arbitrary points s and t in* J_ρ

$$|\phi(s) - \phi(t)| \leq L|s - t|, \quad 0 \leq L < 1. \tag{6.1.16}$$

Then if

$$|x_0 - x_1| \leq (1 - L)\rho, \tag{6.1.17}$$

the equation $x = \phi(x)$ *has a unique solution* α *in the closed interval* $|x - x_0| \leq \rho$. *This solution can be obtained by the convergent iteration process* $x_{k+1} = \phi(x_k)$, $k = 0, 1, \ldots$. *We have the error estimate*

$$|x_k - \alpha| \leq |x_k - x_{k-1}|\frac{L}{1-L} \leq |x_1 - x_0|\frac{L^k}{1-L}. \tag{6.1.18}$$

Proof. We first prove the uniqueness. If there were two solutions x' and x'', we would get $x' - x'' = \phi(x') - \phi(x'')$ so that

$$|x' - x''| = |\phi(x') - \phi(x'')| \leq L|x' - x''|.$$

Since $L < 1$, it follows that $|x' - x''| = 0$, i.e., $x' = x''$.

By (6.1.17) we have $|x_1 - x_0| = |\phi(x_0) - x_0| \le (1 - L)\rho$, and hence $x_1 \in J_\rho$. We now use induction to prove that $x_n \in J_\rho$ for $n < j$, and that

$$|x_j - x_{j-1}| \le L^{j-1}(1 - L)\rho, \qquad |x_j - x_0| \le (1 - L^j)\rho.$$

We already know that these estimates are true for $j = 1$. Using the triangle inequality and (6.1.16) we get

$$\begin{aligned}|x_{j+1} - x_j| &= |\phi(x_j) - \phi(x_{j-1})| \le L|x_j - x_{j-1}| \le L^j(1 - L)\rho,\\ |x_{j+1} - x_0| &\le |x_{j+1} - x_j| + |x_j - x_0| \le L^j(1 - L)\rho + (1 - L^j)\rho\\ &= (1 - L^{j+1})\rho.\end{aligned}$$

This proves the induction step, and it follows that the sequence $\{x_k\}_{k=0}^{\infty}$ stays in J_ρ. We also have for $p > 0$

$$\begin{aligned}|x_{j+p} - x_j| &\le |x_{j+p} - x_{j+p-1}| + \cdots + |x_{j+1} - x_j|\\ &\le (L^{j+p-1} + \cdots + L^j)(1 - L)\rho \le L^j(1 - L^p)\rho \le L^j\rho,\end{aligned}$$

and hence $\lim_{j\to\infty} |x_{j+p} - x_j| = 0$. The sequence $\{x_k\}_{k=0}^{\infty}$ therefore is a Cauchy sequence, and since $\mathbf{R}^n$ is complete has a limit α. Since $x_j \in J_\rho$ for all j it follows that $\alpha \in \{x \mid |x - x_0| \le \rho\}$.

Finally, by (6.1.16) ϕ is continuous, and it follows that $\lim_{k\to\infty} \phi(x_k) = \phi(\alpha) = \alpha$. The demonstration of the error estimates (6.1.18) is left as an exercise for the reader. □

There is an analogue of Theorem 6.1.5 for complex functions $\phi(z)$ analytic in a circle $K = \{z \mid |z - \alpha| \le \rho\}$ containing the initial approximation z_0. Indeed, Theorem 6.1.5 holds in a much more general setting, where $\phi : S_r \to \mathcal{B}$, and $\mathcal{B}$ is a Banach space.[178] The proof goes through with simple modifications. In this form the theorem can be used to prove existence and uniqueness for initial value problems for ordinary differential equations.

An estimate of the error in x_n, which depends only on x_0, x_1 and the Lipschitz constant L, may be derived as follows. For arbitrary positive integers m and n we have

$$x_{m+n} - x_n = (x_{m+n} - x_{m+n-1}) + \cdots + (x_{n+2} - x_{n+1}) + (x_{n+1} - x_n).$$

From the Lipschitz condition we conclude that $|x_{i+1} - x_i| \le C^i|x_1 - x_0|$, and hence

$$|x_{m+n} - x_n| \le (C^{m-1} + \cdots + C + 1)|x_{n+1} - x_n|.$$

Summing the geometric series and letting $m \to \infty$ we obtain

$$|\alpha - x_n| \le \frac{1}{1 - C}|x_{n+1} - x_n| \le \frac{C^n}{1 - C}|x_1 - x_0|. \tag{6.1.19}$$

Note that if C is close to unity, then the error in x_n can be much larger than $|x_{n+1} - x_n|$. Clearly it is not always safe to terminate the iterations when $|x_{n+1} - x_n|$ is less than the required tolerance!

[178] Recall that a Banach space is a normed vector space which is complete, i.e., every Cauchy sequence converges to a point in $\mathcal{B}$; see Dieudonné [100].

For a linearly convergent fixed-point iteration the sequence $\{x_j - \alpha\}$ approximately forms a geometric series. Then, as seen in Sec. 3.4.2, a more rapidly convergent sequence $\{x'_j\}$ can be obtained by Aitken extrapolation,

$$x'_j = x_j - (\nabla x_j)^2/\nabla^2 x_j. \tag{6.1.20}$$

(Note that if the convergence is *not* linear, then the sequence $\{x'_n\}$ will usually converge *slower* than $\{x_n\}$.)

Example 6.1.7.

Let $\phi(x) = 1 - \frac{1}{2}x^2$, and consider the fixed-point iteration $x_{n+1} = \phi(x_n)$, with $x_0 = 0.8$, $n = 0, 1, 2, \ldots$. In this example the theoretical criterion (3.4.2) is satisfied, since one can show that $x_n \to a = \sqrt{3} - 1 \approx 0.73205$,

$$\nabla x_n / \nabla x_{n-1} \to \phi'(a) = -a, \qquad n \to \infty.$$

(See the discussion of iteration in Sec. 1.1.1.)

We obtain $x_1 = 0.68$, $x_2 = 0.7688$, and $x'_2 = 0.7688 - (0.0888)^2/0.2088 = 0.73103$. The error in x'_2 is only about 3% of the error in x_2. Iterated Aitken yields, with $u = 2^{-53} = 1.1 \cdot 10^{-16}$, the results $x'_2 = 0.73103\ldots$, $x''_4 = 0.73211$, etc., and the errors

$$e'_2 = -10 \cdot 10^{-3}, \quad e''_4 = 5.6 \cdot 10^{-5}, \quad e'''_6 = 7.8 \cdot 10^{-7}, \quad e^{(4)}_8 = -2.8 \cdot 10^{-9},$$

after 2, 4, 6, 8 evaluations of the function ϕ, respectively. We shall see (Problem 6.1.13) that the accelerated sequence may converge, even if the basic iteration $x_{n+1} = \phi(x_n)$ diverges.

When the basic sequence $\{x_n\}$, as in the above example, is produced by a convergent iterative process, one can apply Aitken acceleration in a different, usually more efficient way. This can be called **active Aitken acceleration**, since the result of an acceleration is actively used in the basic iterative process. We start as before by computing $x_1 = \phi(x_0)$, $x_2 = \phi(x_1)$ and apply (6.1.20) to compute x'_2. Next we continue the iterations from x'_2, i.e., compute $x_3 = \phi(x'_2)$, $x_4 = \phi(x_3)$. We can now extrapolate from x'_2, x_3, and x_4 to get x'_4, etc. It is easily verified that the sequence $z_n = x'_{2n}$ is generated by the fixed-point iteration

$$z_{n+1} = \psi(z_n), \qquad \psi(z) = z - \frac{(\phi(z) - z)^2}{(\phi(\phi(z)) - \phi(z)) - (\phi(z) - z)}. \tag{6.1.21}$$

This is repeated until some termination criterion is satisfied.

It can be shown that active Aitken extrapolation applied to the fixed point iteration $x_{n+1} = \phi(x_n)$ is equivalent to Steffensen's method applied to the equation $f(x) = x - \phi(x) = 0$; see (6.2.10).

6.1.5 Convergence Order and Efficiency

In general we will be given an equation $f(x) = 0$ to solve and want to construct a rapidly converging fixed-point iteration. Basic concepts to quantify the rate of convergence will now be introduced.

Definition 6.1.6.

Consider a sequence $\{x_n\}_0^\infty$ *with* $\lim_{n\to\infty} x_n = \alpha$, *and* $x_n \neq \alpha$ *for* $n < \infty$. *The sequence is said to have* **convergence order** *equal to* $q \geq 1$ *if for some constant* $0 < C < \infty$,

$$\lim_{n\to\infty} \frac{|x_{n+1} - \alpha|}{|x_n - \alpha|^q} = C, \tag{6.1.22}$$

where q *need not be an integer.* C *is called the* **asymptotic error constant**.

If $q = 1$, *then we require that* $C < 1$ *and then* $\{x_n\}$ *is said to converge* **linearly** *and* C *is the rate of linear convergence. For* $q = 2, 3$ *the convergence is called* quadratic, *and* cubic, *respectively.*

More precisely, the order q in Definition 6.1.6 is called the Q-order of convergence, where Q stands for quotient. The same definitions can be used also for vector-valued sequences. Then absolute values in (6.1.22) are replaced by a vector norm.

There are types of convergence that are not covered by the above definition of order. A sequence may converge more slowly than linear so that (6.1.22) holds with $q = 1$ and $C = 1$. Then convergence is called **sublinear**. If (6.1.22) holds with $q = 1$ and $C = 0$, then convergence is called **superlinear**.

Example 6.1.8.

Examples of sublinear, linear, and superlinear convergence are

$$x_n = 1/n, \quad x_n = 2^{-n}, \quad \text{and} \quad x_n = n^{-n},$$

respectively.

Alternative definitions of convergence order are considered by Ortega and Rheinboldt [278, Chap. 9] and Brent [44, Sec. 3.2]. For example, if

$$\lim_{n\to\infty} \inf (-\log |x_n - \alpha|)^{1/n} = q, \tag{6.1.23}$$

then q is called the **weak order** of convergence for x_n, since (6.1.22) implies (6.1.23), but not vice versa. For example, the sequence $x_n = \exp(-p^n)(2 + (-1)^n)$ converges to 0 with weak order p. But the limit in (6.1.22) does not exist if $q = p$, is zero if $q < p$, and is infinite if $q > p$.

Consider a fixed-point iteration $x_{n+1} = \phi(x_n)$. Assume that $\phi'(x)$ exists and is continuous in a neighborhood of α. It then follows from the proof of Theorem 6.1.5 that if $0 < |\phi'(\alpha)| < 1$ and x_0 is chosen sufficiently close to α, then the sequence x_n generated by $x_{n+1} = \phi(x_n)$ satisfies (6.1.22) with $q = 1$ and $C = |\phi'(\alpha)|$.

The number of accurate decimal places in the approximation x_n is $\delta_n = -\log_{10} |x_n - \alpha|$. Equation (6.1.22) implies that

$$\delta_{n+1} \approx q\delta_n - \log_{10} |C|.$$

Hence, for linear convergence ($q = 1$) as $n \to \infty$ each iteration gives a fixed (fractional) number of additional decimal places. For a method with convergence of order $q > 1$ each iteration increases the number of correct decimal places q-fold as $n \to \infty$. This shows that eventually a method with larger order of convergence will converge faster.

Example 6.1.9.

Consider a sequence x_n with quadratic convergence with $C = 1$. Set $\epsilon_n = |x_n - \alpha|$ and assume that $\epsilon_0 = 0.9$. From $\epsilon_{n+1} \le C\epsilon_n^2$, it follows that ϵ_n, for $n = 2, 3, \dots$, is bounded by

$$0.81, 0.66, 0.43, 0.19, 0.034, 0.0012, 1.4 \cdot 10^{-6}, 1.9 \cdot 10^{-12}, \dots,$$

respectively. For $n \ge 6$ the number of significant digits is approximately doubled at each iteration!

Definition 6.1.7.

Consider an iteration method with convergence order $q \ge 1$. If each iteration requires m units of work (usually the work involved in computing a function value or a value of one of its derivatives) then

$$E = q^{1/m} \tag{6.1.24}$$

is the **efficiency index** *of the iteration.*

The efficiency index gives a basis for comparing the efficiency of iterative methods of different order of superlinear convergence. Assuming that the cost of evaluating $f(x_n)$ and $f'(x_n)$ is two units the efficiency index for Newton's method is $E = 2^{1/2} = \sqrt{2}$. (Methods that converge linearly all have $E = 1$.)

The order of the fixed-point iteration $x_{n+1} = \phi(x_n)$ can be determined if $\phi(x)$ is sufficiently many times continuously differentiable in a neighborhood of α.

Theorem 6.1.8.

Assume that $\phi(x)$ is p times continuously differentiable. Then the iteration method $x_{n+1} = \phi(x_n)$ is of order p for the root α if and only if

$$\phi^{(j)}(\alpha) = 0, \quad j = 1 : p-1, \quad \phi^{(p)}(\alpha) \ne 0. \tag{6.1.25}$$

Proof. If (6.1.25) holds, then according to Taylor's theorem we have

$$x_{n+1} = \phi(x_n) = \alpha + \frac{1}{p!}\phi^{(p)}(\zeta_n)(x_n - \alpha)^p, \qquad \zeta_n \in \operatorname{int}(x_n, \alpha).$$

Hence for a convergent sequence x_n the error $\epsilon_n = x_n - \alpha$ satisfies

$$\lim_{n\to\infty} |\epsilon_{n+1}|/|\epsilon_n|^p = |\phi^{(p)}(\alpha)|/p! \ne 0,$$

and the order of convergence equals p. It also follows that if $\phi^{(j)}(\alpha) \ne 0$ for some j, $1 \le j < p$, or if $\phi^{(p)}(\alpha) = 0$, then the iteration cannot be of order p. $\square$

Example 6.1.10.

We remarked before that to compute a root α of $f(x) = 0$, we can use a fixed-point iteration with $\phi(x) = x - f(x)g(x)$, where $g(x)$ is an arbitrary function such that $g(\alpha) \ne 0$. We evaluate the derivative

$$\phi'(x) = 1 - f'(x)g(x) - f(x)g'(x).$$

To achieve quadratic convergence we take $g(x) = 1/f'(x)$. Assuming that $f'(\alpha) \neq 0$ we find, using $f(\alpha) = 0$, that $\phi'(\alpha) = 1 - f'(\alpha)g(\alpha) = 0$. Hence the iteration

$$x_{n+1} = x_n - f(x_n)/f'(x_n) \tag{6.1.26}$$

achieves quadratic convergence. This is Newton's method, which will be treated at length in Sec. 6.3.

Review Questions

6.1.1 What limits the final accuracy of a root computed by the bisection algorithm?

6.1.2 (a) Given a nonlinear scalar equation $f(x) = 0$ with a simple root α, how can a fixed-point iteration $x_{n+1} = \phi(x_n)$ that converges to α be constructed?

(b) Assume that a fixed point α exists for the mapping $x = \phi(x)$. Give sufficient conditions for convergence of the sequence generated by $x_{n+1} = \phi(x_n)$.

(c) How can the conditions in (b) be modified so that the *existence* of a fixed point can be proved?

6.1.3 (a) Define the concepts "order of convergence" and "asymptotic error constant" for a convergent sequence $\{x_n\}$ with $\lim_{n\to\infty} x_n = \alpha$.

(b) What is meant by sublinear and superlinear convergence? Give examples of sequences with sublinear and superlinear convergence.

6.1.4 (a) Define the efficiency index of a given iterative method of order p and asymptotic error constant $C \neq 0$.

(b) Determine the order of a new iterative method consisting of m consecutive steps of the method in (a). What is the order and error constant of this new method? Show that it has the same efficiency index as the first method.

6.1.5 (a) When can (passive) Aitken extrapolation be applied to speed up the convergence of a sequence?

(b) Describe the difference between active and passive Aitken extrapolation.

6.1.6 What two quantities determine the limiting accuracy of a simple root α to the equation $f(x) = 0$? Give an example of an ill-conditioned root.

6.1.7 Discuss the choice of termination criteria for iterative methods.

Problems and Computer Exercises

6.1.1 Use graphic representation to determine the zeros of the following functions to one correct decimal.

(a) $4\sin x + 1 - x$; (b) $1 - x - e^{-2x}$; (c) $(x+1)e^{x-1} - 1$;
(d) $x^4 - 4x^3 + 2x^2 - 8$; (e) $e^x + x^2 + x$; (f) $e^x - x^2 - 2x - 2$;
(g) $3x^2 + \tan x$.

6.1.2 Show analytically that the equation $xe^{-x} = \gamma$ has exactly two real roots when $\gamma < e^{-1}$.

6.1.3 Plot the functions $f(x) = \cosh x$ and $g(x) = 1/\cos x$ and deduce that the equation $\cosh x \cos x = 1$ has its smallest positive root in the interval $(3\pi/2, 2\pi)$. Determine this root using the bisection method.

6.1.4 The following equations all have a root in the interval $(0, 1.6)$ Determine these with an error less than 10^{-8} using the bisection method.
(a) $x \cos x = \ln x$; (b) $2x = e^{-x}$; (c) $e^{-2x} = 1 - x$.

6.1.5 Locate the real root of the equation

$$e^x(x-1) = e^{-x}(x+1)$$

by graphing both sides. Then compute the root with an error less than 10^{-8} using bisection. How many bisection steps are needed?

6.1.6 Let k be a given nonnegative number and consider the equation $\sin x = -k \cos x$. This equation has infinitely many roots. Separate the roots, i.e., partition the real axis into intervals which contain exactly one root.

6.1.7 The choice of m_k as the *arithmetic* mean of a_{k-1} and b_{k-1} in the bisection method minimizes the worst case maximum *absolute* error. If in the case that $ab > 0$ we take instead the *geometric* mean

$$m_k = \sqrt{a_k b_k},$$

then the worst case *relative* error is minimized. Do Example 6.1.2 using this variation of the bisection method.

6.1.8 In Example 6.1.6 three different fixed-point iterations were suggested for solving the equation $x + \ln x = 0$.
(a) Which of the formulas *can* be used?
(b) Which of the formulas *should* be used?
(c) Give an even better formula.

6.1.9 Investigate if and to what limit the iteration $x_{n+1} = 2^{x_n - 1}$ sequence converges for various choices of x_0.

6.1.10 (Wittmeyer-Koch)
(a) Show that the iteration $x_{n+1} = \phi(x_n)$, where $\phi(x) = x + (x-1)^2$, has a fixed point $\alpha = 1$, and $|\phi'(\alpha)| = 1$. Then verify that it converges by graphing the iteration for $x_0 = 0.6$.
(b) Show that for the iteration in (a) if $x_n = 1 - \epsilon$, then

$$\frac{|x_{n+1} - 1|}{|x_n - 1|} = 1 - \epsilon,$$

i.e., the asymptotic rate of convergence is sublinear.

6.1.11 (a) Consider the fixed-point iteration $x_{n+1} = \phi(x_n)$, where $\phi(x) = x + (x-1)^2$. Show that this has a fixed point for $\alpha = 1$ and that $\phi'(\alpha) = 1$.

(b) Show that the iteration in (a) is convergent for $x_0 < 1$.

6.1.12 Consider the iteration $x_{n+1} = 1 - \lambda x_n^2$. Illustrate graphically how the iteration behaves for $\lambda = 0.7, 0.9, 2$. (For $\lambda = 2$ the iteration is chaotic.)

6.1.13 Apply active Aitken acceleration to the iteration formula $s_{n+1} = \phi(s_n)$, where $\phi(s) = 1 - \frac{1}{2}s^2$, until either you have 10 correct decimals, or there are clear indications that the process is divergent.

(a) with $s_0 = 0.8$; (b) with $s_0 = -2.7$.

6.2 Methods Based on Interpolation

6.2.1 Method of False Position

Given two initial approximations $a_0 = a$ and $b_0 = b$ such that $f(a)f(b) < 0$, a nested sequence of intervals $(a_0, b_0) \supset (a_1, b_1) \supset (a_2, b_2) \supset \cdots$ such that $f(a_n)f(b_n) < 0$, $n = 0, 1, 2, \ldots$, can be generated as follows. Given (a_n, b_n), we take x_{n+1} to be the intersection of the x-axis and the secant through the point $(a_n, f(a_n))$ and $(b_n, f(b_n))$. Then by Newton's interpolation formula x_{n+1} satisfies

$$0 = f(a_n) + (x_{n+1} - a_n)\frac{f(a_n) - f(b_n)}{a_n - b_n},$$

giving

$$x_{n+1} = a_n - f(a_n)\frac{a_n - b_n}{f(a_n) - f(b_n)}, \quad n = 0, 1, 2, \ldots. \tag{6.2.1}$$

If $f(x_{n+1})f(a_n) > 0$, set $a_{n+1} = x_{n+1}$ and $b_{n+1} = b_n$; otherwise set $b_{n+1} = x_{n+1}$ and $a_{n+1} = a_n$. As for bisection, convergence to a root is guaranteed (in exact arithmetic) for a continuous function $f(x)$. This is the **false-position method** or, in Latin, **regula falsi**.[179]

Note that if $f(x)$ is linear we obtain the root in just one step, but sometimes the rate of convergence can be much slower than for bisection.

Suppose now that $f(x)$ is convex on $[a, b]$, $f(a) < 0$, and $f(b) > 0$, as in Figure 6.2.1. Then the secant through $x = a$ and $x = b$ will lie above the curve and hence intersect the x-axis to the left of α. The same is true for all subsequent secants and therefore the right endpoint b will be kept. The approximations $x_1, x_2, x_3, \ldots$ will all lie on the convex side of the curve and cannot go beyond the root α. A similar behavior, with monotone convergence and one of the points a or b fixed, will occur whenever $f''(x)$ exists and has constant sign on $[a, b]$.

[179]Regula falsi is a very old method that originated in fifth century Indian texts and was used in medieval Arabic mathematics. It got its current name from the Italian mathematician Leonardo Pisano, also known as Leonardo Fibonacci (ca 1170–1250). He is considered to be one of the most talented mathematicians of the Middle Ages.

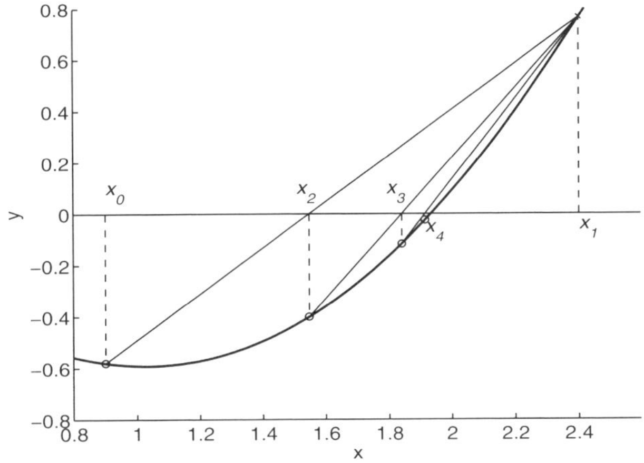

Figure 6.2.1. *The false-position method.*

Example 6.2.1.

We apply the method of false position to the $f(x) = (x/2)^2 - \sin x = 0$ from Example 6.1.2 with initial approximations $a_0 = 1.5$, $b_1 = 2$. We have $f(1.5) = -0.434995 < 0$ and $f(2.0) = +0.090703 > 0$, and successive iterates are as follows.

n	x_n	$f(x_n)$	h_n
1	**1.9**13731 221035	−0.026180060742	−0.019322989205
2	**1.933**054 210240	−0.000924399645	−0.000675397892
3	**1.9337**29 608132	−0.000031930094	−0.000023321005
4	**1.93375**2 929137	−0.000001102069	−0.000000804916
5	**1.933753** 734053		

Note that $f(x_n) < 0$ for all $n \geq 0$ and consequently $b_n = 2$ is fixed. In the limit convergence is linear with rate approximately equal to $C \approx 0.034$.

If f is twice continuously differentiable and $f''(\alpha) \neq 0$, then eventually an interval will be reached on which $f''(x)$ does not change sign. Then, as in the example above, one of the endpoints (say b) will be retained and $a_n = x_n$ in all future steps. By (6.2.1) the successive iterations are

$$x_{n+1} = x_n - f(x_n)\frac{x_n - b}{f(x_n) - f(b)}.$$

To determine the speed of convergence subtract α and divide by $\epsilon_n = x_n - \alpha$ to get

$$\frac{\epsilon_{n+1}}{\epsilon_n} = 1 - \frac{f(x_n)}{x_n - \alpha}\frac{x_n - b}{f(x_n) - f(b)}.$$

Since $\lim_{n\to\infty} x_n = \alpha$ and $f(\alpha) = 0$, it follows that

$$\lim_{n\to\infty} \frac{\epsilon_{n+1}}{\epsilon_n} = C = 1 - (b - \alpha)\frac{f'(\alpha)}{f(b)}, \tag{6.2.2}$$

which shows that convergence is linear. Convergence will be very slow if $f(x)$ is very flat near the root α, $f(b)$ is large, and α near b, since then $(b - \alpha)f'(\alpha) \ll f(b)$ and $C \approx 1$.

6.2.2 The Secant Method

A serious drawback to the method of false position is that ultimately one endpoint of the sequence of bracketing intervals will become fixed and therefore the length $(b_n - a_n)$ will not tend to zero. This can be avoided and convergence substantially improved by always using the secant through *the last two points*, $(x_{n-1}, f(x_{n-1}))$ and $(x_n, f(x_n))$. The next approximation x_{n+1} is determined as the abscissa of the point of intersection between this secant and the x-axis; see Figure 6.2.2.

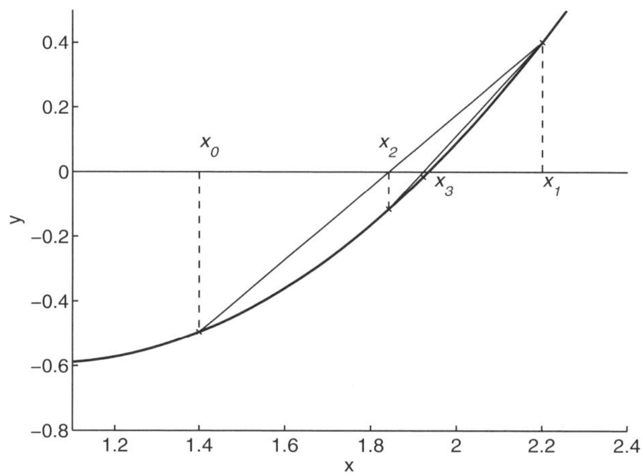

Figure 6.2.2. *The secant method.*

Given initial approximations $x_{-1} = a$ and $x_0 = b$, approximations $x_1, x_2, , x_3, \ldots$ are computed by

$$x_{n+1} = x_n + h_n, \quad h_n = -f(x_n)\frac{x_n - x_{n-1}}{f(x_n) - f(x_{n-1})}, \quad n \geq 1, \tag{6.2.3}$$

assuming that $f(x_n) \neq f(x_{n-1})$. This is the **secant method**, which historically predates Newton's method.

Notice that although regula falsi and the secant method require two initial approximations to the root, only *one function evaluation per step* is needed. The iteration, which is of the form $x_{n+1} = \phi(x_n; x_{n-1})$, is not a fixed-point iteration as defined in Sec. 6.1.4. Sometimes methods of this form, which use old information at points x_{n-k}, $k \geq 1$, are called fixed-point iterations with memory.

When the secant method converges $|x_n - x_{n-1}|$ will eventually become small. The quotient $(x_n - x_{n-1})/(f(x_n) - f(x_{n-1}))$ will then be determined with poor relative accuracy. If x_n and x_{n-1} are both very close to the root α and not bracketing α, then the resulting rounding error in x_{n+1} can become very large. Fortunately, from the error analysis below it follows that the approximations in general are such that $|x_n - x_{n-1}| \gg |x_n - \alpha|$ and the dominant contribution to the roundoff error in x_{n+1} comes from the error in $f(x_n)$. Note that (6.2.3) should *not* be "simplified" to the form

$$x_{n+1} = \frac{x_{n-1}f(x_n) - x_n f(x_{n-1})}{f(x_n) - f(x_{n-1})},$$

since this formula can give rise to severe difficulties with *cancellation* when $x_n \approx x_{n-1}$ and $f(x_n)f(x_{n-1}) > 0$. Even (6.2.3) is not always safe to use. We must take care to avoid overflow or division by zero. Without restriction we can assume that $|f(x_{n-1})| \geq |f(x_n)| > 0$ (otherwise renumber the two points). Then, s_n can be computed without risk of overflow from

$$x_{n+1} = x_n + \frac{s_n}{1 - s_n}(x_n - x_{n-1}), \qquad s_n = \frac{f(x_n)}{f(x_{n-1})}, \tag{6.2.4}$$

where the division with $1 - s_n$ is only carried out if $1 - s_n$ is large enough.

Example 6.2.2.

To illustrate the improved convergence of the secant method we consider once again the equation $f(x) = (x/2)^2 - \sin x = 0$ with initial approximations $x_0 = 1.5$, $x_1 = 2$. The result is shown in the following table.

n	x_n	$f(x_n)$	h_n
−1	1.5	−0.434994 986604	
0	2.0	+0.090702 573174	−0.086268 778965
1	**1.9**13731 221035	−0.026180 060742	+0.019322 989205
2	**1.933**054 210240	−0.000924 399645	+0.000707 253882
3	**1.9337**61 464122	+0.000010 180519	−0.000007 704220
4	**1.933753 7**59902	−0.000000 003867	+0.000000 002925
5	**1.933753 762827**		

Note that the approximations x_1 and x_2 are the same as for the false-position method, but here x_4 is correct to eight decimals and x_5 to twelve decimals. The rapid convergence is partly because $x_1 = 2$ is quite a good initial approximation. But note that although the root is bracketed by the initial intervals $[x_0, x_1]$ and $[x_1, x_2]$ it lies outside the interval $[x_2, x_3]$.

Assume that f is twice continuously differentiable. Then according to Newton's interpolation formula with error term (Theorem 4.3.1) we have

$$f(x) = f(x_n) + (x - x_n)[x_{n-1}, x_n]f + (x - x_{n-1})(x - x_n)\frac{f''(\zeta_n)}{2}, \tag{6.2.5}$$

where $\zeta_n \in \text{int}(x, x_{n-1}, x_n)$ and

$$[x_{n-1}, x_n]f = \frac{f(x_n) - f(x_{n-1})}{x_n - x_{n-1}}.$$

To derive an asymptotic formula for the secant method we put $x = \alpha$ in (6.2.5) and subtract the secant equation $0 = f(x_n) + (x_{n+1} - x_n)[x_{n-1}, x_n]f$. Since $f(\alpha) = 0$ we get

$$(\alpha - x_{n+1})[x_{n-1}, x_n]f + (\alpha - x_{n-1})(\alpha - x_n)f''(\zeta_n)/2 = 0,$$

where $\zeta_n \in \text{int}(\alpha, x_{n-1}, x_n)$. According to the mean value theorem, we have

$$[x_{n-1}, x_n]f = f'(\zeta_n'), \quad \zeta_n' \in \text{int}(x_{n-1}, x_n),$$

and it follows that

$$\epsilon_{n+1} = -\frac{1}{2}\frac{f''(\zeta_n)}{f'(\zeta_n')}\epsilon_n\epsilon_{n-1}. \tag{6.2.6}$$

Example 6.2.3.

The ratios $\epsilon_{n+1}/(\epsilon_n\epsilon_{n-1})$ in Example 6.2.2 are equal to 0.697, 0.527, 0.550, $n = 1 : 3$, which compares well with the limiting value 0.543 of $\frac{1}{2}f''(\alpha)/f'(\alpha)$.

From (6.2.6) it can be deduced that the secant method always converges from starting values x_0, x_1 sufficiently close to α. For this to be true it suffices that the first derivative $f'(x)$ is continuous, since then

$$\epsilon_{n+1} = \left(1 - \frac{f'(\xi_n)}{f'(\zeta_n)}\right)\epsilon_n, \quad \xi_n \in \text{int}(x_{n-1}, \alpha), \quad \zeta_n \in \text{int}(x_n, x_{n-1}).$$

But *in the secant method there is no guarantee that the computed sequence of approximations stays in the initial interval* $[x_0, x_1]$. Unlike the steady convergence of the bisection method things can go seriously wrong using the secant method! A remedy will be discussed in Sec. 6.2.4.

The following theorem gives the order of convergence for the secant method.

Theorem 6.2.1.

Suppose that $f(x)$ is twice continuously differentiable and that in a neighborhood I of the root α, containing $x_0, x_1, x_2, \ldots, x_n$, we have

$$\frac{1}{2}\left|\frac{f''(y)}{f'(x)}\right| \le M, \quad x, y \in I.$$

Denote by ϵ_n the error in the nth iterate of the secant iteration Let q be the unique positive root of the equation $\mu^2 - \mu - 1 = 0$ and set

$$K = \max\left(M|\epsilon_0|, (M|\epsilon_1|)^{1/q}\right). \tag{6.2.7}$$

Then it holds that

$$|\epsilon_n| \le \frac{1}{M}K^{q^n}, \quad n = 0, 1, 2, \ldots; \tag{6.2.8}$$

i.e., the secant iteration has convergence order equal to $q = (1 + \sqrt{5})/2 = 1.618\ldots$.

Proof. The proof is by induction. From the choice of K it follows that (6.2.8) is trivially true for $n = 0, 1$. Suppose that (6.2.8) holds for n and $n-1$. Then since $q^2 = q + 1$ it follows using the assumption and (6.2.6) that

$$|\epsilon_{n+1}| \le M|\epsilon_n|\,|\epsilon_{n-1}| \le \frac{1}{M}K^{q^n}K^{q^{n-1}} = \frac{1}{M}K^{q^n+q^{n-1}} = \frac{1}{M}K^{q^{n+1}}. \quad \square \tag{6.2.9}$$

To compare the efficiency of the secant method and Newton's method, which is quadratically convergent, we use the efficiency index introduced in Sec. 6.1.5. Assume that the work to compute $f'(x)$ is θ times the amount of work required to compute $f(x)$. Then, with the same amount of work we can perform $k(1+\theta)$ iterations with the secant method and k iterations with Newton's method. Equating the errors we get $(m\epsilon_0)^{2^k} = (m\epsilon_0)^{p^{k(1+\theta)}}$, where $q = 1.618\ldots$. Hence the errors are the same for both methods when $p^{k(1+\theta)} = 2^k$ or

$$(1+\theta)\log\left(\frac{1}{2}(1+\sqrt{5})\right) = \log 2,$$

which gives $\theta = 0.4404\ldots$. Thus, from this analysis we conclude that if $\theta > 0.44$, then the secant method is asymptotically more efficient than Newton's method.

In Example 6.2.2 we can observe that the error $\epsilon_n = x_n - \alpha_n$ changes sign at every third step. Hence, in this example,

$$\alpha \in \operatorname{int}(x_{n+1} - x_n), \qquad n = 0, 1, 3, 4, \ldots,$$

i.e., *the root α is bracketed by x_n and x_{n+1} except for every third step.* We shall show that this is no coincidence. Assume that $x_n \in (a, b)$, $n = 0, 1, 2, \ldots$, and that $f'(x) \neq 0$ and $f''(x)$ does not change sign in (a, b). Then from (6.2.6) it follows that the ratio

$$\frac{\epsilon_{n+1}}{\epsilon_n\epsilon_{n-1}}$$

will have constant sign for all n. Then if $\alpha \in \operatorname{int}(x_0, x_1)$ and $\epsilon_0\epsilon_1 < 0$, it follows that the sign of ϵ_n must change every third step according to one of the following two schemes (verify this!):

$$\cdots + - + + - + + - + + \cdots;$$
$$\cdots + - - + - - + - - + \cdots.$$

Hence convergence for the secant method, if it occurs, will take place in a waltz rhythm! This means that at every third step the last two iterates x_{n-1} and x_n will not always bracket the root.

6.2.3 Higher-Order Interpolation Methods

The secant method does not achieve quadratic convergence. Indeed, it can be shown that under very weak restrictions no iterative method using only one function evaluation per step can achieve this. In **Steffensen's method**

$$x_{n+1} = x_n - \frac{f(x_n)}{g(x_n)}, \qquad g(x_n) = \frac{f(x_n + f(x_n)) - f(x_n)}{f(x_n)}, \tag{6.2.10}$$

two function evaluations are used. This method can be viewed as a variant of the secant method, where the step size used to approximate the first derivative goes to zero as $f(x)$.

To show quadratic convergence we put $\beta_n = f(x_n)$ and expand $g(x_n)$ in a Taylor series about x_n; we get

$$g(x_n) = f'(x_n)\left(1 - \frac{1}{2}h_n f''(x_n) + O(\beta_n^2)\right),$$

where $h_n = -f(x_n)/f'(x_n)$ is the Newton correction. Thus,

$$x_{n+1} = x_n + h_n\left(1 + \frac{1}{2}h_n f''(x_n) + O(\beta_n^2)\right).$$

Using the error equation for Newton's method we get

$$\frac{\epsilon_{n+1}}{\epsilon_n^2} \to \frac{1}{2}\frac{f''(\alpha)}{f'(\alpha)}(1 + f'(\alpha)),$$

where $\epsilon_n = x_n - \alpha$. This shows that Steffenson's method is of second order.

Steffensen's method is of particular interest for solving *systems* of nonlinear equations. For a generalization to nonlinear operator equations on a Banach space, see [213].

In the secant method linear interpolation through (x_{n-1}, f_{n-1}) and (x_n, f_n) is used to determine the next approximation to the root. A natural generalization is to use an interpolating method of higher order. Let $x_{n-r}, \dots, x_{n-1}, x_n$ be $r+1$ distinct approximations and determine the (unique) polynomial $p(x)$ of degree r interpolating $(x_{n-j}, f(x_{n-j}))$, $j = 0 : r$. By Newton's interpolation formula (Sec. 4.2.1) the interpolating polynomial is

$$p(x) = f_n + [x_n, x_{n-1}]f \cdot (x - x_n) + \sum_{j=2}^{r}[x_n, x_{n-1}, \dots, x_{n-j}]f\ \Phi_j(x),$$

where

$$\Phi_j(x) = (x - x_n)(x - x_{n-1})\cdots(x - x_{n-j}).$$

The next approximation x_{n+1} is taken as the real root to the equation $p(x) = 0$ closest to x_n and x_{n-r} is deleted. Suppose the interpolation points lie in an interval J, which contains the root α and in which $f'(x) \neq 0$. It can be shown that if there is at least one interpolation point on each side of α, then $p(x) = 0$ has a real root in J. Further, the following formula for the error holds (Traub [354, pp. 67–75]):

$$\epsilon_{n+1} = -\frac{f^{(r+1)}(\zeta_n)}{(r+1)!p'(\eta_n)}\prod_{i=0}^{r}\epsilon_{n-i}, \tag{6.2.11}$$

where $\zeta_n \in \text{int}\,(\alpha, x_{n-1}, x_n)$ and $\eta_n \in \text{int}\,(\alpha, x_{n+1})$. (Recall that by int $(a, b, \dots, w)$ we denote the smallest interval that contains the points $a, b, \dots, w$.) In the special case $r = 2$ we get the quadratic equation

$$p(x) = f_n + (x - x_n)[x_n, x_{n-1}]f + (x - x_n)(x - x_{n-1})[x_n, x_{n-1}, x_{n-2}]f. \tag{6.2.12}$$

We assume that $[x_n, x_{n-1}, x_{n-2}]f \neq 0$ since otherwise the method degenerates into the secant method. Setting $h_n = (x - x_n)$ and writing $(x - x_{n-1}) = h_n + (x_n - x_{n-1})$, this equation becomes

$$h_n^2[x_n, x_{n-1}, x_{n-2}]f + \omega h_n + f_n = 0, \tag{6.2.13}$$

where

$$\omega = [x_n, x_{n-1}]f + (x_n - x_{n-1})[x_n, x_{n-1}, x_{n-2}]f. \tag{6.2.14}$$

The root closest to x_n corresponds to the root h_n of smallest absolute value to (6.2.13). To express this root in a numerically stable way the standard formula for the roots of a quadratic equation should be multiplied by its conjugate quantity (see Example 2.3.3). Using this formula we get

$$x_{n+1} = x_n + h_n, \qquad h_n = -\frac{2f_n}{\omega \pm \sqrt{\omega^2 - 4f_n\,[x_n, x_{n-1}, x_{n-2}]f}}, \tag{6.2.15}$$

where the sign in the denominator should be chosen so as to minimize $|h_n|$. This is the **Muller–Traub method**.

A drawback is that the equation (6.2.13) may not have a real root even if a real zero is being sought. On the other hand, this means that the Muller–Traub method has the useful property that complex roots may be found from real starting approximations.

By (6.2.11) it follows that

$$\epsilon_{n+1} = -\frac{f'''(\zeta_n)}{3!\,p'(\eta_n)}\epsilon_n\epsilon_{n-1}\epsilon_n. \tag{6.2.16}$$

It can be shown that the convergence order for the Muller–Traub method is at least $q = 1.839\ldots$, which equals the largest root of the equation $\mu^3 - \mu^2 - \mu - 1 = 0$ (cf. Theorem 6.2.1). Hence this method does not quite achieve quadratic convergence.

For $r > 2$ there are no useful explicit formulas for determining the zeros of the interpolating polynomial $p(x)$. Then we can proceed as follows. We write the equation $p(x) = 0$ in the form $x = x_n + F(x)$, where

$$F(x) \equiv \frac{-f_n - \sum_{j=2}^{r}[x_n, x_{n-1}, \ldots, x_{n-j}]f\ \Phi_j(x)}{[x_n, x_{n-1}]f}$$

(cf. Sec. 4.2.2). Then a fixed-point *iteration* can be used to solve for x. To get the first guess x^0 we ignore the sum (this means using the secant method) and then iterate, $x^i = x_n + F(x^{i-1})$, $i = 1, 2, \ldots$, until x^i and x^{i-1} are close enough.

Suppose that $x_{n-j} - x_n = O(h)$, $j = 1 : r$, where h is some small parameter in the context (usually some step size). Then $\Phi_j(x) = O(h^j)$, $\Phi_j'(x) = O(h^{j-1})$. The divided differences are $O(1)$, and we assume that $[x_n, x_{n-1}]f$ is bounded away from zero. Then the terms of the sum decrease like h^j. The convergence ratio $F'(x)$ is here approximately

$$\frac{\Phi_2'(x)[x_n, x_{n-1}, x_{n-2}]f}{[x_n, x_{n-1}]f} = O(h).$$

Thus, if h is small enough, the iterations converge rapidly.

A different way to extend the secant method is to use **inverse interpolation**. Assume that $y_n, y_{n-1}, \dots, y_{n-r}$ are distinct and let $q(y)$ be the unique polynomial in y interpolating the values $x_n, x_{n-1}, \dots, x_{n-r}$. Reversing the rule of x and y and using Newton's interpolation formula, this interpolating polynomial is

$$q(y) = x_n + [y_n, y_{n-1}]g \cdot (y - y_n) + \sum_{j=2}^{r} [y_n, y_{n-1}, \dots, y_{n-j}]f \, \Psi_j(y),$$

where $g(y_{n-j}) = x_{n-j}$, $j = 0 : r$,

$$\Psi_j(y) = (y - y_n)(y - y_{n-1}) \cdots (y - y_{n-j}).$$

The next approximation is then taken to be $x_{n+1} = q(0)$, i.e.,

$$x_{n+1} = x_n - y_n \, [y_n, y_{n-1}]g + \sum_{j=2}^{r} [y_n, y_{n-1}, \dots, y_{n-j}]g \, \Psi_j(0).$$

For $r = 1$ there is no difference between direct and inverse interpolation and we recover the secant method. For $r > 1$ inverse interpolation as a rule gives different results. Inverse interpolation has the advantage of not requiring the solution of a polynomial equation. (For other ways of avoiding this see Problems 6.2.3 and 6.2.4.) The case $r = 2$ corresponds to inverse quadratic interpolation:

$$x_{n+1} = x_n - y_n \, [y_n, y_{n-1}]g + y_n y_{n-1} \, [y_n, y_{n-1}, y_{n-2}]g. \tag{6.2.17}$$

Note that it has to be required that y_n, y_{n-1}, and y_{n-2} are distinct. This method has the same order of convergence as the Muller–Traub method.

Even if this is the case it is not always safe to compute x_{n+1} from (6.2.17). Care has to be taken in order to avoid overflow and possibly division by zero. If we assume that $0 \neq |y_n| \leq |y_{n-1}| \leq |y_{n-2}|$, then it is safe to compute

$$s_n = y_n / y_{n-1}, \quad s_{n-1} = y_{n-1} / y_{n-2}, \quad r_n = y_n / y_{n-2} = s_n s_{n-1}.$$

We can rewrite (6.2.17) in the form $x_{n+1} = x_n + p_n / q_n$, where

$$\begin{aligned} p_n &= s_n[(1 - r_n)(x_n - x_{n-1}) - s_{n-1}(s_{n-1} - r_n)(x_n - x_{n-2})], \\ q_n &= (1 - s_n)(1 - s_{n-1})(1 - r_n). \end{aligned}$$

The final division p_n / q_n is only carried out if the correction is sufficiently small.

6.2.4 A Robust Hybrid Method

Efficient and robust root finders can be constructed by combining the secant method (or some higher-order interpolation method) with bisection. A particularly elegant combination of bisection and the secant method was developed in the 1960s by van Wijngaarden, Dekker, and others at the Mathematical Center in Amsterdam; see [61]. It uses a rapidly convergent method for smooth functions and the slow but sure bisection method when necessary. This algorithm was taken up and improved by Brent [44]; see also [123, Section 7.2]. In contrast to Dekker's algorithm, Brent's new algorithm `zeroin` never converges much more slowly than bisection.

In each of the iterations of `zeroin` three approximations a, b, and c to the root are present. Initially we are given an interval $[a, b]$ such that $f(a)$ and $f(b)$ have opposite sign, and set $c = a$. In the following steps a, b, and c are arranged so that

- $f(a)$ and $f(b)$ have opposite sign, and $|f(b)| \leq |f(a)|$.
- b is the latest iteration.
- c is the value of b in the previous step.

At all times a and b bracket the zero and the length $|b - a|$ of the interval shrinks in each iteration.

If $c = a$ a secant step is taken; otherwise inverse quadratic interpolation is used. If the computed step gives an approximation in $[a, b]$ (and not too close to one of the endpoints) take it; otherwise a bisection step is taken. The iterations are terminated if $f(b) = 0$, or

$$|b - a| \leq \text{tol} + 4u|b|,$$

where tol is a user tolerance and u the unit roundoff. The midpoint $r = b + 0.5(a - b)$ is then returned as the root.

Brent claims that his version of `zeroin` will always converge, even in floating-point arithmetic. Further, if f has a continuous second derivative the method will eventually converge at least as fast as the secant method. In this case typically about ten function evaluations are needed. He never found a function requiring more than three times the number of evaluations needed for bisection. On average, using inverse quadratic interpolation when possible saves 0.5 function evaluations over using only the secant method.

The MATLAB function `fzero`, which finds a zero near a given approximation x_0, is based on `zeroin`. A discussion of a slightly simplified version of `fzero` is given in Moler [268, Chap. 4.7].

Review Questions

6.2.1 Sketch a function $f(x)$ with a root in (a, b) such that regula falsi converges very slowly.

6.2.2 Outline how the secant method can be safeguarded by combining it with the bisection method.

6.2.3 What property should the function $f(x)$ have to be unimodal on the interval $[a, b]$?

6.2.4 Discuss root finding methods based on quadratic interpolation (the Muller–Traub method) and inverse quadratic interpolation. What are the merits of these two approaches?

Problems and Computer Exercises

6.2.1 Use the secant method to determine the roots of the following equations to six correct decimals.

(a) $2x = e^{-x}$; (b) $\tan x + \cosh x = 0$.

6.2.2 Assume that we have $f_n f_{n-1} < 0$, and have computed x_{n+1}. If $f_{n+1} f_n < 0$, then in the next step we compute x_{n+2} by a secant step; otherwise we use a line through (x_{n+1}, f_{n+1}) and $(x_{n-1}, \theta f_{n-1})$, where $0 < \theta < 1$. Clearly, $\theta = 1$ corresponds to a step with the method of false position and will usually give $f_{n+2} f_{n+1} > 0$. On the other hand, $\theta = 0$ gives $x_{n+1} = x_n$, and thus $f_{n+1} f_n < 0$. Hence a suitable choice of θ will always give $f_{n+2} f_{n+1} < 0$.

Show that in a modified step $\epsilon_{n+1} \approx -\epsilon_n$ when $\theta = 0.5$. Deduce that the resulting algorithm gives cubic convergence with three function evaluations and hence has efficiency index $E = 3^{1/3} = 1.4422\ldots$.[180]

6.2.3 Another modification of the secant method can be derived by estimating $f'(x_n)$ in Newton's method by quadratic interpolation through the points x_n, x_{n-1}, x_{n-2}. Show that the resulting method can be written $x_{n+1} = x_n - f(x_n)/\omega$, where

$$\omega = f[x_n, x_{n-1}] + (x_n - x_{n-1}) f[x_n, x_{n-1}, x_{n-2}].$$

6.2.4 The Muller–Traub method uses three points to determine the coefficient of an interpolating parabola. The same points can also be interpolated by a rational function of the form

$$g(x) = \frac{x - a}{bx + c}.$$

An iterative method is derived by taking x_{n+1} equal to the root a of $g(x) = 0$.

(a) Show that this is equivalent to calculating x_{n+1} from the "modified secant formula":

$$x_{n+1} = x_n - f_n \frac{x_n - x_{n-2}}{f_n - \tilde{f}_{n-2}}, \qquad \tilde{f}_{n-2} = f_{n-2} \frac{f[x_n, x_{n-1}]}{f[x_{n-1}, x_{n-2}]}.$$

Hint: Use a theorem in projective geometry, according to which the cross ratio of any four values of x is equal to the cross ratio of the corresponding values of $g(x)$ (see Householder [204, p. 159]). Hence

$$\frac{(0 - f_n)/(0 - f_{n-2})}{(y_{n-1} - f_n)/(y_{n-1} - f_{n-2})} = \frac{(x_{n+1} - x_n)/(x_{n+1} - x_{n-2})}{(x_{n-1} - x_n)/(x_{n-1} - x_{n-2})}.$$

(b) Use the result in (a) to show that $x_{n-1} \in \operatorname{int}(x_{n-2}, x_n)$ if

$$\operatorname{sign}(y_n) = -\operatorname{sign}(y_{n-2}), \qquad \operatorname{sign}(y[x_n, x_{n-1}]) = \operatorname{sign}(y[x_{n-1}, x_{n-2}]).$$

[180] The resulting modified rule of false position is often called, after its origin, the **Illinois method**. It is due originally to the staff of the computer center at the University of Illinois in the early 1950s.

6.2.5 The result in Problem 6.2.4 suggests that the Illinois method in Problem 6.2.2 should be modified by taking

$$\beta = f[x_{n+1}, x_n]/f[x_n, x_{n-1}], \qquad \theta = \begin{cases} \beta & \text{if } \beta > 0, \\ \frac{1}{2} & \text{if } \beta \le 0. \end{cases}$$

Implement this modified method. Compare it with the unmodified Illinois method and with the safeguarded secant algorithm. As test equations use the following:

(a) A curve with one inflection point on $[0, 1]$:
$f(x) = x^2 - (1-x)^n$, $a = 25$, $b = 1$, $n = 2, 5, 10$.

(b) A family of curves which lie increasingly close to the x-axis for large n:
$f(x) = e^{-nx}(x-1) + x^n$, $a = 0.25$, $b = 1$, $n = 5, 10, 15$.

(c) A family of curves with the y-axis asymptotic:
$f(x) = (nx-1)/((n-1)x)$, $a = 0.01$, $b = 1$, $n = 2, 5, 10$.

6.3 Methods Using Derivatives

6.3.1 Newton's Method

In Newton's method for solving an equation $f(x) = 0$ the curve $y = f(x)$ is approximated by its tangent at the point $(x_n, f(x_n))$, where x_n is the current approximation to the root. Assuming that $f'(x_n) \neq 0$, the next approximation x_{n+1} is then determined as the abscissa of the point of intersection of the tangent with the x-axis (see Figure 1.1.3).

When $f'(x)$ is available **Newton's method** is usually the method of choice. It is equivalent to replacing the equation $f(x) = 0$ by

$$f(x_n) + (x - x_n)f'(x_n) = 0, \tag{6.3.1}$$

which is obtained by taking the linear part of the Taylor expansion of $f(x)$ at x_n. Hence if $f'(x_n) \neq 0$, then

$$x_{n+1} = x_n + h_n, \qquad h_n = -f(x_n)/f'(x_n). \tag{6.3.2}$$

Clearly Newton's method can also be viewed as the limit of the secant method when the two interpolation points coalesce.

Example 6.3.1.

We want to compute the unique positive root of the equation $f(x) = (x/2)^2 - \sin x = 0$ (cf. Example 6.2.2) for which $f'(x) = x/2 - \cos x$. The following Newton iterates, starting from $x_0 = 1.8$, are given in the table below (correct digits in x_n shown in bold).

n	x_n	$f(x_n)$	$f'(x_n)$	h_n
0	**1.**8	−0.163847 630878	1.127202 094693	−0.145357 812631
1	**1.9**45357 812631	0.015436 106659	1.338543 359427	0.011532 018406
2	**1.933**825 794225	0.000095 223283	1.322020 778469	0.000072 028582
3	**1.933753 7**65643	0.000000 003722	1.3219174 29113	0.000000 002816
4	**1.933753 762827**			

The number of correct digits approximately doubles in each iteration until the limiting precision is reached. Although the initial approximation is not very good, already x_4 is correct to 12 decimals!

If the iterations are broken off when $|h_n| < \delta$ it can be shown (see the error analysis below) that the truncation error is less than δ, provided that $|Kh_n| \leq 1/2$, where K is an upper bound for $|f''/f'|$ in the neighborhood of the root. This restriction is seldom of practical importance. But rounding errors made in computing h_n must also be taken into account.

Note that when the root is approached, the *relative* precision in the computed values of $f(x_n)$ usually decreases. Since $f'(x_n)$ is only used for computing h_n it need not be computed to much greater *relative precision* than $f(x_n)$. In the above example we could have used $f'(x_2)$ instead of $f'(x_n)$ for $n > 2$ without much effect on the accuracy. Such a simplification is of great importance when Newton's method is used on *systems* of nonlinear equations.

We now consider the **local** convergence of Newton's method, i.e., the convergence in a neighborhood of a simple root α.

Theorem 6.3.1.

Assume that α is a simple root of the equation $f(x) = 0$, i.e., $f'(\alpha) \neq 0$. If f' exists and is continuous in a neighborhood of α, then the convergence order of Newton's method is at least equal to two.

Proof. A Taylor expansion of f yields

$$0 = f(\alpha) = f(x_n) + (\alpha - x_n)f'(x_n) + \frac{1}{2}(\alpha - x_n)^2 f''(\zeta_n), \qquad \zeta_n \in \operatorname{int}(x_n, \alpha).$$

Subtracting $f(x_n) + (x_{n+1} - x_n)f'(x_n) = 0$ and solving for $\epsilon_{n+1} = x_{n+1} - \alpha$ we get

$$\epsilon_{n+1} = \frac{1}{2}\epsilon_n^2 \frac{f''(\zeta_n)}{f'(x_n)}, \qquad \zeta_n \in \operatorname{int}(x_n, \alpha). \tag{6.3.3}$$

Provided that $f'(\alpha) \neq 0$, it follows that (6.1.22) is satisfied with $p = 2$ and the asymptotic error constant is

$$C = \frac{1}{2}\left|\frac{f''(\alpha)}{f'(\alpha)}\right|. \tag{6.3.4}$$

If $f''(\alpha) \neq 0$, then $C > 0$ and the rate of convergence is quadratic. $\square$

The following classical theorem gives rigorous conditions for the quadratic convergence of Newton's method. In particular, *it is not necessary to assume the existence of a solution.*

Theorem 6.3.2 (*Ostrowski* [279, Theorem 7.1]).

Let $f(x)$ be a real function, $f(x_0)f'(x_0) \neq 0$, and put $h_0 = f(x_0)/f'(x_0)$. Assume that $f''(x)$ exists in $J_0 = \operatorname{int}[x_0, x_0 + 2h_0]$, and that

$$2M|h_0| \leq |f'(x_0)|, \qquad M = \sup_{x \in J_0} |f''(x)|. \tag{6.3.5}$$

Let the sequence x_k, $k = 1, 2, \dots$, be generated by Newton's method:

$$x_{k+1} = x_k - f(x_k)/f'(x_k), \quad k = 0, 1, \dots. \tag{6.3.6}$$

Then $x_k \in J_0$, and we have $\lim_{k\to\infty} x_k = \alpha$, where α is the only zero of $f(x)$ in J_0. Unless $\alpha = x_0 + 2h_0$, α is a simple zero.[181] *Further, it holds that*

$$|x_{k+1} - x_k| \le \frac{M}{2|f'(x_k)|}|x_k - x_{k-1}|^2, \quad k = 1, 2, \dots, \tag{6.3.7}$$

$$|\alpha - x_{k+1}| \le \frac{M}{2|f'(x_k)|}|x_k - x_{k-1}|^2, \quad k = 1, 2, \dots. \tag{6.3.8}$$

Proof. We have by (6.3.5) that $|f'(x) - f'(x_0)| \le |x - x_0|M$, and

$$|f'(x_1) - f'(x_0)| \le |h_0|M \le \frac{1}{2}|f'(x_0)|.$$

It follows that

$$|f'(x_1)| \ge |f'(x_0)| - |f'(x_1) - f'(x_0)| \ge \tfrac{1}{2}|f'(x_0)|. \tag{6.3.9}$$

Integrating by parts we have

$$\int_{x_0}^{x_1} (x_1 - x)f''(x)\,dx = -(x_1 - x_0)f'(x_0) + f(x_1) - f(x_0) = f(x_1).$$

Introducing a new variable by $x = x_0 + th$, this becomes

$$f(x_1) = h_0^2 \int_0^1 (1 - t)f''(x_0 + th)\,dt,$$

and it now follows that

$$|f(x_1)| = |h_0|^2 \int_0^1 (1 - t)|f''(x_0 + th)|\,dt \le M|h_0|^2 \int_0^1 (1 - t)\,dt = \tfrac{1}{2}M|h_0|^2. \tag{6.3.10}$$

Using (6.3.9) and (6.3.10) gives

$$|h_1| \le \frac{M|h_0|^2}{|f'(x_0)|}, \tag{6.3.11}$$

and applying (6.3.10) gives

$$\frac{2M|h_1|}{|f'(x_1)|} \le \frac{2(M|h_0|)^2}{|f'(x_0)||f'(x_1)|} \le \left(\frac{(2M|h_0|)}{|f'(x_0)|}\right)^2.$$

By (6.3.5) the expression in parenthesis is ≤ 1 and hence

$$2M|h_1| \le |f'(x_1)|, \quad |h_1| \le \tfrac{1}{2}|h_0|. \tag{6.3.12}$$

[181] It can be proved that if $\alpha = x_0 + 2h_0$, then $f(x)$ is a quadratic polynomial with the double root α; see [279, Chapter 40].

This shows that the point x_2 will not get beyond the distance $\frac{1}{2}|h_0|$ from x_1 and will remain in J_0, and $\text{int}\,[x_1, x_1 + 2h_1] \in J_0$.

We can now use induction to prove that the intervals $J_k = \text{int}\,[x_k, x_k + 2h_k]$, $k = 1, 2, \dots$, lie in J_0, that $J_{k+1} \in J_k$, and that the length of J_{k+1} is at most half of the length of J_k. Since J_0 is closed it follows that

$$\lim_{k\to\infty} x_k = \alpha \in J_0.$$

To show that α is a root of $f(x)$ we note that from (6.3.6) it follows that $f'(x_k)(x_k - x_{k+1}) = f(x_k)$. Taking the limit it follows that $\lim_{k\to\infty} f(x_k) = f(\alpha) = 0$.

By (6.3.5) we have for all $x \in J_0$

$$|f'(x) - f'(x_0)| \le |x - x_0|M \le 2M|h_0|.$$

Assume that $|x - x_0| < |h_0|$, i.e., x_0 lies *in the interior* of J_0. Then

$$|f'(x) - f'(x_0)| < 2M|h_0| \le |f'(x_0)|,$$

so that $f'(x) \ne 0$ for all x in the interior of J_0. Therefore, α must be a simple root if it lies in the interior of J_0. Further, since $f'(x)$ does not vanish in J_0, $f(x)$ is strictly monotonic in J_0 and thus has only one root.

The assertion (6.3.8) is equivalent to

$$|h_k| \le \frac{M|h_{k-1}|^2}{2|f'(x_k)|}.$$

Since our starting assumption (6.3.5) is true for all $k \ge 0$, we have $|f(x_k)| \le \frac{1}{2}M|h_{k-1}|^2$, and hence the assertion follows using (6.3.6). To prove (6.3.8) we notice that α lies in an interval with center x_{k+1} and radius $|h_k|$, and use (6.3.8). □

Note that the relation (6.3.8) between the errors holds only as long as the roundoff errors in the calculations can be ignored. As pointed out in Sec. 6.1.3, the accuracy which can be achieved in calculating the root is always limited by the accuracy in the computed values of $f(x)$.

So far we have assumed that α is a simple root. Suppose now that α is a root of multiplicity $q > 1$. Then by Taylor's formula we have (cf. (6.1.9))

$$f'(x) = \frac{1}{(q-1)!}(x - \alpha)^{q-1} f^{(q)}(\xi'), \qquad \xi' \in \text{int}(x, \alpha).$$

It follows that if x_n is close to α, then the Newton correction will satisfy

$$h_n = \frac{f(x_n)}{f'(x_n)} \approx \frac{1}{q}(x_n - \alpha) = \frac{1}{q}\epsilon_n.$$

For the corresponding errors we have

$$\epsilon_{n+1} = \epsilon_n - \epsilon_n/q = (1 - 1/q)\epsilon_n,$$

which shows that for a root of multiplicity $q > 1$ Newton's method only converges *linearly* with rate $C = 1 - 1/q$. (The same is true of other methods which have quadratic or higher rates of convergence for simple roots.) For $q > 2$ this is much slower even than for the bisection method! For a root of multiplicity $q > 1$ convergence is linear with rate $1 - 1/q$. For $q = 20$ it will take 45 iterations to gain one more decimal digit.

Note also that when $x_n \to \alpha$, $f(x_n) \to 0$ and $f'(x_n) \to 0$. Therefore, rounding errors may seriously affect the Newton correction when evaluated close to α, This clearly is related to the lower limiting accuracy for multiple roots as given by (6.1.10).

When the multiplicity q of a root is known a priori the modified Newton's method

$$x_{n+1} = x_n - q\frac{f(x_n)}{f'(x_n)} \tag{6.3.13}$$

is easily shown to have quadratic convergence. For a root of unknown multiplicity we can use the following observation. From (6.1.9) it follows that if the equation $f(x) = 0$ has a multiple root at $x = \alpha$, then α is *a simple root* of the equation

$$u(x) = 0, \quad u(x) = f(x)/f'(x). \tag{6.3.14}$$

Hence, if Newton's method is applied to this equation it will have a quadratic rate of convergence independent of the multiplicity of α as a root to $f(x) = 0$. The Newton iteration for $u(x) = 0$ becomes

$$x_{n+1} = x_n - \frac{f(x_n)}{f'(x_n) - f(x_n)f''(x_n)/f'(x_n)}, \tag{6.3.15}$$

and thus requires the evaluation also of $f''(x_n)$.

Under certain regularity assumptions Newton's method is **locally convergent** from a sufficiently good initial approximation. But in general Newton's method is not **globally convergent**, i.e., it does not converge from an arbitrary starting point. This is not surprising since it is based on a series of local approximations.

It is easy to construct examples where Newton's method converges very slowly or not at all.

Example 6.3.2.

The equation $f(x) = \sin x = 0$ has exactly one root $\alpha = 0$ in the interval $|x| < \pi/2$. Newton's method becomes

$$x_{n+1} = x_n - \tan x_n, \quad n = 0, 1, 2, \dots.$$

If we choose the initial value $x_0 = z$ such that $\tan z = 2z$, then $x_1 = -x_0$, $x_2 = -x_1 = x_0$. Hence the successive approximations show a cyclic behavior.

Newton's method will converge for any starting value such that $|x_0| < z$. The critical value can be shown to be $x_0 = 1.16556\dots$.

Example 6.3.3.

In Example 5.2.5 we encountered the function $u(x)$ defined for $x \geq 0$ by

$$x = u \ln u. \tag{6.3.16}$$

The function $u(x)$ is related to Lambert's W-function $w(x)$ (see Problem 3.1.12 (d) and [370]), which is the inverse of the function $x(w) = we^w$. Clearly $u(x) = e^{w(x)}$. It can be shown that $u(x)$ is a smooth and slowly varying function x.

In Figure 6.3.1 we show a plot of the function $x = u \ln u$. For the inverse function $u(x)$ we have $u^p r(x) = 1/(1 + lnu(x))$. Clearly $u(0) = 1$, $u'(0) = 1$, and for $x > 0$ $u'(x)$, is positive and decreasing, while $u''(x)$ is negative and decreasing in absolute value.

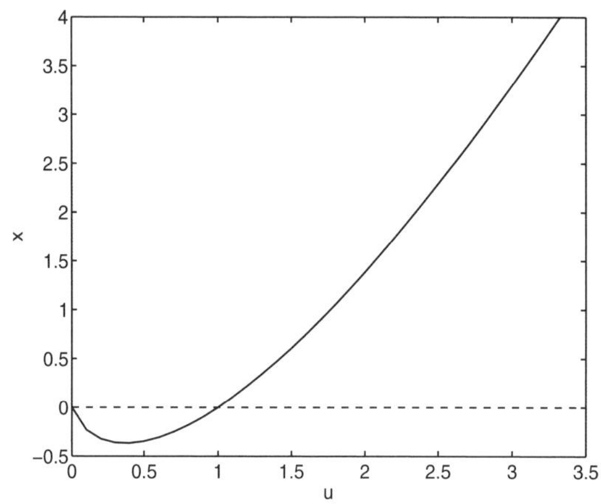

Figure 6.3.1. *The function* $x = u \ln u$.

To compute $u(x)$ to high accuracy for a given value of $x > 0$, we can use Newton's method to solve the equation

$$f(u) = u \ln u - x = 0.$$

This takes the form

$$u_{k+1} = \frac{u_k + x}{1 + \ln u_k}, \qquad k = 0, 1, 2, \dots.$$

For $x > 10$, we use as starting value u_0 the asymptotic approximation $u(x) \sim x/\ln x = u_0$, $x \to \infty$. For $0 < x \leq 10$ the approximation

$$u_0 = 1.0125 + 0.8577x - 0.129013x^2 + 0.208645x^3 - 0.00176148x^4 + 0.000057941x^5,$$

derived by Gautschi from a truncated Chebyshev expansion of $u(x)$, has a maximum error of about 1%. It is left to Problem 6.3.6 to investigate how efficient the outlined Newton method is for computing $u(x)$.

Global Convergence

In some simple cases the global convergence of Newton's method may be easy to verify. Two examples are given in the following theorems.

Theorem 6.3.3.

Suppose that $f'(x)f''(x) \neq 0$ *in an interval* $[a, b]$, *where* $f''(x)$ *is continuous and* $f(a)f(b) < 0$. *Then if*

$$\left|\frac{f(a)}{f'(a)}\right| < b - a, \qquad \left|\frac{f(b)}{f'(b)}\right| < b - a,$$

Newton's method converges from an arbitrary $x_0 \in [a, b]$.

Proof. The theorem follows easily by inspecting Figure 6.3.2. □

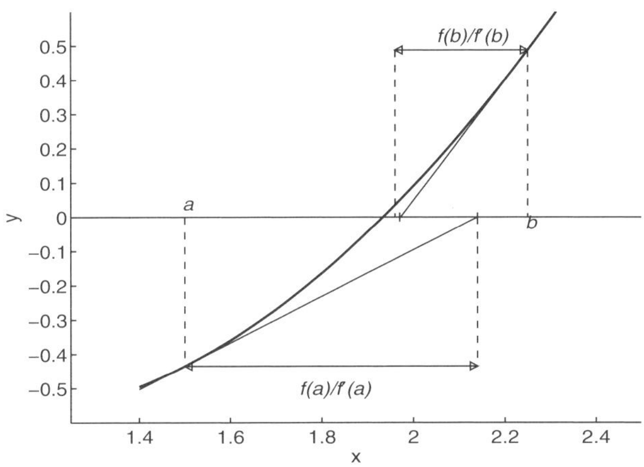

Figure 6.3.2. *A situation where Newton's method converges from any* $x_0 \in [a, b]$.

Lemma 6.3.4.

Let $[a, b]$ *be an interval such that* $f(a)f(b) < 0$. *Assume that the so-called* **Fourier conditions** *are satisfied, i.e.,*

$$f'(x)f''(x) \neq 0, \quad x \in [a, b], \tag{6.3.17}$$

with $f''(x)$ *continuous and* $f(x_0)f''(x_0) > 0$, *for* $x_0 \in [a, b]$. *Then the sequence* $\{x_0, x_1, x_2, \dots\}$ *generated by Newton's method converges monotonically to a root* $\alpha \in [a, b]$.

Proof. We can assume that $f''(x) > 0$; otherwise consider the equation $-f(x) = 0$. Assume first that $f'(x) < 0$ in the interval. Since by assumption $f(x_0) \geq 0$, this corresponds to the situation in Figure 6.3.2, with $b = x_0 > \alpha$. Clearly, it holds that $x_1 > x_0$ and since the curve lies to the left of the tangent in x_0 we also have $x_1 > \alpha$. The case when $f'(x) < 0$ can be treated similarly. The theorem now follows by induction. □

Newton's method can be safeguarded by taking a bisection step whenever a Newton step "fails" in some sense. Assume that initially $a < b$, $f(a)f(b) < 0$, and x is equal to

a or b. At each step a new approximation x' is computed and a, b are updated to a', b' as follows:

- If the Newton iterate $x' = x - f(x)/f'(x)$ lies in (a, b), then accept x'; otherwise take a bisection step, i.e., set $x' = (a + b)/2$.
- Set either $a' = x$, $b' = b$ or $a' = a$, $b' = x$, where the choice is made so that $f(a')f(b') \le 0$.

This ensures that at each step the interval $[a', b']$ contains a root.

When checking if $z \in (a, b)$, it is important to avoid division by $f'(x)$, since this may cause overflow or division by zero. Hence, we note $z \in (a, b)$ if and only if

$$b - z = b - x + f(x)/f'(x) > 0 \quad \text{and} \quad z - a = x - a - f(x)/f'(x) > 0.$$

If $f'(x) > 0$ these two inequalities are equivalent to

$$(x - b)f'(x) < f(x) < (x - a)f'(x).$$

The case when $f'(x) < 0$ is analyzed similarly, giving

$$(x - a)f'(x) < f(x) < (x - b)f'(x).$$

In either case only one of the inequalities will be nontrivial depending on whether $f(x) > 0$.

6.3.2 Newton's Method for Complex Roots

Newton's method is based on approximating f with the linear part of its Taylor expansion. Taylor's theorem is valid for a complex function $f(z)$ around a point of analyticity (see Sec. 3.1.2). Thus Newton's method applies also to an equation $f(z) = 0$, where $f(z)$ is a complex function, analytic in a neighborhood of a root α. An important example is when f is a polynomial; see Sec. 6.5.

The geometry of the complex Newton iteration has been studied by Yao and Ben-Israel [387]. Let $z = x + iy$, $f(z) = u(x, y) + iv(x, y)$, and consider the absolute value of $f(z)$:

$$\phi(x, y) = |f(x + iy)| = \sqrt{u(x, y)^2 + v(x, y)^2}.$$

This is a differentiable function as a function of (x, y), except where $f(z) = 0$. The gradient of $\phi(x, y)$ is

$$\text{grad}\, \phi = (\phi_x, \phi_y) = \frac{1}{\phi}\left(uu_x + vv_x, uu_y + vv_y\right), \qquad (6.3.18)$$

where $u_x = \partial u/\partial x$, $u_y = \partial u/\partial y$, etc. Using the Cauchy–Riemann equations $u_x = v_y$, $u_y = -v_x$, we calculate (see Henrici [196, Sec. 6.1.4])

$$\frac{f(z)}{f'(z)} = \frac{u + iv}{u_x + iv_x} = \frac{(uu_x + vv_x) + i(uu_y + vv_y)}{u_x^2 + v_x^2}.$$

A comparison with (6.3.18) shows that the Newton step

$$z_{k+1} = z_k - f(z_k)/f'(z_k), \quad k = 0, 1, \dots, \qquad (6.3.19)$$

is in the direction of the negative gradient of $|f(z_k)|$, i.e., in the direction of strongest decrease of $|f(z)|$.

Theorem 6.3.5.

Let the function $f(z) = f(x+iy)$ *be analytic and* $z_k = x_k + iy_k$ *be a point such that* $f(z_k)f'(z_k) \neq 0$. *Let* z_{k+1} *be the next iterate of Newton's method* (6.3.19). *Then* $z_{k+1} - z_k$ *is in the direction of the negative gradient of* $\phi(x, y) = |f(x+iy)|$ *and therefore orthogonal to the level set of* $|f|$ *at* (x_k, y_k). *If* T_k *is the tangent plane of* $\phi(x, y)$ *at* (x_k, y_k) *and* L_k *is the line of intersection of* T_k *with the* (x, y)*-plane, then* (x_{k+1}, y_{k+1}) *is the point on* L_k *closest to* (x_k, y_k).

Proof. See [387]. □

Newton's method is very efficient if started from an initial approximation sufficiently close to a simple zero. If this is not the case Newton's method may converge slowly or even diverge. In general, there is no guarantee that z_{n+1} is closer to the root than z_n, and if $f'(z_n) = 0$ the next iterate is not even defined.

It is straightforward to generalize the results in Theorem 6.3.2 to the complex case. In the case of a complex function $f(x)$ of a complex variable the interval $I_0 = \text{int}\,[x_0, x_0 + 2h_0]$ can be replaced by a disk $K_0 : |z - z_1| \leq |h_0|$.

Theorem 6.3.6.

Let $f(z)$ *be a complex function of a complex variable. Let* $f(z_0)f'(z_0) \neq 0$ *and set* $h_0 = -f(z_0)/f'(z_0)$, $x_1 = x_0 + h_0$. *Assume that* $f(z)$ *is twice continuously differentiable in the disk* $K_0 : |z - z_1| \leq |h_0|$, *and that*

$$2\,|h_0| M_2 \leq |f'(z_0)|, \quad M_2 = \max_{z \in K_0} |f''(x)|. \tag{6.3.20}$$

Let z_k *be generated by Newton's method:*

$$z_{k+1} = z_k - \frac{f(z_k)}{f'(z_k)}, \quad k = 1, 2, \ldots.$$

Then $z_k \in K_0$ *and we have* $\lim_{k\to\infty} z_k = \zeta$, *where* ζ *is the only zero of* $f(z)$ *in* K_0. *Unless* ζ *lies on the boundary of* K_0, ζ *is a simple zero. Further, we have the relations*

$$|\zeta - z_{k+1}| \leq \frac{M_2}{2|f'(z_k)|}|z_k - z_{k-1}|^2, \quad k = 1, 2, \ldots. \tag{6.3.21}$$

A generalization of this theorem to systems of nonlinear equations is the famous Newton–Kantorovich theorem; see Volume II, Sec. 11.1.

Since the Newton step is in the direction of the negative gradient of $|f(z)|$ at $z = z_k$, it will necessarily give a decrease in $|f(z_k)|$ if a short enough step in this direction is taken. A modified Newton method based on the descent property and switching to standard Newton when the condition (6.3.20) is satisfied will be described in Sec. 6.5.5.

6.3.3 An Interval Newton Method

Sometimes it is desired to compute a tiny interval that is guaranteed to enclose a real simple root x^* of $f(x)$, even when rounding errors are taken into account. This can be done using an adaptation of Newton's method to interval arithmetic method due to Moore [271].

Suppose that the function $f(x)$ is continuously differentiable. Using the notation in Sec. 2.5.3, let $f'([x_0])$ denote an interval containing $f'(x)$ for all x in a finite interval $[x_0] := [a, b]$. Define the Newton operator on $N[x]$ by

$$N([x]) := m - \frac{f(m)}{f'([x])}, \tag{6.3.22}$$

where $m = \text{mid}\,([x]) = \frac{1}{2}(a + b)$.

Theorem 6.3.7.

If $\alpha \in [x]$ is a zero of $f(x)$, then $\alpha \in N([x])$. If $N([x]) \subseteq [x]$, then $f(x)$ has one and only one zero in $N([x])$.

Proof. Suppose α is a zero of $f(x)$ in $[x]$. If $0 \in f'([x])$, then $N([x]) = [-\infty, \infty]$. Otherwise, by the mean value theorem

$$0 = f(\alpha) = f(m) + f'(\xi)(\alpha - m), \quad \xi \in \text{int}\,[\alpha, m] \subseteq [x].$$

This implies that $\alpha = m - f(m)/f'(\xi) \subseteq N([x])$, which proves the first statement.

If $N([x]) \subseteq [x]$, then $f'([x]) \neq 0$ on $[x]$. Then by the mean value theory there are ξ_1 and ξ_2 in $[x]$ such that

$$\begin{aligned}(m - f(m)/f'(\xi_1)) - a &= -f(a)/f'(\xi_1),\\ b - (m - f(m)/f'(\xi_2)) &= f(b)/f'(\xi_2).\end{aligned}$$

Because $N([x]) \subseteq [a, b]$, the product of the left sides is positive. But since $f'(\xi_1)$ and $f'(\xi_2)$ have the same sign this means that $f(a)f(b) < 0$, and f has therefore a zero in $[x]$.

Finally, there cannot be two or more zeros in $[x]$, because then we would have $f'(c) = 0$ for some $c \in [x]$. $\square$

In the interval Newton method, a starting interval $[x_0]$ is chosen, and we compute for $k = 0, 1, 2, \ldots$ a sequence of intervals $[x_{k+1}]$ given by

$$N([x_k]) = \text{mid}\,([x_k]) - \frac{f(\text{mid}\,[x_k])}{f'([x_k])}.$$

If $N([x_k]) \subset [x_k]$ we set $[x_{k+1}] = N([x_k]) \cap [x_k]$. Otherwise, if $N([x_k]) \cap [x_k]$ is empty, we know that $[x_k]$ does not contain a root and stop. In neither condition holds we stop, subdivide the initial interval, and start again. It can be shown that if $[x_0]$ does not contain a root, then after a finite number of steps the iteration will stop with an empty interval.

If we are close enough to a zero, then the length of the intervals $[x_k]$ will converge quadratically to zero, just as with the standard Newton method.

Example 6.3.4.

Take $f(x) = x^2 - 2$ and $[x_0] = [1, 2]$. Using the interval Newton method

$$N([x_k]) = \text{mid}\,([x_k]) - \frac{(\text{mid}\,[x_k])^2 - 2}{2\,[x_k]}, \qquad [x_{k+1}] = N([x_k]) \cap [x_k],$$

we obtain the sequence of intervals

$$[x_1] = N([x_0]) = 1.5 - \frac{2.25 - 2}{2[1, 2]} = \left[\frac{22}{16}, \frac{23}{16}\right] = [1.375, 1.4375],$$

$$[x_2] = N([x_1]) = \frac{45}{32} - \frac{(45/32)^2 - 2}{2[22/16, 23/16]} = \frac{45}{32} - \frac{(45)^2 - 2(32)^2}{128\,[22, 23]} \subset [1.41406, 1.41442].$$

The quadratic convergence of the radius of the intervals is evident:

$$0.5, 0.03125, 0.00036, \ldots .$$

The interval Newton method is well suited to determine *all zeros in a given interval.* Divide the given interval into subintervals and for each subinterval $[x]$ check whether the condition $N([x]) \subseteq [x]$ in Theorem 6.3.7 holds. If this is the case, we continue the interval Newton iterations, and if we are close enough the iterations converge toward a root. If the condition is not satisfied but $N([x]) \cap [x]$ is empty, then there is no zero in the subinterval and this can be discarded. If the condition fails but $N([x]) \cap [x]$ is not empty, then subdivide the interval and try again. The calculations can be organized so that we have a queue of intervals waiting to be processed. Intervals may be added or removed from the queue. When the queue is empty we are done.

The above procedure may not always work. Its performance will depend on, among other things, the sharpness of the inclusion of the derivative $f'([x])$. The algorithm will fail, for example, in the case of multiple roots where $N([x]) = [-\infty, \infty]$.

6.3.4 Higher-Order Methods

Newton's method has quadratic convergence, which means that the number of significant digits approximately doubles in each iteration. Although there is rarely any practical need for methods of higher order of convergence such methods may be useful in special applications, for example, when higher-order derivatives are easily computed. For the following discussion we assume that α is a simple zero of f and that f has a sufficient number of continuous derivatives in a neighborhood of α.

We first briefly derive some famous methods with cubic convergence for simple roots of $f(x) = 0$. Newton's method was derived by approximating the function $f(x)$ with its linear Taylor approximation. Higher-order iteration methods can be constructed by including more terms from the Taylor expansion. The quadratic Taylor approximation of the equation $f(x_n + h) = 0$ is

$$f(x_n) + hf'(x_n) + \frac{h^2}{2} f''(x_n) = 0, \qquad h = x - x_n. \tag{6.3.23}$$

Assuming that $f'(x_n)^2 \geq 2f(x_n)f''(x_n)$, the solutions of this quadratic equation are real and equal to

$$h_n = -\frac{f'(x_n)}{f''(x_n)}\left(1 \pm \sqrt{1 - 2\frac{f(x_n)f''(x_n)}{(f'(x_n))^2}}\right).$$

Rearranging and taking the solution of smallest absolute value we get

$$x_{n+1} = x_n - u(x_n) \cdot \frac{2}{1 + \sqrt{1 - 2t(x_n)}}, \tag{6.3.24}$$

where we have introduced the notations

$$u(x) = \frac{f(x)}{f'(x)}, \qquad t(x) = \frac{f(x_n)f''(x_n)}{(f'(x_n))^2} = u(x)\frac{f''(x)}{f'(x)}. \tag{6.3.25}$$

The iteration (6.3.24) is **Euler's iteration method**.

Assuming that $|t(x_n)| \ll 1$ and using the approximation

$$\frac{2}{1 + \sqrt{1 - 2t(x_n)}} \approx 1 + \frac{1}{2}t(x_n), \tag{6.3.26}$$

valid for $|t| \ll 1$, we obtain another third order iteration method usually also attributed to Euler:

$$x_{n+1} = x_n - u(x_n)\left(1 + \frac{1}{2}t(x_n)\right). \tag{6.3.27}$$

A different method of cubic convergence is obtained by using a rational approximation of (6.3.26):

$$x_{n+1} = x_n - \cdot \frac{u(x_n)}{1 - \frac{1}{2}t(x_n)}. \tag{6.3.28}$$

This is **Halley's**[182] iteration method [179], which according to Traub [354] has the distinction of being the most frequently rediscovered iteration method. Halley's method has a simple geometric interpretation. Consider a hyperbola

$$h(x) = b + a/(x - c),$$

where a, b, c are determined so that $h(x)$ is tangent to the curve $f(x)$ at the point $x = x_n$, and has the same curvature there. Then x_{n+1} is the intersection of this hyperbola with the x-axis.

The methods (6.3.27) and (6.3.28) correspond to the (1,0) and (0,1) Padé approximations of Euler's method. We now show that both are of third order for simple zeros. Following Gander [129] we consider an iteration function of the general form

$$\phi(x) = x - u(x)H(t(x)), \tag{6.3.29}$$

where $u(x)$ and $t(x)$ are defined by (6.3.25). Differentiating (6.3.29) and using $u'(x) = 1 - t(x)$ we get

$$\phi'(x) = 1 - (1 - t(x))H(t) - u(x)H'(t)t'(x).$$

[182]Edmond Halley (1656–1742), English astronomer, who predicted the periodic reappearance (ca 75 years) of a comet, which is now named after him.

Since $u(\alpha) = t(\alpha) = 0$ it follows that $\phi'(\alpha) = 1 - H(0)$. Hence if $H(0) = 1$, then $\phi'(\alpha) = 0$ and the iteration function (6.3.29) is at least of second order. Differentiating once more and putting $x = \alpha$ we get

$$\phi''(\alpha) = t'(\alpha)H(0) - 2u'(\alpha)H'(0)\,t'(\alpha) = t'(\alpha)(H(0) - 2H'(0)).$$

Hence $\phi'(\alpha) = \phi''(\alpha) = 0$ and the method (6.3.29) yields convergence of at least third order if

$$H(0) = 1, \qquad H'(0) = 1/2. \tag{6.3.30}$$

For Euler's and Halley's method we have

$$H(t) = 2\left(1 + \sqrt{1-2t}\right)^{-1}, \qquad H(t) = \left(1 - \frac{1}{2}t\right)^{-1},$$

respectively, and both these methods satisfy (6.3.30). (Verify this!)

It is not very difficult to construct iteration methods of arbitrarily high order for solving $f(x) = 0$. It can be shown [354, Theorem 5.3] that any iteration function of order p must depend explicitly on the first $p-1$ derivatives of f. Consider the Taylor expansion at x_n:

$$0 = f(x_n + h) = f(x_n) + hf'(x_n) + \sum_{k=2}^{p-1} \frac{h^k}{k!} f^{(k)}(x_n) + O(h^p). \tag{6.3.31}$$

Neglecting the $O(h^p)$-term this is a polynomial equation of degree $p-1$ in h. Assuming that $f'(x_n) \neq 0$ we could solve this by the fixed-point iteration $h_i = F(h_{i-1})$, where

$$F(h) = -\frac{f(x_n) + \sum_{k=2}^{p-1} h^k f^{(k)}(x_n)/k!}{f'(x_n)},$$

taking h_0 to be the Newton correction.

To get an explicit method of order p we set $u = f(x_n)/f'(x_n)$, and write

$$u = -h - \sum_{k=2}^{p-1} a_k h^k, \qquad a_k = \frac{f^{(k)}(x_n)}{k!\,f'(x_n)}, \qquad k = 2 : p-1. \tag{6.3.32}$$

This can be interpreted as a formal power series in h (cf. Sec. 3.1.5). Reversing this series we can express h as a formal power series in u:

$$h = -u - \sum_{k=2}^{p-1} c_k u^k + \cdots, \tag{6.3.33}$$

where the first few coefficients are

$$\begin{aligned} &c_2 = a_2, \qquad c_3 = 2a_2^2 - a_3, \qquad c_4 = 5a_2^3 - 5a_2a_3 + a_4, \\ &c_5 = 14a_2^4 - 21a_2^2a_3 + 6a_2a_4 + 3a_3^2 - a_5, \ldots. \end{aligned} \tag{6.3.34}$$

More coefficients can easily be determined; see Problem 3.1.12. This leads to the family of **Schröder methods** [316]. If f is analytic it can be shown that the method

$$x_{n+1} = x_n - u(x_n) - \sum_{k=2}^{p-1} c_k(x_n)u^k(x_n), \quad p \geq 2, \tag{6.3.35}$$

yields convergence order p for simple roots (see Henrici [196, p. 529]). Setting $p = 2$ gives Newton's method and for $p = 3$ we get the third-order method of Euler (6.3.27).

The Schröder methods make use of *polynomials* in u. As for the case $p = 3$, there are methods which instead use *rational expressions* in u. These can be derived from Padé approximants of the Schröder polynomials; see Traub [354, Sec. 5.2]. There are indications that methods which use rational approximations with about equal degree of numerator and denominator are best. For example, for $p = 4$, the rational approximation

$$1 + c_2u + c_3u^2 = \frac{c_2 + (c_2^2 - c_3)u}{c_2 - c_3u} + \mathcal{O}(u^3)$$

can be used to derive an alternative method of order 4.

Example 6.3.5.

A kth-order iteration method (k odd) for the square root of c is

$$x_{n+1} = x_n \frac{x_n^{k-1} + \binom{k}{2}x_n^{k-3}c + \cdots + kc^{(n-1)/2}}{\binom{k}{1}x_n^{k-1} + \binom{k}{3}x_n^{k-3}c + \cdots + c^{(n-1)/2}}, \tag{6.3.36}$$

where the sums end in a natural way with zero binomial coefficients. For $k = 5$ we obtain the iteration

$$x_{n+1} = \frac{x_n^2 + 10c + 5(c/x_n)^2}{5x_n^2 + 10c + (c/x_n)^2}x_n, \quad n = 0, 1, 2, \ldots, \tag{6.3.37}$$

with order of convergence equal to five. For $k = 3$ we obtain Halley's method

$$x_{n+1} = \frac{x_n + 3c/x_n}{3x_n + c/x_n}x_n, \quad n = 0, 1, 2, \ldots, \tag{6.3.38}$$

which has cubic rate of convergence.

Using Halley's method with $c = 3/4$ and the initial approximation $x_0 = (\sqrt{2} + 2)/4$ (obtained by linear interpolation at $1/2$ and 1) we obtain the following result (correct digits in boldface):

$x_0 = \mathbf{0.8}5355339059327, \qquad x_1 = \mathbf{0.866024}74293290, \qquad x_2 = \mathbf{0.86602540378444}.$

Already two iterations give a result correct to 14 digits. Compared to Newton's method (see Example 1.1.1) we have gained one iteration. Since each iteration is more costly there is no improvement in efficiency.

We now introduce a rational family of iteration methods of arbitrary order, which is very convenient to use. First note that alternative derivation of Halley's method is as follows. Starting from (6.3.23) we get

$$h_n = -f(x_n) \Big/ \left(f'(x_n) + \frac{h_n}{2} f''(x_n) \right).$$

Replacing h_n in the denominator by the Newton correction $-f(x_n)/f'(x_n)$ does not change the order of the method and leads to (6.3.28). We note that this can be written

$$x_{n+1} = x_n + B_2(x_n),$$

where

$$B_2(x) = -f(x)\frac{f'(x)}{\det\begin{pmatrix} f'(x) & \frac{f''(x)}{2!} \\ f(x) & f'(x) \end{pmatrix}}.$$

This method belongs to a rational family of iteration functions of arbitrary order which we now present. Set $D_p(x) = \det(F_p)$, where

$$F_p(x) = \begin{pmatrix} f'(x) & \frac{f''(x)}{2!} & \dots & \frac{f^{(p-1)}(x)}{(p-1)!} & \frac{f^{(p)}(x)}{(p)!} \\ f(x) & f'(x) & \ddots & \ddots & \frac{f^{(p-1)}(x)}{(p-1)!} \\ 0 & f(x) & \ddots & \ddots & \vdots \\ \vdots & \vdots & \ddots & \ddots & \frac{f''(x)}{2!} \\ 0 & 0 & \dots & f(x) & f'(x) \end{pmatrix} \in \mathbf{R}^{p\times p} \tag{6.3.39}$$

is a Toeplitz matrix of upper Hessenberg form whose elements are the normalized derivatives of $f(x)$. (Recall that a square matrix is called Toeplitz if its elements are identical along each diagonal.) The iteration

$$x_{n+1} = x_n + B_p(x_n), \quad B_p(x) = -f(x)\frac{\det(F_{p-2}(x))}{\det(F_{p-1}(x))} \tag{6.3.40}$$

can be shown to be of order p for simple roots. Notice that for $f(x) = x^2 - c$ the matrix F_p becomes tridiagonal, which gives a simple iteration method of arbitrarily high order for the square root.

The determinant formula (6.3.40) is attractive since it leads to a simple implementation. Use Gaussian elimination without pivoting to compute the LU factorization $F_p(x_n) = L_pU_p$, where

$$\text{diag}\,(U_p) = (u_{11}, u_{22}, \dots, u_{pp}).$$

Since L_p is unit lower triangular and U_p upper triangular (see Sec. 1.3.2) we have $\det(F_p) = u_{11}u_{22}\dots u_{pp}$. Hence the ratio

$$\det(F_p(x))/\det(F_{p-1}(x))$$

simply equals the *last diagonal element* u_{pp} in U_p. It follows that

$$x_{n+1}^{(k)} = x_n - f(x_n)/u_{kk}(x_n), \quad k = 1 : p, \tag{6.3.41}$$

gives the result of one iteration step using a sequence of iteration formulas of order $k = 2 : p+1$.

Note that the Gaussian elimination is simplified by the fact that $F_p(x_n)$ is a Hessenberg matrix. Only the $p-1$ subdiagonal elements need to be eliminated, which requires $\frac{1}{2}p^2$

flops. Further, it is possible to implement the computation of diag (U_p) using only two row vectors; see Problem 6.3.16.

Methods using high-order derivatives are useful, especially when seeking zeros of a function satisfying a differential equation (usually of second order). Then higher-order derivatives can be calculated by differentiating the differential equation.

Example 6.3.6.

The Bessel function of the first kind $\mathcal{J}_\nu(x)$ satisfies the differential equation

$$x^2y'' + xy' + (x^2 - \nu^2)y = 0.$$

The smallest zero ξ of $\mathcal{J}_0(x)$ is close to $x_0 = 2.40$, and

$$\mathcal{J}_0(x_0) = 0.00250\,76832\,9724, \qquad \mathcal{J}_0'(x_0) = -\mathcal{J}_1(x_0) = -0.52018\,52681\,8193.$$

Differentiating the differential equation for $\mathcal{J}_0(x)$ we get

$$xy^{(k+1)} + ky^{(k)} + xy^{(k-1)} + (k-1)y = 0, \quad k \geq 1,$$

which gives a recursion for computing higher derivatives. Taking $p = 5$ we compute

$$y''(x_0)/2! = 0.10711\,80892\,2261, \qquad y'''(x_0)/3! = 0.05676\,83752\,3951,$$
$$y^{iv}(x_0)/4! = -0.00860\,46362\,1903,$$

and form the Toeplitz matrix F_4. Computing the diagonal elements of U in its LU factorization, we obtain using (6.3.41) the following sequence of approximations to ξ:

2.40482 07503, **2.40482 55406**, **2.40482 55576** 7553, **2.40482 55576 9573**.

Correct digits are shown in boldface. Hence the fifth-order method gives close to full IEEE double precision accuracy!

Review Questions

6.3.1 (a) Under what assumptions is convergence of Newton's method quadratic?

(b) Construct an example where Newton's method diverges, even though the equation has real roots.

6.3.2 Describe an iteration for the division-free computation of the reciprocal of a positive number c. Determine the largest set of starting values x_0 such that the iterates converge to $1/c$.

6.3.3 The equation $f(x) = \sin x = 0$ has one trivial root $x = 0$ in the interval $(-\pi/2, \pi/2)$. Show that for an initial approximation x_0 chosen so that $\tan x_0 = 2x_0$ Newton's method cycles, and $x_{2k} = x_0$ for all $k \geq 0$!

6.3.4 (a) Assume that f is continuously differentiable in a neighborhood of a double root α of the equation $f(x) = 0$. Describe how the equation can be converted to one with a simple root α.

(b) Discuss the case when $f(x) = 0$ has two distinct roots which *nearly* coincide.

Problems and Computer Exercises

6.3.1 (a) Compute $\epsilon_{n+1}/\epsilon_n^2$ for $n = 0, 1, 2$, and the limit, as $n \to \infty$, in Example 6.3.1.
(b) Treat the equation in Example 6.3.1 using $f'(x_2)$ as a fixed approximation to $f'(x_n)$ for $n > 2$. Compare the convergence of this simplified method with the true Newton method.

6.3.2 The equation $x^3 - 2x - 5 = 0$ is of historical interest because it was the one used by Wallis[183] to present Newton's method to the French Academy. Determine the roots of this equation.
Hint: It has one real and two complex roots.

6.3.3 Use Newton's method to determine the positive root of the equation to six correct decimals: (a) $x = 1 - e^{-2x}$; (b) $x \ln x - 1 = 0$.

6.3.4 Determine the unique positive real root of the equation $x^q - x - 1 = 0$ for $q = 2 : 8$.

6.3.5 (a) Consider the Newton iteration used in Example 6.3.5 for computing square root. Show that the iterations satisfy

$$x_{n+1} - \sqrt{c} = \frac{1}{2x_n}(x_n - \sqrt{c})^2.$$

Use this relation to show that, for all $x_0 > 0$, convergence is monotone $x_1 \geq x_2 \geq x_3 \geq \cdots \geq \sqrt{c}$ and $\lim_{n\to\infty} = \sqrt{c}$ (compare Figure 1.1.2).
(b) In Example 6.3.5 Newton's method was used to compute $\sqrt{c}$ for $1/2 \leq c \leq 1$. Determine the maximum error of the linear initial approximation used there. Then use the expression for the error in (a) to determine the number of iterations that suffices to give $\sqrt{c}$ with an error less than 10^{-14} for *all* c in $[1/2, 1]$ using this initial approximation. Show that the influence of rounding errors is negligible.

6.3.6 (a) Investigate how many Newton iterations are required to compute the inverse function $u(x)$ of $x = u \ln u$, to 12 digits of accuracy in the interval $0 < x \leq 20$. Use the starting approximations given in Example 6.3.3.
(b) Lambert's W-function equals $w(x) = \ln u(x)$, where $u(x)$ is the inverse function in (a). Compute and plot the function $w(x)$ for $0 < x \leq 20$.

6.3.7 Determine p, q, and r so that the order of the iterative method

$$x_{n+1} = px_n + qc/x_n^2 + rc^2/x_n^5$$

for computing $\sqrt[3]{c}$ becomes as high as possible. For this choice of p, q, and r, give a relation between the error in x_{n+1} and the error in x_n.

6.3.8 (Ben-Israel) The function $f(x) = xe^{-x}$ has a unique zero $\alpha = 0$. Show that for any $x_0 > 1$ the Newton iterates move away from the zero.

6.3.9 The Cartesian coordinates of a planet in elliptic orbit at time t are equal to $ea(\sin(x), \cos(x))$, where a is the semimajor axis, and e the eccentricity of the ellipse. Using

[183] John Wallis (1616–1703), the most influential English mathematician before Newton.

Kepler's laws of planetary motion it can be shown that the angle x, called the eccentric anomaly, satisfies Kepler's equation

$$x - e\sin x = M, \quad 0 < |e| < 1,$$

where $M = 2\pi t/T$ is the mean anomaly and T the orbital period.

(a) Newton used his method to solve Kepler's equation. Show that for each e, M there is one unique real solution $x = \alpha$ such that $M - |e| \le \alpha < M + |e|$.

(b) Show that the simple fixed-point iteration method

$$x_{n+1} = e\sin x_n + M, \quad x_0 = 0,$$

is convergent.

(c) Study the convergence of Newton's method:

$$x_{n+1} = x_n + \frac{e\sin x_n - x_n + M}{1 - e\cos x_n}.$$

6.3.10 Determine the multiple root $\alpha = 1$ of the equation $p(x) = (1-x)^5 = 0$, when the function is evaluated using Horner's scheme:

$$p(x) = ((((x-5)x+10)x-10)x+5)x - 1 = 0.$$

(a) Use bisection (cf. Algorithm 6.1) with initial interval $(0.9, 1.1)$ and tolerance $\tau = 10^{-8}$. What final accuracy is achieved?

(b) Use Newton's method, starting from $x_0 = 1.1$ and evaluating $p'(x)$ using Horner's scheme. Terminate the iterations when for the first time $|x_{n+1} - 1| > |x_n - 1|$. How many iterations are performed before termination? Repeat with a couple of other starting values.

(c) Repeat (b), but perform one step of the modified Newton's method (6.3.13) with $x_0 = 1.1$ and $q = 5$. How do you explain that the achieved accuracy is much better than predicted by (6.1.10)?

6.3.11 Show that if Newton's method applied to the equation $u(x) = 0$, where $u(x) = f(x)/f'(x)$, then

$$x_{n+1} = x_n - \frac{u(x_n)}{1 - t(x_n)}, \quad t(x_n) = \frac{f(x_n)f''(x_n)}{(f'(x_n))^2}. \tag{6.3.42}$$

This transformation is most useful if an *analytical simplification* can be done such that $u(x)$ can be evaluated accurately also in a neighborhood of α.

6.3.12 (a) Show that Halley's method can also be derived by applying Newton's method to the equation $f(x)(f'(x))^{-1/2} = 0$.

(b) What are the efficiency indexes of Newton's and Halley's methods, respectively, if it is assumed that evaluating each of f, f', and f'' takes one unit of work?

(c) Show that Halley's method applied to $f(x) = x^2 - c = 0$, $c > 0$, gives rise to the iteration

$$x_{n+1} = x_n \frac{x_n^2 + 3c}{3x_n^2 + c} = x_n - \frac{2x_n(x_n^2 - c)}{3x_n^2 + c}.$$

Apply Halley's method to $f(x) = x^k - c = 0$, $c > 0$.

6.3.13 (Ben-Israel) Consider the **quasi-Halley method**

$$x_{n+1} = x_n - \frac{f(x_k)}{f'(x_k) - \dfrac{f'(x_k) - f'(x_{k-1})}{2(x_k - x_{k-1})f'(x_k)} f(x_k)},$$

where the second derivative $f''(x_k)$ has been approximated by a divided difference. Show that if f'' is Lipschitz continuous near a root α, then

$$|\alpha - x_{k+1}| = O(|\alpha - x_k|^\gamma),$$

where γ satisfies the quadratic equation $\gamma^2 - 2\gamma - 1 = 0$. Conclude that the order of this method is approximately 2.41 as compared to 3 for Halley's method.

6.3.14 (Bailey et al. [11]) In 1976 Brent and Salamin independently discovered the following iteration, which generates a sequence $\{p_k\}$ converging quadratically to π. Set $a_0 = 1$, $b_0 = 1/\sqrt{2}$, and $s_0 = 1/2$. For $k = 1, 2, 3, \ldots$ compute

$$a_k = (a_{k-1} + b_{k-1})/2, \qquad b_k = \sqrt{a_{k-1}b_{k-1}},$$
$$c_k = a_k^2 - b_k^2, \qquad s_k = s_{k-1} - 2^k c_k, \qquad p_k = 2a_k^2/s_k.$$

Perform this iteration in IEEE 754 double precision. Verify the quadratic convergence by listing the errors in $|p_k - \pi|$ in successive iterations. How many iterations can you do before the error starts to grow? What is the best accuracy achieved?

6.3.15 In Example 6.3.4 the first two steps in the interval Newton method for solving the equation $x^2 - 2 = 0$ are shown. Implement this method and carry out the iterations until convergence.

6.3.16 (a) Compute $\det(F_p(x))$ in (6.3.39) for $p = 3$ and write down the corresponding rational fourth-order iteration method in terms of u, a_2, a_3 in (6.3.34).

(b) Implement in MATLAB the iteration method (6.3.40) for arbitrary order p. Input should be an approximation x_n to the root $f(x_n)$ and the row vector of scaled derivatives,

$$\left(f'(x), \ \frac{f''(x)}{2!}, \ \ldots, \ \frac{f^{(p-1)}(x)}{(p-1)!}, \ \frac{f^{(p)}(x)}{p!} \right),$$

evaluated at $x = x_n$. Output should be the diagonal elements of U_p in the LU factorization of $F_p(x_n)$ and the sequence of approximations $x_{n+1,k} = x_n + f(x_n)/u_{kk}(x_n)$, $k = 1 : p$. Try to economize on memory requirement.

6.3.17 Write a program that computes the inverse of the error function erf(x) by solving the equation $\mathrm{erf}(x) - y = 0$, $0 \le y < 1$. Use Newton's method and the series expansion given in Example 1.3.4 to compute values of erf(x) and its derivative. Note that $\mathrm{erf}(x) \approx 1 - 1/(\sqrt{\pi}x)$ for large values of x.

6.3.18 (a) Given $\sigma_1 \ge \sigma_2 \ge \cdots \ge \sigma_n > 0$, $c_1, c_2, \ldots, c_n$, and a parameter $\gamma > 0$, consider the equation

$$f(\lambda) = \left(\sum_{i=1}^{n} \frac{\sigma_i^2 c_i^2}{(\sigma_i^2 + \lambda)^2} \right)^{1/2} = \gamma. \tag{6.3.43}$$

Each term in the sum is a convex and strictly decreasing function of λ for $\lambda > 0$. Show that this also holds for $f(\lambda)$. Further, show that if $f(0) > \gamma$ (6.3.43) has a unique root $\lambda^* > 0$.

(b) Newton's method for solving $f(\lambda) - \gamma = 0$ is

$$\lambda_{k+1} = \lambda_k + h_k, \quad h_k = -\frac{f(\lambda_k) - \gamma}{f'(\lambda_k)}.$$

Show that with $\lambda_0 = 0$, this gives a strictly increasing sequence $\{\lambda_k\}_{k=0}^{\infty}$ converging to the solution. Derive an explicit expression for h_k.

(c) The Newton method suggested in (b) will converge slowly when $\lambda_k \ll \lambda^*$. A more efficient method is obtained by instead applying Newton's method to the equation $h(\lambda) = 1/f(\lambda) = 1/\gamma$. Show that this iteration can be written

$$\lambda_{k+1} = \lambda_k - h_k \frac{f(\lambda_k)}{\gamma},$$

where h_k is the Newton step in (b).

(d) Let

$$\sigma_i = 1/i^2, \qquad c_i = 1/i^2 + 0.001, \quad i = 1 : 20.$$

Plot the function $f(\lambda)$ for $\lambda \in (0, 0.0005)$. Solve the equation $\phi(\lambda) = \alpha = 2.5$, using $\lambda_0 = 0$, comparing the two methods with $p = 1$ and $p = -1$.

6.4 Finding a Minimum of a Function

6.4.1 Introduction

In this section we consider the problem of finding the minimum (maximum) of a real-valued function

$$\min g(x), \quad x \in I = [a, b], \tag{6.4.1}$$

which is closely related to that of solving a scalar equation.

Probably the most common use of unidimensional minimization is in "line search" procedures in multidimensional minimization of a function $\phi(z)$ of n variables, $z \in \mathbf{R}^n$. For example, if z_k is the current approximation to the optimal point, the next approximation is often found by minimizing a function

$$g(\lambda) = \phi(x_k + \lambda d_k),$$

where d_k is a search direction and the step length λ is to be determined. In this application, however, the minimization is rarely performed accurately.

Most algorithms for minimizing a nonlinear function of one (or more) variables find at best a **local minimum**. For a function with several local minima, there is no guarantee that the **global minimum** (i.e., the lowest local minimum) in $[a, b]$ will be found. One obvious remedy is to try several different starting points and hope that the lowest of the local minima found is also the global minimum. However, this approach is neither efficient or safe.

Suppose we want to find a local minimum of $g(x)$ in a given interval $[a, b]$. If g is differentiable in $[a, b]$, a **necessary condition** for an interior point of $\tau \in I$ to be a local minimum is that $g'(\tau) = 0$. If g' does not change sign on I it is also possible that the minimum is at a or b. If this is checked for separately, then it is possible to reduce the problem to finding the zeros of $g'(x)$ in I. However, since g' also vanishes at a point of maximum and inflection, it is necessary to check, for example, the second derivative, to see if the point found really is a minimum.

6.4.2 Unimodal Functions and Golden Section Search

A condition which ensures that a function g has a unique global minimum τ in $[a, b]$ is that $g(x)$ is strictly decreasing for $a \le x < \tau$ and strictly increasing for $\tau < x \le b$. Such a function is called **unimodal**.

Definition 6.4.1.

The function $g(x)$ *is* **unimodal** *on* $[a, b]$ *if there exists a unique* $\tau \in [a, b]$ *such that, given any* $c, d \in [a, b]$ *for which* $c < d$

$$d < \tau \Rightarrow g(c) > g(d), \qquad c > \tau \Rightarrow g(c) < g(d). \tag{6.4.2}$$

This condition does not assume that g is differentiable or even continuous on $[a, b]$. For example, $|x|$ is unimodal on $[-1, 1]$.

We now describe an **interval reduction method** for finding the local minimum of a unimodal function, which only uses function values of g. It is based on the following lemma.

Lemma 6.4.2.

Suppose that g *is unimodal on* $[a, b]$, *and* τ *is the point in Definition* 6.4.1. *Let* c *and* d *be points such that* $a \le c < d \le b$. *If* $g(c) \le g(d)$, *then* $\tau \le d$, *and if* $g(c) \ge g(d)$, *then* $\tau \ge c$.

Proof. If $d < \tau$, then $g(c) > g(d)$. Thus, if $g(c) \le g(d)$, then $\tau \le d$. The other part follows similarly. □

Assume that g is unimodal in $[a, b]$. Then using Lemma 6.4.2 it is possible to find a reduced interval on which g is unimodal by evaluating $g(x)$ at two interior points c and d such that $c < d$. Setting

$$[a', b'] = \begin{cases} [c, b] & \text{if } g(c) > g(d), \\ [a, d] & \text{if } g(c) < g(d), \end{cases}$$

we can enclose x^* in an interval of length at most equal to $\max(b-c, d-a)$. (If $g(c) = g(d)$, then $\tau \in [c, d]$, but we ignore this possibility.) To minimize this length one should take c

and d so that $b - c = d - a$. Hence $c + d = a + b$, and we can write

$$c = a + t(b - a), \qquad d = b - t(b - a), \quad 0 < t < 1/2.$$

Then $d - a = b - c = (1 - t)(b - a)$, and by choosing $t \approx 1/2$ we can almost reduce the length of the interval by a factor 1/2. But $d - c = (1 - 2t)(b - a)$ must not be too small for the available precision in evaluating $g(x)$.

If we only consider one step the above choice would be optimal. Note that this step requires **two** function evaluations. A clever way to save function evaluations is to arrange it so that if $[c, b]$ is the new interval, then d can be used as one of the points in the next step; similarly, if $[a, d]$ is the new interval, then c can be used at the next step. Suppose this can be achieved with a fixed value of t. Since $c + d = a + b$ the points lie symmetric with respect to the midpoint $\frac{1}{2}(a + b)$ and we need only consider the first case. Then t must satisfy the following relation (cf. above and Figure 6.4.1):

$$d - c = (1 - 2t)(b - a) = (1 - t)t(b - a).$$

Hence t should equal the root in the interval $(0, 1/2)$ of the quadratic equation $1 - 3t + t^2 = 0$, which is $t = (3 - \sqrt{5})/2$. With this choice the length of the interval will be reduced by the factor

$$1 - t = 2/(\sqrt{5} + 1) = 0.618034\ldots$$

at each step, which is the **golden section** ratio. For example, 20 steps gives a reduction of the interval with a factor $(0.618034\ldots)^{20} \approx 0.661 \cdot 10^{-5}$.

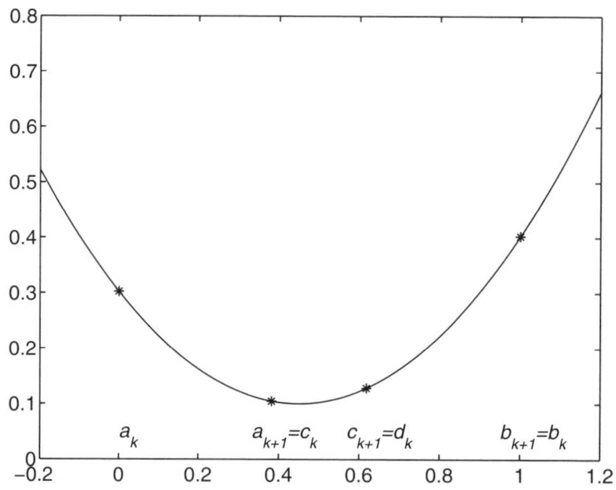

Figure 6.4.1. *One step of interval reduction, $g(c_k) \geq g(d_k)$.*

ALGORITHM 6.2. *Golden Section Search.*

The function `goldsec` computes an approximation $m \in I$ to a local minimum of a given function in an interval $I = [a, b]$, with an error less than a specified tolerance τ.

```
function xmin = goldsec(fname,a,b,delta);
%
t = 2/(3 + sqrt(5));
c = a + t*(b - a);
d = b - t*(b - a);
fc = feval(fname,c);
fd = feval(fname,d);
while (d - c) > delta*max(abs(c),abs(d))
   if fc >= fd     % Keep right endpoint b
      a = c;
      c = d;
      fc = fd;
      d = b - t*(b - a);
      fd = feval(fname, d);
    else             % Keep left endpoint a
      b = d;
      d = c;
      fd = fc;
      c = a + t*(b - a);
      fc = feval(fname, c);
   end;
end;
xmin = (c + d)/2;
```

Rounding errors will interfere when determining the minimum of a scalar function $g(x)$. Because of rounding errors the computed approximation $fl(g(x))$ of a unimodal function $g(x)$ is not in general unimodal. Therefore, we assume that in Definition 6.4.1 the points c and d satisfy $|c-d| > \text{tol}$, where

$$\text{tol} = \epsilon|x| + \tau$$

is chosen as a combination of absolute and relative tolerance. Then the condition (6.4.2) will hold also for the computed function. For any method using only computed values of g there is a fundamental limitation in the accuracy of the computed location of the minimum point τ in $[a,b]$. The best we can hope for is to find $x_k \in [a,b]$ such that

$$g(x_k) \leq g(x^*) + \delta,$$

where δ is an upper bound of rounding and other errors in the computed function values $\bar{g}(x) = fl(g(x))$. If g is twice differentiable in a neighborhood of a minimum point τ, then by Taylor's theorem

$$g(\tau+h) \approx g(\tau) + \tfrac{1}{2}h^2 g''(\tau).$$

This means that there is no difference in the floating-point representation of $g(\tau+h)$ unless h is of the order of $\sqrt{u}$. Hence we can not expect τ to be determined with an error less than

$$\epsilon_\alpha = \sqrt{2\,\delta/|g''(x^*)|}, \tag{6.4.3}$$

unless we can also use values of g' or the function has some special form.

6.4.3 Minimization by Interpolation

For finding the minimum of a unimodal function g, golden section search has the advantage that linear convergence is guaranteed. In that respect it corresponds to the bisection method for finding a zero of a function. If the function is sufficiently smooth and we have a good initial approximation, then a process with superlinear convergence will be much faster. Such methods can be devised using interpolation by a polynomial or rational function, chosen so that its minimum is easy to determine. Since these methods do not always converge they should be combined with golden section search. There is a close analogy with robust methods for solving a nonlinear equation, where a combination of inverse interpolation and bisection can be used; see Sec. 6.2.4.

Since linear functions generally have no minimum the simplest choice is to use a second degree polynomial (a parabola). Suppose that at step n we have three distinct points in u, v, and w. The quadratic polynomial interpolating $g(x)$ at these points is (cf. (6.2.12))

$$p(x) = g(v) + (x - v)[u, v]g + (x - v)(x - u)[u, v, w]g.$$

Setting the derivative of $p(x)$ equal to zero gives

$$0 = [u, v]g + (v - u)[u, v, w]g + 2(x - v)[u, v, w]g,$$

and solving for x gives

$$x = v + d, \qquad d = -\frac{[u, v]g + (v - u)[u, v, w]g}{2[u, v, w]g}. \tag{6.4.4}$$

This is a minimum point of $p(x)$ if $[u, v, w]g > 0$. We assume that of all the points where g has been evaluated v is the one with least function value. Therefore, d should be small, so the effect of rounding errors in computing d is minimized. Initially we can take $u = a$, $w = b$; if $g(c) < g(d)$, then $v = c$, otherwise $v = d$, where c and d are the two golden section points.

Multiplying the numerator and denominator of d by $(v - u)(w - v)(w - u)$, a short calculation shows that $d = -s_1/s_2$, where

$$\begin{aligned} r_1 &= (w - v)(g(v) - g(u)), \qquad & r_2 &= (v - u)(g(w) - g(v)),\\ s_1 &= (w - v)r_1 + (v - u)r_2, \qquad & s_2 &= 2(r_2 - r_1). \end{aligned} \tag{6.4.5}$$

Consider parabolic interpolation at the points x_{i-2}, x_{i-1}, x_i, $i = 2, 3, \dots,$ and let $\epsilon_i = x_i - \tau$. Assuming that $g(x)$ is sufficiently smooth in a neighborhood of τ it can be shown that asymptotically the relation

$$\epsilon_{i+1} \sim \frac{c_3}{2c_2}\epsilon_{i-1}\epsilon_{i-2}, \qquad c_r = \frac{1}{k!}g^{(k)}(\zeta_r), \tag{6.4.6}$$

holds between successive errors. Hence the convergence order equals the real root $p = 1.3247\dots$ of the equation $x^3 - x - 1 = 0$.

If two or more of the points u, v, w coincide, or if the parabola degenerates into a straight line, then $s_2 = 0$. The parabolic interpolation step is only taken if the following inequalities are true:

$$|d| < \frac{1}{2}|e|, \qquad s_2 \neq 0, \qquad v + d \in [a, b],$$

where e is the value of the second last cycle. Otherwise a golden section step is taken, i.e.,

$$x = \begin{cases} (1-t)v + ta & \text{if } v \geq \frac{1}{2}(a+b), \\ (1-t)v + tb & \text{if } v < \frac{1}{2}(a+b), \end{cases}$$

where $1 - t = 2/(\sqrt{5}+1)$.

The combination of inverse quadratic interpolation and golden section search has been suggested by Brent [44, Chapter 5], where the many delicate points to consider in an implementation are discussed. At a typical step there are six significant points a, b, u, v, w, and x, not all distinct. The positions of these points are updated at each step. Initially $[a, b]$ is an interval known to contain a local minimum point. At a later point in the algorithm they have the following significance: A local minimum lies in $[a, b]$; of all the points at which g has been evaluated v is the one with the least value of g; w is the point with the next lowest value of g; u is the previous value of w; and x is the last point at which g has been evaluated.

Review Questions

6.4.1 How many steps are needed in golden section search to reduce an initial interval $[a, b]$ by a factor of 10^{-6}?

6.4.2 Suppose the twice differentiable function $f(x)$ has a local minimum at a point x^*. What approximate limiting accuracy can you expect in a method for computing x^* which uses only function values?

6.4.3 The algorithm FMIN is a standard method for finding the minimum of a function. It uses a combination of two methods. Which?

Problems and Computer Exercises

6.4.1 Use the algorithm `goldsec` to find the minimum of the quadratic function $f(x) = (x - 1/2)^2$ starting from $a = 0.25$, $b = 1$. Plot the successive inclusion intervals.

6.4.2 Modify the algorithm `goldsec` to use parabolic interpolation instead of golden section if this gives a point within the interval.

6.4.3 (a) Plot the function

$$g(x) = \frac{1}{(x-0.3)^2 + 0.01} + \frac{1}{(x-0.9)^2 + 0.04},$$

and show that it has a local minimum in each of the intervals $[0.2, 0.4]$ and $[0.8, 1.0]$.

(b) Use your algorithm from Problem 6.4.2 to determine the location of the two minima of $g(x)$ in (a).

(c) MATLAB includes a function `fminbnd` that also uses a combination of golden section search and parabolic interpolation to find a local minimum. Compare the result using this function with the result from (b).

6.4.4 (Brent [44, Sec. 5.6]) The function

$$g(x) = \sum_{i=1}^{20} \left(\frac{2i-5}{x-i^2} \right)^2$$

has poles at $x = 1^2, 2^2, \ldots, 20^2$. Restricted to the open interval $(i^2, (i+1)^2)$, $i = 1 : 19$, it is unimodal. Determine the minimum points in these intervals and the corresponding values of $g(x)$.

6.5 Algebraic Equations

6.5.1 Some Elementary Results

The problem of solving an algebraic equation

$$p(z) = a_0 z^n + a_1 z^{n-1} + \cdots + a_n = 0, \quad (a_0 \neq 0), \tag{6.5.1}$$

has played a major role in the development of mathematical concepts for many centuries. Even the "high school formula" for solving a quadratic equation requires the introduction of irrational and complex numbers. There is a long history of investigations into algebraic expressions for the zeros of equations of higher degree. In the sixteenth century Cardano published formulas for the roots of a cubic equation (see Problem 2.3.8). Formulas for the roots when $n = 4$ are also known. In 1826 Abel proved that it is not possible to find algebraic expressions for the roots for the class of algebraic equations of degree $n > 4$. But even the existing formulas for $n \leq 4$ are not, in general, suitable for numerical evaluation of the roots. In Sec. 2.3.4, it was shown that care must be taken to avoid cancellation even in the case of a quadratic equation.

Despite the absence of closed solution formulas, the **fundamental theorem of algebra** states that every algebraic equation has at least one root. More precisely, if a polynomial vanishes nowhere in the complex plane, then it is identically constant. The **remainder theorem** states that when a polynomial $p(z)$ is divided by $z - r$ the remainder is $p(r)$, i.e.,

$$p(z) = (z - r)p_1(z) + p(r). \tag{6.5.2}$$

It follows that r_1 is a zero of $p(z)$ if and only if $z - r_1$ divides $p(z)$. If $p(z)$ is of degree n and $p(r_1) = 0$, then $p_1(z)$ in (6.5.2) is of degree $n - 1$. But if $n > 1$, then $p_1(z)$ must vanish for some r_2, and hence has a linear factor $z - r_2$. Continuing, we find that an algebraic equation $p(z) = 0$ of degree n has exactly n roots, counting multiplicities, and it holds that[184]

$$p(z) = a_0(z - r_1)(z - r_2) \cdots (z - r_n). \tag{6.5.3}$$

By this representation it also follows that if the coefficients $a_0, a_1, \ldots, a_n$ are real, then eventual complex roots must occur in conjugate pairs. Hence, if $p(z)$ has the complex zero $s + it$ it also has the zero $s - it$ and therefore contains the quadratic factor $(z - s)^2 + t^2$ which is positive for all real values of z.

[184]Note the corollary that if a polynomial $p(z)$ of degree at most n vanishes in $n + 1$ distinct points, it must be identically zero, a result used repeatedly in Chapters 3 and 4.

Solving algebraic equations of high degree does not play a central role in scientific computing. Usually the applications involve only equations of moderate degree, say 10–20, for which acceptable subroutines exist. Algebraic equations of high degree ($n > 100$) occur in computer algebra. Such problems usually require symbolic, or high multiple-precision computations. Algorithms for such problems are still a subject of research.

Let $p(z)$ be the polynomial (6.5.3) with roots $r_1, r_2, \dots, r_n$, not necessarily equal. Comparing with the coefficients of z^{n-k} in the representations (6.5.1) we find that $(-1)^k a_k/a_0$ is the sum of the products of the roots r_i taken k at a time. Thus we obtain the following relations between the coefficients and zeros of a polynomial:

$$\sum_i r_i = -\frac{a_1}{a_0}, \qquad \sum_{i<j} r_i r_j = \frac{a_2}{a_0}, \qquad \sum_{i<j<k} r_i r_j r_k = -\frac{a_3}{a_0}, \dots,$$

$$r_1 r_2 \cdots r_n = (-1)^n \frac{a_n}{a_0}. \tag{6.5.4}$$

The functions on the left sides of (6.5.4) are called **elementary symmetric functions** of the variables $r_1, r_2, \dots, r_n$, since interchanging any of the variables will not change the functions. A classical result says that any symmetric polynomial in the roots $r_1, r_2, \dots, r_n$ can be expressed as a polynomial in the elementary symmetric functions in (6.5.4).

Other commonly used symmetric functions are the power sums

$$S_k = \sum_{i=1}^{n} r_i^k, \quad k = \pm 1, \pm 2, \dots. \tag{6.5.5}$$

The **Newton formulas** give the connection between the coefficients of the polynomial $p(z)$ (with $a_0 = 1$), and the power sums:

$$\begin{aligned} &S_1 + a_1 = 0, \\ &S_2 + a_1 S_1 + 2a_1 = 0, \\ &S_3 + a_1 S_2 + a_2 S_1 + 3a_3 = 0, \\ &\quad\vdots \\ &S_{n-1} + a_1 S_{n-2} + \cdots + a_{n-2} S_1 + (n-1)a_{n-1} = 0. \end{aligned} \tag{6.5.6}$$

For a proof see Householder [204, Sec. 1.3].

If $a_n \neq 0$, setting $z = 1/y$ in $p(z)$ gives the **reciprocal polynomial**

$$q(y) = y^n p(1/y) = a_n y^n + \cdots + a_1 y + a_0. \tag{6.5.7}$$

The zeros of the reciprocal polynomial are $1/r_1, 1/r_2, \dots, 1/r_n$ and from (6.5.4), giving the Newton formulas,

$$\sum_i 1/r_i = -a_{n-1}/a_n, \quad \text{etc.}$$

Function values of the polynomial $p(z)$ in (6.5.1) at a (real or complex) point w can conveniently be computed by repeated synthetic division of $p(z)$ with $z - w$; cf. Sec. 1.2.2. If we set

$$\begin{aligned} p(z) &= (z - w)q(z) + b_n, \\ q(z) &= b_0 z^{n-1} + b_1 z^{n-2} + \cdots + b_{n-1}, \end{aligned} \tag{6.5.8}$$

then the sequence $\{b_i\}_{i=0}^{n}$ satisfies the recursion

$$b_0 = a_0, \qquad b_i = b_{i-1}w + a_i, \quad i = 1 : n. \tag{6.5.9}$$

Here $p(w) = b_n$ is the remainder and $q(z)$ is the quotient polynomial when dividing $p(z)$ with $(z - w)$. Differentiating (6.5.8) we get

$$p'(z) = (z - w)q'(z) + q(z), \tag{6.5.10}$$

and setting $z = w$ we find that $p'(w) = q(w)$. We can form $q(w)$ by synthetic division of $q(z)$ with $(z - w)$:

$$q(z) = (z - w)r(z) + c_{n-1},$$
$$r(z) = c_0 z^{n-2} + c_1 z^{n-3} + \cdots + c_{n-2}.$$

Now $p'(w) = q(w) = c_{n-1}$, where

$$c_0 = b_0, \qquad c_i = c_{i-1}w + b_i, \quad i = 1 : n - 1.$$

Higher derivatives can be computed in the same fashion. Differentiating once more gives

$$p''(z) = (z - w)q''(z) + 2q'(z),$$

and thus $\frac{1}{2}p''(w) = q'(w) = d_{n-2}$, where

$$d_0 = c_0, \qquad d_i = d_{i-1}w + c_i, \quad i = 1 : n - 2.$$

To compute $p^{(i)}(w)$ using these formulas requires $n - i$ additions and multiplications.

In the important special case where all the coefficients $a_0, a_1, \ldots, a_n$ are real, the above formulas are somewhat inefficient, and one can save operations by performing synthetic division with the quadratic factor

$$(z - w)(z - \bar{w}) = z^2 - 2z Re(w) + |w|^2,$$

which has real coefficients (see Problem 6.5.2 (b)).

Synthetic division can also be used to shift the origin of a polynomial $p(z)$. Given $a_0, a_1, \ldots, a_n$ and s, we then want to find coefficients $c_0, c_1, \ldots, c_n$ so that

$$p(w + s) = q(s) = c_0 s^n + c_1 s^{n-1} + \cdots + c_n. \tag{6.5.11}$$

Clearly this is the Taylor expansion of $p(z)$ at $z = w$. It follows that

$$c_n = p(w), \quad c_{n-1} = p'(w), \quad c_{n-1} = \frac{1}{2}p''(w), \quad \ldots, \quad c_0 = \frac{1}{n!}p^{(n)}(w),$$

and the coefficients c_i can be computed by repeated synthetic division of $p(z)$ by $(z - w)$ as described above in about $n^2/2$ multiplications.

It is often desirable to obtain some preliminary information as to where the zeros of a polynomial $p(z)$ are located. Some information about the location of real roots can be obtained from a simple examination of the sign of the coefficients of the polynomial.

A simple observation is that if $a_i > 0$, $i = 1 : n$, then $p(x)$ can have no real positive zero. A generalization of this result, known as **Descartes' rule of sign**,[185] states that the number of positive roots is either given by the number of variations in sign in the sequence $a_0, a_1, \dots, a_n$, or is *less* than that by an even number. (Multiple roots are counted with their multiplicity.) By considering the sign variations for the polynomial $p(-z)$ we similarly get an upper bound on the number of negative roots. By shifting the origin, we can get bounds on the number of roots larger and smaller than a given number. In Sec. 6.5.6 we give a method to obtain precise information about the number of real roots in any given interval.

Many classical results are known about the number of real or complex roots in a disk or half-plane. For these we refer to surveys in the literature.

6.5.2 Ill-Conditioned Algebraic Equations

Let $a = (a_1, a_2, \dots, a_n)^T \in \mathbf{C}^n$ be the coefficient vector of a monic polynomial $p(z)$ ($a_0 = 1$) and denote the set of zeros of p by $Z(p) = \{z_1, z_2, \dots, z_n\}$. In general, the best we can expect from any numerical zero-finding algorithm is that it computes the zeros of a nearby polynomial $\hat{p}(z)$ with slightly perturbed coefficients $\hat{a}$. Following Trefethen [358] we define the ϵ-pseudo zero set of $p(z)$ by

$$Z_\epsilon = \{z \in \mathbf{C} \mid z \in Z(\hat{p}) \text{ for some } \hat{p} \text{ with } \|\hat{a} - a\| \le \epsilon\}. \tag{6.5.12}$$

These sets describe the conditioning of the zero-finding problem. Depending on the chosen norm they correspond to a coefficientwise or normwise perturbation of the coefficient vector a. The shape of these sets is studied in [353].

The sensitivity of polynomial zeros to perturbations in the coefficients was illustrated in a famous example by Wilkinson in the early 1960s.[186] The paper [379] contains an extensive discussion of numerical problems in determining roots of polynomial equations. Wilkinson considered the polynomial

$$p(z) = (z-1)(z-2)\cdots(z-20) = z^{20} - 210z^{19} + \cdots + 20!,$$

with zeros $1, 2, \dots, 20$. Let $\bar{p}(z)$ be the polynomial which is obtained when the coefficient $a_1 = -210$ in $p(z)$ is replaced by

$$-(210 + 2^{-23}) = -210.000000119\dots,$$

while the rest of the coefficients remain unchanged. Even though the relative perturbation in a_1 is of order 10^{-10}, many of the zeros of the perturbed polynomial $\bar{p}(z)$ deviate greatly

[185] René Descartes (1596–1650), French philosopher and mathematician.

[186] Wilkinson received the Chauvenet Prize of the Mathematical Association of America 1987 for this exposition of the ill-conditioning of polynomial zeros.

from those of $p(z)$. In fact, correct to nine decimal places, the perturbed zeroes are the following.

$$\begin{array}{rl}
1.000000000 & 10.095266145 \pm 0.643500904i \\
2.000000000 & \\
3.000000000 & 11.793633881 \pm 1.652329728i \\
4.000000000 & \\
4.999999928 & 13.992358137 \pm 2.518830070i \\
6.000006944 & \\
6.999697234 & 16.730737466 \pm 2.812624894i \\
8.007267603 & \\
8.917250249 & 19.502439400 \pm 1.940330347i \\
20.846908101 &
\end{array}$$

For example, the two zeros 16, 17 have not only changed substantially but have become a complex pair. It should be emphasized that this behavior is quite typical of polynomials with real coefficients and real roots. Indeed, many polynomials which arise in practice behave much worse than this.

If we assume that the coefficients a_i of a polynomial are given with full machine accuracy, then the error δ in computed values of $p(x)$ (for real x) is bounded by

$$\delta < 1.06u \sum_{i=0}^{n} |(2i+1)a_{n-i}x^i| < \gamma_{2n+1} \sum_{i=0}^{n} |a_{n-i}||x|^i;$$

see Sec. 2.3.2. Hence by (6.1.7) the limiting accuracy of a zero α is equal to

$$\epsilon_\alpha = \frac{\delta}{|p'(\alpha)|} = \frac{\sum_{i=0}^{n} |(2i+1)a_{n-i}\alpha^i|}{|p'(\alpha)|}.$$

In particular, for the root $\alpha = 14$ in the above example we get $\epsilon_\alpha = 1.89 \cdot 10^{16}$. But the changes in this example are so large that this linearized perturbation theory does not apply! For a more detailed discussion of the conditioning of algebraic equations in general and the Wilkinson polynomial, we refer to Gautschi [144, Sec. 4].

It should be emphasized that although a problem may be given in the form (6.5.1), it is often the case that the coefficients of $p(z)$ are not the original data. *Then it may be better to avoid computing them.* An important case is when the polynomial is the **characteristic polynomial** of a matrix $A \in \mathbf{R}^{n\times n}$:

$$p_A(z) = \det(zI - A) = \begin{vmatrix} z - a_{11} & -a_{12} & \cdots & -a_{1n} \\ -a_{21} & z - a_{22} & \cdots & -a_{2n} \\ \vdots & \ddots & \vdots & \vdots \\ -a_{n1} & -a_{n2} & \cdots & z - a_{nn} \end{vmatrix}. \tag{6.5.13}$$

The n roots of $p_A(z) = 0$ are the eigenvalues of A, and *the problem is an eigenvalue problem in disguise.* Then the original data are the elements of the matrix A and numerical values of $p_A(z)$ can then be evaluated much more accurately directly from (6.5.13). It is important to realize that, even when the roots (eigenvalues) are well determined by the elements of A

and well separated, they can be *extraordinarily sensitive to small relative perturbations in the coefficients of* $p_A(z)$.

In classical (i.e., before 1960) methods of linear algebra, the eigenvalues of a matrix are often found by first computing the coefficients of the characteristic polynomial. This is not in general a good idea and one of the highly developed modern eigenvalue algorithms should be used (consult Volume II, and references therein). Note also that if the coefficients of the characteristic polynomial $\det(zI - A)$ are required, these are often best computed by first computing the eigenvalues λ_i of A and then forming

$$p(z) = \prod_{i=1}^{n}(z - \lambda_i). \tag{6.5.14}$$

Example 6.5.1.

Another example where computing the coefficients of the polynomial should be avoided is the following. Suppose the largest positive root of the algebraic equation

$$p(x) = (x + 2)(x^2 - 1)^6 - 3 \cdot 10^{-6} \cdot x^{11} = 0$$

is to be computed. Here $p(z)$ is a polynomial of degree 13. If the coefficients are computed using decimal floating-point arithmetic with seven digits, then the coefficient of x^{11} which is $(12 - 3 \cdot 10^{-6})$ will be rounded to 12.00000. Thus the machine will treat the equation $(x + 2)(x^2 - 1)^6 = 0$, whose exact positive root is one.

This is a poor result. However, by writing the equation in the form

$$x = \phi(x), \quad \phi(x) = 1 + \frac{0.1}{x+1}\left(\frac{3x^{11}}{x+2}\right)^{1/6},$$

and solving this by the iteration $x_0 = 1$, $x_{k+1} = \phi(x_k)$, one can get the root $\alpha = 1.053416973823$ to full accuracy.

Turning the tables, a polynomial zero-finding problem can be transformed into an eigenvalue problem. The **companion matrix** (or Frobenius matrix) to the polynomial $p(z)$ in (6.5.1), normalized so that $a_0 = 1$, is defined as

$$C = \begin{pmatrix} -a_1 & -a_2 & \cdots & -a_{n-1} & -a_n \\ 1 & 0 & \cdots & 0 & 0 \\ 0 & 1 & \cdots & 0 & 0 \\ \vdots & \vdots & \ddots & \vdots & \vdots \\ 0 & 0 & \cdots & 1 & 0 \end{pmatrix}. \tag{6.5.15}$$

(Sometimes the companion matrix is defined slightly differently, for example, with the coefficients of the polynomial in the last column.) Using the definition (1.6.4) it can be verified that the characteristic polynomial of C equals

$$p_C(z) = \det(zI - C) = z^n + a_1 z^{n-1} + \cdots + a_{n-1}z + a_n.$$

Thus the roots of $p(z) = 0$ can be computed by applying an eigenvalue algorithm to C. This trick is used in several of the zero-finding algorithms in current use; see Sec. 6.5.5.

For example, in MATLAB the function `roots(p)` computes the roots of a polynomial $p(z)$ using the QR algorithm to solve the eigenvalue problem for the companion matrix. Although the operation count for this QR algorithm is $\mathcal{O}(n^3)$ and the storage requirement $1.5n^2$, experiments suggest that for small and moderate values of n it is as fast as competing algorithms and can be more accurate. Further problems with overflow or underflow are avoided.

If the coefficients are known (and stored) *exactly*, then by using multiple precision arithmetic the accuracy in the zeros can be increased. It is generally true that the solution of polynomial equations of high degree requires the use of multiple precision floating-point arithmetic in order to achieve high accuracy.

6.5.3 Three Classical Methods

In this section we describe three classical methods which are mainly of theoretical interest, but may be useful in special situations.

The idea in **Graeffe's method** is to replace equation (6.5.1) by an equation whose roots are the square of the roots of (6.5.1). By iterating this procedure roots which are of unequal magnitude become more and more separated and therefore, as we shall see, more easily determined.

If the monic polynomial $p_1(z) = z^n + a_1 z^{n-1} + \cdots + a_n$ has zeros z_i, $i = 1 : n$, then

$$p_1(z) = (z - z_1)(z - z_2)\cdots(z - z_n).$$

Setting $y = z^2$, it follows that

$$p_2(y) = (-1)^n p_1(z) p_1(-z) = (y - z_1^2)(y - z_2^2)\cdots(y - z_n^2);$$

i.e., $p_2(y)$ is a polynomial of degree n whose zeros equal the square of the zeros of $p_1(z)$. The coefficients of

$$p_2(y) = y^n + b_1 y^{n-1} + \cdots + b_n$$

are obtained by convolution of the coefficients of $p_1(z)$ and $p_1(-z)$; see Theorem 3.1.6. This gives

$$b_0 = a_0^2, \qquad (-1)^k b_k = a_k^2 + \sum_{j=1}^{k} (-1)^j 2 a_{k-j} a_{k+j}, \quad k = 1 : n. \tag{6.5.16}$$

Assume now that after repeated squaring we have obtained an equation

$$p_{m+1}(w) = w^n + c_1 w^{n-1} + \cdots + c_n,$$

with zeros $|\alpha_j| = |z_j|^{2m}$, such that $|\alpha_1| \gg |\alpha_2| \gg \cdots \gg |\alpha_n|$. Then we can use the relations (6.5.4) between the coefficients and zeros of a polynomial to deduce that

$$c_1 = -\sum_i \alpha_i \approx -\alpha_1, \qquad c_2 = \sum_{i<j} \alpha_i \alpha_j \approx \alpha_1 \alpha_2, \qquad c_3 = \sum_{i<j<k} \alpha_i \alpha_j \alpha_k \approx \alpha_1 \alpha_2 \alpha_3, \ldots,$$

and therefore

$$|z_1| \approx (-c_1)^{1/2^m}, \qquad |z_2| \approx (-c_2/c_1)r^{1/2^m}, \qquad |z_3| \approx (-c_3/c_2)^{1/2^m}, \dots .$$

Example 6.5.2.
Squaring the polynomial $p_1(z) = z^3 - 8z^2 + 17z - 10$ three times gives

$$p_4(w) = w^3 - 390{,}882w^2 + 100{,}390{,}881w - 10^8.$$

We have $2^m = 8$ and approximations to the roots are

$$\begin{aligned}|z_1| &\approx \sqrt[8]{390{,}882} = 5.00041,\\ |z_2| &\approx \sqrt[8]{100{,}390{,}881/390{,}882} = 2.00081,\\ |z_3| &\approx \sqrt[8]{10^8/100{,}390{,}881} = 0.999512.\end{aligned}$$

The exact roots are 5, 2, and 1.

In **Bernoulli's method** for obtaining the zeros of

$$p(z) \equiv z^n + a_1 z^{n-1} + \cdots + a_n = 0,$$

one considers the related difference equation

$$y_{n+k} + a_1 y_{n+k-1} + \cdots + a_n y_k = 0, \tag{6.5.17}$$

which has $p(z)$ as its characteristic equation. If $p(z)$ only has simple roots $z_1, \dots, z_n$, then by Theorem 3.3.12 the general solution of (6.5.17) is given by

$$y_k = \sum_{j=1}^{n} c_j z_j^k,$$

where $c_1, c_2, \dots, c_n$ are constants which depend on the initial conditions.

We assume in the following that the roots are ordered in magnitude so that

$$|z_1| > |z_2| \geq \cdots \geq |z_n|,$$

where the root of largest magnitude z_1 is distinct. We say that z_1 is a **dominant root**. Assuming that the initial conditions are chosen so that $c_1 \neq 0$, we have

$$y_k = c_1 z_1^k \left[1 + \sum_{j=2}^{n} \frac{c_j}{c_1} \left(\frac{z_j}{z_1} \right)^k \right] = c_1 z_1^k \left[1 + O\left(\frac{|z_2|}{|z_1|} \right)^k \right].$$

If z_1 is real it follows that

$$\lim_{n\to\infty} \frac{y_k}{y_{k-1}} = z_1 \left[1 + O\left(\frac{|z_2|}{|z_1|} \right) \right]. \tag{6.5.18}$$

Note that the rate of convergence is linear and depends on the ratio $|z_2/z_1|$. In practice measures to avoid overflow or underflow must be included. By initially applying a few steps of Graeffe's root-squaring method, convergence can be improved.

The relation (6.5.18) holds also if z_1 is a real multiple root. However, if the root of largest magnitude is complex the method needs to be modified. It is difficult to safeguard against all possible cases.

Bernoulli's method is closely related to the **power method** for computing approximate eigenvalues and eigenvectors of a matrix. In this application a powerful modification is to shift all the eigenvalues so that the desired eigenvalue is close to zero. If we then apply the power method to the inverse matrix rapid convergence is assured.

In **Laguerre's method**[187] the polynomial $p(z)$ of degree n is approximated in the neighborhood of the point z_k by a special polynomial of the form

$$r(z) = a(z - w_1)(z - w_2)^{n-1},$$

where the parameters a, w_1, and w_2 are determined so that

$$p(z_k) = r(z_k), \qquad p'(z_k) = r'(z_k), \qquad p''(z_k) = r''(z_k). \tag{6.5.19}$$

If z_k is an approximation to a simple zero α, then the simple zero w_1 of $r(z)$ is taken as the new approximation z_{k+1} of α. Laguerre's method has very good global convergence properties for polynomial equations, and with *cubic* convergence for simple roots (real or complex). For multiple roots convergence is only linear.

In order to derive Laguerre's method we note that the logarithmic derivative of $p(z) = (z - \alpha_1)\cdots(z - \alpha_n)$ is

$$S_1(z) = \frac{p'(z)}{p(z)} = \sum_{i=1}^{n} \frac{1}{z - \alpha_i}.$$

Taking the derivative of this expression we obtain

$$-\frac{dS_1(z)}{dz} = S_2(z) = \left(\frac{p'(z)}{p(z)}\right)^2 - \frac{p''(z)}{p(z)} = \sum_{i=1}^{n} \frac{1}{(z - \alpha_i)^2}.$$

Using (6.5.19) to determine the parameters of the approximating polynomial r(z) we obtain the equations

$$S_1(z_k) = \frac{1}{z_k - w_1} + \frac{(n-1)}{z_k - w_2}, \qquad S_2(z_k) = \frac{1}{(z_k - w_1)^2} + \frac{(n-1)}{(z_k - w_2)^2}.$$

Eliminating $z_k - w_2$ gives a quadratic equation for the correction $z_k - w_1 = z_k - z_{k+1}$. After some algebra we obtain (check this!)

$$z_{k+1} = z_k - \frac{np(z_k)}{p'(z_k) \pm \sqrt{H(z_k)}}, \tag{6.5.20}$$

where

$$H(z_k) = (n-1)^2[p'(z_k)]^2 - n(n-1)p(z_k)p''(z_k).$$

[187]Edmond Nicolas Laguerre (1834–1886), French mathematician at École Polytechnique, Paris and best known for his work on orthogonal polynomials.

The sign in the denominator in (6.5.20) should be chosen so that the magnitude of the correction $|z_{k+1} - z_k|$ becomes as small as possible.

For polynomial equations with only real roots, Laguerre's method is globally convergent, i.e., it converges for *every* choice of real initial estimate z_0. Suppose the roots are ordered such that $\alpha_1 \le \alpha_2 \le \cdots \le \alpha_n$. If $z_0 \in (\alpha_{j-1}, \alpha_j)$, $j = 2 : n$, then Laguerre's method converges to one of the roots α_{j-1}, α_j; if $z_0 < \alpha_1$ or $z_0 > \alpha_n$, then convergence is to α_1 or α_n, respectively.

For polynomial equations with complex roots, Laguerre's method no longer converges for every choice of initial estimate. But experience has shown that the global convergence properties are good also in this case. In particular, if we take $z_0 = 0$, then Laguerre's method will usually converge to the root of smallest modulus. We finally remark that, as might be expected, for multiple roots convergence of Laguerre's method is only linear.

Consider the polynomial equation $p(z) = 0$ and assume that $a_n \ne 0$ so that $\alpha = 0$ is not a root. Now suppose that $a_{n-2}a_{n-1} \ne 0$, and take $z_0 = 0$ in Laguerre's method. A simple calculation gives

$$z_1 = \frac{-na_n}{a_{n-1} \pm \sqrt{H(z_0)}}, \qquad H(z_0) = (n-1)^2 a_{n-1}^2 - 2n(n-1)a_n a_{n-2}, \tag{6.5.21}$$

where the sign is to be chosen so the $|z_1|$ is minimized. In particular, for $n = 2$, $H(z_0)$ is the discriminant of $p(z)$ and z_1 is the root of smallest modulus.

Example 6.5.3.

If there are complex roots, then there may be several distinct roots of smallest modulus. For example, the equation

$$p(z) = z^3 - 2z^2 + z - 2$$

has roots $\pm i$ and 2. Using the above formula (6.5.21) for z_1 with $n = 3$, we get

$$z_1 = \frac{6}{1 \pm 2i\sqrt{11}} = \frac{2}{15} \pm i\frac{4\sqrt{11}}{15} = 0.06666666667 \pm 0.88443327743i.$$

Continuing the iterations with Newton's method we get convergence to one of the two roots $\pm i$,

$$z_2 = -0.00849761051 + 1.01435422762i, \quad z_3 = -0.00011503062 + 1.00018804502i,$$
$$z_4 = -0.00000002143 + 1.00000003279i, \quad z_5 = -0.00000000000 + 1.00000000000i.$$

6.5.4 Deflation and Simultaneous Determination of Roots

Suppose we have found a root α to the equation $p(z) = 0$. Then taking $z_k = \alpha$ in (6.5.9)–(6.5.8) we have $b_n = p(\alpha) = 0$ and the remaining roots of $p(z)$ are also roots of the polynomial equation

$$q(z) = \frac{p(z)}{z - \alpha} = 0.$$

Hence we can continue the iterations with the quotient polynomial $q(z)$ of degree $n-1$. This process is called **deflation** and can be repeated; as soon as a root has been found it is factored out. Proceeding like this, all roots are eventually found. Since we work with polynomials of lower and lower degree, deflation saves arithmetic operations. More important is that it prevents the iterations from converging to the same simple root more than once.

So far we have ignored that roots which are factored out are only known with finite accuracy. Also, rounding errors occur in the computation of the coefficients of the quotient polynomial $q(x)$. Clearly there is a risk that both these types of errors can have the effect that the zeros of the successive quotient polynomials deviate more and more from those of $p(z)$. Indeed, deflation is not unconditionally a stable numerical process. A closer analysis performed by Wilkinson [379] shows that if the coefficients of the quotient polynomials are computed by the recursion (6.5.9), then errors resulting from deflation are negligible provided that

1. the roots are determined in order of *increasing* magnitude;
2. each root is determined to its limiting accuracy.

Note that if the above procedure is applied to the reciprocal polynomial $z^n p(1/z)$ we obtain the zeros of $p(z)$ in order of *decreasing* magnitude.

With Laguerre's method it is quite probable that we get convergence to the root of smallest magnitude from the initial value $z_0 = 0$. But this cannot be guaranteed and therefore one often proceeds in two steps. First, all n roots are determined using deflation in the process. Next, each root found in the first step is refined by doing one or several Newton iterations using the *original* polynomial $p(z)$.

Deflation can be avoided by using a **zero suppression** technique suggested by Maehly [255]. He notes that the derivative of the reduced polynomial $q(z) = p(z)/(z - \xi_1)$ can be expressed as

$$q'(z) = \frac{p'(z)}{z - \xi_1} - \frac{p(z)}{(z - \xi_1)^2}.$$

More generally, assume that we have determined approximations $\xi_1, \dots, \xi_j$ to j roots of $p(z) = 0$. Then the first derivative of the reduced polynomial $q_j(z) = p(z)/[(z - \xi_1) \cdots (z - \xi_j)]$ can be expressed as

$$q_j'(z) = \frac{p'(z)}{(z - \xi_1) \cdots (z - \xi_j)} - \frac{p(z)}{(z - \xi_1) \cdots (z - \xi_j)} \sum_{i=1}^{j} \frac{1}{z - \xi_i}.$$

Hence Newton's method applied to $q_j(z)$ can be written

$$z_{k+1} = z_k - \frac{p(z_k)}{p'(z_k) - \sum_{i=1}^{j} p(z_k)/(z_k - \xi_i)}, \tag{6.5.22}$$

which is the **Newton–Maehly method**. This iteration has the advantage that it is not sensitive to the accuracy in the approximations of the previous roots $\xi_1, \dots, \xi_j$. Indeed, the iteration (6.5.22) is locally quadratically convergent to simple zeros of $p(z)$ for *arbitrary* values of $\xi_1, \dots, \xi_j$.

For the removal of a linear factor by deflation it is necessary that the zero be computed to full working accuracy, since otherwise the remaining approximative zeros can be meaningless. This is a disadvantage if only low accuracy is required. An alternative to deflation is to use an iterative method that, under appropriate separation assumptions, allows for the *simultaneous* determination of all the roots of a polynomial equation. Suppose that the numbers $\xi_i^{(k)}$, $i = 1 : n$, are a set of n distinct approximations of $p(z)$. A new set of approximations are then computed from

$$\xi_i^{(k+1)} = \xi_i^{(k)} - p(\xi_i^{(k)}) \Big/ \left[a_0 \prod_{\substack{j=1 \\ j \neq i}}^{n} (\xi_i^{(k)} - \xi_j^{(k)}) \right], \quad i = 1 : n. \tag{6.5.23}$$

This is **Weierstrass' method**, introduced in 1891 in connection with a new constructive proof of the fundamental theorem of algebra. The method was rediscovered and analyzed in the 1960s by Durand and is also known as the **Durand–Kerner method**.

With $q(z) = (z - \xi_1^{(k)})(z - \xi_2^{(k)}) \cdots (z - \xi_n^{(k)})$ the formula may also be written

$$\xi_i^{(k+1)} = \xi_i^{(k)} - p(\xi_i^{(k)})/q'(\xi_i^{(k)}),$$

which shows that to first approximation the method is identical to Newton's method. This relation can be used to prove that for simple (real or complex) zeros the asymptotic order of convergence of the Weierstrass method equals 2. (For multiple zeros the method will only converge linearly.) The relation

$$\sum_{i=1}^{n} \xi_i^{(k)} = \sum_{i=1}^{n} \alpha_i = -a_1, \quad k \geq 1,$$

which holds independent of the initial approximations, can be used as a control; see Kjellberg [227].

It is possible to accelerate Weierstrass' method by using the new approximations of the roots in (6.5.23) as they become available. This leads to the iteration

$$\xi_i^{(k+1)} = \xi_i^{(k)} - p(\xi_i^{(k)}) \Big/ \left[a_0 \prod_{j<i} (\xi_i^{(k)} - \xi_j^{(k+1)}) \prod_{j>i} (\xi_i^{(k)} - \xi_j^{(k)}) \right], \quad i = 1 : n.$$

This *serial* version of the Weierstrass method can be shown to have an order of convergence at least $1 + \sigma_n$, where $1 < \sigma_n < 2$ is the unique positive root to $\sigma^n - \sigma - 1 = 0$.

If no a priori information about the roots is available, then the initial approximations $\xi_i^{(0)}$ can be chosen equidistantly on a circle $|z| = \rho$, centered at the origin, which encloses all the zeros of $p(z)$. Such a circle can be found by using the result that all the roots of the polynomial $p(z)$ lie in the disk $|z| \leq \rho$, where

$$\rho = \max_{1 \leq k \leq n} 2 \left(\frac{|a_k|}{|a_0|} \right)^{1/k}.$$

Note that this is (6.5.25) applied to the reciprocal polynomial.

The zeros z_i, $i = 1 : n$, of a polynomial of degree n,

$$p(z) = b_0 + b_1 z + \cdots + b_n z^n, \quad (b_0 b_n \neq 0), \tag{6.5.24}$$

are the poles of the rational function $r(z) = 1/p(z)$. Hence the progressive form of the qd algorithm, described in Sec. 3.5.5, can be used to simultaneously determine the zeros. This method can be considered as a modern, more powerful version of Bernoulli's method.

In the qd scheme for $r(z)$ the values in the first two rows can be expressed in terms of the coefficients of $p(z)$ as

$$q_1^{(0)} = -b_1/b_0, \qquad q_k^{(1-k)} = 0, \quad k > 1,$$

$$e_k^{(1-k)} = b_{k-1}/b_k, \quad k = 1 : n-1.$$

Further, since r has a zero of order n at infinity,

$$e_n^{(m)} = 0, \qquad m \geq -n.$$

Thus the qd scheme is flanked on both sides by a column of zeros. This allows the extended qd scheme for $r(z)$ to be constructed row by row. It can be shown that a sufficient condition for the qd scheme to exist is that the zeros are positive and simple. Then the kth q-column tends to z_k^{-1}. By reversing the order of the coefficients, i.e., by considering the polynomial $z^k p(z^{-1})$ instead, we can also construct a scheme where the q-columns tends to the zeros z_m.

Example 6.5.4.

The Laguerre polynomial of degree four is

$$L_4(z) = \frac{1}{24}(24 - 96z + 72z^2 - 16z^3 + z^4).$$

The corresponding qd scheme is shown in Table 6.5.1.

Table 6.5.1. *The qd scheme for computing the zeros of $L_y(z)$.*

	16.00000000		0		0		0	
0		−4.50000000		−1.33333333		−0.25000000		0
	11.50000000		3.16666667		1.08333333		0.25000000	
0		−1.23913043		−0.45614035		−0.05769231		0
	10.26086957		3.94965675		1.48178138		0.30769231	
0		−0.47697126		−0.17112886		−0.01197982		0
	9.78389831		4.25549915		1.64093042		0.31967213	
0		−0.20745829		−0.06598769		−0.00233381		0
	9.57644002		4.39696974		1.70458430		0.32200594	
0		−0.09525333		−0.02558161		−0.00044087		0
	9.48118669		4.46664146		1.72972504		0.32244681	
	⋮		⋮		⋮		⋮	
	9.39507749		4.53661379		1.74576103		0.32254769	
0		−0.00000340		−0.00000004		0.00000000		0
	9.39507409		**4.53661**715		**1.74576108**		**0.32254769**	

The zeros of $L_4(z)$ are correctly rounded to eight decimals:

$$9.39507091, \quad 4.53662030, \quad 1.74576110, \quad 0.32254769.$$

Note that the convergence of $e_i^{(n)}$ is linear with rate (z_{i+1}/z_i), $i = 1 : 3$, and can be accelerated with Aitken extrapolation.

6.5.5 A Modified Newton Method

There are three competing methods in current use for determining all zeros of a given polynomial. The Jenkins–Traub method [211], used in the IMSL library, is equivalent to a so-called variable-shift Rayleigh quotient iteration for finding the eigenvalues and eigenvectors of the companion matrix. By taking advantage of the matrix structure the work per iteration can be reduced to $O(n)$. A three stage procedure is used, each stage being characterized by the type of shift used. The code CPOLY (see [212]) is available via Netlib; see Sec. C.7 in Online Appendix C. For a description of the Jenkins–Traub method we refer to Ralston and Rabinowitz [296, Sec. 8.11].

The MATLAB zero-finding code `roots` applies the QR algorithm, which is a standard method for solving eigenvalue problems, to a balanced companion matrix. The balancing involves a diagonal similarity transformation,

$$\tilde{A} = DAD^{-1}, \qquad D = \mathrm{diag}\,(d_1, d_2, \dots, d_n),$$

which preserves the eigenvalues. The aim is to reduce the norm of A and thereby reduce the condition number of its eigenvalue problem. The balancing algorithm used is that of Parlett and Reinsch [286].

Another excellent algorithm is the modified Newton algorithm PA16, due to Madsen and Reid [253, 254], used by NAG Library. In its first stage it uses the Newton formula to find a search direction for minimizing $|p(z)|$. Once the iterates are close to a zero it enters stage 2 and switches to standard Newton. This will be described in more detail below.

Theoretical and experimental comparisons of the three algorithms above are given in [353]. It is shown that the Jenkins–Traub, Madsen–Reid, and QR algorithms all have roughly the same stability properties. The highest accuracy is typically achieved by PA16 and the next best by `roots`. A possible drawback with the QR algorithm is that it requires $\mathcal{O}(n^3)$ work and $\mathcal{O}(n^2)$ memory. Both the Madsen–Reid and Jenkins–Traub require only $\mathcal{O}(n^2)$ work $\mathcal{O}(n)$ memory. Since one is rarely interested in solving polynomial equations of high degree this is usually not important.

We now describe the modified Newton method due to Madsen [253]. By including a one-dimensional search along the Newton direction this method achieves good global convergence properties and is also effective for multiple roots.

To initialize let $z_0 = 0$,

$$\delta z_0 = \begin{cases} -p(0)/p'(0) = -a_n/a_{n-1} & \text{if } a_{n-1} \neq 0, \\ 1 & \text{otherwise,} \end{cases}$$

and take

$$z_1 = \frac{1}{2\rho} \frac{\delta z_0}{|\delta z_0|}, \qquad \rho = \max_{1 \le k \le n} \left(\frac{|a_{n-k}|}{|a_n|} \right)^{1/k}. \tag{6.5.25}$$

This assures that $|z_1|$ is less than the modulus of any zero of $p(z)$ (see [204, Exercise 2.2.11]). Further, if $p'(0) \neq 0$, it is in the direction of steepest descent of $|p(z)|$ from the origin (see Sec. 6.3.2). This choice makes it likely that convergence will take place to a root of near minimal modulus.

The general idea of the algorithm is that given z_k, a tentative step h_k is computed by Newton's method. The next iterate is found by taking the best point (in terms of minimizing $|f(z)|$) found by a short search along the line through z_k and $z_k + h_k$. When the search yields no better value than at z_k we take $z_{k+1} = z_k$ and make sure that the next search is shorter and in a different direction. Since the line searches will be wasteful if we are near a simple root, we then switch to the standard Newton's method.

In the first stage of the algorithm, when searches are being performed, new iterates z_{k+1} are computed as follows.

1. If the last iteration was successful ($z_k \neq z_{k-1}$), then the Newton correction

$$h_k = -p(z_k)/p'(z_k) \tag{6.5.26}$$

is computed. The next tentative step is taken as

$$\delta z_k = \begin{cases} h_k & \text{if } |h_k| \leq 3|z_k - z_{k-1}|, \\ 3|z_k - z_{k-1}|e^{i\theta} h_k/|h_k| & \text{otherwise,} \end{cases}$$

where θ is chosen rather arbitrarily as $\arctan(3/4)$. This change of direction is included because if a saddle point is being approached, the direction h_k may be a bad choice.

2. If the last step was unsuccessful ($z_k = z_{k-1}$) the search direction is changed and the step size reduced. In this case the tentative step is chosen to be

$$\delta z_k = -\frac{1}{2} e^{i\theta} \delta z_{k-1}.$$

Repeated use of this is sure to yield a good search direction.

3. Once the tentative step δz_k has been found the inequality $|p(z_k + \delta z_k)| < |p(z_k)|$ is tested. If this is satisfied the numbers

$$|p(z_k + p\,\delta z_k)|, \quad p = 1, 2, \dots, n,$$

are calculated as long as these are strictly decreasing. Note that if we are close to a multiple root of multiplicity m we will find the estimate $z_k + mh_k$, which gives quadratic convergence to this root. A similar situation will hold if we are at a fair distance from a cluster of m zeros and other zeros are further away.

If $|p(z_k + \delta z_k)| \geq |p(z_k)|$, we calculate the numbers

$$|p(z_k + 2^{-p} \delta z_k)|, \quad p = 0, 1, 2,$$

again continuing until the sequence ceases to decrease.

A switch to standard Newton is made if in the previous iteration a standard Newton step $z_{k+1} = z_k + h_k$ was taken, and Theorem 6.3.3 ensures the convergence of Newton's method with initial value z_{k+1}, i.e., when $f(z_k)f'(z_k) \neq 0$ and

$$2\,|f(z_k)| \max_{z\in K_k} |f''(z)| \leq |f'(z_k)|^2, \quad K_k : |z - z_k| \leq |h_k|,$$

is satisfied; cf. (6.3.20). This inequality can be approximated using already computed quantities by

$$2\,|f(z_k)||f'(z_k) - f'(z_{k-1})| \leq |f'(z_k)|^2 |z_{k-1} - z_k|. \tag{6.5.27}$$

The iterations are terminated and z_{k+1} accepted as a root whenever $z_{k+1} \neq z_k$ and

$$|z_{k+1} - z_k| < u|z_k|$$

holds, where u is the unit roundoff. The iterations are also terminated if

$$|p(z_{k+1})| = |p(z_k)| < 16nu|a_n|,$$

where the right-hand side is a generous overestimate of the final roundoff made in computing $p(z)$ at the root of the smallest magnitude. The polynomial is then deflated as described in the previous section.

More details about this algorithm and methods for computing error bounds can be found in [253, 254].

6.5.6 Sturm Sequences

Precise information about the number of real zeros of a polynomial $p(z)$ in an interval $[a, b]$, $-\infty \leq a < b \leq \infty$, can be obtained from a **Sturm sequence**[188] for $p(z)$.

Definition 6.5.1.

A sequence of real polynomials $p_0(x), p_1(x), \dots, p_m(x)$ is a strict Sturm sequence for $p(x) = p_0(x)$ on the interval $[a, b]$ if the following conditions hold:

(i) *No two consecutive polynomials in the sequence vanish simultaneously on the interval $[a, b]$.*

(ii) *If $p_j(r) = 0$ for some $j < m$, then $p_{j-1}(r)p_{j+1}(r) < 0$.*

(iii) *Throughout the interval $[a, b]$, $p_m(x) \neq 0$.*

(iv) *If $p_0(r) = 0$, then $p_0'(r)p_1(r) > 0$.*

Given a polynomial $p_1(x)$ of degree not greater than that of $p_0(x)$ a Sturm sequence can be constructed by the Euclidean algorithm as follows. Let $q_1(x)$ be the quotient polynomial and $-p_2(x)$ the remainder in the quotient $p_0(x)/p_1(x)$, i.e., $p_0(x) = q_1(x)p_1(x) - p_2(x)$,

[188] J. C. F. Sturm (1803–1855), a Swiss mathematician best known for his theorem on Sturm sequences, discovered in 1829, and his theory of Sturm–Liouville differential equations. In 1839 he succeeded Poisson in the chair of mechanics at the École Polytechnique, Paris.

where the degree of $p_2(x)$ is strictly less than that of $p_1(x)$. Continuing in this way, we compute $p_2(x), \dots, p_m(x)$ by

$$p_{k+1}(x) = q_k(x)p_k(x) - p_{k-1}(x), \quad k = 1 : m-1, \tag{6.5.28}$$

where $q_k(x)$ is the quotient and $-p_{k+1}$ the remainder in the quotient $p_{k-1}(x)/p_k(x)$. We stop when $p_m(x)$ nowhere vanishes on the interval $[a, b]$. Clearly, if $p_j(r) = 0$, then $p_{j+1}(r) = -p_{j-1}(r) < 0$, so condition (ii) is satisfied.

Let $V(x)$ denote the number of variations in sign in the Sturm sequence at x. If $p_0(x)$ and $p_1(x)$ have only simple zeros that separate each other, then it can be shown that the number of zeros of $p_0(x)$ on $[a, b]$ is equal to $|V(a) - V(b)|$.

Theorem 6.5.2.

Take $p_1(x) = p_0'(x)$ and define $p_2(x), \dots, p_m(x)$ by (6.5.28), where $p_m(x)$ has a fixed sign on the interval $[a, b]$ ($p_0(a) \neq 0$ and $p_0(b) \neq 0$). Let V(r) denote the number of variations of sign in the sequence of values

$$p_0(r), p_1(r), \dots, p_m(r),$$

vanishing terms not being counted. Then the number of roots of $p_0(x)$ in $[a, b]$, each multiple root being counted once, is exactly equal to $|V(a) - V(b)|$.

Note that if all real zeros of $p_0(x)$ are simple and $p_1(x) = p_0'(x)$, then (6.5.28) generates a Sturm sequence. If $p_0(x)$ has multiple zeros, then $p_0(x)$ and $p_0'(x)$ have a common divisor, which divides every $p_i(x)$ in the sequence, and this will not affect $V(r)$.

Example 6.5.5.

The equation $p(x) = p_0 = x^5 - 3x - 1 = 0$ has three real roots, $z_1 = -1.21465$, $z_2 = -0.33473$, and $z_3 = 1.38879$, and two complex roots. The derivative equals $p'(x) = p_1 = 5x^4 - 3$, and the rest of the Sturm chain is given by

$$p_2 = \frac{12}{5}x + 1, \qquad p_3 = \frac{59083}{20736}.$$

Here p_2 is a polynomial of degree one and the Sturm chain ends with $s = 3 < n$.

We denote by $[l_k, u_k]$ an interval containing the zero x_k. Evaluating the sign changes of the Sturm sequence at $x = -2$ and $x = 2$ shows that there are $3 - 0 = 3$ roots x_k, $k = 1, 2, 3$, in the interval $[-2, 2]$. Counting the number of sign changes at the midpoint $x = 0$ allows us to deduce that $u_k = 0$, $k = 1, 2$, and $l_3 = 0$; see Table 6.5.2. The interval $[-2, 0]$ contains two roots so we determine next the number of sign changes at the midpoint $x = -1$.

At this point we have determined three disjoint intervals $[-2, -1]$, $[-1, 0]$, and $[0, 2]$, which each contain one root. We continue bisecting each of these intervals, which can be performed in parallel.

Methods based on Sturm sequences can be competitive, when only a relatively small number of real roots in a given interval are of interest. Consider a real symmetric tridiagonal

Table 6.5.2. *Left: Sign variations in the Sturm sequence. Right: Intervals $[l_k, u_k]$ containing the zero x_k.*

x	p_0	p_1	p_2	p_3	δ
-2	$-$	$+$	$-$	$+$	3
$+2$	$+$	$+$	$+$	$+$	0
0	$-$	$-$	$+$	$+$	1
-1	$+$	$+$	$-$	$+$	2
1	$-$	$+$	$+$	$+$	1

l_1	u_1	l_2	u_2	l_3	u_3
-2	2	-2	2	-2	2
	0		0	0	
	-1	-1			
				1	

matrix,

$$A = \begin{pmatrix} \alpha_1 & \beta_2 & & & \\ \beta_2 & \alpha_2 & \beta_3 & & \\ & \ddots & \ddots & \ddots & \\ & & \beta_{n-1} & \alpha_{n-1} & \beta_n \\ & & & \beta_n & \alpha_n \end{pmatrix},$$

such that $\beta_k \neq 0$, $k = 2 : n$, has only simple eigenvalues. Let $p_k(\lambda)$ be the characteristic polynomial of the kth leading principal minor of $(A - \lambda I)$. Define $p_0(\lambda) = 1$, and $p_k(\lambda)$ by the three-term recursion

$$p_1(\lambda) = \alpha_1 - \lambda, \quad p_k(\lambda) = (\alpha_k - \lambda)p_{k-1}(\lambda) - \beta_k^2 p_{k-2}(\lambda), \quad k = 2 : n. \tag{6.5.29}$$

Then the sequence

$$1,\ p_0(\lambda), \dots, p_n(\lambda) = \det(A - \lambda I)$$

is known to form a Sturm sequence. Combined with the bisection method, this recursion can be used to develop an efficient numerical method for determining the eigenvalues of a symmetric tridiagonal matrix A in a given interval $[a, b]$ without reference to any of the others. It can also be used for determining the singular values of a bidiagonal matrix in a given interval; see Volume II.

The Sturm sequence algorithm only works when $f(x)$ is a real function of a real variable. To determine complex zeros an algorithm that performs a search and exclusion of the complex plane can be used. The **quadtree** exclusion algorithm, due to H. Weyl [372] and illustrated in Figure 6.5.1, is such a "two-dimensional bisection algorithm."[189] It was one of the first algorithms with guaranteed convergence to all n zeros of a polynomial of degree n. The algorithm is based on an **exclusion test** applied to squares in the complex plane. Assume that $f(z)$ is analytic in K and that

$$|f'(z)| \leq M \quad \forall\ z \in K.$$

Then if $|f(z_0)| > \eta M$ there can be no zero of $f(z)$ in K. Any square that does not pass this exclusion test may contain a root and is called suspect. (Note that it is not required that a suspect square actually contains a root.)

The computations begin with an initial suspect square S containing all the zeros of $p(x)$. This square can be found from an upper bound on the absolute value of the zeros of

[189] In general a quadtree is a tree where each node is split along d dimensions giving 2^d children.

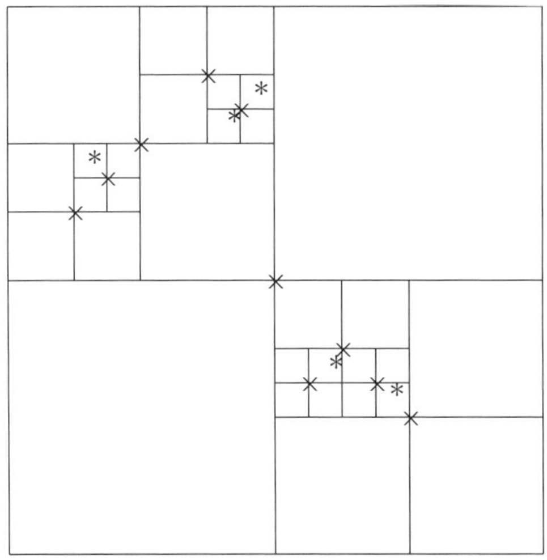

Figure 6.5.1. *Suspect squares computed by Weyl's quadtree method. Their centers (marked by $\times$) approximate the five zeros marked by $*$.*

$p(x)$. In the algorithm, as soon as we have a suspect square, this is partitioned into four congruent subsquares. At the center of each of them a test estimating the distance to the closest zero of $p(x)$ is performed. (A relative error within, say, 40% will suffice.) If the test guarantees that this distance exceeds half of the length of the diagonal of the square, then the square cannot contain any zero and is discarded. Each remaining suspect square undergoes the same recursive partitioning into four subsquares and the test. The zeros lying in a suspect square are approximated by its center with errors bounded by half the length of its diagonal. Each iteration step decreases the diagonal of the remaining squares by a factor of two so the errors will decrease by a factor of $1/2$.

6.5.7 Finding Greatest Common Divisors

A problem that arises in many applications, including computer graphics, geometric modelling, and control theory, is the computation of the **greatest common divisor** (GCD) of two polynomials:

$$p(z) = a_0 \prod_{j=1}^{m} (z - \alpha_i) = a_0 z^n + a_1 z^{n-1} + \cdots + a_n, \quad a_0 \neq 0,$$

$$q(z) = b_0 \prod_{j=1}^{m} (z - \beta_j) = b_0 z^m + b_1 z^{n-1} + \cdots + b_m, \quad b_0 \neq 0.$$

The GCD can in principle be determined by the Euclidean algorithm; see Problem 1.2.6. Assume that $n > m$ and perform the divisions

$$p(z) = q(z)s(z) + r_1(z),$$
$$q(z) = r_1(z)s_1(z) + r_2(z),$$

$$q(z) = r_1(z)s_1(z) + r_2(z), \tag{6.5.30}$$
$$\vdots$$
$$r_{s-2}(z) = r_{s-1}(z)s_{s-1}(z) + r_s(z).$$

The degrees of the polynomials r_i, $i = 1 : s$, decrease at each step and we stop when r_s is a constant. If $p(z)$ and $q(z)$ have a common factor, then that factor is a divisor of r_i, $i = 1 : s$, as well. This means that if $r_s \neq 0$, then $p(z)$ and $q(z)$ have no common divisor; otherwise r_{s-1} is the greatest common divisor.

From the representation of $r_s(z)$ in (6.5.30) it follows that if $p(z)$ and $q(z)$ are two polynomials, then there always exist polynomials $x(z)$ and $y(z)$ such that $xp - yq$ equals a constant. If $p(z)$ and $q(z)$ have a common divisor, then $xp(z) - yq(z) = 0$.

The Euclidean algorithm is suitable for symbolic computation, but not effective in finite precision arithmetic because of roundoff. We now describe a suitable algorithm for determining the GCD when the coefficients of $p(z)$ and $q(z)$ are given as floating-point numbers. Note that in this case we can only hope to determine "almost common factors."

Consider the product of the mn factors,

$$\bar{R}(p, q) = \prod_{i=1}^{n}\prod_{j=1}^{m}(\beta_j - \alpha_i), \tag{6.5.31}$$

of all possible differences between the two sets of roots. This is a symmetric function of the α's and β's and hence expressible as a polynomial in the quotients a_i/a_0 and b_j/b_0. Clearly $\bar{R}(p, q)$ is zero if and only if $p(z)$ and $q(z)$ have a common zero, and no polynomial of lower degree in the α's and β's can have this property. The polynomial $R(p, q) = a_0^m b_0^n \bar{R}(p, q)$ is called the **resultant** of $p(z)$ and $q(z)$ and can also be expressed as

$$R(p, q) = a_0^m b_0^n \bar{R}(p, q) \equiv a_0^m \prod_{i=1}^{n} q(\alpha_i) \equiv b_0^n \prod_{j=1}^{m} p(\beta_j). \tag{6.5.32}$$

From the coefficients of $p(z)$ and $p(z)$ we form the **Sylvester matrix**[190]

$$S(p, q) = \begin{pmatrix} T_m(p) \\ T_n(q) \end{pmatrix} \in \mathbf{R}^{(m+n)\times(m+n)}, \tag{6.5.33}$$

where

$$T_m(p) = \begin{pmatrix} a_0 & a_1 & \cdots & a_{n-1} & a_n & & & \\ & a_0 & a_1 & \cdots & a_{n-1} & a_n & & \\ & & \ddots & \ddots & \ddots & \ddots & & \\ & & & a_0 & a_1 & \cdots & a_{n-1} & a_n \end{pmatrix} \in \mathbf{R}^{m\times(m+n)},$$

$$T_n(q) = \begin{pmatrix} b_0 & b_1 & \cdots & b_{m-1} & b_m & & & \\ & b_0 & b_1 & \cdots & b_{m-1} & b_m & & \\ & & \ddots & \ddots & \ddots & \ddots & & \\ & & & b_0 & b_1 & \cdots & b_{m-1} & b_m \end{pmatrix} \in \mathbf{R}^{n\times(m+n)}.$$

190 The Sylvester matrix appeared in [346].

Note the Toeplitz structure of the two blocks of rows. It can be shown that *the resultant* $R(p, q)$ *equals the determinant of* $S(p, q)$ *except for a possible multiplier* -1. Thus $p(z)$ and $q(z)$ have a common zero if and only if the Sylvester matrix is singular.

If we form $m+n$ polynomials of degree less than or equal to $m+n-1$ by multiplying $p(z)$ with z^k, $k = 1 : m$, and multiplying $q(z)$ with z^k, $k = 1 : n$, this can be written in matrix notation as

$$S(p, q)\,(\, z^{m+n-1} \quad \cdots \quad z \quad 1\,)^T = \begin{pmatrix} f_p \\ f_q \end{pmatrix},$$

where

$$f_p = p(z)\,(\, z^{m-1} \quad \cdots \quad z \quad 1\,)^T, \qquad f_q = q(z)\,(\, z^{n-1} \quad \cdots \quad z \quad 1\,)^T.$$

If γ is a common root to the two polynomials $p(z)$ and $q(z)$, then

$$S(p, q)(\gamma^{m+n-1}, \dots, \gamma, 1)^T = 0.$$

In other words this vector belongs to the nullspace of $S(p, q)$. The degree of the GCD of $p(z)$ and $q(z)$ equals the dimension of the nullspace of the Sylvester matrix, i.e.,

$$\deg\,(\gcd\,(p, q)) = m + n - r, \qquad r = \operatorname{rank}\,(S(p, q)).$$

The best way to estimate the rank is by computing the SVD of the Sylvester matrix; see Theorem 1.4.4. The coefficients of the GCD can also be obtained from the decomposition.

Review Questions

6.5.1 Describe the method of iterated successive synthetic division for computing function values and derivatives of a polynomial.

6.5.2 Consider the polynomial $p(z) = z^4 - 2z^3 - 4z^2 + z + 1$. Using Descartes's rule of sign, what can you deduce about the number of real positive roots?

6.5.3 Suppose that all roots of a polynomial equation are to be determined. Describe two methods to avoid the problem of repeatedly converging to roots already found.

6.5.4 Discuss the ill-conditioning of roots of polynomial equations. What famous polynomial did Wilkinson use as an example?

6.5.5 (a) What is the companion matrix of a polynomial $p(x) = x^n + a_1x^{n-1} + \cdots + a_{n-1}x + a_n$?

(b) One approach to computing the eigenvalues of a matrix A is to find the coefficients of the characteristic polynomial $p_A(\lambda) = \det(\lambda I - A)$, and then solve the algebraic equation $p_A(\lambda) = 0$. Why should such a method usually be avoided?

6.5.6 What properties are satisfied by a Sturm sequence of real polynomials $p_0(x), p_1(x), \dots, p_m(x)$? Describe one way of generating a Sturm sequence using the Euclidean algorithm.

Problems and Computer Exercises

6.5.1 Apply Newton's method to determine one of the complex roots of the equation $z^2+1 = 0$. Start with $z_0 = 1 + i$.

6.5.2 Consider a polynomial with real coefficients

$$p(z) = a_0 z^n + a_1 z^{n-1} + \cdots + a_n, \quad a_i \neq 0, \quad i = 0 : n.$$

(a) Count the number of (real) additions and multiplications needed to compute a value $p(z_0)$ by synthetic division of $p(z)$ by $(z - z_0)$, when z_0 is a real and complex number, respectively.

(b) For a complex number $z_0 = x_0 + iy_0$, $p(z_0)$ can also be computed by performing the synthetic division of $p(z)$ with the real quadratic factor

$$d(z) = (z - z_0)(z - \bar{z}_0) = z^2 - 2x_0 z + (x_0^2 + y_0^2).$$

Derive a recursion for computing the quotient polynomial $q(z)$ and $p(z_0) = b_{n-1}z_0 + b_n$, where

$$q(z) = b_0 z^{n-2} + b_1 z^{n-3} + \cdots + b_{n-2},$$
$$p(z) = q(z)d(z) + b_{n-1}z + b_n.$$

Count the number of real additions and multiplications needed to compute $p(z_0)$ and also show how to compute $p'(z_0)$.

6.5.3 (a) Using the Cardano–Tartaglia formula the real root α to the equation $x^3 = x + 4$ can be written in the form

$$\alpha = \sqrt[3]{2 + \frac{1}{9}\sqrt{321}} + \sqrt[3]{2 - \frac{1}{9}\sqrt{321}}.$$

Use this expression to compute α. Discuss the loss of accuracy due to cancellation.

(b) Compute α to the same accuracy by Newton's method using the initial approximation $x_0 = 2$.

6.5.4 Let C be the companion matrix of $p(x)$. Show that in Bernoulli's method the vector sequence $m_n = (y_{n+k}, \ldots, y_{n+1}, y_n)^T$ can be generated by forming successive matrix-vector products $m_n = Cm_{n-1}$. Conclude that $m_k = C^k m_0$.

6.5.5 (a) Assume that the zeros of a polynomial satisfy the stronger conditions $|u_1| > |u_2| > |u_3| \geq \cdots \geq |u_k|$, where u_1 and u_2 are real and simple. Show that in Bernoulli's method

$$\lim_{n\to\infty} D_n/D_{n-1} = u_1 u_2, \quad D_n = \begin{vmatrix} y_n & y_{n+1} \\ \Delta y_n & \Delta y_{n+1} \end{vmatrix}. \tag{6.5.34}$$

(b) The result in (a) can be combined with (6.5.18) to also compute u_2. What is the rate of convergence in this process?

(c) Apply Bernoulli's method to the polynomial p_4 in Example 6.5.2 to improve the approximations computed by Graeffe's method.

6.5.6 (a) Use the MATLAB command `x = roots(poly(1:20))` to compute the roots of the classical Wilkinson polynomial $\prod_{k=1}^{20}(x-k)$. Use IEEE double precision. What is the maximum error in the roots?

(b) Using the command `poly(x)` to compute the coefficients of the polynomial with the erroneous roots from (a). What is the maximum error in these coefficients?

6.5.7 (a) Generate the coefficients of the Legendre polynomials $P_n(x)$, $n = 2 : 24$, using the three-term recursion $P_0(x) = 1$, $P_1(x) = x$,

$$P_{n+1}(x) = \frac{2n+1}{n+1} x P_n(x) - \frac{n}{n+1} P_{n-1}(x), \quad n \geq 1,$$

where the recursion coefficients are exact rational numbers. Compute using IEEE double precision the roots of $P_{24}(x)$ as in Problem 6.5.6. Verify that the errors in some computed roots λ_k^* close to 1 exceed $10^6 u$, where $u = 1.11\,10^{-16}$ is the unit roundoff.

(b) Compute the coefficients of the polynomial $\prod_{k=1}^{24}(x-\lambda^*)$. Show that the relative errors in the computed coefficients are all less than $16u$.

6.5.8 Consider the iteration $z_{n+1} = z_n^2 + c$, where $c = p + iq$ is a fixed complex number. For a given z_0 the sequence of iterates $z_n = x_n + iy_n$, $n = 0, 1, 2, \dots$, may either converge to one of the two roots of the quadratic equation $z^2 - z + c = 0$ or diverge to infinity. Consider z_0 chosen, for example, in the unit square of the complex plane. The boundary separating the region of convergence from other points in the plane is a very complex **fractal curve** known as the **Julia set**. The **Mandelbrot set** is obtained by fixing $z_0 = 0$ and sweeping over values of c in a region of the complex plane.

(a) Picture the Julia set as follows. Set $c = 0.27334 + 0.000742i$. Sweep over points of z_0 in the region $-1 \leq \Re z_0 \leq 1$, $-1.3 \leq \Im z_0 \leq 1.3$. If $|z_N| < R$, for $N = 100$ and $R = 10$, color the point z_0 black; otherwise color the point from hot (red) to cool (blue) according to how fast the iteration is diverging, i.e., according to how fast the inequality $|z_n| > R$ becomes satisfied.

(b) Picture the Mandelbrot set in a similar way. Sweep over values of c in the region $-2.25 \leq \Re c \leq 0.75$, $-1.5 \leq \Im c \leq 1.5$.

6.5.9 The following MATLAB function, `sylvester`, generates the Sylvester matrix of two polynomials whose coefficients are given in the row vectors a and b:

```
function S = sylvest(a,b);
% SYLVEST computes
n = length(a) - 1;
m = length(b) - 1;
r = [a,zeros(1,m-1)];
c = [a(1),zeros(1,m-1)];
T1 = toeplitz(c,r);
r = [b,zeros(1,n-1)];
c = [b(1),zeros(1,n-1)];
T2 = toeplitz(c,r);
S = [T1; T2];
```

Generate $S(p, q)$ for the two polynomials

$$p(x) = x^5 + x^4 + x^3 + x^2 + x + 1,$$
$$q(x) = x^4 - 2x^3 + 3x^2 - x - 71.$$

The GCD of these polynomials is $x + 1$. How is this revealed by the SVD of $S(p, q)$?
(b) Try some more difficult examples with several common zeros.

Notes and References

The general idea of solving an equation by repeatedly improving an estimate of the solution has been used in many cultures for thousands of years. Examples are ancient Greek and Babylonian methods as well as Arabic algebraists from the eleventh century.

An interesting historical account of Newton's method is given in Ypma [388]. Newton's method is contained in his book *Method of Fluxions*, written in 1671 but not published until 1736. Joseph Raphson was allowed to see Newton's work and Newton's method was first published in a book by Raphson 1690. This is why the method in English literature is often called the Newton–Raphson method. The first to give a modern description of Newton's method using derivatives seems to have been Thomas Simpson 1740. Edmond Halley was a contemporary of Isaac Newton and his third-order method was published more than 300 years ago [179].

Several comprehensive monographs dealing with methods for solving scalar nonlinear equations are available. Traub [354] gives an exhaustive enumeration of iteration methods with and without memory, with their order of convergence and efficiency index. Much classical material is also found in Ostrowski [279]. The recently reprinted book by Brent [44] deals exclusively with methods which only use function values for finding zeros and minima of functions of a single variable. It is unique in its careful treatment of algorithmic details which are crucial when developing reliable computer codes.

Fortran and C versions of some of the zero-finding and minimization routines given in [44] are available from Netlib. The MATLAB function `fminbnd` is based on the Fortran implementation FMIN of Brent's algorithm given in Forsythe, Malcolm, and Moler [123, pp.184–187].

Halley's method has been rediscovered by J. H. Lambert [234] and numerous other people; see Traub [354, Sec. 5.22]. A nice exposition of this and other third-order methods is given by Gander [129]. Families of iteration methods of arbitrary order are studied in a remarkable paper by Schröder [316], which has been translated into English by G. W. Stewart [317]. The determinant family of iteration functions $B_p(x)$ is a special case of a parametrized family of iteration functions for polynomials given by Traub [356]; see also [204, Sec. 4.4]. This family was derived independently by Kalantari, Kalantari, and Zaare-Nahandi [221].

There is a vast literature on methods for the classical problem of solving algebraic equations. Indeed, the nonexistence of an algebraic formula for equations of fifth and higher degree gave rise to modern algebra. Householder [204] gives an elegant treatment and is an excellent source for classical results. Detailed surveys are found also in Durand [102] (in French) and Sendov, Andreev, and Kjurkchiev [320]. Numerical polynomial algebra is an emerging area that falls between numerical analysis and computer algebra. A comprehensive survey of this area is given by Stetter [334]. Further studies of the speed and accuracy of computing polynomial zeros by the MATLAB method of solving the eigenvalue problem for the companion matrix are given in [157, 104].

The theory of Sturm sequences is treated in [204, Sec. 2.5]. The quadtree method was used by Weyl [372] to give a constructive proof of the fundamental theorem of algebra. An interesting analysis of the efficient implementation of this method is given by Pan [282], who also gives a brief account of the history of algorithms for solving polynomial equations. Some background on algorithms for computing the GCD is found in [2].

Bibliography

[1] Milton Abramowitz and Irene A. Stegun (eds.). *Handbook of Mathematical Functions*. Dover, Mineola, NY, 1965. Republication of work published in 1964 by National Bureau of Standards (Cited on pp. xix, 67, 77, 164, 167, 168, 189, 208, 210, 212, 260, 260, 269, 269, 270, 270, 291, 301, 301, 315, 315, 316, 316, 318, 318, 320, 328, 349, 467, 529, 535, 545, 546, 555, 572, 572, 574, 587, 593, 606, 607, 607.)

[2] A. G. Akritas. A new method for computing polynomial greatest common divisor and remainder sequences. *Numer. Math.*, 52:119–127, 1988. (Cited on p. 686.)

[3] Götz Alefeld and Jürgen Herzberger. *Introduction to Interval Computation*. Academic Press, New York, 1983. Translated from German by Jon Rokne. (Cited on p. 147.)

[4] Götz Alefeld and Günter Mayer. Interval analysis: theory and applications. *J. Comput. Appl. Math.*, 121:421–464, 2000. (Cited on p. 147.)

[5] G. S. Ammar, W. B. Gragg, and Lothar Reichel. An analogue for Szegő polynomials of the Clenshaw algorithm. *J. Comput. Appl. Math.*, 46:211–216, 1993. (Cited on p. 466.)

[6] Edward Anderson, Zhaojun Bai, Christian Bischof, S. Blackford, James W. Demmel, Jack J. Dongarra, Jeremy Du Croz, Anne Greenbaum, Sven Hammarling, A. McKenney, and D. C. Sorensen, editors. *LAPACK Users' Guide*. SIAM, Philadelphia, PA, third edition, 1999. (Cited on p. 42.)

[7] Tom M. Apostol. *Mathematical Analysis*. Addison Wesley, Reading, MA, second edition, 1974. (Cited on p. 84.)

[8] Mario Arioli, Hans Z. Munthe-Kaas, and L. Valdettaro. Componentwise error analysis for FFTs with applications to fast Helmholtz solvers. *Numer. Algorithms*, 12:65–88, 1996. (Cited on p. 519.)

[9] David H. Bailey. Algorithm 719: Multiprecision translation and execution of FORTRAN programs. *ACM Trans. Math. Software*, 19(3):288–319, 1993. (Cited on p. 105.)

[10] David H. Bailey. A Fortran 90-based multiprecision system. *ACM Trans. Math. Software*, 21(4):379–387, 1995. (Cited on p. 105.)

[11] David H. Bailey, Jon M. Borwein, Peter B. Borwein, and Simon Plouffe. The quest for pi. *Notices Amer. Math. Soc.*, 19(1):50–57, 1975. (Cited on pp. 104, 655.)

[12] N. T. J. Bailey. A study of queues and appointment systems in hospital outpatient departments, with special reference to waiting times. *J. Roy. Statist. Soc. Ser. B*, 14:185–199, 1952. (Cited on p. 79.)

[13] G. A. Baker, Jr. *Essentials of Padé Approximants*. Academic Press, New York, 1975. (Cited on p. 349.)

[14] G. A. Baker, Jr. and P. Graves-Morris. *Padé Approximants*. Encyclopedia Math. Appl. 59, Cambridge University Press, Cambridge, UK, second edition, 1996. (Cited on p. 349.)

[15] Jesse L. Barlow and E. H. Bareiss. On roundoff error distributions in floating point and logarithmic arithmetic. *Computing*, 34:325–347, 1985. (Cited on p. 156.)

[16] R. W. Barnard, Germund Dahlquist, K. Pearce, Lothar Reichel, and K. C. Richards. Gram polynomials and the Kummer function. *J. Approx. Theory*, 94:128–143, 1998. (Cited on p. 464.)

[17] R. H. Bartels, A. R. Conn, and C. Charalambous. On Cline's direct method for solving overdetermined linear systems in the l_∞ sense. *SIAM J. Numer. Anal.*, 15:255–270, 1978. (Cited on p. 472.)

[18] R. H. Bartels, A. R. Conn, and J. W. Sinclair. Minimization techniques for piecewise differentiable functions: The l_1 solution to an overdetermined linear system. *SIAM J. Numer. Anal.*, 15:224–241, 1978. (Cited on p. 473.)

[19] F. L. Bauer. Genauigkeitsfragen bei der Lösung linearer Gleichungssysteme. *Z. Angew. Math. Mech.*, 46:7:409–421, 1966. (Cited on p. 156.)

[20] F. L. Bauer, Heinz Rutishauser, and E. Stiefel. New aspects in numerical quadrature. In *Proc. of Symposia in Appl. Math.*, volume 15, pages 199–218. Amer. Math. Soc., Providence, RI, 1963. (Cited on pp. 548, 549, 550.)

[21] B. J. C. Baxter and Arieh Iserles. On the foundations of computational mathematics. In P. G. Ciarlet and F. Cucker, editors, *Handbook of Numerical Analysis*, volume XI, pages 3–34. North–Holland Elsevier, Amsterdam, 2003. (Cited on p. 84.)

[22] B. K. Beatson. On the convergence of some cubic spline interpolation schemes. *SIAM J. Numer. Anal.*, 23:903–912, 1986. (Cited on pp. 422, 425.)

[23] Jean-Paul Berrut and Hans D. Mittelmann. Matrices for the direct determination of the barycentric weights of rational interpolation. *J. Comput. Appl. Math.*, 78:355–370, 1997. (Cited on pp. 394, 395.)

[24] Jean-Paul Berrut and Lloyd N. Trefethen. Barycentric Lagrange interpolation. *SIAM Review*, 46(3):501–517, 2004. (Cited on pp. 367, 518.)

[25] G. W. Bickley. Difference and associated operators, with some applications. *J. Math. Phys. MIT*, 27:183–192, 1948. (Cited on p. 231.)

[26] O. Biermann. *Volesungen über Mathematische Näherungsmetoden*. Vieweg, Braunschweig, 1905. (Cited on p. 397.)

[27] Garret Birkhoff. Piecewise polynomial interpolation and approximation. In Henry L. Garabedian, editor, *Approximation of Functions*, pages 164–190. Elsevier, New York, 1965. (Cited on p. 518.)

[28] Garret Birkhoff and Henry L. Garabedian. Smooth surface interpolation. *J. Math. Phys.*, 39:164–190, 1960. (Cited on p. 411.)

[29] Åke Björck and Tommy Elfving. Algorithms for confluent Vandermonde systems. *Numer. Math.*, 21:130–137, 1973. (Cited on p. 382.)

[30] Åke Björck and Gene H. Golub. Eigenproblems for matrices associated with periodic boundary problems. *SIAM Review.*, 19(1):5–16, 1977. (Cited on p. 424.)

[31] Åke Björck and Victor Pereyra. Solution of Vandermonde systems of equations. *Math. Comp.*, 24:893–903, 1970. (Cited on pp. 376, 518.)

[32] Harry Björk. Contribution to the problem of least squares approximations. Tech. Report TRITA-NA-7137, Dept. Comp. Sci., Royal Institute of Technology, Stockholm, 1971. (Cited on p. 471.)

[33] Petter Bjørstad, Germund Dahlquist, and Eric Grosse. Extrapolation of asymptotic expansions by a modified Aitken δ^2-formula. *BIT*, 21(1):56–65, 1981. (Cited on pp. 276, 312, 312.)

[34] G. Blanche. Numerical evaluation of continued fractions. *SIAM Review*, 6(4):383–421, 1964. (Cited on p. 349.)

[35] G. A. Bliss. *Algebraic Functions*. Amer. Math. Soc., New York, 1933. Republished in 1947 by Edwards Brothers, Ann Arbor, MI. (Cited on p. 204.)

[36] Carl de Boor. On calculating with B-splines. *J. Approx. Theory*, 6:50–62, 1972. (Cited on pp. 418, 435.)

[37] Carl de Boor. *A Practical Guide to Splines*. Springer-Verlag, Berlin, 1991. Revision of first edition from 1978. (Cited on pp. 421, 423, 431, 432, 434, 435, 518, 518.)

[38] Carl de Boor. Divided differences. *Surveys in Approximation Theory*, 1:49–69, 2005. (Cited on p. 518.)

[39] Carl de Boor, Klaus Höllig, and Sherman Riemenschneider. *Box Splines*. Applied Mathematical Sciences, Volume 98. Springer-Verlag, New York, 1993. (Cited on p. 519.)

[40] Carl de Boor and Allan Pinkus. Backward error analysis for totally positive linear systems. *Numer. Math.*, 27:485–490, 1977. (Cited on p. 434.)

[41] Folkmar Bornemann, Dirk Laurie, Stan Wagon, and Jürg Waldvogel. *The SIAM 100-digit Challenge. A Study in High-Accuracy Numerical Computing*. SIAM, Philadelphia, PA, 2004. (Cited on p. 558.)

[42] Tibor Boros, Thomas Kailath, and Vadim Olshevsky. A fast parallel Björck–Pereyra-type algorithm for parallel solutions of Cauchy linear systems. *Linear Algebra Appl.*, 302–303:265–293, 1999. (Cited on pp. 376, 377.)

[43] Jon M. Borwein, Peter B. Borwein, and K. Dilcher. Pi, Euler numbers and asymptotic expansions. *Amer. Math. Monthly*, 96:681–687, 1989. (Cited on p. 321.)

[44] Richard P. Brent. *Algorithms for Minimization without Derivatives*. Prentice-Hall, Englewood Cliffs, NJ, 1973. Republished by Dover Publications, Mineola, NY, 2002. (Cited on pp. 478, 622, 635, 661, 662, 685, 685.)

[45] Richard P. Brent. Algorithm 524: A Fortran multiple-precision arithmetic package. *ACM Trans. Math. Software*, 4(1):71–81, 1978. (Cited on p. 105.)

[46] Richard P. Brent. A Fortran multiple-precision arithmetic package. *ACM Trans. Math. Software*, 4(1):57–70, 1978. (Cited on p. 105.)

[47] Richard P. Brent and H. T. Kung. Fast algorithms for manipulating formal power series. *J. Assoc. Comput. Mach.*, 25:581–595, 1978. (Cited on p. 179.)

[48] Richard P. Brent, Colin Percival, and Paul Zimmermann. Error bounds on complex floating-point multiplication. *Math. Comp.*, 76:1469–1481, 2007. (Cited on p. 112.)

[49] Claude Brezinski. *Accélération de la Convergence en Analyse Numérique*. Number 584 in Lecture Notes in Mathematics. Springer Verlag, Berlin, 1977. (Cited on p. 307.)

[50] Claude Brezinski. *Padé-Type Approximations and General Orthogonal Polynomials*. Birkhäuser Verlag, Basel, 1980. (Cited on p. 332.)

[51] Claude Brezinski. *History of Continued Fractions and Padé Approximants*. Springer-Verlag, Berlin, 1991. (Cited on pp. 332, 349.)

[52] Claude Brezinski. *Projection Methods for Systems of Equations*. Elsevier, Amsterdam, 1997. (Cited on p. 349.)

[53] Claude Brezinski. Convergence acceleration during the 20th century. *J. Comput. Appl. Math.*, 122:1–21, 2000. (Cited on p. 349.)

[54] Claude Brezinski and J. Van Iseghem. A taste of Padé approximation. *Acta Numerica*, 4:53–103, 1995. (Cited on p. 336.)

[55] Claude Brezinski and Michela Redivo-Zaglia. *Extrapolation Methods. Theory and Practice*. North-Holland, Amsterdam, 1991. (Cited on p. 349.)

[56] Claude Brezinski and Luc Wuytack. Numerical analysis in the twentieth century. In Claude Brezinski and Luc Wuytack, editors, *Numerical Analysis: Historical Developments in the 20th Century*, pages 1–40. North Holland Elsevier, Amsterdam, 2001. (Cited on p. 84.)

[57] W. L. Briggs and Van Emden Henson. *The DFT. An Owner's Manual for the Discrete Fourier Transform*. SIAM, Philadelphia, PA, 1995. (Cited on p. 519.)

[58] E. O. Brigham. *The Fast Fourier Transform and Its Application*. Prentice-Hall, Englewood Cliffs, NJ, 1988. (Cited on p. 519.)

[59] M. Buhmann. Radial basis functions. *Acta Numerica*, 9:1–38, 2000. (Cited on p. 519.)

[60] Hans-Joachim Bungartz and Michael Griebel. Sparse grids. *Acta Numerica*, 13:147–270, 2004. (Cited on p. 607.)

[61] J. C. P. Bus and T. J. Dekker. Two efficient algorithms with guaranteed convergence for finding a zero of a function. *ACM Trans. Math. Software*, 1:330–345, 1975. (Cited on p. 634.)

[62] Daniela Calvetti, Gene H. Golub, W. B. Gragg, and Lothar Reichel. Computation of Gauss–Kronrod quadrature rules. *Math. Comp.*, 69:1035–1052, 2000. (Cited on p. 607.)

[63] Françoise Chaitin-Chatelin and Valerie Frayssé. *Lectures on Finite Precision Computations*. SIAM, Philadelphia, PA, 1996. (Cited on p. 156.)

[64] Françoise Chaitin-Chatelin and Elisabeth Traviesas-Cassan. PRECISE and the quality of reliable numerical software. In Bo Einarsson, editor, *Accuracy and Reliability in Scientific Computing*, pages 95–108. SIAM, Philadelphia, 2005. (Cited on p. 156.)

[65] B. W. Char, K. O. Geddes, G. H. Gonnet, B. L. Leong, M. B. Monagan, and S. M. Watt. *Maple V Library Reference Manual*. Springer-Verlag, Berlin, 1991. (Cited on pp. 103, 105.)

[66] E. W. Cheney. *Introduction to Approximation Theory*. McGraw-Hill, New York, NY, 1966. (Cited on pp. 323, 326, 332, 410, 473.)

[67] E. W. Cheney and W. Light. *A Course in Approximation Theory*. Brooks/Cole, Pacific Grove, CA, 2000. (Cited on p. 356.)

[68] Ward Cheney and David Kincaid. *Numerical Mathematics and Computing*. Brooks/Cole, Pacific Grove, CA, fourth edition, 1999. (Cited on p. 85.)

[69] E. B. Christoffel. Über die Gaussische Quadratur und einige Verallgemenierung derselbern. *J. Reine Angew. Math.*, 55:61–82, 1858. (Cited on p. 568.)

[70] Philippe G. Ciarlet and Jacques Louis Lions. *Handbook of Numerical Analysis*, volume I–XIII. North-Holland, Amsterdam, 1990–2005. (Cited on p. 85.)

[71] C. W. Clenshaw. A note on summation of Chebyshev series. *Math. Tables Aids Comput.*, 9:118–120, 1955. (Cited on p. 467.)

[72] C. W. Clenshaw and A. R. Curtis. A method for numerical integration on an automatic computer. *Numer. Math.*, 2:197–205, 1960. (Cited on p. 539.)

[73] W. J. Cody. Implementation and testing of function software. In P. C. Messina and A. Murli, editors, *Problems and Methodologies in Mathematical Software*, pages 24–47. Springer-Verlag, Berlin, 1982. (Cited on pp. 104, 121.)

[74] W. J. Cody. Algorithm 714: CELEFUNT: A portable test package for complex elementary functions. *ACM Trans. Math. Software*, 14(4):1–21, 1993. (Cited on p. 104.)

[75] W. J. Cody and W. Waite. *Software Manual for the Elementary Functions*. Prentice-Hall, Englewood Cliffs, NJ, 1980. (Cited on pp. 103, 123.)

[76] James W. Cooley. The re-discovery of the fast Fourier transform algorithm. *Microchimica Acta*, 3:33–45, 1987. (Cited on p. 519.)

[77] James W. Cooley, Peter A. W. Lewis, and Peter D. Welsh. The fast Fourier transform and its application. *IEEE Trans. Education*, E–12:27–34, 1969. (Cited on p. 519.)

[78] James W. Cooley and John W. Tukey. An algorithm for machine calculation of complex Fourier series. *Math. Comp.*, 19:297–301, 1965. (Cited on p. 503.)

[79] D. Coppersmith and S. Winograd. Matrix multiplication via arithmetic progressions. *J. Symbolic Comput.*, 9:251–280, 1990. (Cited on p. 28.)

[80] G. Corliss, C. Faure, A. Griewank, and L. Hascoet. *Automatic Differentiation of Algorithms. From Simulation to Optimization.* Springer-Verlag, Berlin, 2002. (Cited on p. 349.)

[81] R. M. Corless, Gaston H. Gonnet, D. E. G. Hare, D. J. Jeffrey, and Donald E. Knuth. On the Lambert W function. *Adv. Comput. Math.*, 5:329–359, 1996. (Cited on p. 190.)

[82] R. Courant. *Differential and Integral Calculus*, volume I. Blackie & Son, London, 1934. Republished 1988 in Classics Library, John Wiley. (Cited on p. 170.)

[83] R. Courant and D. Hilbert. *Methods of Mathematical Physics*, volume I. Interscience, New York, 1953. (Cited on pp. 485, 495.)

[84] Maurice G. Cox. The numerical evaluation of B-splines. *J. Inst. Math. Appl.*, 10:134–149, 1972. (Cited on p. 418.)

[85] H. S. M. Coxeter. *Introduction to Geometry*. Wiley, New York, 1969. (Cited on p. 595.)

[86] Germund Dahlquist. Preliminär rapport om premieobligationsdragning med datamaskin. (In Swedish), Riksgäldskontoret, Stockholm, 1962. (Cited on p. 67.)

[87] Germund Dahlquist. On summation formulas due to Plana, Lindelöf and Abel, and related Gauss–Christoffel rules II. *BIT*, 37(4):804–832, 1997. (Cited on pp. 212, 286, 320, 349.)

[88] Germund Dahlquist. On summation formulas due to Plana, Lindelöf and Abel, and related Gauss–Christoffel rules I. *BIT*, 37(2):256–295, 1997. (Cited on pp. 582, 606.)

[89] Germund Dahlquist and Åke Björck. *Numerical Methods*. Prentice-Hall, Englewood Cliffs, NJ., 1974. Republished in 2003 by Dover, Mineola, NY. (Cited on pp. 42, 85.)

[90] G. Danielson and Cornelius Lanczos. Some improvements in practical Fourier analysis and their applications to X-ray scattering from liquids. *J. Franklin Inst.*, 233:365–380, 435–452, 1942. (Cited on pp. 505, 519.)

[91] I. Daubechies. *Ten Lectures on Wavelets*. SIAM, Philadelphia, 1992. (Cited on p. 519.)

[92] Philip J. Davis. *Interpolation and Approximation*. Blaisdell, New York, 1963. Republished in 1975 by Dover, Mineola, NY. (Cited on pp. 450, 456, 457, 457, 458, 476.)

[93] Philip J. Davis and Philip Rabinowitz. *Methods of Numerical Integration*. Academic Press, Orlando, FL, second edition, 1984. (Cited on pp. 539, 606, 606, 606, 607, 607.)

[94] James W. Demmel. Underflow and the reliability of numerical software. *SIAM J. Sci. Stat. Comput.*, 5(4):887–919, 1984. (Cited on p. 101.)

[95] James Demmel and Plamen Koev. The accurate and efficient solution of a totally positive generalized Vandermonde linear system. *SIAM J. Matrix Anal. Appl.*, 27:142–152, 2005. (Cited on p. 376.)

[96] Peter Deuflhard and Andreas Hohmann. *Numerical Analysis in Modern Scientific Computing*. Springer, Berlin, second edition, 2003. (Cited on pp. 85, 260.)

[97] P. Dierckx. FITPACK User Guide Part I: Curve fitting routines. TW Report 89, Department of Computer Science, Katholieke Universiteit, Leuven, Belgium, 1987. (Cited on p. 518.)

[98] P. Dierckx. FITPACK User Guide Part II: Surface fitting routines. TW Report 122, Department of Computer Science, Katholieke Universiteit, Leuven, Belgium, 1989. (Cited on p. 518.)

[99] P. Dierckx. *Curve and Surface Fitting with Splines*. Oxford University Press, New York, 1993. (Cited on p. 436.)

[100] J. Dieudonné. *Foundations of Modern Analysis*. Academic Press, New York, NY, 1960. Republished by Hesperides Press, 2006. (Cited on p. 620.)

[101] J. J. Dongarra, James R. Bunch, Cleve B. Moler, and G. W. Stewart. *LINPACK Users' Guide*. SIAM, Philadelphia, PA, 1979. (Cited on p. 42.)

[102] E. Durand. *Solutions Numériques des Équations Algébriques. Tom I: Équations du Type* $F(x) = 0$*, Racines d'un Polynôme*. Masson et Cie, Paris, 1960. (Cited on p. 685.)

[103] Alan Edelman. The mathematics of the Pentium division bug. *SIAM Review*, 39(1):54–67, 1997. (Cited on p. 89.)

[104] Alan Edelman and H. Murakami. Polynomial roots from companion matrix eigenvalues. *Math. Comp.*, 64:763–776, 1995. (Cited on p. 686.)

[105] Bo Einarsson. Numerical calculation of Fourier integrals with cubic splines. *BIT*, 8(4):279–286, 1968. (Cited on p. 555.)

[106] Bo Einarsson. On the calculation of Fourier integrals. In C. V. Freiman, editor, *Information Processing* 71, volume 8, pages 1346–1350, Amsterdam, 1972. North–Holland. (Cited on pp. 555, 556, 557.)

[107] Bo Einarsson. Use of Richardson extrapolation for the numerical calculation of Fourier transforms. *J. Comput. Phys.*, 21(3):365–370, 1976. (Cited on p. 555.)

[108] Bo Einarsson, editor. *Accuracy and Reliability in Scientific Computing*. SIAM, Philadelphia, 2005. (Cited on p. 156.)

[109] Lars Eldén, Linde Wittmeyer-Koch, and Hans Bruun Nielsen. *Introduction to Numerical Computation*. Studentlitteratur, Lund, Sweden, 2004. (Cited on p. 85.)

[110] Sylvan Elhay and Jaroslav Kautsky. Algorithm 655. IQPACK: FORTRAN subroutines for the weights of interpolatory quadratures. *ACM Trans. Math. Software*, 13(4):399–415, 1987. (Cited on p. 607.)

[111] Erik Elmroth, F. G. Gustavson, Isak Jonsson, and Bo Kågström. Recursive blocked algorithms and hybrid data structures for dense matrix library software. *SIAM Review*, 46:3–45, 2004. (Cited on p. 22.)

[112] H. Engels. *Numerical Quadrature and Cubature*. Computational Mathematics and Applications. Academic Press, New York, 1980. (Cited on pp. 606, 607.)

[113] K. Eriksson, D. Estep, P. Hansbo, and C. Johnson. *Computational Differential Equations*. Cambridge University Press, Cambridge, UK, 1996. (Cited on p. 594.)

[114] Terje O. Espelid. On floating-point summation. *SIAM Review*, 37:603–607, 1995. (Cited on p. 21.)

[115] Terje O. Espelid. Doubly adaptive quadrature routines based on Newton–Cotes rules. *BIT*, 43(2):319–337, 2003. (Cited on p. 606.)

[116] Gerald Farin. *Curves and Surfaces for Computer Aided Geometric Design: A Practical Guide*. Academic Press, New York, 1988. (Cited on p. 518.)

[117] L. Fejér. Mechanische Quadraturen mit positiven Cotesschen Zahlen. *Math. Z.*, 37:287–309, 1933. (Cited on p. 539.)

[118] Louis Napoleon George Filon. On a quadrature formula for trigonometric integrals. *Proc. Roy. Soc. Edinburgh*, 49:38–47, 1928–1929. (Cited on p. 555.)

[119] George E. Forsythe. Generation and use of orthogonal polynomials for data-fitting with a digital computer. *J. Soc. Indust. Appl. Math.*, 5(2):74–88, 1957. (Cited on p. 518.)

[120] George E. Forsythe. Algorithms for scientific computation. *Comm. ACM*, 9:255–256, 1966. (Cited on p. 127.)

[121] George E. Forsythe. What is a satisfactory quadratic equation solver? In B. Dejon and P. Henrici, editors, *Constructive Aspects of the Fundamental Theorem of Algebra*, pages 53–61. Wiley-Interscience, London, 1969. (Cited on p. 16.)

[122] George E. Forsythe. Pitfalls in computation, or why a math book isn't enough. *Amer. Math. Monthly*, 77:931–956, 1970. (Cited on p. 156.)

[123] George E. Forsythe, Michael A. Malcolm, and Cleve B. Moler. *Computer Methods for Mathematical Computations*. Prentice-Hall, Englewood Cliffs, NJ, 1977. (Cited on pp. 85, 635, 685.)

[124] George E. Forsythe and Cleve B. Moler. *Computer Solution of Linear Algebraic Systems*. Prentice-Hall, Englewood Cliffs, NJ, 1967. (Cited on p. 33.)

[125] Leslie Fox. How to get meaningless answers in scientific computation (and what to do about it). *IMA Bulletin*, 7(10):296–302, 1971. (Cited on p. 156.)

[126] G. Freud. *Orthogonale Polynome*. Birkhäuser, Basel, 1969. (Cited on p. 573.)

[127] F. G. Frobenius. Über Relationen zwischen den Näherungsbrüchen von Potenzreihen. *J. für Math.*, 90:1–17, 1881. (Cited on p. 349.)

[128] Carl-Erik Fröberg. *Lärobok i Numerisk Analys*. Svenska Bokförlaget/Bonniers, Stockholm, 1962. In Swedish. (Cited on p. 314.)

[129] Walter Gander. On Halley's iteration method. *Amer. Math. Monthly*, 92:131–134, 1985. (Cited on pp. 648, 685.)

[130] Walter Gander. Change of basis in polynomial interpolation. *Numer. Linear Algebra Appl.*, 12(8):769–778, 2005. (Cited on p. 353.)

[131] Walter Gander and Walter Gautschi. Adaptive quadrature—Revisited. *BIT*, 40(1):84–101, 2000. (Cited on pp. 562, 576.)

[132] Walter Gander and Dominik Gruntz. Derivation of numerical methods using computer algebra. *SIAM Review*, 41(3):577–593, 1999. (Cited on p. 606.)

[133] B. S. Garbow, J. M. Boyle, J. J. Dongarra, and G. W. Stewart. *Matrix Eigensystems Routines: EISPACK Guide Extension*. Springer-Verlag, New York, 1977. (Cited on p. 42.)

[134] Mariano Gasca and Thomas Sauer. On the history of multivariate polynomial interpolation. *J. Comput. Appl. Math.*, 122:23–35, 2000. (Cited on p. 518.)

[135] M. Gasca and J. M. Peña. On factorizations of totally positive matrices. In M. Gasca and C. A. Micchelli, editors, *Total Positivity*, pages 109–130. Kluwer Academic Publ., Dordrecht, 1996. (Cited on p. 376.)

[136] C. F. Gauss. Methodus nova integralium valores per approximationem inveniendi. *Comm. Soc. R. Sci. Göttingen Recens*, 3:39–76, 1814. (Cited on p. 568.)

[137] C. F. Gauss. Theoria interpolationis methodo nova tractata. In *Carl Friedrich Gauss, Werke*, volume Band 3, pages 265–303. Königlichen Gesellschaft der Wissenschaften, Göttingen, 1866. (Cited on p. 519.)

[138] Carl Friedrish Gauss. *Theory of the Motion of of the Heavenly Bodies Moving about the Sun in Conic Sections*. Dover, New York, (1963), C. H. Davis, Translation, 1809. (Cited on p. 84.)

[139] Walter Gautschi. Computational aspects of three-term recurrence relations. *SIAM Review*, 9(1):24–82, 1967. (Cited on p. 18.)

[140] Walter Gautschi. On the construction of Gaussian quadrature rules from modified moments. *Math. Comp.*, 24:245–260, 1970. (Cited on p. 606.)

[141] Walter Gautschi. Attenuation factors in practical Fourier analysis. *Numer. Math.*, 18(5):373–400, 1972. (Cited on p. 490.)

[142] Walter Gautschi. Anomalous convergence of a continued fraction for ratios of Kummer functions. *Math. Comp.*, 31(140):994–999, 1977. (Cited on p. 329.)

[143] Walter Gautschi. A survey of Gauss–Christoffel quadrature formulae. In P. L. Butzer and F. Fehér, editors, *E. B. Christoffel: The influence of his work on mathematics and the physical sciences*, pages 72–147. Birkhäuser, Basel, 1981. (Cited on p. 606.)

[144] Walter Gautschi. Questions of numerical condition related to polynomials. In Gene H. Golub, editor, *Studies in Numerical Analysis*, pages 140–177. The Mathematical Association of America, Washington, DC, 1984. (Cited on pp. 356, 666.)

[145] Walter Gautschi. Algorithm 726: ORTHPOL—a package of routines for generating orthogonal polynomials and Gauss-type quadrature rules. *ACM Trans. Math. Software*, 20:21–62, 1994. (Cited on p. 607.)

[146] Walter Gautschi. Orthogonal polynomials: applications and computation. *Acta Numerica*, 5:45–119, 1996. (Cited on p. 606.)

[147] Walter Gautschi. *Numerical Analysis. An Introduction*. Birkhäuser, Boston, MA, 1997. (Cited on pp. 85, 135, 356.)

[148] Walter Gautschi. *Orthogonal Polynomials: Computation and Approximation*. Numerical Mathematics and Scientific Computation. Oxford University Press, New York, 2004. (Cited on pp. 468, 606.)

[149] Walter Gautschi. The numerical evaluation of a challenging integral. Tech. Report, Department of Computer Science, Purdue University, West Lafayette, IN, 2005. (Cited on p. 558.)

[150] Walter Gautschi. Orthogonal polynomials (in MATLAB). *J. Comput. Appl. Math.*, 178(1–2):215–234, 2005. (Cited on p. 607.)

[151] A. O. Gelfond. *Calculus of Finite Differences*. Hindustan Publishing Corporation, Dehli, 1971. Translation of third Russian edition. (Cited on p. 518.)

[152] James E. Gentle. *Random Number generation and Monte Carlo Methods*. Springer-Verlag, New York, second edition, 2003. (Cited on p. 84.)

[153] W. M. Gentleman. Algorithm 424: Clenshaw–Curtis quadrature. *Comm. Assoc. Comput. Mach.*, 15(5):353–355, 1972. (Cited on p. 540.)

[154] W. M. Gentleman. Implementing Clenshaw–Curtis quadrature I and II. *Comm. Assoc. Comput. Mach.*, 15(5):337–346, 1972. (Cited on p. 540.)

[155] W. M. Gentleman and G. Sande. Fast Fourier transforms—for fun and profit. In *Proceedings AFIPS 1966 Fall Joint Computer Conference*, pages 503–578. Spartan Books, Washington, D.C., 1966. (Cited on p. 508.)

[156] John R. Gilbert, Cleve Moler, and Robert Schreiber. Sparse matrices in MATLAB: Design and implementation. *SIAM J. Matrix Anal. Appl.*, 13:333–356, 1992. (Cited on p. 40.)

[157] S. Goedecker. Remark on algorithms to find roots of polynomials. *SIAM J. Sci. Comput.*, 15:1059–1063, 1994. (Cited on p. 686.)

[158] David Goldberg. What every computer scientist should know about floating point arithmetic. *ACM Computing Surveys*, 23:5–48, 1991. (Cited on pp. 101, 102, 106, 106, 110, 116, 156.)

[159] Herman H. Goldstine. *The Computer from Pascal to von Neumann*. Princeton University Press, Princeton, NJ, 1972. (Cited on pp. 245, 293, 349.)

[160] Herman H. Goldstine. *A History of Numerical Analysis from the 16th through the 19th Century*. Stud. Hist. Math. Phys. Sci., Vol. 2. Springer-Verlag, New York, 1977. (Cited on p. 84.)

[161] Herman H. Goldstine and John von Neumann. Numerical inverting of matrices of high order II. *Proceedings Amer. Math. Soc.*, 2:188–202, 1951. (Cited on p. 156.)

[162] Gene H. Golub. Some modified matrix eigenvalue problems. *SIAM Review*, 15:318–334, 1973. (Cited on pp. 581, 606.)

[163] Gene H. Golub. Matrix computations and the theory of moments. In S. D. Chatterji, editor, *Proceedings of the International Congress of Mathematicians, Zürich* 1994, volume 2, pages 1440–1448, Basel, Switzerland, 1995. Birkhäuser-Verlag. (Cited on p. 574.)

[164] Gene H. Golub and J. Kautsky. Calculation of Gauss quadratures with multiple free and fixed knots. *Numer. Math.*, 41:147–163, 1983. (Cited on p. 606.)

[165] Gene H. Golub and Gérard Meurant. Matrices, moments and quadrature. In D. F. Griffiths and G. A. Watson, editors, *Numerical Analysis 1993: Proceedings of the 13th Dundee Biennial Conference*, Pitman Research Notes in mathematics, pages 105–156. Longman Scientific and Technical, Harlow, Essex, UK, 1994. (Cited on p. 582.)

[166] Gene H. Golub and James M. Ortega. *Scientific Computing and Differential Equations. An Introduction to Numerical Methods*. Academic Press, San Diego, CA, 1992. (Cited on p. 85.)

[167] Gene H. Golub and Christian Reinsch. Singular value decomposition and least squares solutions. *Numer. Math*, 14:403–420, 1970. (Cited on p. 50.)

[168] Gene H. Golub and Lyle B. Smith. Chebyshev approximation of continuous functions by a Chebyshev system of functions. Algorithm 414. *Comm. ACM.*, 14(11):737–746, 1971. (Cited on p. 478.)

[169] Gene H. Golub and Charles F. Van Loan. *Matrix Computations*. Johns Hopkins University Press, Baltimore, MD, third edition, 1996. (Cited on p. 27.)

[170] Gene H. Golub and J. H. Welsch. Calculation of Gauss quadrature rules. *Math. Comp.*, 23:221–230, 1969. (Cited on p. 606.)

[171] I. S. Gradshteyn and I. M. Ryzhik. *Table of Integrals, Series and Products*. Academic Press, London, UK, fifth edition, 1993. (Cited on p. 521.)

[172] W. B. Gragg. The Padé table and its relation to certain algorithms of numerical analysis. *SIAM Review*, 14:1–62, 1972. (Cited on pp. 333, 349.)

[173] P. R. Graves-Morris, D. E. Roberts, and A. Salam. The epsilon algorithm and related topics. *J. Comput. Appl. Math.*, 122:51–80, 2000. (Cited on p. 349.)

[174] Ulf Grenander and G. Szegő. *Toeplitz Forms and their Applications*. Chelsea, New York, 1984. (Cited on p. 466.)

[175] A. Griewank. *Evaluating Derivatives; Principles and Techniques of Algorithmic Differentiation*. Frontiers in Applied Mathematics, volume 19. SIAM, Philadelphia, PA, 2000. (Cited on p. 349.)

[176] Sven-Åke Gustafson. Control and estimation of computational errors in the evaluation of interpolation formulae and quadrature rules. *Math. Comp.*, 24:847–854, 1970. (Cited on p. 357.)

[177] Sven-Åke Gustafson. Convergence acceleration on a general class of power series. *Computing*, 21:53–69, 1978. (Cited on p. 292.)

[178] Ernst Hairer and Gerhard Wanner. *Analysis by Its History*. Springer-Verlag, Berlin, 2000. (Cited on p. 443.)

[179] Edmond Halley. A new and general method of finding the roots of equations. *Philos. Trans. Roy. Soc. London*, 18:136, 1694. (Cited on pp. 648, 685.)

[180] John Halton. On the efficiency of certain quasi-random sequences of points in evaluating multidimensional integrals. *Numer. Math.*, 2:84–90, 1960. (Cited on p. 603.)

[181] John Halton and G. B. Smith. Algorithm 247: Radical-inverse quasi-random point sequence. *ACM Trans. Math. Software*, 7:701–702, 1964. (Cited on p. 606.)

[182] Günther Hämmerlin and Karl-Heinz Hoffmann. *Numerical Mathematics*. Springer-Verlag, Berlin, 1991. (Cited on p. 85.)

[183] J. M. Hammersley and D. C. Handscomb. *Monte Carlo Methods*. Methuen, London, UK, 1964. (Cited on pp. 81, 84, 603.)

[184] Richard W. Hamming. *Numerical Methods for Scientists and Engineers*. McGraw-Hill, New York, second edition, 1974. Republished in 1986 by Dover, Mineola, NY. (Cited on p. 85.)

[185] Hermann Hankel. *Über eine besondere Klasse der symmetrischen Determinanten*. Inaugural Diss., Universität Göttingen, Germany, 1861. (Cited on p. 337.)

[186] G. I. Hargreaves. Interval analysis in MATLAB. Numer. anal. report 416, Department of Mathematics, University of Manchester, 2002. (Cited on p. 147.)

[187] J. F. Hart, E. W. Cheney, C. L. Lawson, H. J. Maehly, C. K. Mesztenyi, J. F. Rice, Jr. H. G. Thacher, and C. Witzgall. *Computer Approximations*. John Wiley, New York, New York, 1968. (Cited on pp. 102, 103.)

[188] T. Håvie. Generalized Neville type extrapolation schemes. *BIT*, 19(2):204–213, 1979. (Cited on p. 349.)

[189] J. G. Hayes. Fitting data in more than one variable. In J. G. Hayes, editor, *Numerical Approximation to Functions and Data*, pages 84–97. The Athlon Press, London, 1970. (Cited on p. 518.)

[190] Michael T. Heath. *Scientific Computing. An Introductory Survey*. McGraw-Hill, Boston, MA, second edition, 2002. (Cited on p. 85.)

[191] Peter Hellekalek. Good random number generators are (not so) easy to find. *Math. Comput. Simulation*, 46:485–505, 1998. (Cited on p. 84.)

[192] Peter Henrici. *Discrete Variable Methods in Ordinary Differential Equations*. John Wiley, New York, 1962. (Cited on p. 254.)

[193] Peter Henrici. *Elements of Numerical Analysis*. John Wiley, New York, 1964. (Cited on pp. 273, 462.)

[194] Peter Henrici. Fast Fourier methods in computational complex analysis. *SIAM Review*, 21(4):481–527, 1979. (Cited on p. 195.)

[195] Peter Henrici. *Essentials of Numerical Analysis*. John Wiley, New York, 1982. (Cited on p. 368.)

[196] Peter Henrici. *Applied and Computational Complex Analysis. Volume 1: Power Series, Integration, Conformal Mapping, Location of Zeros*. Wiley Classics Library, New York, 1988. Reprint of the 1974 original. (Cited on pp. 181, 183, 191, 326, 340, 341, 342, 342, 349, 644, 650.)

[197] Peter Henrici. *Applied and Computational Complex Analysis. Volume 2: Special Functions, Integral Transforms, Asymptotics, Continued Fractions*. Wiley Classics Library, New York, 1991. Reprint of the 1977 original. (Cited on pp. 301, 326, 344.)

[198] Nicholas J. Higham. Error analysis of the Björck–Pereyra algorithm for solving Vandermonde systems. *Numer. Math.*, 50:613–632, 1987. (Cited on p. 376.)

[199] Nicholas J. Higham. *Accuracy and Stability of Numerical Algorithms*. SIAM, Philadelphia, PA, second edition, 2002. (Cited on pp. 97, 111, 112, 156, 324, 356, 377, 518.)

[200] Nicholas J. Higham. The numerical stability of barycentric Lagrange interpolation. *IMA J. Numer. Anal.*, 24:547–556, 2004. (Cited on p. 369.)

[201] F. B. Hildebrand. *Introduction to Numerical Analysis*. McGraw-Hill, New York, second edition, 1956. Republished in 1988 by Dover, Mineola, NY. (Cited on pp. 85, 393.)

[202] C. A. R. Hoare. Quicksort. *Computer J.*, 5(1):10–15, 1962. (Cited on p. 21.)

[203] Roger A. Horn and Charles R. Johnson. *Topics in Matrix Analysis*. Cambridge University Press, Cambridge, 1991. (Cited on p. 518.)

[204] Alston S. Householder. *The Numerical Treatment of a Single Nonlinear Equation*. McGraw-Hill, New York, 1970. (Cited on pp. 636, 663, 676, 685, 685, 686.)

[205] IEEE Standard for Binary Floating-Point Arithmetic. ANSI/IEEE Standard 754-1985. *SIGPLAN Notices*, 22(2):9–25, 1987. (Cited on pp. 99, 156.)

[206] IEEE Standard for Radix-Independent Floating Point Arithmetic. ANSI/IEEE Standard 854-1987. Technical report, IEEE Computer Society, New York, 1987. (Cited on p. 99.)

[207] J. P. Imhof. On the method for numerical integration of Clenshaw and Curtis. *Numer. Math.*, 5:138–141, 1963. (Cited on p. 539.)

[208] Eugene Isaacson and Herbert B. Keller. *Analysis of Numerical Methods*. John Wiley, New York, 1966. Republished in 1994 by Dover, Mineola, NY. (Cited on pp. 85, 372, 396, 518.)

[209] Arieh Iserles. On the numerical quadrature of highly-oscillating integrals I: Fourier transforms. *IMA J. Numer. Anal.*, 24:365–391, 2004. (Cited on p. 557.)

[210] Glenn James and Robert C. James, editors. *James & James Mathematics Dictionary*. Van Nostrand, Princeton, NJ, fifth edition, 1992. (Cited on p. 84.)

[211] M. A. Jenkins and J. F. Traub. A three-stage variable shift iteration for polynomial zeros and its relation to generalized Rayleigh quotient iteration. *Numer. Math.*, 14:252–263, 1970. (Cited on p. 675.)

[212] M. A. Jenkins and J. F. Traub. Algorithm 419—Zeros of a complex polynomial. *Comm. ACM.*, 15:97–99, 1972. (Cited on p. 675.)

[213] L. W. Johnson and D. R. Scholz. On Steffensen's method. *SIAM J. Numer. Anal.*, 5:296–302, 1968. (Cited on p. 632.)

[214] W. B. Jones and W. J. Thron. *Continued Fractions. Analytic Theory and Applications*, volume 11 of *Encyclopedia of Mathematics and its Application*. Addison-Wesley, Reading, MA, 1980. (Cited on pp. 334, 349.)

[215] D. C. Joyce. Survey of extrapolation processes in numerical analysis. *SIAM Review*, 13(4):435–490, 1971. (Cited on p. 349.)

[216] W. Kahan. A survey of error analysis. In B. Dejon and P. Henrici, editors, *Proceedings IFIP Congress Ljubljana, Information Processing 1971*, pages 1214–1239. North-Holland, Amsterdam, 1971. (Cited on pp. 121, 156.)

[217] W. Kahan. Branch cuts for elementary functions or much ado about nothing's sign bit. In Arieh Iserles and M. J. D. Powell, editors, *The State of the Art in Numerical Analysis*, pages 165–211. Clarendon Press, Oxford, UK, 1987. (Cited on p. 104.)

[218] W. Kahan. Miscalculating Area and Angles of a Needle-like Triangle. Work in Progress, Department of Computer Science, University of California, Berkeley, CA, 2000. (Cited on p. 124.)

[219] W. M. Kahan. Numerical linear algebra. *Canad. Math. Bull.*, 9:757–801, 1966. (Cited on p. 37.)

[220] David Kahaner, Cleve B. Moler, and Stephen G. Nash. *Numerical Methods and Software*. Prentice-Hall, Englewood Cliffs, NJ, 1989. (Cited on pp. 76, 85, 576.)

[221] Bahman Kalantari, Iraj Kalantari, and Rahim Zaare-Nahandi. A basic family of iteration functions for polynomial root finding and its characterizations. *J. Comput. Appl. Math.*, 80:209–226, 1997. (Cited on p. 685.)

[222] S. Karlin and W. J. Studden. *Tchebysheff Systems with Applications in Analysis and Statistics*. Interscience, New York, 1966. (Cited on p. 518.)

[223] Alan H. Karp. Bit reversal on uniprocessors. *SIAM Review*, 38(1):1–26, 1996. (Cited on p. 519.)

[224] Jaroslav Kautsky and Sylvan Elhay. Calculation of the weights of interpolatory quadratures. *Numer. Math.*, 40:407–422, 1982. (Cited on p. 607.)

[225] R. Kearfott. Interval computations: Introduction, uses, and resources. *Euromath. Bulletin*, 2(1):95–112, 1996. (Cited on p. 147.)

[226] David Kincaid and Ward Cheney. *Numerical Analysis*. Brooks/Cole, Pacific Grove, CA, third edition, 2002. (Cited on p. 85.)

[227] Göran Kjellberg. Two observations on Durand–Kerner's root finding method. *BIT*, 24(4):556–559, 1984. (Cited on p. 673.)

[228] Felix Klein. *Elementary Mathematics from an Advanced Standpoint*. Dover, Mineola, NY, 1945. Translation from German original, 1924. (Cited on p. 324.)

[229] Konrad Knopp. *Infinite Sequences and Series*. Dover, Mineola, NY, 1956. (Cited on p. 349.)

[230] Donald E. Knuth. *The Art of Computer Programming, Volume 2: Seminumerical Algorithms*. Addison-Wesley, Reading, MA, third edition, 1998. (Cited on pp. 68, 69, 70, 72, 73, 74, 76, 77, 77, 84, 103, 155, 174, 179.)

[231] Donald E. Knuth. *The Art of Computer Programming, Volume 3: Sorting and Searching*. Addison-Wesley, Reading, MA, second edition, 1998. (Cited on p. 83.)

[232] A. S. Kronrod. Integration with control of accuracy. *Dokl. Akad. Nauk. SSSR*, 154:283–286, 1964. (In Russian.) (Cited on p. 575.)

[233] A. S. Kronrod. *Nodes and Weights for Quadrature Formulae. Sixteen-place tables*. Izdat. Nauka, Moscow, 1964. (In Russian.) English translation in: Consultants Bureau, New York, 1965. (Cited on pp. 575, 576.)

[234] Johann H. Lambert. *Beiträge zum Gebrauche der Mathematik und deren Anwendungen. Zweiter Theil*. 1770. (Cited on p. 685.)

[235] Cornelius Lanczos. *Applied Analysis*. Prentice-Hall, Englewood Cliffs, NJ, 1956. Republished in 1988 by Dover, Mineola, NY. (Cited on pp. 487, 519.)

[236] Cornelius Lanczos. *Linear Differential Operators*. D. Van Nostrand, London, UK, 1961. Republished in 1974 by Dover, Mineola, NY. (Cited on p. 583.)

[237] D. P. Laurie. Calculation of Gauss–Kronrod quadrature rules. *Math. Comp.*, 66:1133–1145, 1997. (Cited on p. 607.)

[238] D. P. Laurie. Computation of Gauss-type quadrature formulas. *J. Comput. Appl. Math.*, 127:201–217, 2001. Numerical Analysis 2000. Vol. V. Quadrature and Orthogonal Polynomials. (Cited on p. 607.)

[239] Charles L. Lawson, Richard J. Hanson, David R. Kincaid, and Fred T. Krogh. Basic Linear Algebra Subprograms for Fortran usage. *ACM Trans. Math. Software*, 5:308–323, 1979. (Cited on p. 42.)

[240] N. N. Lebedev. *Special Functions and Their Applications*. Dover, Mineola, NY, 1972. Translation from Russian original. (Cited on pp. 210, 212, 219, 301, 348.)

[241] H. Lebesgue. Sur l'approximation des fonctions. *Bull. Sciences Math.*, 22:278–287, 1898. (Cited on p. 449.)

[242] Pierre L'Ecuyer. Uniform random number generation. *Ann. Oper. Res.*, 53:77–120, 1994. (Cited on p. 84.)

[243] Pierre L'Ecuyer. Random number generation. In Jerry Banks, editor, *The Handbook of Simulation*, pages 93–137. John Wiley, New York, 1998. (Cited on p. 84.)

[244] Pierre L'Ecuyer. Software for uniform random number generation: Distinguishing the good and bad. In *Proc. 2001 Winter Simulation Conference*, pages 95–105. IEEE Press, Piscataway, NJ, 2001. (Cited on pp. 70, 73.)

[245] Adrien-Marie Legendre. *Nouvelles méthodes pour la détermination des orbites des comètes*. Courcier, Paris, 1805. (Cited on p. 84.)

[246] C. C. Lin and L. A. Segel. *Mathematics Applied to Deterministic Problems in the Natural Sciences*. Macmillan, New York, 1974. (Cited on p. 204.)

[247] Xiaoye S. Li, James W. Demmel, David H. Bailey, Greg Henry, Y. Hida, Jimmy Iskandar, William Kahan, Suh Y. Kang, Anil Kapur, Michael C. Martin, Brandon J. Thompson, Teresa Tung, and Daniel J. Yoo. Design, implementation and testing of extended and mixed precision BLAS. *ACM Trans. Math. Software*, 28(2):152–205, 2002. (Cited on p. 115.)

[248] L. Lorenzen and H. Waadeland. *Continued Fractions with Applications*. North-Holland, Amsterdam, 1992. (Cited on p. 349.)

[249] Daniel W. Lozier. NIST digital library of mathematical functions. *Annals of Mathematics and Artificial Intelligence*, 38:105–119, 2003. (Cited on p. 348.)

[250] D. W. Lozier and F. W. J. Olver. Numerical evaluation of special functions. In W. Gautschi, editor, *Mathematics of Computation 1943–1993: A Half-Century of Computational Mathematics*, volume 48 of *Proc. Sympos. Appl. Math.*, pages 79–125, Providence, RI, 1994. Amer. Math. Soc. (Cited on p. 348.)

[251] James N. Lyness. Notes on the adaptive Simpson quadrature routine. *J. Assoc. Comput. Mach.*, 16:483–495, 1969. (Cited on p. 606.)

[252] James N. Lyness and Cleve B. Moler. Numerical differentiation of analytic functions. *SIAM J. Numer. Anal.*, 4:202–210, 1967. (Cited on p. 195.)

[253] Kaj Madsen. A root-finding algorithm based on Newton's method. *BIT*, 13(1):71–75, 1973. (Cited on pp. 675, 677.)

[254] Kaj Madsen and John K. Reid. Fortran subroutines for finding polynomial zeros. Tech. Report A.E.R.E. R.7986, Computer Science and Systems Division, A.E.R.E. Harwell, 1975. (Cited on pp. 675, 677.)

[255] H. J. Maehly. Zur iterativen Auflösung algebraischer Gleichungen. *Z. angew. Math. Phys.*, 5:260–263, 1954. (Cited on p. 672.)

[256] Michael A. Malcolm. On the computation of nonlinear spline functions. *SIAM J. Numer. Anal.*, 14(2):254–282, 1977. (Cited on p. 417.)

[257] George Marsaglia. Expressing a random variable in terms of uniform random variables. *Ann. Math. Statist.*, 32:894–898, 1961. (Cited on p. 77.)

[258] George Marsaglia. Random numbers falls mainly in the planes. *Proc. Nat. Acad. Sci. USA*, 61(5):25–28, 1968. (Cited on p. 69.)

[259] George Marsaglia. A current view of random number generators. In L. Billard, editor, *Computer Science and Statistics: The Interface*, pages 3–10. Elsevier Science Publishers, Amsterdam, 1985. (Cited on p. 84.)

[260] George Marsaglia and W. W. Tsang. A fast, easily implemented method for sampling from decreasing or symmetric unimodal density functions. *SIAM J. Sci. Stat. Comput.*, 5(2):349–359, 1984. (Cited on p. 76.)

[261] M. Matsumoto and T. Nishimura. Mersenne twister: A 623-dimensionally equidistributed uniform pseudo-random number generator. *ACM Trans. Modeling Comput. Software*, 8(1):3–30, 1998. (Cited on p. 70.)

[262] Urs von Matt. Gauss quadrature. In Walter Gander and Jiři Hřebiček, editors, *Solving Problems in Scientific Computing using Maple and MATLAB*, pages 251–279. Springer-Verlag, Berlin, fourth edition, 2004. (Cited on p. 607.)

[263] J. McNamee and Frank Stenger. Construction of fully symmetric numerical integration formulas. *Numer. Math.*, 10:327–344, 1967. (Cited on p. 607.)

[264] Günter Meinardus. *Approximation of Functions: Theory and Numerical Methods*, volume 16 of *Springer Tracts in Natural Philosophy*. Springer-Verlag, Berlin, 1967. (Cited on p. 478.)

[265] L. M. Milne-Thomson. *Calculus of Finite Differences*. Macmillan & Co., London, 1933. (Cited on p. 518.)

[266] Cleve Moler. Random thoughts, 10^{435} years is a very long time. *MATLAB News and Notes*, Fall, 1995. (Cited on p. 70.)

[267] Cleve Moler. Normal behavior. *MATLAB News and Notes*, Spring, 2001. (Cited on p. 76.)

[268] Cleve B. Moler. *Numerical Computing with MATLAB*. SIAM, Philadelphia, PA, 2004. (Cited on p. 635.)

[269] Cleve Moler and Steve Eddins. Fast finite Fourier transforms. *MATLAB News and Notes*, pages 14–15, Winter, 2001. (Cited on p. 517.)

[270] Giovanni Monegato. An overview of results and questions related to Kronrod schemes. In G. Hämmerlin, editor, *Numerische Integration*, pages 231–240. Birkhäuser, Numer. Math. 45, Basel, 1979. (Cited on p. 607.)

[271] Ramon E. Moore. *Interval Analysis*. Prentice-Hall, Englewood Cliffs, NJ, 1966. (Cited on pp. 147, 646.)

[272] J.-M. Muller. *Elementary Functions: Algorithm and Implementation*. Birkhäuser, Boston, MA, 1997. (Cited on pp. 103, 121.)

[273] I. P. Mysovskih. On the construction of cubature formulas with the smallest number of nodes. *Soviet Math. Dokl.*, 9:277–280, 1968. (Cited on p. 606.)

[274] J. C. Nash. *Compact Numerical Methods for Computers: Linear Algebra and Function Minimization*. American Institute of Physics, New York, second edition, 1990. (Cited on p. 84.)

[275] H. Niederreiter. *Random Number Generation and Quasi-Monte Carlo Methods*. SIAM, Philadelphia, PA, 1992. (Cited on pp. 84, 602.)

[276] N. E. Nörlund. *Vorlesungen über Differenzenrechnung*. Springer-Verlag, Berlin, 1924. Republished by Chelsea, New York, 1954. (Cited on p. 315.)

[277] W. Oettli and W. Prager. Compatibility of approximate solution of linear equations with given error bounds for coefficients and right-hand sides. *Numer. Math.*, 6:404–409, 1964. (Cited on p. 138.)

[278] James M. Ortega and Werner C. Rheinboldt. *Iterative Solution of Nonlinear Equations in Several Variables*. Academic Press, New York, 1970. (Cited on p. 622.)

[279] A. M. Ostrowski. *Solution of Equations in Euclidian and Banach Spaces*. Academic Press, New York, third edition, 1973. (Cited on pp. 615, 638, 639, 685.)

[280] Michael L. Overton. *Numerical Computing with IEEE Floating Point Arithmetic*. SIAM, Philadelphia, PA, 2001. (Cited on p. 156.)

[281] H. Padé. Sur la représentation approcheée d'un fonction par des fraction rationelles. *Thesis Anal. Ecole Norm. Sup.*, 3:1–93, supplement, 1892. (Cited on p. 349.)

[282] Victor Y. Pan. Solving a polynomial equation: Some history and recent progress. *SIAM Review*, 39:187–220, 1997. (Cited on p. 686.)

[283] S. K. Park and K. W. Miller. Random number generators: Good ones are hard to find. *Comm. Assoc. Comput. Mach.*, 31:1192–1201, 1988. (Cited on p. 69.)

[284] T. W. Parks and J. J. McClellan. Chebyshev approximation for nonrecursive filters with linear phase. *IEEE Trans. Circuit Theory*, 19:189–194, 1972. (Cited on p. 518.)

[285] Beresford N. Parlett. The new qd algorithm. *Acta Numerica*, 4:459–491, 1995. (Cited on p. 349.)

[286] Beresford N. Parlett and Christian Reinsch. Balancing a matrix for calculation of eigenvalues and eigenvectors. *Numer. Math.*, 13:293–304, 1969. (Cited on p. 675.)

[287] K. Pearson. On the criterion that a given system of deviations from the probable in the case of a correlated system of variables is such that it can be reasonably supposed to have arisen from random sampling. *Phil. Mag. Series 5*, 50:157–175, 1900. (Cited on p. 71.)

[288] R. Penrose. A generalized inverse for matrices. *Proc. Cambridge Philos. Soc.*, 51:406–413, 1955. (Cited on p. 52.)

[289] O. Perron. *Die Lehre von den Kettenbrüchen. Volume II.* Teubner, Stuttgart, third edition, 1957. (Cited on p. 326.)

[290] R. Piessens, E. de Doncker, C. W. Überhuber, and David K. Kahaner. *QUADPACK, A Subroutine Package for Automatic Integration.* Springer-Verlag, Berlin, 1983. (Cited on p. 607.)

[291] Allan Pinkus. Weierstrass and approximation theory. *J. Approx. Theory*, 107:1–66, 2000. (Cited on p. 449.)

[292] M. J. D. Powell. *Approximation Theory and Methods.* Cambridge University Press, Cambridge, UK, 1981. (Cited on p. 430.)

[293] M. J. D. Powell. A Review of Methods for Multivariable Interpolation of Scattered Data Points. In I. S. Duff and G. A. Watson, editors, *The State of the Art in Numerical Analysis*, pages 283–309. Clarendon Press, Oxford, UK, 1997. (Cited on p. 518.)

[294] William H. Press, Brian P. Flannery, Saul A. Teukolsky, and William T. Vetterling. *Numerical Recipes in Fortran 77. The Art of Scientific Computing.* Cambridge University Press, Cambridge, UK, second edition, 1992. (Cited on pp. 70, 77, 77, 83, 84, 85, 105, 219, 270, 349, 349, 481, 513, 515.)

[295] RAND Corporation. *A Million Random Digits with 100,000 Normal Deviates.* Free Press, Glencoe, IL, 1955. (Cited on p. 67.)

[296] Anthony Ralston and Philip Rabinowitz. *A First Course in Numerical Analysis.* McGraw-Hill, New York, second edition, 1978. Republished in 2001 by Dover, Mineola, NY. (Cited on pp. 85, 675.)

[297] L.B. Rall. Automatic Differentiation. Techniques and Applications. Number 120 in Lecture Notes in Computer Science. Springer-Verlag, Berlin, 1981. (Cited on p. 349.)

[298] Lothar Reichel. On polynomial approximation in the uniform norm by the discrete least squares method. *BIT*, 26(2):349–366, 1986. (Cited on p. 471.)

[299] Lothar Reichel. Newton interpolation at Leja points. *BIT*, 30(2):332–346, 1990. (Cited on pp. 366, 518.)

[300] Evgeny Yakolevich Remez. Sur le calcul effectif des polynômes d'approximation des Tchebyscheff. *C. R. Acad. Sci. Paris*, 199:337–340, 1934. (Cited on p. 478.)

[301] John R. Rice. A theory of condition. *SIAM J. Numer. Anal.*, 3(2):287–310, 1966. (Cited on p. 156.)

[302] Lewis F. Richardson. The approximate arithmetical solution by finite differences of physical problems involving differential equations, with an application to the stresses in a masonry dam. *Proc. Roy. Soc. London Ser. A*, 210:307–357, 1910. (Cited on p. 40.)

[303] Hans Riesel. *Prime Numbers and Computer Methods for Factorization.* Progr. Math. 126, Birkhäuser, Boston, MA, second edition, 1994. (Cited on p. 349.)

[304] Friedrich Riesz and Béla Sz.-Nagy. *Vorlesungen über Funktionalanalysis.* VEB Deutscher Verlag der Wissenschaften, Berlin, 1956. (Cited on p. 451.)

[305] J. L. Rigal and J. Gaches. On the compatibility of a given solution with the data of a linear system. *J. Assoc. Comput. Mach.*, 14(3):543–548, 1967. (Cited on p. 137.)

[306] Sara Robinson. Toward an optimal algorithm for matrix multiplication. *SIAM News*, 38(9), Nov. 2005. (Cited on p. 28.)

[307] Sigfried M. Rump. Fast and parallel interval arithmetic. *BIT*, 39(3):534–554, 1999. (Cited on pp. 150, 154.)

[308] Sigfried M. Rump. INTLAB—INTerval LABoratory. In T. Csendes, editor, *Developments in Reliable Computing*, pages 77–104. Kluwer Academic Publishers, Dordrecht, 1999. (Cited on p. 154.)

[309] C. Runge and H. König. *Vorlesungen über Numerisches Rechnen. Band XI*. Die Grundlehren der Matematischen Wissenachaften. Verlag von Julius Springer, Berlin, 1924. (Cited on p. 519.)

[310] Heinz Rutishauser. Der Quotienten-Differenzen-Algoritmus. *Z. Angew. Math. Phys.*, 5:233–251, 1954. (Cited on p. 339.)

[311] Heinz Rutishauser. Solution of eigenvalue problems with the LR-transformation. *Nat. Bureau of Standards, Appl. Math. Ser.*, 49:47–81, 1958. (Cited on p. 349.)

[312] Heinz Rutishauser. *Lectures on Numerical Mathematics. Volume I*. Birkhäuser, Boston, 1990, 1990. English translation by W. Gautschi of *Vorlesungen über numerische Mathematik*, Birkhäuser, Basel–Stuttgart, 1976. (Cited on pp. 85, 367.)

[313] Robert Schaback and Helmut Werner. *Numerische Mathematik*. Springer, Berlin, fourth edition, 1991. (Cited on p. 391.)

[314] I. J. Schoenberg. Contributions to the problem of approximation of equidistant data by analytic functions. *Quart. Appl. Math.*, 4:45–99 and 112–141, 1946. (Cited on pp. 418, 426.)

[315] I. J. Schoenberg and A. Whitney. On Pólya frequency functions III: The positivity of translation determinants with an application to the interpolation problem by spline curves. *Trans. Amer. Math. Soc.*, 74:246–259, 1953. (Cited on p. 434.)

[316] Ernst Schröder. Über unendlich viele Algorithmen zur Auflösung der Gleichungen. *Mathematische Annalen*, 2:317–365, 1870. (Cited on pp. 650, 685.)

[317] Ernst Schröder. On infinitely many algorithms for solving equations (translated by G. W. Stewart). Tech. Report TR-92-121, Department of Computer Science, University of Maryland, College Park, MD, 1992. (Cited on p. 685.)

[318] H.-R. Schwarz. *Numerical Analysis: A Comprehensive Introduction*. John Wiley, New York, 1989. English translation of *Numerische Mathematik*: Teubner, Stuttgart, 1986. (Cited on pp. 85, 369.)

[319] Hubert Schwetlick and Torsten Schütze. Least squares approximation by splines with free knots. *BIT*, 35(3):361–384, 1995. (Cited on p. 436.)

[320] Bl. Sendov, A. Andreev, and N. Kjurkchiev. Numerical solution of polynomial equations. In Philippe G. Ciarlet and Jacques-Louis Lions, editors, *Handbook of Numerical Analysis*, volume III, pages 629–778. Elsevier Science, Cambridge, UK, 1994. (Cited on p. 685.)

[321] Lawrence F. Shampine. Discrete least squares polynomial fits. *Comm. Assoc. Comput. Mach.*, 18:179–180, 1975. (Cited on p. 518.)

[322] D. Shanks. Nonlinear transformations of divergent and slowly convergent sequences. *J. Math. Phys.*, 34:1–42, 1955. (Cited on p. 337.)

[323] Avram Sidi. A user-friendly extrapolation method for oscillatory infinite integrals. *Math. Comp.*, 51:249–266, 1988. (Cited on p. 558.)

[324] R. C. Singleton. On computing the fast Fourier transform. *Comm. Assoc. Comput. Mach.*, 10:647–654, 1967. (Cited on p. 519.)

[325] R. C. Singleton. Algorithm 338: Algol procedure for the fast Fourier transform with arbitrary factors. *Comm. Assoc. Comput. Mach.*, 11:773–779, 1968. (Cited on p. 519.)

[326] Robert D. Skeel. Roundoff error and the patriot missile. *SIAM News*, 25(4):11, Jul. 1992. (Cited on p. 94.)

[327] Ian H. Sloan and S. Joe. *Lattice Methods for Multiple Integration*. Clarendon Press, Oxford, UK, 1994. (Cited on p. 607.)

[328] Ian H. Sloan and Henryk Wozniakowski. When are Quasi-Monte Carlo algorithms efficient for high dimensional integrals? *J. Complexity*, 14:1–33, 1998. (Cited on p. 607.)

[329] A. Smoktunowicz. Backward stability of Clenshaw's algorithm. *BIT*, 42(3):600–610, 2002. (Cited on p. 468.)

[330] J. F. Steffensen. *Interpolation*. Chelsea, New York, second edition, 1950. Republished by Dover Publication, Mineola, NY. (Cited on pp. 349, 367, 518, 535.)

[331] Irene A. Stegun and Milton Abramowitz. Pitfalls in computation. *J. Soc. Indust. Appl. Math.*, 4:207–219, 1956. (Cited on p. 156.)

[332] Frank Stenger. *Numerical Methods Based on Sinc and Analytic Functions*. Springer-Verlag, Berlin, 1993. (Cited on p. 171.)

[333] P. H. Sterbenz. *Floating Point Computation*. Prentice-Hall, Englewood Cliffs, NJ, 1974. (Cited on pp. 109, 156.)

[334] Hans J. Stetter. *Numerical Polynomial Algebra*. SIAM, Philadelphia, 2004. (Cited on p. 686.)

[335] George W. Stewart. *Matrix Algorithms Volume* I: *Basic Decompositions*. SIAM, Philadelphia, PA, 1998. (Cited on p. 27.)

[336] E. Stiefel. Altes und Neues über numerische Quadratur. *Z. Angew. Math. Mech.*, 41:408–413, 1961. (Cited on p. 548.)

[337] S. M. Stigler. Gauss and the invention of least squares. *Ann. Statist.*, 9:465–474, 1981. (Cited on p. 84.)

[338] Joseph Stoer and Roland Bulirsch. *Introduction to Numerical Analysis*. Springer-Verlag, New York, third edition, 2002. (Cited on pp. 85, 394, 518, 531.)

[339] Gilbert Strang. *Introduction to Applied Mathematics*. Wellesley-Cambridge Press, Wellesley, MA, 1986. (Cited on pp. 217, 256, 497, 497, 502.)

[340] Gilbert Strang. The discrete cosine transform. *SIAM Rev.*, 41(1):135–147, 1999. (Cited on p. 514.)

[341] Volker Strassen. Gaussian elimination is not optimal. *Numer. Math.*, 13:354–356, 1969. (Cited on p. 28.)

[342] A. H. Stroud. *Approximate Calculation of Multiple Integrals*. Prentice-Hall, Englewood Cliffs, NJ, 1972. (Cited on p. 607.)

[343] A. H. Stroud and D. Secrest. *Gaussian Quadrature Formulas*. Prentice-Hall, Englewood Cliffs, NJ, 1966. (Cited on p. 607.)

[344] Endre Süli and David F. Mayers. *Introduction to Numerical Analysis*. Cambridge University Press, Cambridge, UK, 2003. (Cited on p. 85.)

[345] Paul N. Swarztrauber. On computing the points and weights for Gauss–Legendre quadrature. *SIAM J. Sci. Comput.*, 24:945–954, 2002. (Cited on p. 584.)

[346] J. J. Sylvester. On a theory of the syzygetic relations of two rational integral functions, comprising an application to the theory of Sturm's functions, and that of the greatest algebraic common measure. *Philos. Trans. Roy. Soc. London*, 143:407–548, 1853. (Cited on p. 681.)

[347] Gabor Szegő. *Orthogonal Polynomials*. Colloq. Publ., Volume 23. Amer. Math. Soc., Providence, RI, fourth edition, 1975. (Cited on p. 407.)

[348] Eihan Tadmor. Filters, mollifiers and the computation of the Gibbs phenomenon. *Acta Numer.*, 16:305–378, 2007. (Cited on p. 519.)

[349] Nico M. Temme. Numerical aspects of special functions. *Acta Numerica*, 16:379–478, 2007. (Cited on p. 348.)

[350] T. N. Thiele. *Interpolationsrechnung*. B. G. Teubner, Leipzig, 1909. (Cited on p. 393.)

[351] E. C. Titchmarsh. *The Theory of Functions*. Oxford University Press, London, second edition, 1939. (Cited on pp. 172, 193, 450.)

[352] John Todd. Notes on numerical analysis I, solution of differential equations by recurrence relations. *Math. Tables Aids Comput.*, 4:39–44, 1950. (Cited on p. 257.)

[353] Kim-Chuan Toh and Lloyd N. Trefethen. Pseudozeros of polynomials and pseudospectra of companion matrices. *Numer. Math.*, 68(3):403–425, 1994. (Cited on pp. 665, 675.)

[354] J. F. Traub. *Iterative Methods for the Solution of Equations*. Prentice-Hall, Englewood Cliffs, NJ, 1964. Republished in 1997 by Amer. Math. Soc., Providence, RI. (Cited on pp. 632, 648, 649, 650, 685, 685.)

[355] J. F. Traub. Associated polynomials and uniform methods for the solution of linear problems. *SIAM Rev.*, 8(3):277–301, 1966. (Cited on p. 518.)

[356] J. F. Traub. A class of globally convergent iteration functions for the solution of polynomial equations. *Math. Comp.*, 20:113–138, 1966. (Cited on p. 685.)

[357] Lloyd N. Trefethen. The definition of numerical nalysis. *SIAM News*, 25, Nov. 1992. (Cited on p. 87.)

[358] Lloyd N. Trefethen. Pseudospectra of linear operators. *SIAM Rev.*, 39:383–406, 1997. (Cited on p. 665.)

[359] Lloyd N. Trefethen. Is Gauss quadrature better than Clenshaw–Curtis? *SIAM Rev.*, 50:67–87, 2008. (Cited on pp. 540, 606.)

[360] Lloyd N. Trefethen and Robert S. Schreiber. Average-case stability of Gaussian elimination. *SIAM J. Matrix Anal. Appl.*, 11:335–360, 1990. (Cited on p. 37.)

[361] W. Tucker. A rigorous ODE solver and Smale's 14th problem. *Found. Comput. Math.*, 2(1):53–117, 2002. (Cited on p. 147.)

[362] A. M. Turing. Rounding-off errors in matrix processes. *Quart. J. Mech. Appl. Math.*, 1:287–308, 1948. (Cited on p. 84.)

[363] E. Tyrtychnikov. *A Brief Introduction to Numerical Analysis*. Birkhäuser, Boston, 1997. (Cited on p. 85.)

[364] Christoph W. Ueberhuber. *Numerical Computation: Methods, Software, and Analysis. Volumes* 1 & 2. Springer-Verlag, Berlin, 1997. (Cited on p. 607.)

[365] J. G. van der Corput. Verteilungsfunktionen I & II. *Nederl. Akad. Wetensch. Proc.*, 38:813–820, 1058–1066, 1935. (Cited on p. 602.)

[366] Charles F. Van Loan. *Computational Framework for the Fast Fourier Transform*. SIAM, Philadelphia, 1992. (Cited on pp. 507, 508, 513, 513, 516, 519.)

[367] Charles F. Van Loan. *Introduction to Scientific Computing*. Prentice-Hall, Upper Saddle River, NJ, second edition, 2000. (Cited on p. 85.)

[368] Jürg Waldvogel. Fast construction of thr Fejér and Clenshaw–Curtis quadrature rules. *BIT*, 46:195–202, 2006. (Cited on p. 541.)

[369] H. S. Wall. *Analytic Theory of Continued Fractions*. Van Nostrand, Princeton, NJ, 1948. (Cited on p. 326.)

[370] Eric W. Weisstein, editor. *CRC Concise Encyclopedia of Mathematics*. CRC Press, Boca Raton, FL, 1999. (Cited on pp. 84, 642.)

[371] Wilhelm Werner. Polynomial interpolation: Lagrange versus Newton. *Math. Comp.*, 43(167):205–217, 1984. (Cited on p. 518.)

[372] Hermann Weyl. Randbemerkungen zu Hauptproblemen der Mathematik, II, Fundamentalsatz der Algebra und Grundlagen der Mathematik. *Math. Z.*, 20:131–151, 1924. (Cited on pp. 679, 686.)

[373] D. V. Widder. *The Laplace Transform*. Princeton University Press, Princeton, NJ, 1941. (Cited on p. 286.)

[374] D. V. Widder. *An Introduction to Transform Theory*. Academic Press, New York, 1971. (Cited on p. 286.)

[375] James H. Wilkinson. Error analysis of direct methods of matrix inversion. *J. Assos. Comput. Mach.*, 8:281–330, 1961. (Cited on p. 37.)

[376] James H. Wilkinson. *Rounding Errors in Algebraic Processes*. Notes on Applied Science No. 32. Her Majesty's Stationery Office, London, UK, 1963. Republished in 1994 by Dover, Mineola, NY. (Cited on pp. 117, 156, 156, 156.)

[377] James H. Wilkinson. *The Algebraic Eigenvalue Problem*. Clarendon Press, Oxford, 1965. (Cited on pp. 114, 141.)

[378] James H. Wilkinson. A priori error analysis of algebraic processes. In *Proceedings International Congress Math.*, pages 629–639. Izdat. Mir, Moscow, 1968. (Cited on p. 139.)

[379] James H Wilkinson. The perfidious polynomial. In G. H. Golub, editor, *Studies in Numerical Analysis*, pages 1–28. American Mathematical Society, Providence, RI, 1984. (Cited on pp. 665, 672.)

[380] James H. Wilkinson. Error analysis revisited. *IMA Bull.*, 22:192–200, 1986. (Cited on pp. 145, 156.)

[381] James H. Wilkinson and C. Reinsch, editors. *Handbook for Automatic Computation. Vol.* II. *Linear Algebra*. Springer-Verlag, New York, 1971. (Cited on pp. 38, 42.)

[382] Stephen Wolfram. *The Mathematica Book*. Wolfram Media, Champaign, IL, fifth edition, 2003. (Cited on p. 105.)

[383] Henryk Wozniakowski. Average case complexity of multivariate integration. *Bull. Amer. Math. Soc.*, 24(1):185–194, 1991. (Cited on p. 603.)

[384] Peter Wynn. On a device for computing the $e_m(s_n)$ transformation. *Math. Tables Aids Comput.*, 10:91–96, 1956. (Cited on p. 338.)

[385] Peter Wynn. On a Procrustean technique for the numerical transformation of slowly convergent sequences and series. *Proc. Cambridge Phil. Soc.*, 52:663–671, 1956. (Cited on p. 518.)

[386] Peter Wynn. On the convergence and stability of the epsilon algorithm. *J. SIAM Numer. Anal.*, 3(1):91–122, 1966. (Cited on p. 338.)

[387] L. Yao and Adi Ben-Israel. The Newton and Halley methods for complex roots. *Amer. Math. Monthly*, 105:806–818, 1998. (Cited on pp. 644, 645.)

[388] Tjalling J. Ypma. Historical development of the Newton–Raphson method. *SIAM Rev.*, 37:531–551, 1995. (Cited on p. 685.)

[389] M. Zelen. Linear estimation and related topics. In John Todd, editor, *Survey of Numerical Analysis*, pages 558–584. McGraw-Hill, New York, 1962. (Cited on p. 47.)

Index